Optics in Four Dimensions—1980
(ICO, Ensenada)

ICO Hymn, J. Blabla 1975

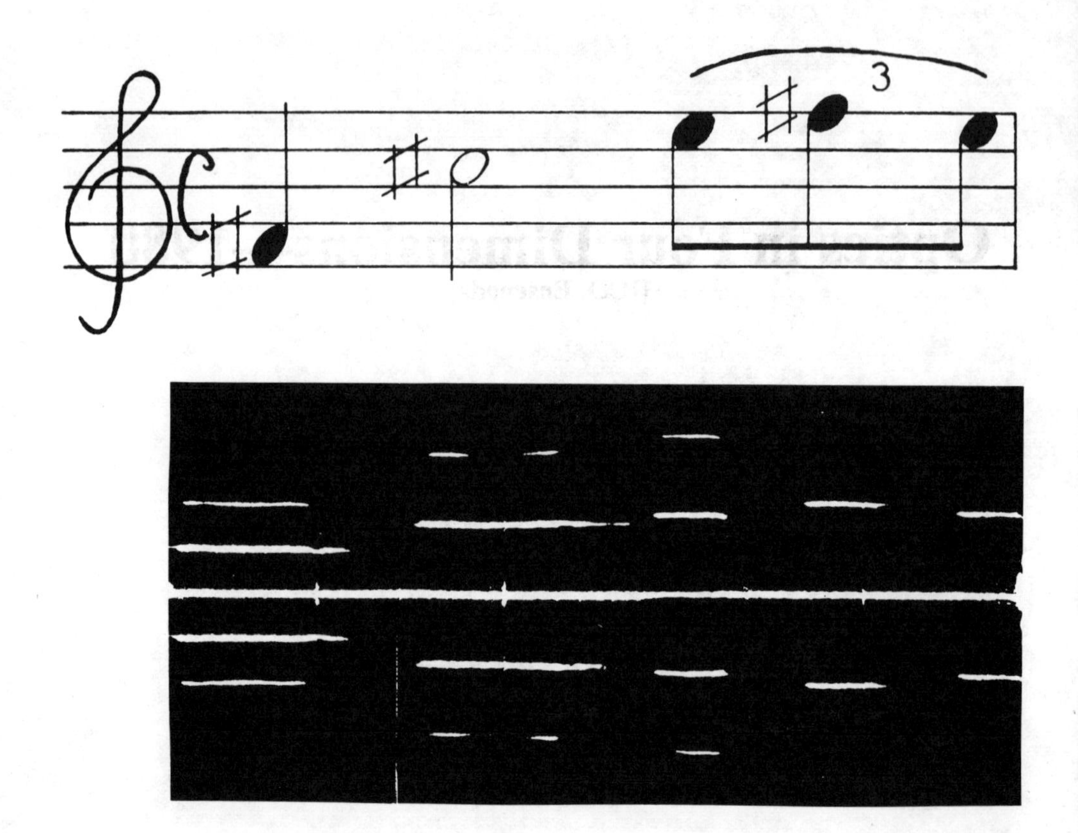

AIP Conference Proceedings
Series Editor: Hugh E. Wolfe
Number 65
Subseries on Optical Science and Engineering No. 1

Optics in Four Dimensions—1980
(ICO, Ensenada)

Edited by

M.A. Machado
CICESE, Ensenada, BC, Mexico
and
L.M. Narducci
Drexel University, Philadelphia, PA.

American Institute of Physics
New York 1981

L.C. Catalog Card No. 80-70771
ISBN 0-88318-164-9
DOE CONF- 800880

PREFACE

This volume contains most of the papers that were presented at the Conference, Optics in Four Dimensions, which was held at the Convention Center of Ensenada, B.C., Mexico, during the period August 4-8, 1980.

The Conference represents the end product of Professor Adolf Lohmann's urging, and it was organized under the auspices and with the partial support of the International Commission for Optics (ICO), of which Professor Lohmann is the current President. Still, the meeting would have had little chance of success without the generous sponsorship of the Consejo Nacional de Ciencia y Technologia (CONACYT) and the enthusiastic support of the Centro de Investigacion Cientifica y de Educacion Superior de Ensenada (CICESE) and of its staff.

The Conference was attended by about 130 participants. A total of 85 papers were presented in 18 sessions.

The Conference was organized by a Program Committee consisting of:

A. Lohmann, ICO President	Federal Republic of Germany
E. Wolf, Program Chairman	USA
L.M. Narducci, Program Secretary	USA
N. Abramson	Sweden
H.H. Barrett	USA
D.J. Bradley	Great Britain
Y.N. Denisyuk	USSR
A.A. Friesem	Israel
G. Gavrilow	USSR
A.K. Ghatak	India
F. Gori	Italy
E.N. Leith	USA
W. Lukosz	Switzerland
H.M. Nussenzveig	Brazil
T. Okoshi	Japan
E.R. Pike	Great Britain
G. Schulz	German Democratic Republic
J. Ch. Vienot	France

The local organization was handled by:

H.J. Frankena, ICO Secretary	The Netherlands
M.A. Machado, Chairman	Mexico
L.E. Celaya, Secretary	Mexico
M. Celaya	Mexico
D. Malacara	Mexico
R. Noble	Mexico
J. Ojeda Castaneda	Mexico

The purpose of the meeting was to provide a forum for discussion of several areas in Modern Optics where the space-time or space-frequency descriptions of optical fields are finding growing theoretical and practical applications. The titles of the various sessions provide evidence of a strong current interest in space-time optics, phase conjugation, phase-space methods, holography, coding and encoding of optical signals and optical reconstruction, to mention only a few of the areas that were covered in the Conference. The papers in this Volume reveal a concerted effort on the part of the participants to generalize many of the traditional ideas and to move along the lines that were originally proposed by Lohmann for discussion and review.

The Conference organizers owe a special thanks to the many people who devoted months of their time to the preparation of the innumerable details that make an international gathering a pleasant and successful one. In particular, we would like to acknowledge the interest and support of E. Landgrave, J. Signeiros, D. Tentori and R. Machorro, of the Optics Group at CICESE. A special acknowledgement is due to Mrs. Debra DeLise-Hughes for her competent and enthusiastic handling of innumerable organizational tasks. Without the help of these valued collaborators, our efforts would have met with far more limited success.

M.A. Machado
CICESE, Ensenada, Mexico

L.M. Narducci
Drexel University
Philadelphia, PA (USA)

CONTENTS

Image Formation & Holographic Interferometry

Phase Space Methods & Fourier Optics

OPTICS IN FOUR DIMENSIONS - WHY?

A. W. Lohmann
Physikalisches Institut der Universität
Erwin-Rommel-Straße 1, D-8520 Erlangen

ABSTRACT

"Optics in four dimensions" is the title of this conference. Like any other conference, it hopes to stimulate progress. How is that possible with a conference title, that links optics with a central term out of geometry: dimensions? Thinking about optics by using geometrical concepts is fruitful, because our mind's computer language employs many geometrical "words" when communicating with the subconsciousness. That is where much of our creativity resides.

OPTICS AND GEOMETRY IN FOUR STEPS

Step 1: Optics in two dimensions (x,y), which might also be called "classical image science", is not yet exhausted but sometimes in the style of a routine. What to do? Look for problems with more than two dimensions such as (x,y,z), (x,y,t) or (x,y,z,t). More about step (1) in the second chapter.

Step 2: The dimension "time" in optics can be tricky, because we do not observe the field u(t) in first order, instead $|u(t)|^2$ or sometimes $\langle u(t+\tau)u^*(t)\rangle$, which is related to the spectrum $S(\lambda)$. This step 2 is well covered at this conference.

Step 3: There are other domains besides the "direct geometrical domain" in (x,y,z). Very popular is the Fourier domain and recently also the WIGNER domain, which is a special case of a "phase space". The need for signal representation in "phase space" is very old, as the musical score shows us. Einstein's "world lines", Feynman's graphs, flow charts etc. are other cases of useful pseudo-geometrical domains, that will be discussed in one of the coming chapters.

Step 4: During the previous steps we have experienced how much optics is penetrated by geometrical concepts. This has prepared us for discussing the central question "how do we think creatively?" The hypothesis presented in our final chapter is that much of creative thinking is performed with "pseudo-geometrical concepts". If that is so, we should cultivate the penetration of optics with geometrical concepts.

ISSN:0094-243X/81/650001-08$1.50

OPTICS IN MORE THAN TWO DIMENSIONS

Optics in higher dimensions such as (x,y,z), (x,y,t) or (x,y,z,t) is by no means new as the examples on figure 1 indicate. Curiously, only an illustrous minority of optical scientists went beyond (x,y). Three-dimensional holography is an exception, but even there the number of fundamental contributors is fairly small.

The Talbot-effect (= self imaging), the Talbot bands and Einstein's world lines are known well enough. But the "forerunner" of Sommerfeld and Brillouin /1/ needs some comments. When an otherwise monochromatic wave with a sharp onset enters a prism the electrons of the glass material will be at rest at first. Hence, the front end of the wave train slips straight through the prism at velocity c. Only later, when the forced oscillations of the electrons are at steady state, will ordinary refraction occur. The week "forerunners" have never been observed experimentally. Their merit is of more conceptual nature, since the transient behaviour of refraction is elucidated, similarly as the transient behaviour of diffraction gratings in the case of Talbot's bands.

PSEUDO-GEOMETRICAL DOMAINS

What was said about "step 3" is illustrated in fig. 2. The point we want to make here is that certain situations in physics are complicated in the genuine or direct space (x,y,z) but simple in a pseudo-space such as the Fourier transform domain. In other situations, for example when a sequence of sounds (= music) is to be described, neither the direct time domain nor the frequency or Fourier domain satisfies the needs of the musician. Hence, the musical score was invented. The musical score is illegal from the point of view of the uncertainty principle, since both time and frequency are specified with unlimited accuracy. Wigner's distribution function resolved in 1931 the dilemma of the musicians, who can now perform music in good conscience.

There exist many more useful pseudo-geometrical domains as sketched in fig. 3. Some of these examples may be called "geometryfied thoughts". This might be also a suitable description of Bohr's model of the atom, that became populary almost instantly, partially because it was in numerical agreement with spectroscopic observations. Taken literally, Bohr's model is nonsense because the circulating electron would have to radiate. This was not a lasting obstacle against Bohr's success.

OPTICS IN >2 DIMENSIONS

(x, y, z) TALBOT-EFFECT = SELF-IMAGING

GRATING

etc.

1834

$$Z_T = 2d^2/\lambda$$

X-RAYS + CRYSTALS, VON LAUE, BRAGG 1912

3D HOLOGRAPHY, GABOR, LEITH, DENISYUK

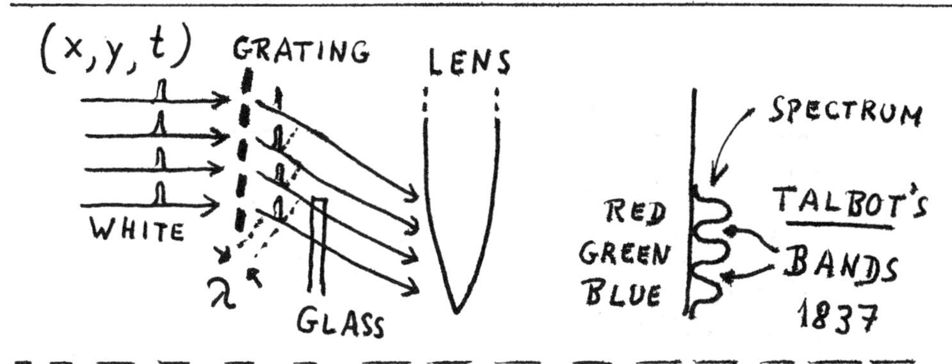

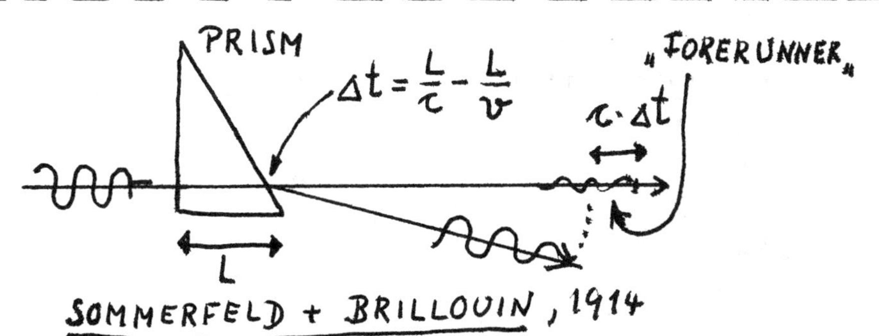

SOMMERFELD + BRILLOUIN, 1914

(x, y, z, t) EINSTEIN's WORLD LINES 1905

PSEUDO-GEOMETRICAL DOMAINS

WAVE PROPAGATION

IN (x, y, z): $$\frac{\partial^2 u}{\partial x^2} + \frac{\partial^2 u}{\partial y^2} + \frac{\partial^2 u}{\partial z^2} + K^2 u = 0$$

IN (ν_x, ν_y, ν_z): $\tilde{u}(\nu_x, \nu_y, \nu_z) \neq 0$ ONLY ON EWALD's SPHERE

3D-FOURIER DOMAIN

ν_x, $\frac{1}{\lambda}$, ν_z

$|\vec{\nu}| = 1/\lambda$

4 REPRESENTATIONS OF SOUND

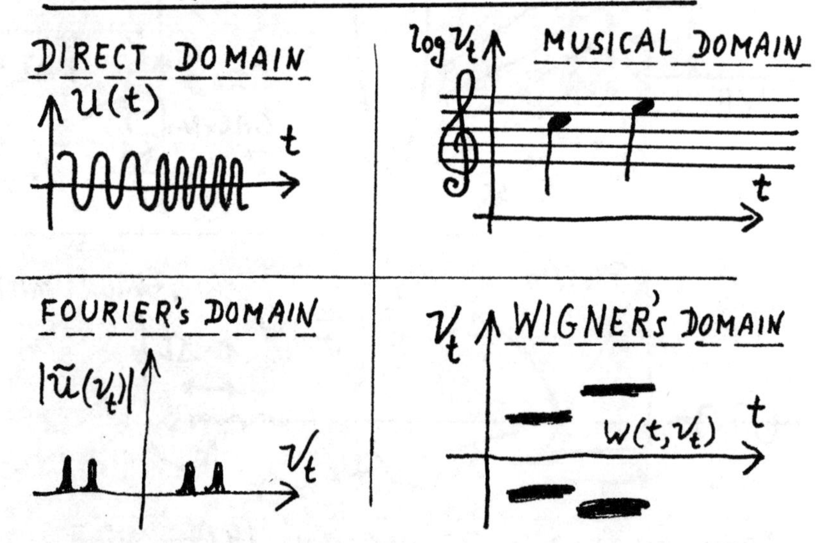

Fig. 2

OTHER PSEUDO-GEOMETRIC DOMAINS

GRAPHIC TOOLS FOR PERSUASIVE COMMUNICATION

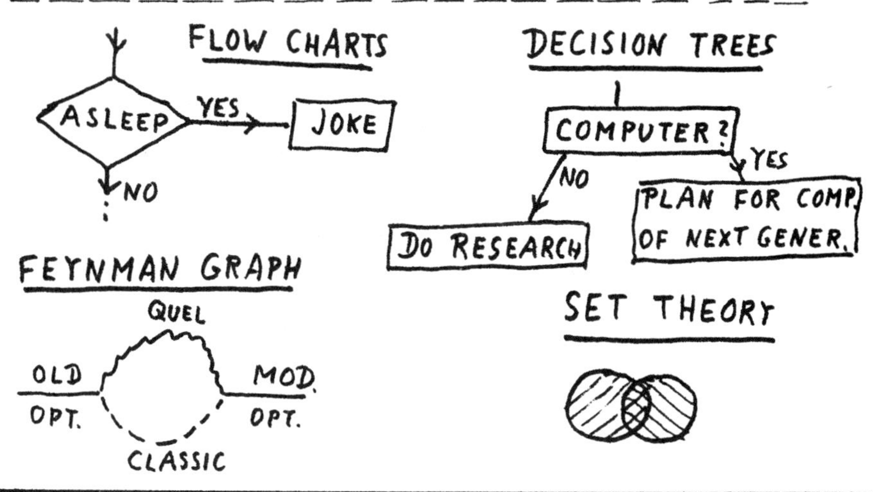

ISOMORPHISMS

1596 - 1650

ANALYT. GEOM. + LIN. ALGEBRA, DESCARTES

HILBERT's SPACE , NOETHER's THEOREMS
(ISOTROPY ≈ CONSERVATION)

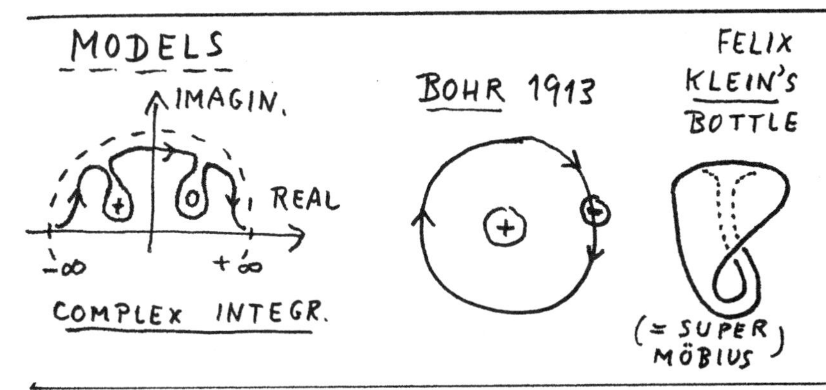

TRANSCENDENT MODEL

MORPHOLOG. INVENTIONS
F. ZWICKY

Fig. 3

My guess is that the simple and beautiful visualisation of Bohr's postulates helped by persuading in a somewhat irrational manner via the subconsciousness.

THINKING AND GEOMETRY

Now we begin with the central part of this essay. We want to explain what we meant in the abstract, when we recommended to use geometrical concepts for creative thinking. The next few lines contain the essence of this chapter in telegram style, that will be elaborated by subsequent comments.

Hypothesis (1): The "computer language" of thinking contains (not only) geometrical concepts.

Supporting arguments

(A1) The mind is not an arithmetic computer /2/.

(A2) The mind enjoys input in pseudo-geometrical format, because the mind's CPU and memory can absorb input in that format directly.

(A3) The newly born mind is not empty, It contains already the "operating system" of linguistics /3/, transmitted by "linguistic genes".

Hypothesis (2): The mind inherits geometrical concepts via "geometrical genes".

Hypothesis (3): The subconscious part of the mind likes to communicate in geometrical symbols.

Hypothesis (4): Genuine creativity takes place in our subconscious mind.

(A4) Some inventions appear like a flash, without the help of the conscious mind.

In our comments on these hypothesis and arguments we will continue to use the terminology of computer science for describing certain aspects of the human mind. When doing so, one is tempted to ask the question "what type of intelligence is better, human or artificial?" We will resist and state only that both types of "intelligence" are very useful and quite different from each other.

Hypothesis (1) and arguments (1) and (2) were prepared already in the chapter on "pseudo-geometrical domains" with the help of figures 2 and 3. Some additional comments give more credibility. Many persons had probably an enjoyable experience, when set theory suddenly became simple while looking at the two overlapping circles. Two linear equations with three unknowns are difficult to digest until they are visualised as two intersecting planes in 3D space. Not only learning is often much easier, when "images" go along with intellectual input. Also memory re-call is greatly facilitated since "images" are apparently like "addresses" in the human information storage system.

We remember our friends by their faces, not by their telephone numbers. In a way, that is surprising because the face may occur in many projections or under various circumstances. Our pseudo-geometrical computer language must be quite sophisticated. Just think how you remember the concept "chair", for example.

The argument (3) about "linguistic genes" is quite generally accepted among modern linguists. In part it is based on abservations on language learning of babies and chimpanzees. Even if some special chimpanzees were tought with considerably more effort than human children ever get, the apes never got very far.

Argument (3) encourages us to postulate in hypothesis (2) "geometrical genes", that may be considered as a revival of one of Descartes' "innate ideas". Experiences shows that babies learn easily how to orient themselves in the 4D world without much instruction. Behaviouristic pyschology alone cannot explain this achievement. There would be much more bloody noses and black eyes, if behaviourism were right, exclusively. The same is true for animals, I believe. Hence, animals are probably our equals in terms of "geometrical genes".

In hypotheses (3) we claim that communications with the subconscious part of our mind is performed largely with "geometrical words" of our mind's computer language. Maybe that is why psychologists can take a look at our innermost self with the help of the phantastic pattern of the Rohrschach test.

For supporting the last hypotheses (4) we will re-tell a story about an amazing act of creativity of Poincaré, as reported by Hadamard /4/. Poincaré had been trying to solve a problem for over two weeks, but in vain. This was quite uncommon for a man, who published quite regularly a new result every week. Poincaré had to interrupt his attempts in order to participate in a mineralogic excursion. One day, while discussing mineralogy intensively, the solution to the mathematical problem came upon Poincaré unexpectedly like a flash. He did not even interrupt the mineralogic sentence he was speaking at that instance. Poincaré continued to be engulfed in mineralogy until he was by himself in his hotel room, where he wrote without any hesitation his newly discovered results about "fuchsian functions".

What had happened? Apparently, while trying hard but without success, Poincaré had impregnated his subconscious mind with the problem and with the urgent wish to solve it. His subconscious mind must have worked autonomously on the problem until it was solved. At that instance, a red light was flashing at the console of the conscious mind-computer, indicating that the back-up computer for tough problems (i.e. the subconscious mind) had found out what the fuchsian functions were all about.

CONCLUSION

The story about Poincaré's subconscious mind, that was solving for him difficult problems, while Poincaré in his actual life did something quite different under the guidance of his conscious mind, makes us envious. If we only knew how to delegate difficult problems to our own subconscious mind. Actually, the effect of subconscious productivity itself is probably widely known, although the results are not so dramatic as for Poincaré. Many persons could report similar events that had occured to them. School children try something of this sort when putting their school books underneeth their pillows before falling asleep.

Can we use Poincaré's method for creating progress in optics? Everybody has to find out for himself, in what "language" he is able to tell his subconscious mind what to do. Maybe some will be lucky when using a "pseudo-geometric language" while thinking about optics in N dimensions.

REFERENCES

1. A. Sommerfeld, L. Brillouin, Ann der Physik (4) 44, 177, 203 (1914)
2. J. von Neumann, The computer and the brain, Yale Univ. Press, 1958
3. N. Chomsky, Information and Control 2, 137 (1959)
4. J. Hadamard, The psychology of inventions in the mathematical field, Princeton Univ. Press, 1948

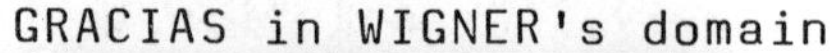
GRACIAS in WIGNER's domain

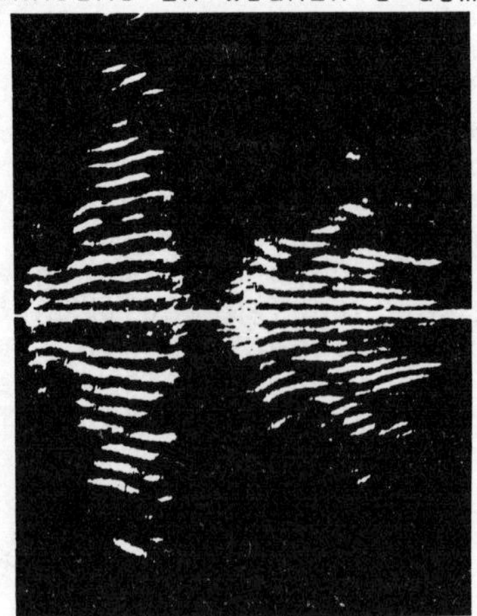

CAUSALITY AND ANALYTICITY IN OPTICS

H. M. Nussenzveig
Instituto de Física, Universidade de São Paulo, Brazil

ABSTRACT

In order to provide an overall picture of the broad range of optical phenomena that are directly linked with the concepts of causality and analyticity, the following topics are briefly reviewed, emphasizing recent developments: 1) Derivation of dispersion relations for the optical constants of general linear media from causality. Application to the theory of natural optical activity. 2) Derivation of sum rules for the optical constants from causality and from the short-time response function (asymptotic high-frequency behavior). Average spectral behavior of optical media. Applications. 3) Role of spectral conditions. Analytic properties of coherence functions in quantum optics. Reconstruction theorem. 4) Phase retrieval problems. 5) Inverse scattering problems. 6) Solution of nonlinear evolution equations in optics by inverse scattering methods. Application to self-induced transparency. Causality in nonlinear wave propagation. 7) Analytic continuation in frequency and angular momentum. Complex singularities. Resonances and natural-mode expansions. Regge poles. 8) Wigner's causal inequality. Time delay. Spatial displacements in total reflection. 9) Analyticity in diffraction theory. Complex angular momentum theory of Mie scattering. Diffraction as a barrier tunnelling effect. Complex trajectories in optics.

INTRODUCTION

"On revient toujours à ses premiers amours"

It is often useful and illuminating to survey a broad field of research from the vantage point of some general unifying principle, recognizing common features in apparently very diverse subjects. The present work is devoted to the task of discussing the relevance to optics of the principle of causality and of the closely related concept of analyticity. This is done by very briefly reviewing a number of illustrative examples, both in classical and in quantum optics.

1. DERIVATION OF DISPERSION RELATIONS

It was in optics that dispersion relations were first applied to physics and that their connection with causality was first realized[1]. Typically, in the propaga-

ISSN:0094-243X/81/650009-22$1.50

tion of light through a linear medium, the medium response $r(t)$ (we omit all tensor indices) is linked with the excitation $e(t')$ through a response function $g(t,t')$, so that, for a time-invariant medium,

$$r(t) = \int_{-\infty}^{\infty} g(t-t')\, e(t')\, dt' , \qquad (1.1)$$

corresponding to a constitutive relation for the Fourier transforms,

$$R(\omega) = G(\omega)\, E(\omega) , \qquad (1.2)$$

where the material constant $G(\omega)$ is a function of the circular frequency ω (dispersion).

According to the "primitive causality condition", the response at time t cannot depend on the excitation at later times, i.e.,

$$\delta r(t)/\delta e(t') = 0 , \qquad t' > t , \qquad (1.3)$$

where $\delta r/\delta e$ denotes the functional derivative. Thus,

$$g(\tau) = 0 , \qquad \tau < 0 , \qquad (1.4)$$

so that

$$G(\omega) = \int_{0}^{\infty} g(\tau) \exp(i\omega\tau)\, d\tau . \qquad (1.5)$$

Since this is a half-range Fourier transform, $G(\omega)$ is the boundary value of an analytic function, regular in the half-plane $\mathrm{Im}\ \omega > 0$.

The derivation of dispersion relations for $G(\omega)$ is usually based on additional information concerning high-frequency behavior. Thus, if $G(\omega)$ is the complex dielectric susceptibility $\chi'(\omega)$, the expected behavior at frequencies well above optical absorption bands is free electron-like,

$$\chi'(\omega) \approx -\mathcal{N} e^2/m\omega^2 , \qquad \omega \to \infty , \qquad (1.6)$$

where $\mathcal{N}$ is the electron density. If, e.g., $G(\omega)$ is square integrable on the real axis (as far as the behavior at infinity is concerned, (1.6) is more than enough to guarantee this for $\chi'(\omega)$) , it follows[2] that

$$G(\omega) = \frac{1}{\pi i} P\int_{-\infty}^{\infty} \frac{G(\omega')}{\omega'-\omega}\, d\omega' , \qquad (1.7)$$

where P denotes the Cauchy principal value. Thus, the real (dispersive) and imaginary (absorptive) parts of $G(\omega)$ obey the Kramers-Kronig (dispersion) relations, i.e., they are each other's Hilbert transforms.

The asymptotic high-frequency behavior is not always as readily determined as for (1.6); it may happen, for instance, that the macroscopic description associated with the constitutive relation (1.2) breaks down at high frequencies. However, it may still be possible to derive dispersion relations by an alternative procedure, based on the properties of a passive linear system[3,4].

A passive system is one that can absorb or store energy, but cannot generate it. Every passive system is causal, but the converse need not be true. What is important is that the response function of a passive linear system not only has a regular analytic continuation in a half-plane, but it also has a positive semi-definite imaginary part in it, i.e., it is a Herglotz function[2]. This turns out to be a strong constraint on the high-frequency growth properties, leading to the dispersion relations in a slightly generalized form.

If one assumes that the system is not just passive, but that it is actually dissipative, i.e., that it has strictly positive absorption at _all_ positive frequencies, one gets even stronger constraints on the asymptotic behavior, as well as the absence of zeros of $G(\omega)$ in $\mathrm{Im}\,\omega > 0$. This stronger assumption is usually made in the work of the Russian school[5,6].

The dielectric susceptibility is associated with a local response function of the type (1.1). This is not true, however, for the important case of the complex refractive index $N'(\omega) = n(\omega) + i\kappa(\omega)$, which is connected with the propagation factor $\exp(i\,N'\omega z/c)$ of a monochromatic plane wave. Although $N'(\omega) = \sqrt{\varepsilon'(\omega)}$ (taking $\mu = 1$), the analyticity of ε' does not entail that of N', unless one accepts[5] the strong dissipativity assumption to exclude branch points due to eventual zeros of ε' in $\mathrm{Im}\omega > 0$. This is unnecessary, however,[2] if one employs the "relativistic causality condition", according to which _no signal can propagate with velocity greater than c_. Even in the presence of conductivity, which gives rise to a $\omega^{-1/2}$ singularity at $\omega = 0$, this condition allows one to derive the dispersion relations for the refractive index.[7]

The derivation of dispersion relations for more general linear media, including the effects of spatial dispersion, has also been discussed.[6,8] The response function in such media goes beyond the usual electric dipole approximation, bringing in higher multipoles.

The theory of natural optical activity provides one of the simplest illustrations of these effects. The normal modes of propagation in an optically active medium are circularly polarized waves. Besides circular double refraction, such a medium also shows circular dichroism, i.e., both the phase velocity and the extinction are different for right and left circular polarization, leading to different complex refractive indices $N'_+(\omega)$ and $N'_-(\omega)$, respectively.

As a consequence of these features, the medium rotates the plane of polarization of linearly polarized light and also converts it into elliptically polarized light. The rotatory power $\rho(\omega)$ and the ellipticity $\phi(\omega)$ per unit length of the medium are related with the complex refractive indices by [9]

$$\rho(\omega) + i\phi(\omega) = \omega \left[N'_-(\omega) - N'_+(\omega)\right]/2c \;. \qquad (1.8)$$

The usual derivation of dispersion relations does not go through[10] for $N'_\pm$, because one cannot build up a real field using only one of them; we need both, in view of the crossing relation

$$\left[N'_+(\omega)\right]^* = N'_-(-\omega) \quad (\omega \text{ real}). \qquad (1.9)$$

However, one can derive[11] dispersion relations between ρ and ϕ from primitive causality, by applying it to the constitutive equation[9]

$$\vec{B}(\omega) - \vec{H}(\omega) = i\omega\, g(\omega)\, \vec{E}(\omega) \;, \qquad (1.10)$$

where $g(\omega)$ is related with ρ and ϕ by

$$\rho(\omega) - i\phi(\omega) = \omega^2 g(\omega)/c \;. \qquad (1.11)$$

The dispersion relations between ρ and ϕ follow from those between Reg and Img . Physically, (1.10) describes the induced magnetic dipole moment that arises from a time-varying electric field, due to the helical structure that is associated with the natural optical activity of the medium. Thus, this procedure captures the basic causal relationship that is responsible for the effect.

2. SUM RULES FOR THE OPTICAL CONSTANTS

We usually know a bit more about the response function besides its causal behavior represented by the dispersion relations. We can take advantage of this extra knowledge to get additional constraints.

Suppose, for instance, that the excitation takes the

form of a step function in time, switched on at $t = 0$. Causality requires that the response must vanish for $t < 0$, but it does not prevent a jump at $t = 0$, whereas we actually expect that the response will rise continuously from zero at $t = 0$. The short-time behavior of the response is related with the high-frequency asymptotic behavior of $G(\omega)$. For the dielectric susceptibility $\chi'(\omega)$, this is given by (1.6): at high frequencies, we get just the inertial free-electron response, which prevents a sudden jump.

Instead of (1.6), a slower decrease at infinity, e.g., like $\omega^{-1-\varepsilon}$, $\varepsilon > 0$, would suffice for square integrability. Thus, in this case, the dispersion integrals converge faster than they "need" to at the high-frequency end; this is called "superconvergence". By comparing the actual high-frequency behavior with the high-frequency limit of the Kramers-Kronig relations, one gets[12,13] additional constraints in the form of sum rules.

Besides the well-known "f-sum rules", one finds a whole set of new ones, which may be called "inertial sum rules", because of their connection with the inertial response. Thus, for the dielectric constant of an isotropic nonconducting medium, one finds

$$\int_0^\infty \left[\mathrm{Re}\ \varepsilon'(\omega) - 1\right] d\omega = 0 , \qquad (2.1)$$

which is equivalent[13] to the condition $g(0) = 0$ for the associated response function $g(\tau)$ of (1.4).

Similarly, if $n(\omega) = \mathrm{Re}\ N'(\omega)$ is the real refractive index of an isotropic medium, conducting or not, one has

$$\int_0^\infty \left[n(\omega) - 1\right] d\omega = 0 . \qquad (2.2)$$

For the optical rotatory power of a medium with natural optical activity, one finds[11,14]

$$\int_0^\infty \rho(\omega) d\omega = 0 . \qquad (2.3)$$

The inertial sum rules (2.2) and (2.3) may be interpreted in terms of the average behavior of optical media over the whole frequency spectrum. According to (2.2), frequency ranges where $n(\omega) > 1$ must be compensated by other ones where $n(\omega) < 1$, in such a way that the total area under the curve of $n(\omega) - 1$ vanishes. Thus, the phase velocity is sometimes greater than c and sometimes smaller, but, on the average over the whole spectrum, in the sense of (2.2), it is just equal to c.

Similarly, according to (2.3), optically active media are "middle-of-the-road": "leftist" and "rightist" frequency ranges compensate each other on the average.

These results might be summed up by saying that, on the average over the whole frequency spectrum, optical media tend to behave like the vacuum!

From (1.6) follows the well-known fact that $n(\omega) < 1$ in the asymptotic high-frequency domain, and it might be thought that the whole negative contribution to (2.2) arises from this domain. That this is not so follows from the fact[12] that $n(\omega) - 1$ satisfies a sum rule analogous to (2.2), but with the extra weighting factor $\omega\kappa(\omega)$ (where $\kappa(\omega) = \mathrm{Im}\, N'(\omega)$), which reduces the high-frequency contribution and enhances that of the absorption bands. Similarly,[11] $\rho(\omega)$ satisfies sum rules analogous to (2.3), but with extra weighting factors $\omega\phi(\omega)$ and $\phi(\omega)/\omega$.

In terms of a Lorentz-like dispersion formula, the compensation of positive and negative contributions in the inertial sum rules can be physically related with the change from in-phase to out-of-phase character of the response of a forced damped oscillator as it goes across a resonance.

The new sum rules for the refractive index and dielectric constant have been verified experimentally for a variety of substances, including both metals and insulators.[13,15-18] For metals, the refractive index sum rule (2.2) is particularly sensitive to the low-frequency behavior, because of the $\omega^{-1/2}$ singularity of $n(\omega)$ at $\omega = 0$; this property has been employed[13] to choose between different sets of optical constants, based on different extrapolation procedures in the far infrared. The sum rules with extra weighting factors play a similar role, emphasizing the contributions of different frequency ranges.

3. ANALYTIC PROPERTIES OF COHERENCE FUNCTIONS

Coherence functions in quantum optics are generally defined by[19]

$$G^{(n,m)}(x_1,\ldots,x_n;\, x_{n+1},\ldots x_{n+m})$$
$$= \mathrm{Tr}\left[\rho E^-(x_1)\ldots E^-(x_n)E^+(x_{n+1})\ldots E^+(x_{n+m})\right], \qquad (3.1)$$

where x_i are spacetime points, ρ is the density operator, and E^- and E^+ are, respectively, creation and annihilation operators associated with the quantized radiation field (tensor indices are omitted). For $m = n$ and $x_{n+i} = x_i$, the function $G^{(n,n)}$ can be interpreted in terms of an n-fold delayed coincidence counting rate involving n photodetectors. For $m = n = 1$, (3.1) corresponds to the mutual coherence function between the spacetime points x_1 and x_2.

The normal ordering in (3.1) arises from the fact that ordinary photon detectors operate by absorption.[20] There are certain parallels[19] between the set of coherence functions (3.1) in quantum optics and the set of Wightman functions in quantum field theory. In particular, the analogue of the "spectral condition" is the fact that the Fourier transforms of $G^{(n,m)}$ with respect to time contain only positive frequencies in the m annihilation operators, and only negative frequencies in the n creation operators.

It follows, by analogy with (1.5) (but note that the roles of frequency and time are reversed!), that $G^{(n,m)}$ is the boundary value of an analytic function of the <u>time</u> variables, regular in $\mathrm{Im}(x_i^o) > 0$ for $1 \leqslant i \leqslant n$ and in $\mathrm{Im}(x_i^o) < 0$ for $n+1 \leqslant i \leqslant n+m$, where $x_i^o = ct_i$ is the time component of the 4-vector x_i. In the particular case of a time-stationary field, $G^{(1,1)}(\vec{x}_1,t_1; \vec{x}_2,t_2)$ depends only on $\tau = t_2-t_1$, and

$$G^{(1,1)}(\vec{x}_1,\vec{x}_2, \tau = t_2-t_1) = \int_0^\infty W(\vec{x}_1,\vec{x}_2,\omega)\, e^{-i\omega\tau} d\omega, \quad (3.2)$$

where W is the cross-spectral density, is regular in $\mathrm{Im}\,\tau < 0$. Note that this analyticity in time does not follow from causality, but rather from the normal ordering in (3.1), which expresses the privileged role of absorption as opposed to emission in photodetection. This may be likened to the passivity condition in Sect. 1.

Since this may look like a purely accidental feature of the detection process[20], one may wonder whether, by employing (3.1) to define the set of coherence functions, we do not lose some statistical information about the field. Given the set of coherence functions (3.1) for all values of m and n, can one reconstruct from it the density operator ρ associated with the field? This is analogous to reconstructing a field theory from the set of Wightman functions.

A <u>sufficient</u> condition for the reconstruction to be possible has been given[21,19] making use of the coherent state representation.[22] This representation, which emphasizes the correspondence with classical coherence theory and has played a central role in quantum optics, can be regarded as a one-to-one mapping between Hilbert space vectors and entire analytic functions[23], thus establishing another close link between analyticity and optics.

For the reconstruction problem, it suffices to consider a single mode of the radiation field (the general case corresponds to a direct product involving all the modes). Let $|z>$ be a coherent state associated with this mode (z is a complex eigenvalue of the annihilation operator). The above-mentioned sufficient condition then is

$$< z|\,\rho\,|z> = \mathcal{O}\left[\exp(-\,\varepsilon|z|^2)\right]\ ,\quad |z| \to \infty \qquad (\varepsilon > 0)\ . \qquad (3.3)$$

Within limits set by the uncertainty principle, the left-hand side of (3.3) can be interpreted as a sort of "probability of finding the field in a coherent state $|z>$ ". According to (3.3), this "probability" should decrease at least exponentially with the average number of photons $\bar{n} = |z|^2$ in $|z>$, as $\bar{n} \to \infty$. The proof of the "reconstruction theorem"[21,19] makes use of the analytic properties of the normally-ordered characteristic function in the coherent-state representation.

The result is expected to remain valid under considerably more general conditions, but (3.3) is satisfied by all of the most widely employed models in quantum optics. The reconstruction theorem implies that the set of coherence functions (3.1) indeed contains all the statistical information about the field.

4. PHASE RETRIEVAL PROBLEMS

Applying (3.2) for $\vec{x}_1 = \vec{x}_2$, we see that the complex degree of temporal coherence $\gamma(\tau)$ and the spectral density are Fourier transforms of each other. Michelson's method of interference spectroscopy is based on this relationship. In practice, however, while $|\gamma(\tau)|$ (which is proportional to the visibility of the interference fringes) is readily measurable, it is very difficult to measure $\arg\gamma(\tau) = \phi(\tau)$, the phase of the degree of coherence. The problem of determining ϕ is known as the <u>phase problem</u> of coherence theory. Similar phase retrieval problems occur in x-ray crystallography and in image reconstruction.

According to (3.2), $\gamma(\tau)$ has a regular analytic continuation in $\mathrm{Im}\,\tau < 0$. Is this analyticity helpful in connection with the phase problem? We have

$$\ell n\ \gamma(\tau) = \ell n|\gamma(\tau)| + i\phi(\tau)\ , \qquad (4.1)$$

so that, if $\ell n\ \gamma$ has similar analytic properties, yielding a dispersion relation between ϕ and $\ell n|\gamma|$, we would have the solution of the phase problem. Such "logarithmic Hilbert transforms" have been discussed by Toll;[24] an important practical application is the determination of optical constants by reflectivity measurements.[25]

If there are zeros of $\gamma(\tau)$ in $\mathrm{Im}\,\tau < 0$, they are branch points for $\ell n\,\gamma(\tau)$, so that the phase ϕ contains, besides the contribution from the "minimal phase", a Hilbert transform involving $\ell n|\gamma(\tau)|$, an additional contribution from the zeros, known as the "Blaschke phase". In some cases, including that of blackbody radiation, the zeros are absent, and the solution of the phase problem is given by the minimal phase.[26]

In general, however, this is no longer true[27]: for a variety of practically important spectra, including natural (Lorentzian) and Doppler-broadened line shapes, as well as for band-limited spectra, one finds not only that $\gamma(\tau)$ has a large number of zeros, but also that their contribution to the phase is essential; the minimal phase is a very poor approximation.

Therefore, in order to solve the phase problem, one needs additional information. It is interesting to note that holography, the most successful practical method of image reconstruction, can be regarded as a device to solve the phase problem by generating a zero-free half-plane.[28]

Indeed, if we consider a band-limited function (one-dimensional case)

$$f(z) = \int_a^b F(k)\ \exp(ikz)\ dk\ , \qquad (4.2)$$

where k is a spatial frequency, then $f(z)$ is [29] an entire function of exponential type. By Rouché's theorem,[29] the function

$$f(z) + A\ \exp(icz)\ , \qquad (4.3)$$

where $|A| > |f(x)|$ (x real) and c lies outside the support of $F(k)$, i.e.,

$$c \notin [a,\ b], \qquad (4.4)$$

is still entire but has a zero-free half-plane, so that its phase may be reconstructed from its modulus by a logarithmic Hilbert transform.

The added plane wave term in (4.3) is a "reference function"[28] that plays exactly the same role as the reference beam in off-axis holography.[30] The off-axis condition corresponds to (4.4); this condition can also be enforced by spatial filtering, as in single-sideband holography.[31]

The proposal of adding a reference source to solve the phase problem of coherence theory[32,33] has been experimentally demonstrated[34,35] as a practically feasible method of solution.

5. INVERSE SCATTERING

The inverse scattering problem, in general terms, is that of reconstructing the scatterer given a set of scattering data. Depending on the choice of this set, there is a variety of different inverse problems, not only in optics, but also in many other fields.[36-38]

Analyticity plays an important role in the solution of most such problems. The physical reasons for this can be illustrated by a couple of examples.

Consider first the problem of diffraction by an aperture, employing the representation in terms of an angular spectrum of plane waves.[39-40] The angular distribution in Fraunhofer diffraction (scattering data) and the aperture distribution are related to each other by Fourier transformation. However, only the homogeneous plane wave components in the angular spectrum, which carry information about details larger than the wavelength, will survive in the Fraunhofer region. In order to recover the aperture distribution by an inverse Fourier transform, one needs also the inhomogeneous (evanescent) wave components, and one may try to relate them to the homogeneous ones by analytic continuation.

Consider next scattering by a three-dimensional scatterer. In the first Born approximation, the scattering amplitude $A(\vec{k}, \vec{k}')$ depends only on the difference $\vec{K} = \vec{k} - \vec{k}'$ between the incident and scattered wave vectors, and it is in fact just the Fourier transform of the scattering distribution with respect to $\vec{K}$. It then follows from an extension of the Paley-Wiener theorem that A is an entire function in the components of $\vec{K}$. For a nonspherically symmetric scatterer, already in this approximation, the analyticity of A suffices to show that the inverse scattering problem cannot be uniquely solved from scattering data obtained in a single experiment. This remains true for the exact amplitude.[41]

The most successful approach to the solution of the inverse scattering problem is the Gelfand-Levitan-Marchenko method[42-45] in nonrelativistic quantum scattering by a spherically symmetric potential $V(r)$, which has also been applied to optics, as will be seen in the next Section. The set of scattering data consists of the phase shift for a given partial wave, assumed known for all energies, plus the energies and normalization constants of the bound states associated with this partial wave. In the Gelfand-Levitan version (the Marchenko method is similar), the solution is given by

$$V(r) = 2 \frac{d}{dr} K(r,r) , \tag{5.1}$$

where $K(r, r')$, for each r, is the solution of the Gelfand-Levitan integral equation, a linear integral equation of the Fredholm type,

$$K(r,r') + G(r,r') + \int_0^r K(r,r'') G(r'',r')dr'' = 0, \qquad (5.2)$$

whose kernel G is constructed entirely from the set of scattering data.

The essential feature of K for the success of the method is that it is a triangular kernel, i.e., that

$$K(r,r') = 0 \quad \text{for} \quad r' > r . \qquad (5.3)$$

This looks like a causal property. Indeed, in a classical application to the plasma inverse problem, this property of the Gelfand-Levitan kernel has been shown[46] to follow from causality. For a cutoff potential in quantum scattering, it can also be traced back to primitive causality.[2]

The quantum inverse scattering problem has also been solved in one dimension, i.e., for scattering on the line[47-49]. However, for the three-dimensional problem, where the scattering amplitude $A(\vec{k}, \vec{k}')$ is given, the proposed solutions[50-51] are still incomplete.

6. NONLINEAR OPTICS

One of the most remarkable theoretical advances in the treatment of nonlinear waves was the discovery[52] of a totally unexpected connection between the inverse scattering problem and a number of nonlinear evolution equations of physical interest, allowing them to be exactly solved[53]. This new method of solution represents, in a sense, a nonlinear generalization of the Fourier transform.[54]

The physical basis of the new method appears to be best understood in its application to the phenomenon of self-induced transparency[55,19] in nonlinear optics, so that we shall illustrate it in this connection.

The Mc Call-Hahn equations of self-induced transparency, for an inhomogeneously broadened medium of two-level atoms, are a set of coupled Maxwell-Bloch equations. By suitable choice of variables, they may be reduced to the form[56]

$$\partial\Lambda/\partial\tau - i\gamma\Lambda = -eW , \qquad (6.1a)$$

$$\partial W/\partial\tau = \frac{1}{2}(e^*\Lambda + e\Lambda^*) , \qquad (6.1b)$$

$$\partial e/\partial\zeta = -\langle\Lambda\rangle , \qquad (6.2)$$

where $e(\tau,\zeta)$ is a slowly-varying complex electric field amplitude propagating in the z direction, ζ is proportional to z, and $\tau = t - z/U$, with U representing the phase velocity in the host medium; $\Lambda(\gamma,\tau,\zeta)$ is proportional to the slowly-varying complex polarization amplitude at the detuning $\gamma = \omega - \omega_0$ (ω_0 = resonance frequency), and $W(\gamma,\tau,\zeta)$ is proportional to the excitation energy density in the active medium; finally, $\langle\Lambda\rangle$ denotes the average of Λ over the inhomogeneously broadened line shape.

The Bloch equations correspond to (6.1), with the Bloch vector normalized by $|\Lambda|^2 + W^2 = 1$. Maxwell's equations reduce to (6.2), describing $\langle\Lambda\rangle$ as a source for e. Asymptotically, for $\tau \to -\infty$, the field and the polarization vanish. The problem is to determine $e(\tau,\zeta)$, given $e(\tau, 0)$.

The solution by the inverse scattering method in the presence of inhomogeneous broadening was given by Ablowitz, Kaup and Newell[57] and, under less restrictive assumptions, by de Castro.[58] A physical interpretation of the inverse scattering method in this problem was proposed by Haus.[59] However, a different interpretation,[58] in which the roles of space and time variables are interchanged, turns out to be more appropriate, and it will be adopted here.

The linear "scattering" problem associated with (6.1) is simply the time evolution of a two-level atom driven by the electric field, as described by the Schrödinger equation. "Scattering" in time means asymptotic excitation, and the "scattering data" correspond to the asymptotic excitation state in which the atom is left by the passage of the electric field pulse (there is no spontaneous emission in this semiclassical description).

Given $e(\tau,0)$, the determination of the associated set of "scattering data" by solving the direct scattering problem is the analogue of the Fourier transform. The evolution in ζ of this set is obtained so as to make it compatible with the McCall-Hahn equations. Finally, the analogue of the inverse Fourier transform is the solution of the inverse scattering problem from the set of "scattering data" at ζ, yielding $e(\tau,\zeta)$.

The inverse problem is solved by a generalization of the Marchenko method due to Zakharov and Shabat.[60] The analytic properties of scattering solutions play a basic role in the solution; in particular, the reduction to an integral equation of the Marchenko type depends crucially on causality[58]: the state of a two-level atom at a given time can depend only on the values taken by the electric field at earlier times. Self-transparent pulses correspond to solitons.[53]

In the unrealistic (but semiclassically allowed) case where the medium is initially fully inverted, there exist "fast" self-transparent solutions which may travel faster than c . However, this does not constitute a violation of causality[61]: the solution at a given spacetime point still depends only on data contained within the backward light cone with vertex at that point. Though there may exist a preserved pulse shape travelling faster than c , the energy does not travel within the pulse: rather, the excitation energy stored in the active medium ahead of the pulse is triggered by its weak leading edge, through stimulated emission. This is an interesting example of how causality still applies to the case of an active system.

The derivation of dispersion relations from causality can also be extended to nonlinear response.[62] E.g., for quadratic nonlinearity, (1.1) is replaced by

$$r(t) = \int_{-\infty}^{\infty} dt' \int_{-\infty}^{\infty} dt'' \; g(t-t', t-t'') \; e(t')e(t''), \qquad (6.3)$$

where, from causality (cf.(1.4)),

$$g(\tau_1, \tau_2) = 0 \quad \text{for} \quad \tau_1 < 0 \quad \text{or} \quad \tau_2 < 0 \; . \qquad (6.4)$$

It follows that the double Fourier transform $G(\omega_1, \omega_2)$ of $g(\tau_1, \tau_2)$ satisfies dispersion relations in each frequency variable when the other one is kept fixed, or, what is more appropriate, dispersion relations in terms of the sum and difference frequencies $\omega_1 \pm \omega_2$. These relations connect susceptibilities for sum and difference frequency generation, showing that they represent different aspects of the same analytic function. However, in contrast with the linear case, the real and imaginary parts of the nonlinear susceptibilities do not correspond to different physical processes like dispersion and absorption, so that the Kramers-Kronig relations are apparently less useful here.

7. NATURAL MODES

So far, we have considered the role of analytic properties only within a regularity domain. However, one may often extend the domain of analyticity by analytic continuation, including simple types of singularities such as poles. This procedure has long been employed in electric circuit theory, where poles in the frequency domain are associated with natural modes of oscillation, that play an important role in transient response.

In scattering by a bounded scatterer, one can consider the analytic continuation of the scattering amplitude in the frequency plane. For a spherically symmetric scatterer of finite radius, each partial-wave amplitude is meromorphic in ω or, equivalently, in the wave number k , so that its only singularities are poles. Probably the earliest example was Thomson's treatment[63] of the natural modes of oscillation of the eletromagnetic field outside a perfectly conducting sphere. The modes were defined as purely outgoing solutions of the boundary-value problem. This is equivalent to associating them with complex poles of the scattering matrix. In quantum mechanics, "complex-energy eigenfunctions" were introduced by Gamow,[64] in his treatment of alpha decay.

The imaginary part of a complex pole in the frequency plane yields exponential decay in time. By the same token, it leads to exponential growth in space, since distant waves were emitted a long time ago. It follows that "complex-frequency eigenfunctions" cannot be legitimate solutions for all time, i.e., a satisfactory formulation of the decay process also involves considering the excitation process. This can be done[2] by looking for the general solution of the initial-value problem in terms of natural modes.

For the initial-value problem in the exterior of an arbitrary spherically symmetric scatterer of finite radius, the natural mode expansion is obtained[2] through a partial-fraction decomposition of the scattering matrix in terms of its poles. In view of the infinite number of poles , this is a Mittag-Leffler or Cauchy expansion. In the cross section, the contribution from a complex pole close to the real axis and well separated from other poles has a typical resonance shape, with a width given by the imaginary part of the pole. In this sense, natural modes correspond to resonances.

The contribution from such a pole to Green's function is a "transient-mode" term, corresponding to a "complex-frequency eigenfunction", but with a temporal cutoff due to the excitation at the initial time. This removes the exponential growth problem at sufficiently large distances. Besides the pole contributions, the Green's function also contains terms associated with direct reflection from the surface of the scatterer, arising from the entire function contribution in the Cauchy expansion.[2]

Natural mode expansions have also been investigated for the interior problem, both in quantum mechanics[65] and in optics[66], where they have been related with the Ewald-Oseen extinction theorem.[67]

For a spherically symmetric scatterer, the elements of the scattering matrix $S_{\ell}(k)$ are functions of k and of the multipole or partial-wave order ℓ , which is associated with angular momentum. For a scatterer of finite range, one can introduce an interpolating function $S(\lambda,k)$ of a continuous variable λ , which reduces to $S_\ell(k)$ at the discrete integer values ℓ , and one can also extend the definition of S to complex values of λ ("complex angular momentum"). This was done at the beginning of the century by Poincaré and Watson[68], in connection with the propagation of radio waves around the Earth, giving rise to the well-known Watson transform method.

Poles of $S(\lambda,k)$ in the λ plane have become known as Regge poles, in view of their application by Regge to quantum scattering.[2] Poles in the λ plane for fixed k and poles in the k plane for fixed λ (in particular, for the physical values ℓ) are different aspects of the same singularity surface in both variables.

The counterpart[2] of the conjugate variable pair frequency and time is the pair angular momentum and angle. Thus, the imaginary part of a Regge pole leads to angular damping, and one can associate it with a "life-angle". Narrow resonances correspond to large "life-angles", i.e., to a "capture" situation involving many turns around the scattering center. For a scatterer with finite radius, Regge poles can also describe surface waves travelling around the surface of the scatterer, with an angular damping due to radiation. These are the well-known "creeping modes" found in the Watson transformation.

8. TIME DELAY AND SPATIAL DISPLACEMENTS

Another implication of causality was first pointed out by Wigner[70] in quantum scattering. It is based on the concept of time delay of a spherical multipole wave packet due to the scattering process.[71]

Consider an incoming spherical wave packet, formed by the superposition of a narrow range of frequencies. The effect of the scattering on the corresponding outgoing wave packet is to shift the phase of each frequency component, through multiplication by the S-matrix element

$$S(\omega) = \exp\left[2i\eta(\omega)\right] , \qquad (8.1)$$

where η is the scattering phase shift.

Under suitable conditions, the "center" of the outgoing wave packet can be determined by applying the principle of stationary phase. This involves differentiating the total phase, including, among others, the contribution from the time factor $\exp(-i\omega t)$, with respect to ω . It then follows from (8.1) that the effect of the scattering is to shift the center of the outgoing wave packet, relative to the situation when the scatterer is not present, by introducing a time delay

$$\Delta t = 2\, d\eta/d\omega \ , \qquad (8.2)$$

which corresponds to a spatial displacement

$$\Delta r = 2\, d\eta/dk \ . \qquad (8.3)$$

For classical scattering by a spherically symmetric scatterer of radius a, it follows[2] from the analytic properties of S that

$$d\eta/dk \geqslant -a \ . \qquad (8.4)$$

This result has an immediate interpretation in terms of causality. According to (8.3), it implies that the maximum possible time advance of the outgoing wave packet corresponds to a spatial displacement of $-2a$, i.e., to its appearance as soon as (but no sooner than) the incoming wave packet reaches the surface of the scatterer.

On the other hand, causality allows an arbitrarily large positive time delay. Thus, if S, within the frequency spectrum of the wave packet, is dominated by an isolated narrow resonance, it is readily found from (8.2) that the associated time delay is just the lifetime of the resonance.

The extension of (8.4) to quantum scattering, known as Wigner's causal inequality,[70] contains an extra term related with the uncertainty principle.

The concept of time delay can be extended to the scattering of plane wave packets.[72] In terms of the scattering amplitude $A(\vec{k},\vec{k}')$, the time delay of the scattered wave packet in the direction of $\vec{k}'$ is given by

$$\Delta t = \frac{\partial}{\partial \omega} \arg A(\vec{k},\vec{k}') \qquad (\vec{k}' \neq \vec{k}) . \qquad (8.5)$$

This result is not valid in the forward direction, where there is an extra term[73] due to interference with the incident wave. For light propagation in a medium, the forward time delay is related with the real refractive index.

The corresponding spatial displacement of a bounded light beam in total reflection is the well-known Goos-Hänchen effect.[74] Light penetrates into the rarer medium at a beam boundary and travels along the interface as a surface wave before exiting again at the boundary of the reflected beam.

For the conjugate variable pair angular momentum and angle, there is a corresponding concept of angular deflection due to scattering.[75] This is given by

$$\Theta = 2\, d\eta/d\lambda \ , \qquad (8.6)$$

where λ is the continuous (interpolating) angular momentum variable and Θ is the deflection angle for a wave packet formed by superposing a range of angular momenta (impact parameters), i.e., a cylindrical pencil of rays. The total reflection of such a beam on a homogeneous (optically rarer) sphere gives rise to an angular beam displacement[75] that can be derived from (8.6) and represents the spherical analogue of the Goos-Hänchen effect.

9. ANALYTICITY IN DIFFRACTION THEORY

Sommerfeld's celebrated exact solution[76] of the half-plane diffraction problem was an early application of analytic function theory to diffraction. Several other problems, besides this one, were exactly solved by extensions of the Wiener-Hopf technique.[39]

Other exact solutions of diffraction problems have been known for a long time in the form of eigenfunction expansions (partial wave series). A classic example is the Mie solution[77,68] for the scattering of a monochromatic plane wave by a homogeneous sphere. The trouble with such expansions is that they converge extremely slowly when the wavelength is much smaller than the dimensions of the diffracting object, as usually happens for visible light. The remedy is again found in analytic continuation: the complex angular momentum method was invented for this purpose.

The original Watson transformation had several shortcomings; it could only be applied in a few disjoint spatial regions. During the past few years, however, a modified version of the Watson transformation has been developed, allowing one to derive the asymptotic short-wavelength behavior of the partial wave amplitude in any region of space. We give only a very brief outline of the method, referring to previous reviews[78,79] for a more detailed survey.

The passage to the λ plane proceeds by first applying to the partial wave series the Poisson sum formula,

$$\sum_{\ell=0}^{\infty} \phi\left(\ell + \frac{1}{2}, \vec{r}\right) = \sum_{m=-\infty}^{\infty} (-)^m \int_0^{\infty} \phi(\lambda,\vec{r}) \exp(2im\pi\lambda)\, d\lambda \,. \tag{9.1}$$

Analiticity in λ is then employed to deform the path of integration in the λ plane. The freedom thus gained is the main advantage of the method.

Slow convergence of the left-hand side of (9.1) at short wavelengths means that significant contributions to the result are spread over many terms of the partial-wave series. The object of the path deformation on the right-hand side is to concentrate the dominant contributions in a relatively small number of "critical points" in the λ plane. This requires different path deformations in different spatial regions. When Regge poles are swept through, we also get their residue contributions besides the deformed path integrals.

Thus, one finds two types of critical points: saddle points (real and/or complex), which dominate the contributions to the deformed path integrals, and poles.

Real saddle points correspond to the usual geometric-optic rays; their contributions lead to the well-known WKB series. Complex saddle points are associated with complex rays, a kind of analytic continuation of real ones. Such rays occur in total reflection, where they describe the evanescent waves penetrating into the rarer medium; the quantum analogue is the barrier tunnelling effect. They also occur on the shadow side of caustics.[80]

Regge pole contributions (cf. Sect. 7) represent both the effects of resonances and of surface waves, i.e., "creeping modes", which can also be described in terms of diffracted rays.[80]

Near the boundary between spatial regions where different path deformations are required, several interesting diffraction effects occur. One of them, involving transitions between saddle points and poles, is a penumbra region, described by a generalization of Fock's theory of diffraction.[81] A collision between real saddle points, followed by their transformation into complex ones, corresponds to a rainbow.[79] Both complex saddle points and complex poles give the dominant contributions to the glory,[79] an effect that was first explained by complex angular momentum theory.

Diffraction, by definition, is the penetration of light into regions that are inaccessible to the real rays of geometrical optics. In terms of the Hamiltonian analogy between geometrical optics and classical mechanics, these are "classically forbidden" regions. Thus, in wave-mechanical terms, diffraction effects may be regarded as a kind of tunnelling through potential barriers. These barriers represent "inertial forces" related with the geometry of the diffracting objects;[82] e.g., in the angular momentum representation, we get the centrifugal barrier.

The strongest effects are found near the top of such a potential barrier, where it is most easily penetrable. In the Mie problem, this is the "edge domain", corresponding to rays near tangential incidence. The dominant contributions to the glory arise from this domain, where sur face waves are launched. It also yields significant contributions to the Mie efficiency factors;[83] such contributions represent an extension of the geometrical-optic ones to complex angles of incidence and refraction.

Diffraction effects may therefore be explained by a kind of analytic continuation of geometrical optics to complex values of ray parameters (complex trajectories). This is a vast new domain, which we are just beginning to explore.

REFERENCES

1. H. A. Kramers, Atti Congr. Intern. Fis. Como 2, 545 (1927); Phys. Z. 30, 522 (1929); R. de L. Kronig, J. Opt. Soc. Am. 12, 547 (1926); Ned. T. Natuurk. 9, 402 (1942).
2. H. M. Nussenzveig, Causality and Dispersion Relations (Academic, N.Y., 1972).
3. V. Vladimirov, Distributions en Physique Mathématique (Mir, Moscow, 1979), and references therein.
4. R. L. Weaver and Y.-H. Pao, to appear in J. Math. Phys.
5. L. D. Landau and E. M. Lifshitz, Statistical Physics (Pergamon, London, 1958), p. 394; Electrodynamics of Continuous Media (Pergamon, London, 1960), p. 256.
6. V. M. Agranovich and V. L. Ginzburg, Spatial Dispersion in Crystal Optics and the Theory of Excitons (Interscience, London, 1966), pp. 9,49.
7. M. Altarelli, D. L. Dexter, H. M. Nussenzveig and D. Y. Smith, Phys. Rev. B6, 4502 (1972).
8. P. C. Martin, Phys. Rev. 161, 143 (1967).
9. E. U. Condon, Rev. Mod. Phys. 9, 432 (1937); J. P. Mathieu, Handb. d. Physik 28 (Springer, Berlin, 1957), p. 333.
10. D. Y. Smith, J. Opt. Soc. Am. 66, 454 (1976).
11. M. T. C. S. Thomaz, Master's Thesis (Univ. of São Paulo, 1979); M. T. C. S. Thomaz and H. M. Nussenzveig, to be published; W. P. Healy, J. Phys. B9, 2499 (1976).
12. M. Altarelli, D. L. Dexter, H. M. Nussenzveig and D. Y. Smith, Phys. Rev. B6, 4502 (1972).
13. M. Altarelli and D. Y. Smith, Phys. Rev. B9, 1290 (1974).
14. D. Y. Smith, Phys. Rev. B13, 5303 (1976).
15. H. J. Hagemann, W. Gudat and C. Kunz, DESY SR-74/7 (Hamburg, 1974), J. Opt. Soc. Am. 65, 742 (1975).
16. B. W. Veal and A. P. Paulikas, Phys. Rev. B10, 1280 (1974).
17. T. Inagaki, E. T. Arakawa, R. N. Hamm and M. W. Williams, Phys. Rev. B15, 3243 (1977).

18. C. Kunz,, ed., Synchrotron Radiation, Techniques and Applications (Springer, Berlin, 1979).
19. Cf., e.g., H. M. Nussenzveig, Introduction to Quantum Optics (Gordon and Breach, N.Y., 1973).
20. Photon detectors that would operate by emission, leading to antinormal ordering, are conceivable in principle; cf. L. Mandel, Phys. Rev. 152, 438 (1966).
21. H. M. Nussenzveig, Proc. II Brazilian Symp. Theor. Phys. (Rio de Janeiro, 1969).
22. R. J. Glauber, Phys. Rev. 131, 2766 (1963).
23. V. Bargmann, Commun. Pure Appl. Math. 14, 187 (1961).
24. J. S. Toll, Phys. Rev. 104, 1760 (1956).
25. T. Inagaki, H. Kuwata and A. Ueda, Phys. Rev. B19, 2400 (1979).
26. E. Wolf, Proc. Phys. Soc. (London) 80, 1269 (1962); Y. Kano and E. Wolf, ibid. 80, 1273 (1962).
27. H. M. Nussenzveig, J. Math. Phys. 8, 561 (1967).
28. R. E. Burge, M. A. Fiddy, A. H. Greenaway and G.Ross, Proc. Roy. Soc. (London) A350, 191 (1976).
29. E. C. Titchmarsh, The Theory of Functions, 2nd ed. (Oxford, 1939).
30. E. N. Leith and J. Upatnieks, J. Opt. Soc. Am. 54 , 1295 (1964).
31. O. Bryngdahl and A. Lohmann, J. Opt. Soc. Am. 58 , 620 (1968).
32. H. Gamo, in Electromagnetic Theory and Antennas, ed. E. C.Jordan (Pergamon, Oxford 1963), p. 801.
33. C. L. Mehta, J. Opt. Soc. Am. 58, 1233 (1968).
34. T. D. Beard, Appl. Phys. Lett. 15, 227 (1969).
35. D. Kohler and L. Mandel, J. Opt. Soc. Am. 63, 126 (1973).
36. K. Chadan and P. S. Sabatier, Inverse Problems in Quantum Scattering Theory (Springer, N.Y., 1977).
37. P. C. Sabatier, ed., Applied Inverse Problems (Springer, N.Y., 1978).
38. H. P. Baltes, ed., Inverse Source Problems in Optics (Springer, N.Y., 1978).
39. C. J. Bouwkamp, in Rep. Progr. Phys. 17, 41 (1954).
40. J. R. Shewell and E. Wolf, J. Opt. Soc. Am. 58, 1596 (1968).
41. A. J. Devaney, J. Math. Phys. 19, 1526 (1978).
42. I. M. Gelfand and B. M. Levitan, Izv. Akad. Nauk SSSR, Ser. Mat. 15, 309 (1951) [Am. Math. Soc. Transl. (2) 1, 253 (1955).]
43. V. A. Marchenko, Doklad. Akad. Nauk SSSR, 72, 457 (1950); 104, 695 (1955).
44. R. Jost and W. Kohn, Kgl. Danske Vid. Selsk. Mat.-fys. Medd. 27, No. 9 (1953).
45. L. D. Faddeev, J. Math. Phys. 4, 72 (1963).
46. G. N. Balanis, J. Math. Phys. 13, 1001 (1972).

47. L. D. Faddeev, Trudy Mat. Inst. Steklov 73, 314 (1964) [Am. Math. Soc. Transl. 2, 139 (1965)].
48. P. Deift and E. Trubowitz, Commun. Pure Appl. Math. 32, 121 (1979).
49. R. G. Newton, J. Math. Phys. 21, 493 (1980).
50. L. D. Faddeev, Preprint Kiev ITP 71.106 E; J. Sov. Math. 5, 334 (1976).
51. R. G. Newton, in Scattering Theory in Mathematical Physics, eds. J. A. Lavita and J. P. Marchand (D. Reidel, Dordrecht, 1974), p. 193; Phys. Rev. Lett. 43, 541 (1979).
52. C. S. Gardner, J. M. Greene, M. D. Kruskal and R. M. Miura, Phys. Rev. Lett. 19, 1095 (1967).
53. A. C. Scott, F. Y. F. Chu and D. W. McLaughlin, Proc. IEEE 61, 1443 (1973).
54. M. J. Ablowitz, D. J. Kaup, A. C. Newell and H. Segur, Studies Appl. Math. 53, 249 (1974).
55. S. L. Mc Call and E. L. Hahn, Phys. Rev. 183, 457 (1969).
56. G. L. Lamb, Jr., Phys. Rev. Lett. 31, 196 (1973); Phys. Rev. A9, 422 (1974).
57. M. J. Ablowitz, D. J. Kaup and A. C. Newell, J. Math. Phys. 15, 1852 (1974); D. J. Kaup, Phys. Rev. A16, 704 (1977).
58. H. M. A. de Castro, Master's Thesis (Univ. of São Paulo, 1979); H. M. A. de Castro and H. M. Nussenzveig, to be published.
59. H. A. Haus, Rev. Mod. Phys. 51, 331 (1979).
60. V. E. Zakharov and A. L. Shabat, Sov. Phys. JETP 34, 62 (1972).
61. R. C. T. da Costa, J. Math. Phys. 11, 2799 (1970).
62. F. L. Ridener, Jr. and R. H. Good, Jr., Phys. Rev. B10, 4980 (1974), and references therein.
63. J. J. Thomson, Proc. London Math. Soc. 15 (1), 197 (1884).
64. G. Gamow, Z. Phys. 51, 204; 52, 510 (1928).
65. W. J. Romo, J. Math. Phys. 21, 311 (1980), and references therein.
66. B. J. Hoenders, in Coherence and Quantum Optics, eds. L. Mandel and E. Wolf (Plenum, N.Y., 1978), p. 221; J. Phys. A11, 1815 (1978).
67. E. Wolf and D. N. Pattanayak, in Symposia Mathematica, vol. 18 (Academic, N.Y., 1976); Opt. Commun. 6, 217 (1972).
68. Cf. N. A. Logan, Proc. IEEE 53, 773 (1965).
69. W. Franz, Theorie der Beugung Elektromagnetischer Wellen (Springer, Berlin, 1957).
70. E. P. Wigner, Phys. Rev. 98, 145 (1955).
71. L. Eisenbud, Ph.D. Thesis (Princeton Univ., 1948, unpublished).

72. M. Froissart, M. L. Goldberger and K. M. Watson, Phys. Rev. 131, 2820 (1963).
73. H. M. Nussenzveig, Phys. Rev. D6, 1534 (1972).
74. F. Goos and H. Hänchen, Ann. Physik (6) 1, 333 (1947); 5, 251 (1949); K. Artmann, ibid. 2, 87 (1948).
75. N. F. Ferrari Jr. and H. M. Nussenzveig, to be published.
76. A. Sommerfeld, Math. Ann. 47, 317 (1896).
77. G. Mie, Ann. Physik 25, 377 (1908).
78. H. M. Nussenzveig, in Methods and Problems of Theoretical Physics, ed. J. E. Bowcock (North-Holland, Amsterdam, 1970), p. 203.
79. H. M. Nussenzveig, J. Opt. Soc. Am. 69, 1068 (1979) and references therein.
80. J. B. Keller, in Calculus of Variations and its Applications, ed. L. M. Graves (McGraw-Hill, N.Y., 1958), p. 27.
81. V. A. Fock, Electromagnetic Diffraction and Propagation Problems (Pergamon, Oxford, 1965).
82. G. Beck and H. M. Nussenzveig, Nuovo Cimento 16, 416 (1960).
83. H. M. Nussenzveig and W. J. Wiscombe, to be published.

OPTICAL PHASE CONJUGATION FOR THE STUDY OF THE ANALYTIC PROPERTIES OF OPTICAL FIELDS

R.E. Burge, M.A. Fiddy, T.J. Hall and G. Ross

Physics Department, Queen Elizabeth College,
University of London, Campden Hill Road,
London W8 7AH, U.K.

ABSTRACT

The importance of the analytic character of optical fields is discussed, in particular the encoding of information by complex zeros. The properties of the real part of the complex field are outlined and its usefulness in determining the zero distribution for a field from intensity data is presented. The real part of an optical field may be formed by optical phase conjugation.

INTRODUCTION

In the study of the structure of materials by light (or X-ray, electron or neutron) scattering, let us assume that a Fourier transform relates the object wave, $f(\vec{t})$, which contains information about the structure of interest, and the scattered field, $F(\vec{x})$. We assume also that a Fourier transform relates the scattered field to the image. Since the object wave $f(\vec{t})$ is always confined to a finite volume it can be shown that $F(\vec{x})$ may be regarded as a function in real space which is entire and of exponential type (class E) in the associated complex space.[1,2] Thus, in one real dimension, we may write

$$F(x+iy)= \int_a^b f(t)\exp(i(x+iy)t)\,dt \qquad (1)$$

ISSN:0094-243X/81/650031-11$1.50

where $0 \leq |a| < b < \infty$

Being analytic, the real and imaginary parts of F(x) are related. Moreover, being of class E, F(x) is completely determined by the location of its complex zeros at the points z_n, i.e.

$$F(x+iy) = F(z) = F(0)e^{i(b+a)z} \prod_{n=-\infty}^{\infty} (1 - z/z_n) \qquad (2)$$

This Hadamard product representation of the field highlights the difficulties associated with inverse scattering problems in optics. When only intensity data are available, the intensity function has the zeros of F(z) and, in conjugate positions, those of $F^*(z^*)$; thus the zeros of F(x) cannot be separated from those of $F^*(x)$[3].

Methods for phase retrieval have been proposed but not all are based upon zero location (see for example[4,5]). However, the location of zeros does provide a logical solution to this problem and also offers a convenient means for encoding the complex scattered field in terms of geometrical measurements [6].

Of great interest recently [7,8,9], has been the use of nonlinear optical systems that create the conjugate wavefront of an arbitrary incident wave. One manifestation of this phenomenon is based on a phase-matched degenerate four wave mixing in a non linear medium. By interference of the phase conjugate wave, $F^*(x)$, with the input field, F(x), one can form the real or imaginary part of F(x). It will be shown in the following sections how measuring or observing $|\mathrm{Re}F(x)|$ enables one to determine the z_n of equation (2) and thus reconstruct F(x).

THE ANALYTIC OPTICAL FIELD

For convenience only the one dimensional situation will be considered but the conclusions reached are valid in general. As indicated in equations (2) and (3) the scattered field in an entire function of exponential type being characterised in the complex z plane by the location of its zeros. If in equation (1), $a \geq 0$, $F(x)$ is known as a causal transform [10] and $F(z)$ will be regular in the upper half of the complex plane (uhp). An equivalent statement to the latter is that the real and imaginary parts of $F(x)$ are related by a Hilbert transformation [11], i.e.

$$\mathrm{Re}F(x) = 1/\pi \, P\int_{-\infty}^{\infty} \frac{\mathrm{Im}F(x')\,dx'}{x'-x} \qquad (3)$$

where P denotes that the Cauchy principal value is taken. As the support in t space is moved, the asymptotic growth properties and domain of regularity of $F(z)$ are altered; if $b \leq 0$ $F(z)$ is regular in the lower half plane (lhp) and if $b = -a$, $F(z)$ is no longer regular in either half plane but increases exponentially. Despite changes in the direction of maximum growth by redefining the t origin, the distribution of zeros of $F(z)$ remains unchanged.

Clearly if we could measure $\mathrm{Re}F(x)$ rather than $|F(x)|^2$ there would be no difficulty in determining $f(t)$: one could apply the Hilbert relation directly. Unfortunately phase conjugation would provide only $|\mathrm{Re}F(x)|$ and it is not obvious that one can always specify the negative excursions of this function, but in some cases this may be possible. However, the modulus of the real part or the imaginary part of an analytic function does provide useful information. $\mathrm{Re}F(x)$ and $\mathrm{Im}F(x)$ are real functions with continous second partial derivatives which each satisfy Laplace's equation. Since $F(z)$ is analytic, $\mathrm{Re}F(x)$ and $\mathrm{Im}F(x)$ are termed conjugate harmonic functions[12]. From the theorem of maximum modulus[12], the function $|F(z)|$ can have no maxima, or minima other than zeros, within any region of analyticity of $F(z)$. It is easy to show that the same constraint must apply to a harmonic function: the maximum modulus of these functions necessarily lies on the boundary delimiting the region of analyticity. The zeros of $F(z)$ occur at the intersection of the level curves $\mathrm{Re}F(z) = 0$ and $\mathrm{Im}F(z) = 0$, and at a simple zero, it follows from the Cauchy-Riemann equations that these level curves cross at right angles. These properties lead to a theorem given by Titchmarsh[12], namely that if C is a simple closed contour inside and on which $F(z)$ is analytic, then if $\mathrm{Re}F(z)$ vanishes at 2k distinct points on C, $F(z)$ has at most k zeros inside C.

This information about the zeros of $F(z)$ is equally well provided by $|\mathrm{Re}F(z)|$ on C.

DETERMINING WHICH COMPLEX ZEROS ARE IN EACH HALF PLANE

If the scattered field $F(x)$ is a causal transform, then its continuation into the uhp will be regular, tending to zero as $|z| \to \infty$. The real and imaginary parts of such a function, having a constant $f(t)$, are shown in figure 1; there are two zero crossings of $\mathrm{Re}F(x)$ for each zero of $F(x)$. These real zeros are inside the contour C referred to in Titchmarsh's theorem, C being the real axis closed by a semicircle of infinite radius in the uhp. In figure 2 the change in the number of zero crossings of $|\mathrm{Re}F(x)|$ can be seen to confirm this theorem when the pair of zeros nearest to the z origin are either in the lhp (fig 2 a,b) or in the uhp (fig 2 c,d). The usual observable, namely $|F(x)|^2$ is also shown and is the same in each case since it has this pair of complex zeros in both half planes, symmetric about the x axis. The non zero minima of $|F(x)|^2$ which lie between real zeros of $|\mathrm{Re}F(x)|$ are associated with zeros of $F(z)$ belonging, in this case, to the uhp.

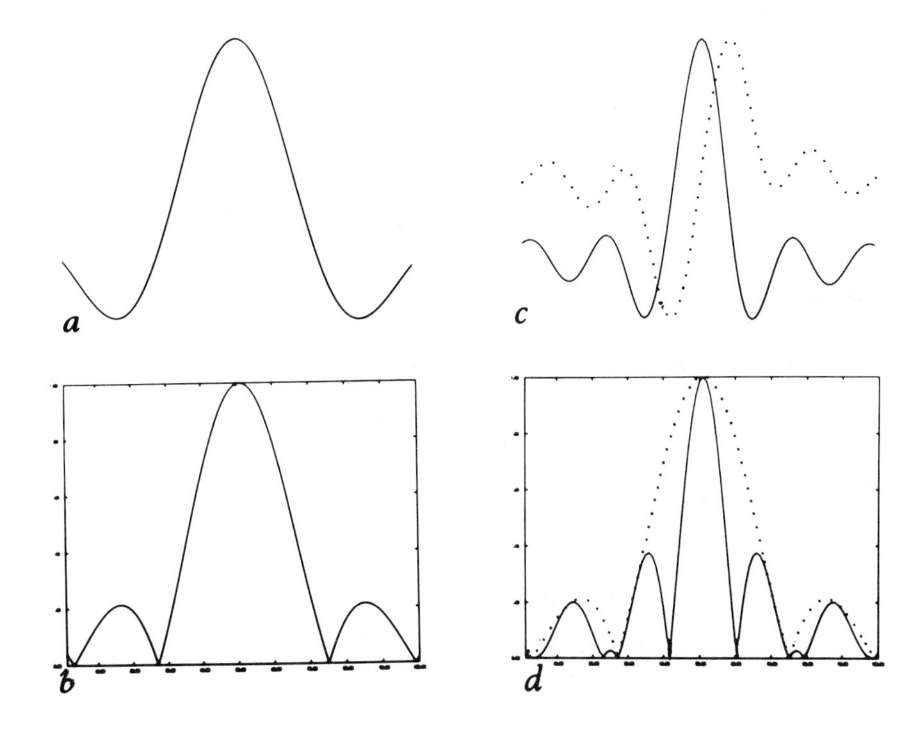

Figure 1

a) The central part of F(x) corresponding to an even constant f(t) and b) the modulus of ReF(x) (= |F(x)| in this case)

c) The complex F(x) corresponding to a causal f(t), (real (solid) and imaginary (dotted) parts are now related by H.T. to within an arbitrary constant)

d) |ReF(x)| from c) with |F(x)| represented by dotted curve.

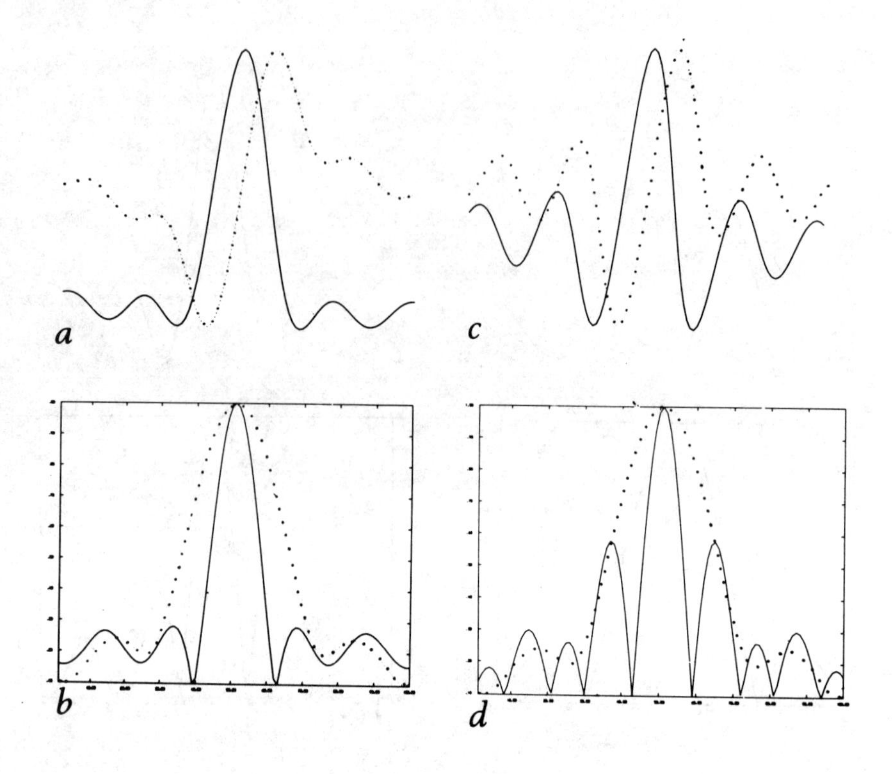

Figure 2

a) The central part of F(x), (real (solid) and imaginary (dotted) parts shown), corresponding to all zeros real and equidistant except for the pair nearest the origin which are moved into the lhp.

b) |ReF(x)| corresponding to a) with |F(x)| represented by dotted curve.

c) ,d) as for a), b) respectively but with zero pair moved into the uhp.

If one has located the co-ordinates of those complex zeros of $|F(x)|^2$ in, say, the uhp it now becomes possible to determine which of these zeros belongs to the lower half plane in order to encode $F(z)$ using equation (2). The positions of the zeros of $|F(x)|^2$ may be determined either optically[6] or digitally by double Fourier transformation and the multiplication in conjugate space by a variable exponential function as indicated by equation (1). Illustrations of $\mathrm{Re}F(z)$ and $\mathrm{Im}F(z)$ are given in Fig. 3.

It is interesting to note that monitoring the zero crossings of $|\mathrm{Re}F(x)|$ also enables an optimum reference wave to be chosen for a holographic experiment. The addition of a suitable reference function to $F(x)$ removes zeros from the uhp by implicity satisfying Rouché's theorem[3]. Only when all zeros are removed from one half plane will the corresponding $|F(x)|^2$ encode uniquely the phase of $F(x)$. A zero free half plane occurs when $|\mathrm{Re}F(x)| \neq 0$. Since $|\mathrm{Re}F(x)| = \mathrm{Re}F(x)$ in this case the Hilbert transform may be applied to yield $\mathrm{Im}F(x)$. Only under these conditions will replaying the hologram or applying a logarithmic Hilbert transform[3] yield the true phase distribution. By observing the zeros of $|\mathrm{Re}F(x)|$ and locating optically the zeros of $|F(z)|^2$ it is possible to determine, in principle, $F(x)$ without any measurement of intensity, performing on geometrical measurements.

OPTICAL PHASE CONJUGATION

The use of phase conjugation for phase retrieval has already been suggested [13]. The set up we envisage is shown schematically in figure 4 using DFWM in ruby[9] or BSO[14]. Two counter propagating pump beams assumed to be phase conjugates of each other with wave vectors $\pm \vec{k}_p$ interact with a weaker input field having wave vector $\vec{k}_F$. The output field propagates in direction $-\vec{k}_F$ having an amplitude proportional to the complex conjugate of the amplitude of the input wave. The input field in split by a high quality beam splitter onto the conjugating medium. The output field is then combined as shown with the input field in order to generate $\mathrm{Re}F(x)$.

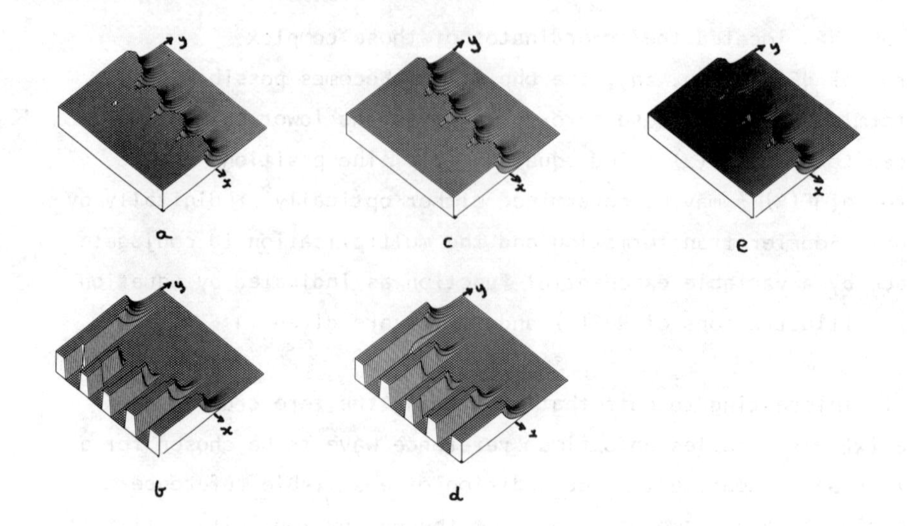

Figure 3

a) |ReF(x+iy)| corresponding to case a) of Figure 2.

c) |ReF(x+iy)| corresponding to case c) in Figure 2.

b) and d) show contours of |ReF(z)| = 0

e) |F(z)| corresponding to the cases in Figure 1.

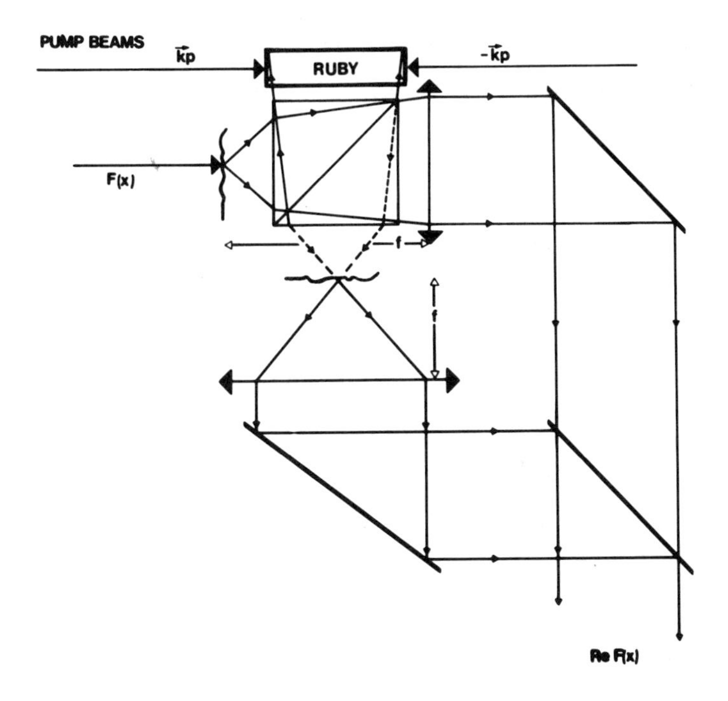

Figure 4

Schematic wave front conjugation set up to provide |ReF(x)|.

ACKNOWLEDGEMENTS

The authors are grateful to the Procurement Executive (U.K., M.O.D) and the Science Research Council (U.K.) for financial support for this work.

REFERENCES

1. R.E.A.C. Paley and N.Wiener,'Fourier transforms in the complex domain' Amer. Math. Soc. Colloq., Pub. XIX Providence, Rhode Island (1934)

2. B.A. Fuks, 'Introduction to the theory of analytic functions of several variables' Translations of Math. Monographs. 8 Amer. Math. Soc. (1963)

3. R.E. Burge, M.A. Fiddy, A.H. Greenaway and G. Ross, 'The phase problem' Proc. Roy. Soc. Lond. A.350 191 (1976)

4. H.A. Ferwerda, ch 2 in 'Inverse Source Problems in Optics' 9 of Topics in Current Physics, Ed. H.P. Baltes, Springer-Verlag, Heidelberg (1979)

5. G. Ross, M.A. Fiddy and M. Nieto-Vesperinas, ch. 2. in 'Inverse Scattering Problems in Optics' 20 of Topics in Current Physics, Ed. H.P. Baltes, Springer-Verlag, Heidelberg (1980)

6. M.A. Fiddy, G. Ross and M. Nieto-Vesperinas, 'Optical processing for the encoding of information in inverse problems' see this volume of proceedings 'Optics in Four Dimensions'

7. A. Yariv, 'Phase conjugate optics and real-time holography'. IEEE QE - 14 , 650 (1978)

8. P.F. Liao and D.M. Bloom, 'Continuous-wave backward-wave generation by DFWM' Opt. Lett. 3 4 (1978)

9. M.D. Levenson, 'High resolution imaging by wave-front conjugation'. Opt. Lett. 5 182 (1980)

10. J.S. Toll, 'Causality and dispersion relations: logical foundations', Phys. Rev. 104 1760 (1956)

11. E.C. Titchmarsh 'The Theory of the Fourier Integral' O.U.P. (1948)

12. E.C. Titchmarsh 'The Theory of Functions' O.U.P. (1939)

13. G.P. Agrawal 'Phase determination by conjugate wave-front generation' J.Opt. Soc. Amer. 68 1135 (1978)

14. J.P. Huignard, J.P. Herriau and G. Rivet 'Phase conjugation and spatial frequency dependence of wave-front reflectivity in $Bi_{12}SiO_{20}$ crystals'. Opt. Lett. 5 102 (1980)

A NEW DESCRIPTION OF SECOND-ORDER COHERENCE PHENOMENA IN THE SPACE-FREQUENCY DOMAIN*

E. Wolf
Department of Physics and Astronomy and The Institute of Optics
University of Rochester, Rochester, NY, 14627, USA

ABSTRACT

A new representation is obtained for the cross-spectral density function of stationary optical wavefields. We show that under very general conditions it is possible to construct an ensemble of monochromatic wave functions, all of the same frequency ω, with the property that the cross-spectral density function $W(\underline{r}_1,\underline{r}_2,\omega)$ of the field is expressible as a correlation between the values of the wave functions at the points $\underline{r}_1$ and $\underline{r}_2$. The representation is intimately related to a new mode expansion of the field and has a number of potential applications to various problems that involve wavefields of any state of coherence.

1. INTRODUCTION

The study of many coherence phenomena in stationary optical fields is based on the concept of correlation of the field vibrations at two space-time points. If $V(\underline{r},t)$ is the complex analytic signal representation[1] of the field at a point specified by the position vector $\underline{r}$, at time t, the correlation between the field vibrations at space-time points $(\underline{r}_1,t)$ and $(\underline{r}_2,t+\tau)$ is characterized by the mutual coherence function

$$\Gamma(\underline{r}_1,\underline{r}_2,\tau) = \langle V^*(\underline{r}_1,t)V(\underline{r}_2,t+\tau)\rangle_t . \tag{1.1}$$

In (1.1) the asterisk denotes the complex conjugate and the sharp brackets denote an ensemble average.[2] We will be dealing with two ensembles, one consisting of time-dependent sample functions, the other consisting of sample functions that are frequency-dependent. For the sake of clarity we distinguish averaging over the respective ensembles by subscripts of t or ω, attached to the sharp brackets.

Under usual circumstances the mutual coherence function may be assumed to be square-integrable with respect to τ, so that it will possess a Fourier frequency representation

$$\Gamma(\underline{r}_1,\underline{r}_2,\tau) = \int_0^\infty W(\underline{r}_1,\underline{r}_2,\omega)e^{-i\omega\tau}d\omega, \tag{1.2a}$$

and the Fourier inversion formula

$$W(\underline{r}_1,\underline{r}_2,\omega) = \frac{1}{2\pi}\int_{-\infty}^{\infty} \Gamma(\underline{r}_1,\underline{r}_2,\tau)e^{i\omega\tau}d\tau \tag{1.2b}$$

ISSN:0094-243X/81/650042-07$1.50

will then hold. The function $W(\underline{r}_1,\underline{r}_2,\omega)$ is the cross-spectral density of the field and we will assume it to be a continuous function of ω. In the special case when the two points $\underline{r}_1$ and $\underline{r}_2$ coincide (in which case we drop the common subscript), Eq. (1.2b) reduces to

$$W(\underline{r},\underline{r},\omega) = \frac{1}{2\pi}\int_{-\infty}^{\infty} \Gamma(\underline{r},\underline{r},\tau)e^{i\omega\tau}d\tau. \qquad (1.2c)$$

Within the accuracy of the scalar model that we use, $W(\underline{r},\underline{r},\omega)$ represents the spectrum of the light at the point $\underline{r}$.

Although mathematically rigorous, the introduction of the spectrum of light via the formula (1.2c) is not entirely satisfactory from a physicist's point of view. It expresses the spectrum via a correlation in the space-time domain $(\underline{r},t)$ rather than via a field representation in the space-frequency domain $(\underline{r},\omega)$, to which it seems to be more intimately related. Several different mathematical techniques actually do exist that justify - via a space-frequency description of the field - the identification of $W(\underline{r},\underline{r},\omega)$ with the spectrum of the light; however, all of them require a considerable degree of mathematical sophistication. In fact a classic solution to this problem, due to Norbert Wiener[3], necessitated the development of a new branch of mathematics - the generalized harmonic analysis. Alternative approaches employ the Fourier-Stieltjes integral[4] and the theory of generalized functions.[5]

The difficulty that one is faced with when one attempts to introduce the spectrum of light via a space-frequency representation of the field is the following: the probability distributions that govern the statistical properties of a stationary random process are invariant with respect to the translation of the origin of time and hence the sample functions of such a process cannot tend to zero as the time $t \to \pm\infty$. This implies that the sample functions $V(\underline{r},t)$ of a stationary optical field do not possess a Fourier frequency transform with respect to the temporal variable.

There are numerous problems in optics for which a space-frequency description of the field in terms of ordinary functions would nevertheless appear to be natural and also highly desirable. Such problems are encountered, for example, in the analysis of speckle patterns, in the theory of laser modes, in problems related to the propagation of laser light through the atmosphere and in spectral radiometry with partially coherent sources.

In the present paper, we show that a mathematically rigorous space-frequency description of stationary optical fields within the framework of ordinary function theory is possible, which provides a representation of the cross-spectral density and of the ordinary spectral density (the spectrum of the light) as averages, taken over an ensemble of realizations in the space-frequency domain. We will also show that this description is closely related to an interesting new mode representation of stationary optical fields.

2. THE CROSS-SPECTRAL DENSITY FUNCTION AS A CORRELATION FUNCTION IN THE SPACE-FREQUENCY DOMAIN

Let us consider a stationary optical field $V(\underline{r},t)$ in some finite region D of space and let $\Gamma(\underline{r}_1,\underline{r}_2,\tau)$ be its mutual coherence function, defined by Eq. (1.1). We assume that $\Gamma(\underline{r}_1,\underline{r}_2,\tau)$ admits the Fourier frequency representation (1.2a), that the inversion formula (1.2b) also applies, and that the cross-spectral density $W(\underline{r}_1,\underline{r}_2,\omega)$ is a continuous function[6] of ω. If the total energy of the field in the volume D is assumed to be finite, the cross-spectral density may be shown to be square integrable with respect to each of the two spatial variables, and hence

$$\iint_{DD} |W(\underline{r}_1,\underline{r}_2,\omega)|^2 d^3r_1 d^3r_2 < \infty. \tag{2.1}$$

It follows from the definition (1.2b) and from the properties of the mutual coherence function, defined by Eq. (1.1), that the cross-spectral density satisfies the requirements

$$W(\underline{r}_2,\underline{r}_1,\omega) = W^*(\underline{r}_1,\underline{r}_2,\omega) \tag{2.2}$$

and that[7]

$$\iint_{DD} W(\underline{r}_1,\underline{r}_2,\omega)f^*(\underline{r}_1)f(\underline{r}_2)d^3r_1 d^3r_2 \geqslant 0, \tag{2.3}$$

where $f(\underline{r})$ is any function that is square-integrable in D. The formulae (2.1) - (2.3) imply that $W(\underline{r}_1,\underline{r}_2,\omega)$ is a Hermitian, non-negative definite Hilbert-Schmidt kernel and hence by Mercer's theorem[8] admits a uniformly and absolutely convergent expansion

$$W(\underline{r}_1,\underline{r}_2,\omega) = \sum_n \lambda_n(\omega)\psi_n^*(\underline{r}_1,\omega)\psi_n(\underline{r}_2,\omega), \tag{2.4}$$

where the ψ_n's are the eigenfunctions and the λ_n's are the eigenvalues of the Fredholm integral equation

$$\int_D W(\underline{r}_1,\underline{r}_2,\omega)\psi_n(\underline{r}_1,\omega)d^3r_1 = \lambda_n(\omega)\psi_n(\underline{r}_2,\omega). \tag{2.5}$$

Let us now introduce an ensemble of functions $\{U(\underline{r},\omega)\}$, each member of which is a linear combination of the eigenfunctions $\psi_n(\underline{r},\omega)$,

$$U(\underline{r},\omega) = \sum_n a_n(\omega)\psi_n(\underline{r},\omega), \tag{2.6}$$

with random coefficients $a_n(\omega)$, whose properties we will specify shortly. It follows from Eq. (2.6) that

$$\langle U^*(\underline{r}_1,\omega)U(\underline{r}_2,\omega)\rangle_\omega = \sum_n \sum_m \langle a_n^*(\omega)a_m(\omega)\rangle_\omega \psi_n^*(\underline{r}_1,\omega)\psi_m(\underline{r}_2,\omega), \tag{2.7}$$

where the sharp brackets denote the average over the ensemble of the U's. We attached the suffix ω to the sharp brackets to stress that the averaging is now over an ensemble of frequency-dependent sample functions. If we choose the random coefficients $a_n(\omega)$ so that they satisfy the requirement

$$<a_n^*(\omega)a_m(\omega)>_\omega \;=\; \lambda_n(\omega)\delta_{nm}, \tag{2.8}$$

where δ_{nm} is the Kronecker symbol, then Eq. (2.7) reduces to

$$<U^*(\underline{r}_1,\omega)U(\underline{r}_2,\omega)>_\omega \;=\; \sum_n \lambda_n(\omega)\psi_n^*(\underline{r}_1,\omega)\psi_n(\underline{r}_2,\omega). \tag{2.9}$$

Comparison of Eq. (2.9) with Eq. (2.4) shows that we have succeeded in constructing an ensemble of sample functions $U(\underline{r},\omega)$ such that

$$W(\underline{r}_1,\underline{r}_2,\omega) \;=\; <U^*(\underline{r}_1,\omega)U(\underline{r}_2,\omega)>_\omega\,, \tag{2.10}$$

i.e., such that <u>the cross-spectral density function of the field is expressed as a correlation function in the space-frequency domain.</u>

Before proceeding further we might mention that the above analysis has much in common with that encountered in connection with the well-known Karhunen-Loéve orthogonal expansion[9] of a random process. However, our expansion leads to a representation of the cross-spectral density of a stationary field, rather than of the autocorrelation of a random process, in terms of a set of orthogonal functions.

3. PHYSICAL SIGNIFICANCE OF THE NEW REPRESENTATIONS (2.4) AND (2.10) OF THE CROSS-SPECTRAL DENSITY

The new representations (2.4) and (2.10) for the cross-spectral density have interesting interpretations. To see this, let us recall that in free space the cross-spectral density satisfies the two Helmholtz equations[10]

$$\nabla_1^2\, W(\underline{r}_1,\underline{r}_2,\omega) + k^2 W(\underline{r}_1,\underline{r}_2,\omega) \;=\; 0, \tag{3.1a}$$

$$\nabla_2^2\, W(\underline{r}_1,\underline{r}_2,\omega) + k^2 W(\underline{r}_1,\underline{r}_2,\omega) \;=\; 0. \tag{3.1b}$$

Here ∇_1^2 and ∇_2^2 represent the Laplacian operator acting with respect to the coordinates of $\underline{r}_1$ and $\underline{r}_2$, respectively, and

$$k \;=\; \omega/c \tag{3.2}$$

is the wave number associated with the frequency ω, c being the speed of light in vacuo.

If we substitute into Eqs. (3.1) from (2.4) and make use of the fact that the eigenfunctions $\psi_n(\underline{r},\omega)$ can be taken to be mutually orthogonal, one readily finds that each eigenfunction also satisfies the Helmholtz equation, i.e., that

$$\nabla^2\psi_n(\underline{r},\omega) + k^2\psi_n(\underline{r},\omega) \;=\; 0. \tag{3.3}$$

Hence, if we rewrite Eq. (2.4) in the form

$$W(\underline{r}_1,\underline{r}_2,\omega) = \sum_n \lambda_n(\omega) W_n(\underline{r}_1,\underline{r}_2,\omega), \tag{3.4}$$

where

$$W_n(\underline{r}_1,\underline{r}_2,\omega) = \psi_n^*(\underline{r}_1,\omega)\psi_n(\underline{r}_2,\omega), \tag{3.5}$$

we find at once that the cross-spectral density $W(\underline{r}_1,\underline{r}_2,\omega)$ is expressed as a linear combination of cross-spectral densities $W_n(\underline{r}_1,\underline{r}_2,\omega)$ of wave fields all of which obey the same equations, namely (3.1), as does the cross-spectral density of the total field. Thus our new expansion (3.4) is a mode representation of the cross-spectral density $W(\underline{r}_1,\underline{r}_2,\omega)$. Moreover, since $W_n(\underline{r}_1,\underline{r}_2,\omega)$ is, according to (3.4), a product of a function of $\underline{r}_1$ and a function of $\underline{r}_2$, each mode represents a wavefield that is spatially completely coherent.[11]

Let us now consider the other new representation, (2.10), of the cross-spectral density. It follows from Eqs. (2.6) and (3.3) that each member of the ensemble of $U(\underline{r},\omega)$ satisfies the Helmholtz equation with the wave number k, viz.,

$$\nabla^2 U(\underline{r},\omega) + k^2 U(\underline{r},\omega) = 0. \tag{3.6}$$

Hence our formula (2.10) expresses the cross-spectral density as a correlation between the values of the monochromatic wave functions, all of the same frequency ω, at the points $\underline{r}_1$ and $\underline{r}_2$. This result provides a rigorous mathematical framework for the treatment of many problems of statistical optics, some of which were briefly mentioned in the introduction. In particular, the present approach can be shown to resolve certain conceptual difficulties that arise in the usual treatments of speckle patterns.[12] We might also mention that this new approach was recently used to provide a new mode representation of partially coherent sources and of the fields that they generate.[13]

4. DISCUSSION

We will now briefly compare the approach outlined in this paper with the usual one.

In the usual formulation, one starts with an ensemble $\{V(\underline{r},t)\}$ of time-dependent realizations of the stationary random field that satisfies the free-space wave equation

$$\nabla^2 V = \frac{1}{c^2}\frac{\partial^2 V}{\partial t^2}, \tag{4.1}$$

and one defines, via the average (1.1), the mutual coherence function $\Gamma(\underline{r}_1,\underline{r}_2,\tau)$. The cross-spectral density $W(\underline{r}_1,\underline{r}_2,\omega)$ is then obtained as its Fourier transform [Eq. (1.2b)]. These steps are indicated by arrows on the diagram (a) in Table I. A step from $V(\underline{r},t)$ to a frequency representation of the field [indicated symbolically as $\hat{V}(\underline{r},\omega)$] is not possible within the framework of ordinary function theory, for reasons that were explained earlier.

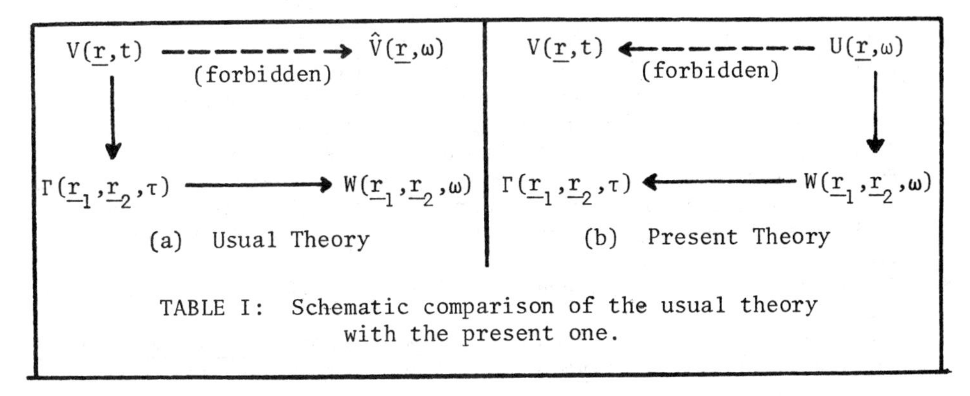

TABLE I: Schematic comparison of the usual theory with the present one.

In the present theory we deal with an ensemble $\{U(\underline{r},\omega)\}$ of the time-independent parts of monochromatic wave functions $U(\underline{r},\omega)\exp(-i\omega t)$. All of them are of the same frequency ω, and satisfy the Helmholtz Eq. (3.6). The cross-spectral density $W(\underline{r}_1,\underline{r}_2,\omega)$ is then obtained via the average indicated by Eq. (2.10). The mutual coherence function can then be obtained as its Fourier transform [Eq. (1.2a)]. These steps are indicated by arrows on the diagram (b) in Table I. In this theory a step from $U(\underline{r},\omega)$ to $V(\underline{r},t)$ is not possible via an ordinary Fourier transform because, as we noted earlier, the sample functions of a stationary random process do not admit a Fourier integral representation within the framework of ordinary function theory.

Apart from introducing greater symmetry between the space-time and space-frequency descriptions, the new theory is likely to provide new insights into the properties of stationary optical wavefields of any state of coherence.

Finally we mention that although the present paper deals only with second-order properties, the theory can readily be extended to provide space-frequency representations (in terms of ensembles of well-behaved monochromatic wavefields) of cross-spectral densities of all orders. This extension, as well as various implications of our analysis, will be described in other publications.

REFERENCES AND FOOTNOTES

*Research supported by the U. S. Army Research Office.

1. For the definition of the complex analytic signal, as well as for an account of some of the other concepts of optical coherence theory that are used in the present paper, see, for example, M. Born and E. Wolf, Principles of Optics, [Pergamon Press, Oxford and New York, 5th ed., 1975] or L. Mandel and E. Wolf, Rev. Mod. Phys., 37, 231 (1965).
2. We implicitly assume that the ensemble is ergodic, so that the ensemble average in Eq. (1.1) is equal to the corresponding time average.
3. N. Wiener, Acta Math., 55, 117 (1930).

4. A.M. Yaglom, An Introduction to the Theory of Stationary Random Functions (Prentice-Hall, Englewood Cliffs, NJ, 1962), Secs. 9 and 10.
5. J.L. Lumley, Stochastic Tools in Turbulence (Academic Press, New York, 1970), Sec. 3.12 [see also Sec. 3.15].
6. It is not difficult to show that all these requirements follow from the single assumption that the mutual coherence function $\Gamma(\underline{r}_1,\underline{r}_2,\tau)$ is absolutely integrable with respect to τ.
7. The requirement (2.3) can be established by a similar argument as used in the derivation of Bochner's theorem [S. Bochner, Lectures on Fourier Integrals (Princeton University Press, Princeton, NJ, 1959), p. 325 et seq.
8. F. Riesz and B.Sz.-Nagy, Functional Analysis (F. Ungar, New York, 1955), Secs. 97 and 98.
9. See, for example, W.B. Davenport and W.L. Root, An Introduction to the Theory of Random Signals and Noise (McGraw Hill, New York, 1958), p. 96 et seq.
10. E. Wolf, J. Opt. Soc. Amer., 68, 6 (1978), Eqs. (5.3).
11. The fact that such a factorization implies complete spatial coherence (at frequency ω) can be seen at once by considering the degree of spatial coherence $\mu(\underline{r}_1,\underline{r}_2,\omega)$ of a mode W_n [cf. L. Mandel and E. Wolf, J. Opt. Soc. Amer., 66, 529 (1976), Eq. (2.10)]. One then finds that for all points $\underline{r}_1$ and $\underline{r}_2$ in the domain D that contains the field, $|\mu(\underline{r}_1,\underline{r}_2,\omega)|$ has the value unity for each mode W_n; this is precisely the condition for complete spatial coherence of W_n.
12. J.W. Goodman, "Role of Coherence Concepts in the Study of Speckle" in Applications of Optical Coherence, Proc. Soc. Photo-Opt. Instr. Eng., ed. W.H. Carter, 194, 86-94 (1979).
13. E. Wolf, submitted to Phys. Rev. Lett.

CONJUGATION OF SPACE AND TIME VARIABLES IN OPTICS

J.Ch. Viénot, R. Ferrière, J.P. Goedgebuer
Lab. d'Optique, Ass. CNRS, U. de Franche-Comté, 25030 Besançon, France

ABSTRACT

A review of the main landmarks which proceed the fundamentals of space-time optics is given. It includes a comparison of the roles of space and time variables in optics as far as information transfer is concerned.

Illustrations deal with interference and diffraction phenomena, the light beam being considered as a carrier whose capacity is linked to the width of the spectral band. Experimental examples are reported in metrology, information processing, spectral analysis and wavelength coding, holography, transmission by spectral modulation, etc.

INTRODUCTION

A representation of a 4D optical signal at time t is given in Fig. 1 ($\vec{r}$ stands for the three geometrical dimensions of a body at rest). The wavefront W describes the state of vibration of the disturbance propagating along the radius $\vec{r}$. It is a function of the passing time t. In a way the curve drawn along the direction of propagation is also a representation of the disturbance. If the object is time-varying its three dimensions are functions of time t individually. In short, the three spatial variables x, y and z are linked to

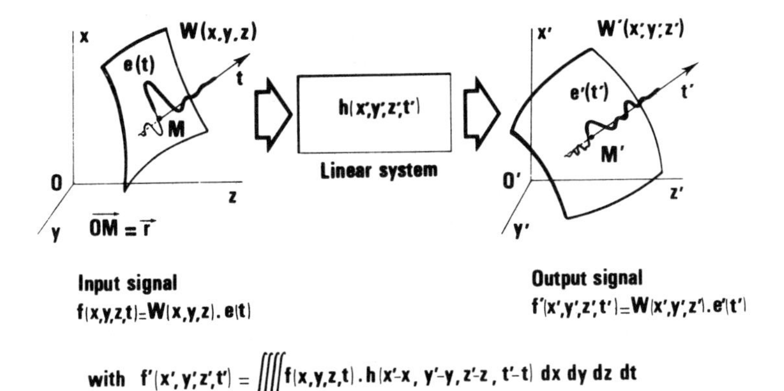

Fig. 1. Input/output approach of a 4D system. The transfer process concerns the three geometrical variables x, y, z and time t.

ISSN:0094-243X/81/650049-14$1.50

the geometrical structure of the wavefront $W(x,y,z)$ of the optical signal. The fourth degree of freedom corresponding to the temporal structure of the signal, i.e. to the carrier, and to the possible change of the signal with time, is expressed through the fourth variable t. Both following cases ought to be considered : (i) stable source and time-varying object, (ii) time-varying source and stable object. In this context the general relationship holds between the input and output signals as transfer is assumed through a linear system characterized by a space-time impulse response $h(x',y',z',t')$. The latter describes a spatial and chromatic (or temporal) filtering in which the reciprocal variables of x,y,z,t are u,v,σ,ν respectively (σ: wavenumber, ν: temporal frequency). This is not so awkward as it seems. In conventional optics, with coherent illumination, the object is often regarded as the signal itself. In the description of a 3D object, z is an implicit function of x and y. Also, since information is transmitted by means of light, there is a coding of the longitudinal dimension as a function of time through the relation $z = ct$. Would it be sensible to say that the situation is quite different if the signal is taken in the most general case ? The longitudinal structure of the signal, function of z, also depends on the chromatic components of wavelengths λ in the signal, which is in turn a function of the four variables x,y,z,t. As an illustration, if the transmission system is dispersive one observes a modification of the longitudinal profile of the signal, that can be interpreted as a chromatic filtering, function of time; the analysis of it would proceed with the determination of a "local" spectrum. In the following we are going to play with the four dimensions defined so far, without considering the state of polarization which represents a fifth degree of freedom. Fig. 2 reminds us that if the information carrier is monochromatic,

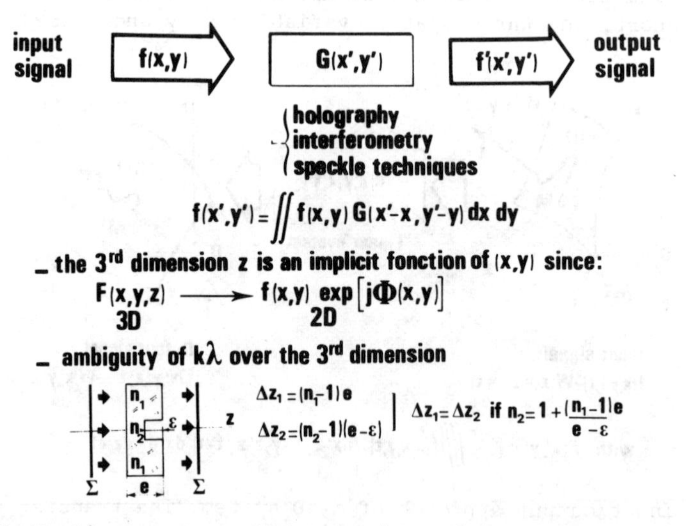

Fig. 2. Particular case of the previous figure : large observation time, sinusoidal carrier (conventional spatial optics).

one deals with spatial domain only. Such a scheme is quite valid for holography, interferometry and speckle techniques. In the holographic image the function $z(x,y)$ that describes the outer shape of the 3D reconstructed image is included in the phase term of the wavefront scattered by the object. The 3D appearance and parallax effects are linked to the scanning of the reconstructed scene by the observer who moves around. Double exposure holographic interferometry provides means of freezing the fourth dimension in the sense that the evolutive waves recorded at times t_1 and t_2 are made to interfere when reconstructed : so beside the 3D aspect, information is given along its temporal evolution (e.g. a displacement, unfortunately given with an ambiguity of $k\lambda$). In real-time holographic interferometry, changes in the object are observed continuously. In speckle techniques a well-known manner to remove the ambiguity is to use polychromatic speckle. The correlation between various monochromatic patterns allows absolute measurements. Despite the previous examples, from the general point of view of the experimentalist, assessing the four degrees of freedom simultaneously is not obvious. Following the access to the first two variables, a coding or any other trick is necessary to reach the third and the fourth ones.

RE-INTRODUCTION OF WHITE LIGHT, CORNER-STONE OF MODERN OPTICS

There was a long search for generating monochromatic light that simplifies the analysis of phenomena by eliminating time. Working with coherent beams seemed easier. The narrow frequency band of laser light is perfectly defined but very soon it was felt that the capacity of transmission of a system could be increased by a huge amount (several orders of magnitude) if the carrier was white light. Each spectral element behaves as an individual channel and the overall system becomes a multiplexing processor when the power-spectra, modulated by the information to be processed, are suitably analyzed.

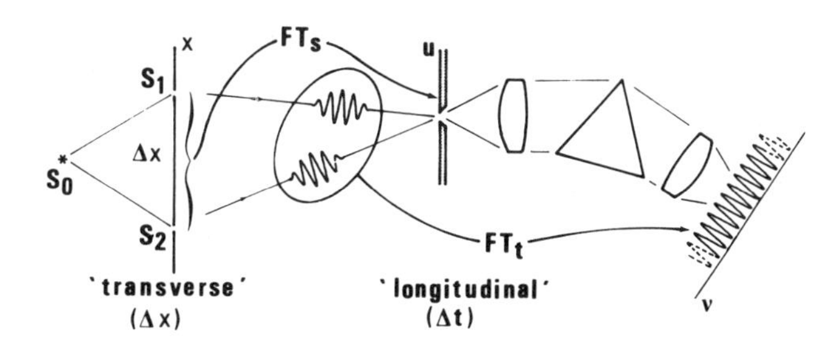

Fig. 3. Simple device for producing a channelled spectrum. It shows the two Fourier Transforms (Young's slits / interference patterns, entrance / exit of the spectroscope).

The principle is given in Fig. 3 above; between the transverse pupil S_1S_2 (x-dimension) and the interference plane (u-dimension) is a spatial Fourier Transform, x and u being reciprocal. The x-plane being now illuminated with white light, the entrance slit of the spectroscope can be set in a region where the path-difference between the two longitudinal signals is $(2k+1)\lambda/2$: the intensity is then minimum for this wavelength. The analysis performed by the spectroscope is a temporal Fourier Transform applied to the longitudinal information at the input, i.e. a function of time. The result of the analysis of the coherent superposition of the temporal spectra is a channelled spectrum in which chromatic frequencies have been filtered out (¶). It can be noted that one of the components being considered as the reference for the other, the pattern must behave as a hologram in the usual meaning. Precisely it is a temporal Fourier hologram lying along a ν-axis (replacing the u-axis for spatial Fourier holograms).

Another type of filtering is given in various works sketched in Fig. 4. The carrier is the chromatic spectrum modulated by the message when passing through it. This message plays role of a geometrical amplitude filter. The spectroscopes at the end of the chain perform the analysis of the power-spectrum of the light filtered in the (x,σ) plane (¶¶). The information being chromatically coded, it can be transmitted at a distance without being disturbed by any accident along the path. In the lower device the spectral combination inserted in front of the fiber allows gathering all the spectral components and the system works then in 2D (instead of 1D as the previous ones).

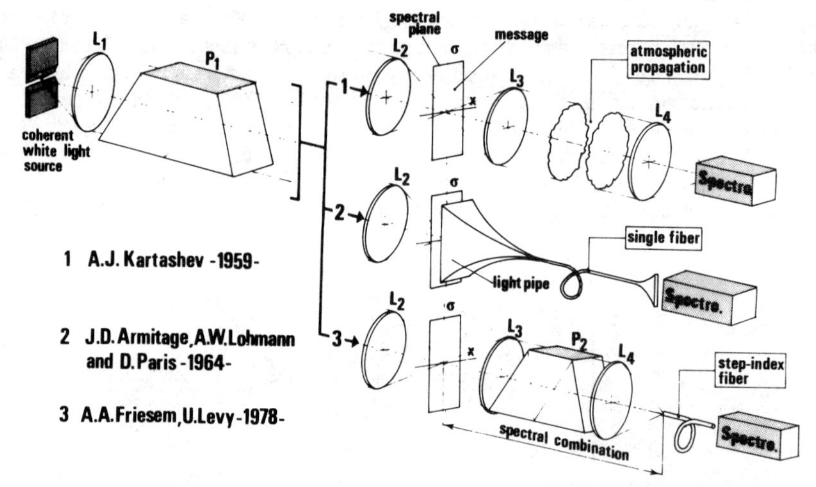

Fig. 4. Examples of passive chromatic filtering applied to communications.

(¶) Active filtering since it results from an interference process linked to the complex information inserted along the path of the carrier.
(¶¶) This is a passive filtering.

Once more let's come back to fundamentals, keeping in mind the essential aspect of coding with its two possibilities, spatial and temporal, that expresses similar filtering properties (Fig. 5). In both situations one needs the help of a modulation (the spectrum of a single impulse is a constant). The spectrum resulting from the addition of a reference to the first impulse is a sinusoidal signal. There is a slight difference between the roles of space and time variables, due to the causality principle, since there is no symmetry for the time domain. Moreover the Fourier spectrum of any function of time is necessarily limited within a band of the positive part of the frequency axis. A concrete case is shown in Fig. 6 through the determination of the shape of a wave-group of white light (top and left center diagrams) where $\Delta\lambda$ = 700nm - 400nm = 300nm, or, in ν, $\Delta\nu$ = 747THz - 428THz = 320THz, centered at λ_0 = 550nm, or ν_0 = 598THz. As the emission of a set of impulses occurs at random the result is given by the convolution of the wave-group with noise *n(t)*.

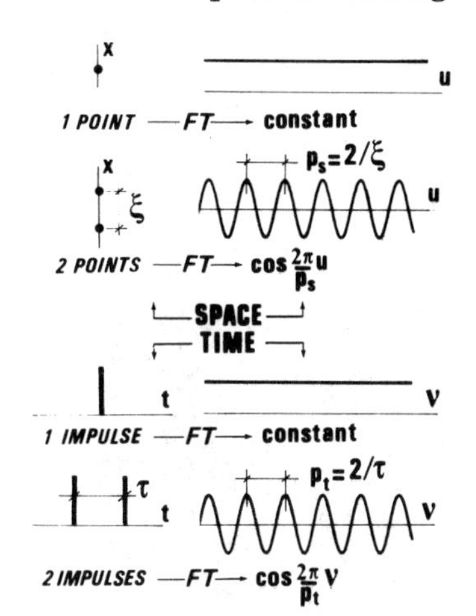

Fig. 5. Elementary approach to compare frequencies ν, u.

Further arguments for comparison between space and time variables in optics concern Abbe's theory of optical imaging applied to time signals, the transmitting system being assumed to be linear, time-invariant (see ref. 18 and 26). The diffracted amplitude in the far-field can be derived from the Fresnel-Kirchhoff integral, and a consequence is illustrated in Fig. 7 : the time response attached to a given geometrical pupil is the convolution of the temporal input function with the time derivative of the projection of the pupil along the direction of diffraction.

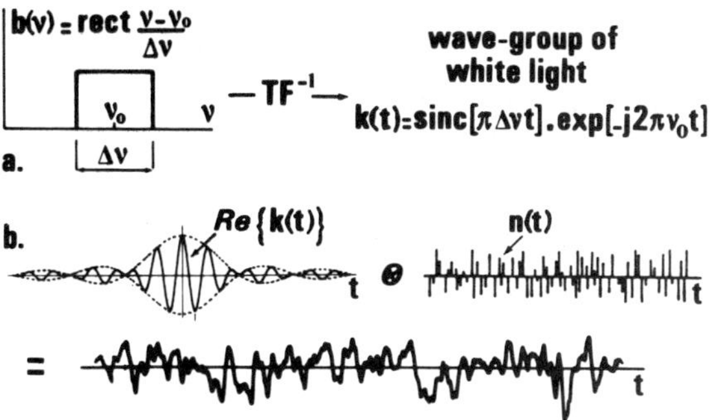

Fig. 6. Representation of white light.

Fig. 7. →

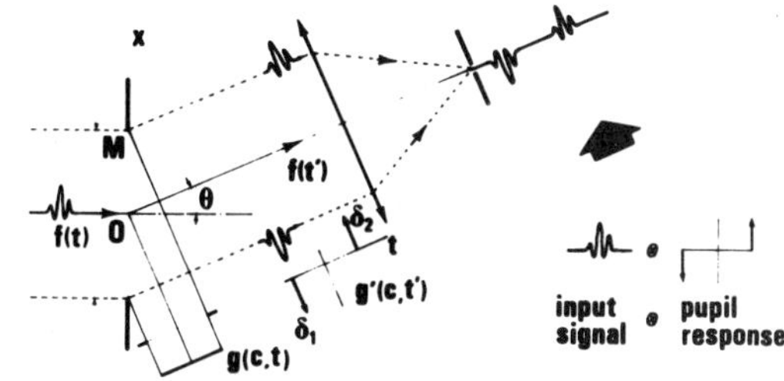

An example of practical application of space-time optics is dimensional metrology in white light. As seen before the uncertainty about the phase of $2k\pi$ is overcome by using white light, the achromatic fringe standing as a reference. Since a channelled spectrum behaves as a hologram does, spectral analysis makes the transposition from the non-accessible longitudinal dimension to a transverse one which is coded with temporal frequencies. Suppose then that a part of one mirror of a two-beam interferometer be a rough surface, the other mirror being of good quality. At the output of a spectroscope, set in line with the interferometer, a channelled spectrum is observed (Fig. 8 a).

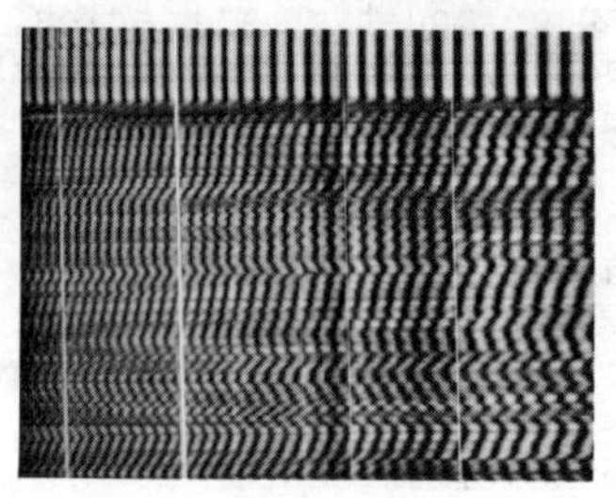

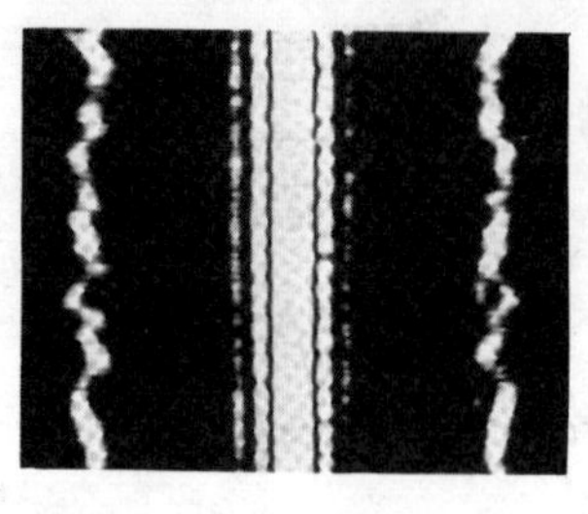

Fig. 8 (a, left). Channelled spectrum obtained from a rough surface, the upper part corresponds to a region where the surface is optically plane; (b, right) reconstructed profile of the surface.

The spacing of the fringes is reciprocal to the time delays separating the various local longitudinal signals. Such delays are then determined either by measuring the fringe spacing, or through a diffraction process as explained in Fig. 9. The distance from the diffracted terms ± 1 to the zero (or central term) is linearly related to the local path difference and the overall picture, like that shown in Fig. 8 b), is the actual image of the profile along one direction.

Direct access to $\Delta(t)$ through channelled spectra

$|F'(\nu)|^2$ spectrogram = channelled spectrum

temporal frequency domain (transverse)

DIRECT ACCESS TO $\Delta(t)$:

- by measuring $\Delta\nu = 1/\Delta(t)$
- by diffraction i.e. Fourier analysis of the spectrogram

SPECTRO $\mathcal{F}_{t\to\nu}$

temporal domain (longitudinal) NO DIRECT ACCESS TO $\Delta(t)$

$z = ct \longrightarrow \sigma = \frac{\nu}{c}$ $(\mathcal{F}_{z\to\sigma})$

$|F'(\sigma)|^2 = |F(\sigma)|^2 [1 + \cos 2\pi\sigma\Delta(z)]$

σ — also geometrical transverse coordinate (ξ)

$\Delta\sigma = \frac{1}{\Delta(z)}$ DIFFRACTION

Fig. 9

CHROMATIC CODING OF 3D METROLOGICAL INFORMATION

In Fig. 10 (see next page) a 3D object, transparent or not, is set along one arm of a two-beam interferometer illuminated in white light. For any λ it is always possible to find couples of homologous "slices" or corresponding scattered wavelets, stretched along x-direction, that interfere in a given region conjugated of the object through the lens L. The energy in the interferogram is the product of three terms as shown in the figure.

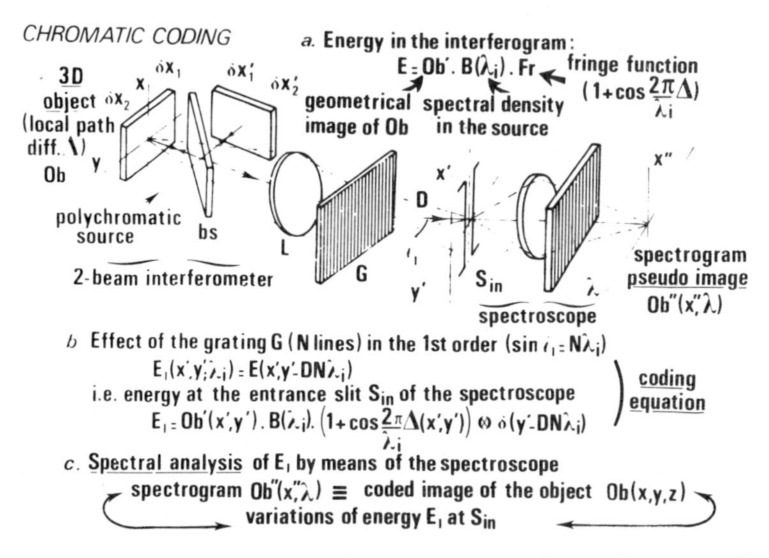

Fig. 10. Landmarks for working out chromatic coding relations.

Now the grating G, placed behind the lens, dispatches various monochromatic patterns. The power-spectrum displayed at the output of the spectroscope "sees" every slice of the object chromatically coded. The power-spectrum pattern is a 2D image of the object striped

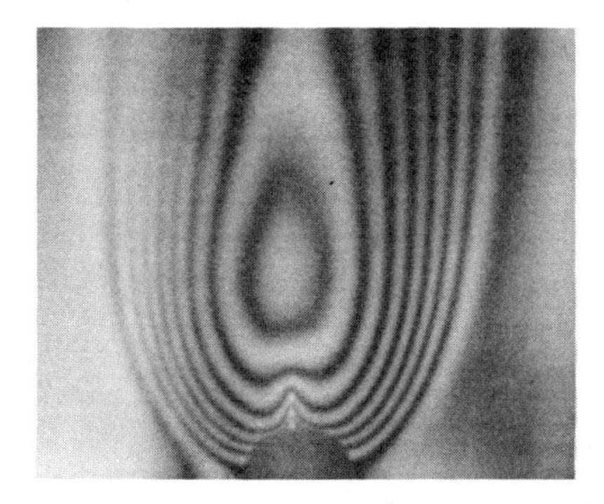

Fig. 11. Image-spectrogram of a flame for the study of gradients of temperature.

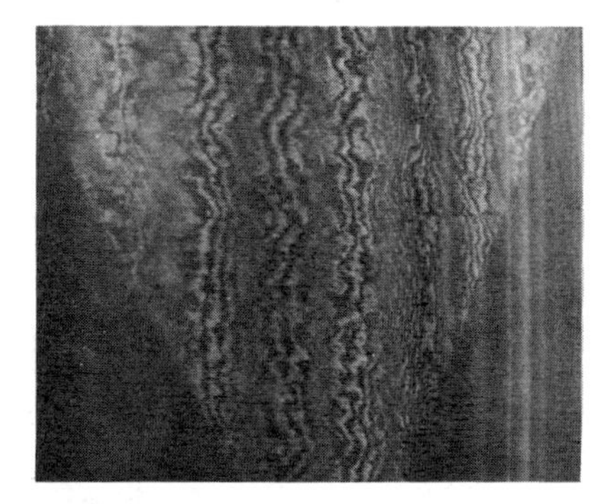

Fig. 12. Moiré pattern with equal chromatic order fringes for measuring absolute longitudinal displacements of the object.

by equal order chromatic fringes.

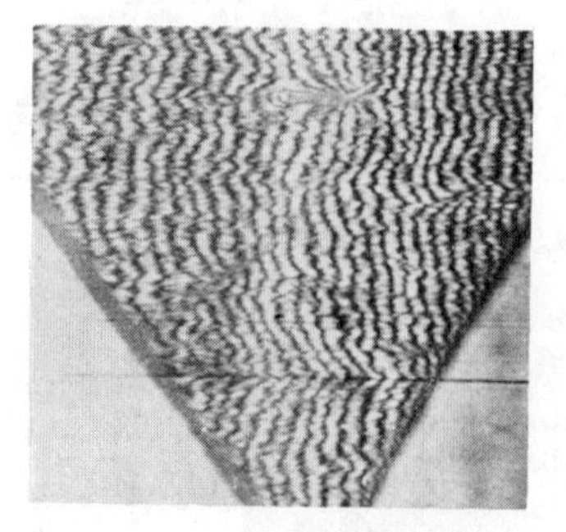

Fig. 13. Overall topography of a triangle-shaped rough surface.

The advantages of these methods over conventional interferometry are, in particular, the absolute measurement of path-differences from several millimeters to a fraction of one micron with two-beam interferometers (possibly down to a few angstroms with multiple-beam devices), and the direct discrimination between peaks and valleys in surface testing. Automatic apparatus can be developed for industrial uses.

CHROMATIC OR TEMPORAL HOLOGRAPHY

Here again the conjugation of space and time variables, transverse and longitudinal respectively, is evidenced. The principle of holographic recording is given in Fig. 14 where longitudinal wavetrains of white light (information and reference) are dispersed by a grating, then a part of the energy is analyzed by a second dispersive device. The recorded power-spectrum results from the coherent superposition of the temporal spectra of the information and reference, yielding a hologram in the temporal domain just as it would be in the spatial domain, now with dimension ν instead of u. Fig. 15 (a) shows

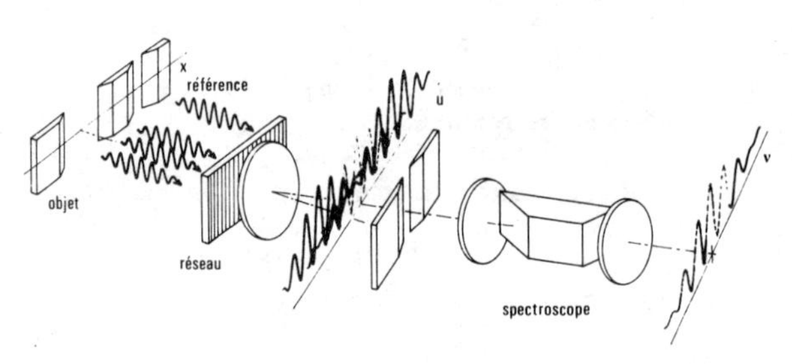

Fig. 14. Holography in the chromatic domain.

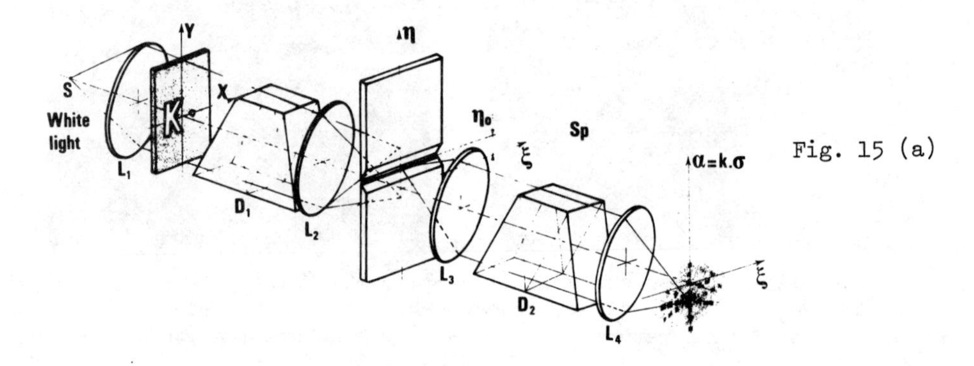

Fig. 15 (a)

the experimental set-up for recording a 2D (and simultaneously chromatic) hologram. The temporal hologram of the letter *K* and reconstruction illustrate the whole process (Fig. 15 (b) and 15 (c)).

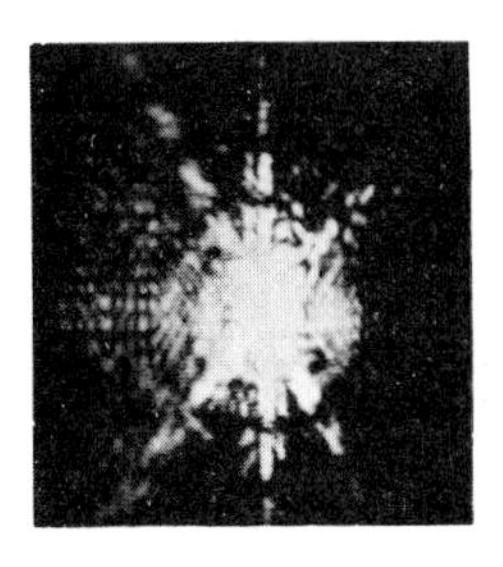

Fig. 15 (b). Hologram : it should be imagined in colours.

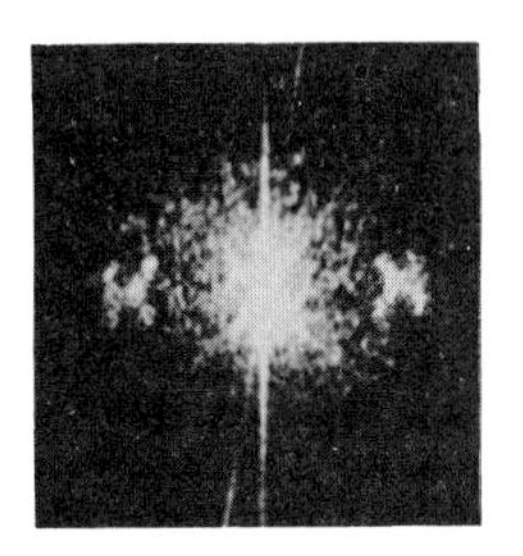

Fig. 15 (c). The letter *K* is retrieved in the side-bands.

Alternative techniques of temporal holography were studied at our laboratory. It was shown that holograms could be recorded in white light together with extended sources or self-luminous objects, i.e. spatially and temporally incoherent (see for example ref. 16).

TRANSPOSITION OF CONVENTIONAL CORRELATORS IN THE TEMPORAL DOMAIN

The arrangement in Fig. 16 is a multiplex incoherent correlator. The two parts, before and after the filtering plane, would be symmetrical if the various parameters or variables could be inserted at that plane at once. Since one axis is occupied by λ there is room for one spatial variable only. This is why a selection of the information must be done along one single direction at the entrance. It is carried out by means of the telescopic system L_1, L_2 with the selection slit suitably adjusted in between. Now the input and output planes are conjugated and the dispersors matched so that their dispersion laws cancel each other. The message to be processed being illuminated in white light the output is also displayed in white light.

Fig. 16.

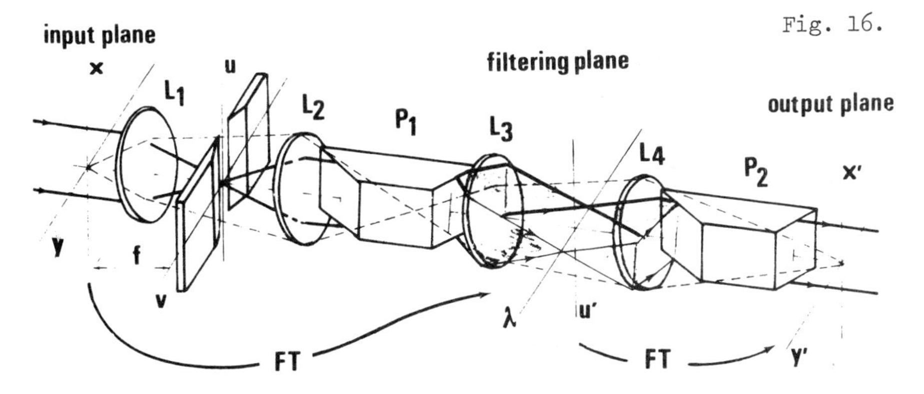

Then, how to make a matched filter which ought to be a temporal hologram ? Fig. 17 shows it in the case of a rectangular aperture : the various spectral components are dispersed by a spectroscope in (u',λ)-plane where the hologram is recorded. Such a hologram being placed in the filtering plane of the correlator with the word ZIGZAG

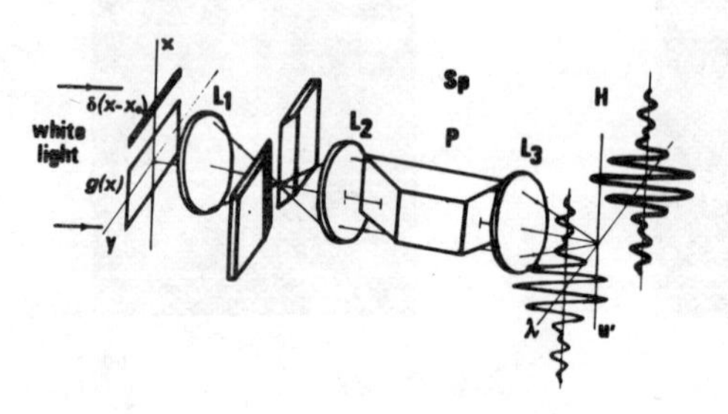

Fig. 17. Recording of the chromatic matched filter of the function $g(x)$.

Fig. 18. Correlation between rectangular pupil and ZIGZAG.

as input, a picture of the input is given in Fig. 18. It should be noted that the input is actually displayed in black and white; this is an advantage in many practical cases. The dedoubling is due to the saturation of the matched filter with low frequencies : the relative importance of the remaining high frequencies underlines the contours, unfortunately in one direction only, the processed direction.

A somewhat different approach is shown in Fig. 19. It essentially works with incoherent white light. Such an arrangement is hardly comparable with coherent correlators. The correlation between the object 1 (X) and the object 2 (Y) is two-step : (i) the dispersor

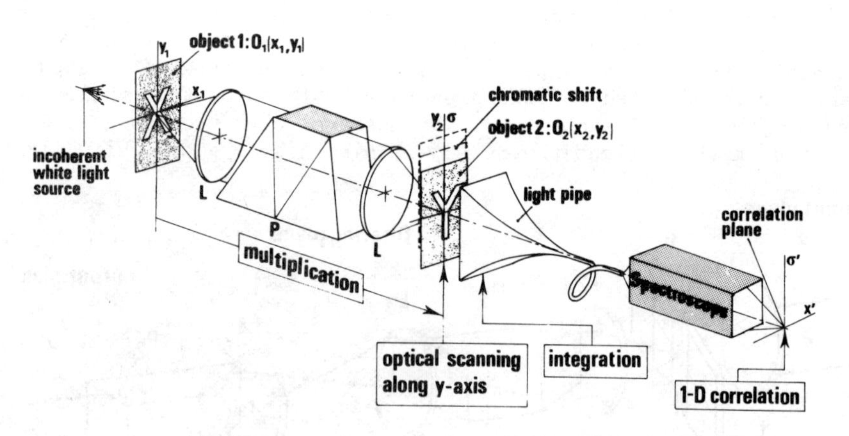

Fig. 19. Incoherent white light correlator.

shifts the various components of X which are multiplied by Y, (ii) the light pipe integrates the elementary products then the spectroscope analyzes the energy at the end of the light pipe.

TEMPORAL OR CHROMATIC SPECKLE

The analogy between spatial and temporal speckles was stressed many times in the past (see ref. 28). Elementary considerations are presented in Fig. 20 : the power-spectrum of a random time signal limited within a given period shows a grainyness whose average size is reciprocal of that period. As a source of white light illuminates a rough surface, its chromatic distribution is spatially modulated by the profile function of the surface under test. Adding a uniform background to a temporal speckle leads to a temporal Fourier hologram. In Fig. 21 the tiny wavelets reflected back from the various grains of the surface along one longitudinal direction are delayed respectively by a temporal distance which is a coding of the surface profile. These wavelets interfere with the reference. The spectroscope performs chromatic analysis of that coded information whose power-spectrum is recorded (Fig. 22 a). The demodulation is made in coherent light. In the display the ±*1* orders represent the probability density function of the accidents of the surface, considered along one line parallel with the temporal axis. A microphotometric section of one side-band (Fig. 22 c)

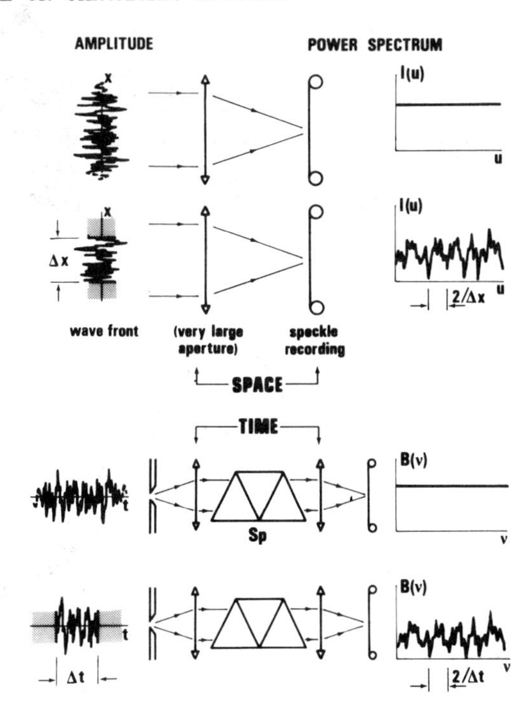

Fig. 20. Comparison between spatial and temporal speckles.

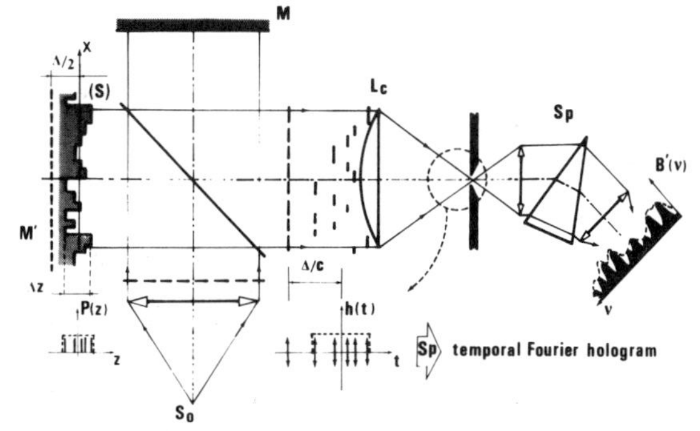

Fig. 21. Holographic recording of temporal speckle.

shows that the r.m.s. value of the roughness is about five microns in the present situation. The main advantage of this method is that no model of surface and no particular assumption are needed a priori.

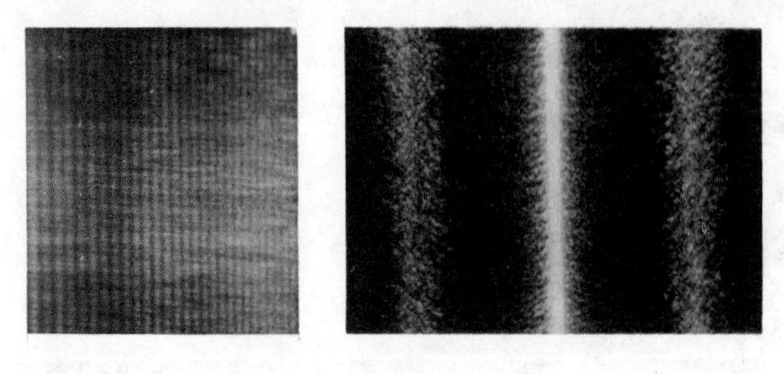

Fig. 22 a (left). Temporal hologram of the surface histogram.

Fig. 22 b (right). Reconstruction : statistical image of the surface profile.

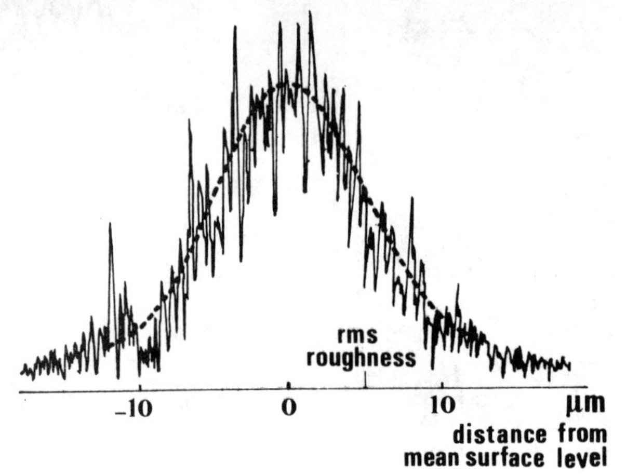

Fig. 22 c. Histogram of the surface along one given line.

CONCLUSION

What can be said in conclusion? To say that the parallelism between space and time variables and its application to practical problems is entirely satisfactory, appears a bit premature. In optics, we are almost always faced with all four dimensions, without knowing with certainty that in actual practice the time dimension is reliable.

We have dealt with the temporal propagation of optical signals in free space and with the time evolution of the signal itself. The time variable must also be taken into account at the detection stage, and then in the averaging process. At any rate, an optical device is rarely four dimensional. At least at the present time, a photocathode behaves as a two dimensional surface. Thick holograms are three dimensional. Optical feedback and adaptive systems need the time as a parameter... These few examples show that it is rather difficult to work with four independently chosen variables. In addition, problems arise when trying to translate Fourier Optics into the time domain because the Fourier Transform, a basic tool, appears to be noncausal. In fact, should it be causal, the notion of Wiener spectrum would lose its meaning because time dependent signals are not stationary, and this spectrum might have to be replaced by a time-dependent instantaneous quantity.

At any rate, Old Father Time is awkward, as well as variable when hidden in a spectrum. If we went along with such philosophy, the reader might be inclined to think that he has not put his time to good use.

REFERENCES

1 - R.A. Michelson, Phil. Mag. 531, (1891), 338.

2 - C.H. Page, J.Appl. Phys. 23, (1952), 103.

3 - D.G. Lampart, J.Appl. Phys. 25, (1954), 803.

4 - M. Born, E.Wolf, Principle of Optics, Pergamon Press (1959).

5 - A.J. Kartashev, Opt. Spect. 9, (1960), 204.

6 - E. Leith, J. Upatnieks, J.Opt.Soc.Am, 52, (1962), 1123.

7 - E. Leith, J. Upatnieks, J.Opt.Soc.Am, 53, (1963), 1377.

8 - E. Leith, J. Upatnieks, J.Opt.Soc.Am, 54, (1964), 1285.

9 - J.D. Armitage, A. Lohmann, and D. Paris, Jap. J.Appl.Phys.4, (1965), 273.

10 - E. Leith, J. Upatnieks, J.Opt.Soc.Am, 57, (1967), 975.

11 - P.A. Flournoy, R.W. Mc Clure, G. Wintjes, Applied Optics 11, (1972), 1907.

12 - C. Froehly, A. Lacourt and J.Ch. Viénot, Nouv.Revue Opt.4 (4), (1973), 183.

13 - G. Parry, Optics Commun. 12, (1975), 75.

14 - C. Roychoudhuri, J.Opt.Soc.Am, 65, 12 (1975), 1418.

15 - J.Ch. Viénot, A. Lacourt and J.P. Goedgebuer, Jap.J. of Appl. Phys., 14, Suppl. 14-1, (1975).

16 - J.Ch. Viénot, A. Lacourt and J.P. Goedgebuer, Proc.Conf.Optical Computing (Washington DC 1975), Ed. IEEE, pp. 133-136.

17 - H.M. Pedersen, Optica Acta 22, 1, (1975) pp. 15-24.

18 - J.Ch. Viénot, J.P. Goedgebuer and A. Lacourt, Proc. ICO X, Prague (August 1976).

19 - J.P. Goedgebuer, A. Lacourt and J.Ch. Viénot, Optics Commun. 16 (1976), 99.

20 - J.P. Goedgebuer, J.Ch. Viénot, Optics Commun. 19, (1976), 229.

21 - J. Calatroni, Optics Commun. 19, (1976), 49.

22 - C. Froehly, R. Desailly, Optics Commun. 21, (1977), 258.

23 - A. Lacourt, J.P. Goedgebuer, Optica Acta 24, 8 (1977), pp. 827, 835.

24 - E. Leith, J. Roth, Applied Optics 16, 9 (1977), pp.2565, 2567.

25 - J.H. Eberly, K. Wodkiewicz, J.Opt.Soc.Am, 67, 9, (1977), 1252.

26 - J.Ch. Viénot, J.P. Goedgebuer and A. Lacourt, Applied Optics 16, 2, (1977), 454.

27 - N. Aebischer, J.P. Goedgebuer, Optics Commun. 22, (1977), 103.

28 - J.Ch. Viénot, J.P. Goedgebuer, Applications of Holography and Optical Data Processing. Israël (1976) and Pergamon Press Ed. 1977.

29 - J.P. Goedgebuer, J.Ch. Viénot, Coherent Optical Engineering. North Holland Publishing Company (1977).

30 - J.P. Goedgebuer, A. Lacourt and M. Guignard, Optics and Laser Technology 8, (1978), 193.

31 - J.P. Goedgebuer, R. Gazeu and J.Ch. Viénot, Proc. ICO XI, Madrid (1978).

32 - A.A. Friesem, U. Levy, Opt. Lett. 2, (1978), 133.

33 - H.O. Bartelt, Optics Commun. 27, (1978), 365.

34 - J.P. Goedgebuer, R. Gazeu, Optics Commun. 27 (1978), 53.

35 - A. Lacourt, Optics Commun. 27 (1978), 47.

36 - A. Lacourt, P. Boni, Optics Commun. 27 (1978), 57.

37 - H.J. Caulfield, Optics Commun. 26 (1978), 322.

38 - G. Crosta, Optics Commun. 26 (1978), 141.

39 - H.O. Bartelt, Optics Commun. 28 (1979), 45.

40 - H.O. Bartelt, Optics Commun. 29 (1979), 37.

41 - E. Leith, J. Roth, Applied Optics 18, 16 (1979), 2803.

42 - R. Ferrière, J.P. Goedgebuer and J.Ch. Viénot, Optics Commun. 31, (1979), 285.

43 - R. Ferrière, J.P. Goedgebuer and J.Ch. Viénot, Proc. Conférence Européenne d'Optique : "Horizon de l'Optique" Pont-à-Mousson (1980).

44 - G. Morris, N. George, Opt. Lett. 5 (1980), 202.

HOLOGRAPHIC ANALYSIS OF DISPERSIVE PUPILS IN SPACE-TIME OPTICS(*)

J. Calatroni
Universidad Simón Bolívar.Dpto. Física, A.P. 80659, Caracas, Venezuela

J. Ch. Viénot
Laboratoire d'Optique, Université de Besançon, Fac. des Sciences, 25030 Besançon cedex, France.

ABSTRACT

Extension of Space-Time Optics to objects whose transparency is a function of the temporal frequency $\nu = c/\lambda$ is examined. Considering the effects of such stationary pupils on white light waves, they are called "temporal pupils". It is shown that simultaneous encoding, both in the space and time frequency domains, is required in order to record pupil parameters. The space-time impulse response and transfer functions are calculated for a dispersive non-absorbing material.

An experimental method providing holographic recording of the dispersion curve of any transparent material is presented.

INTRODUCTION

Space Time Optics has been developed in the last years resting upon a parallelism between space and time variables in optics[1]. The basic idea of Space Time Optics (STO) is to consider any optical pupil not only as a spatial frequency filter, but also as a time frequency filter. As a consequence of the filtering process in the time domain, the wide spectrum of a white light wave has been used to carry information about geometrical pupils.

In order to illustrate the above concepts, let us recall two classical experiments: (*i*) a transparent non-dispersive stationary pupil of variable thickness, $e(x)$, and refraction index, η, introduces a delay, $\eta\, e(x)/c$, for any white light wave; the wave is then superposed to a white light reference wave, the combination is analyzed by a spectroscopic device. The resulting spectrum is modulated by the $e(x)$ function. This channelled spectrum behaves as a Temporal Fourier Hologram (*i.e.* in the temporal frequency domain) of a longitudinal spatial pupil[2,3]; (*ii*) a transverse amplitude pupil $g(x)$ diffracts a white light wave ; for each direction of diffraction the temporal spectrum is

(*) *A part of the preliminary experiments was presented at the ICO XI Meeting, Madrid, 1978, and at the Int. Congress on High Speed Photography and Photonics, Tokyo, 1978.*

ISSN:0094-243X/81/650063-09$1.50

filtered according to a characteristic law which depends on dg/dx. A reference wave being added, this filtered spectrum becomes a Temporal Fourier Hologram distribution correspon ding to the spatial transverse pupil[1,2]. In both cases (*i*) and (*ii*) a single message - $e(x)$ or $g(x)$ - of spatial natu re, is transfered into the temporal domain. However STO is well adapted to analyse pupils which are of "temporal" nature. "Temporal pupils" means the objects (time invariant) whose complex transparency function are explicit functions of the temporal frequency ν ($\nu = c/\lambda$) *i. e.* either frequency dependent absorbing or dispersive objects.

Such "temporal pupils" modify the temporal profile of the illuminating wave.

The temporal objects can be properly analyzed and recorded by taking advantage of the high information capacity of white light beams. This can be performed by assigning one monochromatic carrier to each individual message. As a result, temporal and spatial characteristics of the ob - ject can be recorded. Along this way, Fourier Holograms of one-dimensional coloured pupils (temporal amplitude pupils) have been obtained[4].

This work deals with the extension of simultaneous fil tering (temporal and spatial) to the analysis of temporal phase pupils, *i. e.* transparent and dispersive objects. In particular an experimental method is discussed; it enables one to record the dispersive curve of a transparent mate — rial in a white light Fourier Hologram.

SPACE-TIME IMPULSE RESPONSE OF A DISPERSIVE PUPIL

The transparency of any 1-D pupil can be characteri - zed by a complex function:

$$T\,(x\,;\,\nu) = b(x\,;\,\nu)\,exp\,\{j\,\phi\,(x\,;\,\nu)\} \qquad (1)$$

where ν is the temporal frequency of the illuminating wave and x a transverse coordinate of the object. We shall refer to a phase pupil for which $b\,(x\,;\nu) = 1$. The phase term may be written as :

$$\phi\,(x\,;\,\nu) = \frac{2\pi}{c}\nu\,\eta\,(\nu)\,.\,e(x) \qquad (2)$$

$e(x)$ being the thickness of the pupil and $\eta\,(\nu)$ its refrac tive index; the refractive index is considered independent of x-coordinate.

Let us calculate the temporal impulse response and the temporal transfer function of this pupil. The luminous amplitude in the Fourier plane is:

$$A\,(\xi;\nu) = j\frac{\nu}{c}\,F(\nu)\,\exp\{j\,2\,\pi\frac{\nu}{c}\,e_0\}\int_{\text{pupil}}\exp j\,2\,\pi\frac{\nu}{c}\left[\eta\,(\nu) - 1\right]e(x).\exp\{-j2\pi\frac{\nu\xi}{cF}x\}dx \qquad (3)$$

when the object is illuminated by a plane wave of amplitude $F(\boldsymbol{\nu})$; e_0 is the maximum thickness of the pupil; ξ is the conjugate coordinate of x.

When $e(x)$ is a general function, the expression (3) is difficult to calculate; but it becomes a simple integral when $e(x)$ is a linear function

$$e(x) = \alpha x \tag{4}$$

In this case expression (3) reduces to

$$A(u\ ;\ \boldsymbol{\nu}) = \exp\{j\ 2\pi \frac{\boldsymbol{\nu}}{c} e_0\}\ F(\boldsymbol{\nu})\delta\left[u - \alpha\ (\eta - 1)\right], \tag{5}$$

where $u = \xi/\lambda$; for convenience we ignore a phase factor and we neglect diffraction effects.

Expression (5) is the spectral distribution in the Fourier plane of the pupil; in turns one obtains the transfer function (frequency response for a constant spectrum input) of the refractive prism of index η and angle α *i.e.* that of an elementary spectroscopic component:

$$H\ (u\ ;\ \boldsymbol{\nu}) = \exp\{j\ 2\pi \frac{\boldsymbol{\nu}}{c} e_0\}.\delta\left[u - \alpha\ (\eta - 1)\right] \tag{6}$$

It is important to note that for a prism, the distinction between temporal transfer function and spatial transfer function has no meaning, since the dispersive pupil calls for a description in terms of both spatial and temporal frequencies, even in the absence of diffraction effects. With geometrical apertures, the spectrum of the pupil can be described by a transfer function of the spatial frequencies as monochromatic illumination is used ($\boldsymbol{\nu}$ constant). This is the classical picture in monochromatic Fourier Optics. But also a transfer function of the temporal frequencies can be defined in order to describe the same geometrical pupil as before, but now fixing one direction of diffraction (u= constant); this is space-time approach.

This double picture of the filtering action of a geometrical pupil was mentioned in previous works[1,2,3], in order to specify the extensions and limits of the symmetry between space and time variables in Optics. In those papers the interest was mainly the analysis of the mechanism by which the spatial behaviour of an optical system was transposed into the temporal domain. The present interest is reciprocal and plans to transpose the temporal behaviour of a dispersive phase pupil -η ($\boldsymbol{\nu}$) function- into the space domain. With temporal pupils the situation is different from that of geometrical pupils. Whatever the amplitude or phase temporal pupils, both domains must be simultaneously considered in order to get meaningful results. In other words, if $\boldsymbol{\nu}$ or u are kept constant, the pupil reduces to a simple geometrical aperture.

The impulse response of the prism is the inverse Fourier

Transform (FT^{-1}) of $H(u\ ;\ \mathbf{v})$ in the temporal domain:

$$h\ (u\ ;\ t) \ = \ \exp\ \{j2\pi \frac{1}{\eta}\ (\frac{u}{\alpha} + 1)\ (t - e_0/c)\} \qquad (7)$$

where $\frac{1}{\eta}\ (\frac{u}{\alpha} + 1)$ stands for the value of $\mathbf{v}$ which satisfies the equation $u = \alpha\ (\eta\ (\mathbf{v}) - 1)$.

Coming back to the physical situation we realize that the prism has transposed the η ($\mathbf{v}$) function into the space domain, because each frequency necessarily travels associated with a particular spatial frequency $u = \alpha\ (\eta\ (\mathbf{v}) - 1)$. This is somewhat reciprocal to the situation in refs. (1), (2), (3) in which the spatial information is coded in the temporal domain.

Let us emphasize the importance of the transposition carried out by the prism when one has to analyze the phase term η ($\mathbf{v}$) of the pupil. For dispersive objects, the longitudinal variable z, -along the main axis of the optical system- is no longer simply related to time, due to the dispersive relation, $dz/dt = c/\eta\ (\mathbf{v})$, of the wave train. This is the reason why temporal encoding (channelled spectrum) of the message is not fairly adapted. Indeed, one is able to record the η ($\mathbf{v}$) function when encoded in space and time domains simultaneously. In other words it remains impossible to encode both thickness and dispersion parameters along the same temporal frequency axis. The above arguments again illustrate the advantage of interchanging spatial and temporal variablesin optics.

SPACE TIME FOURIER HOLOGRAM OF A PRISM

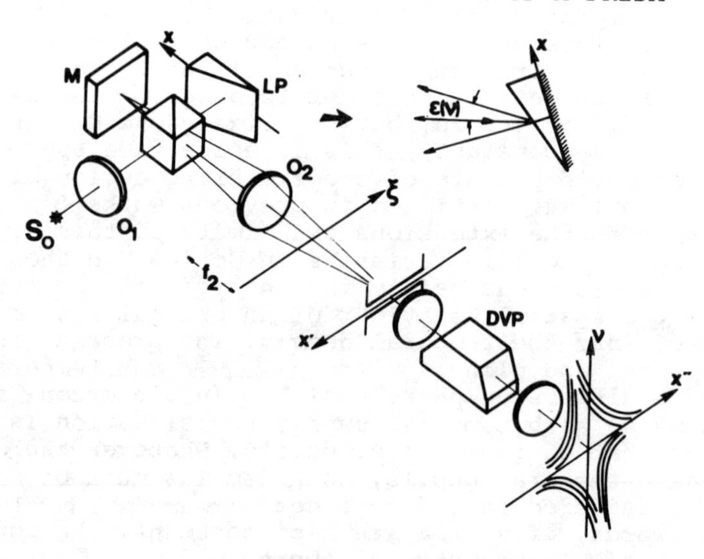

Fig.1: Experimental set-up for recording dispersion curve of a prism.

Figure 1 shows the experimental set-up which has been developed to analyze the dispersion of a transparent material, namely here a Littrow prism with a small deviation angle ε (ν); this quantity is then a linear function of the refractive index

$$\varepsilon\ (\nu) = 2\ \alpha\ \{\ \eta\ (\nu) - \eta_0\ (\nu_0)\} \tag{8}$$

where ν_0 stands for some arbitrary frequency for which $\eta\ (\nu_0) = \eta_0$ is know.

The prism is set along one arm of a two-beam interferometer, the back face is a mirror. The interferometer is illuminated by a white light point source.

A spectroscope placed at the exit of the interferometer analyzes the white light interference pattern, spread along the ν - axis in a series of monochromatic fringe systems. Thus a 2 - D interferogram is diplayed along the space (x'') and time (ν) frequency axes.

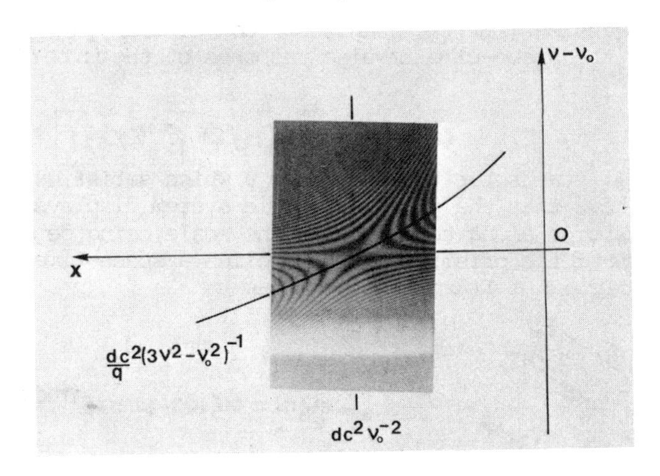

Fig.2: White light interferogram;it corresponds to a hollow prism of 7°23' filled with distilled water at room temperature.The drawing curve shows the relation of the pattern with involved parameters.

Figure 2 shows the recorded interferogram. The recorded intensity distribution of the 2 - D interferogram is given by

$$I\ (x''\ ;\ \nu) = 2|\ F(\nu)\,|^2\{1 + \cos \frac{2\pi}{c} \nu\, \Delta(\nu\ ;\ x'')\} \tag{9}$$

where $|F\ (\nu)|^2$ is the power spectrum of the source and $\Delta(x''\ ;\ \nu)$ the difference between optical paths.For each frequency ν, the last expression represents the Fourier hologram of a pair of point sources placed along the ξ - axis and whose amplitude is

$$A(\xi\,;\,\boldsymbol{\nu}) \;=\; F(\boldsymbol{\nu})\left[\delta\;(\xi)\exp\{j\;2\pi\frac{\boldsymbol{\nu}}{c}\,2d\} \;+\; \delta\,(\xi-f_2\,\varepsilon\,(\boldsymbol{\nu}))\right], \qquad (10)$$

$2d$ stands for the arbitrary delay introduced at $\boldsymbol{\nu}_0$ by the interferometer. This is just as if a series of pairs of coherent point sources of different frequencies were placed at the back focal plane of O_2; the distance between the elements of each pair being proportional to $\eta\;(\boldsymbol{\nu}) - \eta\;(\boldsymbol{\nu}_0)$. Going a bit further into formalism, the overall interferogram can be considered as a single Fourier Hologram, simultaneously recorded in space and time domains. Indeed, the interferogram comes from a double FT in both domains

$$I\;(x'';\boldsymbol{\nu}) \;=\; \left|FT_{\xi;t}\{\delta(\xi)\;\;f(t-\frac{2d}{c}) + FT^{-1}\left[F(\boldsymbol{\nu})\,\delta\,(\xi-f_2\varepsilon)\right]\}\right|^2 \;=$$

$$=\;\left|FT_{\xi;t}\{\;f\;(t)\otimes \mathrm{k}\;\;(\xi\;;t)\}\right|^2 \qquad (11)$$

In expression (11), $\otimes$ stands for convolution product, and $\mathrm{k}(\xi;t)$ for the space-time impulse response of the interferometric system:

$$k\,(\xi;t) \;=\; \delta\,(\xi)\;\;\delta\;\;(t-\frac{2d}{c})+\exp\{-j\,2\pi\left[\varepsilon^{-1}(\xi/f_2)\right]t\}, \qquad (12)$$

where $\varepsilon^{-1}(\xi/f_2)$ stands for the frequency $\boldsymbol{\nu}$ which satisfies $\xi/f_2 = \varepsilon(\boldsymbol{\nu})$

This shows that the interferometric system displays the spectrum of the source along the ξ-axis, the scale being determined by the dispersion of the prism; it also supplies a space-time reference source to record a Fourier Hologram

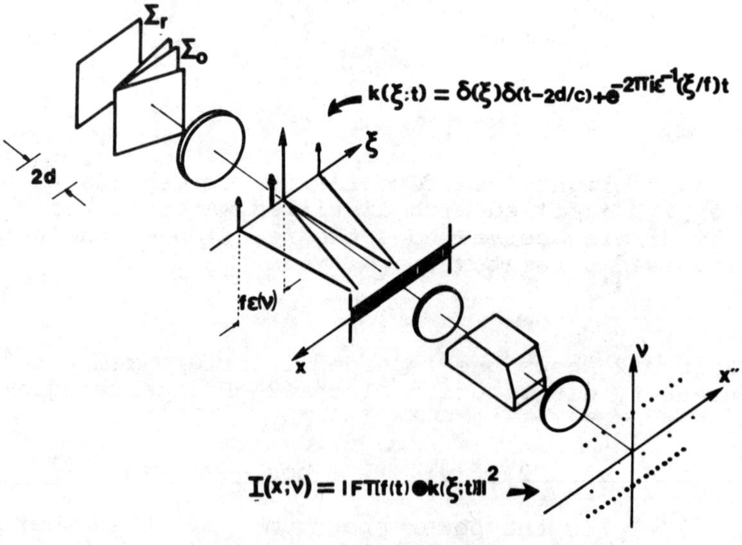

Fig.3: Impulse response and recorded intensity for the arrangement of Fig.1

Briefly, the interferometric device behaves as a spectroscope with a reference source, providing "holographic phase spectroscopy" of the dispersive pupil. Figure 3 summarizes the above discussion.

RECONSTRUCTION PROCESS

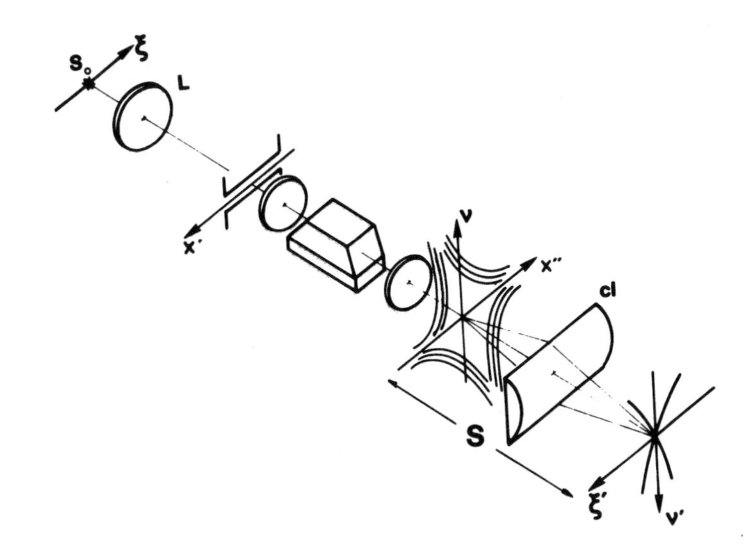

Fig.4: Experimental setting for hologram decoding

Figure 4 schematically shows the reconstruction setting. S_0 is a white light point source. The axes ξ and ξ' are conjugated through the lenses L and $c\ell$. The hologram is replaced at the exit plane of the spectroscope in the same position as that of recording. The cylindrical lens $c\ell$ images the ν-axis into ν' so as to perform the FT^{-1} in the space domain only.

Figure 5 shows the reconstructed image of the dispersion function of water. The central line is the zero order term and the other two curves correspond to

$$\xi' = \pm 2 \alpha s \{\eta(\nu) - \eta(\nu_0)\} \quad (13)$$

It must be noted that a spectroscope is not necessary to observe any particular dispersion law; whether or not it is linear in λ or ν, its dispersion function merely fixes the variable and scale in which the dispersion curve will be displayed.

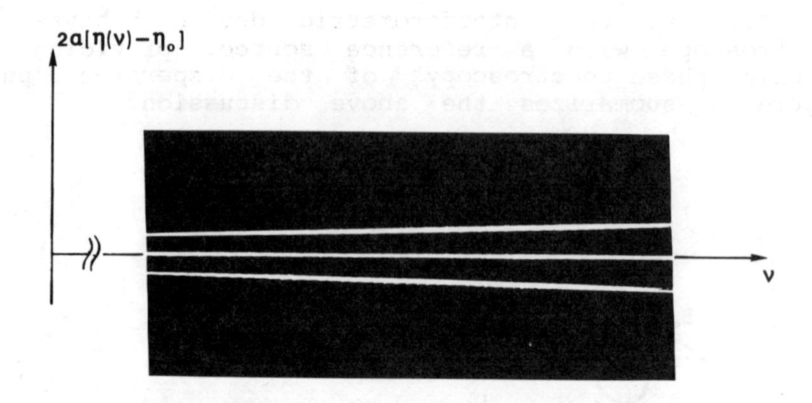

Fig.5: Reconstructed image corresponding to dispersion curve of water at room temperature.This image comes from a hologram for which ν_0 was chosen near the violet limit of the visible spectrum in order to get high spatial frequencies;the hologram is not shown in Fig.2.

CONCLUSION

This work proves that STO is well suited to the analysis of pupils which modify the shape of the temporal profile of light waves. The term "temporal pupils" fits their properties as far as the amplitude and/or phase of each monochromatic component are concerned.

In STO "temporal pupils" operate somewhat differently from "spatial pupils". For "spatial pupil" (diffracting apertures or frequency independent time delays) the message is transferred as a modulating function of the time spectrum of a white light beam. On the other hand, a separate message is assigned to each monochromatic carrier by modulating its amplitude (coloured objects) or its phase (dispersive objects) when temporal pupils are involved.

The above concepts and distinctions play an important role when the analysis of various approaches of STO is intended. In particular in the method developed along this work,the refractive index function η ($\boldsymbol{\nu}$) of a prism is holographically recorded in a 2 - D interferogram which is simultaneously displayed in the space and time frequency domains. The reconstruction process requires a single diffraction step but through a tuning spectroscope. It may be thought that the overall system applies to recording the temporal evolution of η ($\boldsymbol{\nu}$) functions of non—stationary samples, by means of succesive holograms recorded on a running film; in the reconstruction process the movement of the film along one direction (x"-axis in figure 4) will not affect the image because of the well known in-plane translation properties (shift theorem) of Fourier Holograms.

It is interesting to point out that the interferometric set-up used when recording behaves like a spectroscope with reference beam. A second spectroscope in cascade follows. The overall operation may be called "Phase Holographic Spectroscopy".

Many systems which allow highly precise measurements of refraction index exist in Optics; moreover through the very old method of "crossed-spectra" the dispersion can be obtained directly. The experimental technique described above is neither more accurate than earlier methods, nor simpler than the "crossed-spectra" method. Yet it was important to show how STO can be well extended to dispersive pupil analysis. Furthermore, in spite of its inherent experimental difficulties and lack of precision "Phase Holographic Spectroscopy" opens the possibility of ν-dependent filtering on the phase term of a wave and synthesis of dispersive elements in an equivalent grating.

REFERENCES

1. J.Ch. Viénot, J.P. Goedgebuer and A. Lacourt, Appl. Opt. 16, 2, pp 454-461 (Febr. 1977).
2. Cl. Frohely, A. Laocurt, and J. Ch. Viénot. Nouv.Rev. Opt. 4, 4, pp 183-196 (1973).
3. J.P. Goedgebuer, A. Lacourt and J. Ch. Viénot. Opt. Comm. 16, 1, pp 99-103 (Jan. 1976).
4. J. Calatroni. Opt. Comm. 19, 1, pp 49-53 (Oct. 1976).

PROPAGATION OF NON MONOCHROMATIC SIGNALS ATTACHED TO HIGH DISTORTION WAVEFRONT

(NON DISPERSIVE HOMOGENEOUS MEDIUM)

J.P. Goedgebuer, J.Ch. Viénot, R. Ferrière
Lab. d'Optique, Ass. CNRS, U. de Franche-Comté, 25030 Besançon, France

ABSTRACT

A description of diffraction phenomena of a white light beam passing through a complex pupil, i.e. corresponding to a very irregular wavefront profile, demands the use of a local obliquity factor. The general form of the resulting signals is derived from the Kirchhoff integral. It depends on the transit time, and its evolution law, here studied with one transverse and with one longitudinal coordinates, can be extended to *x, y, z, t*. An illustration is given in the field of rough surface testing.

INTRODUCTION AND PROCEDURE

The object of the paper is the determination of the evolution of a non monochromatic disturbance, function of time, as it propagates in a non dispersive medium. The signal can be a laser pulse or a wave group of white light whose amplitude and wavefront have been modulated by a complex pupil introducing various optical delays. The general solution to this problem is well-known. It is given by the Helmholtz-Kirchhoff integral, *as far as monochromatic waves are concerned*. Fig. 1 briefly recalls the fundamentals.

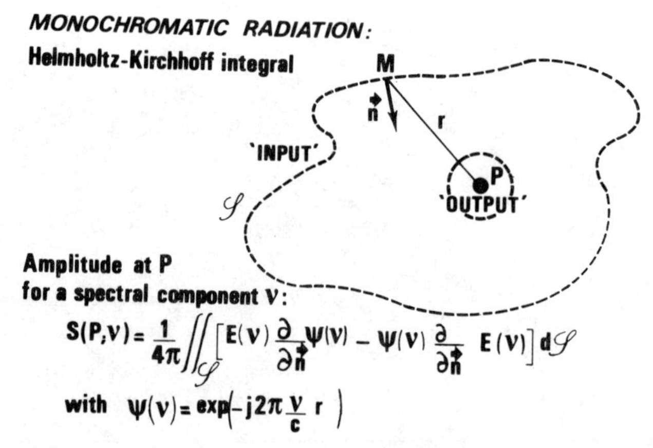

Fig. 1. Derivation of the Helmholtz-Kirchhoff integral.

ISSN:0094-243X/81/650072-07$1.50

The amplitude $S(P,\nu)$ at point P for a monochromatic radiation of temporal frequency ν is [1] :

$$S(P,\nu)= \frac{1}{4\pi} \iint_{\mathcal{S}} \{E(\nu) \frac{\partial}{\partial \vec{n}} \psi(\nu) - \psi(\nu) \frac{\partial}{\partial \vec{n}} E(\nu)\} d\mathcal{S} \qquad (1)$$

where $E(\nu)$ and $\frac{\partial}{\partial \vec{n}} E(\nu)$ are the values of the amplitude of the incident wave and normal derivative at every point M taken on the closed surface $\mathcal{S}$ surrounding the point P ; $\psi(\nu)$ is the Kirchhoff auxiliary function :

$$\psi(\nu)= exp \{-j2\pi(\nu/c)r\} \qquad (2)$$

where r is the distance from M to P. In fact, extension to non monochromatic disturbances derives from eq.(1) as each individual spectral component ν of the signal may be considered independently.

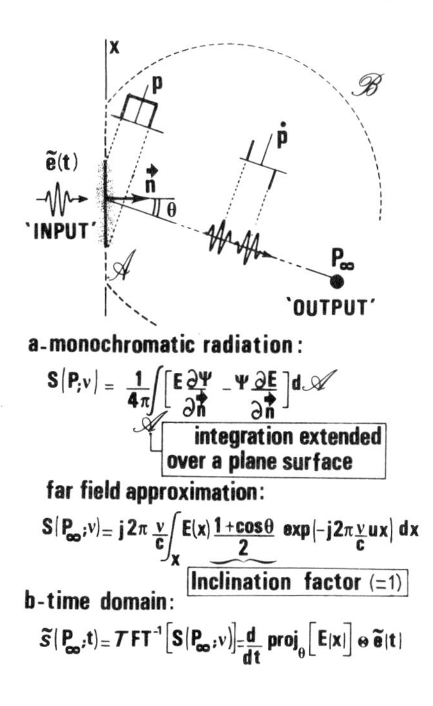

Fig. 2. Recalling transposition of the general relation (a) into the time domain (b) : the result is shown in the diagram.

Fig. 2 illustrates the procedure in the particular case of an initial plane wavefront disturbance $e(x,t)$, called "input" signal in Fig. 2 , propagating at infinity. The light amplitude $S(P,\nu)$ at point P is ruled by the Helmholtz-Kirchhoff integral, the integration being taken over the closed surface $\mathcal{A} + \mathcal{B}$ surrounding P. It reduces to an integration over the plane surface $\mathcal{A}$, yielding the spatial Fourier Transform of the amplitude $E(\nu)$ of the input signal. The time signal $\tilde{s}(P,t)$ at point P, called "output" signal in the figure, is the inverse temporal Fourier Transform of $S(P,\nu)$. It has been demonstrated that the output signal $\tilde{s}(P,t)$ at infinity is given by the first derivative (*vs* time) of the projection of W along the z-direction of observation convoluted with the input disturbance [2, 3, 4]. It is worth noting that this result only concerns the time signal at infinity arising from an original *plane-wavefront* disturbance. Next section shows what happens for any disturbance as it propagates in a non dispersive medium.

GENERALIZATION TO ANY DISTURBANCE

Assessement of the Helmholtz-Kirchhoff integral :

The general case of a distorted wavefront is more complicated ; it involves considerations on the local slopes. From the previous statements and usual reasoning in spatial optics, one may think that the fluctuations of the distorted wavefront can be expressed through relative phase shifts $\phi(x)=(2\pi/\lambda)\,\Delta(x)$, where $\Delta(x)$ is the departure between W and a reference plane X (Fig. 3). The next step is the evalua-

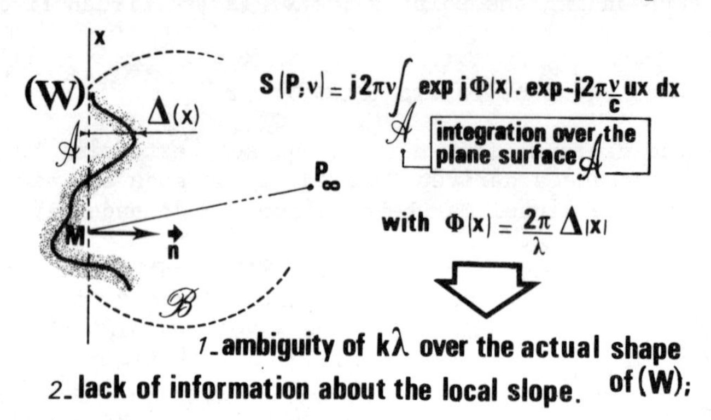

Fig. 3. Calculation of the scattered amplitude implying phase shifts only.

tion of the Helmholtz-Kirchhoff integral over the plane surface $\mathcal{A}$. However, the method suffers from an ambiguity : the actual shape of the wavefront is not explicited, namely as far as local slopes are concerned. This is important as one deals with polychromatic light or signals functions of time. It can be overcome by taking another surface of integration in place of $\mathcal{A}$, for instance the wavefront W itself. (Fig. 4) Under these conditions, the light amplitude

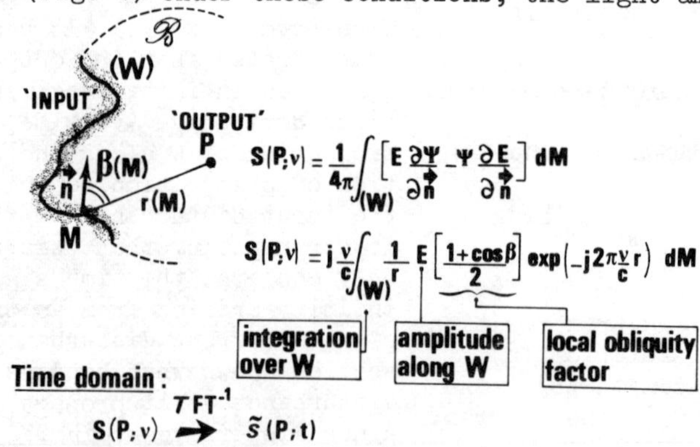

Fig. 4. Rigorous calculation of the amplitude taking the local shape of W into account.

$S(P,\nu)$ at point P reads :

$$S(P,\nu)=\frac{1}{4\pi}\iint_W\{E(\partial\psi/\partial\vec{n}) - \psi(\partial E/\partial\vec{n})\}dM \qquad (3)$$

yielding [5, 6] :

$$S(P,\nu)=j(\nu/c)\iint_W(1/r)E(1+\cos\beta)/2.\exp\{-j2\pi(\nu/c)r\}dM \qquad (4)$$

where E : amplitude along W ;
β : local angle between the inward normal $\vec{n}$ and $\vec{r}(M)$;
ν : temporal frequency and c : velocity of light.

Eq. (4) is an alternative version of the Fresnel-Kirchhoff diffraction formula [1], that offers several advantages : no ambiguity occurs on the shape of the wavefront, the integration being taken over W itself ; the local slopes of the wavefront are taken into account through a local inclination factor $(1+\cos\beta)/2$ which may vary between 0.5 and 1 as the element of integration dM explores the domain of integration W, the normal $\vec{n}$ being not constant in direction.

Time domain :

Eq. (4) rules the amplitude of a monochromatic radiation at point P. The time signal $\tilde{s}(P,t)$ arriving at P is given by the inverse temporal Fourier Transform of $S(P,\nu)$:

$$\tilde{s}(P,t)= TFT^{-1}\{S(P,\nu)\}=\int_{-\infty}^{+\infty} S(P,\nu)\exp(j2\pi\nu t)d\nu \qquad (5)$$

yielding [6] :

$$\tilde{s}(P,t)=(1/2\pi c)\iint_W (1/r)(1+\cos\beta)/2.\frac{d}{dt}\{\tilde{e}(x;t-\frac{r}{c})\}dM \qquad (6)$$

Eq. (6) is the general relationship governing the time signal $\tilde{s}(P,t)$ arriving at any point P and generated by any disturbance(¶) $\tilde{e}(x,t)$. An observer at P receives a sequence of signals proportional to the time derivative of the disturbance $\tilde{e}(x;t-r/c)$ arising from every point M of the wavefront W, weighted by a local obliquity factor $(1+\cos\beta)/2$ and with delayed times r/c depending on the distance of propagation r (Eq.(6) is a generalization of the relationship derived by other authors [2, 3, 4] in the case of non-distorted wavefront disturbance).

(¶) In fact, Eq.(6) holds for any disturbance whose wavefront is locally flat [5, 6], that involves radii of curvature of the irregularities of the wavefront much larger than the wavelength.

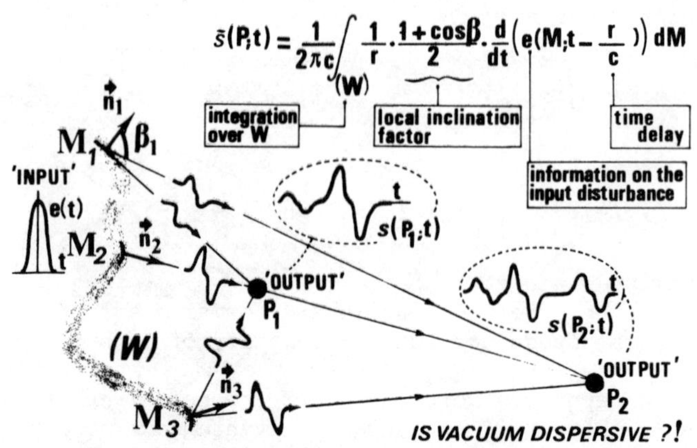

Fig. 5. Evolution of a time signal as it propagates. Since the resulting signal at a given point *P* varies with the distance to *P* from the contributing elements of the wavefront, the modification of profile of the signal could be felt as a dispersive effect. This is not the case, in fact the relative distance from M_i to P_i varies as well as the angle β_i when P_i is chosen in various regions of the output space.

Fig. 5 roughly represents signals $\tilde{s}(P_1,t)$ and $\tilde{s}(P_2,t)$ at two points P_1 and P_2, called "outputs" and resulting from the contributions of three elements of the wavefront isolated in the vicinity of points M_1, M_2, M_3 (called "inputs").

To summarize the situation, eq.(6) describes the space-time evolution of any wave or signal function of time as it propagates in a non dispersive medium. In other words, eq.(6) can be considered as the mathematical formulation for a generalized Huygens theorem that might be expressed as :"a disturbance propagates as if every element *dM* of its wavefront emitted a secondary disturbance whose temporal structure is given by the time derivative of the original signal weighted by an inclination factor, function of the local slope of the element *dM* of the wavefront".

DETERMINATION OF THE TIME IMPULSE RESPONSE OF COMPLEX PUPILS WITH HIGH PHASE GRADIENTS

Fig. 6 represents the situation in the far field equivalent to Fraunhofer diffraction in the spatial domain. Then the surface integral (6) reduces to the convolution product [6] :

$$\tilde{s}(P_\infty,t)=(1/2\pi cd)\int\left(\sum_{i=1}^{n} E_i\frac{\cos\beta_{i+1}}{2\sin\beta_i}\right).\frac{d}{dt}\,\tilde{e}(x;t-d/c+z'/c)dz'$$

$$=1/2\pi cd\left[\frac{d}{dt}\sum_{i=1}^{n} E_i\frac{\cos\beta_{i+1}}{2\sin\beta_i}\right]\otimes\tilde{e}(x;t) \qquad (7)$$

where E_i and θ_i are the local amplitude and inclination angle of the wavelet at $z'=z_i$.

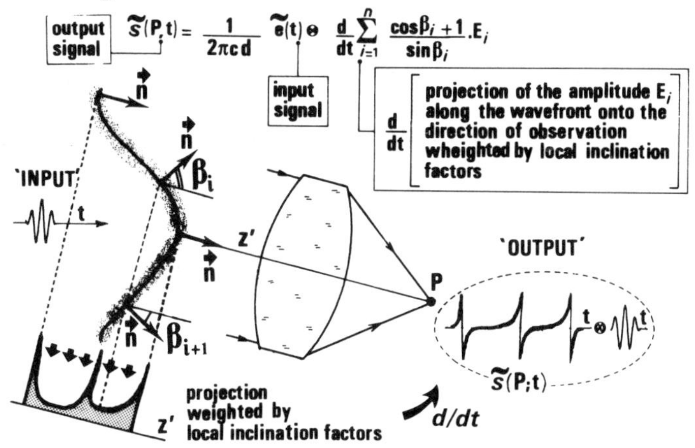

Fig. 6. Far field observation. The situation should be the same if P was at the infinity or at the focus of an optical signal. But physically the transit time is finite in the second case ; d : distance from the elements of the considered wavefront to P. The temporal output signal is the convolution of the input one with the derivative of the projection of the wavefront suitably weighted.

A straightforward application is sketched in Fig. 7, dealing with disturbances whose wavefront is altered by irregularities randomly distributed. Such a situation is encountered when illuminating a rough surface in white light [7]. For abrupt slopes, the correlation

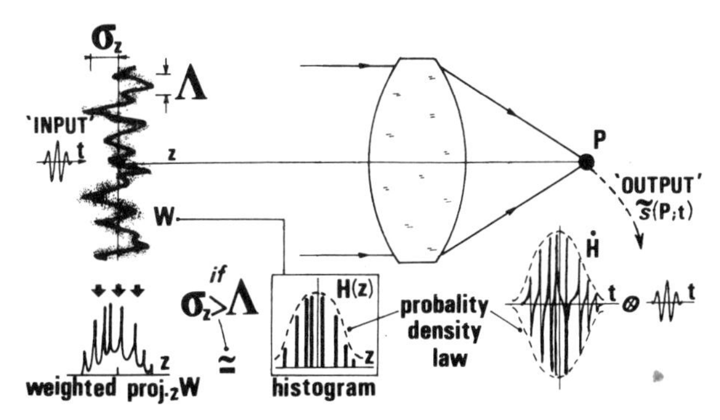

Fig. 7. Interpretation of temporal speckle phenomena as a rough surface is the concrete image of a disturbance with rapid variations of phase randomly distributed. The resulting signal is observed in the far field (other explanations in the text). This also illustrates, as an application, the general considerations developed in this paper.

lenghth Λ is much smaller than r.m.s. roughness σ_z, then the main contributions to the projection of the wavefront W on *z-axis* come from peaks and valleys only. As a result, the projection of the wavefront W weighted by the local inclination factors can be approximated to the histogram $H(z)$ of the surface. Its derivative leads to the time signal $\tilde{s}(P_\infty,t)$ at P_∞. The envelope of $H(z)$ being a slowly varying function, the time signal $\tilde{s}(P_\infty,t)$ is modulated by the envelope of $H(z)$, the latter being nothing but the probability density law of the in-depth accidents of the surface along *z-axis*. It means that the time signal $\tilde{s}(P_\infty,t)$ stores and carries the information about the statistics of the surface. Such a statement is useful for metrological applications dealing with surface testing through "temporal speckle" phenomena [7], and justifies, if need be, some assumptions encountered in a method of determination of the statistics of rough surfaces developed elsewhere.

REFERENCES

1 - e.g. M.Born, E.Wolf, Principles of Optics, Pergamon Press Ed. 1975, pp. 371-386.

2 - J.W.Goodman, Introduction à l'Optique de Fourier (Mc Graw Hill Ed. 1968) p.42-44.

3 - C.Froehly, A.Lacourt, J.Ch.Viénot, Nouvelle Revue d'Optique 4, (4), 1973, 183.

4 - G.Bonnet, Annales des Télécommunications, 30, 7-8, 1975.

5 - P.Beckmann, A.Spizzichino, The Scattering of Electromagnetic Waves from Rough Surfaces.(Pergamon Press Ed.) 1963. pp.17-28.

6 - A complete formulation of the approach and results presented here is to be published.

7 - J.P.Goedgebuer, J.Ch.Viénot, Optics Commun. 19, 2 (1976), 229.

PICTURE PROCESSING IN (x,y,t)-DOMAIN WITH TV-OPTICAL FEEDBACK METHODS

G.Ferrano
G.Häusler
Physikalisches Institut der Universität
Erlangen-Nürnberg
Erwin-Rommel-Str.1,D-8520 Erlangen

H.Maître
Ecole Nationale Supérieure des Télécommunications
36 Rue Barrault, 75013 Paris

ABSTRACT

This paper shows some examples of time-dependent picture processing by means of a hybrid optical processing system. It consists of electronical and optical parts. The advantages of both are combined in our TV-optical feedback system.

INTRODUCTION

We want to report three applications of our TV-optical feedback system. This system allows us to translate electronical feedback concepts and digital image processing methods into optics. The advantages of optical processing like a fast two-dimensional convolution can be fully used. Combined with TV technology it is possible to implement "real-time" picture processing.

The first application we want to report is the van Cittert filtering method, a concept for digital image enhancement.

The other two examples shall show, how to use TV-components, in order to build optical storage elements.

We begin by explaining our TV-optical feedback system.

TV-OPTICAL FEEDBACK SYSTEM

In such a system a TV-camera looks onto its own synchronously running monitor(fig.1).

A two-dimensional (x,y)-signal circles around and during each cycle it can be processed either one-dimensionally in the electronical part or two-dimensionally in the optical part of the system.

ISSN:0094-243X/81/650079-08$1.50

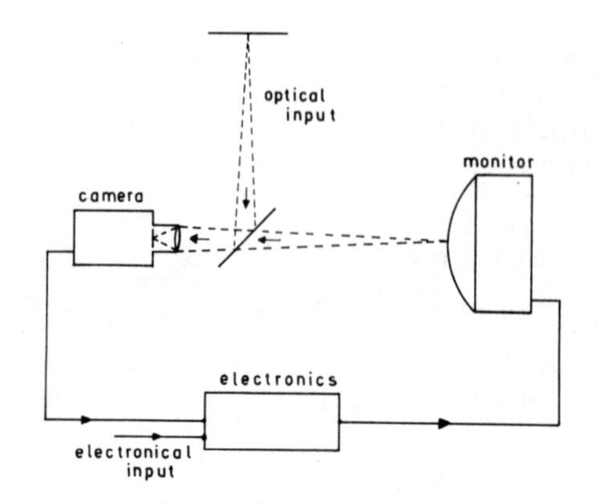

Fig.1. TV optical feedback set-up

For a more detailed description see references 1,2.

VAN CITTERT FILTERING

This is an iterative process[3] for picture enhancement. So far the concept of van Cittert has been implemented only digitally. Here we want to execute it by means of our TV optical system[4].

Mathematically this process has the following form. If, for example u(x,y) represents a picture and H(x,y) describes a defect, the blurred picture can be written as:

$$v(x,y)=u(x,y)*H(x,y)$$

This function v(x,y) shall be enhanced iteratively by a process according to which the k+1st iteration has the form:

$$v_{k+1}(x,y)=v_k(x,y)+v(x,y)-v_k(x,y)*H(x,y)$$

for $k=0$, $v_k=v$

With growing k, v_k converges towards u.

$$v_k(x,y) \rightarrow u(x,y) \text{ for } k \rightarrow \infty$$

The optical translation of these formulas are shown in the next figure(fig.2)

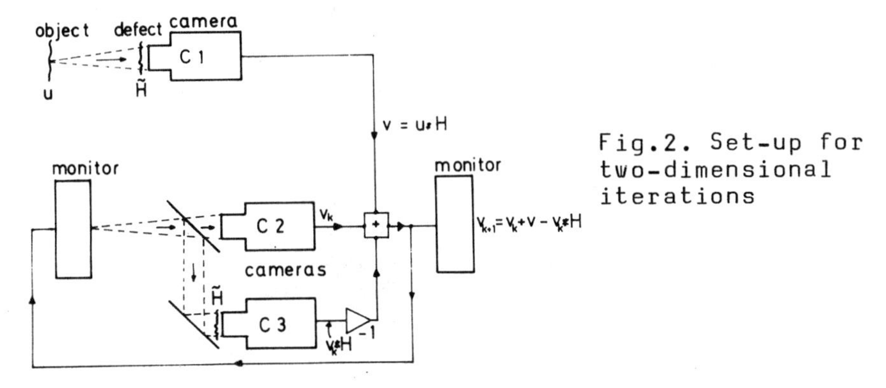

Fig.2. Set-up for two-dimensional iterations

In this set-up camera C1 picks up the input, the defective picture.The cameras C2 and C3 introduce a time delay of one frame time. This time delay increases the iteration index from k to k+1. The rays entering camera C3 are affected by the same blurring defect H as the object had encountered originally.

This set-up allows a two-dimensional restoration of defective pictures. But in practice it is nearly impossible to adjust the two cameras C2,C3 pixel by pixel.

Therefore we first worked with an one-dimensional defect. Hence we needed only one feedback camera. The set-up for one-dimensional filtering is shown in fig.3.

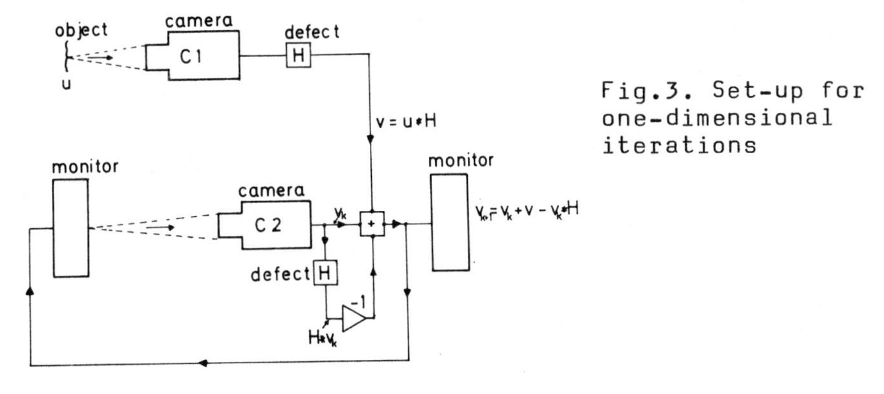

Fig.3. Set-up for one-dimensional iterations

With this set-up we got rather good results. They are presented in the next figures. Figs.5,6 show a digital simulation of a low pass filter and the corresponding restoration. Figs.7,8 show the optical analogues.

Fig.4. Original picture

Fig.5. Digital blurred picture

Fig.7. Optical blurred picture

Fig.6. Digital restored picture

Fig.8. Optical restored picture

If we look again into fig.2, we easily can see that camera C2 is only used to introduce a time delay. This also can be done by a TV-storage device. When this device will be available to us we will attempt two-dimensional restoration according to fig.9.

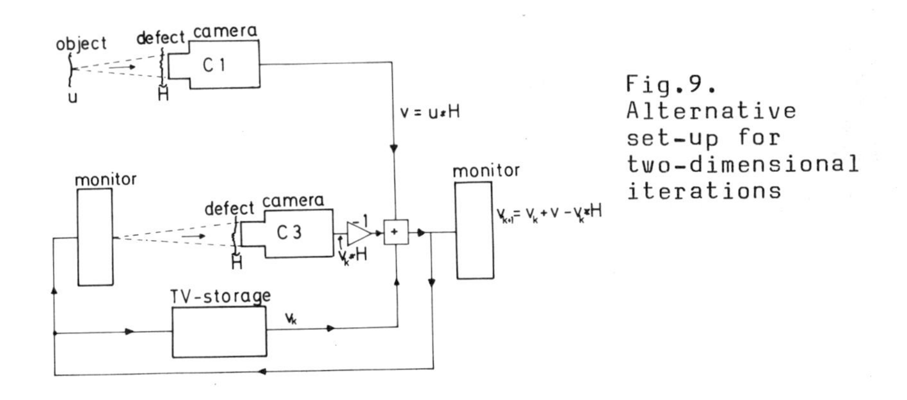

Fig.9.
Alternative set-up for two-dimensional iterations

TV-LOOP AS PICTURE MEMORY

As mentioned already, a camera which looks onto a monitor produces a time delay of one frame period (40msec,European standard). This can be used to build a picture storage system or a picture buffer. It allows us to subtract two subsequent frames according to fig. 10.

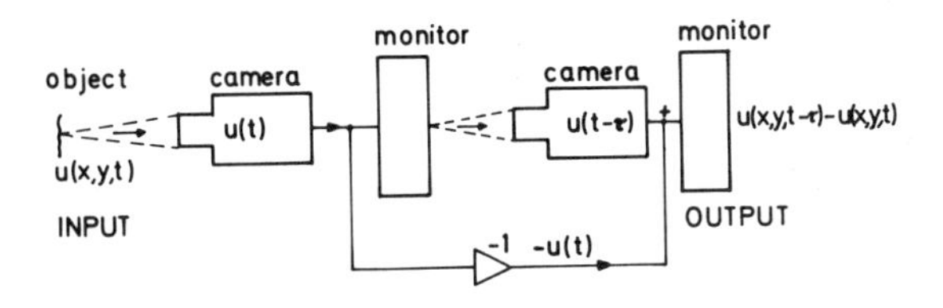

Fig.10. TV optical "movement detector"

This set-up may serve to detect moving object details. A result is shown in fig.11.

Fig.11.
Difference of two frames. The bars on the right side are moved. The movement causes the black and white edges.
(Arrow=direction of the movement)

The system of fig.10 could store the picture only for one TV frame time. For longer storage times we may use a system that will be described now.

TV-OPTICAL FLIP-FLOPS

A feedback system with suitable nonlinearity allows bistabilities. We used this for a TV optical flip-flop system[5],which allows us to store binary pictures. It also enables to connect these pictures by logical operations.

Fig.12 shows a typical output of our system. The TV-screen is divided into two halves. A pair of small patches, situated symmetrically with respect to the vertical middle line, forms a flip-flop. The sizes and the shapes of these patches are determined by the distortion correction, that is necessary.

Fig. 12. Output of our TV optical flip-flop system.

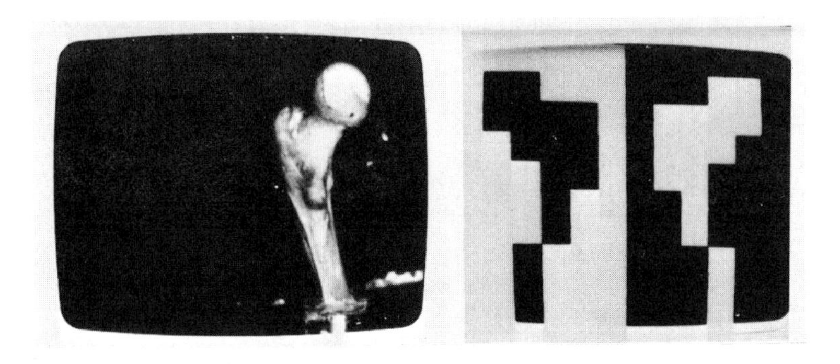

Fig.13. Greytone picture (input) with its output produced by the TV optical flip-flop system

These pictures can be stored forever, if no changes are introduced. We may perform a logical OR operation pixel by pixel, if two binary pictures are written sequentially into the same half of the picture field. This can be seen in fig.14.

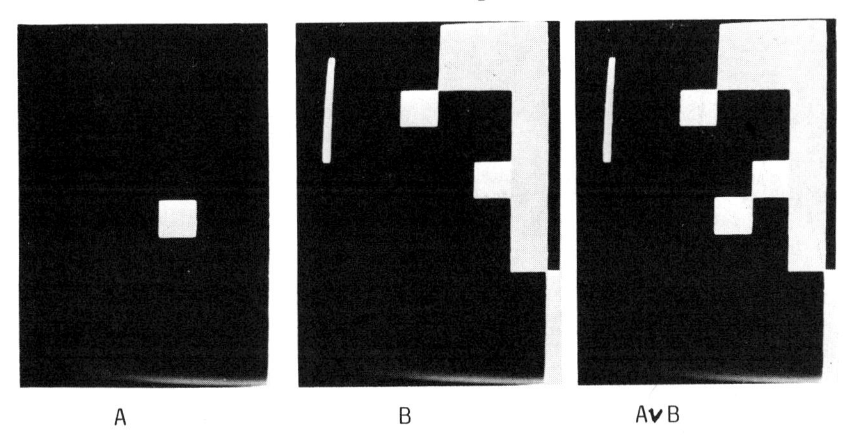

Fig.14. OR relation of two pictures

CONCLUSION

These experiments show that the combination of electronical TV-systems and optics provides a suitable set of tools for implementing concepts of space-time optics.

REFERENCES

1. S.H.Lee,Opt.Eng. 13,198(1974)
   T.Sato,K.Sasaki,R.Yamamoto,Appl.Opt.,17,717 (1978)
2. G.Ferrano,G.Häusler,to be published in Opt Eng.
3. P.A.Jansson,J.Opt.Am. 32,1980(375)
4. H.Maitre, to be published in Comp.Graph.Im.Proc.
5. G.Ferrano,G.Häusler,Opt.Comm. 32,1980(375)

CONCEPTS OF SPECTROSCOPY OF PULSED LIGHT

Chandrasekhar Roychoudhuri*, Jesús M. Siqueiros and E. Landgrave.

C I C E S E
Ave. Espinoza No. 843, A.P. 2732. Ensenada, B. C. MEXICO

ABSTRACT

We derive a generalized but simple expression for the interferograms generated due to the propagation of a pulse through classical interference spectrometers. All multiple beam interferograms (Fabry-Perots, gratings) can be expressed as the summation of many two-beam interferograms. The time-integrated interferograms can be expressed as due to the Fourier spectrum of the signal through the autocorrelation theorem. The time resolved interferogram helps separate the effects of carrier frequency from that due to the amplitude envelope. The point is illustrated with some time-resolved 3-D computer plots.

Any radiative source has a finite lifetime, either by virtue of its natural properties or by design of the experiment. Thus any spectroscopy has to deal with pulsed radiation whether long or short. The established concept in classical spectroscopy is that the measured spectrum (energy distribution as a function of frequency) is given by the Fourier spectrum of the time pulse. The Fourier frequency distribution is centered around the carrier frequency of the pulse. We assume the definition of carrier frequency as that at which the electric and magnetic fields of the radiation pulse are oscillating. Although there may be some controversy or misconception as to what is the physical spectrum of a radiation, [1,2,3] it is generally assumed that there is no basic problem regarding the measurability of the spectrum of long pulses. Then there should not be any fundamental problem in measuring the spectrum of a short pulse either, at least, not until the size of the amplitude envelope becomes comparable to the period of oscillation of the field.

*Current address: TRW Systems, 1 Space Park, Redondo Beach, CA.90278 U.S.A.

ISSN:0094-243X/81/650087-08$1.50

In this paper we will consider pulses of duration that is many times the period of the carrier frequency. The objective is to determine the response of classical spectrometers (like Fabry-Perots and gratings) to pulsed radiation and explore the possibility of obtaining information on the carrier frequency distribution and the pulse envelope.

Our view of interference spectrometers like Fabry-Perots and gratings is as follows.[4,5,6,7] These instruments replicate the incident pulse (by amplitude or wavefront division) into a train of pulses with regular characteristic delay $\tau = m/\nu$, where m is the order of interference and ν is the carrier frequency. If the instrumental arrangement is such that the replicated pulses are physically superposed, the dispersive effect, i.e. the separation of energies corresponding to different carrier frequencies, becomes measurable. Thus the dispersive property is actually due to the superposition of fields only and arises when superposed pulses bring different phase information through propagation delay at the plane of interference. For the same path delay τ, the phase of the wave $\omega\tau$ is different for different carrier frequency, and hence the measurable intensity distribution (interference fringes) is different for different carrier frequency. (A prism can be thought of as a blazed grating where the layers of atoms with their characteristic energy levels introduce different propagation delays for different carrier frequencies of the incident wave). Now, since the intensity distribution of the interference fringes depend simultaneously on the amplitude and the total effective phase $(\omega\tau - \phi(\tau))$ of the wave, the information on carrier frequency distribution will, in general, remain masked due to the variation of the amplitude and the phase. Thus an exact knowledge of the carrier frequency can not be obtained by a classical spectrometer, under a detection scheme of long time averaging, unless an exact knowledge of the amplitude and the total phase of the wave is determinable through some other independent experimentation. However, under a very short time (over a few cycles) averaging detection scheme, the carrier frequency distribution can be easily determined if the amplitude and the phase of the wave do not vary significantly.

We represent the input signal $V_i(t)$ as,

$$V_i(t) = A_i(t) \exp i(\omega_o \tau - \phi(t)). \tag{1}$$

Then the output signal from a spectrometer is,

$$V_o(t,\tau) = \sum_{n=0}^{n-1} TR^n A_i(t-n\tau) \exp (\omega_o(t-n\tau) - \phi(t-n\tau)). \tag{2}$$

For a Fabry-Perot, T and R are the transmittance and reflectance of the mirrors and N is a few times its finesse number. For a grating, T = R =1 and N is the number of slits. The single slit diffraction envelope is neglected in this analysis. For a Michelson T = R = 1 and N = 2. Assuming that the phase is constant and zero, and the amplitude is slowly varying, the "instantaneous" intensity (short time, a few cycles average) is

$$I(t,\tau)= \sum_n T^2 R^{2n} A_i(t-n\tau) + 2\sum_{n<m} T^2 R^{n+m} A_i(t-n\tau) A_i(t-m\tau) \cos\omega_o(m-n)\tau. \tag{3}$$

If the intensity is recorded by a long-time integrating device like a photographic plate, the normalized intensity is given by

$$i(\tau) = \sum_n T^2 R^{2n} + 2 \sum_{n<m} T^2 R^{n+m} C\{(m-n)\tau\} \cos\omega_o (m-n)\tau, \tag{4}$$

where $C(\tau)$ is the normalized time autocorrelation of the amplitude envelope,

$$C(p\tau) = \int_{-\infty}^{\infty} A_i(t-n\tau) A_i(t-m\tau)\, dt / \{\int_{-\infty}^{\infty} A_i^2(t-m\tau)\, dt\}^{1/2} \{\int_{-\infty}^{\infty} A_i^2(t-m\tau)\, dt\},^{1/2} \tag{5}$$

where, $p = m-n$.

For a grating, Eq. (4) can be simplified to,

$$i(\tau) = N + 2\sum_{p=1}^{N-1} (N-p) C(p\tau) \cos\omega_o p\tau, \tag{6}$$

and for a Michelson, one gets,

$$i(\tau) = 2\{1+C(\tau)\cos\omega_o\tau\}. \qquad (7)$$

It is now clear from the equations (4), (6) and (7) that for all interference spectrometers, when the output is integrated over the entire duration of the pulse train, the interferogram is represented by the summation of many two-beam interferograms multiplied by appropriately weighted pulse autocorrelation factors.

For a continuous wave, the equations (4), (6) and (7) reduce respectively to the standard Airy function for Fabry-Perots, $\sin^2/\sin^2$ function for gratings and (1+cos) function for Michelson interferometers.

For two-beam interferometers (Eq. 7), the normal procedure is to extract the oscillatory part of the interferogram,

$$i_{osc}(\tau) = C(\tau)\cos\omega_o\tau. \qquad (8)$$

The carrier frequency ω_o can be determined by counting the number of fringes with τ, and the autocorrelation of the pulse will be the visibility of the fringes. Using the autocorrelation theorem, Eq. (8) can be represented as

$$i_{osc}(\tau) = \int_{-\infty}^{\infty} W(\omega)\cos\omega\tau d\omega \qquad (9)$$

Where $W(\omega)$ is the Fourier energy spectrum of the signal. Thus, the time integrated interferogram of a pulse can be thought of as formed by the superposition of fringes produced by all the Fourier component frequencies of the pulse, as if the mutual interferences between the Fourier frequencies averaged to zero. A substitution similar to that of Eq. (9) in Eq. (4) and (6) formalizes the customary assumption that a pulsed light shows an instrumental broadening that is equivalent to the Fourier spectrum of the pulse. However, the equivalence is only a formal one through the autocorrelation theorem. It should not be considered as the physical spectrum of the signal. The case where the pulse contains many carrier frequencies requires a summation over all ω in Eq. (2), and will be presented elsewhere.

That the Fourier spectrum should not be considered as the physispectrum of a pulse can be appreciated in two ways. First, from the operational standpoint, the time Fourier transform integral is non-causal because to produce the Fourier spectrum, an instrument would require to respond to information that is distributed from past to the future.[1,2] Second, the equivalence of the broadening of the Fourier spectrum is true only if the interferometer output is integrated over the entire train of replicated pulses. If the output is recorded by a fast detector device like a streak camera one would record a fringe broadening (visibility or contrast) that would evolve with time. The time-evolving fringe broadening certainly can not be interpreted as due to time-varying Fourier spectrum. Besides, in spite of time-evolving fringe broadening, the frequency of the interference fringes would remain the same all the time and would correspond to the carrier frequency only.

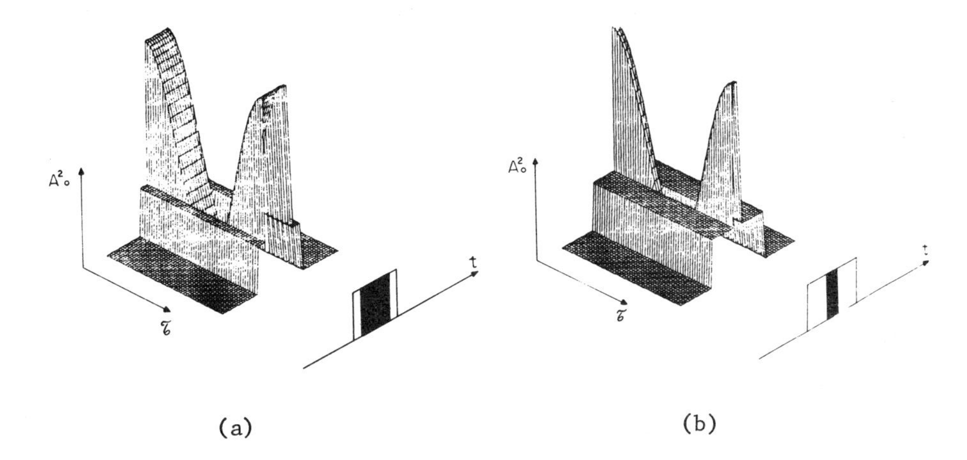

FIG. 1 Time evolution of a fringe formed by a rectangular pulse in a Michelson interferometer. In Fig. (a) the delay between the interfering pulses is $\tau = 0.25\ \delta t$, where δt is the width of the pulse, and in Fig. (b) $\tau = 0.75\ \delta t$. The plot on the right hand side of each figure represents the superposed intensity envelopes of the pulses.

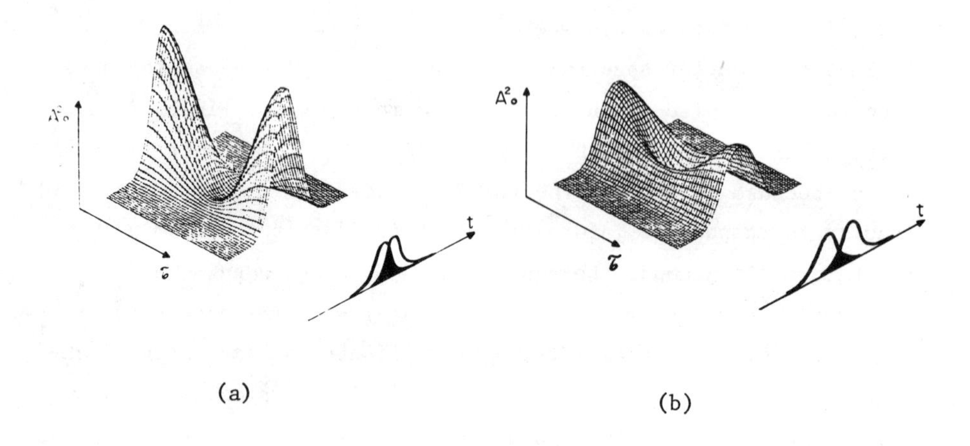

FIG. 2 Time evolution of a fringe formed by a gaussian pulse in a Michelson interferometer with (a) $\tau = 0.25\ \delta t$ and (b) $\tau = 0.75\ \delta t$.

In figures 1 and 2 we show the time-evolving fringes due to a Michelson interferometer for two different interpulse delays and for two different kinds of amplitude envelopes: rectangular and gaussian. For rectangular pulses, it is clear that during the interval when there is no superposition between the pulses, one has uniform energy (no fringes; white light spectrum?); during the interval when parts of the two pulses are superposed with equal amplitude, one has perfect visibility fringes (no fringe broadening; monochromatic light?). For gaussian pulses, the time-evolving fringe broadening is, of course, smoother than due to rectangular pulses. Figure 3 shows the time varying fringe broadening for a Fabry-Perot interferometer due to an incident gaussian pulse. The fringe broadening is much faster compared to Michelson interferometer for the same interpulse separation. As in Michelson, fringe counting with τ would give the value of the carrier frequency ω_o, but the computation of the pulse autocorrelation is substantially more complicated compared to the two-beam case because of the involved relation of the Eq.(4) or (6).

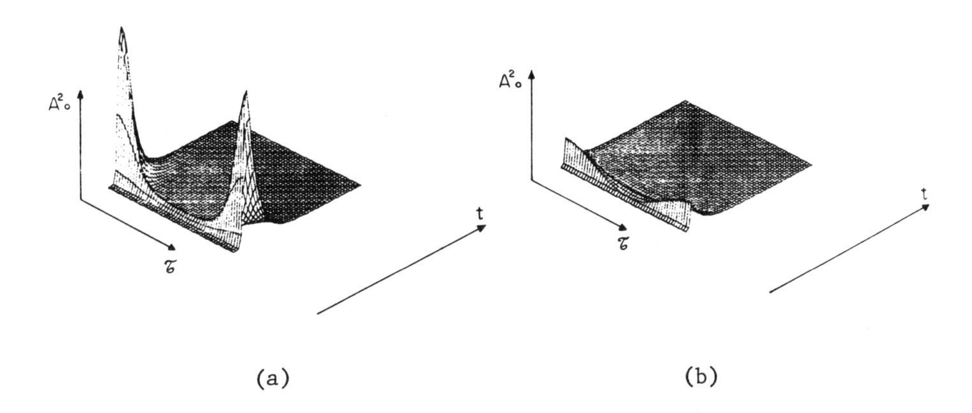

FIG. 3 Time evolution of a fringe formed by a gaussian pulse in a Fabry-Perot interferometer for (a) τ = o.25 δt and (b) = τ = 0.75 δt. Ao^2 has been scaled up 40 times with respect to Figs. 1 and 2, due to the fast decay in the values of Ao^2 with increasing delay.

Let us consider the case of a radiation consisting of such a large number of random pulses that one has a stationary random process. In such a case, even a fast detector would see all the phases of a time pulse simultaneously and one can reproduce Eq.(4) where $C(\tau)$ now would be an ensemble average quantity instead of a time average quantity. However, this is a significant point in classical spectroscopy. If the thermal emission of an ensemble of atoms or molecules is considered to be emitting random classical pulses, even the time resolved spectroscopy would show a fringe broadening corresponding to $C(\tau)$ of Fourier spectrum. One can not separate the effect of pulse autocorrelation and the carrier frequency as in the case of a single pulse.

In summary we have shown from the basic principle of interference, how the dispersive property of interference spectrometers arise. The time-resolved interferogram of a single pulse does not lend itself to Fourier spectrum interpretation. However, the interferogram of stationary random pulses, whether time-resolved or integrated, and that of a single pulse when time-integrated, can be interpreted as due to the Fourier frequencies of the amplitude envelope.

We wish to thank S. Castañeda, F. Reyes-Spíndola and A. Valenzuela for their collaboration in the computer work in this paper.

REFERENCES

1.- J. H. Eberly and K. Wodkiewicz, J. Opt. Soc. Am. 67, 1252 (1977).
2.- M. S. Gupta, Am. J. Phys. 43, 1087 (1975).
3.- L. Mandel, Am. J. Phys. 42, 840 (1974).
4.- G. Cesini, G. Guattari, G. Lucarini and C. Palma, Opt. Acta. 24 1217 (1977).
5.- A. Kastler, Nuov. Rev. Opt. 5, 133 (1974).
6.- C. Roychoudhuri, S. Calixto, Boletín Inst. Tonantzintla, 2, 165 and 187 (Dec. 1977).
7.- C. Roychoudhuri, J. Opt. Soc. Am. 65, 1418 (1975).

GEOMETRY OF WAVE ELECTROMAGNETICS*

E. C. G. Sudarshan†

Center for Particle Theory and Physics Department
University of Texas, Austin
Austin, TX 78712

and

Lyman Laboratory of Physics
Harvard University
Cambridge, MA 02138

Geometry of Wave Electromagnetics

It has been known now for over a century that light is an electromagnetic phenomenon; and that light propagates in empty space as a wave according to Maxwell's equations. In direct confirmation of this wave nature are the observation and quantitative explanation of interference and diffraction; and, in evidence of its transverse nature, of polarization and double refraction. There is no substitute for the wave theory of light.

On the other hand, there are a number of simple situations where a ray picture of light is useful. These include the formation of shadows of images by the pinhole camera, and reflection and refraction at simple surfaces like planes and conics. We have an adequate idea of the conditions under which this obtains as an approximation: it is most simply seen as a consequence of the extremely small wavelength of light.[1] We even know how to compute corrections to the ray approximations. We have, for example, the typical fluting of the shadow in diffraction at a straight edge. It has therefore been thought that the ray picture is always an approximation and should be transcended. Since we know the wave theory to be accurate and adequate albeit clumsy.

We present a challenge to this commonly held view and construct an exact realization of light as generalized pencils of rays.[2] It is important to stress that we must use pencils of rays rather than single rays. We shall display exact equations of motion for the rays in the pencil, and we shall see that these rays tend to travel in straight lines in empty space (not too near the edge of a beam).

The same kind of question had arisen once before in wave optics but this was in relation to the question to what extent quantum optics had classical pictures.[3] It was known that the correspondence

*Work supported by the Department of Energy and †by the National Science Foundation under Grant No. PHY77-22864.

ISSN:0094-243X/81/650095-11$1.50

principle suggested an approximation to quantum theory by classical phase space distributions: high photon number states could be related to intense classical wave fields. But we had shown that there was an exact correspondence between the classical and quantum pictures.[4] The complete expression is embodied in the first and second fundamental theorems of quantum optics.[5]

While the problem of ray picture of wave electromagnetics is distinct from the above, it is logically of the same type. What we wish to show is that the exact correspondence reproduces all the known features of wave and ray optics but provides a clear insight into the scaling of interference and diffraction patterns. These patterns in the Franhoufer region scale as if they are shadows cast by rectilinear rays.[2]

My interest in this work stems from a discussion I had with Emil Wolf exactly two decades ago! It was rekindled by Jacob Kuriyan who pointed out the immediate relevance of the work of Wolf[6] and the previous work of Walther[7] on radiative transfer. The astute reader will see the germs of the ideas discussed here in their work, but also see that there are definite points of departure.

Rays in Wave Fields

We shall deal with statistical wave fields exclusively; the pure wave field can be considered as a special case. We may realize the wave field in terms of the two-point correlations of the electric field over the ensemble:

$$\Gamma_{jk}(r_1,r_2,t_1t_2) = \langle E_j^{\dagger}(r_1,t_1)E_k(r_2,t_2)\rangle \tag{1}$$

The ensemble averages so obtained obey a wave equation in the variables r_1,t_1 as well as in the variables r_2,t_2. It is known that quite rigorously we may show complete one-to-one correspondence between the quantum mechanical expectation values for second quantized field operators and the classical ensemble averages for classical (analytic signal) wave fields: this is the second fundamental theorem of quantum optics.[5] We shall therefore not distinguish the two theories, but deal directly with the correlation functions themselves. For a stationary ensemble the correlation function does not depend individually on the times t_1,t_2 but only on their differences. In this case we can define the specific correlation function:

$$\Gamma_{jk}(r_1,r_2;\nu) = (1/2\pi)\int_{-\infty}^{\infty} d\tau\, e^{-i\nu\tau}\Gamma_{jk}(r_1,r_2,t+\tau,t) \tag{2}$$

which is independent of time. To avoid inessential complications, we shall deal exclusively with stationary ensembles and time-independent specific correlation functions.

Various physical quantities of interest can be expressed in terms of the specific correlation tensors. The energy density $u(\underline{r})$ is given by

$$u(\underline{r}) = (1/4\pi) \sum_j \int d\nu \, \Gamma_{jj}(\underline{r},\underline{r};\nu). \tag{3}$$

We now define the Wolf tensor

$$W_{jk}(\underline{r},\underline{p};\nu) = (1/2\pi)^3 \nu^{-1} \int d^3q \, e^{+i\underline{p}\cdot\underline{q}} \, \Gamma_{jk}(r + \tfrac{1}{2}q, r - \tfrac{1}{2}q;\nu) \tag{4}$$

which will be presently associated with the density of polarizable rays with wave number $\underline{p}$ in the direction of $\underline{p}$ at the location $\underline{r}$. We note that while the sum on the right hand side of (3) runs over

three indices, by virtue of Gauss's theorem contained in Maxwells' equations it really reduces to two polarizations. One may go to the helicity basis appropriate for description of circularly polarized photons and introduce the two valued helicity index σ:

$$u(\underline{r}) = (1/4\pi) \sum_{\sigma=1}^{2} \int d\nu \, \Gamma(\underline{r},\underline{r};\nu,\sigma). \tag{5}$$

The Wolf tensor W is not diagonal in the helicity σ in general but we may introduce in place of (4) the Wolf matrix

$$W(\underline{r},\underline{p};\nu,\sigma_1,\sigma_2) \tag{6}$$

adapted to the helicity basis.

We now identify $W(\underline{r},\underline{p})$ as the density of light rays in the pencil passing through the point $\underline{r}$ in the direction $\underline{p}$. For the model of stationary ensembles that we choose to discuss we note that there is a specific frequency ν for each ray; and there is a wave number which is $(p^2)^{1/2}$, not necessarily coinciding with ν. There is also a polarization label for the rays; one may take the diagonal elements of the Wolf matrix $W(\underline{r},\underline{p};\nu,\sigma,\sigma)$ as giving the density of rays with polarization σ at $\underline{r}$ in the direction $\underline{p}$ with wave number $(p^2)^{1/2}$ and frequency ν. We observe that W so defined is real (which is not true of the correlation tensor) but not necessarily positive.

Light Rays: Bright and Dark

Since a negative density of rays might appear unsatisfactory and unfamiliar we digress to point out that it is not unphysical. The measurable quantities of the theory are dynamical quantities like the energy density in the form of the photometric intensity of illumination at any location; the total flux of light from a source; or the flux of momentum across a surface as determined by the light pressure. Wave theory of light cautions as against the attempt to make measurements on localized pencils of light which are too narrow despite their formal use in consideration of radiative transfer. On the other hand, a little reflection shows that the redistribution of light implicit in the phenomenon of "light added to light producing darkness" in the double slit interference pattern most clearly shows that the ray density W cannot always be nonnegative. In fact, it is precisely the indefinite sign of the ray density that makes it possible to have an exact ray picture for arbitrary wave fields. We shall, therefore, not be surprised to have indefinite signs for the ray density in a number of contexts where wave optical effects make their appearance. (See, however (17) and the following discussion.)

We note, in passing, that Babinet's Principle is automatically satisfied.

Since $W(\underline{r},\underline{p};\nu)$ is interpreted as the density of rays, the density of energy and the density of momentum are, respectively

$$\varepsilon(\underline{r}) = \int d\nu\ \nu \int d^3p\ W(\underline{r},\underline{p};\nu) \tag{7}$$

and

$$p_j(\underline{r}) = \int d\nu \int d^3p\ p\ W(\underline{r},\underline{p};\nu). \tag{8}$$

The first of these terms coincide with the energy density (5) but the second does not quite coincide with the Poynting vector, which in our present notation may be written

$$S_j(\underline{r}) = \nu^{-1}\left\{+\frac{\partial}{\partial q_k}\Gamma_{jk}(\underline{r}+\tfrac{1}{2}\underline{q},\underline{r}-\tfrac{1}{2}\underline{q}) - \frac{\partial}{\partial q_j}\Gamma_{kk}(\underline{r}+\tfrac{1}{2}\underline{q},\underline{r}-\tfrac{1}{2}\underline{q})\right\}_{\underline{q}=0}. \tag{9}$$

The second term coincides with (8) but there is an additional term which may be expressed as a curl. Consequently, it does not contribute to net fluxes. We conclude that the Poynting vector includes, in addition to the density of momentum associated with the pencil of rays a circulating local momentum current. In general, it is a very small quantity and can be estimated to be of the order of

$$c/\nu L$$

where c is the velocity of light and L the linear dimension of the beam across which there is an appreciable variation of the intensity. For most practical purposes this difference may be neglected.

We still have to deal with the problem of both wave number $(p^2)^{1/2}$ and frequency ν being defined for the pencil of rays without a relation of the form

$$\nu = c(p^2)^{1/2}.$$

This happens in those cases we have great departure from locally plane waves; and when we have many waves, the wave numbers are averages of pairs of wave numbers of the waves. Accordingly the wave numbers satisfy the inequality

$$(p^2) \lesssim \nu^2/c^2. \tag{10}$$

It is as if we have "dispersion in free space" but remember that this is only appearance since the wave number we define is to be interpreted as the "momentum of a localized photon".

Propagation of Pencils of Rays

The two wave equations satisfied by $\Gamma_{jk}(r_1,r_2;\nu)$ can be displayed in a manner reminiscent of the eikonal equation for a classical wave. To start with, we have

$$(\nabla_1^2 + \nu^2/c^2)\ \Gamma_{jk}(\underline{r}_1,\underline{r}_2;\nu) = 0;$$

$$(\nabla_2^2 + \nu^2/c^2)\ \Gamma_{jk}(\underline{r}_1,\underline{r}_2;\nu) = 0; \tag{11}$$

which can be rewritten

$$\left\{\left(\frac{1}{2}\nabla \pm i\underline{p}\right)^2 + \nu^2/c^2\right\} W_{jk}(\underline{r},\underline{p};\nu) = 0. \tag{12}$$

Consequently,

$$\underline{p}\cdot\underline{\nabla}\ \ W_{jk}(\underline{r},\underline{p};\nu) = 0 \tag{13}$$

and

$$\left(p^2 - \frac{1}{4}\nabla^2\right) W_{jk}(\underline{r},\underline{p};\nu) = \frac{\nu^2}{c^2} W_{jk}(\underline{r},\underline{p},\nu). \tag{14}$$

The inequality of p^2 and ν^2/c^2 is evident from (14); they are equal in those locations where there is no local variation of the Wolf tensor and hence of the energy density. But in the edge domain of a light beam, we have bending of the light rays and the wave number p then no longer equals ν/c. The wave vector $\underline{p}$ must bend according to (13). So as to be along the level surfaces of $W_{jk}(\underline{r})$. One notes that while the specific correlation tensors were divergence-free, this translates into the polarization being transverse to the wave vector only in the region of local constancy. More generally we have

$$\nabla_{1j}\,\Gamma_{jk}(\underline{r}_1,\underline{r}_2;\ ) = \nabla_{2k}\,\Gamma_{jk}(\underline{r}_1,\underline{r}_2;\nu) = 0$$

and hence

$$\left(p_j - \frac{i}{2}\nabla_j\right) W_{jk}(\underline{r},\underline{p};\nu) = 0.$$

$$\left(p_k + \frac{i}{2}\nabla_k\right) W_{jk}(\underline{r},\underline{p};\nu) = 0. \tag{15}$$

Hence the polarization will be purely transverse as long as the variation of the Wolf tensor vanishes locally. These are immediate consequences of the equations of propagation (13) and (14).

Rewriting (14) in the form of a wave equation for the Wolf tensor

$$\left\{\nabla^2 + 4\left(\frac{\nu^2}{c^2} - p^2\right)\right\} W_{jk}(\underline{r},\underline{p};\nu) = 0. \tag{16}$$

We see that the "wave effects" in the propagation of the pencil of rays is confined to the departure of the wave number p from ν/c (apart from the possibility of the Wolf tensor!). By virtue of the inequality (10) we see that those components with

$$\nu^2/c^2 - p^2 = \mu^2/4 > 0$$

alternate rapidly with a typical Yukawa falloff $e^{-\mu r/r}$ from the source. At large distances only the "mass shell" contributions with $\nu = cp$ are available.

Pencils of Rays Forming Interference and Diffraction Patterns

To illustrate the use of the ray pencil picture of wave electromagnetics and display the scaling law, we simplify life (for the author and for the reader!) by replacing transverse vector light by "scalar light". The wave equations (11) are now scalar equations for a scalar two-point correlation function $\Gamma(r_1,r_2;\nu)$ and the Wolf function (4) is now a scalar function $W(\underline{r},\underline{p};\nu)$. It is still not necessarily point wise positive, but instead satisfies the positivity condition

$$\int d^3r \int d^3p\; W(\underline{r},\underline{p};\nu)\; M(\underline{r},\underline{p}) \geq 0 \tag{17}$$

where $M(\underline{r},p)$ is any "testing phase space density"

$$M(\underline{r},\underline{p}) = \int d^3q\ f^*(\underline{r} + \frac{1}{2}\underline{q})\ f(\underline{r} - \frac{1}{2}\underline{q}). \tag{18}$$

So we may say that while the ray pencil contains <u>tamasic</u> (dark) rays and <u>pradipa</u> (shining) rays[8], any reasonable test will produce non-negative illumination.

To simplify life further, we shall work in the Huyghen's approximation. It is possible to carry out the more elaborate exact calculations with transverse waves and without the use of the Huyghen's approximation. But here we present the simplified version in the Franhoufer region for the specific amplitude:

$$\phi_\nu(\underline{r}) \sim \frac{i\nu}{|\gamma-\gamma'|} \iint dx'\,dy'\,|\underline{r}-\underline{r}'|^{-1}\cdot \exp\left(i\frac{\nu}{c}|\underline{r}-\underline{r}'|\right)\phi_\nu(\underline{r}'). \tag{19}$$

For a single slit at x'=a we obtain the two-point function

$$\Gamma\left(x+\frac{\xi}{2}, y+\frac{\eta}{2}, z+\frac{\zeta}{2}; x-\frac{\xi}{2}, y-\frac{\eta}{2}, z-\frac{\zeta}{2}\right) \sim \exp -i\left(\frac{\nu\zeta}{c}\right)\exp\left\{i\frac{\nu}{c}(x-a)\xi/z\right\}$$

and the scalar Wolf function

$$W(x,y,z,p_1,p_2,p_3;\nu)\delta(p_2)\delta(p_3-\nu)\cdot\delta\left(p_1-\frac{\nu(x-a)}{c}\right) \tag{20}$$

apart from inessential constants. This expression is easily interpreted. It describes an elementary pencil originating at the slit x'=a and propagating geometrically to the point x on the "screen"; we have been careless about quadratic terms and so the fact that $p^2 \neq \nu^2/c^2$ is not to be taken seriously. There is an inverse square law attenuation contained in (19), but apart from that there is local isotropy. The ray density given by the scalar Wolf function (20) is clearly nonnegative.

We now turn to the two-slit interference pattern. In this case we can still calculate by elementary methods the two-point function

$$\Gamma \sim \exp\left(-\frac{i\nu}{c}\right)\left\{\exp\frac{i\nu(x-a)\xi}{cz} + \exp i\frac{\nu(x+a)\xi}{cz} + 2\exp\frac{i\nu\zeta}{cz}\cos\left(2\frac{\nu ax}{cz}+\phi\right)\right\} \tag{21}$$

and the Wolf function

$$W \sim \frac{c^2}{\nu^2 z^2} \; \delta(p_2)\delta(p_3-\nu)$$

$$\cdot \left\{ \delta\left[p_1 - \frac{\nu(x-a)}{cz}\right] + \delta\left[p_2 - \frac{\nu(x+a)}{cz}\right] \right.$$

$$\left. - 2 \cos \left[\frac{\nu ax}{cz} + \phi\right] \delta\left[p_1 - \frac{\nu x}{cz}\right] \right\}. \qquad (22)$$

Here ϕ is the phase difference between the two slits at $x' = \pm a$. The pattern corresponds to three distinct pencils of rays. One from each slit with elementary geometric propagation with only pradīpa rays but the third pencil apparently originating at a location between the two slits contains both tamasic and pradipa rays. For the $\phi=0$ case the central ray from the third beam is pradipa, while for the $\phi=\pi$ case it is tāmasic. The angular separation between the pradipa and tamasic rays is given by

$$\Delta\theta = \pi e/\nu a = \pi\lambda/a \qquad (23)$$

in accordance with the elementary result. The scaling law for the fringe pattern

$$\Delta x = z\Delta\theta = \pi\lambda z/a \qquad (24)$$

is an immediate consequence of the geometric propagation law and (22). The pattern scales since light rays travel in straight lines even in the formation of interference patterns!

It is of interest to point out that the need for three rather than two beams for two (coherent) sources is an essential component of the present exact ray picture and originates in the linearity of ray patterns in the two-point function. We recognize that the third pencil with both tamasic and pradipa rays is directly dependent on the coherence of the two (virtual) sources given by the two slits as evidenced by the appearance of the phase angle ϕ. For incoherent virtual sources this pencil of rays vanishes: this can be most easily seen as an average over the phase angle.

As an additional bonus we now recognize that most traditional descriptions of photon interference, for example, by the Copenhagen school are incomplete and inadequate. They all depend on two pencils of photon rays.

As a second illustration we consider the diffraction of (scalar) light by a thin slit. The two-point function and the Wolf function can be calculated by elementary techniques. The latter has the form

$$W \sim \frac{c^2}{\nu^2 z^2}\,\delta(p_2)\delta(p_3-\nu)$$

$$\cdot \left\{\delta\left(p_1 - \frac{\nu(x-a)}{cz}\right) + \delta\left(p_1 - \frac{\nu(x+a)}{cz}\right)\right.$$

$$\left. - 2\cos\frac{\nu a x}{cz}\,\delta\left(p_1 - \frac{x\nu}{cz}\right)\right\}. \tag{25}$$

In this case, again, we see three pencils of rays, the first two are very similar to the case of two slits but the third one originates in the center of the slit. The third pencil contains both kinds of rays.

We note, that in all these cases the total contribution of the three pencils at any field point add up to a nonnegative value by virtue of the Schwartz inequality. Hence the photometric intensity at any field point is nonnegative.

As a third example of some interest in its own right, let us consider the van Cittert-Zernike-Wolf theorem[9] on the propagation of light emerging from a primary radiating surface and the development of partial coherence by propagation. A primary radiating surface is a surface from which light emerges outward, no two distinct points of the surface having a mutual coherence. In accordance with an obvious generalization of (22) we then see that the illumination field will be simply represented in the ray pictures. Every point on the primary radiating surface becomes the source of rectilinear rays. Consequently in the "far zone" the rays of light will all be practically parallel over areas which subtend small solid angles at the source. By reading (4) from right to left, we see this to imply a sizable coherence patch in the far field. This is the van Cittert-Zernike-Wolf theorem.

Discussion

In this paper we have discussed the construction of pencils of rays which exactly represent any given wave field. This is not a high frequency limit in the sense of the geometric theory of Keller and others[1]; nor is it an asymptotic treatment. It is an exact picture, the wave aspects making themselves felt in the possible "dispersion in free space" and the curved light rays in empty space (13) and (14); and in the possible nonpositivity of the Wolf function (4). We consider these effects to be real physical effects rather than defects of the theory.

The phase space description of quantum theory was introduced into physics by Wigner[10] using the characteristic function defined as the expectation values of the Weyl unitary operators. Moyal[11] showed that quantum dynamics can be reformulated in terms of it using the Moyal sine brackets. Related distribution functions were introduced by Bargman and by Segal from the point of view of the

Hilbert space of entire functions[12] and by Klauder from the point of view of continuous representations.[13] The distribution functions which correctly give expectation values of normal ordered quantum electrodynamical variables was introduced by Glauber and by Sudarshan.[3,4] But the application to radiative transfer and rays in electromagnetic wave fields began with the work of Walther[7] and of Wolf.[6] The present theory was worked out by the author during the past few years; and we note that Bastiaans[14] has made several applications of the theory. Lohman[15] has pointed out that the traditional musical notation is an early application of such phase space distributions!

It may not be amiss to clarify a point of some confusion in the literature. The Wigner-Moyal density function is appropriate to a normalized quantum mechanical state vector (or more generally, a density matrix). As such the double integral of this distribution function is always unity; and all eigenvalues of the Wigner-Moyal distribution considered as a kernel would be of absolute value less than unity but summing to untiy. The Wolf tensors, on the other hand, do not have any such constraints. This stems from their being expectation values in a field, and it could have any non-negative trace whatever. The modes[5] are identifiable with wave functions but they have nontrivial multiplier factors. This is, of course, immediately obvious from elementary optics!

As a matter of some historical interest, it may not be amiss to recall two earlier works. About three decades ago Bohm[16] published two papers on "Quantum Theory in Terms of Hidden Variables" which made use of a "quantum-mechanical force". This work which gave birth to a considerable amount of discussion was modeled on the eikonal theory[1]; but the quantum-mechanical force is reminiscent of the curved light rays in empty space of the present theory. Bohm did not examine the ray density function but instead concentrated on the discussion of the equation for the path and, consequently, the possible nonpositivity of the ray density function was not recognized. Secondly, we have shown that dynamics in terms of phase space pictures obtained by Moyal[11] is essentially the only one available for any linear phase space formulation.[17]

Finally, since the picture in terms of ray pencils is an exact reformulation of the wave theory it follows that given the pencils of rays, the wave field can be obtained by inverting (4). This correspondence obtains for more general situations involving arbitrary passive linear media[18] which could be suitably nonuniform.

References and Footnotes

1. Morris Kline and Irvin W. Kay, "Electromagnetic Theory and Geometrical Optics", Interscience, New York (1965).

2. E. C. G. Sudarshan, Physica 96A, 315 (1979).

3. R. J. Glauber, Phys. Rev. Letts. 10, 84 (1963); E. C. G. Sudarshan, Phys. Rev. Lett. 10, 277 (1963), Proceedings of the Symposium on Optical Masers, edited by J. Fox, Wiley, New York (1963).

4. R. J. Glauber, Phys. Rev. 130, 2529 (1963); C. L. Mehta and E. C. G. Sudarshan, Phys. Rev. 138B, 274 (1965).

5. E. C. G. Sudarshan, J. Math. Phys. Sciences 3, 121 (1969).

6. E. Wolf, Phys. Rev. D 13, 869 (1976).

7. A. Walther, J. Opt. Soc. Amer. 63, 1622 (1973).

8. From sanskrit = tamas = darkness, pradīp = shining.

9. P. H. van Cittert, Physica 6, 1129 (1939); F. Zernike, Proc. Phys. Soc. (Lond) 61, 158 (1948); E. Wolf, Proc. Roy. Soc. A230, 246 (1955)

10. E. P. Wigner, Phys. Rev. 40, 749 (1932).

11. J. E. Moyal, Proc. Cambridge Phil. Soc. 45, 99 (1949).

12. V. Bargman, Comm. Pure Appl. Math. 14, 187 (1961).

13. J. R. Klauder, Ann. Phys. (N.Y.) 11, 123 (1960); S. S. Schweber, Math. Phys. 3, 831 (1962).

14. M. J. Bastiaans, "The Wigner Distribution Function and Its Applications to Optics", These Proceedings.

15. A Lohman, Introductory Lecture, These Proceedings.

16. D. Bohm, Phys. Rev. 85, 166; 180 (1952).

17. A. Simoni, F. Zaccaria, and E. C. G. Sudarshan, Nuovo Cimento 5B, 134 (1971).

18. B. Kagali, Ph.D. Thesis, University of Texas at Austin (1980).

A GENERALIZED RECIPROCITY THEOREM FOR ELECTROMAGNETIC OPTICS IN THE MOVING MEDIA

Hideya Gamo
University of California, Irvine
School of Engineering
Irvine, California 92717

ABSTRACT

A general reciprocity relation for electromagnetic waves in the bianisotropic medium is formulated based on the time-reversal operations, conditions of coefficients of constitutive relations for non-dissipative bianisotropic medium. The wave propogation in the uniformly moving isotropic and non-dispersive medium is treated as an example. Potential elaboration on the reciprocity relation for dissipative medium, unilateral operation of a ring dye laser by flowing active media and light scattering in laser transmission through turbulent flow are briefly discussed.

INTRODUCTION

The violation of reciprocity theorem in gyromagnetic medium has been well recognized in both microwave and optical regions but the non-reciprocal effect in moving medium seems to have been given less attention in optics. This is perhaps due to the fact that the velocity of the moving medium in our laboratory is much smaller than the speed of light. Since the laser has been introduced to various instrumentation, the non-reciprocal effect of the moving medium plays an important role in some laser systems. For instance, the DC flow of plasma in a ring laser gyroscope causes oscillation frequency difference between clock-wise and counter-clock-wise propagation directions without rotational motion.[1] Furthermore, the unilateral oscillation in a ring dye laser[2] was obtained by flowing dye solution in one direction very rapidly without an optical isolator[3] including a Faraday rotator.[4] In the optical communication system utilizing the laser transmission through turbulent atmosphere, the non-reciprocal wave propagation in the turbulent flow may become noticeable.[5]

With the objective of treating these non-reciprocal phenomena in the moving medium, a general formulation of reciprocity theorem is developed starting from the real electromagnetic fields and currents for the bianisotropic medium. The time-reversal reciprocity is described using the time-reversal transforms of electromagnetic fields and currents. The latter is especially useful for characterizing the non-dissipative bianisotropic medium.

Simple examples of non-reciprocal effects in non-dispersive moving dielectric medium such as Fresnel-Fizeau effect will be described.

ISSN:0094-243X/81/650106-17$1.50

BASIC EQUATIONS

The Maxwell equations are given by

$$\nabla \times E = -\frac{\partial B}{\partial t} - M \tag{1}$$

$$\nabla \times H = \frac{\partial D}{\partial t} + J \tag{2}$$

$$\nabla \cdot D = \rho \tag{3}$$

$$\nabla \cdot B = 0 \tag{4}$$

A fictitious magnetic conduction current density M is included for the convenience of calculating field distributions based on the induction and equivalence theorems in engineering electromagnetics.[6] The induced or scattered field can be calculated by using the electric and magnetic current sheets over a surface S. The current densities of these sheets are give by

$$J_s = \hat{n} \times (H_t^t - H_t^r) = \hat{n} \times H^i \tag{5}$$

$$M_s = -\hat{n} \times (E_t^t - E_t^r) = -\hat{n} \times E^i \tag{6}$$

where subscript t indicates the components tangential to the surface and superscripts t, r and i indicate transmission, reflection and incident wave, respectively. The special case of homogeneous medium in which the reflected wave is zero can be treated by the equivalence theorem:

$$J_s = \hat{n} \times H^i \tag{7}$$

$$M_s = -\hat{n} \times E^i \tag{8}$$

In case of dispersive media, such as the dispersive dielectrics, constitutive relations are given by

$$D(t) = \int_0^\infty \bar{\varepsilon}(\tau) \cdot E(t-\tau)d\tau + \int_0^\infty \bar{\xi}(\tau) \cdot H(t-\tau)d\tau . \tag{9}$$

$$B(t) = \int_0^\infty \bar{\mu}(\tau) \cdot H(t-\tau)d\tau + \int_0^\infty \bar{\zeta}(\tau) \cdot E(t-\tau)d\tau . \tag{10}$$

These constitutive relations can be re-written in terms of Fourier components $\bar{\varepsilon}(\omega)$, $\bar{\mu}(\omega)$, $E(\omega)$, $H(\omega)$:

$$D(t) = \frac{1}{2\pi}\int_0^\infty e^{j\omega t}\left[\bar{\varepsilon}(\omega)\cdot E(\omega) + \bar{\xi}(\omega)\cdot H(\omega)\right] d\omega \tag{11}$$

$$B(t) = \frac{1}{2\pi}\int_0^\infty e^{j\omega t}\left[\bar{\varepsilon}(\omega)\cdot H(\omega) + \bar{\zeta}(\omega)\cdot E(\omega)\right] d\omega \tag{12}$$

where complex tensors $\bar{\varepsilon}(\omega)$, $\bar{\mu}(\omega)$, and $\bar{\zeta}(\omega)$ generally satisfy:

$$\bar{\varepsilon}(-\omega) = \bar{\varepsilon}^*(\omega) \tag{13a}$$

$$\bar{\mu}(-\omega) = \bar{\mu}^*(\omega) \tag{13b}$$

$$\bar{\xi}(-\omega) = \bar{\xi}^*(\omega) \tag{13c}$$

$$\bar{\zeta}(-\omega) = \bar{\zeta}^*(\omega). \tag{13d}$$

In accordance with the causality principle, real and imaginary parts of $\varepsilon(\omega)$, $\xi(\omega)$ and $\zeta(\omega)$ satisfy the Kramers-Kronig relations, respectively.

When the incident electromagnetic radiation is monochromatic, we can express the fields as follows:

$$E(t) = \tfrac{1}{2}\left[e^{j\omega t}E(\omega) + e^{-j\omega t}E^*(\omega)\right], \tag{14}$$

$$H(t) = \tfrac{1}{2}\left[e^{j\omega t}H(\omega) + e^{-j\omega t}H^*(\omega)\right], \tag{15}$$

$$D(t) = \tfrac{1}{2}e^{j\omega t}\left[\bar{\varepsilon}(\omega)\cdot E(\omega) + \bar{\xi}(\omega)\cdot H(\omega)\right] + \text{c.c.} \tag{16}$$

and

$$B(t) = \tfrac{1}{2}e^{j\omega t}\left[\bar{\mu}(\omega)\cdot H(\omega) + \bar{\zeta}(\omega)\cdot E(\omega)\right] + \text{c.c.} \tag{17}$$

The constitutive relations of the complex time-harmonic field $D(\omega)$ and $B(\omega)$ are give by

$$D(\omega) = \bar{\varepsilon}(\omega)\cdot E(\omega) + \bar{\xi}(\omega)\cdot H(\omega), \tag{18}$$

and

$$B(\omega) = \bar{\mu}(\omega)\cdot H(\omega) + \bar{\zeta}(\omega)\cdot E(\omega), \tag{19}$$

where the complex tensors $\bar{\varepsilon}(\omega)$, $\bar{\mu}(\omega)$, $\bar{\xi}(\omega)$ and $\bar{\zeta}(\omega)$ of the non-dissipative or loss-less bianisotropic medium satisfy the following relations:

$$\overline{\varepsilon}(\omega) = \overline{\varepsilon}(\omega)^{\dagger}, \tag{20a}$$

$$\overline{\mu}(\omega) = \overline{\mu}(\omega)^{\dagger}, \tag{20b}$$

$$\overline{\zeta}(\omega) = \overline{\xi}(\omega)^{\dagger}, \tag{20c}$$

$$\overline{\xi}(\omega) = \overline{\zeta}(\omega)^{\dagger}. \tag{20d}$$

The proof of these relations is described in the Appendix.

TIME-REVERSAL TRANSFORMS

In order to keep the Maxwell equations (Eqs. 1,2,3,4) invariant under the time-reversal operation (T), the real fields E, D, B, H and real current densities J and M must satisfy:

$$TE = E, \tag{21a}$$

$$TD = D, \tag{21b}$$

$$TB = -B, \tag{21c}$$

$$TH = -H, \tag{21d}$$

$$TJ = -J, \tag{21e}$$

$$TM = M. \tag{21f}$$

In case of constitutive relations of the non-dispersive bianisotropic medium given by

$$D = \overline{\varepsilon} \cdot E + \overline{\xi} \cdot H , \tag{22}$$

$$B = \overline{\mu} \cdot H + \overline{\zeta} \cdot E . \tag{23}$$

we obtain the following relations:

$$T\overline{\varepsilon} = \overline{\varepsilon}, \tag{24a}$$

$$T\overline{\mu} = \overline{\mu}, \tag{24b}$$

$$T\overline{\xi} = -\overline{\xi}, \tag{24c}$$

$$T\overline{\zeta} = -\overline{\zeta} , \tag{24d}$$

where use has been made of Eqs. (21a)-(21f).

The complex time-harmonic fields introduced above will satisfy the following Maxwell equations:

$$\nabla \times E(\omega) = -j\omega B(\omega) - M(\omega) \tag{25}$$

$$\nabla \times H(\omega) = j\omega D(\omega) + J(\omega) \tag{26}$$

The fields under the time-reversal transform satifsy the equations which are obtained as the complex conjugate of the above equations:

$$\nabla \times E^*(\omega) = + j\omega B^*(\omega) - M^*(\omega) \tag{27}$$

$$\nabla \times H^*(\omega) = - j\omega D^*(\omega) + J^*(\omega) \tag{28}$$

The time reversal transforms of these complex time-harmonic fields must satisfy the following relations:

$$TE(\omega) = E^*(\omega)\ , \tag{29a}$$

$$TD(\omega) = D^*(\omega)\ , \tag{29b}$$

$$TH(\omega) = -H^*(\omega)\ , \tag{29c}$$

$$TB(\omega) = -B^*(\omega)\ , \tag{29d}$$

$$TJ(\omega) = -J^*(\omega)\ , \tag{29e}$$

$$TM(\omega) = M^*(\omega)\ . \tag{29f}$$

The time-reversal transforms of the coefficients of constitutive relations for complex time-harmonic fields satisfy

$$T\ \bar{\varepsilon}(\omega) = \bar{\varepsilon}^*(\omega)\ , \tag{30a}$$

$$T\ \bar{\mu}(\omega) = \bar{\mu}^*(\omega)\ , \tag{30b}$$

$$T\ \bar{\xi}(\omega) = -\bar{\xi}^*(\omega)\ , \tag{30c}$$

$$T\ \bar{\zeta}(\omega) = -\bar{\zeta}^*(\omega)\ . \tag{30d}$$

When the dispersive medium is under the magnostatic field H, the tensors $\bar{\varepsilon}$, $\bar{\mu}$, $\bar{\xi}$ and $\bar{\zeta}$ are generally functions of H. In this case, we must reverse the sign of the bias field for time reversal operation:

$$T\bar{\varepsilon}(\omega;H) = \bar{\varepsilon}^*(\omega,-H), \tag{31a}$$

$$T\bar{\mu}(\omega;H) = \bar{\mu}^*(\omega,-H), \tag{31b}$$

$$T\bar{\xi}(\omega;H) = -\bar{\xi}^*(\omega,-H), \tag{31c}$$

$$T\bar{\zeta}(\omega;H) = -\bar{\zeta}^*(\omega,-H). \tag{31d}$$

The time reversal of the real electric field in Eq. (14) is given by

$$TE(t) = \tfrac{1}{2}\left[e^{-j\omega t}E^*(\omega) + e^{j\omega t}E(\omega)\right] = E(t), \tag{32}$$

where Eq. (29a) is applied to Eq. (14).
This agrees with Eq. (21a). On the other hand, the real magnetic field H(t) in Eq. (15) is transformed under the time reversal into

$$TH(t) = \tfrac{1}{2}\left[e^{-j\omega t}\, H(\omega) - e^{+j\omega t}H^*(\omega)\right] = -H(t) \tag{33}$$

in accordance with Eq. (21d).

The time reversals of D(t) and B(t) in Eqa. (16) and (17) are respectively given using Eqs. (30 a,b,c,d) by

$$TD(t) = \tfrac{1}{2}e^{-j\omega t}\left[\bar{\varepsilon}^*(\omega)\cdot E^*(\omega) + \bar{\xi}^*(\omega)\cdot H^*(\omega)\right] + \text{c.c.} \tag{34}$$

$$= D(t),$$

and

$$TB(t) = \tfrac{1}{2}e^{-j\omega t}\left[-\bar{\mu}^*(\omega)\cdot H^*(\omega) - \bar{\zeta}^*(\omega)\cdot E^*(\omega)\right] + \text{c.c.} \tag{35}$$

$$= -B(t).$$

THE NORMAL RECIPROCITY THEOREM

The reciprocity theorem of electromagnetic radiation has been usually formulated in terms of the complex time-harmonic fields instead of real time signal. However, physical meaning of the formulation can be more clearly defined by using the real electromagnetic fields by Eqs. (14), (15), (16) and (17). Assuming that the bianisotropic medium and radiation sources are limited within a finite region, one can show that the following surface integral will vanish in the limit of infinitely large distance from the sources:

$$\int (E_1 \times H_2 - E_2 \times H_1)\cdot dS = 0. \tag{36}$$

The reason for this is that at very large distances from the sources the electromagnetic waves can be approximated by the following electric dipole radiation:

$$H_1 = \hat{n} \times E_1/\eta\,, \tag{37}$$

$$H_2 = \hat{n} \times E_2/\eta\,. \tag{38}$$

Thus, using these equations we obtain

$$E_1 \times H_2 = E_1 \times (\hat{n} \times E_2) = E_2(E_1 \cdot \hat{n})/\eta - \hat{n}\,(E_1 \cdot E_2)/\eta \tag{39}$$

$$E_2 \times H_1 = E_2 \times (\hat{n} \times E_1) = E_1(E_2 \cdot \hat{n})/\eta - \hat{n}\,(E_2 \cdot E_1)/\eta \tag{40}$$

where $\eta = \sqrt{\mu/\varepsilon}$ is the wave impedance.

Since the first terms of the above equations satisfy in the plane wave limit

$$(E_1 \cdot \hat{n}) = (E_2 \cdot \hat{n}) = 0. \tag{41}$$

and the second terms will cancel each other on the surface mentioned above. Thus, we obtain the normal reciprocity theorem, Eq. (36).

According to the divergence theorem, the surface integral is reduced to the volume integral of

$$\nabla \cdot (E_1 \times H_2 - E_2 \times H_1).$$

By using the vector formula

$$\nabla \cdot (E_1 \times H_2) = H_2 \cdot (\nabla \times E_1) - E_1 \cdot (\nabla \times H_2) \tag{42}$$

and substituting $\nabla \times E$ and $\nabla \times H$ by the right hand side of the Maxwell equations, we obtain

$$\iint \left\{ E_2 \cdot (\dot{D}_1 + J_1) - H_2 \cdot (\dot{B}_1 + M_1) - E_1 \cdot (\dot{D}_2 + J_2) + H_1 \cdot (\dot{B}_2 + M_2) \right\} dV$$

$$= 0. \tag{43}$$

First, regarding J_1, J_2, M_1, M_2 and displacement and induction currents $\dot{D}_1$, $\dot{D}_2$ and $\dot{B}_1$, $\dot{B}_2$ as delta functions of spatial coordinates, respectively, we obtain

$$E_2 \cdot (\dot{D}_1 + J_1) - H_2 \cdot (\dot{B}_1 + M_1) - E_1 \cdot (\dot{D}_2 + J_2) +$$

$$H_1 \cdot (\dot{B}_2 + M_2) = 0. \tag{44}$$

By inserting

$$D = \tfrac{1}{2}(j\omega D(\omega)e^{j\omega t} - j\omega D^*(\omega)e^{-j\omega t}), \tag{45a}$$

$$\dot{B} = \tfrac{1}{2}(j\omega B(\omega)e^{j\omega t} - j\omega B^*(\omega)e^{-j\omega t}), \tag{45b}$$

$$J = \tfrac{1}{2}(J(\omega)e^{j\omega t} + J^*(\omega)e^{-j\omega t}), \tag{45c}$$

$$M = \tfrac{1}{2}(M(\omega)e^{j\omega t} + M^*(\omega)e^{-j\omega t}), \tag{45d}$$

into Eq. (44), we obtain the term with $\exp(2j\omega t)$ and time-independent term, respectively:

$$E_2\cdot (J_1+ j\omega D_1) - H_2\cdot(M_1+j\omega B_1) - E_1\cdot(J_2+ j\omega D_2)$$
$$+ H_1\cdot (M_2+ j\omega B_2) = 0, \tag{46}$$

and

$$E_2{}^*\cdot(J_1+j\omega D_1) - H_2{}^*\cdot(M_1+j\omega B_1) - E_1{}^*\cdot(J_2+ j\omega D_2)$$
$$+ H_1{}^*\cdot(M_2+ j\omega B_2) + E_2\cdot(J_1{}^* - j\omega D_1{}^*) -H_2\cdot(M_1{}^*-j\omega B_1{}^*)$$
$$- E_1\cdot(J_2{}^* - j\omega D_2{}^*) + H_1\cdot(M_2{}^*-j\omega B_2{}^*) = 0. \tag{47}$$

By substituting the constitutive equations, Eqs. (18) and (19) in these equations and utilizing the following relations, for instance,

$$E_1\cdot\bar{\varepsilon}\cdot E_2 = E_2\cdot\bar{\varepsilon}'\cdot E_1 \tag{48}$$

and

$$E_1\cdot\varepsilon^*\cdot E_2{}^* = E_2{}^*\cdot\bar{\varepsilon}^\dagger\cdot E_1 \tag{49}$$

we obtain for the $\exp(2j\omega t)$ and time-independent terms, respectively:

$$E_2\cdot J_1 - H_2\cdot M_1 - E_1\cdot J_2 + H_1\cdot M_2+ j\omega\Big[E_2\cdot(\bar{\varepsilon} - \bar{\varepsilon}')\cdot E_1+ E_2\cdot(\bar{\xi} + \bar{\zeta}')\cdot H_1 -$$
$$H_2\cdot(\bar{\mu}-\bar{\mu}')\cdot H_1 - H_2\cdot(\zeta+\bar{\xi}')\cdot E_1\Big] = 0 \tag{50}$$

and

$$E_2\cdot J_1{}^* - H_2\cdot M_1{}^* - E_1\cdot J_2{}^* + H_1\cdot M_2{}^* + \text{ c.c.t } j\omega\Big[E_2{}^*\cdot(\bar{\varepsilon}+\bar{\varepsilon}^\dagger)\cdot E_1$$
$$- H_2{}^*\cdot(\bar{\mu} + \mu^\dagger)\cdot H_1+ E_2{}^*\cdot(\bar{\xi} - \bar{\zeta}^\dagger)\cdot H_1 - H_2{}^*\cdot(\bar{\zeta} - \bar{\xi}^\dagger)\cdot E_1 - \text{c.c.}\Big]$$
$$= 0. \tag{51}$$

Note that $\bar{\xi}'$ and $\bar{\varepsilon}^\dagger$ denote the transposed and Hermitian conjugate of $\bar{\xi}$ and $\bar{\varepsilon}$, respectively.

Eq. (50) is equal to the result usually obtained by using the complex electromagnetic fields and the Maxwell equation for complex time-harmonic fields, Eq. (25) and (26).

In case of reciprocal media in which both the dielectric permittivity and magnetic permeability tensors are symmetric,

$$\overline{\varepsilon} = \overline{\varepsilon}', \tag{52a}$$

$$\overline{\mu} = \overline{\mu}', \tag{52b}$$

$$\overline{\zeta} = -\overline{\xi}' \tag{52c}$$

$$\overline{\xi} = -\overline{\zeta}'. \tag{52d}$$

Substituting these relations into Eq. (51) we obtain the well-known reciprocity relation:

$$E_2 \cdot J_1 - H_2 \cdot M_1 = E_1 \cdot J_2 - H_1 \cdot M_2. \tag{53}$$

By introducing the transfer-impedance tensors Z_{12} and Z_{21} by

$$E_1 = Z_{21} \cdot J_1 , \tag{54a}$$

$$E_2 = Z_{12} \cdot J_1 , \tag{54b}$$

and the transfer-admittance tensors Y_{21} and Y_{12} by

$$H_1 = Y_{21} \cdot M_1 , \tag{55a}$$

$$H_2 = Y_{12} \cdot M_2 , \tag{55b}$$

and substituting them into Eq. (53), we finally obtain the reciprocity relations,

$$Z_{12} = Z_{21} , \tag{56}$$

and

$$Y_{12} = Y_{21} , \tag{57}$$

for the reciprocal medium.

However, the above equation does not hold in case of the non-reciprocal media such as gyrotropic medium for which the off-diagonal elements of permeability and permittivity tensors are anti-symmetric:

$$\overline{\mu}_{ij} = -\overline{\mu}_{ji} \tag{58a}$$

$$\overline{\varepsilon}_{ij} = -\overline{\varepsilon}_{ji} \tag{58b}$$

The equations (46) and (47) may be respectively regarded as a generalized reciprocity relation between electric and magnetic fields, and the generalized currents which include displacement and magnetic induction currents, respectively

$$C = J + j\omega D, \tag{59a}$$

$$G = M + j\ B. \tag{59b}$$

Thus we obtain

$$E_2 \cdot C_1 - H_2 \cdot G_1 = E_1 \cdot C_2 - H_1 \cdot G_2\ , \tag{60}$$

and

$$E_2 \cdot C_1^* - H_2 \cdot G_1^* - E_1 \cdot C_2^* - H_1 \cdot G_2^* + c.c. = 0. \tag{61}$$

THE TIME-REVERSAL RECIPROCITY

In acase of gyromagnetic medium the reciprocity relation, Eq. (53), holds when the direction of static magnetic field is reversed because

$$\bar{\varepsilon}'(-H) = \bar{\varepsilon}(H), \tag{62a}$$

$$\bar{\mu}'(-H) = \bar{\mu}(H), \tag{62b}$$

$$\bar{\xi}'(-H) = \bar{\xi}(H), \tag{62c}$$

$$\bar{\zeta}'(-H) = \zeta(H), \tag{62d}$$

and the terms with $j\omega$ in Eq. (50) vanish. According to Eq. (21d) the reversal of magnetic field is equivalent to the time reversal. Thus, we shall investigate the reciprocal relation between E_1, H_1 and TE_2, TH_2.

By noting that $TE_2 = E_2$ and $TH_2 = -H_2$, we obtain from Eq. (36) the following integral:

$$\iint \left[E_1 \times (TH_2) + (TE_2) \times H_1\right] \cdot dS = 0. \tag{63}$$

Here we assumed that the radiation sources and the bianisotropic medium are restricted within a finite region and the integrating surface is located infinitely distant from sources. Thus, the electric dipole radiation approximation holds in these fields on the surface.

We shall denote the time-reversal fields and currents by $\bar{E}$, $\bar{D}$, $\bar{H}$, $\bar{B}$, $\bar{J}$ and $\bar{M}$. The fields and currents E, D, H, B, J, M, $\dot{D}$, and $\dot{B}$ are replaced by $\bar{E}$, $\bar{D}$, $-\bar{H}$, $-\bar{B}$, $-\bar{J}$, $\bar{M}$, $-\dot{\bar{D}}$, and $\dot{\bar{B}}$, respectively, in accordance with Eqs. (21a-f). From Eq. (44), we obtain

$$\bar{E}_2 \cdot (D_1 + J_1) + \bar{H}_2 \cdot (B_1 + M_1) + E_1 \cdot (\bar{D}_2 + \bar{J}_2) + H_1 \cdot (\bar{B}_2 + \bar{M}_2)$$
$$= 0. \tag{64}$$

By substituting Eqa. (45a-d) and their time-reversals, we obtain the following two equations for the term with exp $(2j\omega t)$ and the time-independent term, respectively:

$$E_2 \cdot (J_1 + j\omega D_1) + \overline{H}_2 \cdot (M_1 + j\omega B_1) + E_1 \cdot (\overline{J}_2 + j\omega \overline{D}_2)$$
$$+ H_1 \cdot (\overline{M}_2 + j\omega \overline{B}_2) = 0, \tag{65}$$

and

$$\overline{E}_2^{\,*} \cdot (J_1 + j\omega D_1) + \overline{H}_2^{\,*} \cdot (M_1 + j\omega B_1) + E_1^{\,*} \cdot (\overline{J}_2 + j\omega \overline{D}_2) + H_1^{\,*} \cdot (\overline{M}_2 + j\omega \overline{B}_2)$$
$$+ \overline{E}_2 \cdot (J_1^{\,*} - j\omega D_1^{\,*}) + H_2 \cdot (M_1^{\,*} - j\omega B_1^{\,*}) + E_1 \cdot (J_2^{\,*} - j\omega \overline{D}_2^{\,*})$$
$$+ H_1 \cdot (\overline{M}_2^{\,*} - j\omega B_2^{\,*}) = 0. \tag{66}$$

By inserting the constitutive relations, Eqa. (18) and (19) into these equations, we obtain

$$\overline{E}_2 \cdot J_1 + \overline{H}_2 \cdot M_1 + E_1 \cdot \overline{J}_2 + H_1 \cdot M_2 + j\omega \Big[\overline{E}_2 \cdot (\overline{\varepsilon} + \overline{\varepsilon}') \cdot E_1 + \overline{H}_2 \cdot (\overline{\mu} + \overline{\mu}')$$
$$\cdot H_1 + \overline{E}_2 \cdot (\overline{\xi} + \overline{\zeta}') \cdot H_1 + \overline{H}_2 \cdot (\overline{\zeta} + \overline{\xi}') \cdot E_1 \Big] = 0 \tag{67}$$

and

$$\overline{E}_2 \cdot J_1^{\,*} + \overline{H}_2 \cdot M_1^{\,*} + E_1 \cdot \overline{J}_2^{\,*} + H_1 \cdot \overline{M}_2^{\,*} + \text{c.c.}$$
$$+ j\omega \Big[\overline{E}_2^{\,*} \cdot (\overline{\varepsilon} - \overline{\varepsilon}^{\dagger}) \cdot E_1 + H_2^{\,*} \cdot (\overline{\mu} - \overline{\mu}^{\dagger}) \cdot H_1$$
$$+ \overline{E}_2^{\,*} \cdot (\overline{\xi} - \overline{\zeta}^{\dagger}) \cdot H_1 + \overline{H}_2^{\,*} \cdot (\overline{\zeta} - \overline{\xi}^{\dagger}) \cdot E_1 - \text{c.c.} \Big] = 0. \tag{68}$$

The equation (68) is especially useful for the non-dissipative medium because the term with $j\omega$ will vanish owing to Eqs.(20a--d). Then, the following reciprocity relation must hold for the non-dissipative medium:

$$\overline{E}_2 \cdot J_1^{\,*} + \overline{H}_2 \cdot M_1^{\,*} + E_1 \cdot \overline{J}_2^{\,*} + H_1 \cdot M_2^{\,*}$$
$$+ \overline{E}_2^{\,*} \cdot J_1 + \overline{H}_2^{\,*} \cdot M_1 + E_1^{\,*} \cdot \overline{J}_2 + H_1^{\,*} \cdot M_2 = 0. \tag{69}$$

By substituting Eqs. (54 a,b) and Eqs.(55 a,b) into the above equation, we find that

$$Z_{12} + Z_{21}^{\,*} = 0, \tag{70}$$
$$Y_{12} + Y_{21}^{\,*} = 0, \tag{71}$$

for the non-dissipative medium.

It might be interesting to note that Eqs. (67) and (69) can be formally derived from Eqs. (51) and (50) by using the time-reversal transforms for complex time-harmonic fields, Eqs. (29,30), respectively. However, this does not mean these equations are equivalent.

WAVE PROPAGATION IN MOVING MEDIUM

We shall examine the reciprocity relation for the waves propagating in the moving medium by using the constitutive relations of a uniformly moving isotropic and non-dispersive medium:

$$D = \varepsilon A \cdot E + \Omega \times H \;, \tag{72}$$

$$B = \mu A \cdot H - \Omega \times E \;, \tag{73}$$

where

$$A = \frac{1 - \beta^2}{1 - n^2\beta^2}\left(I - \frac{n^2 - 1}{1 - \beta^2}\beta\beta \right) , \tag{74}$$

$$\Omega = \frac{n^2 - 1}{1 - n^2\beta^2} \; \frac{\beta}{c} , \tag{75}$$

$$\beta = \frac{v}{c} , \tag{76}$$

and the index of refraction

$$n = \sqrt{\varepsilon\mu/\varepsilon_o\mu_o} \;. \tag{77}$$

These constitutive relations were derived by Minkowski[14] under the[15] postulate that macroscopic Maxwell's equations are Lorentz covariant.[16] [17] By introducing a 3 × 3 dyad $\overline{k}$ defined by $\overline{k} \cdot A \equiv k \times A$ and represented by a skew symmetric matrix,

$$\overline{k} = \begin{pmatrix} 0 & -k_z & k_y \\ k_z & 0 & -k_x \\ -k_y & k_x & 0 \end{pmatrix} , \tag{78}$$

the above constitutive relations can be reformulated as follows:

$$D = \overline{\varepsilon} \cdot E + \overline{\Omega} \cdot H \;, \tag{79}$$

$$B = \overline{\mu} \cdot H - \overline{\Omega} \cdot E \;, \tag{80}$$

where

$$\overline{\varepsilon} = \varepsilon\overline{\alpha} \ , \tag{81}$$

$$\overline{\mu} = \varepsilon\overline{\alpha} \ , \tag{82}$$

and

$$\overline{\alpha} = \begin{pmatrix} a & 0 & 0 \\ 0 & a & 0 \\ 0 & 0 & 1 \end{pmatrix}, \tag{83}$$

and

$$a = (1 - \beta^2)/(1 - n^2\beta^2), \tag{84}$$

$$\overline{\Omega} = \frac{n^2 - 1}{1 - n^2\beta^2} \ \frac{1}{c^2} \begin{pmatrix} 0 & -v_z & v_y \\ v_z & 0 & -v_x \\ -v_y & v_x & 0 \end{pmatrix}, \tag{85}$$

and the velocity of light in the vacuum:

$$c = 1/\sqrt{\varepsilon_o \mu_o} \ . \tag{86}$$

Note that $\overline{\varepsilon}$ and $\overline{\mu}$ defined above are symmetric tensors but $\overline{\Omega}$ is anti-symmetric or skew symmetric and satisfies the relations $\zeta = \overline{\xi}\dagger$ and $\overline{\zeta} = \overline{\zeta}\dagger$ (Eqs.20c and d) because $\overline{\xi} = \overline{\Omega}, = \overline{\zeta} = - \overline{\Omega}$ and, $\overline{\Omega}' = - \overline{\Omega}$.

As examples we shall treat the case in which the electromagnetic wave propagating in the same direction as the moving medium: Assuming the z component of velocity $v_z = v$ and $v_x = v_y = 0$, we can easily obtain the propagation constant

$$\gamma_{\pm} = j\omega\sqrt{\mu\varepsilon}\ [1 \pm (n^2-1)\ v/c] \tag{87}$$

The phase velocity $v_+ = \omega/\beta_+$ in the same direction as the moving medium is greater than the phase velocity $v_- = \omega/\beta_-$ for the wave propagating in the opposite direction. This represents the Fresnel-Fizeau drag effect. The wave impedance for waves in positive and negative z direction satisfies

$$\eta^+ = -\eta^- = \sqrt{\mu/\varepsilon} \ .$$

In the case where the electromagnetic wave propagates in the direction perpendicular to the direction of moving medium; $v_x = v$ and $v_y = v_z = 0$. Then, the phase constants $\beta_{\parallel}$ and $\beta_{\perp}$ satisfy

$$\beta_{\parallel} = \beta_{\perp} = \omega\sqrt{\mu\varepsilon}\ \sqrt{1 - (n^2-1)\ v^2/c^2} \ , \tag{89}$$

but the wave impedances are given by

$$\eta_{\parallel} = \eta \sqrt{1 - (n^2-1)\, v^2/c^2} \tag{90}$$

and

$$\eta_{\perp} = \eta \sqrt{1 - (n^2-1)\, v^2/c^2} \tag{91}$$

where $\eta = \sqrt{\mu/\varepsilon}$

DISCUSSION

In the derivation of reciprocity relation in Eq. (56), we have assumed the symmetry of tensors $\overline{\varepsilon}$, $\overline{\mu}$, $\overline{\xi}$ and $\overline{\zeta}$ in the constitutive relations, Eqs. (52a-d). In order to justify these symmetry relations of dissipative medium we must follow the procedures used by Landau and Lifshitz[11] based on the Onsager priniple of microscopic reversibility.[12 13] The statistical theory of thermal radiation in dissipative bianisotropic medium should be formulated using the procedures by Landau and Lifshitz including correlation functions of electromagnetic fields.

The transfer impedance tensor Z_{12} defined by Eqs. (54a) and (54b) may be regarded as a special case of dyadic Green's function . The reciprocity relation can be reduced as the symmetric properties of the dyadic Green's function. With regards to the dyadic Green's function for the moving medium, refer to Ref. 18. It would be interesting to clarify the symmetric properties of dyadic Green's function from the standpoint of space time reversal operations.

In order to treat practical systems such as the moving dye, we have to elaborate on the above treatment by including the dispersive property of active medium and also treat the reflection and transmission at the boundary of the moving medium. The theory which has been developed for the moving medium contains the basic material necessary for further development although it treats only the idealized problem; for instance, reflection and transmission at the interface of moving medium[19] and cut-off nature of waveguide modes in moving materials[20] and the dispersion effect.[21 22]

In the practical laser communication system through turbulent atmosphere, the uniform and turbulent air flow introduces unique disturbances because of the multi-path transmission in non-reciprocal media. For instance, there are significant differences between optical transmission using reflection by a flat mirror and a corner-cube reflector.[4] Since the velocity of air flow is very small compared to the speed of light, we may expect the reciprocity theorem of Helmholtz[23] can explain the corner-cube case very well. This is not necessarily the case. It would be meaningful to generalize the reciprocity relation in terms of scattering matrix[24] using the constitutive relations of the moving medium.

APPENDIX

POYNTING THEOREM FOR BIANISOTROPIC MEDIUM

Substituting Eq. (1) and (2) into

$$\nabla\cdot(E\times H) = H\cdot(\nabla\times E) - E\cdot(\nabla\times H) \tag{A.1}$$

we obtain

$$\nabla\cdot(E\times H) = -H\cdot\frac{\partial B}{\partial t} - E\cdot\frac{\partial B}{\partial t} - H\cdot M - E\cdot J \tag{A.2}$$

Inserting Eqs. (14) (15) (16) and (17) into Eqs. (A.2), and taking the time average

$$\begin{aligned}\langle\nabla\cdot(E\times H)\rangle_t &= \mathrm{Re}\ \nabla\cdot[\tfrac{1}{2}E(\omega)\times H^*(\omega)]\\ &= -\mathrm{Re}[\frac{j\omega}{2} H^*(\omega)\cdot(\overline{\mu}-\overline{\mu}^{\dagger})\cdot H(\omega)\\ &\quad+\frac{j\omega}{2} E^*(\omega)\cdot(\overline{\varepsilon}-\overline{\varepsilon}^{\dagger})\cdot E(\omega)\\ &\quad+\frac{j\omega}{2} H^*(\omega)\cdot(\overline{\zeta}-\overline{\xi}^{\dagger})\cdot E(\omega)\\ &\quad+\frac{j\omega}{2} E^*(\omega)\cdot(\overline{\xi}-\overline{\zeta}^{\dagger})\cdot H(\omega)]\\ &\quad-\tfrac{1}{2}\mathrm{Re}[H^*(\omega)\cdot M(\omega)+E^*(\omega)\cdot J(\omega)]\ .\end{aligned} \tag{A.4}$$

In the non-dissipative medium the outgoing power and power induced by current sources must be balanced.

$$\begin{aligned}\langle\nabla\cdot(E\times H)\rangle_t &+ \tfrac{1}{2}\mathrm{Re}[H^*(\omega)\cdot M(\omega)+E^*(\omega)\cdot J(\omega)]\\ &= \mathrm{Re}[j\frac{\omega}{2}\{H^*(\omega)\cdot(\overline{\mu}-\overline{\mu}^{\dagger})\cdot H(\omega)\\ &\quad+E^*(\omega)\cdot(\overline{\varepsilon}-\overline{\varepsilon}^{\dagger})\cdot E(\omega)\\ &\quad+E^*(\omega)\cdot(\overline{\xi}-\overline{\zeta}^{\dagger})\cdot H(\omega)\\ &\quad+H^*(\omega)\cdot(\overline{\zeta}-\overline{\xi}^{\dagger})\cdot E(\omega)\}] = 0\ .\end{aligned} \tag{A.5}$$

For any arbitrary sources and fields the tensor $\overline{\mu} - \overline{\mu}^{\dagger}$, $\overline{\varepsilon} - \overline{\varepsilon}^{\dagger}$, $\overline{\xi} - \overline{\zeta}^{\dagger}$ and $\overline{\zeta} - \overline{\xi}^{\dagger}$ must vanish. Thus, Eqs. (20a,b,c,d) result.

REFERENCES

1. Podgorski, T.J. and Aronowitz, F., IEEE J. Quant. Elec. 4, 11 (1968).
2. R. Tansey and P. Slaymaker, "Longitudinal Mode Tuning of Forward and Reverse Travelling Waves in Unstable Ring Resonators with a Fast Flowing Gain Medium," 1979 Opt. Soc. Am. Conf. Rochester, N.Y., Oct. 9-12, 1979; J. Opt. Soc. Am. 64, 1475 (1979).
3. H. Gamo, S.S. Chuang and R.E. Grace, IEEE J. Quant. Elec. QE-3, 243 (1967).
4. H. Gamo and N. Takahashi, "Statistical Model of the CW Unilateral Partially Homogeneously Broadened Ring Laser," 1979 Opt. Soc. Am. Conf., Rochester, N.Y., Oct. 9-12, 1979; J. Opt. Soc. Am. 69, 1475 (1979).
5. H. Gamo, "Comparison of a Corner-Cube Reflector and a Plane Mirror in Folded Path and Direct Transmission Through Atmospheric Turbulence," App. II, Optical Transmission Through the Turbulent Atmosphere, Final Scientific Report AFOSR 76-3097, Oct. 1, 1976 to Jan. 31, 1978.
6. E. C. Jordan and K.G. Balmain, Electromagnetic Waves and Radiating Systems, 2nd Ed., Prentice-Hall, 1968.
7. J.A. Kong, Theory of Electromagnetic Waves, John Wiley and Sons, N.Y., Chap. 2 (1975).
8. T.H. O'Dell, Electrodynamics of Magneto-Electric Media, North-Holland Publ. Co. (1970).
9. W. Lukocz, Symmetries and Time-Reversal in Optics, Proc. of ICO Conf. on Optics in Four Dimensions, Ensenada, Mexico, Aug. 4-8, 1980.
10. R.F. Harrington, Time Harmonic Electromagnetic Fields, McGraw-Hill (1961).
11. L.D. Landau and E.M. Lifshitz, Electrodynamics of Continuous Media, Pergamon Press, Chapt. 13 (1960).
12. L. Onsager, Phys. Rev. 37, 405 (1931) and 38, 2265 (1931).
13. H.B.G. Casimir, Rev. Mod. Phys. 17, 343 (1945).
14. H. Minkowsi, Göttingen Nachrihten, p. 53-116 (1908).
15. A. Sommerfeld, Electrodynamics, Academic Press (1952).
16. E.J.Post, Formal Structure of Electromagnetics, North-Holland Publ. Co., Amsterdam, Chap. 6 (1962).
17. R. Becker, Electromagnetic Fields and Interactions, Vol. 1, Electromagnetic Theory and Relativity, Blaisdell Publ. Co., N.Y. p. 374.
18. Chen-To Tai, Dyadic Green's Functions in Electromagnetic Theory, Intext Educational Publ. 1971; see also, M.J. van Weert, J. Opt. Soc. Am. 68, 1275 (1978).
19. B. Podolsky and K.S. Kunz, Fundamentals of Electrodynamics, Marcel Dekker, N.Y., Chap. 8, Sec. 29 (1969).
20. L.J. Du and R.T. Compton, IEEE Trans. Microwave Theory and Techniques MTT-14, 358 (1966).

21. C.H. Papas, Theory of Electromagnetic Wave Propagation, McGraw-Hill, N.Y., Chap. 7 (1965).
22. I.M. Frank, J. Phys. USSR 7(2), p. 49-67 (1943).
23. M. Born and E. Wolf, Principles of Optics, Pergamon Press, 4th Ed., p. 381 (1970).
24. D.S. Saxon, Lecture on the Scattering of Light, Proc. UCLA Int. Conf. on Radiation and Remote Probing of the Atmosphere, UCLA Dept. Atmospheric Science, Ed. by J.G. Kuriyan, p. 227-308 (1974).

ON A RELATIONSHIP BETWEEN SPATIAL COHERENCE AND TEMPORAL COHERENCE IN FREE FIELDS

E. Wolf
Department of Physics and Astronomy and The Institute of Optics
University of Rochester, Rochester, NY, 14627, USA

A. J. Devaney
Schlumberger-Doll Research, P.O. Box 307, Ridgefield, CT, 06877, USA

J. T. Foley
Physics Department, Mississippi State University
Mississippi State, MS, 39762, USA

ABSTRACT

An expression is derived that describes the temporal development of the second-order correlation function of free fields. When specialized to free fields that are stationary (at least in the wide sense), the solution provides an expression for the mutual coherence function of the field at any two points in space in terms of the mutual intensity, specified at all pairs of points. As an immediate consequence of this result, a formula is obtained that expresses the temporal coherence of the field in terms of its spatial coherence. The results confirm quantitatively certain qualitative predictions of E.C.G. Sudarshan, regarding correlation properties of statistical ensembles of free fields.

1. INTRODUCTION

In a remarkable but little known paper[1], Sudarshan presented a number of new basic results relating to coherence properties of fluctuating free fields. In particular, he showed that the second-order correlation function of the field obeys two equations that are first-order in the temporal variables, but are non-local in the spatial variables. Sudarshan discussed, without proofs, some implications of these equations. In particular, it appears from his discussion that there must exist a close relationship between spatial coherence and temporal coherence in any fluctuating free field that is statistically stationary.

In this paper, we derive an expression that describes the temporal development of the second-order correlation function. When specialized to stationary free fields, the solution expresses the mutual coherence function at any two points in space in terms of the mutual intensity at all pairs of points. An immediate consequence of this result is an expression that shows explicitly how the temporal coherence of the field can be determined from the knowledge of its spatial coherence.

ISSN:00940243X/81/650123-08$1.50

2. SUDARSHAN'S EQUATION FOR THE SECOND-ORDER CORRELATION FUNCTION OF AN ENSEMBLE OF FREE FIELDS.

Let $V(\underline{r},t)$ be the analytic signal[2] representing a scalar wavefield in free space. Here $\underline{r}$ denotes the position vector of a typical point and t denotes the time. $V(\underline{r},t)$ satisfies the wave equation

$$\nabla^2 V = \frac{1}{c^2}\frac{\partial^2 V}{\partial t^2}, \tag{2.1}$$

(with c denoting the speed of light in vacuum), valid throughout the whole space, for all values of t. The general solution of Eq. (2.1) may be expressed as superposition of plane wave modes, viz.,

$$V(\underline{r},t) = \int a(\underline{K})e^{i(\underline{K}\cdot\underline{r} - Kct)}d^3K, \tag{2.2}$$

where $K = |\underline{K}|$ and the integration extends over the whole $\underline{K}$-space.

If we differentiate Eq. (2.2) with respect to t, we obtain

$$\frac{\partial V}{\partial t} = -ic\int Ka(\underline{K})e^{i(\underline{K}\cdot\underline{r} - Kct)}d^3K. \tag{2.3}$$

Following Sudarshan, we will express Eq. (2.3) in a form that involves a non-local operator, denoted by $\sqrt{-\nabla^2}$, defined as follows: Let $F(\underline{r})$ be any scalar function of $\underline{r}$ that admits a three-dimensional Fourier representation:

$$F(\underline{r}) = \int f(\underline{K})e^{i\underline{K}\cdot\underline{r}}d^3K. \tag{2.4}$$

The operator $\sqrt{-\nabla^2}$, acting on $F(\underline{r})$ is then defined by the formula[3]

$$\sqrt{-\nabla^2}\,F(\underline{r}) = \int Kf(\underline{K})e^{i\underline{K}\cdot\underline{r}}d^3K. \tag{2.5}$$

With this definition, Eq. (2.3) may be expressed in the form

$$\sqrt{-\nabla^2}\,V = -\frac{1}{ic}\frac{\partial V}{\partial t}. \tag{2.6}$$

Next let us consider an ensemble (not necessarily stationary) of free fields. We define the second-order correlation function of the ensemble by the usual formula

$$\Gamma(\underline{r}_1,t_1;\underline{r}_2,t_2) = \langle V^*(\underline{r}_1,t_1)V(\underline{r}_2,t_2)\rangle, \tag{2.7}$$

where the sharp brackets denote an ensemble average. Let $\sqrt{-\nabla_1^2}$ and $\sqrt{-\nabla_2^2}$ denote the operator $\sqrt{-\nabla^2}$ when acting with respect to the spatial variable $\underline{r}_1$ and $\underline{r}_2$, respectively. Let us apply either of these two operators to Eq. (2.7) and interchange the order of the operations on the right-hand side of the resulting equation. This gives

$$\sqrt{-\nabla_j^2}\,\Gamma(\underline{r}_1,t_1;\underline{r}_2,t_2) = \langle\sqrt{-\nabla_j^2}\,V^*(\underline{r}_1,t_1)V(\underline{r}_2,t_2)\rangle, \quad (j=1,2). \tag{2.8}$$

If we make use of Eq. (2.6) on the right-hand side of Eq. (2.8) and again interchange the order of operations, we find that

$$\sqrt{-\nabla_1^2}\,\Gamma(\underline{r}_1,t_1;\underline{r}_2,t_2) = \frac{1}{ic}\frac{\partial}{\partial t_1}\Gamma(\underline{r}_1,t_1;\underline{r}_2,t_2), \tag{2.9a}$$

$$\sqrt{-\nabla_2^2}\,\Gamma(\underline{r}_1,t_1;\underline{r}_2,t_2) = -\frac{1}{ic}\frac{\partial}{\partial t_2}\Gamma(\underline{r}_1,t_1;\underline{r}_2,t_2). \tag{2.9b}$$

Equations (2.9) are Sudarshan's equations[1] which must be satisfied by the second-order correlation function $\Gamma(\underline{r}_1,t_1;\underline{r}_2,t_2)$ of any ensemble of free fields.[4] We note that, unlike the well-known wave equations that the correlation function is known to satisfy (ref. 2, Sec. 10.7.1), Sudarshan's equations are first-order equations in time and are non-local in space. The fact that Eqs. (2.9) are first-order in time implies, as Sudarshan pointed out, that if $\Gamma(\underline{r}_1,0;\underline{r}_2,0)$ is specified for all values of $\underline{r}_1$ and $\underline{r}_2$, $\Gamma(\underline{r}_1,t_1;\underline{r}_2,t_2)$ may be determined. In other words, the spatial dependence of Γ completely specifies its temporal dependence. We will now derive an explicit formula for the temporal dependence of the correlation function.

3. TEMPORAL DEPENDENCE OF THE CORRELATION FUNCTION

If we substitute in Eq. (2.7) for the field variable $V(\underline{r},t)$, the mode representation (2.2) and interchange the order of the ensemble averaging and integration, we obtain the following expression for Γ:

$$\Gamma(\underline{r}_1,t_1;\underline{r}_2,t_2) = \iint A(\underline{K}_1,\underline{K}_2)e^{i(\underline{K}_2\cdot\underline{r}_2-\underline{K}_1\cdot\underline{r}_1)}e^{-ic(K_2t_2-K_1t_1)}d^3K_1d^3K_2, \tag{3.1}$$

where

$$A(\underline{K}_1,\underline{K}_2) = \langle a^*(\underline{K}_1)a(\underline{K}_2)\rangle\ . \tag{3.2}$$

The function $A(\underline{K}_1,\underline{K}_2)$, defined by Eq. (3.2), represents the correlation of the field in wave-vector space. It is analogous to the so-called angular correlation function that was introduced previously for another class of fields.[5]

If we apply the Fourier inversion formula to Eq. (3.1), we obtain the following expression for $A(\underline{K}_1,\underline{K}_2)$ in terms of $\Gamma(\underline{r}_1,t_1;\underline{r}_2,t_2)$:

$$A(\underline{K}_1,\underline{K}_2) = e^{ic(K_2t_2-K_1t_1)}\frac{1}{(2\pi)^6}\iint\Gamma(\underline{r}_1,t_1;\underline{r}_2,t_2)e^{-i(\underline{K}_2\cdot\underline{r}_2-\underline{K}_1\cdot\underline{r}_1)}d^3r_1d^3r_2. \tag{3.3}$$

Next we replace in Eq. (3.3), $\underline{r}_1, t_1, \underline{r}_2, t_2$ by $\underline{r}_1', t_1', \underline{r}_2', t_2'$ respectively and substitute for $A(\underline{K}_1, \underline{K}_2)$ into Eq. (3.1). After a straightforward calculation, we then obtain the formula

$$\Gamma(\underline{r}_1,t_1;\underline{r}_2,t_2) = \iint \Gamma(\underline{r}_1',t_1';\underline{r}_2',t_2') G^*(\underline{r}_1-\underline{r}_1',t_1-t_1') G(\underline{r}_2-\underline{r}_2',t_2-t_2') d^3r_1' d^3r_2', \tag{3.4}$$

where

$$G(\underline{R},T) = \frac{1}{(2\pi)^3} \int e^{i(\underline{K}\cdot\underline{R}-KcT)} d^3K. \tag{3.5}$$

The formula (3.4) is valid for all values of t_1 and t_2 ($t_2 \gtrless t_1$) and implies that $\Gamma(\underline{r}_1,t_1;\underline{r}_2,t_2)$ is uniquely specified for all values of its arguments by $\Gamma(\underline{r}_1',t_1';\underline{r}_2',t_2')$, where $\underline{r}_1'$ and $\underline{r}_2'$ take on all possible values, but t_1' and t_2' are any two fixed temporal arguments.

It is shown in Appendix A [Eq. (A3)] that $G(\underline{R},T)$ may be expressed in the form

$$G(\underline{R},T) = -\frac{1}{2\pi R}\frac{\partial}{\partial R}\left[\delta^{(+)}(R-cT) + \delta^{(-)}(R+cT)\right]. \tag{3.6}$$

Here $\delta^{(+)}(\xi)$ and $\delta^{(-)}(\xi)$ are the positive and the negative frequency parts of the Dirac delta function, $\delta(\xi)$, viz.[6],

$$\delta^{(+)}(\xi) = \frac{1}{2\pi}\int_0^\infty e^{iK\xi}dK, \quad \delta^{(-)}(\xi) = \frac{1}{2\pi}\int_0^\infty e^{-iK\xi}dK. \tag{3.7}$$

These functions may be represented symbolically as

$$\delta^{(\pm)}(\xi) = \pm\frac{i}{2\pi} P\frac{1}{\xi} + \frac{1}{2}\delta(\xi), \tag{3.8}$$

where P represents the Cauchy principal value.

4. STATIONARY FREE FIELDS

We will now specialize our results to the important case when the ensemble of the free fields is stationary, at least in the wide sense.[7] In this case the correlation function $\Gamma(\underline{r}_1,t_1;\underline{r}_2,t_2)$ will depend on its two time arguments only through the difference $t_2 - t_1$. We set

$$t_2 - t_1 = \tau \tag{4.1}$$

and write[8]

$$\Gamma(\underline{r}_1,t_1;\underline{r}_2,t_2) \equiv \Gamma(\underline{r}_1,\underline{r}_2,\tau). \tag{4.2}$$

The function $\Gamma(\underline{r}_1,\underline{r}_2,\tau)$ is the well-known mutual coherence function (ref. 2, Sec. 10.3.1).

Making use of Eq. (4.2) and of the fact that, in view of Eq. (4.1),

$$-\frac{\partial}{\partial t_1} = \frac{\partial}{\partial t_2} = \frac{\partial}{\partial \tau}, \tag{4.3}$$

Sudarshan's equations (2.9) reduce to

$$\sqrt{-\nabla_1^2}\,\Gamma(\underline{r}_1,\underline{r}_2,\tau) = -\frac{1}{ic}\frac{\partial}{\partial\tau}\Gamma(\underline{r}_1,\underline{r}_2,\tau), \tag{4.4a}$$

$$\sqrt{-\nabla_2^2}\,\Gamma(\underline{r}_1,\underline{r}_2,\tau) = -\frac{1}{ic}\frac{\partial}{\partial\tau}\Gamma(\underline{r}_1,\underline{r}_2,\tau). \tag{4.4b}$$

Let us now examine the form that the "time-evolution" Eq. (3.4) takes when the ensemble is stationary. For this purpose we set, in analogy to Eq. (4.1),

$$t_2' - t_1' = \tau' \tag{4.5}$$

and we eliminate t_2 and t_2' from Eq. (3.4) by the use of Eqs. (4.5) and (4.1). We then obtain the formula

$$\begin{aligned}\Gamma(\underline{r}_1,t_1;\underline{r}_2,t_1+\tau) = \iint &\Gamma(\underline{r}_1',t_1';\underline{r}_2',t_1'+\tau')G^*(\underline{r}_1-\underline{r}_1',t_1-t_1') \\ &\times G(\underline{r}_2-r_2',t_1+\tau-t_1'-\tau')d^3r_1'd^3r_2'.\end{aligned} \tag{4.6}$$

If we now use Eq. (4.2) and set $t_1 = t_1' = 0$, Eq. (4.6) reduces to

$$\Gamma(\underline{r}_1,\underline{r}_2,\tau) = \iint\Gamma(\underline{r}_1',\underline{r}_2',\tau')G^*(\underline{r}_1-\underline{r}_1',0)G(\underline{r}_2-\underline{r}_2',\tau-\tau')d^3r_1'd^3r_2'. \tag{4.7}$$

It follows from Eq. (3.5) and the Fourier representation of the three-dimensional Dirac delta function that

$$G^*(\underline{R}_1,0) = \delta^{(3)}(\underline{R}_1) \tag{4.8}$$

and we have, according to Eq. (3.6)

$$G(\underline{R}_2,T) = -\frac{1}{2\pi R_2}\frac{\partial}{\partial R_2}\left[\delta^{(+)}(R_2-cT) + \delta^{(-)}(R_2+cT)\right]. \tag{4.9}$$

If we now substitute from Eq. (4.8) into Eq. (4.7) and carry out the trivial integration with respect to $\underline{r}_1'$, we obtain for $\Gamma(\underline{r}_1,\underline{r}_2,\tau)$ the expression

$$\Gamma(\underline{r}_1,\underline{r}_2,\tau) = \int\Gamma(\underline{r}_1,\underline{r}_2',\tau')G(\underline{r}_2-\underline{r}_2',\tau-\tau')d^3r_2'. \tag{4.10}$$

The formula (4.10) is the simplified form that the "time-evolution" equation (3.4) takes in the special case when the ensemble of the free fields is stationary.

5. A RELATIONSHIP BETWEEN TEMPORAL COHERENCE AND SPATIAL COHERENCE IN FREE STATIONARY FIELDS

In Eq. (4.10) the parameter τ' is arbitrary. Let us now choose $\tau' = 0$ and let us denote by $J(\underline{r}_1,\underline{r}_2)$ the mutual intensity function (ref. 1, Sec. 10.4.1):

$$J(\underline{r}_1,\underline{r}_2) = \Gamma(\underline{r}_1,\underline{r}_2,0) = \langle V^*(\underline{r}_1,t)V(\underline{r}_2,t)\rangle. \tag{5.1}$$

This function, which may also be called the "equal-time" coherence function, is a measure of spatial coherence at the points specified by the position vectors $\underline{r}_1$ and $\underline{r}_2$, at the same instant of time. Eq. (4.10) then gives,

$$\Gamma(\underline{r}_1,\underline{r}_2,\tau) = \int J(\underline{r}_1,\underline{r}_2')G(\underline{r}_2-\underline{r}_2',\tau)d^3r_2'. \tag{5.2}$$

This formula expresses the mutual coherence function in terms of the mutual intensity.

Next let us take $\underline{r}_2 = \underline{r}_1$ (= $\underline{r}$ say). Then Eq. (5.2) gives, if we drop the subscript 2 on $\underline{r}_2'$,

$$\Gamma(\underline{r},\underline{r},\tau) = \int J(\underline{r},\underline{r}')G(\underline{r}-\underline{r}',\tau)d^3r'. \tag{5.3}$$

The expression on the left is the self-coherence function (ref. 1, Sec. 10.3.1) and is a measure of the temporal coherence at the point specified by the position vector $\underline{r}$. Thus the formula (5.3) expresses the temporal coherence throughout the field in terms of its spatial coherence, [represented by the mutual intensity $J(\underline{r},\underline{r}')$].

APPENDIX A: EVALUATION OF THE INTEGRAL (3.5) FOR $G(\underline{R},T)$

The Green's function $G(\underline{R},T)$ is defined by Eq. (3.5) as

$$G(\underline{R},T) = \frac{1}{(2\pi)^3}\int e^{i(\underline{K}\cdot\underline{R}-KcT)}d^3K. \tag{A1}$$

Let us introduce spherical polar coordinates in $\underline{K}$-space, with the polar axis $\Theta = 0$ along $\underline{R}$. Eq. (A1) then becomes

$$G(\underline{R},T) = \frac{1}{(2\pi)^3}\int_0^\infty dKK^2\int_0^{2\pi} d\phi\int_0^\pi d\Theta\sin\Theta\, e^{i[KR\cos\Theta-KcT]}. \tag{A2}$$

The integration with respect to ϕ gives 2π and if we set $x = \cos\Theta$, Eq. (A2) then reduces to

$$G(\underline{R},T) = \frac{1}{(2\pi)^2} \int_0^\infty dK K^2 \int_{-1}^{1} dx\, e^{iK(Rx-cT)}$$

$$= \frac{1}{(2\pi)^2} \int_0^\infty K^2 e^{-iKcT} \left[\frac{e^{iKR} - e^{-iKR}}{iKR} \right] dK$$

$$= -\frac{1}{2\pi R} \left[\frac{1}{2\pi} \int_0^\infty iK\, e^{iK(R-cT)} dK - \frac{1}{2\pi} \int_0^\infty iK\, e^{-iK(R+cT)} dK \right]$$

$$= -\frac{1}{2\pi R} \frac{\partial}{\partial R} \left[\delta^{(+)}(R-cT) + \delta^{(-)}(R+cT) \right], \qquad \text{(A3)}$$

where

$$\delta^{(+)}(\xi) = \frac{1}{2\pi} \int_0^\infty e^{iK\xi} dK, \qquad \delta^{(-)}(\xi) = \frac{1}{2\pi} \int_0^\infty e^{-iK\xi} dK. \qquad \text{(A4)}$$

The functions $\delta^{(+)}$ and $\delta^{(-)}$ are the positive and the negative frequency parts of the Dirac delta function $\delta(\xi)$, which may be represented by the symbolic formulae[6]

$$\delta^{(\pm)}(\xi) = \pm \frac{i}{2\pi} \mathcal{P} \frac{1}{\xi} + \frac{1}{2} \delta(\xi) , \qquad \text{(A5)}$$

where $\mathcal{P}$ represents the Cauchy principal value.

FOOTNOTES AND REFERENCES

1. E.C.G. Sudarshan, J. Math. Phys. Sci., 3, 121 (1969).
2. M. Born and E. Wolf, Principles of Optics, 5th ed., (Pergamon Press, Oxford and New York, 1975), Sec. 10.2.

3. The symbol $\sqrt{-\nabla^2}$ is used, because applying this operator twice in succession to the right-hand side of Eq. (2.4) is equivalent to operating on $F(\underline{r})$ by $-\nabla^2$.
4. An error in sign in the equation corresponding to Eq. (2.9a) in ref. 1 is corrected here.

5. E.W. Marchand and E. Wolf, J. Opt. Soc. Amer., 62, 379 (1972).
6. W. Heitler, The Quantum Theory of Radiation, 3rd ed., (Clarendon Press, Oxford, 1954), p. 69.

7. Throughout this paper the term "stationary" is used in this sense. For the definition of wide-sense stationarity see, for example, W.B. Davenport and W.L. Root, *An Introduction to the Theory of Random Signals and Noise* (McGraw Hill, New York, 1958), p. 60.
8. No confusion should arise from the use of the same symbol (Γ), for the two correlation functions, since the variables are shown explicitly in each case.

GENERALIZATION OF THE DEBROGLIE RELATIONS FOR PROPAGATION IN ANISOTROPIC MEDIA

Ronald G. Newburgh
Deputy for Electronic Technology
Rome Air Development Center
Electromagnetic Sciences Division
Hanscom AFB, MA 01731

ABSTRACT

By assuming wave-particle duality for those phenomena in which the directions of energy transport and wave propagation are co-parallel, one can show the co-variant nature of the de Broglie energy-frequency and momentum - wave vector relations. If the energy transport and wave propagation are not parallel, as is the case for light propagation in anisotropic media, it follows necessarily that the de Broglie relations are no longer valid as written. In this paper the relations are generalized for such non-parallel situations by defining a new wave vector ℓ in the direction of energy transport and distinct from the propagation wave vector. The introduction of ℓ follows the method by which de Broglie associated a wave with an electron. The magnitude of ℓ is $\omega\, u/c^2$ where ω is the angular frequency and u the velocity of energy transport. The physical meaning of ℓ is discussed including its interpretation as the wave vector for the photon wave function.

The covariant nature of the de Broglie relations has far reaching implications for the propagation of light in anisotropic media. Examination of their covariance shows that Planck's constant is a Lorentz invariant as is the velocity of light in vacuo. However, a simple derivation based on two assumptions only, makes clear that the relations as normally written cannot hold for propagation in anisotropic media. The object of this paper is to generalize the relations so that they do apply in such media and to provide a physical interpretation of the new formulation.

The first step is to derive the relations

$$E = \hbar\,\omega \tag{1}$$

and

$$p = \hbar\, k \tag{2}$$

in such a way as to demonstrate their Lorentz invariance. In Eqs (1) and (2) E is the energy, $\hbar$ Planck's constant

ISSN:0094-243X/81/650131-07$1.50

divided by 2Π, ω the circular frequency, $\underset{\sim}{p}$ the 3-momentum, and $\underset{\sim}{k}$ the 3-wave vector. The first assumption used in the derivation is that wave particle duality exists for both light and matter so that each may be considered both as wave and particle. The second is that the direction of energy transport and that of wave propagation are identical. The basis of the derivation is a property of 4-vectors pointed out by Landau and Lifshitz [1]. If two 4-vectors are parallel in an inertial frame, the ratio of corresponding components is a Lorentz invariant in all inertial frames. Therefore, if the 4-momentum $\underset{\sim}{P}$,

$$\underset{\sim}{P} = (\underset{\sim}{p},\ iE/c), \tag{3}$$

and the 4-wave vector $\underset{\sim}{K}$,

$$\underset{\sim}{K} = (\underset{\sim}{k},\ i\omega/c), \tag{4}$$

are parallel, the ratios p/k and E/ω are Lorentz invariant by the application of the Landau and Lifshitz rule. (The constant c is the velocity of light in vacuo). The value of the invariant can be determined by experiment and is $\hbar$.

This derivation of the two relations differs from those of Planck, Einstein, and de Broglie. Planck postulated Eq. (1) for material oscillators in the walls comprising a black body cavity but did not apply it to electromagnetic radiation. This was done by Einstein who indeed discovered the quantum nature of light by asserting the validity of Eq. (1) for radiation. De Broglie some twenty years later then postulated the wave nature of electrons. His thesis, as given by Ruark and Urey[2], considered relativistic time changes in the frequency of periodic motions of electrons within atoms. From these time changes he argued that the phase of such motion must agree with the phase of the wave at a given point. From this he showed that the possible paths of the electrons are rays of the phase waves. In de Broglie's words "the motion can be stable only if the phase wave is tuned with the length of the path".

In 1924 before publication of de Broglie's ideas Dirac[3] had followed a line of thought very similar to the derivation of this paper in discussing the Bohr frequency equation for atomic transitions,

$$\delta E = \hbar\ \omega \tag{5}$$

Here δE is the loss (gain) of energy of an atom which emits (absorbs) a photon of frequency ω. Dirac, too concluded that the direction of energy transport (really that of the energy - momentum 4-vector) and that of light propagation must be identical for Eq. (5) to hold in all inertial frames.

If one looks at Eqs. (1) and (2) from the view point of Lorentz invariants, the interpretation is both simple and far reaching. Neither energy nor frequency is invariant under a Lorentz transformation. Yet if the assumptions of duality and parallelism hold, their ratio is invariant in all inertial frames. For photons this invariance encompasses such diverse phenomena as aberration and Doppler shifts. It is clear that the derivation reproduced here has the obvious merit of bringing relativity into the very foundations of quantum theory.

However, it is equally evident that the relations cannot hold if the assumption of parallelism fails. But this is precisely the situation for light propagation in an anisotropic medium. There are many crystals as well as plasmas in which energy transport is not in the direction of the wave normal. Born and Wolf[4] discuss this case at length. Therefore on the basis of the necessity of parallelism we conclude that we must reexamine Eqs. (1) and (2) for propagation in anisotropic media.

We begin by assuming the universality of the de Broglie relations. Therefore we must generalize them for the case of anisotropy. Given the non-parallelism of energy transport and wave propagation, we know that we can no longer associate the 3-wave vector k with the 3 - momentum p. There exists, however, a ray in the anisotropic medium whose direction and velocity are those of the energy transport. We also know that the wave momentum and energy form a 4-vector in agreement with the definition of Eq. (3). We therefore ask if we can define a new 4-vector $\underset{\sim}{L}$ consonant with the de Broglie relations of Eqs. (1) and (2) and with physical significance.

We define $\underset{\sim}{L}$ through the expression

$$\underset{\sim}{P} = \hbar \underset{\sim}{L} \tag{6}$$

where $\underset{\sim}{L}$ is equivalent to $(\underset{\sim}{\ell}, i\omega/c)$. The circular frequency ω is that associated with the wave itself. The 3-vector $\underset{\sim}{\ell}$ is a new propagation vector parallel to the 3 - momentum $\underset{\sim}{p}$ and distinct from the 3-wave vector $\underset{\sim}{k}$. We will deal with the physical meaning of $\underset{\sim}{L}$ and $\underset{\sim}{\ell}$ shortly, but first let us look at some of their properties.

From Eq. (6) we now write the de Broglie relations as

$$E = \hbar\omega \tag{7}$$

and

$$\underset{\sim}{p} = \hbar\underset{\sim}{\ell} \tag{8}$$

Equation (7) is identical with Eq. (1); the linear relation between energy and frequency holds. Equation (8) is quite different from Eq. (2) however, for $\underset{\sim}{\ell}$ is a new wave vector, quite different from $\underset{\sim}{k}$ which governs the propagation of the electric and magnetic field vectors.

Since P is a 4-vector, the defining equation for $\underset{\sim}{L}$, Eq. (6), shows that $\underset{\sim}{L}$ must also be a 4-vector. If we consider two inertial frames Σ and Σ' such that Σ' moves with relative velocity V along the x-axis with respect to Σ, the standard Lorentz transformations give the relations for $\underset{\sim}{p}$, E, $\underset{\sim}{\ell}$, and ω measured in the two frames. These transformations are

$$\begin{aligned} p_x' &= \gamma\,(p_x - VE/c^2) \\ p_y' &= p_y \\ p_z' &= p_z \\ E' &= \gamma\,(E - p_x V) \end{aligned} \tag{9}$$

and

$$\begin{aligned} \ell_x' &= \gamma\,(\ell_x - V\omega/c^2) \\ \ell_y' &= \ell_y \\ \ell_z' &= \ell_z \\ \omega &= \gamma\,(\omega - \ell_x V), \end{aligned} \tag{10}$$

where γ equals $(1-V^2/c^2)^{-1/2}$ and c is the velocity of light in vacuo.

To determine the physical significance of $\underset{\sim}{\ell}$ consider the quantity $\underset{\sim}{w}$ defined as a vector in the direction of $\underset{\sim}{\ell}$ with magnitude $\omega/|\ell|$. Its dimensions are those of a velocity. There is a second velocity $\underset{\sim}{u}$ which is the velocity of energy transport and in the direction of $\underset{\sim}{p}$, therefore by definition also in the direction of $\underset{\sim}{\ell}$. We now do precisely what de Broglie[5] did in treating an electron. To a particle of velocity $\underset{\sim}{u}$ moving in a given direction he ascribed a plane wave propagating in the same direction with phase velocity c^2/u which we call $\underset{\sim}{w}$. We conclude that the quantity $\underset{\sim}{\ell}$ defined in Eq. (8) is the wave vector for a plane wave with phase velocity c^2/u propagating in the direction of energy transport. This plane wave is not the same as the waves describing the electric and magnetic field vectors which are functions of $\underset{\sim}{k}$ and ω. It is the wave we associate directly with the light quantum. Introducing a function ψ, we write

$$\psi \sim \exp i(\omega t \pm \underset{\sim}{\ell} \cdot \underset{\sim}{r}) \tag{11}$$

This ψ is the wave function of the photon and we interpret $\psi\psi^*$ as the energy density. (For the relation among the various velocities and wave vectors see Fig. 1)

When propagation occurs in free space or in isotropic media, the introduction of $\underset{\sim}{\ell}$ and ψ would be an unnecessary complication. Because of the co-parallelism of propagation and energy transport, classical electromagnetic wave theory works perfectly well. We see though that the electromagnetic wave, even in this simple case, cannot be looked on directly as the wave associated with the light quantum. In the classical electromagnetic formulation the product of the group velocity $\underset{\sim}{u}$ and the phase velocity $\underset{\sim}{v}$ derived from $\underset{\sim}{k}$ does not equal c^2 unlike the product of $\underset{\sim}{u}$ and $\underset{\sim}{w}$. In fact this phase velocity $\underset{\sim}{v}$, apart from some cases for x-rays, is always less than or equal to c. In crystal optics a ray index of refraction N_R is defined as

$$N_R = c/u \tag{12}$$

[See, for example reference (4)] Since u, the group velocity, can never exceed c, N_R is always greater than or equal to unity. Therefore the phase velocity $\underset{\sim}{w}$ of the wave function may be written as

$$w = \omega/\ell = c^2/u = N_R c, \tag{13}$$

a velocity always greater than or equal to c. The result of this paper, which generalizes the de Broglie relations for anisotropic media is not surprising when one considers that the group velocity, not the phase velocity, obeys the Einstein velocity addition theorem. When the two are no longer parallel, it is clear that the direction of the group velocity, hence of energy transport, is the one to which we look for proper relativistic relations. Our measurements, after all, always require energy transfer

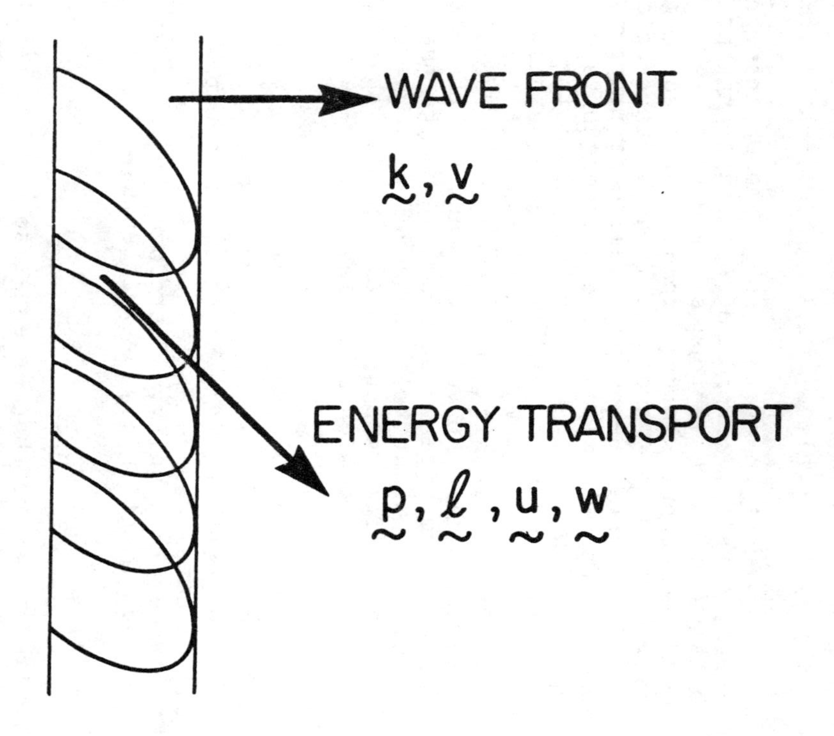

FIGURE 1 PROPAGATION IN AN ANISOTROPIC MEDIUM

from wave or particle to detector. Ko and Chuang[6] have made this point in discussing flux in a moving dispersive dielectric.

We conclude that the introduction of a new 4-vector $\underset{\sim}{L}$ associated with direction of energy flow rather than the wave normal preserves the Lorentz invariance of the de Broglie relations in anisotropic media and provides a consistent relativistic description. It also gives a single wave function to describe the photon. In short the vector $\underset{\sim}{L}$ completes the wave particle description for light.

REFERENCES

1. L.D. Landau and E.M. Lifshitz, The Classical Theory of Fields, Rev. 2nd Edition (Pergamon, Oxford, 1962), Sec 10., p. 31.
2. A.B. Ruark and H.C. Urey, Atoms, Molecules, and Quanta (McGraw-Hill, New York, 1930) pp. 516-520.
3. P.A.M. Dirac, Proc. Cambridge Phil. Soc. 22, 432 (1924).
4. M. Born and E. Wolf, Principles of Optics, 2nd Rev. Edition (Pergamon, Oxford, 1964), Ch. 14, p. 665.
5. L. de Broglie, Thesis, University of Paris (Masson et cie, Paris, 1924).
6. H.C. Ko and C.W. Chuang, Radio Science, 12, 337 (1977).

ON A LAGRANGIAN FORMULATION OF OPTICS AND GRAVITATION

O.R. Mata and A. Aburto
Instituto Politécnico Nacional
Escuela Superior de Física y Matemáticas
México, D.F. Lindavista

ABSTRACT

We propose a Lagrangian that indicates the existence of an intimate relation between Optics and Gravitation under certain conditions.

INTRODUCTION

The gravitational field and refractive media have traditionally been considered as physical entities whithout any connection, even if occasionally some authors (1) have dealt with problems in gravitation using techniques borrowed from optics (2). It appears, however that no formal equivalence has been drawn between these two sciences. In this paper we suggest that, under certain conditions, the gravitational field and a refractive medium can be considered to be formally equivalent, and we provide the mathematical basis for a theory that describes the behavior of photons in these two systems.

The theory presented in this paper is a first approximation to a more rigorous formulation devoted to proving the main premises. We suggest that a scalar potential ϕ be associated to a refractive medium. The potential must be only a function of the index of refraction because this determines completely the behavior of light. In the same way one can associate a refractive index n depending only on ϕ to all gravitational systems described by such potential. To make this equivalence more precise we postulate the existence of the action

$$S = \int_{t_1}^{t_2} L \; dt = \int_{t_1}^{t_2} h\nu \left[\frac{1}{2} \ln v^2 + \ln \frac{1}{c}\left(1+\frac{2\phi}{c}\right)^{-\frac{1}{2}} \right] dt \qquad (1)$$

where h is Planck's constant, ν is the frequency, v and c are the velocities of light in the medium and in vacuum, respectively. We will analyze the Lagrangian associated to the action of eq.(1) by means of conservation theorems that can be deduced from the symmetries of the systems. For an intuitive justification of eq. (1), see Ref. (3).

ISSN:0094-243X/81/650138-07$1.50

ENERGY

From eq. (1) we see that the Lagrangian function is time independent and therefore the Hamiltonian function is a constant of the motion. We use cartesian coordinates to write the Lagrangian in the form:

$$L = h\nu\left[\frac{1}{2}\ln(v_x^2+v_y^2+v_z^2) + \ln\frac{1}{c}\left(1 + \frac{2\phi(x,y,z)}{c^2}\right)^{-\frac{1}{2}}\right] \qquad (2)$$

It follows that the Hamiltonian is given by:

$$H = h\nu - h\nu\left[\frac{1}{2}\ln\frac{h^2\nu^2}{p^2} + \ln\frac{1}{c}\left(1+\frac{2\phi(x,y,z)}{c^2}\right)^{-\frac{1}{2}}\right] \qquad (3)$$

Where $p^2 = p_x^2 + p_y^2 + p_z^2$. Here P_i, $i = x,y,z$ is the canonical momentum conjugate to the generalized coordinate i.

Because we are dealing with photons it is not unreasonable to postulate that:

$$H = h\nu \qquad (4)$$

as a consequence of Planck's relation. From eqs. (3) and (4) we obtain:

$$\frac{c}{v} = n = \left(1 + \frac{2\phi}{c^2}\right)^{-\frac{1}{2}} \qquad (5)$$

and the inverse relation:

$$\phi = \frac{c^2}{2n^2} - \frac{c^2}{2} \qquad (6)$$

The last two equations show how the equivalence between optics and gravitation is realized. If we know ϕ, from eq. (5) we can calculate the index n associated with this potential; if n is known, on the other hand, the corresponding potential ϕ can be calculated.

Consider a photon in the gravitational field of a star or a planet. From this system it is generally true that $\left|\frac{\phi}{c^2}\right| \ll 1$, so that from eq. (5) we have:

$$v = c\left(1 + \frac{\phi}{c^2}\right) \qquad (7)$$

Equation (7) coincides with a result obtained by

Einstein (1) and Scott (2). From eq. (7) we have obtained the deflection of light to be one half of that predicted by general relativity.

LINEAR MOMENTUM

It is interesting to express the momentum in terms of the energy $E = h\nu$:

$$p_x = \frac{Ev_x}{v^2} \qquad p_y = \frac{Ev_y}{v^2} \qquad p_z = \frac{Ev_z}{v^2} \tag{8}$$

In this way we obtain $E = pv$ for the momentum of the photon. In vacuum we have $E = pc$, in agreement with special relativity.

ANGULAR MOMENTUM

A system possessing spherical symmetry must be described by $\phi = \phi(r)$ or $n = n(r)$, where r is the radial coordinate. For such system we have found:

$$\vec{L} = \vec{r} \times \vec{p} = \vec{const} \tag{9}$$

From eq. (8) we find that:

$$\vec{p} = \frac{h\nu}{c} n\hat{t} \tag{10}$$

where $\hat{t}$ is a unit vector tangent to the trajectory in the direction of motion. From eq. (9) and (10) we obtain the well known result:

$$\vec{r} \times n\hat{t} = \vec{const} \tag{11}$$

In view of this, if we use polar coordinates (r,θ) we obtain:

$$n^2 r^2 \mathring{\theta} = const \tag{12}$$

If we substitute eq. (5) into eq. (12) we find:

$$\left(1 + \frac{2\phi}{c^2}\right)^{-1} r^2 \mathring{\theta} = const. \tag{13}$$

For the case of a star of mass M, eq. (13) reduces to:

$$\left(1 - \frac{2GM}{rc^2}\right)^{-1} r^2 \mathring{\theta} = const. \tag{14}$$

This result is very similar to the one obtained in

General Relativity (4) for a homogeneous static star in a Schwarzchild metric:

$$(1 - \frac{GM}{rc})^{-1} r^2\theta = \text{const.} \tag{15}$$

In the case of axial symmetry, it is useful to express the Lagrangian in cylindrical coordinates ρ,θ and z, where the z axis coincides with the symmetry axis. Because n is a function of r and z, we expect the potential to be a function of the same variables. As the Lagrangian is independent of θ, the associated canonical momentum is a conserved quantity, i.e.:

$$nr^2 \frac{d\theta}{ds} = \text{const.} \tag{16}$$

In principle, eq. (16) allows us to reduce an optical problem containing s variables to one with s-1 variables. The latter problem should be easier to solve than the original one.

LAGRANGE'S EQUATIONS

In analogy with analytical mechanics we define the force in cartesian coordinates as:

$$\vec{F} = \frac{\partial L}{\partial x}\hat{i} + \frac{\partial L}{\partial y}\hat{j} + \frac{\partial L}{\partial z}\hat{k} \tag{17}$$

Using eqs. (1), (5) in eq. (17) we obtain:

$$\vec{F} = -\frac{h\nu}{c^2} n^2\nabla\phi = h\nu\nabla\ln n \tag{18}$$

This result was derived in 1968 by Tangherlini (5) using a different procedure based on the assumption that a photon has a mass proportional to the square of the refractive index.

Using eqs. 10, 18 and the Lagrange's equations:

$$\vec{F} = \frac{d\vec{p}}{dt}$$

we have found:

$$\frac{dn\hat{t}}{ds} = \nabla n \tag{19}$$

which is the ray equation.

HAMILTON-JACOBI EQUATION

Let S be the action obtained from our Lagrangian. It must satisfy the Hamilton-Jacobi equation:

$$\frac{\partial S}{\partial t} + H\ (q_i, \frac{\partial S}{\partial t}, t) = 0 \qquad (20)$$

From $H = h\nu$ and eq. 10, we find in cartesian coordinates:

$$H = \frac{cp}{n} = \frac{c}{n}\sqrt{p_x{}^2 + p_y{}^2 + p_z{}^2} \qquad (21)$$

From eqs. (20) and (21), we can derive the following equation:

$$\frac{\partial S}{\partial t} = \frac{c}{n}\ |\nabla S| \qquad (22)$$

In addition, because $\frac{\partial S}{\partial t} = -H$ we have that:

$$|\nabla S| = \frac{h\nu n}{c} \qquad (23)$$

and:

$$|\nabla L| = n \qquad (24)$$

where $L = \frac{Sc}{h\nu}$. Equation (24) can be recognized as the Eikonal equation of optics. Finally from Maupertuis principle we can derive Fermat's principle, as one would expect.

RED SHIFT

We now calculate the space-time metric which is consistent with the postulated Lagrangian. A small point mass M will bend the associated space-time. We expect that the metric determining the geometrical properties should be close to the Galileian metric. In spherical polar coordinates we assume the following form:

$$ds^2 = -a(r)dt^2 + b(r)\ dr^2 + r^2(d\theta^2 + sen^2\theta d\phi^2) \qquad (25)$$

where a and b are function of r only, and $a(r) \simeq c^2$, $b(r) \simeq 1$. We then look for a correction in the temporal coordinate of eq. (25), with $b(r) = 1$, taking into account that $ds = 0$ for photons, we have:

$$a(r) = \frac{dr^2 + r^2(d\theta^2 + sen^2\theta\ d\phi^2)}{dt^2} \quad (25')$$

On the basis of eqs. (5) and (25') we have:

$$a(r) = v^2 = c^2(1 + \frac{2\theta}{c^2})$$

Thus, the metric that describes the properties of space-time is approximately given by:

$$ds^2 = -c^2(1+\frac{2\phi}{c^2})\ dt^2 + dr^2 + r^2\ (d\theta^2 + sen^2\theta d\phi^2) \quad (26)$$

Its temporal part coincides with the Schwarzschild metric of general relativity. From eq. (26) the red shift gravitational formula follows at once.

A GAS

From our Lagrangian we can also derive an important result that allows us to apply techniques from Celestial Mechanics to the study of gases with spherical symmetry, that is to say, gases with a density that is only a function of r. We consider the case of a weak gravitational field or an optical system with a refractive index close to unity. In this case we have approximately:

$$L = \frac{h\nu}{2}\ (\frac{v^2}{c^2} - 1)\ - \frac{h\nu}{4}\ (\frac{v^2}{c^2} - 1)\frac{h\nu}{c^2} + h\nu(\frac{\phi}{c^2})^2 \quad (27)$$

If we define $m = \frac{E}{c}$, we obtain as a first approximation:

$$L = \frac{mv^2}{2} - m\phi \quad (28)$$

to within an additive constant. The last equation has the same form as the classical Lagrangian of a massive particle of mass M in a field ϕ.

ACKNOWLEDGEMENTS

We thank G. Rojas, A. Cisneros, F. Angulo and J. L. Bonilla for a critical review of the material contained in this paper, and A. Queijeiro for preparing the manuscript.

REFERENCES

1.- A. Einstein, On the influence of Gravitation on the propagation of light, The principle of Relativity (Dover, 1950), p. 60.

2.- J. C. Scott, Canad. J. Phys. 44, 100, 1966.

3.- A. Aburto, Thesis, Escuela Superior de Física y Matemáticas, México.

4.- Misner, Thorne, Wheeler, Gravitation (Freeman 1973) 1200.

5.- F. Tangherlini, Am. J. Phys. 36, 1001 (1968).

SELECTIVE PROPERTIES OF THE NONLINEAR FOUR-WAVE FILTER

A.L. Gulamiryan, A.V. Mamaev, N.F. Pilipetsky,
V.V. Ragul'sky, V.V. Shkunov
Institute for Problems in Mechanics AS USSR
117526, Moscow, Prospect Vernadskogo 101.

ABSTRACT

Nondegenerate four-wave nonlinear mixing as a reflecting frequency filter is considered. Several simple methods to tune the central frequency, angle of reflection and selective properties of this filter are proposed. The narrowband variant of such a filter is considered. The reflection of the weak signal wave, which has a frequency shift with respect to the strong reference waves in four-wave mixing in liquid carbon disulfide (CS_2) was experimentally obtained. The signal wave reflection coefficient versus the frequency shift was measured. The experimental results are in a good accordance with theory. These results confirm also that for the pulses of about 50 nsec. duration the main nonlinear mechanism in liquid CS_2 is the orientational (Kerr) one. The relaxation time of this nonlinearity is small enough. It permits construction of filters with a wide frequency band right up to 10^{12} hz.

THEORY

It is well known that every dynamic hologram and, in particular, nearly degenerate four-wave mixing reflects a signal in a frequency selective fashion.[1-3] For counterpropagating reference waves with a frequency ω_0, four-wave reflection of an arbitrarily directed signal is effective only in the narrow frequency band $\omega_o - \delta\omega < \omega_s < \omega_o + \delta\omega$ shown in Fig. 1. The spectral width $\delta\omega$ is determined from the minimum of the following two expressions:

$$\delta\omega \sim \omega_o \frac{\pi}{K_o \ell}$$

and

$$\delta\omega \sim 1/\tau,$$

where

$$K_o = \frac{\omega_o}{C} \eta$$

is the wave vector, ℓ is the length of the nonlinear

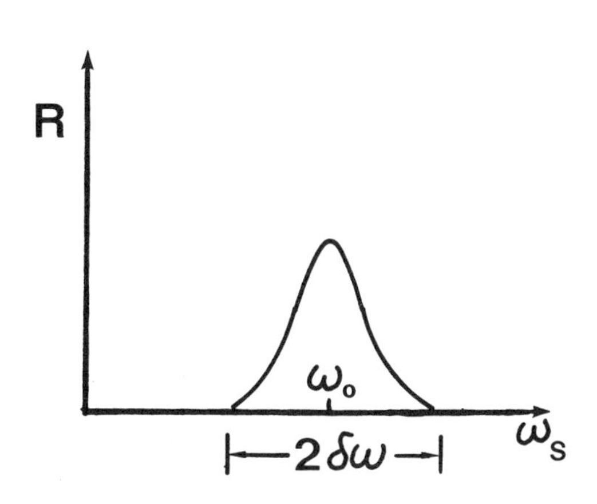

Fig. 1 Four-wave mixing reflection coefficient R versus the signal wave frequency.

ISSN:0094-243X/81/650145-09$1.50

medium, and τ is the relaxation time of the nonlinearity used for mixing.

(1) Let us consider the case when two reference waves

$$E_1\exp(-i\omega_1 t + i\omega\vec{K}_1\cdot\vec{R}) \text{ and } E_2\exp(-i\omega_2 t + i\vec{K}_2\cdot\vec{R})$$

propagate in a medium with third-order nonlinearity $\chi^{(3)}$. The signal wave

$$E_s\exp(-i\omega_s t + i\vec{K}_s\cdot\vec{R})$$

is mixed with the two reference waves inducing a nonlinear polarization in the medium, which contains the component with the spatial structure $P^{N\ell}$ defined by

$$P^{N\ell} = 2\chi^{(3)} E_1E_2E_s^* \exp(-i\omega t + i\vec{Q}\cdot\vec{R}). \quad (1)$$

Here $\omega = \omega_1 + \omega_2 - \omega_s$, $\vec{Q} = \vec{K}_1 + \vec{K}_2 - \vec{K}_s$, $|\vec{K}_i| = \eta(\omega_i)\frac{\omega_i}{C}$, and $\eta(\omega_i)$ is the medium refractive index at frequency ω_i. The medium will radiate a light at a frequency ω, with wave vector $\vec{Q}$, if the phase matching condition is satisfied:

$$|\vec{K}_1 + \vec{K}_2 - \vec{K}_s| = |\vec{Q}| = \eta(\omega)\frac{\omega}{C}. \quad (2)$$

Deviation from this condition decreases the radiated intensity.

Let the mixing occur in a nonlinear medium with Z-axis chosen perpendicular to the surface H as shown in Fig. 2. Then one has to attribute variations of the slowly varying signal and reflected wave amplitudes within the layer to Z-dependence. The transverse component of the wave vector $\vec{K}$ of the radiating field is related to the vector $\vec{Q}$ by the condition $K_H = Q_H$, equating the projections on the layer boundary surface H, through which the signal wave is induced. A longitudinal component K_Z is defined in terms of the frequency ω by

$$K_Z = \sqrt{\left(\frac{\eta\omega}{C}\right) - \vec{Q}_H^{\,2}}$$

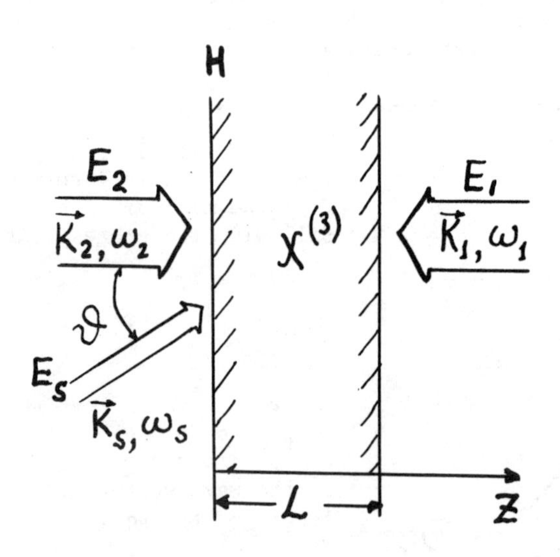

Fig. 2 Four-wave mixing scheme.

The coefficient R of reflection from the signal wave E_s into radiated one, in a first Born approximation, when $R << 1$, is defined by[4]

$$R(\omega_s) = \left|\frac{4\pi\omega}{C\eta}\ \chi^{(3)}\ E_1E_2\ell\ /\cos\theta\right|^2 \frac{\sin^2\left(\frac{\delta K\ell}{2}\right)}{\left(\frac{\delta K\ell}{2}\right)^2}\ , \tag{3}$$

where $\delta K = |\vec{Q} - \vec{K}|$ and ℓ - is the nonlinear layer thickness.

(2) Let us fix the frequencies ω_1 and ω_2 and a mutual orientation $\vec{K}_1 + \vec{K}_2$ of the reference waves E_1 and E_2, and find in the space of wave vectors the surface of strict phase matching $\delta K = 0$, defined by the phase matching condition (2). In media without dispersion, this surface is an ellipsoid of revolution (Fig. 3). Two focuses of the ellipsoid lie at the beginning and end points of the vector $\vec{K}_1+\vec{K}_2$. The sum of the vectors' $\vec{K}_s$ and $\vec{K}$ lengths is constant--the well known property of an ellipse:

$$\frac{(\omega_s + \omega)\eta}{C} = \frac{(\omega_1 + \omega_2)\eta}{C}\ .$$

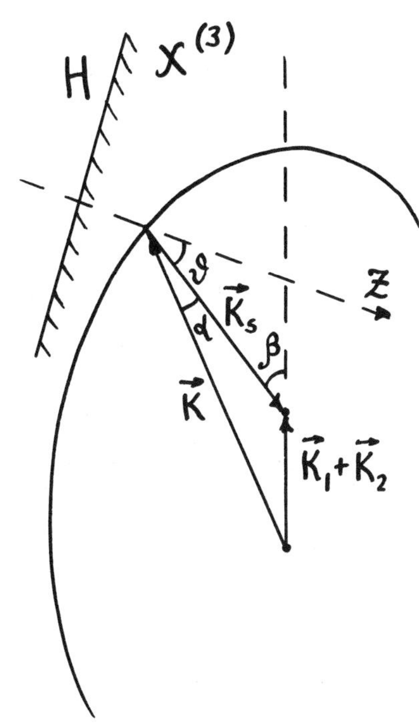

Fig. 3 The surface of strict synchronism.

Thus, for every signal direction there exists an optimal signal frequency ω_s^{opt}, corresponding to the maximal value of the reflection coefficient

$$R\left(\omega_s^{opt}\right) = \left|\frac{4\pi\omega}{C\eta}\ \chi^{(3)}\ E_1E_2\ell/\cos\theta\right|^2$$

On the other hand, if the signal frequency lies within the limits defined by

$$\frac{C}{2\eta}\ (|\vec{K}_1| - |\vec{K}_1+\vec{K}_2| + |\vec{K}_2|) \le \omega_s \le \frac{C}{2\eta}\ \cdot (|\vec{K}_1| + |\vec{K}_1+\vec{K}_2| + |\vec{K}_2|), \tag{4}$$

for every ω_s one can find the cone of directions[5,6] for optimal reflection around the $\vec{K}_1 + \vec{K}_2$ direction (Fig. 3) tilted an angle β. The reflected wave direction $\vec{K}$ is tilted an angle α in the plane of the $\vec{K}_s$ and $\vec{K}_1+\vec{K}_2$ vectors with respect to the reversed direction of the signal propagation. One may discuss several particular variants of the general geometry considered[5-7].

(3) Now we consider in detail the most interesting case, when the reference waves have the same frequency ω_0 and are nearly counter-propagating so that the parameter

$\gamma = \frac{|\vec{K}_1 + \vec{K}_2|}{K_o} \ll 1$. The signal and reflected waves in this case are nearly counterpropagating too, and are reversed in spatial structure with respect to each other[5-7].

An optimal frequency of a signal ω_s^{opt} is related to the angle β by

$$\omega_s^{opt} \simeq \omega_o \left(1 - \frac{1}{2} \gamma \cos \beta \right), \tag{5}$$

and could be effectively tuned by the variation of two parameters β and γ. Both these parameters are defined completely by the directions of the reference waves and are changed with the angular tilting of one of them. For example to tune this filter one could slightly tilt the mirror, which gives the second reference wave E_2 by reflection of the first reference wave E_1.

The width of the frequency band of a signal reflection near the central frequency ω_s^{opt} for a fixed direction of a signal could be derived from the expression for the reflection coefficient (3). If θ is the angle between the $\vec{K}_s$ vector and the Z-axis, which is normal to the medium surface H, the frequency shift of the signal Δ from the optimal value ω_s^{opt} gives the detuning from the phase matching condition $\delta K \simeq 2\eta(\Delta/C)\cos\theta$. Hence, the halfwidth of the band $\delta\omega_{HW}$ defined by the first zero level, is given by

$$\delta\omega_{HW} = \frac{\pi C}{\eta \ell} \cos\theta = \omega_o \frac{\pi}{K_o \ell} \cos\theta . \tag{6}$$

On the other hand, when the signal frequency ω_s is fixed within the interval (4), the four-wave mixing provides an angular filtering of a signal radiation[5-6]. Maximal reflection occurs when the signal angular component is tilted an angle $\beta^{opt} = \arccos\,[2(\omega_o - \omega_s)/\gamma\omega_o]$ with respect to the direction of the vector $\vec{K}_1 + \vec{K}_2$. The reflection of the signal wave E_s occurs with small angular shift $\alpha \simeq \gamma \sin\beta$ in the plane of the vectors $\vec{K}_s$ and $\vec{K}_1 + \vec{K}_2$. An angular shift could be easily tuned too by tilting the mirror that reflects the first reference wave E_1 into the second one E_2.

An angular deviation ψ from this optimal value gives the detuning from the phase-matching condition $\delta K = 2 K_o \psi\gamma \sin\beta$. Hence, the halfwidth of the band of angles around β^{opt}, for which four-wave mixing reflection is effective, defined by the first zero level, is given by

$$\delta\beta_{HW} \simeq \frac{\pi \cos\theta}{K_o \ell \gamma \sin\beta} \tag{7}$$

There are no restrictions on the angular deviation without angle β variations[5].

(4) The halfwidth of the reflection bands for the four-wave filter, as for every parametric interaction, is defined by the value $\ell/\cos\theta$, that is, by the signal-wave path in the interaction volume. Hence, to narrow these bands, one should increase the path.

For this purpose, it is very attractive to utilize multiple reflections of the signal from mirrors surrounding a nonlinear substance (see Fig. 4). Reference waves propagate along the mirrors without reflection and have relatively small paths through the interaction volume. Such a scheme was proposed and realized in

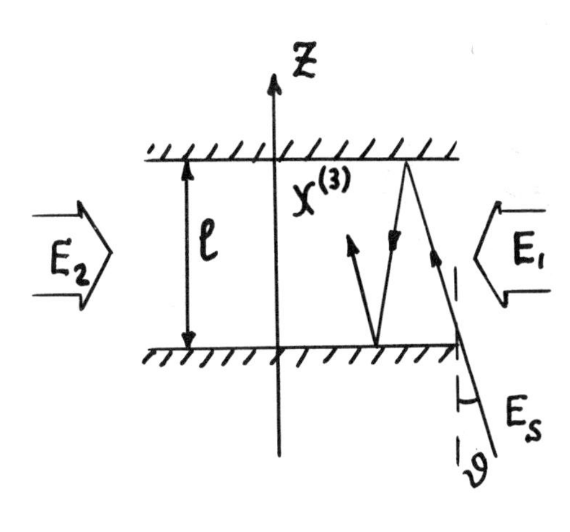

Fig. 4 The multiple reflection scheme of four-wave mixing.

Ref. 8 obtaining effective four-wave mixing in a small volume without a self-action of the reference waves and attaining the fully spatially-polarized wave-front reversal of a signal[8,9]. A similar geometry was also proposed for frequency filtering of signals[7].

If in this scheme two plane mirrors are parallel and separated by a nonlinear layer thickness ℓ, the full signal mixing path is $L = N\ell/\cos\ \theta$, where N-1 is the number of reflections.

For reference waves which are uniform in space between mirrors and have identical frequency ω_0, the system works as a narrowband filter with a maximal reflection at the frequency ω_0. However this filter's properties are substantially changed if the nonlinear polarization that radiates a beam, which in this case is reversed relative to the signal, is modulated in the space between the mirrors. In particular, one can choose a suitable distribution of the reference waves' intensity along the Z-direction so that maximal reflection of the signal wave occurs at a frequency differing from the reference waves frequency ω_0.

It is necessary to emphasize that for successful realization of a four-wave mixing for waves at different frequencies, the nonlinear medium should be resonant with one or both of the moving interference gratings E_1E_s* and E_2E_s*. Hence, the mechanism of nonlinearity should be sufficiently rapid, so that its relaxation time should be less than one of the inverse values of the frequency differences, i.e., $(\omega_s-\omega_1)^{-1}$ or $(\omega_s-\omega_2)^{-1}$. In the opposite case, the interaction effectiveness degrades as $[\tau(\omega_s - \omega_{1,2})]^{-2}$.[2,3]

EXPERIMENT

For the realization of one of the four-wave filter variations and an experimental investigation of its selective properties, the setup in Fig. 5 was implemented. A Nd-glass laser with a passive Q-switch is used to obtain a coherent radiation pulse. By means of a pinhole and plane-parallel glass plate inside a laser resonator, a single frequency is selected. The single-frequency regime of the laser radiation is verified by means of a Fabry-Perot interferometer with a 1 cm base and also by noting the absence of neighboring mode beatings on a high speed oscilloscope, which registers the signal from the photodetector. The laser radiation then is amplified by

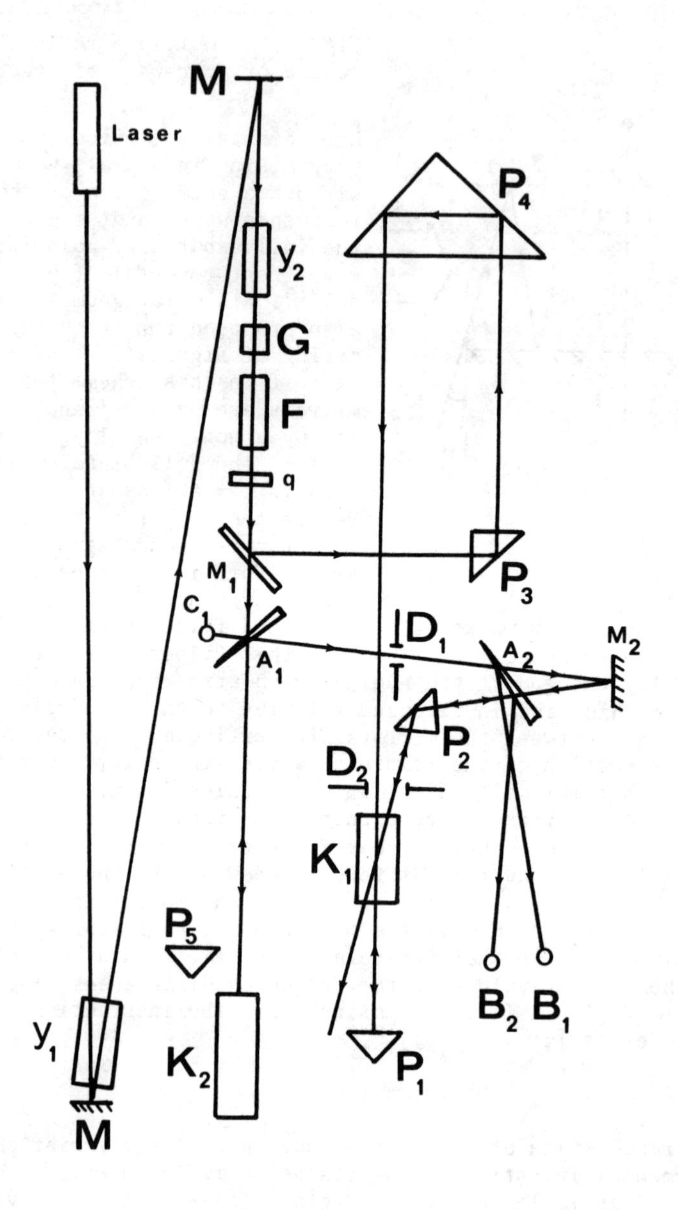

Fig. 5
EXPERIMENTAL SET-UP

means of two amplifiers y_1 and y_2 up to the power 20 MW. The time evolution of the amplified laser radiation represents an approximately Gaussian pulse of about 35 nsec duration full width at half height. After the amplifiers, the laser radiation is transmitted by the Glan-prism polarizer G, and is directed into an optical Faraday cell isolator. A plate of monochristalline quartz q with an optical axis parallel to the incident light rotates the plane of polarization an angle 45°, so that the plane of polarization of the radiation retransmitted through the Faraday cell and quartz plate is perpendicular to the planes of incidence of the light on the surfaces of prisms and mirrors. This allows one to minimize the influence of the light reflection from this surface on the state of polarization. After the quartz plate, the laser radiation is directed onto the semitransparent mirror M_1 with a reflection coefficient of 50%. The wave reflected from mirror M_1 with a power of $\sim$10 MW, is directed by means of prisms P_3 and P_4 into the cell K_1 with a nonlinear liquid of CS_2. This wave serves as a first reference wave E_1. The length of this cell is 4 cm. The relaxation time of the orientational (or Kerr) nonlinearity is less than the reciprocal of the frequencies' maximal difference $\Delta_{max} = (\omega_o - \omega_s)_{max}$ of the interacted waves ($\tau << 1/\Delta_{max}$) in the nonlinear medium, so that the four-wave interaction coefficient in fact does <u>not</u> depend on the frequency difference. The counterpropagating reference wave E_2 is formed by means of the prism P_1 that reflects the first reference wave exactly into the backward direction. The angle between the two reference waves is less than 1 mrad. For the long laser pulses, the thermal nonlinearity of the medium also contributes four-wave mixing because of the weak absorption of the 1.06 μm radiation. But for the 35 nsec duration of the light pulses, such a nonlinearity, which has a longer relaxation time than the orientational one, has no time to influence the four-wave mixing significantly. The other part of the laser radiation, transmitted through the semitransparent mirror M_1 is focused into an auxiliary cell K_2 where the stimulated Brillouin Scattering (SBS) is excited. The Stokes wave scattered into the backward direction from the auxiliary cell is partially reflected from the glass wedge A_1, and is directed by means of mirror M_2 and prism P_2 into the cell K_1 with a nonlinear medium. This wave serves as the signal. The angle between the reference and signal waves is about 20 mrad. The signal wave's frequency differs from the reference wave's by the value of the Stokes shift in the SBS from the auxiliary cell K_2. The discrete frequency shifts are obtained in SBS from the sulfur hexafluoride (SF_6), acetone and polystirole. The minimal frequency shift, obtained in SBS from the gaseous sulfur hexafluoride is 220 MHz. The Stokes shift from the liquid acetone is 3.2×10^9 hz, and from the polystirole is 6.5×10^9 hz. By means of the calorimeter C_1, the energy of the reference waves is measured.

Due to the nearly degenerate four-wave mixing on the orientational CS_2 nonlinearity, a counterpropagating wave conjugate to the signal wave is excited. By means of the glass wedge A_2, parts of the signal and conjugate waves are directed onto the photodetectors B_1 and B_2 to measure their energy. These photodetectors were calibrated initially. The distance between the prisms P_3 and P_4 is chosen so that the optical path difference of the reference and signal waves is less than

15 cm, which corresponds to a light-pulse travel time of about 0.5 nsec. The pinhole D_1 with a 4 mm diameter and the 8 mm gap D_2 selects the central parts of the waves with a more uniform distribution of the intensity along the section. The nonuniform intensity distribution (both for the reference and signal waves) affects the dependence of the reflection coefficient on the frequency shift, and also decreases the effective length of the four-wave mixing $\ell_{eff} \leq \ell$.

In order to compare these measurements with the results of degenerate four-wave mixing when all the frequencies of the reference and signal waves coincide, the auxiliary cell K_2 is replaced by a prism P_5 which retroreflects the light falling on it. The signal wave intensity is about 100 times smaller than the intensities of the reference waves.

We measured the energies of the signal and conjugate waves, and constructed a curve for the reflection coefficient versus the frequency shift Δ as shown in Fig. 6. The absolute reflection coefficient of the signal wave in the degenerate four-wave mixing, when the frequency shift $\Delta = 0$, is about 6%.

If the pulses' time-dependence is taken into account, the expression (3) which gives the reflection coefficient dependence on the frequency shift is changed slightly by a factor $U = T_1^2/(T_1^2 + 2T_2^2)$, where T_1 is the reference wave's duration and T_2 is the signal wave's duration. Here we consider a Gaussian time-dependence of the pulses: $E_1 = E_{10}\, e^{-(t/T_1)^2}$; $E_2 = E_{20}\, e^{-(t/T_1)^2}$; $E_s = E_{so}\, e^{-(t/T_2)^2}$, where $T_2 \leq T_1$. If $T_1 >> T_2$ then $U = 1$. If $T_1 \approx T_2$, then $U \approx 1/3$. In SBS from the sulfur hexafluoride, acetone and polystirole, the duration of the Stokes pulses is decreased relative to the pump-wave duration; the decreases are not the same. Therefore, we correct the results obtained by including the coefficient U above. Figure 6 shows a good agreement between the experimental points and the theoretical curve (3) for the effective length of the nonlinear interaction $\ell_{eff} \simeq 3.6$ cm. An experimental error is created by the background noise and the dispersion of the results from pulse to pulse.

Fig. 6 The four-wave mixing reflection coefficient versus the frequency shift Δ.

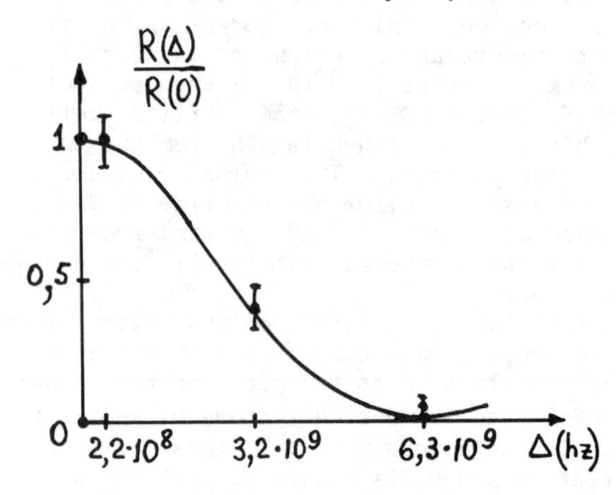

CONCLUSIONS

The conjugate wave emission, when the signal has a frequency shifted relative to the reference waves in four-wave interaction is obtained first. For small frequency shifts the reflection efficiency is approximately the same as in the degenerate four-wave mixing. When the frequency shift increases, the reflection efficiency is decreased by a small amount. Thus, the results obtained show that for pulses of up to 35 nsec duration, the main contribution to the four-wave mixing in carbon disulfide results from the fast orientational nonlinearity. This permits us to vary the line-width of the nonlinear filter and to tune it over a range of 10^{12} hz.

REFERENCES

1. D.M. Pepper, R.L. Abrams, Opt. Lett., 3, 212 (1978).
2. J. Nilsen, A. Yariv, Appl. Opt., 18, 143 (1979).
3. TaoYi Fu, M. Sargent III, Opt. Lett. 4, 366 (1979).
4. O.W. Hellwarth, JOSA, 66, 301 (1976).
5. B.YA. Zel'dovich, V.V. Shkunov, in the book "Wavefront reversal of the light in a nonlinear mediums", p.22-44, Gorkii, 1979.
6. V.N. Balschuk, A.V. Mamaev, N.F. Filipetsky, V.V. Shkunov, B.Ya. Zel'dovich, Opt. Comm., 31,383 (1979).
7. V.N. Balschuk, N.F. Pilipetsky, V.V. Shkunov, Soviet Academy Doklady 251, 70 (1980).
8. V.N. Blaschuk, B.Ya. Zel'dovich, A.V. Mamaev, N.F. Pilipetsky, V.V. Shkunov, Kvantovaia Electron., 7, (3) 627 (1980).
9. B.Ya. Zel'dovich, V.V. Shkunov, Kvantovaia Electron., 6, 629 (1979).

SQUARE ROOT OPERATION BY DYNAMIC SAMPLING

J. Santamaría*, J. Ojeda-Castañeda and L. R. Berriel-Valdos
Instituto Nacional de Astrofísica, Optica y Electrónica
A. P. 216, Puebla, Pue., Mexico

ABSTRACT

It is shown that if the Fourier plane, in a conventional coherent optical processor, is sampled by a time varying filter, then the complex amplitude of the object is mapped as image irradiance variations. Therefore, a square root operation is performed. The mathematical treatment of the method is presented as well as some computer simulated results.

INTRODUCTION

A conventional coherent optical processor can be described as a linear transfer device in complex amplitude. However, since the detectors are quadratic devices, the whole process is a nonlinear process, and only in very specific cases it is possible to speak of linearity[1]. To avoid this nonlinearity due to the detector, we propose a method to implement a square root operation. The method employs a dynamical filter formed by two holes. One of the holes is located at the center of the Fourier spectrum, while the other scans the rest of the Fourier plane. For each position of the mobile hole, an irradiance cosinusoidal grating is recorded at the image plane. In this manner, an on-axis Fourier hologram of Fourier spectrum is recorded. Consequently, the image irradiance variations are proportional to the complex amplitude of the object. Thus, all the linear operations which can be performed in complex amplitude will be displayed as image irradiance. Specifically, the methods employed to render thick phase structures visible can be described as linear filters, in a similar way as it happens in thin phase imagery[2]. In this manner, a dynamic phase contrast filter or dynamic Schlieren method can be implemented.

THEORY

The present mathematical treatment uses suitably normalized coordinates[3]. An object with a complex amplitude transmittance, $A'(u')$, has a Fourier spectrum given by

$$a(x) = \int_{-\infty}^{\infty} A'(u') \exp(-j2\pi u' x) \, du'. \tag{1}$$

*Permanent address: Instituto de Optica, Serrano 121, Madrid, España

ISSN:0094-243X/81/650154-06$1.50

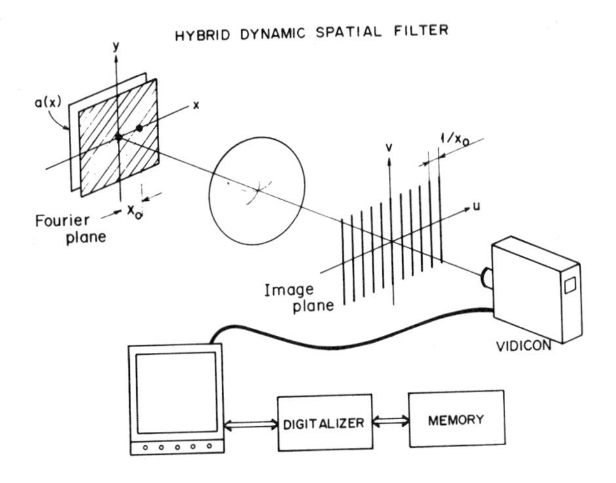

Fig. 1 Optical set-up.

If at a given instance of time t, we sampled only two points of the Fourier spectrum,the zero order and another point chosen by moving a hole with constant speed **v**, see Fig. 1, then the Fourier spectrum of the image is given by

$$a'(x;\ x_o) = a(x)\left[\delta(x) + K\delta(x - x_o)\right], \tag{2}$$

where $x_o = vt$ and K denotes any complex amplitude variation introduced between the two points. In this case,the image irradiance is given by

$$B(u;\ x_o) = |a(0)|^2 + |Ka(x_o)|^2$$
$$+ 2\text{Re}\{Ka^*(0)\ a(x_o)\ \exp(j2\pi u x_o)\}, \tag{3}$$

as shown in Fig. 1. When several of the image irradiance patterns described by Eq. (3) are stored with a real weighting function $f(x_o)$, the total image irradiance is given by

$$B(u) = \int_{-\infty}^{\infty} f(x_o)\ B(u;\ x_o)\ dx_o$$
$$= c_o + 2\text{Re}\{Ka^*(0) \int_{-\infty}^{\infty} f(x_o)\ a(x_o)\ \exp(j2\pi u x_o)\ dx_o\}. \tag{4}$$

This can be conveniently written, invoking the convolution theorem, as

$$B(u) = c_o + 2\text{Re}\{Ka^*(0)\ F(u) \otimes A(u)\}, \tag{5}$$

where F(u) is the Fourier transform of $f(x_o)$.

For the particular case in which $f(x_o) = K = 1$ and the object is real, it is clear that a square root operation has been implemented, that is

$$B(u) = c_o + c\,A(u); \tag{6}$$

see Figs. 2(a) and 2(b).

In the general case, Eq. (5) can be explicitly written as

$$B(u) = c_o + 2\mathrm{Re}\{Ka^*(0)\,F(u)\} \otimes |A(u)|\cos\Phi(u)$$
$$- 2\mathrm{Im}\{Ka^*(0)\,F(u)\} \otimes |A(u)|\sin\Phi(u). \tag{7}$$

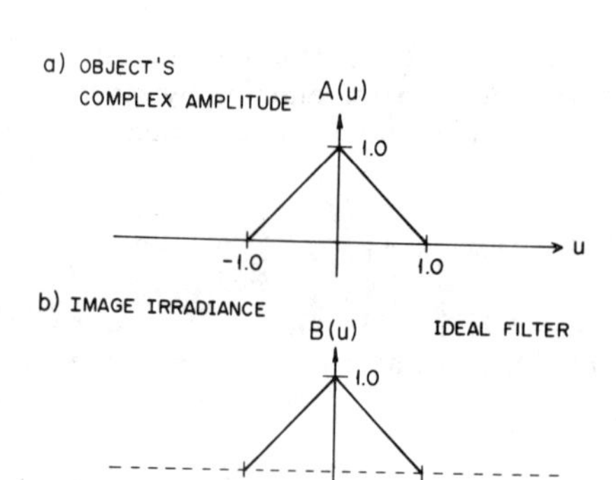

c) IMAGE IRRADIANCE B(u) LOW PASS FILTER

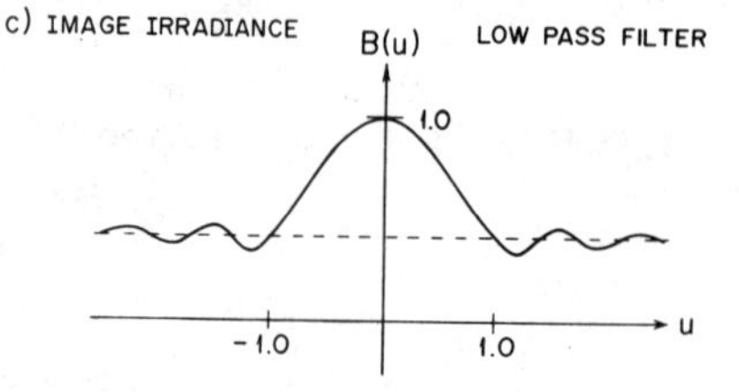

Fig. 2 Square root operation applied to a real object.

From Eq. (7) the following points are clear.

i) Since the interference between the nonzero orders is eliminated, then the image irradiance shows only the interference between the zero and the nonzero orders, as should be expected.

ii) This interference term can be considered as a "true" on-axis Fourier hologram of the Fourier spectrum, in the sense that the Fourier spectrum is mapped as cosinusoidal irradiance gratings. It is to be noted that when recording the image irradiance, the second lens is unnecessary since a lensless Fourier hologram can be recorded[4].

iii) In general, for complex amplitude transmittance objects, the image irradiance variation is related to both the real and the imaginary parts of the object. However, by choosing suitably the constant K and the function F(u), it is possible to isolate each part. This can be very useful to determine the modulus and the phase of the object. For a phase object, $|A(u)| = 1$, only the third term in Eq. (7) would be sufficient to determine the magnitude and the sign of the phase structure, $\Phi(u)$. This can be done by a dynamic phase contrast or by dynamic Schlieren filters. All these filters would map linearly thick phase object, in a similar way as it happens with thin phase imagery[2].

COMPUTER SIMULATION

The following cases have been simulated digitally. Except in one case, Fig. 2(b), the function $f(x_o)$ describes an even lowpass function.

1. The object has an even and purely real transmittance, see Fig. 2(a). For $K = 1$ Eq. (5) becomes

$$B(u) = c_o + 2a(0)\ F(u) \otimes A(u), \tag{8a}$$

as illustrated in Fig. 2(c).

2. The object is a phase object with a hermitian complex transmittance, as shown in Fig. 3(a). Then Eq. (5) for $K = 1$ becomes

$$B(u) = c_o + 2a(0)\ F(u) \otimes \mathrm{Re}\{A(u)\}, \tag{8b}$$

see Fig. 3(b); for $K = j$ we have that

$$B(u) = c_o + 2a(0)\ F(u) \otimes \mathrm{Im}\{A(u)\}, \tag{8c}$$

as shown in Fig. 3(c). This case can be thought of as a dynamical phase contrast method.

3. The object is a nonhermitian phase object, $A(u) = \exp(j\pi\cos 2\pi s u)$, as indicated in Fig. 4(a). Then, for $K = a(0)$, Eq. (5) becomes

$$B(u) = c_o + 2|a(0)|^2\ F(u) \otimes \cos\Phi(u), \tag{9a}$$

see Fig. 4(b), and for $K = j\ a(0)$, we have that

$$B(u) = c_o + 2|a(0)|^2\ F(u) \otimes \sin\Phi(u), \tag{9b}$$

as shown in Fig. 4(c).

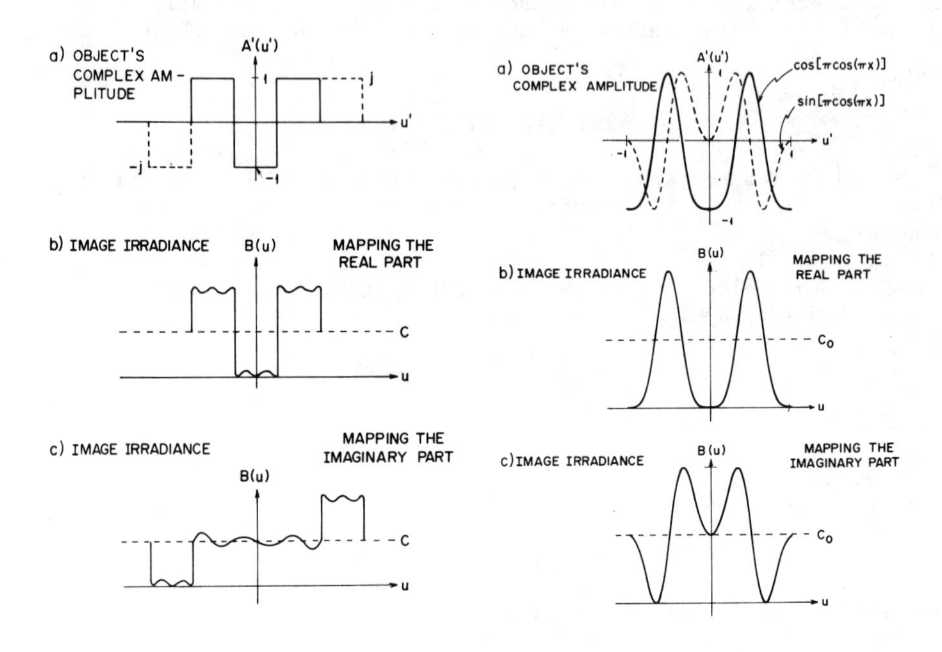

Figs. (3) and (4). Square root operation applied to a phase object.

CONCLUSIONS

We have shown that by eliminating the interference between the nonzero orders of the Fourier spectrum, a square root operation can be implemented optically. This operation makes it possible to consider any conventional optical processor as a linear filter for mapping complex amplitude into image irradiance. In the above process, both the real and the imaginary parts of the complex amplitude are mapped simultaneously. However, it is possible to choose to map either the real or the imaginary part, and in this way to determine the modulus and the phase of the object. For the particular case of phase objects, it is sufficient to map the imaginary part of the object in order to obtain the sign and the magnitude of the phase structure. This can be done either by a dynamic phase contrast filter or by a dynamic Hilbert transform.

REFERENCES

1. J. Ojeda-Castañeda, E. Jara, J. Opt. Soc. Am. 70 458 (1980).
2. J. Ojeda-Castañeda, Optica Acta (accepted for publication).
3. H. H. Hopkins, Jpn. J. Appl. Phys. 4, 31 (1965).
4. G. W. Stroke, Appl. Phys. Lett. 6, 10 (1965).

SOME CONTRIBUTIONS TO ACTIVE OPTICAL FEEDBACK SYSTEMS

T. Tschudi, F. Laeri, B. Schneeberger, F. Heiniger
University of Bern, CH-3012 Bern, Switzerland

ABSTRACT

Optical feedback techniques will extend the region of possible applications of optics in data processing. Only a few different transfer functions can be obtained in a passive optical feedback system, but a wide variety of new possibilities can be suggested by introducing active elements (e.g. amplifiers and spatial phase modulators). A flashlamp pumped dye amplifier adequate for optical feedback is presented and the resulting consequences for data processing systems are discussed.

INTRODUCTION

Optical feedback techniques will extend the region of possible applications of optics in data processing. Especially the realization of optical multidimensional operational amplifiers would be a very important step towards an analog-optical data processing system. Multidimensional operation, together with the possibilities of integration and very high processing speed, are the promising advantages of optical processing in comparison with electronical systems. Recently, important experiments with coherent [1,2] and incoherent [3,4] optical feedback systems have been reported. A review is given in Ref. 5.

Only a few different transfer functions can be obtained in a passive optical feedback system, but a vide variety of new possibilities can be suggested by introducing active elements (e.g. amplifiers and spatial phase modulators). The possibility of varying the transfer function of a coherent optical feedback system simply by changing its loop length can be used for the purpose of realizing filter operations which normally would require different filter transparencies in the optical processor. The first experiments with coherent amplifiers have been reported e.g. by J.W. Hänsch et al [6] and R. Akins and and S. Lee [7].

We report on results of a simple flashlamp pumped dye amplifier suitable for coherent optical feedback systems. A wall-ablating flashlamp system pumps a Coumarin 519 dye with a fluorescent maximum at 514 nm. An argon ion laser, the wavelength of which matches the peak fluorescence of the dye, is used as a signal source. The flashlamp pumped amplifier has the advantage of relative long amplification time - typically several µsec - which is essential to get enough roundtrips of the light in our feedback loop. In addi-

ISSN:0094-243X/81/650160-06$1.50

tion, flashlamp pumped cells will give an isotropic amplification factor over an image field of about 1 cm^2 by using diffuse reflectors around the flash tubes.

DYE AMPLIFIER CELL

A pulsed dye system is well suited for single pass light amplification because of its high gain factor. The broad absorption band facilitates optical pumping [8]. Due to the short lifetime of the excited levels, the broadband fluorescence emission is rather intense but can be reduced with narrowband filters. The gain and the overall efficiency of flash pumped dyes are generally higher than those of laser pumped dyes [9], and pulses are still short enough to get high amplification if one uses wall-ablating flash tubes.

Fig. 1 shows the simple arrangement of the dye amplifier. The dye cell consists of a 45 mm long quartz tube with 5 mm internal diameter. The two optical windows are tilted and the inner surface of the cell is roughend to prevent internal self-oscillation. For the amplification of the 514 nm Argon line (signal source), Coumarin 519 (Coumarin 343) saturated in Methanol ($\sim 10^{-3}$ m) is used.

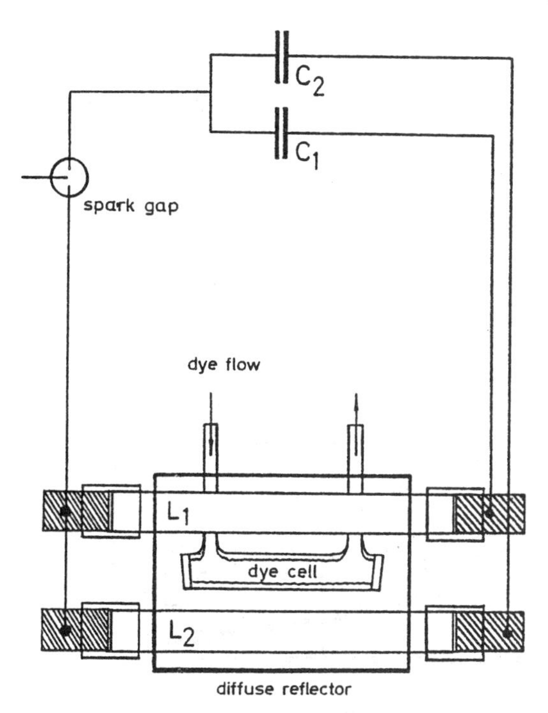

Fig. 1. Arrangement of the dye amplifier.

The transversal pump source consists of four, 18 KV driven, wall-ablating flashlamps with an emission in the wavelength region corresponding to the absorption band of the dye [10]. The total maximum electrical pump power is about 600 Joule which results in an usable light power of about 100 Joule within the absorption spectra of the dye. Lamps and dye cell are mounted close together, surrounded by a MgO_2-diffuser to ensure homogeneous pump light distribution over the image field of about 25 mm^2.

MEASUREMENT OF THE AMPLIFICATION FACTOR

The experimental set up for the amplification measurement is schematically shown in fig. 2. A cw-Argon laser was used as signal source.

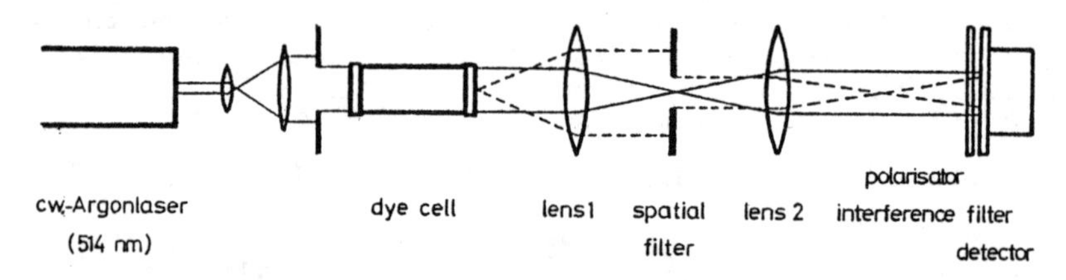

Fig. 2. Experimental set up for the amplification measurement.

Fig. 3 shows typical oscilloscope traces of the spontaneous dye emission (trace a) as well as the amplified laser signal superposed to the spontaneous emission (trace b).

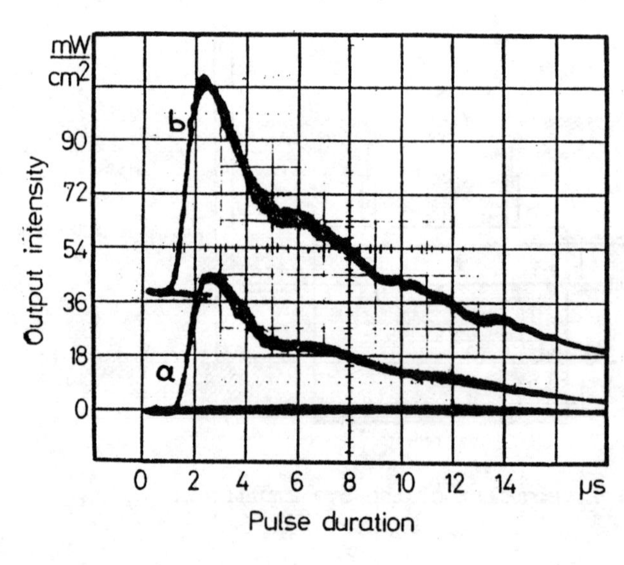

Fig. 3. Oscilloscope trace a) of the spontaneous dye emission b) of the amplified laser signal.

For various input intensities the resulting output signal is shown in fig. 4.

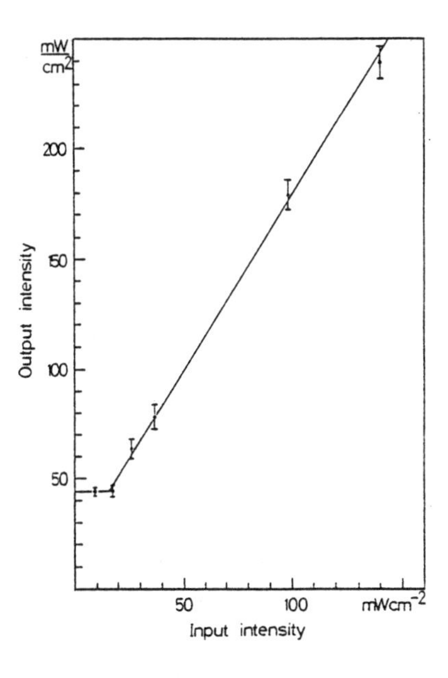

Fig. 4. Determination of the gain factor. The spontaneous emission intensity is about 45 mW cm^{-2}.

The maximum amplification factor for our arrangement is tgα = 1.7 over the 4.5 cm cell, e.g. 1.125 cm^{-1}; enough to compensate losses in the feedback loop and to obtain a net gain.
The optical homogeneity of the dye during the usable pulse width of about 2 μsec was tested in a Michelson interferometer. Thermal distortions were found smaller than one tenth of the signal wavelength. For image amplifier it is useful to express the fluorescent noise intensity in mW per spatial frequency. In our arrangement, the noise intensity is 0.1 mW $(line/mm)^{-2}$ which means a typical value of about 45 mW cm^{-2} for the optics used in the feedback loop (f = 0.4 m). The consequences of this arrangement, from the viewpoint of information processing, are discussed using a slit as the image: if at least the first three maxima of the slit-spectrum

$$\left(\frac{\sin \pi x}{\pi x}\right)^2$$

have to be transmitted and amplified, the third maximum should be just more intense than the background noise; this implies that the zero order of the slit spectrum should be one hundred times more intense than the noise level just before the amplifier cell. In a feedback system this means that the signal intensity should be approximately 1 KW cm^{-2}. Such intensities can be supplied by pulsed lasers (dyes or solid state) as well as injection locked dye lasers [7].

CONCLUSIONS

Coherent amplification of monochromatic light beams in image processing systems will open the possibility to realize a multi-dimensional optical operational amplifier. Dye amplifiers are particularly well suited for this purpose due to their high gain, broad tunability and homogeneous broadening on a nanosecond time scale. In many optical processing systems, not only high gain but also preservation of signal frequency bandwidth is important. Furthermore, active times of sufficient duration to allow passage of light through macroscopic systems are needed. In optical feedback systems, µsec pulses are necessary when a reasonable number of passes through the system is essential to reach steady state. We showed that a flashlamp pumped dye amplifier which is well suited for such applications, is easy realizable. The experimental parameters are summarized in table 1.

Active medium	Coumarin 519 saturated in MeOH ($\sim 10^{-3}$ m)
Gain	1.125 cm^{-1}
Active area for image processing	25 mm^2
Active volume	$\sim$ 1 cm^3
Pump power	5 . 10^4 W cm^{-3}
Usable pulse duration	2 - 3 µsec
Thermal phase distortion	$< \lambda/10$
Possible operation rate	200 Hz
Noise	0.1 mW / $(Lp/mm)^2$

Table 1.

The relatively high noise level due to the spontaneous emission of the dye requires a high power signal beam, e.g. solid state lasers or injection locked dye laser systems [5].

Coherent image amplifiers in combination with spatial phase modulators (e.g. described in Ref. 11) will give a high versatility of optical feedback systems. Corresponding experiments are in progress.

ACKNOWLEDGEMENT

We thank P. Anliker for his helpful discussions and A. Friedrich for his essential contributions in preparing the hardware.

REFERENCES

1. P.P. Jablonowski and S.H. Lee, Appl. Phys. 8, 51 (1975).
2. E. Händler and U. Röder, Opt. Commun. 23, 352 (1977).
3. G. Häusler and A. Lohmann, Opt. Commun. 21, 365 (1977).
4. T. Sato, K. Sasaki and R. Yamamoto, Appl. Opt. 17, 717 (1978).
5. R.P. Akins, R.A. Ahale and S.H. Lee, Opt. Eng. 19, 347 (1980).
6. T.W. Hänsch, F. Varzanyi and A.L. Schawlow, Appl. Phys. Lett. 18, 108 (1971).
7. R. Akins and S. Lee, Appl. Phys. Lett. 35, 660 (1979).
8. G.A. Reynolds and K.H. Drexhage, Opt. Commun. 13, 222 (1975).
9. B.B. Snavely, Proc. of IEEE 57, 1374 (1969).
10. T. Okada, K. Fuiiwara, M. Maeda and Y. Miyazoe, Appl. Phys. 15, 191 (1978).
11. F. Laeri, B. Schneeberger and T. Tschudi, Opt. Commun. 34, 23 (1980).

PHASE CONJUGATION IN 3, 4 AND INFINITE DIMENSIONS

Murray Sargent III
Optical Sciences Center, University of Arizona, Tucson, AZ 85721

ABSTRACT

The theory of resonant four-wave mixing using one and two-photon transitions is described in terms of saturation spectroscopy. Detuning the signal wave from the pump wave in the one-photon case leads to a narrow-band reflection spectrum, with width given by $1/T_1$, where T_1 is the lifetime of the population difference. In the corresponding time-domain picture (the fourth dimension), this produces a reflected pulse at least as long as T_1, regardless of the signal. This time-domain picture also leads to a novel kind of coherent transient spectroscopy. In the two-photon case, an infinite number of levels are involved in principle, i.e., phase conjugation in infinite dimensions! In this case, two kinds of mechanisms compete in generating conjugate waves leading to two phenomena that do not occur in two-level media: 1) a double-peaked reflection spectrum and 2) coupled-mode oscillation in absorbing media. In addition, due to the dynamic-Stark shifted line center, the reflection coefficient approaches a constant value at large intensities, unlike the two-level coefficient which bleaches to zero.

INTRODUCTION

The theme of this conference, optics in four dimensions, fits beautifully into the discussion of phase conjugation given in this paper. First I consider cw four-wave mixing, which takes place in the three spacial dimensions. Then the signal frequency is varied, offering a fourth dimension. The Fourier transform can be applied with care to treat pulsed signals, so that time becomes the fourth dimension. Finally I consider two-photon phase conjugation, which involves an infinity of atomic levels. As such, one might call it phase conjugation in infinite dimensions. In a related way, the name "optics in four dimensions" invites connecting up ideas from various origins, and this paper discusses phase conjugation from the point of view of saturation spectroscopy. People involved in holography think of phase conjugation as real-time holography. But in terms of the interaction of light with matter, phase conjugation can be thought of as an extension of saturation spectroscopy. In this connection, phase conjugation offers new spectroscopic methods for studying the nature of matter, particularly the relaxation processes of excited systems.

So what is phase conjugation? Phase conjugation is an important process[1] that inverts a phase front in space so that it retraces the path through which it came. This has useful applications in propagation through turbulant media, through bad optics and through optical fibers. As such the area is important for subjects as varied as astronomy, military weapons, laser induced fusion, and optical communications. In addition it provides an alternative and potentially very effective way to study properties of the phase conjugating medium itself. Section 1 describes the physical origin of the phase conjugation process in terms of spacial-hole burning (SHB), a nonlinear response of a medium to an electromagnetic field fringe pattern. SHB is used in a simpler problem, grating-dip spectroscopy, to measure relaxation rates in the medium by studying the spectrum of transmitted signal light. Section 2 defines and

ISSN:0094-243X/81/650166-12$1.50

discusses phase conjugation itself as an extension of grating-dip spectroscopy, and shows how the finite lifetimes of the nonlinear medium lead to a narrow-band reflection spectrum. Section 3 uses the Fourier transform to apply the spectral responses to pulsed signal operation in both saturation-spectroscopy and phase conjugation cases. The former leads to a novel kind of coherent transient spectroscopy that combines aspects of the traditional coherent transient spectroscopy with saturation spectroscopy. Section 4 presents two-photon phase conjugation and summarizes three exciting predictions to challange the experimentalists.

1. Saturation spectroscopic origins of phase conjugation

The basic physical process underlying phase conjugation is the nonlinear response of a medium to an electromagnetic fringe pattern. The first appearance of such responses

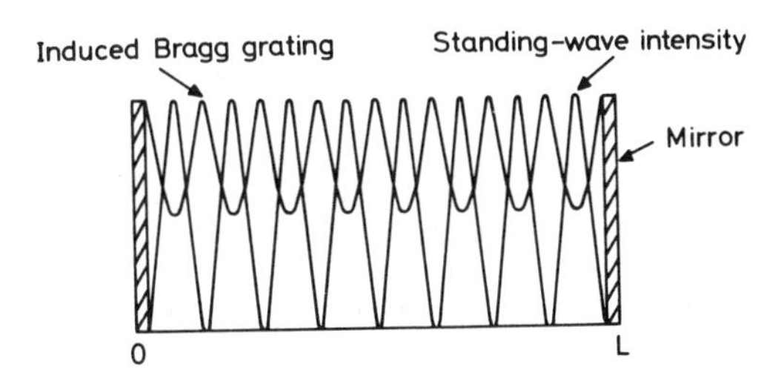

Fig. 1. Spacial hole burning (SHB) in a laser cavity. The standing-wave fringe pattern in a laser saturates the atomic response at the fringe peaks, while ignoring the atoms at fringe nodes. This spacially-varying saturation acts as a grating that backscatters the two running waves comprising the standing wave into one another.

was observed in spacial hole burning (SHB) in lasers (Fig. 1). The laser field consists of two counterpropagating running waves that are constrained by the mirrors to be a standing wave with fringe spacing equal to half a wavelength.[2] This fringe pattern induces spacial holes into the polarization of the nonlinear medium. These holes act, in turn, as a grating to backscatter the running waves into one another.

More recently such gratings have been used in a number of time-varying contexts to study relaxation processes in media. Eichler watched the reflection of Ar^+ laser light off spacial holes burned in ruby.[3] The temporal decay of the reflected light after ruby pump was turned off yields a measure of the relaxation rate of the spacial holes. Siegman and coworkers[4] have considered scattering off transient gratings of this nature. Perhaps the simplest approach is that of grating-dip spectroscopy.[5,6] For this a strong running wave is used to saturate the medium and a weak probe wave is used to test the response (Fig. 2). Although the weak wave is assumed not to saturate the medium itself, it adds to the strong wave to yield a small fringe component on top of the predominantly strong running wave. This small fringe induces a small grating that scatters part of the strong wave into the path of the weak wave. The amount of the scattering can have the same order of magnitude as the weak wave itself. By tuning the weak wave frequency away from the strong wave's, the fringe pattern "walks" increasing rapidly. The nonlinear medium attempts to follow this moving fringe pattern, but ultimately gives up, eliminating the scattered component. Hence by studying the transmitted weak-wave intensity spectrum, one can measure the temporal response of

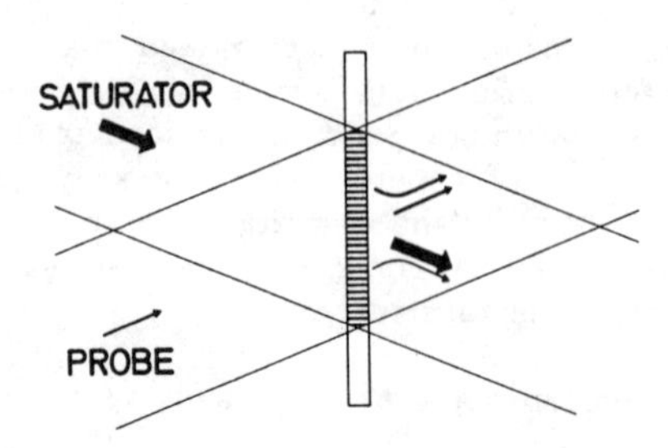

Fig. 2. Grating-dip spectroscopy principle. The superposition of a strong and a weak wave induces a small grating in a nonlinear medium that scatters the strong wave into the path of the weak wave. By varying the weak wave frequency relative to the strong wave's, the fringe pattern moves, modifying the ability of the medium to respond, and hence the magnitude of the scattered component.

the medium in an alternative way to the time-domain approaches of Eichler[3] and Siegman[4]. Which method is best depends on the medium and tools available to the experimenter.

So far my discussion has treated a general nonlinear medium. It could be purely classical, it could be resonant or nonresonant. Let us now consider a specific simple

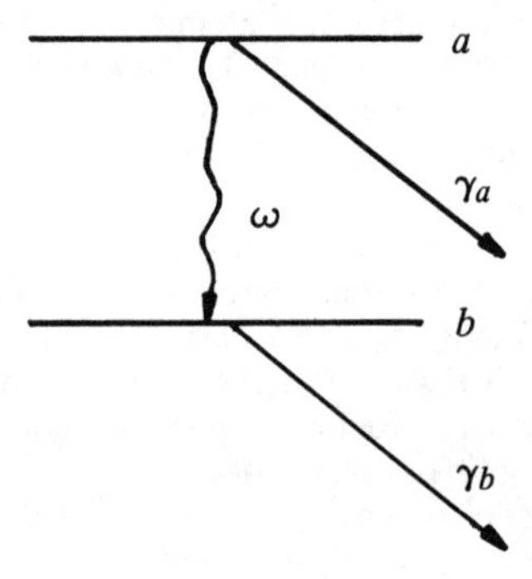

Fig. 3. Two-level medium energy diagram. This medium provides a simple case to illustrate resonant nonlinear interactions leading to grating dips and to phase conjugation

medium consisting of two-level atoms (Fig. 3). If the population difference has a lifetime (T_1) much longer than the induced dipole lifetime (T_2), the weak wave absorption coefficient reduces to

$$\alpha = \alpha_o\left[\frac{1}{1+I} - \frac{I}{(1+I)^2 + (\nu_s-\nu_w)^2 T_1^2}\right] \tag{1}$$

where α_o is the linear absorption coefficient, ν_s is and ν_W are the strong and weak wave frequencies, respectively, and I is the strong wave intensity in units of the

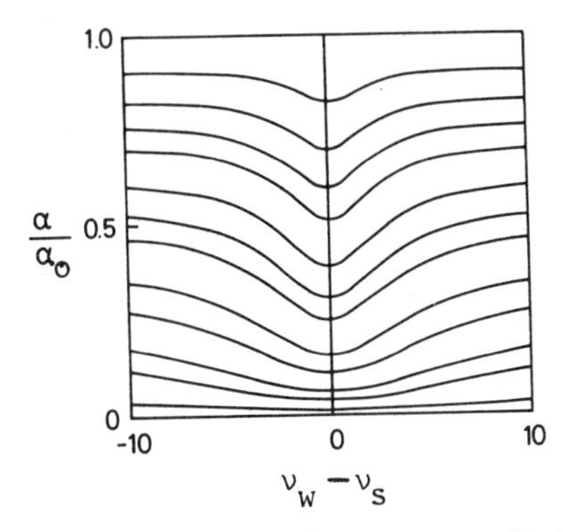

Fig. 4. Grating dip in weak-wave absorption coefficient plotted vs strong-weak wave beat frequency.

saturation intensity. This simple power-broadened Lorentzian is plotted in Fig. 4, and allows one to read off the lifetime T_1 easily.

The theory can be extended to arbitrary relations between the dipole and

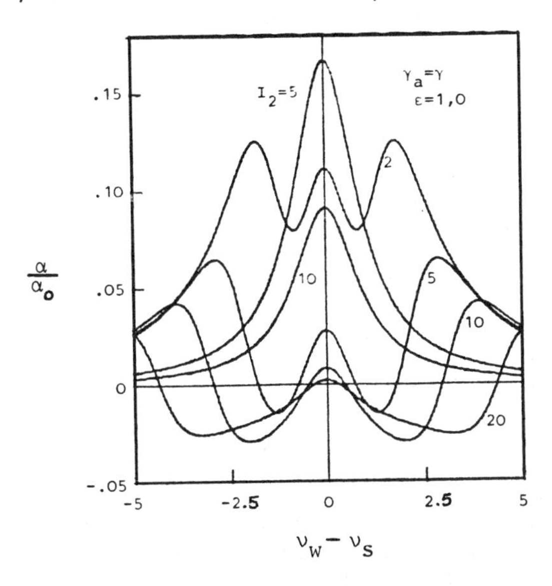

Fig. 5. Weak-wave absorption coefficient α vs strong-weak wave beat frequency for equal dipole and population lifetimes ($T_2=T_1$).

population lifetimes.[7,8] Figure 5 reveals that the scattering contribution reshapes the absorption spectrum in the form of a dynamic Stark effect. In fact, Fig. 5 can be derived alternatively by diagonalizing the Hamiltonian that includes the strong wave interaction, but not the weak wave. The weak wave then interacts with atomic levels split by the dynamic Stark effect. The theory can also be extended to treat inhomogeneously-broadened media. Such an approach has been used by Keilmann[9] to determine the picosecond T_1 in p-type Ge.

Note that in all cases, the moving grating Doppler shifts the scattered wave back into resonance with the transmitted weak wave. Alternatively, one can understand this frequncy shift in terms of a Raman effect. The medium responds to the beat frequency between the strong and weak waves with induced pulsations in its populations. The population pulsations act as a "Raman shifter" or modulator to put two sidebands on the strong wave, one of which has the same frequency as the weak wave, although usually different phase. The other sideband is not phase matched except for collinear weak and strong waves, and hence doesn't build up in general.

2. Four-wave Phase conjugation

Now consider the addition of a second strong wave propagating exactly opposite to

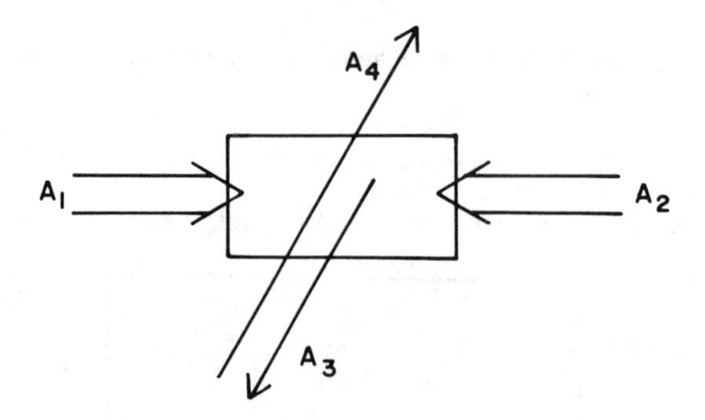

Fig. 6. Four-wave mixing resulting in phase conjugation. A second counterpropagating strong wave scatters off the small grating induced by the weak wave with the first strong wave. The scattered wave propagates in the direction opposite to the weak wave and is called the conjugate wave.

the first as shown in Fig. 6.[10,11] This wave also scatters off the weak-strong wave grating, but in the opposite direction. The scattered wave is called the conjugate wave as can be seen from the following notation. Suppose the weak wave has a propagation factor $\exp[-i\mathbf{K}\cdot\mathbf{r}-i\nu_4 t]$. Then the retroreflected wave has the propagation factor $\exp[+i\mathbf{K}\cdot\mathbf{r}-i\nu_3 t]$, i.e., the spacial factor is conjugated. While considering notation, let's note that the strong waves are called the pump waves in phase conjugation, while the incident weak wave is called the signal wave. These names lead, unfortunately, to confusing subscripts, since the Probe and Saturator waves of saturation spectroscopy are replaced by the Signal and Pump waves of phase conjugation, respectively. We sidestep this problem, by calling the two pump waves 1 and 2, the conjugate wave 3 and the signal wave 4, in agreement with Yariv and Pepper[11] and the degenerate two-level theory of Abrams and Lind.[12]

The theory is obviously complicated by the addition of the second pump and the resulting conjugate wave, but the analysis proceeds along similar lines. In particular, a second fringe and associated grating exist due to the interference between the signal and the second pump wave. This second grating scatters the first pump wave, and contributes to the conjugate-wave amplitude. Another feature of the theory is that the spacial holes resulting from the interference of the two pump waves do not contribute directly to the scattering. They do modify the absorption and scattering coefficients, which have to include averages over the pump intensity. The signal and conjugate waves are again assumed not to saturate the atomic response. Their amplitudes obey the propagation equations

$$dA_3/dz = \alpha A_3 + i\,\kappa^* A_4^* \tag{2}$$

$$dA_4^*/dz = -\,\alpha^* A_4^* + i\,\kappa A_3 \tag{3}$$

familiar in coupled-mode theory and in the distributed feedback laser.[13] These equations can be solved analytically for the amplitude reflection coefficient

$$r = A_3/A_4^* \tag{4}$$

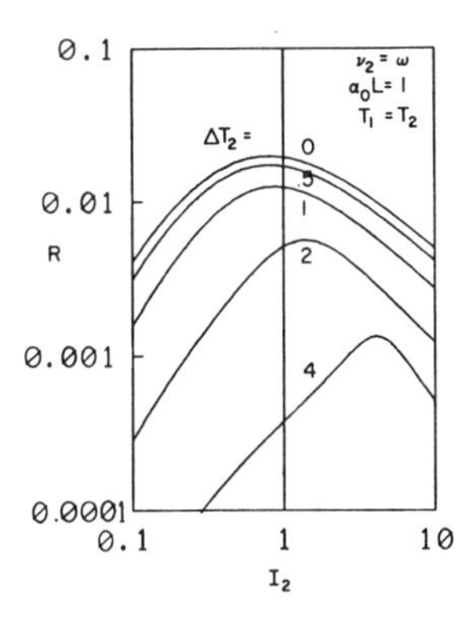

Fig. 7. Intensity reflection coefficient vs pump intensity for various signal-pump detunings. As the detuning increases, R falls off in a fashion similar to the reduction of scattering in grating-dip spectroscopy.

and the intensity reflection coefficient $R=|r|^2$. Figure 7 illustrates the results in the form of curves of R vs pump intensity. Consider first the zero-detuning case (top curve).[12] As the pump intensity increases, R increases up to a point. The increase is due to the larger grating induced, which scatters a more intense pump wave. After a point, however, the two-level medium starts to bleach, ultimately failing to interact with the fields at all. In this limit, the strong pumps have equilibrated the upper and lower level populations, leading to as much stimulated absorption as stimulated emission. In the two-photon problem, this doesn't occur, as we see in Sec. 4 below.

Secondly, note that R falls off as the signal is detuned[14] from the pumps. This is due to the fact that the fringe patterns move, ultimately too rapidly for the medium to follow, just as in the grating-dip spectroscopy case. The results in reduced grating size and a correspondingly reduced reflectivity. The degree to which the reflection is reduced depends on the lifetimes of the medium. Figure 8 shows how a medium with long T_1 has a narrow reflection band, leading to a bandpass filter. The mechanism here differs from that considered by Pepper and Abrams[15] and by Gulamiryan et al.[16] This latter version uses the fact that the conjugate wave is not phase-matched for significant signal-pump detunings. In fact, while the moving gratings Doppler shift the pump frequency to the transmitted signal frequency, they Doppler shift the conjugate

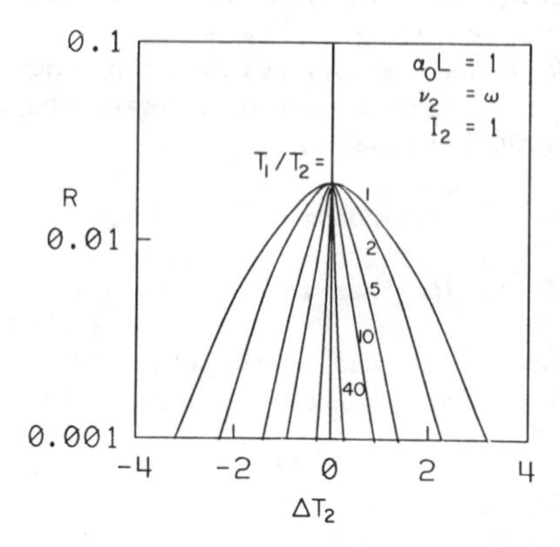

Fig. 8. Intensity reflection coefficient R vs signal-pump detuning for various values of the population difference lifetime T_1. The reflection bandwidth provides a measure of T_1.

wave frequency in the opposite way. Hence the conjugate wave frequency is symmetrically placed on the other side of the pump wave frequency with respect to the signal. Nevertheless, the signal and conjugate wave vectors are equal in magnitude and opposite in sign, thereby leading to a phase mismatch. Which mechanism dominates the reflection bandpass depends on the characteristics of the medium. In general, long lifetime media are dominated by the T_1 effect, while off-resonant χ^3 type media are dominated by the phase mismatch effect.

3. Pulsed signal operation

Now let's use the cw detuning theory to treat pulsed signal operation, first in the two-wave interactions of saturation spectrosocpy and then in four-wave phase conjugation. In the theories of saturation spectroscopy both waves are cw (at least during the relevant medium lifetimes) and typically one wave doesn't saturate the medium. Using the resulting spectral information, we can study the transmission of a nonsaturating pulse through a medium subjected to a cw saturating field. The technique is related to the χ^3 phase-conjugation discussions of Yariv et al[17] and Feldman et al,[18] and relevant four-wave experiments have been carried out by Mandelberg.[19] The pulse is Fourier analysed, multiplied by the amplitude spectral transmission function and resummed, yielding the transmitted pulse. The spectral transmission function is determined from the signal absorption coefficients. At first glance, this looks to be fraught with difficulty, since we are using Fourier analysis in a nonlinear problem. However, we are considering linear deviations about a steady-state that is not in thermal equilibrium. Fourier analysis applies, and we discover information about how the medium attempts to return to thermal equilibrium, namely relaxation times, like T_1 and T_2. The transmission functions look simple enough that we may be able to derive analytic formulas for simple pulse shapes, like delta functions and step functions.

Qualitatively, we can predict the effect of the medium on the transmitted pulse. A delta function signal (i.e., probe wave) induces a grating with the help of the pump wave. The cw pump wave then continues to read out this grating for as long as it lasts. The effect is much like a bell: the sound lasts much longer than the blow that created it. Hence the transmitted pulse consists of something looking like the incident signal plus a long pulse with length about equal to T_1. This approach leads to a new kind of coherent transient spectroscopy, which combines some features of the traditional coherent transient spectroscopy with some features of saturation spectroscopy. A notable difference relative to the former is that the pulse cannot saturate the medium (if the theory is to apply).

Now consider the four-wave problem in which a pulsed signal is reflected off a medium. We know the reflection coefficients for a single frequency from Ref. 14. Using the same Fourier analysis appropriate for the two-wave case we derive the reflected pulse amplitude spectrum. The cw pump waves then proceed to scatter off of the induced gratings (or holograms) for the length of time that the holograms exist, as

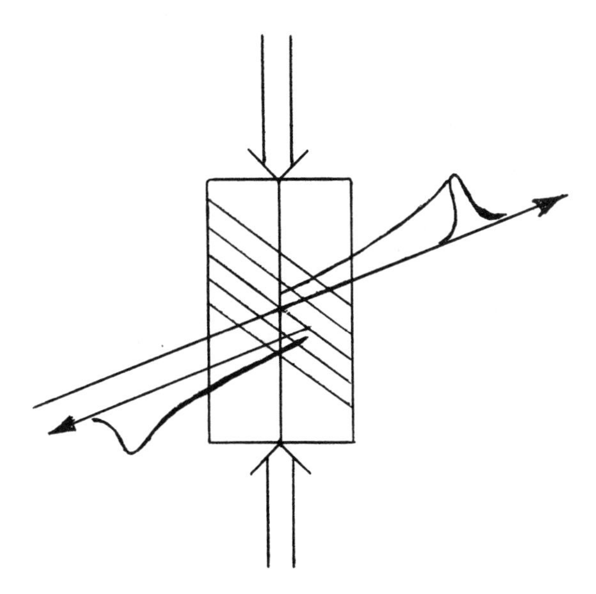

Fig. 9. Short pulsed signal induces gratings (holograms) in a nonlinear medium, that scatter the cw pumps for the lifetime of the gratings.

illustrated in Fig. 9. In a two-level medium the relevant time is once again that for the population difference, namely T_1. The early experiments of Liao and Bloom[20] demonstrated that such effects do occur in ruby, which has a long (about 1 msec) T_1. The Fourier approach is currently being investigated by Barry Feldman, Bob Fisher, Jerry Sudam and myself, and preliminary results show the bell effect as predicted. Understanding the response of a medium to a pulse is very important in a number of applications for phase conjugation and specifically with regard to laser induced fusion. It also provides a new and potentially very important spectroscopic technique.

4. Two-photon Phase Conjugation

Two photon phase conjugation occurs in a medium with an energy level diagram depicted in Fig. 10. Pairs of off-resonant virtual transitions to dipole allowed states occur that transfer population from the lower level b to the upper level a. The corresponding two-photon coherence (which doesn't have an electric dipole moment) ρ_{ab} interacts with the field to produce a polarization. Two photon phase conjugation created by these mechanisms has been observed by Liao, Economu and Freeman[21] and

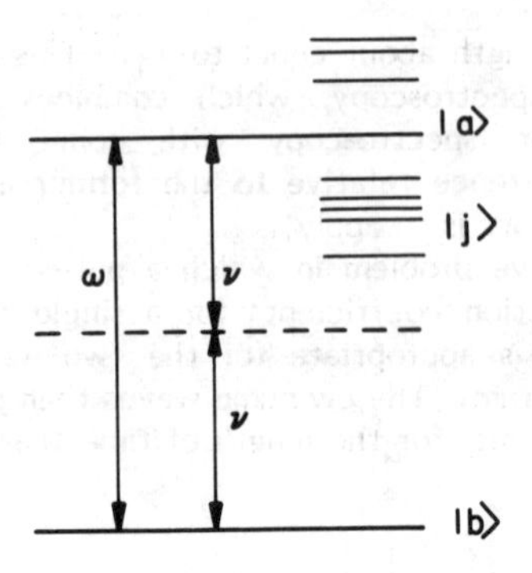

Fig. 10. Energy-level diagram for two-photon transitions an active atom.

also by Steel et al.[22]

The first theory of two-photon phase conjugation has been given by Fu and Sargent,[23] which combines the Abrams-Lind phase conjugation theory[12] with the two-photon theory of Narducci et al.[24] The signal, pumps and hence the conjugate are all chosen to have the same frequency, and the medium is assumed to be homogeneously-broadened. The theory predicts three significant effects that are absent in the two-level medium case. To understand the origin of these effects, consider the polarization expression of Narducci et al[24]:

$$\mathscr{P}(\vec{r}) = \frac{N}{2}(k_{aa} + k_{bb})\mathscr{E} + Nk_{ab}\sqrt{T_1T_2}\,\omega_s \frac{\mathscr{E}}{1+I^2\mathscr{L}} + \frac{Nk_{ab}}{\sqrt{T_1T_2}}\mathscr{D}\frac{\mathscr{E}\mathscr{E}\mathscr{E}^*}{1+I^2\mathscr{L}} . \tag{5}$$

This polarization is averaged and projected onto the appropriate field modes to yield the signal and conjugate absorption and coupling coefficients. The first term does not lead to phase conjugation. The second can yield phase conjugation and results from off-resonant single-photon transitions that are saturated by two-photon transitions. The term is real causing an index change with no absorption. The corresponding phase-conjugation contribution involves an induced grating or hologram as does that in the two-level atom case. Specifically a pump interferes with the signal forming a grating that is then read out as a conjugate by the other pump wave. The final term in Eq. (5) is the dominant conjugate generator for low pump intensities, and has both index and absorptive terms. The primary phase-conjugate contribution results from the conjugated signal interacting with a two-photon coherence term whose pump wave-vector dependence cancels out. Unlike the usual two-level phase conjugation, no population grating is involved in this process.

For sufficiently large pump intensities, three phenomena occur in the two-photon case that do not occur in the single-photon case. First, the pure-index phase-conjugation contribution (second term in Eq. (5)), interferes with the index contribution from the third term in (5), leading to increased reflection on one side of the effective line center ($\omega + \omega_s I$, where ω_s is the Stark shift parameter) and decreased reflection on the other. This leads to the double-peaked spectrum in Fig. 11. Secondly, the pure-index conjugation term increases the signal-conjugate coupling enough to dominate the absorption and produce coupled-mode oscillation even in an absorbing medium (see Fig. 12). In contrast, the resonant two-level theory[12] predicts oscillation only in inverted (gain) media, since the absorption can dominate the signal-conjugate coupling. Thirdly,

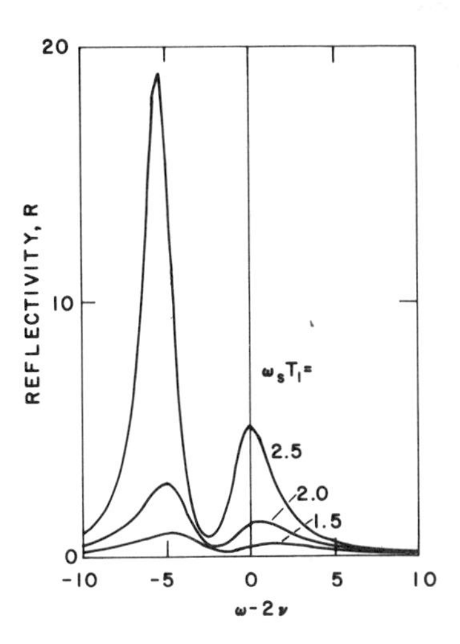

Fig. 11. Reflectivity R versus frequency detuning, normalized to T_2. $T_2=T_1$, $I_2=1.0$.

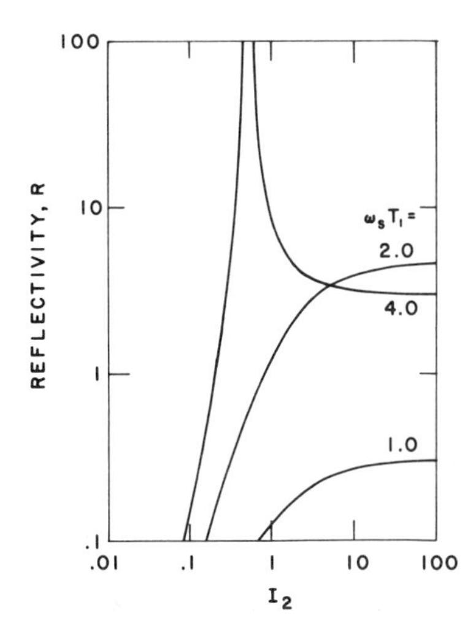

Fig. 12. Reflectivity R versus pump intensity I_2 in an absorbing medium, $\alpha_o L=1.0$, and $T_1=T_2$. Notice that coupled-mode oscillation occurs for $\omega_s T_1=4$.

whereas the two-level reflection coefficient bleaches to zero at large intensities, the Stark-shifted line center of the two-photon case prevents bleaching and leads to an intensity independent reflectivity at large intensities as shown in Fig. 12.

Plans for further research in this area include generalizing the theory in two important ways: to include Doppler broadening and signal detuning. Two-photon work has traditionally be carried out in Doppler broadened media because the Doppler effect cancels out. The four wave case is more complicated and needs to be considered carefully. The signal detuning generalization is prerequisite to the pulsed response analysis.

Two of the interesting predictions above involved "one-photon" processes. Here off-resonant transitions treatable with first-order perturbation theory acquire a nonlinear character by virtue that the initial or final populations are affected by two-photon transitions. This kind of contribution is also significant in two-photon optical bistability.[25]

ACKNOWLEDGEMENTS

It is a pleasure to thank Tao Yi Fu, Bob Fisher, Barry Feldman and Jerry Suydam for numerous helpful discussions. This work was supported in part by the U. S. Air Force Office of Scientific Research (AFSC), U. S. Air Force, and in part by the U S. Army Research Office, U. S. Army.

References

1. For reviews, see A. Yariv, IEEE J. Quantum Electron, **QE-14**, 650 (1978); and B. Ya. Zel'dovich, N. F. Pilipetskii, V. V. Reful'skii, and V. V. Shkunov, Kvantovaya Electron. Moscow **5**, 1800 (1978) [Sov. J. Quantum Electron. **8**, 8 (1978)]; R. W. Hellwarth, IEEE J. Quantum Electron. **QE-15**, 101 (1979).

2. See, for example, Chap. 8 in M. Sargent III, M. O. Scully and W. E. Lamb, Jr., Laser Physics, Addison-Wesley, Reading, Mass (1974).

3. H. J. Eichler and H. Stahl, J. Appl. Phys. **44**, 3429 (1973); **44**, 5383 (1973); H. J. Eichler and J. Eichler, J. Appl. Phys **45**, 4950 (1974).

4. D. W. Phillion, D. J. Kuizenga and A. E. Siegman, Appl. Phys. Letters **27**, 85 (1975).

5. M. Sargent III, Appl. Phys. **9**, 127 (1976).

6. For a review of this subject, see M. Sargent III, Phys. Rep. **43**, 223 (1978).

7. M. Sargent III, P. E. Toschek and H. G. Danielmeyer, Appl. Phys. **11**, 55 (1976).

8. M. Sargent III and P. E. Toschek, Appl. Phys. **11**, 107 (1976).

9. F. Keilmann, Appl. Phys. **14**, 29 (1977).

10. R. W. Hellwarth, J. Opt. Soc. Am. **67**, 1 (1977).

11. A. Yariv and D. M. Pepper, Opt. Lett. **1**, 16 (1977).

12. R. L. Abrams and R. C. Lind, Opt. Lett. **2**, 94 (1978); **3**, 205 (1978).

13. M. Sargent III, W. H. Swantner and J. D. Thomas, IEEE J. Quantum Electron. **QE-16**, 465 (1980).

14. T. Fu and M. Sargent III, Opt. Letters **4**, 366 (1979).

15. D. M. Pepper and R. L. Abrams, Opt. Lett. **3**, 212 (1978).

16. A. L. Gulamiryan, A. V. Mamaev, N. F. Pilipetskii, N. F. Rugulskii and V. V. Shkunov, contribution to this volume.

17. A. Yariv, D. Fekete and D. M. Pepper, Opt. Lett. **4**, 52 (1979).

18. R. A. Fisher, B. R. Suydam, W. W. Rigrod and B. J. Feldman, XIth Int. Quant. Electronics Conf., Boston, June (1980).

19. H. I. Mandelberg, Opt. Lett. **5**, 258 (1980).

20. P. F. Liao and D. M. Bloom, Opt. Lett. **3**, 4 (1978).

21. P. F. Liao, N. P. Economou, and R. R. Freeman, Phys. Rev. Lett. **39**, 1473 (1977).

22. D. G. Steel and J. F. Lam, Phys. Rev. Letters **43**, 1588 (1979).

23. T. Fu and M. Sargent III, Opt. Lett. **6**, (1980).

24. L. M. Narducci, W. W. Eidson, P. Furcinitti, and P. C. Eteson, Phys. Rev. A**16**, 1665 (1977).

25. E. Giacobino, M. DeVaud, F. Biraben, G. Grynberg, Phys. Rev. Letters **45**, 434 (1980).

PHOTOELECTRIC CORRELATIONS AND FOURTH ORDER COHERENCE PROPERTIES OF OPTICAL FIELDS*

L. Mandel
University of Rochester, Rochester, N.Y. 14627

ABSTRACT

Photoelectric correlation measurements of optical fields are now becoming commonplace and, as is well known, they require fourth order correlation functions of the field for a description. Nevertheless, the fourth order theory of optical coherence has not received a great deal of attention. This is the subject of the present paper. When interference effects are combined with intensity correlations, we encounter correlation functions of the general form

$$\Gamma^{(2,2)} \equiv \langle V^*(\underline{r}_1,t_1)V^*(\underline{r}_2,t_2)V(\underline{r}_3,t_3)V(\underline{r}_4,t_4)\rangle \quad ,$$

which are functions of four space-time arguments. Some properties of such fourth order correlations, as they relate to interference and intensity correlation effects, are derived. The space arguments are generally suppressed in the calculation. The correlation functions are expressible as multiple Fourier transforms of certain fourth order spectral densities, which are functions of three frequencies in the stationary state.

The superposition law for fourth order spectral densities is derived for independent light beams, and it is shown that the fourth order spectral densities exhibit the reproducing property under superposition only when the light obeys thermal statistics. Next the superposition law is found for the spectral densities of the photoelectric signals generated by light beams falling on photodetectors. Investigation of the photoelectric signal spectrum yields information on the spectrum of the light intensity fluctuations, and removes the measurement from the domain of optics to the domain of electronics. It is shown that the photoelectric spectral densities reproduce under superposition when the two light beams in question obey a certain factorization condition, that is analogous to the cross-spectral purity condition for second order correlations. Finally, it is shown that, under certain circumstances, all the relevant fourth order correlation properties of the optical field are expressible in terms of simpler two-frequency spectral functions.

INTRODUCTION

When several light waves are superposed, and the expectation value of the resultant light intensity is calculated, one encounters second order correlation functions of the electromagnetic field of the form $\langle \underline{V}^*(\underline{r}_1,t_1)\cdot\underline{V}(\underline{r}_2,t_2)\rangle$. Here $\underline{V}(\underline{r},t)$ is the complex amplitude

*This work was supported in part by the National Science Foundation, and by the Air Force Office of Scientific Research.

ISSN:0094-243X/81/650178-18$1.50

of the optical field at the space-time point $\underline{r},t$ and < > denotes the average over the ensemble. If the two space-time points coincide, then $\underline{V}^*(\underline{r},t)\cdot\underline{V}(\underline{r},t)$ reduces to the light intensity $I(\underline{r},t)$ at $\underline{r},t$. It is well known that the second order correlation function is closely related to the interference fringes that result from the superposition,[1] and in the absence of any correlations between the waves, there is no stationary interference pattern. The measurement of interference patterns therefore represents an important class of experiments for the investigation of the second order correlation properties of optical fields.

A different class of optical correlation functions can be derived from photoelectric correlation measurements. If the light at $\underline{r}_1,t_1$ and $\underline{r}_2,t_2$ is allowed to fall on two photodetectors, and the correlation between the resulting photoelectric signals is measured, one encounters intensity or fourth order correlation functions of the electromagnetic field, in the form

$$\langle I(\underline{r}_1,t_1)I(\underline{r}_2,t_2)\rangle = \langle V_i^*(\underline{r}_1,t_1)V_j^*(\underline{r}_2,t_2)V_j(\underline{r}_2,t_2)V_i(\underline{r}_1,t_1)\rangle \quad .$$

The existence of such photoelectric correlations was first demonstrated by Hanbury Brown and Twiss in some classic experiments.[2] It is easy to show that still higher order correlation functions are associated with measurements involving three or more photodetectors. This has led to the development of the coherence formalism for higher order correlations and higher order cross-spectral densities,[3-7] and several generalizations of theorems from the second order theory, like the Wiener-Khintchine theorem, were discovered.

In recent years photoelectric correlation measurements either with two photodetectors, or with one photodetector at two different times, have become relatively commonplace in optics. Nevertheless, the fourth order coherence theory relating to intensity correlations has not received a great deal of attention. In the following we examine some fourth order coherence properties of optical fields when interference effects and intensity correlation effects are both involved. We derive the superposition law for fourth order spectral densities, and determine the conditions under which this spectral density reproduces. We find that the reproducing property is closely associated with light having thermal statistics. One spectral density of the field is easily derivable from photoelectric measurements of the intensity fluctuations of light. Superposition of the signals from two photoelectric detectors leads to a reproducibility condition that is the fourth order analog of the condition for cross-spectral purity. Finally, we show that many important fourth order properties of the optical field are expressible in terms of three simpler spectral densities derived from the general fourth order spectral density.

THE FORMALISM

Although the electromagnetic field is a vector function of both space and time coordinates, we shall find it convenient to simplify our treatment to scalar functions of time only. We shall describe the field by a complex function $V(t)$, which may be regarded as the

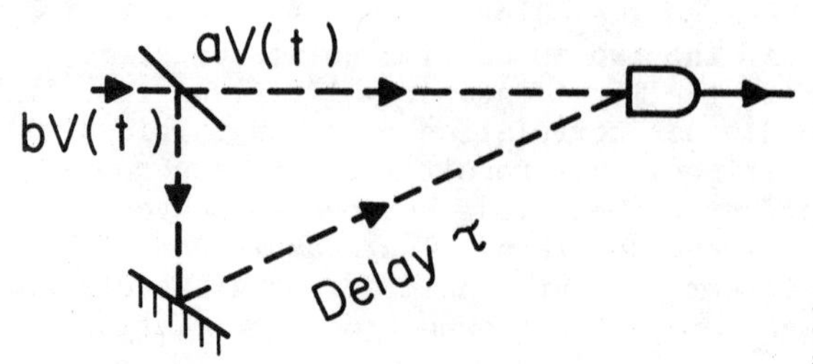

Fig. 1. The geometry for an interference experiment.

amplitude of either a polarized light beam, or a Cartesian component of the vector field. For example, when such a beam is passed through a beam splitter, as shown in Fig. 1, and the two emerging beams are recombined after a time delay τ is inserted in one path, the resultant complex amplitude is related to the incident complex amplitude $V(t)$ by

$$aV(t-T) + bV(t-T-\tau) \quad . \tag{1}$$

Here T is the transit time along the shorter path, and a,b are constants. In this way we are able to describe simple superposition or interference experiments without explicitly introducing space coordinates. When it becomes necessary to distinguish between two light beams in other ways we shall introduce suffices, and label them $V_1(t)$ and $V_2(t)$. We shall confine our attention to stationary fields. An auto-correlation function of order (N,M) is then defined by

$$\Gamma^{(N,M)}(\tau_2,\tau_3,\ldots\tau_N;\tau_M',\ldots\tau_1') \equiv$$

$$\langle V^*(t)V^*(t+\tau_2)\ldots V^*(t+\tau_N)V(t+\tau_M')\ldots V(t+\tau_1')\rangle \quad , \tag{2}$$

and if the field is stationary, this is a function only of N+M-1 variables, which are the differences between the time arguments. In particular, the second order correlation $\Gamma^{(1,1)}(\tau)$ is a function of only one time argument, and

$$\Gamma^{(1,1)}(0) = \langle V^*(t)V(t)\rangle = \langle I\rangle \quad , \tag{3}$$

where $I(t) \equiv V^*(t)V(t)$ is the instantaneous light intensity. It is sometimes convenient to use a normalized second order auto-correlation function also, defined by

$$\gamma(\tau) \equiv \Gamma^{(1,1)}(\tau)/\langle I\rangle \quad . \tag{4}$$

The intensity auto-correlation is an example of a fourth order correlation function, and in our notation

$$\langle I(t)I(t+\tau)\rangle = \Gamma^{(2,2)}(\tau;\tau,0) \quad . \tag{5}$$

It is also a function of only one time argument. More generally, the cross-correlation function of several different light beams distinguished by suffices 1,2,...etc. is defined by

$$\Gamma^{(N,M)}_{a\ldots k;\ell\ldots s}(\tau_2,\ldots\tau_N;\tau'_M,\ldots\tau'_1) =$$
$$\langle V^*_a(t)V^*_b(t+\tau_2)\ldots V^*_k(t+\tau_N)V_\ell(t+\tau'_M)\ldots V_s(t+\tau'_1)\rangle \quad . \tag{6}$$

With the auto-correlation function $\Gamma^{(N,M)}$ we associate a spectral density $\Phi^{(N,M)}$ that is the (N+M-1)'th order Fourier transform of $\Gamma^{(N,M)}$ with respect to the N+M-1 variables $\tau_2,\ldots\tau_N,\tau'_M,\ldots\tau'_1$. Thus, we write

$$\Phi^{(N,M)}(\omega_2,\ldots\omega_N;\omega'_M,\ldots\omega'_1) = \int\ldots\int \Gamma^{(N,M)}(\tau_2,\ldots\tau_N;\tau'_M,\ldots\tau'_1)$$
$$\times \exp i(\omega_2\tau_2+\ldots+\omega_N\tau_N-\omega'_M\tau'_M-\ldots-\omega'_1\tau'_1)d\tau_2\ldots d\tau_N d\tau'_M\ldots d\tau'_1 \quad , \tag{7}$$

so that $\Phi^{(N,M)}$ is a function of N+M-1 frequencies. In a strictly analogous manner we define the cross-spectral density

$$\Phi^{(N,M)}_{a\ldots k;\ell\ldots s}(\omega_2,\ldots\omega_N;\omega'_M,\ldots\omega'_1)$$

to be the (N+M-1)'th order Fourier transform of the cross-correlation function $\Gamma^{(N,M)}_{a\ldots k;\ell\ldots s}(\tau_2,\ldots\tau_N;\tau'_M,\ldots\tau'_1)$.

Although we have introduced correlations and spectral densities of arbitrary order, our interest in this article will focus on the second order and the fourth order quantities, which can be determined from interference and photoelectric correlation measurements. An important difference between the second order and the fourth order correlations is that, whereas $\Gamma^{(1,1)}(\tau)$ tends to zero for large argument τ for any physically realizable field, the same is not necessarily true of $\Gamma^{(2,2)}(\tau_2;\tau'_2,\tau'_1)$, at least when $\tau_2 = \tau'_2$ and $\tau'_1 = 0$. The reason is that V(t) has a zero mean for a stationary field, but $|V(t)|^2$ does not.

Because of the importance of the intensity correlation defined by Eq. (5), we shall find it convenient to associate a normalized correlation function $\lambda(\tau)$ and a corresponding normalized spectral density $\psi(\omega)$ with it. We define $\lambda(\tau)$ in such a way that the contribution of the non-zero mean of I(t) is subtracted out, and we write

$$\begin{aligned}\lambda(\tau) &\equiv \langle\Delta I(t)\Delta I(t+\tau)\rangle/\langle I\rangle^2 \\ &= \langle I(t)I(t+\tau)\rangle/\langle I\rangle^2 - 1 \\ &= \Gamma^{(2,2)}(\tau;\tau,0)/[\Gamma^{(1,1)}(0,0)]^2 - 1 \quad .\end{aligned} \tag{8}$$

This definition makes $\lambda(\tau)$ dimensionless, but $\lambda(0)$ is not necessarily equal to unity. However, for any physically realizable (ergodic) field we have $\lambda(\infty) = 0$. The Fourier transform of $\lambda(\tau)$ with respect to τ is the normalized spectral density $\psi(\omega)$ of the intensity fluctuations, so that we write

$$\psi(\omega) = \int_{-\infty}^{\infty} \lambda(\tau)\, e^{i\omega\tau} d\tau \quad , \tag{9}$$

and by definition

$$\lambda(0) = \frac{1}{2\pi} \int_{-\infty}^{\infty} \psi(\omega)\, d\omega \quad . \tag{10}$$

$\psi(\omega)$ is, of course, related to the more general spectral density $\Phi^{(2,2)}(\omega_2;\omega_2',\omega_1')$. Thus, from the Fourier inverse of Eq. (7) we have

$$\Gamma^{(2,2)}(\tau;\tau,0) = \frac{1}{(2\pi)^3} \int \Phi^{(2,2)}(\omega_2;\omega_2',\omega_1') e^{-i(\omega_2-\omega_2')\tau}\, d\omega_2 d\omega_2' d\omega_1'$$

$$= \frac{1}{(2\pi)^3} \int \Phi^{(2,2)}(\omega_2'+\omega;\omega_2',\omega_1') e^{-i\omega\tau}\, d\omega d\omega_2' d\omega_1' \quad ,$$

and with the help of Eqs. (8) and (9) this leads immediately to

$$\psi(\omega) = \frac{1}{\langle I\rangle^2} \frac{1}{(2\pi)^2} \int \Phi^{(2,2)}(\omega+\omega_2';\omega_2',\omega_1')\, d\omega_2' d\omega_1' - 2\pi\delta(\omega) \quad . \tag{11}$$

The term $2\pi\delta(\omega)$ on the right merely serves to subtract out the corresponding δ-function contribution from $\Phi^{(2,2)}$, and to make $\psi(\omega)$ nonsingular for an ergodic process. In a strictly analogous manner we can define a normalized cross-correlation function $\lambda_{12}(\tau)$, and a cross-spectral density $\psi_{12}(\omega)$ that is the Fourier transform of $\lambda_{12}(\tau)$, and this is related to $\Phi^{(2,2)}_{12;21}(\omega_2;\omega_2',\omega_1')$ through an integral as in Eq. (11).

SUPERPOSITION LAW FOR SPECTRAL DENSITIES AND THE REPRODUCING PROPERTY

Consider two light beams described by complex amplitudes $V_1(t)$ and $V_2(t)$ which are superposed, and result in a new complex amplitude

$$V(t) = V_1(t) + V_2(t) \quad . \tag{12}$$

The second order auto-correlation function of $V(t)$ is then given by

$$\Gamma^{(1,1)}(\tau) = \langle V^*(t)V(t+\tau)\rangle = \Gamma^{(1,1)}_{11}(\tau)+\Gamma^{(1,1)}_{22}(\tau)+\Gamma^{(1,1)}_{12}(\tau)+\Gamma^{(1,1)}_{21}(\tau), \tag{13}$$

and the spectral density $\Phi^{(1,1)}(\omega)$ is obtained by taking Fourier transforms of each term,

$$\Phi^{(1,1)}(\omega) = \Phi_{11}^{(1,1)}(\omega) + \Phi_{22}^{(1,1)}(\omega) + \Phi_{12}^{(1,1)}(\omega) + \Phi_{21}^{(1,1)}(\omega). \quad (14)$$

It has been shown that even when the component spectral distributions $\Phi_{11}^{(1,1)}(\omega)$ and $\Phi_{21}^{(1,1)}(\omega)$ are identical except for a scale factor, the resultant spectral distribution $\Phi^{(1,1)}(\omega)$ will in general be different.[8] Only in the special case when $\Phi_{12}^{(1,1)}(\omega)$ and $\Phi_{21}^{(1,1)*}(\omega)$ are both proportional to $\Phi_{11}^{(1,1)}(\omega)e^{-i\omega\eta}$, where η is a constant, will the original spectral distribution be reproduced. This condition implies a similar degree of correlation between all the Fourier components of $V_1(t)$ and $V_2(t)$, and is known as the condition for cross-spectral purity.[8,9] In the special case in which $V_1(t)$ and $V_2(t)$ are statistically independent, $\Phi_{12}^{(1,1)}(\omega)=0=\Phi_{21}^{(1,1)}(\omega)$ and Eq. (14) simplifies to

$$\Phi^{(1,1)}(\omega) = \Phi_{11}^{(1,1)}(\omega) + \Phi_{22}^{(1,1)}(\omega) \quad . \quad (15)$$

The condition for cross-spectral purity is then satisfied automatically, and the original spectral distribution is reproduced in the superposition.

However, when we come to the fourth order spectral density, the situation is less simple, even when $V_1(t)$ and $V_2(t)$ are independent. From Eq. (12) we find for the fourth order auto-correlation function

$$\Gamma^{(2,2)}(\tau_2;\tau_2',\tau_1') = \langle [V_1^*(t)+V_2^*(t)] [V_1^*(t+\tau_2)+V_2^*(t+\tau_2)]$$

$$\times [V_1(t+\tau_2')+V_2(t+\tau_2')] [V_1(t+\tau_1')+V_2(t+\tau_1')] \rangle \quad .$$

When $V_1(t)$ and $V_2(t)$ are independent, the only non-vanishing terms on the right are those in which factors V_1^* are paired with V_1 and V_2^* with V_2. Ten out of the 16 terms on the right therefore vanish, and we find

$$\Gamma^{(2,2)}(\tau_2;\tau_2',\tau_1') =$$

$$\Gamma_{11;11}^{(2,2)}(\tau_2;\tau_2',\tau_1')+\Gamma_{22;22}^{(2,2)}(\tau_2;\tau_2',\tau_1')+\Gamma_{11}^{(1,1)}(\tau_2')\Gamma_{22}^{(1,1)}(\tau_1'-\tau_2)$$

$$+ \Gamma_{11}^{(1,1)}(\tau_1')\Gamma_{22}^{(1,1)}(\tau_2'-\tau_2)+\Gamma_{11}^{(1,1)}(\tau_1'-\tau_2)\Gamma_{22}^{(1,1)}(\tau_2')$$

$$+ \Gamma_{11}^{(1,1)}(\tau_2'-\tau_2)\Gamma_{22}^{(1,1)}(\tau_1') \quad . \quad (16)$$

If we take the three-dimensional Fourier transform of each term in this equation, and use Eq. (7), we obtain the following relation between the spectral densities,

$$\Phi^{(2,2)}(\omega_2;\omega_2',\omega_1') = \Phi^{(2,2)}_{11;11}(\omega_2;\omega_2',\omega_1')+\Phi^{(2,2)}_{22;22}(\omega_2;\omega_2',\omega_1')$$

$$+ 2\pi[\delta(\omega_1'-\omega_2)+\delta(\omega_2'-\omega_2)][\Phi^{(1,1)}_{11}(\omega_1')\Phi^{(1,1)}_{22}(\omega_2')+\Phi^{(1,1)}_{11}(\omega_2')\Phi^{(1,1)}_{22}(\omega_1')]. \tag{17}$$

This is the superposition law for the fourth order spectral densities when independent beams are superposed. It will be seen that the component spectral densities do not simply add, as they do in Eq. (14) for the second order quantities, because of the δ-function contributions.

Let us now use these results to relate the corresponding normalized correlations and spectral densities of the intensity fluctuations. From Eqs. (13) and (16), and the definitions (4) and (8), we obtain with an obvious extension of the notation, on putting $\tau_2=\tau=\tau_2',\tau_1'=0$,

$$\lambda(\tau) = \frac{\langle I_1\rangle^2}{\langle I\rangle^2}\lambda_{11}(\tau)+\frac{\langle I_2\rangle^2}{\langle I\rangle^2}\lambda_{22}(\tau)+\frac{\langle I_1\rangle\langle I_2\rangle}{\langle I\rangle^2}[\gamma_{11}(\tau)\gamma^*_{22}(\tau)+\text{c.c.}]. \tag{18}$$

This is the superposition law for intensity correlations, and it involves both the intensity correlations and the amplitude correlations of the component beams. In the special case in which the component beams have similar fluctuation properties

$$\lambda_{11}(\tau) = \lambda_{22}(\tau)$$
$$\gamma_{11}(\tau) = \gamma_{22}(\tau) \quad , \tag{19}$$

Equation (18) yields

$$\lambda(\tau) = [(\langle I_1\rangle^2+\langle I_2\rangle^2)\lambda_{11}(\tau)+2\langle I_1\rangle\langle I_2\rangle|\gamma_{11}(\tau)|^2]/\langle I\rangle^2 \quad . \tag{20}$$

On taking the Fourier transform of each term in Eq. (18), or alternatively, from Eqs. (11) and (17), we find the superposition law for the corresponding normalized spectral densities

$$\psi(\omega) = \frac{\langle I_1\rangle^2}{\langle I\rangle^2}\psi_{11}(\omega)+\frac{\langle I_2\rangle^2}{\langle I\rangle^2}\psi_{22}(\omega)$$

$$+\frac{1}{\langle I\rangle^2}\frac{1}{2\pi}\int[\Phi^{(1,1)}_{11}(\omega')\Phi^{(1,1)}_{22}(\omega'+\omega)+\Phi^{(1,1)}_{11}(\omega'+\omega)\Phi^{(1,1)}_{22}(\omega')]d\omega' \quad . \tag{21}$$

In the special case when Eq. (19) holds and the two component spectral distributions are similar,

$$\psi_{11}(\omega) = \psi_{22}(\omega)$$
$$\Phi^{(1,1)}_{11}(\omega)/\langle I_1\rangle = \Phi^{(1,1)}_{22}(\omega)/\langle I_2\rangle \quad , \tag{22}$$

this simplifies to

$$\psi(\omega) = \frac{<I_1>^2+<I_2>^2}{<I>^2}\psi_{11}(\omega) + \frac{2}{<I>^2}\frac{1}{2\pi}\int\Phi_{11}^{(1,1)}(\omega')\Phi_{22}^{(1,1)}(\omega'+\omega)d\omega' \quad . \tag{23}$$

In general, the spectrum of the intensity fluctuations of the superposed beam therefore differs from the corresponding spectrum of the component beams, even when the components are independent. This is particularly obvious in the case of two single-mode laser beams, for which the component intensities are nearly constant and $\lambda_{11}(\tau)$ and $\psi_{11}(\omega)$ are close to zero, whereas the resultant intensity fluctuates because of interference effects.

However, an exception occurs when

$$\lambda_{11}(\tau) = |\gamma_{11}(\tau)|^2 \quad , \tag{24}$$

or, in the frequency domain, when

$$\psi_{11}(\omega) = \frac{1}{2\pi<I_1><I_2>}\int\Phi_{11}^{(1,1)}(\omega')\Phi_{22}^{(1,1)}(\omega'+\omega)d\omega' \quad , \tag{25}$$

which holds true whenever the light obeys thermal statistics.[3-7] In that case we have immediately from Eq. (20)

$$\lambda(\tau) = \lambda_{11}(\tau) \tag{26}$$

and from Eq. (23)

$$\psi(\omega) = \psi_{11}(\omega) \quad . \tag{27}$$

The spectral distribution therefore reproduces on superposition. Why thermal light beams obey the reproducibility property, whereas other kinds of light beam, such as those from a laser, do not, can be understood when we recall that superposition results in interference between various Fourier components. This causes phase fluctuations to be converted to amplitude fluctuations, and generally leads to fluctuations of the resultant amplitude, even when the components that are being superposed are free from amplitude fluctuations. But thermal light has fluctuating Fourier components that obey Gaussian statistics, and the superposition merely preserves the Gaussian nature of these fluctuations.

SUPERPOSITION LAW FOR PHOTOELECTRIC SIGNALS - REPRODUCING PROPERTY

When fluctuating light falls on a photoelectric detector, the fluctuations are impressed on the photoelectric signal emerging from the detector. A spectral analysis of the photoelectric current therefore yields information about the spectrum of the intensity fluctuations of the incident optical field. The theory of this phenomenon has been treated by a number of workers[2,10-13] and it has become the basis for a practical spectroscopic technique.[13-22]

If the release of an electron at the cathode of a photodetector at time t_i results in a photoelectric current pulse $k(t-t_i)$ of standard form, then the photoelectric signal $S(t)$ resulting from the release of electrons at various times $t_1, t_2, \ldots$ is given by

$$S(t) = \sum_i k(t-t_i) \quad . \tag{28}$$

From this we may readily show[2,10,13] that the average signal is

$$\langle S \rangle = \alpha A \langle I \rangle Q = RQ \quad , \tag{29}$$

where

$$Q = \int_0^\infty k(t')dt' \tag{30}$$

is the total charge delivered by a single current pulse, $\langle I \rangle$ is the mean intensity of the light beam falling on the detector expressed in photons per unit area per second, A is the effective detector area, and α is the quantum efficiency of the detector. The product $\alpha A \langle I \rangle$ is just the average counting rate R of the illuminated detector. From Eq. (28) we may readily obtain[2,10,13] for the auto-correlation function of the signal S,

$$\langle \Delta S(t) \Delta S(t+\tau) \rangle = R\int_0^\infty k(t')k(t'+\tau)dt' + R^2 \iint_0^\infty k(t')k(t'')\lambda(t'-t''+\tau)dt'dt'', \tag{31}$$

where $\lambda(\tau)$ is the normalized intensity auto-correlation function of the light given by Eq. (8). Similarly, we may show from Eq. (28) that when two photodetectors labelled 1 and 2 are illuminated by partially coherent light beams, the cross-correlation of the two photoelectric signals is given by[2,10,13]

$$\langle \Delta S_1(t) \Delta S_2(t+\tau) \rangle = R_1 R_2 \iint_0^\infty k_1(t')k_2(t'')\lambda_{12}(t'-t''+\tau)dt'dt'' \quad , \tag{32}$$

where

$$\lambda_{12}(\tau) \equiv \langle \Delta I_1(t) \Delta I_2(t+\tau) \rangle / \langle I_1 \rangle \langle I_2 \rangle = \Gamma^{(2,2)}_{12;21}(\tau;\tau,0)/\langle I_1 \rangle \langle I_2 \rangle - 1 \tag{33}$$

describes the time dependence of the cross-correlation of the two intensity fluctuations, and the other symbols have obvious meanings.

On taking the Fourier transform of each term in Eq. (31) we obtain the spectral density $\chi(\omega)$ of the signal fluctuations, which can be measured with a spectral analyzer. We then find[10,11,13]

$$\chi(\omega) = \int_{-\infty}^{\infty} \langle\Delta S(t)\Delta S(t+\tau)\rangle e^{i\omega\tau} d\tau$$

$$= R|K(\omega)|^2\left[1+R\psi(\omega)\right] \quad , \tag{34}$$

where we have written

$$K(\omega) \equiv \int_0^{\infty} k(t)e^{i\omega t}dt \tag{35}$$

for the Fourier transform of a current pulse. $K(\omega)$ is of course a known function characteristic of the detector. The first term on the right of Eq. (34) is present even in the absence of any intensity fluctuations of the incident light, and represents the shot noise fluctuations of the photoelectric signal. This term, which is connected with the discreteness of the photocurrent, is not of great interest. However, the second term in Eq. (34) clearly carries information about the optical field. Indeed, we note that the spectral density $\psi(\omega)$ of the intensity fluctuations of the incident light is derivable from a spectral analysis of the photoelectric signal, and the measurement can therefore be transformed from the domain of optics to the domain of electronics. It is convenient to subtract the uninteresting shot noise contribution in Eq. (34), and to define a new, more relevant spectral density by

$$\tilde{\chi}(\omega) = R^2|K(\omega)|^2 \psi(\omega) \quad . \tag{36}$$

Similarly, on taking the Fourier transform of the cross-correlation given by Eq. (32), we obtain the following relation between the cross-spectral densities $\chi_{12}(\omega)$ and $\psi_{12}(\omega)$ of the two photoelectric signals and of the two light intensities, respectively,

$$\chi_{12}(\omega) = R_1R_2K_1(\omega)K_2^*(\omega)\psi_{12}(\omega) \quad . \tag{37}$$

This time there is no explicit shot noise contribution, because both cross-spectral densities vanish in the absence of any intensity correlations between the two incident light beams.

Let us now consider the situation illustrated in Fig. 2, in which two light beams fall on two photodetectors and the photoelectric signals $S_1(t)$ and $S_2(t)$ are added to produce a new signal

$$S(t) = S_1(t)+S_2(t) \quad . \tag{38}$$

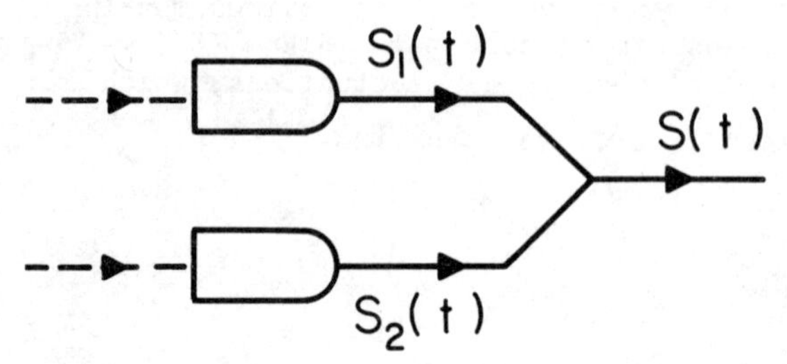

Fig. 2. An experiment in which two photoelectric signals are combined.

Suppose $K_1(\omega) = K_2(\omega)$ for the two detectors, and $\psi_{11}(\omega) = \psi_{22}(\omega)$ for the two light beams. Then the spectral densities $\tilde{\chi}_{11}(\omega)$ and $\tilde{\chi}_{22}(\omega)$ of the signals emerging from the two detectors, after shot noise contributions are subtracted, are also equal, except for a possible scale factor. We now inquire under what conditions the spectral distribution $\tilde{\chi}(\omega)$ of the resultant photoelectric signal S(t) reproduces the original spectral distribution $\tilde{\chi}_{11}(\omega)$. This problem is reminiscent of the problem posed at the beginning of Section 3, except that two photoelectric signals, rather than two light beams, are now being added. Nevertheless, we shall see that the answer is again expressible as a condition on the correlation properties of the light.

From Eq. (38) we have immediately

$$\langle S\rangle = \langle S_1\rangle+\langle S_2\rangle \quad ,$$

and with the help of Eq. (29)

$$\langle S\rangle = (R_1+R_2)Q_1 \quad . \tag{39}$$

For the auto-correlation function of $\Delta S(t)$ we find

$$\begin{aligned}\langle\Delta S(t)\Delta S(t+\tau)\rangle &= \langle\Delta S_1(t)\Delta S_1(t+\tau)\rangle+\langle\Delta S_2(t)\Delta S_2(t+\tau)\rangle \\ &+ \langle\Delta S_1(t)\Delta S_2(t+\tau)\rangle+\langle\Delta S_2(t)\Delta S_1(t+\tau)\rangle \quad ,\end{aligned}$$

and on taking Fourier transforms, and making use of Eqs. (34) and (37), we arrive at

$$\tilde{\chi}(\omega) = |K_1(\omega)|^2\Big[(R_1^2+R_2^2)\psi_{11}(\omega)+R_1R_2\big(\psi_{12}(\omega)+\text{c.c.}\big)\Big] \quad . \tag{40}$$

In the special case in which the two photoelectric signals $S_1(t)$ and $S_2(t)$ are uncorrelated, which implies the absence of any correlations between the intensity fluctuations of two light beams in question, $\psi_{12}(\omega) = 0$ and $\tilde{\chi}(\omega)$ coincides with $\tilde{\chi}_{11}(\omega)$ or $\tilde{\chi}_{22}(\omega)$, except for a scale factor. But in general, $\tilde{\chi}(\omega)$ will be different from $\tilde{\chi}_{11}(\omega)$, and the spectral density does not reproduce. However, in the special case in which

$$\psi_{12}(\omega) = \lambda_{12}(0)\psi_{11}(\omega) \quad , \tag{41}$$

Equation (40) reduces to

$$\tilde{\chi}(\omega) = |K_1(\omega)|^2\psi_{11}(\omega)\left[R_1^2+R_2^2+2R_1R_2\lambda_{12}(0)\right] \quad . \tag{42}$$

$\tilde{\chi}(\omega)$ is then proportional to $\tilde{\chi}_{11}(\omega)$ or $\tilde{\chi}_{22}(\omega)$, and the reproducing property holds. The condition (41) is strongly reminiscent of the condition for cross-spectral purity,[8,9] except that it is imposed on the cross-spectral density of the intensity fluctuations, rather than on the more usual second order cross-spectral density $\Phi_{12}^{(1,1)}(\omega)$. However, the implications are similar. The spectral distribution reproduces only when the degree of correlation between corresponding Fourier components of the intensity fluctuations of the two light beams is the same for all frequencies.

REDUCTION TO SIMPLER FOURTH ORDER SPECTRAL DENSITIES

In the previous sections we studied the intensity correlations of one or two light beams, and we showed that several important reproducing properties of the field are expressible in terms of the one-parameter correlation function $\Gamma^{(2,2)}(\tau;\tau,0)$, or the associated spectral density $\psi(\omega)$. In this way we effectively bypassed the need to treat the full fourth order correlations $\Gamma^{(2,2)}(\tau_2;\tau_2',\tau_1')$ or the fourth order spectral densities $\Phi^{(2,2)}(\omega_2;\omega_2',\omega_1')$. However, when interference effects with substantial time shifts as well as intensity correlations are involved, the situation is more complicated, and correlation functions of one argument no longer suffice to describe the field. Although the full spectral density $\Phi^{(2,2)}(\omega_2;\omega_2',\omega_1')$ carries all the necessary information about fourth order correlations, we now show that three simpler spectral densities of two arguments are sufficient to describe the relevant properties.

Consider the situation illustrated in Fig. 3, in which an incoming light beam with complex amplitude V_o is split into two parts aV_o and bV_o with the aid of a partly silvered mirror. The two parts are recombined after a differential delay τ_1, is inserted, and the combined field, after encountering another beam splitter, is allowed to

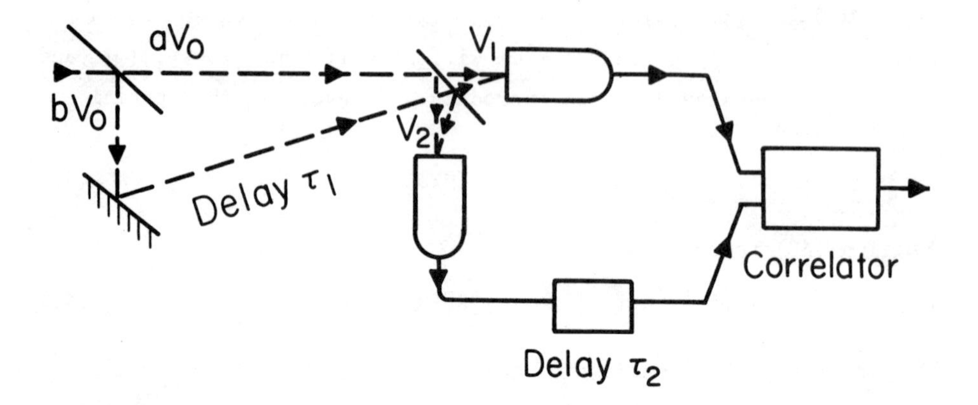

Fig. 3. The geometry for an intensity correlation measurement with interference.

fall on two photodetectors. The complex amplitudes $V_1(t)$ and $V_2(t)$ at the two detectors can then be expressed in the form

$$\begin{aligned} V_1(t) &= c\left[aV_o(t)+bV_o(t+\tau_1)\right] \\ V_2(t) &= d\left[aV_o(t)+bV_o(t+\tau_1)\right] \ . \end{aligned} \tag{43}$$

We now investigate correlations in the signals $S_1(t)$ and $S_2(t)$ delivered by the two photodetectors, after a further delay τ_2 is inserted in one channel. It is clear that both interference and intensity correlation effects play a role in such measurements. From Eq. (32) it follows that

$$\langle S_1(t)S_2(t+\tau)\rangle = \frac{R_1R_2}{\langle I_1\rangle\langle I_2\rangle}\iint_o^\infty k_1(t')k_2(t'')\langle I_1(t)I_2(t+t''-t'+\tau_2)\rangle dt'dt'' \ ,$$

with

$$\langle I_1\rangle = |c|^2\left[\left(|a|^2+|b|^2\right)\langle I_o\rangle+a^*b\Gamma_{oo}^{(1,1)}(\tau_1)+c.c.\right]$$

$$\langle I_2\rangle = |d|^2\left[\left(|a|^2+|b|^2\right)\langle I_o\rangle+a^*b\Gamma_{oo}^{(1,1)}(\tau_1)+c.c.\right] \ ,$$

from Eq. (43). The intensity cross-correlation under the integral is of the general form $\langle I_1(t)I_2(t+\tau_2)\rangle$, and it can be broken into the sum of 16 terms with the help of Eqs. (43),

$$\begin{aligned}<I_1(t)I_2(t+\tau_2)> = |c|^2|d|^2\Big[&|a|^4\Gamma^{(2,2)}(\tau_2;\tau_2,0)+|b|^4\Gamma^{(2,2)}(\tau_2;\tau_2,0)\\
&+ a^{*2}b^2\Gamma^{(2,2)}(\tau_2;\tau_1+\tau_2,\tau_1) + \text{c.c.}\\
&+ |a|^2|b|^2\{\Gamma^{(2,2)}(\tau_1+\tau_2;\tau_1+\tau_2,0)+\Gamma^{(2,2)}(\tau_1-\tau_2;\tau_1-\tau_2,0)\\
&+ \Gamma^{(2,2)}(\tau_1+\tau_2;\tau_2,\tau_1) + \text{c.c.}\}\\
&+ |a|^2\{a^*b\Gamma^{(2,2)}(\tau_2;\tau_1+\tau_2,0)+\text{c.c.}+a^*b\Gamma^{(2,2)}(\tau_2;\tau_1,0) + \text{c.c}\}\\
&+ |b|^2\{a^*b\Gamma^{(2,2)}(\tau_2-\tau_1;\tau_2,0) + \text{c.c.}\\
&+ a^*b\Gamma^{(2,2)}(\tau_2+\tau_1;\tau_2+\tau_1,\tau_1) + \text{c.c.}\}\Big] \quad . \end{aligned} \tag{44}$$

Close examination of Eq. (44) shows that all the auto-correlation functions on the right-hand side are either functions of one time argument of the general form $\Gamma^{(2,2)}(\tau;\tau,0)$, or they are functions of two time arguments. Among the latter we distinguish the following three general types, together with their complex conjugates:

$$\text{(A)} \quad \Gamma^{(2,2)}(T_1;T_2,0)$$

$$\text{(B)} \quad \Gamma^{(2,2)}(T_1+T_2;T_2,T_1)$$

$$\text{(C)} \quad \Gamma^{(2,2)}(T_1-T_2;T_2,T_1) \quad .$$

With the help of the Fourier inverse of Eq. (7) we can express these three correlation functions in terms of the fourth order spectral densities $\Phi^{(2,2)}$ as follows,

$$\text{(A)} \quad \Gamma^{(2,2)}(T_1;T_2,0) = \frac{1}{(2\pi)^3}\int\Phi^{(2,2)}(\omega_2;\omega_2',\omega_1')e^{i(-\omega_2T_1+\omega_2'T_2)}d\omega_2d\omega_2'd\omega_1' \tag{45}$$

$$\begin{aligned}\text{(B)} \quad \Gamma^{(2,2)}(T_1+T_2;T_2,T_1) &= \frac{1}{(2\pi)^3}\int\Phi^{(2,2)}(\omega_2;\omega_2',\omega_1')e^{i(-\omega_2+\omega_2')T_2}\\
&\quad \times e^{-i(\omega_2-\omega_1')T_1}\,d\omega_2d\omega_2'd\omega_1'\\
&= \frac{1}{(2\pi)^3}\int\Phi^{(2,2)}(\omega_2;\omega_2-\omega_2'',\omega_2+\omega_1'')\\
&\quad \times e^{i(\omega_1''T_1-\omega_2''T_2)}\,d\omega_2d\omega_2''d\omega_1''\end{aligned} \tag{46}$$

$$\text{(C)}\quad \Gamma^{(2,2)}(T_1-T_2,T_2,T_1) = \frac{1}{(2\pi)^3}\int \Phi^{(2,2)}(\omega_2;\omega_2',\omega_1')e^{i(\omega_2+\omega_2')T_2} \times e^{i(\omega_1'-\omega_2)T_1}\,d\omega_2 d\omega_2' d\omega_1'$$

$$= \frac{1}{(2\pi)^3}\int \Phi^{(2,2)}(\omega_2;-\omega_2+\omega_2'',\omega_2+\omega_1'') \times e^{i(\omega_1''T_1+\omega_2''T_2)}\,d\omega_2 d\omega_2'' d\omega_1'' \,. \tag{47}$$

It follows that all the terms appearing in Eq. (44) are derivable by Fourier inversion from $\psi(\omega)$ given by Eq. (9), together with the following three spectral densities, all of which are of course expressible in terms of $\Phi^{(2,2)}(\omega_2;\omega_2',\omega_1')$:

$$\Theta_A(\omega_2,\omega_1) \equiv \frac{1}{2\pi}\int \Phi^{(2,2)}(\omega_2;\omega_1,\omega_1')d\omega_1' \tag{48}$$

$$\Theta_B(\omega_2,\omega_1) \equiv \frac{1}{2\pi}\int \Phi^{(2,2)}(\omega_2';\omega_2'-\omega_2,\omega_2'+\omega_1)d\omega_2' \tag{49}$$

$$\Theta_C(\omega_2,\omega_1) \equiv \frac{1}{2\pi}\int \Phi^{(2,2)}(\omega_2';\omega_2-\omega_2',\ \omega_2'+\omega_1)d\omega_2' \quad . \tag{50}$$

So long as we are dealing with correlation experiments in which the fields arise from delayed superpositions of the general type described by Fig. 3, these three spectral densities suffice for a complete description, and being functions of only two frequencies, they are simpler objects than the full fourth order spectral density $\Phi^{(2,2)}$. Cross-correlations and cross-spectral densities can of course be introduced in a strictly analogous manner. In the special case in which the two time delays in Fig. 3 are equal, and $\tau_1=\tau=\tau_2$, all the correlation functions in Eq. (44) become functions of one argument, and they can be expressed in terms of $\psi(\omega)$ and two simpler spectral densities derivable from Θ_A and Θ_B alone. However, $\Phi^{(2,2)}$ and $\Gamma^{(2,2)}$ remain the fundamental quantities for the description of fourth order correlation properties of the optical field under all conditions.

DERIVATION OF LOWER ORDER FROM HIGHER ORDER AUTO-CORRELATIONS

The fourth and higher order auto-correlation functions of the optical field contain progressively more detailed information about the fluctuation properties of the field. In the case of ergodic processes for which correlations always die out in time, the structure is hierarchical; the higher order auto-correlations contain the lower order auto-correlations. We start by illustrating this property for correlations of the fourth order.

Let us consider the intensity auto-correlation function

$$\Gamma^{(2,2)}(T;T,0) = <I(t)I(t+T)> \quad .$$

As the time interval T becomes longer and longer, it is to be expected that the light intensity at time t+T eventually ceases to be correlated with that at time t, so that

$$<I(t)I(t+T)> \rightarrow <I(t)><I(t+T)> \quad ,$$

and for a stationary field each average is $<I>$. It follows that

$$\lim_{T\to\infty}\Gamma^{(2,2)}(T,T,0) = <I>^2 \quad , \tag{51}$$

so that $<I>$ is derivable from $\Gamma^{(2,2)}$. In a similar manner we may derive $\Gamma^{(1,1)}$ from $\Gamma^{(2,2)}$. Thus, if we consider

$$\Gamma^{(2,2)}(T;T,\tau) = <V^*(t)V^*(t+T)V(t+T)V(t+\tau)> \quad ,$$

and let $T \rightarrow \infty$, the factors depending on T become uncorrelated from those that do not. It follows that

$$\lim_{T\to\infty}\Gamma^{(2,2)}(T;T;\tau) \rightarrow <V^*(t)V(t+\tau)><V^*(t+T)V(t+T)>$$

$$= <I>\Gamma^{(1,1)}(\tau) \quad , \tag{52}$$

and this is a prescription for deriving $\Gamma^{(1,1)}$ from $\Gamma^{(2,2)}$. By taking the Fourier transform we can also express the second order spectral density $\Phi^{(1,1)}$ in terms of $\Phi^{(2,2)}$.

More generally, a similar argument can be used to derive auto-correlations of order N,M from those of order N+n, M+n. Thus, if we introduce

$$\Gamma^{N+n,M+n}(\tau_2,\ldots,\tau_N,\tau_N+T,\ldots,\tau_N+nT;\tau_N+nT,\ldots,\tau_N+T,\tau_M',\ldots,\tau_1')$$

$$= <V^*(t)V^*(t+\tau_2)\ldots V^*(t+\tau_N)V^*(t+\tau_N+T)\ldots V^*(t+\tau_N+nT)$$

$$\times\ V(t+\tau_N+nT)\ldots V(t+\tau_N+T)V(t+\tau_M')\ldots V(t+\tau_1')>$$

and again let $T \rightarrow \infty$, then the terms of argument without T become uncorrelated from those containing the argument T, which in turn become uncorrelated from those containing the argument 2T, etc. We therefore have in the limit

$$\lim_{T\to\infty}\Gamma^{N+n,M+n}(\tau_2,\ldots,\tau_N,\tau_M+T,\ldots,\tau_N+nT;\tau_N+nT,\ldots,\tau_N+T,\tau_M',\ldots,\tau_1')$$

$$= \langle V^*(t)V^*(t+\tau_2)\ldots V^*(t+\tau_N)$$

$$\times V(t+\tau_M')\ldots V(t+\tau_1')\rangle\langle I\rangle^n$$

$$= \langle I\rangle^n \, \Gamma^{(N,M)}(\tau_2,\ldots,\tau_N;\tau_M',\ldots,\tau_1') \quad . \tag{53}$$

This establishes the hierarchical structure of the higher order auto-correlation functions.

REFERENCES

1. See for example, Born and Wolf, PRINCIPLES OF OPTICS (Pergamon Press, Oxford, 1975) 5th ed., Ch. 10.
2. R. Hanbury Brown and R.Q. Twiss, Nature 177, 27 (1956) and Proc. Roy Soc. A 243, 291 (1957).
3. R.J. Glauber, Phys. Rev. 130, 2529 (1963).
4. L. Mandel, in QUANTUM ELECTRONICS III, ed. P. Grivet and N. Bloembergen (Dunod, Paris, 1964) p. 101; and in OPTICAL INSTRUMENTS AND TECHNIQUES, ed. J. Home Dickson (Oriel Press, London, 1970) p. 8; C.L. Mehta and L. Mandel, in ELECTROMAGNETIC WAVE THEORY, ed. J. Brown (Pergamon Press, Oxford, 1967) p. 1069; L. Mandel and C.L. Mehta, Il Nuov. Cim., Ser. X, 61B, 149 (1969).
5. C.L. Mehta, Il Nuov. Cim. 45B, 280 (1966) and J. Math. Phys. 8, 1798 (1967).
6. E. Wolf, Jap. J. Appl. Phys. 4 (Suppl. 1), 1 (1965).
7. J.R. Klauder and E.C.G. Sudarshan, FUNDAMENTALS OF QUANTUM OPTICS (W.A. Benjamin, New York, 1968); J. Perina, COHERENCE OF LIGHT (Van Nostrand Reinhold, London, 1972); see also L. Mandel and E. Wolf, Rev. Mod. Phys. 37, 231 (1965).
8. L. Mandel, J. Opt. Soc. Am. 51, 1342 (1961).
9. L. Mandel and E. Wolf, J. Opt. Soc. Am. 66, 529 (1976).
10. L. Mandel, in PROGRESS IN OPTICS, Vol. 2, ed. E. Wolf (North-Holland Publishing, Amsterdam, 1963) p. 181.
11. L. Mandel, in MODERN OPTICS (Polytechnic Press, New York, 1967) p. 143.
12. H. Gamo, J. Appl. Phys. 34, 875 (1963).
13. C. Freed and H.A. Haus, Phys. Rev. 141, 287 (1966).
14. A. Javan, E.A. Ballik and W.L. Bond, J. Opt. Soc. Am. 52, 96 (1962).
15. M.S. Lipsett and L. Mandel, Nature 199, 553 (1963).
16. B.L. Morgan and L. Mandel, Phys. Rev. Lett. 16, 1012 (1966).
17. D.B. Scarl, Phys. Rev. Lett. 17, 663 (1966).
18. D.T. Phillips, H. Kleiman and S.P. Davis, Phys. Rev. 153, 113 (1967).

19. F.T. Arecchi, E. Gatti and A. Sona, Phys. Lett. 20, 27 (1966).
20. R.L. Bailey and J.H. Sanders, Phys. Lett. 10, 295 (1964).
21. J.A. Bellisio, C. Freed and H.A. Haus, Appl. Phys. Lett. 4, 5 (1964).
22. L.R. Prescott and A. Van der Ziel, Phys. Lett. 12, 317 (1964).

LASER DOPPLER TECHNIQUES FOR SIZING PARTICULATE POLLUTANTS AND THERAPEUTIC AEROSOLS

M. K. Mazumder, F. C. Hiller,[1] R.E. Ware,
J. D. Wilson, and P. C. McLeod
University of Arkansas Graduate Institute of Technology
P. O. Box 3017, Little Rock, Arkansas 72203

ABSTRACT

A particle size analyzer based on laser Doppler velocimetry has been developed for measuring, in real time, the aerodynamic size distribution of aerosol particulates in the respirable range 0.1 to 10.0 μm in diameter. The instrument, a single particle aerodynamic relaxation time analyzer, measures the aerodynamic relaxation time of individual suspended particles and droplets without removing the particulates from their aerosol phase. Measurements can be made at a maximum count rate of 200 particles/sec, although coincidence error restricts the count rate to a lower limit. The size resolution is within ±5% of the measured aerodynamic diameter. Currently, three prototype SPART analyzers are being used for: (1) studying aerodynamic size distribution and lung retention of therapeutic aerosols, (2) measuring fractional efficiency of electrostatic precipitators as a function of size and electrical resistivity of flyash particles, and (3) characterizing atmospheric and household aerosols.

INTRODUCTION

The deposition of inhaled particulate materials within the lung depends primarily upon their aerodynamic diameter[1] which is defined as the diameter of a unit-density spherical particle having the same aerodynamic properties as the particle in question. The site and quantity of lung deposition and, consequently, the possible health hazard from inhaled particles can be estimated from measurement of the aerodynamic size distribution of the aerosol.[2] While there are numerous methods of sizing airborne particulates, none of the commercially available instruments can measure the aerodynamic diameter of individual suspended particles and droplets in real time in the size range 0.1 to 10.0 μm in diameter. This size range contains the major mass fraction of respirable aerosols.[3]

The geometric size distribution of particles can be measured by microscopic techniques, but this method is tedious and conversion of geometric size to aerodynamic diameter is difficult

[1]University of Arkansas Medical Sciences Campus, 4301 West Markham, Little Rock, Arkansas 72205.

ISSN:0094-243X/81/650196-17$1.50

for nonspherical particles. The optical diameters can be measured in real time by optical particle counters[4] if the intensity of light scattered from particles of unknown geometrical size and refractive index is correlated with the optical diameter of a standard spherical particle of known optical properties. Also, it is difficult to correlate the optical diameter to aerodynamic diameters.

The aerodynamic diameter can be determined by using a technique where site of particle deposition is dependent upon the aerodynamic size. Cascade impactors[5] are most widely used for determining aerodynamic size distributions. Although able to provide much useful information, various technical limitations impede their ability to determine the aerodynamic size distribution in many applications as reviewed by several workers.[6,7]

Any method requiring sample collection and subsequent analysis suffers the delay between deposition and measurement. Evaporation or condensation after collection may alter the mass, thereby altering the unstable particles. When the size distribution of particulates changes with time because of condensation, evaporation, or coagulation or when there are changes in the generation, sedimentation, or transportation processes, measurements are required in real time. For example, inhaled therapeutic aerosols exposed to the high humidity in the respiratory tract are likely to grow due to the condensation process. Consequently, a size distribution measurement in real time at airway humidity and temperature is required. Real time measurements also are necessary for efficient in-line monitoring of air filtration in devices where critical control adjustments are required.

This paper describes the development and applications of a laser-based instrument that can be used to monitor the size distribution of atmospheric aerosols within the respirable fraction 0.1 to 10.0 μm in aerodynamic diameter. Measurements are made in real time and on a single particle basis,and the process of measurement is not influenced by the nonsize-related properties such as refractive index, electrical charge, and chemical composition. The instrument, called the single particle aerodynamic relaxation time (SPART) analyzer, employs a frequency-biased differential laser Doppler velocimeter (LDV) and a microphone to determine the aerodynamic relaxation[8] time τ_p of individual particles suspended in air. The relaxation time is the time during which the velocity of a particle,moving with respect to a surrounding fluid medium,decays by a factor of 1/e because of fluid resistance. It is the time in which a particle, when suspended in still air, reaches 63.2 percent of its terminal settling velocity. The aerodynamic diameter d_a is obtained from the value of τ_p. Thus, the measurement of size distribution can be made depending upon the aerodynamic properties of the particulates.

The SPART analyzer is currently used in several measuring applications: (1) the size distribution of atmospheric aerosols, (2) the size distribution of therapeutic aerosols at a temperature of 37°C and at a relative humidity 95% or higher to simulate the ambient condition of the lung, and (3) the fractional efficiency of air pollution control devices as a function of aerodynamic diameter.

MEASUREMENT OF AERODYNAMIC DIAMETER

A particle suspended in air will oscillate in an acoustic field. The amplitude and phase of oscillation depends upon the aerodynamic relaxation time τ_p of the particle. The equation of motion of a particle of mass m_p can be written as

$$m_p(dv_p/dt) = F_{drag} - F_{ext}, \qquad (1)$$

where v_p is the velocity of the particle, F_{drag} is the drag force acting on the particle, and F_{ext} is the sum of the external forces on the particle. The external forces may be a combination of gravity, Brownian motion, electrostatic charges, thermophoresis, etc. If the Reynolds number of particle motion is less than 0.1, F_{drag} can be written from Stokes' Law for a spherical particle of diameter d_p as

$$F_{drag} = -3\pi\eta d_p(v_p - u_g)/C_c, \qquad (2)$$

where η is the viscosity of the surrounding fluid medium, u_g is the fluid velocity, and C_c is the Cunningham correction factor for molecular slip. If F_{ext} is negligible, equation (1) can be written as

$$m_p(dv_p/dt) + 3\pi\eta d_p(v_p - u_g)/C_c = 0. \qquad (3)$$

If the fluid is subjected to an oscillatory motion caused by an acoustic excitation in a standing wave of angular frequency ω, and amplitude u_g

$$u_g(t) = U_{g\pm} \sin \omega t, \qquad (4)$$

equation (3) can be written as for $v_p(t)$, the motion of a suspended particle as a function of time,

$$(m_pC_c/3\pi\eta d_p)(dv_p/dt) + v_p = U_{g\pm} \sin \omega t \qquad (5)$$

or

$$\tau_p(dv_p/dt) + v_p = U_{g\pm} \sin \omega t \qquad (6)$$

where $\tau_p = m_pZ$ is the relaxation time, and where

$$Z = C_c/3\pi\eta d_p \qquad (7)$$

is the mechanical mobility of the particle. For a spherical particle,

$$\tau_p = \rho_p d_p^2 C_c / 18\eta \qquad (8)$$

The dimensionless Cunningham correction factor C_c is unity for particles with diameters much larger than the mean free path, λ_g, of air molecules. For particles with diameters less than or comparable with λ_g, C_c can be written as

$$C_c = 1 + (2\lambda_g/d_p)[A + B \exp(-Cd_p/2\lambda g)] \qquad (9)$$

where A = 1.257, B = 0.4, and C = 1.1, determined experimentally. For a nonspherical particle, d_p must be replaced by an equivalent diameter.

The speed of an individual particle $v_p(t)$ in the direction of motion of the acoustic wave can be written from the steady state solution of equation (6),

$$v_p(t) = [U_g / (1 + \omega^2 \tau_p^2)] \sin(\omega t - \phi) , \qquad (10)$$

where ϕ is the phase lag of the motion of a particle with respect to the motion of the element of air surrounding the particle undergoing acoustic excitation, and U_g is the amplitude of the fluid velocity. The quantity ϕ is related to τ_p by

$$\phi = \tan^{-1}(\omega \tau_p). \qquad (11)$$

For any particle, irrespective of its shape and density, τ_p can be written as

$$\tau_p = \rho_0 d_a^2 C_{ca} / 18\eta \qquad (12)$$

where d_a is the aerodynamic diameter, ρ_0 represents unit density, C_{ca} is the Cunningham molecular slip correction factor referred to d_a, and η is the viscosity of air. The Cunningham correction factor is strictly a function of the geometric diameter of a particle. But, by regarding C_{ca} as a function of aerodynamic diameter, equations (3) and (4) can be combined to yield a unique relationship between ϕ and d_a (Fig. 1) permitting the value of d_a to be determined from the measurement of ϕ.

Figure 2 shows the relative waveforms of an acoustic field of excitation frequency 5 kHz and particle velocity as a function of time for 2.0, 4.0, and 6.0 μm in aerodynamic diameter.

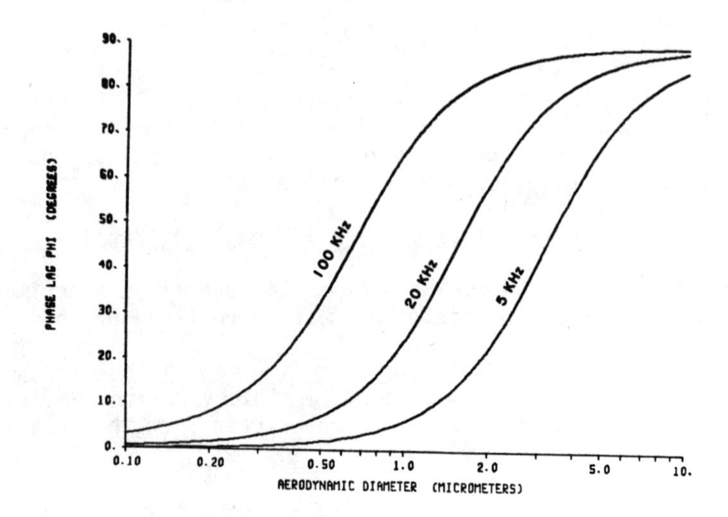

Fig. 1 Relative phase lag (ϕ) of particle velocity with respect to the acoustic excitation as functions of the frequency of excitation and aerodynamic diameter of particles.

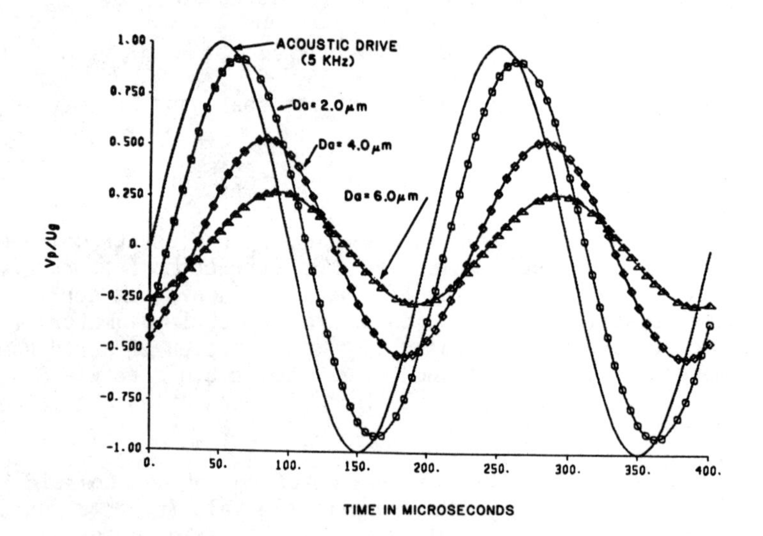

Fig. 2 Relative waveforms of particle velocity in an acoustic field of excitation frequency 5 kHz plotted for three aerodynamic diameters.

DESCRIPTION OF THE SPART ANALYZER

Basic principles of the SPART analyzer have been discussed in detail previously[10,11] and therefore will be only briefly discussed here. The SPART analyzer (Fig. 3) incorporates a laser Doppler velocimeter[12] (LDV), a relaxation cell, and a microcomputer-based data processor.

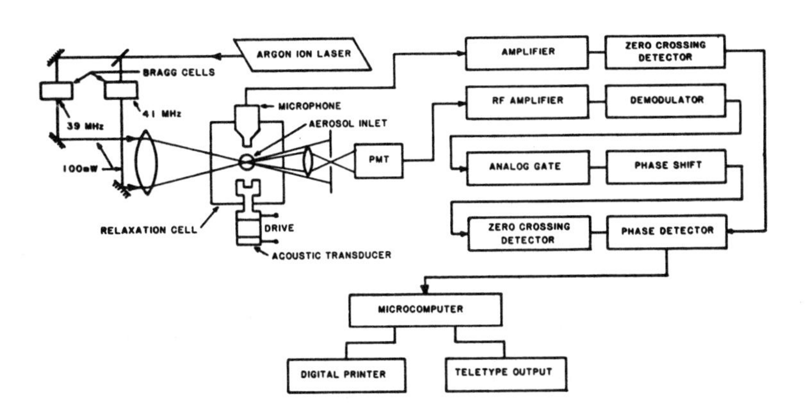

Fig. 3 Schematic of the SPART analyzer showing the laser Doppler velocimeter, relaxation cell, and the signal and data processing circuits.

The sensing volume of the LDV is located at an antinode of an acoustic standing-wave pattern. The wave pattern is generated by using two acoustic transducers, an acoustic driver and a microphone placed inside the relaxation cell where the particle motion $v_p(t)$ relative to the acoustic drive $u_g(t)$ is measured. An excitation frequency (f) of 27.0 kHz was chosen to obtain a high size resolution in the range 0.3 to 6.0 μm. The LDV measures $v_p(t)$ by detecting the Doppler shift of light scattered by an oscillating particle passing through the sensing volume. The microphone measures $u_g(t)$. The relative phase lag ϕ between $u_g(t)$ and $v_p(t)$ is measured by the data-processing circuit. A microprocessor is then used to determine d_a as a function of ϕ. The measurement process takes approximately 2.5 msec for each particle.

A frequency-shifted[13] dual-beam LDV is used to measure $v_p(t)$. The transmission optics of the LDV comprises an argon ion laser, a beam splitter, two Bragg cells, three mirrors, and a focusing lens. The receiving system of the LDV comprises a receiving lens, a pinhole, and a photomultiplier tube (PMT). The sensing volume of the LDV is formed by the intersection of two laser beams with slightly different frequencies. The Bragg cells generate frequency shifts.

When a particle transits the sensing volume, its oscillatory motion along the direction of the acoustic excitation is detected by the Doppler-shifted scattered light. The output of the detector is a frequency-modulated (FM) signal. The FM signal, when demodulated, represents the particle motion within the acoustic field.

The signal-processing circuitry generates a square wave with a time interval Δt beginning at the zero crossing of the microphone signal and ending at the zero crossing of the phase-compensated demodulated LDV signal. Phase compensation is needed to correct for the phase difference between the acoustic excitation measured by the microphone located at a distance d from the LDV sensing volume and the acoustic excitation that the particles experience at the sensing volume. Because there are two zero-crossings per cycle of the acoustic and demodulated signals, there are two square wave pulses per period of the acoustic wave. The time interval Δt is measured by a time interval counter that averages Δt over 8 such pulses. The time interval counter uses a clock oscillator whose frequency (f_s) is approximately 13.8 MHz and the number of clock pulses in Δt is proportional to Δt.

A KIM-1 microprocessor is programmed to compute the phase delay corresponding to Δt and registers a count in a memory location, or channel number n_c, which relates to that phase shift. There are 128 such channels, each relating to a range of phase shifts. Because a given phase shift corresponds to a unique aerodynamic diameter, each channel corresponds to a known interval of aerodynamic diameter. The number of counts in a channel equals the number of particles counted which have aerodynamic diameters within the given interval. The maximum number of pulses is 128 for a maximum possible phase lag of 90°. The microcomputer sums the pulses for a given particle and assigns the particle count to the size channel with the corresponding number in the 128-channel data-storage system. Table I shows the relationships between the aerodynamic diameter (d_a), Cunningham correction factor (C_{ca}), phase lag (ϕ) with respect to the acoustic drive of 27.0 kHz, time interval (Δt), and channel number (n_c).

The number of particle counts (ΔN per channel), the size distribution $dN/d(\log d_a)$, and the volume distribution $dV/d(\log d_a)$ are computed and stored. The total sampling time t in seconds and the total number of counts N are also recorded. The data are printed by a line printer, and a Teletype generates a punched paper tape listing the number of counts in each channel in order of increasing channel number. The data from the paper tape is processed by a minicomputer to generate graphs showing the size, volume, and cumulative distributions as a function of aerodynamic diameter. The minicomputer program also computes the mass median aerodynamic diameter (MMAD), count median diameter (CMD), and the geometric standard deviation (σ_g).

TABLE I

C_{ca}, τ_p, ϕ, Δt, and n_c as a function of d_a for

$f = 27.0$ kHz, $f_s = 13.824$ MHz.

d_a µm	C_{ca}	τ_p µsec	ϕ Degrees	Δt µsec	n_c (Channel Number)
0.08	3.38	0.067	0.65	0.069	1
0.09	3.09	0.077	0.75	0.077	2
0.10	2.87	0.084	0.86	0.088	2
0.20	1.87	0.23	2.24	0.23	4
0.30	1.56	0.43	4.21	0.43	6
0.40	1.41	0.70	6.76	0.79	10
0.50	1.33	1.03	9.88	1.02	15
0.60	1.27	1.42	13.50	1.39	20
0.70	1.23	1.87	17.58	1.81	25
0.80	1.21	2.38	21.99	2.26	32
0.90	1.18	2.96	26.63	2.74	38
1.00	1.16	3.59	31.36	3.23	45
2.00	1.08	13.36	66.19	6.81	95
3.00	1.05	29.30	78.62	8.09	112
4.00	1.04	51.41	83.46	8.59	119
5.00	1.03	79.69	85.77	8.82	122
6.00	1.02	114.20	87.04	8.96	124
7.00	1.02	154.78	87.82	9.03	125
8.00	1.02	201.58	88.33	9.09	126
9.00	1.01	254.56	88.67	9.12	127
10.00	1.01	313.70	88.92	9.15	127
20.00	1.01	1244.70	89.73	9.23	128

The size range and resolution of the SPART analyzer depends on the acoustic frequency. Operating at 27 kHz, the instrument can measure within the so-called respirable size range 0.1 to 10.0 µm in aerodynamic diameter; but its resolution and effective sampling rate were calibrated in the range of 0.3 to 6.0 µm (channel number 6 to 124) using test aerosols containing monodisperse polystyrene latex spheres (PLS). The size resolution outside this range is reduced. A typical size distribution for different PLS aerosols is shown in Figure 4.

The instrument can count and size aerosol particulates at a maximum rate of 200 particles/sec. The sensing volume of the instrument is approximately 10^{-5} cc, which allows measurement of the size distribution of an aerosol with a relatively high particulate concentration without an appreciable coincidence loss. The analyzer samples an aerosol at a rate of 550 cc/min, of which 500 cc/min is used as sheath air. The effective sampling rate, that is, the flow rate of aerosol passing through the LDV sensing volume per unit time, is approximately 6 cc/min.

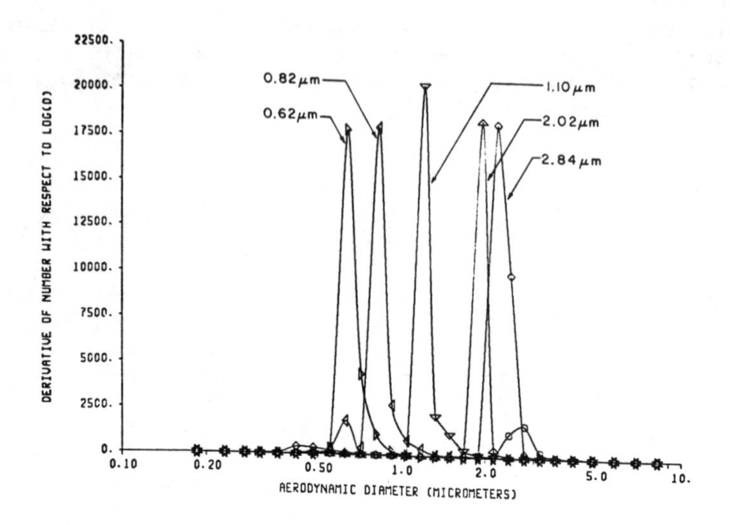

Fig. 4 Aerodynamic size distributions of monodisperse polystyrene latex spheres (PLS) aerosols as measured by the SPART analyzer.

APPLICATIONS OF THE SPART ANALYZER

Therapeutic Aerosols

Therapeutic aerosols are widely used to treat obstructive lung disease. Table II shows some of the uses for medical aerosols. The aerosol particulates contain either solid particles or liquid droplets or both. The efficacy of these medications is determined by their amount and site of deposition inside the lung, which depends primarily on the aerodynamic behavior of the particles, the inhalation technique, and the airway patency. Sympathomimetic drugs are used to dilate obstructed airways. These drugs are commonly inhaled from an inert gas driven atomizer which delivers a fixed dose. Cromolyn sodium is dispensed by a turbine activated inhaler in the form of a finely divided powder. Other therapeutic aerosols, including sodium chloride aerosols, are generated by ultrasonic and hand-held pneumatic nebulizers. These aerosols are being analyzed for lung retention.

Size distributions of the therapeutic aerosols[11,14-16] were studied in the following manner. The test aerosol was introduced into a 200 1-capacity aerosol chamber which was maintained at 37° C and continuously stirred by a fan. The aerosol was sampled from the chamber at this temperature and at ambient relative humidity, approximately 18 to 20 percent. The size distribution of the

TABLE II

COMMON USES FOR MEDICAL AEROSOLS

DISEASES TREATED	DRUGS USED
(1) Emphysema	Sympathomimetic (Adrenalin Like Drugs (1),(2),(3),(4)
(2) Asthma	Cromolyn Sodium (2)
(3) Chronic Bronchitis	Mucolytic Agents (4)
(4) Cystic Fibrosis	Antibiotics (4)
(5) Migraine Headaches	Ergot Alkaloids (5)

particulates was determined. Relative humidity of the chamber was then increased to approximately 95% while maintaining the temperature of 37° C to simulate the ambient condition inside the human lung. The size distribution of the particulate was determined again. Most of the aerosols tested showed significant growth under high humditiy. Typical size distributions of a bronchodilator aerosol (sympathomimetic drug) at ambient and lung humidities are shown in Figure 5. The number of particles per dose for this aerosol increased four-fold when humidified. This increase indicates that particles which existed below the resolution limit (0.3 μm) of the analyzer grew to a measureable size range at high humidity resulting in the higher number of particles above 0.3 μm. The mass per dose also increased at high humidity. A size shift similar to the one shown in Figure 5 will possibly occur when these aerosols are inhaled into the human respiratory tract. The probability of lung deposition of the inhaled bronchodilator aerosol calculated for two humidity conditions as a function of aerodynamic diameter is shown in Table III.

The first three columns represent the number of particles that will be deposited in zones when the aerosol is inhaled in nose breathing. The last column indicates the number of particles that will be deposited in the pulmonary region via mouth breathing -- in a manner recommended for the use of metered-dose bronchodilators. These calculations are based on a lung model developed by the

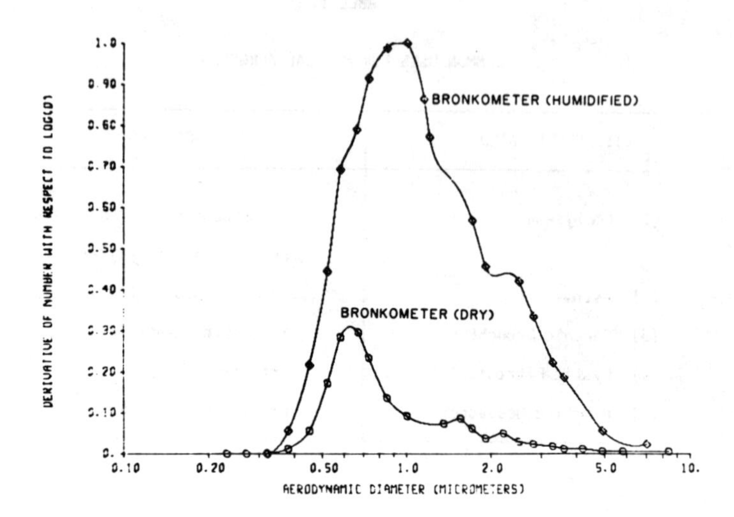

Fig. 5 Growth of aerosol particulates of a metered-dose bronchodilator when exposed to airway humidity (RH 95 percent) and at 37°C. The lower curve represents the size distribution when measured at ambient humidity (RH 18 percent)and at 37°C.

TABLE III

Estimated lung deposition of inhaled bronchodilator aerosol from the measured aerodynamic size distributions in the range 0.3 to 6.0 µm at ambient and airway humidities. The apparent increase in number of particles suggest size shifts because of the condensation process at high humidity.

Human Respiratory Tract	Calculated number of particles likely to be deposited per dose in size range 0.3 to 6.0 µm	
	RH 18%	RH 95%
Nasopharyngeal	2.11×10^7	1.17×10^8
Tracheo-Bronchial	4.98×10^6	2.28×10^7
Pulmonary	2.73×10^7	1.16×10^8
Pulmonary (via mouth)	2.42×10^7	1.24×10^8

International Commission of Radiological Protection (ICRP).[3] The lung model predicts the fraction of particles that will be deposited in the human respiratory tract when inhaled in a standard breathing cycle. Table IV gives the approximate values of the deposition fraction in three regions as functions of aerodynamic diameter.

TABLE IV

DEPOSITION FRACTION
(Based on the ICRP Lung Model, Ref. 3)

Aerodynamic Diameter μm	Pulmonary	Tracheo-Bronchial	Nasopharyngeal
0.01	.60	.29	--
0.03	.57	.14	--
0.05	.54	.10	--
0.10	.44	.07	--
0.30	.32	.04	.02
0.50	.28	.03	.06
1.00	.27	.05	.12
3.00	.25	.07	.51
5.00	.21	.07	.73
10.0	.12	.05	.86
30.0	.05	.03	.94
50.0	.03	.02	.97

This table indicates that while the small particles contribute little to the total mass of aerosol particulates, they particularly are able to penetrate distal lung zones. By controlling the inhalation technique and the size distribution of the therapeutic aerosols, it is possible to maximize the deposition of the aerosol in the desired locations of the respiratory tract.

To perform experimental studies on the total retention of inhaled aerosol in the human lung, the aerosol chamber is equipped with an inhalation system that allows respiratory flow control and monitoring. A desired mouth breathing cycle is first established which allows inhalation and exhalation of aerosol at a preset respiratory cycle. The aerosol is inhaled directly from the chamber and exhaled to a sample collection bag located inside the chamber. The particulate size distributions of the inhaled and the exhaled aerosol are recorded to determine the retention properties as a function of aerodynamic diameter for both normal volunteers and patients with chronic obstructive pulmonary disease.

Flyash Aerosols

Over the next decade an increased usage of coal is predicted to supply the demand for power generation. Emission of SO_2, resulting from the presence of sulphur in the coal,is one of the primary air-pollution problems associated with coal-fired electric generating stations. Use of coal with low sulphur content appears to be a viable means of staying within the strict SO_2 emission standard. However, while the use of low-sulphur coal and lignite will help reduce the emission of SO_2, it also creates a problem in that the flyash - the particulate pollutant that results from burning pulverized coal - acquires high electrical resistivity due to the lack of sulfur content. It is difficult to precipitate high resistivity flyash from the effulent gas stream by an electrostatic precipitator.[17] Because of high resistivity, the flyash is also difficult to charge electrically and, when charged and collected, it builds a highly charged surface layer on the collection plates of the electrostatic precipitator. The high resistivity flyash discharges through the collection plate very slowly, generating a back corona which reduces efficiency and impedes further collection efforts.

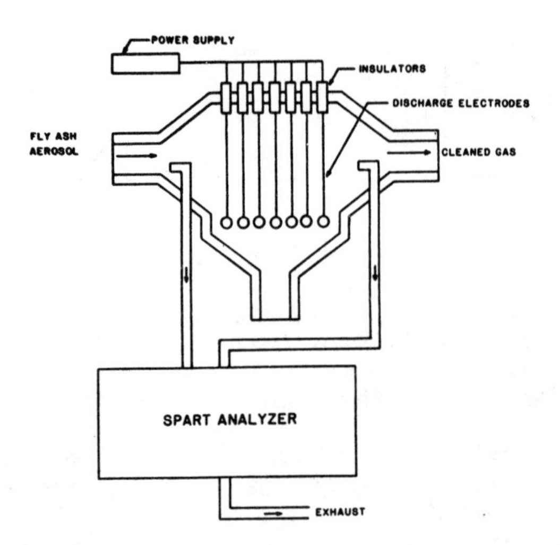

Fig. 6 Experimental setup for measuring fractional collection efficiency of an electrostatic precipitator as a function of particle size using the SPART analyzer.

Due to the complexity of the process of electrostatic precipitation of high resistivity flyash, its analysis and possible remedy require measurement of the fractional collection efficiency of flyash as a function of aerodynamic diameters. Currently, SPART analyzers are being used to measure the particle size distribution of flyash at the inlet and the outlet of a pilot plant precipitator (Fig. 6). The size distribution data are stored within the analyzer, and the fractional efficiency of the precipitator as a function of aerodynamic diameter is computed. The operational parameters of the precipitator can be varied to examine its effect on the collection efficiency. Figure 7 shows a typical plot of collection efficiency as a function of aerodynamic diameter.

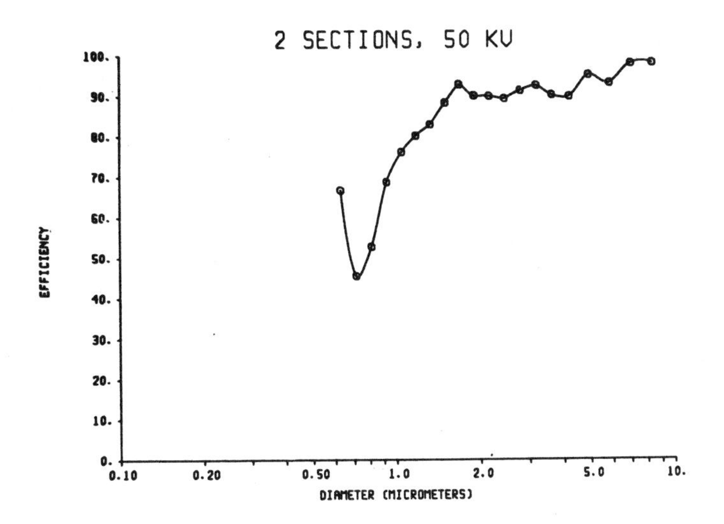

Fig. 7 Fractional collection efficiency of pilot plant electrostatic precipitator as a function of aerodynamic size of flyash particle size using the SPART analyzer.

The SPART analyzer is used to measure size distribution of atmospheric aerosols as well as aerosols generated by commercial pressurized products within the respirable fraction. An example of the size distribution of atmospheric aerosol is shown in Figure 8. Figure 9 shows the size distribution of particles generated in a hair spray within the respirable fraction.

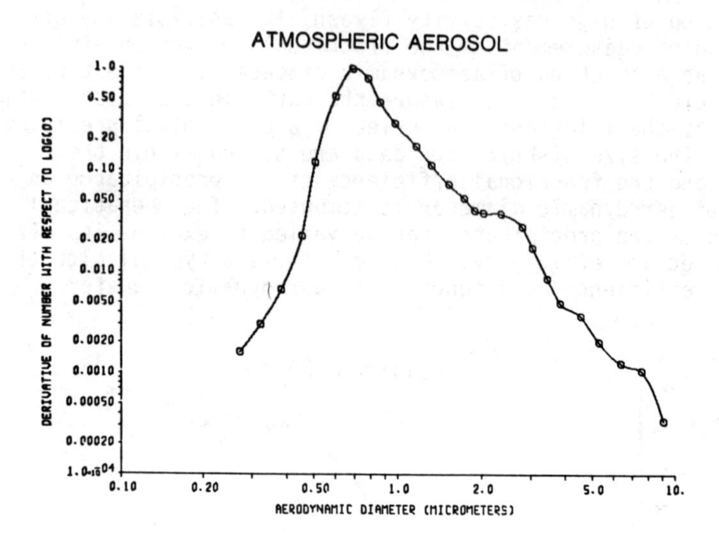

Fig. 8 Normalized size distribution of atmospheric aerosol as measured by SPART analyzer. The effective sampling rate of the analyzer reduces below 0.6 μm resulting in a lower indicated value of particulate concentration.

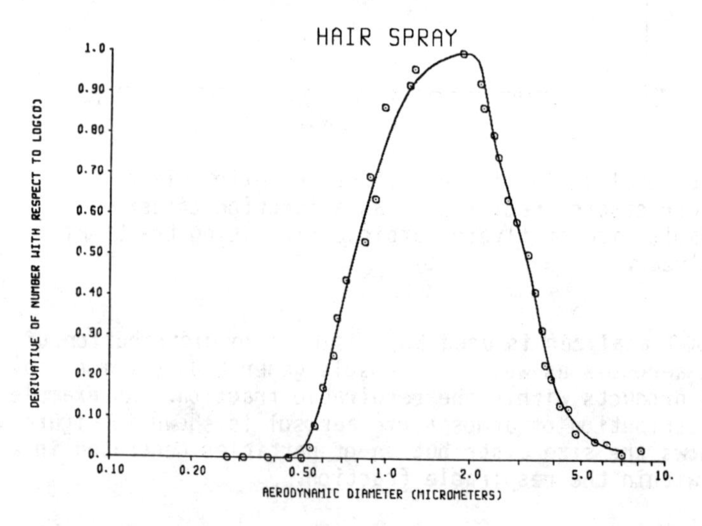

Fig. 9 Normalized size distribution of particles and droplets in a hair spray aerosol within the respirable fractions.

CONCLUSION

An application of laser Doppler velocimetry in sizing liquid droplets and solid particles in therapeutic and potentially harmful aerosols within the respirable fraction is described. Some of its current applications are also included.

ACKNOWLEDGEMENTS

The development of the SPART analyzer and construction of the first prototype instrument were financed in part under grant No. R804429 from the Environmental Protection Agency. Construction of a second analyzer and its application to therapeutic aerosols were supported in part by funds from the National Heart, Lung and Blood Institute under grant No. HL20024. Construction of the third prototype SPART analyzer and its application to electrostatic precipitators were made possible through research grant No. R806192 from the Environmental Protection Agency. The authors thank R. G. Renninger, W. Hood, R. A. Sims, R. W. Raible, P. Roberson and L. Neidhardt for their technical assistance and discussions in various phases of this research program and P. T. Archer, M. Elms, and D. Watson for their assistance in the preparation of the manuscript.

REFERENCES

1. T. T. Mercer, Aerosol Technology in Hazard Evaluation, (New York: Academic Press, [1973]), p. 31.
2. P. E. Morrow, "Aerosol Characterization and Deposition," Am. Rev. Respir. Dis., 110, 88 (1974).
3. Task Group on Lung Dynamics, "Deposition and Retention Models for Internal Dosimetry of the Human Respiratory Tract," Health Phy. 12, 173 (1966).
4. B. Y. H. Liu, ed. Fine Particles: Aerosol Generation, Measurement, Sampling and Analysis, (New York: Academic Press, [1976]), p. 40.
5. H. T. Mercer, "On the Calibration of Cascade Impactors," Ann. Occup. Hyg., 6, 1 (1963).
6. A. K. Rao and K. T. Whitby, Am. Ind. Hyg. Assoc. 5, 38, 174 (1977).
7. V. A. Marple and K. Willeke, Atoms Env. 10, 891 (1976).
8. N. A. Fuchs, The Mechanics of Aerosols, (New York: Pergamon Press [1964]), p. 77.
9. L. L. Beranek, Acoustics (New York: McGraw-Hill [1954]), p. 286.
10. M. Mazumder, R. Ware, J. Wilson, R. Renninger, C. Hiller, P. McLeod, R. Raible and M. Testerman, J. Aerosol Sci. 10, 561, (1979).
11. C. Hiller, M. Mazumder, D. Wilson and R. Bone, Am. Rev. Respir. Dis., 118, 311, (1978).

12. F. Durst, A. Melling and J. H. Whitelaw, Principles and Practice of Laser-Doppler Anemometry, (New York: Academic Press [1976]).
13. M. K. Mazumder, Appl. Phys. Letts., 16, (1970).
14. C. Hiller, M. Mazumder, D. Wilson and R. Bone, Am. Rev. Respir. Dis., 118, 311 (1978).
15. C. Hiller, M. Mazumder, D. Wilson and R. Bone, Journal of Pharmecutical Sciences 69 (March 1980): 3.
16. G. Smith, C. Hiller, M. Mazumder and R. Bone, Am. Rev. Res. Dis., 121, 513 (1980).
17. S. Oglesby, Jr., and G. B. Nichols, Electrostatic Precipitation, (New York: Marcel Dekker, Inc. [1978]).

LASER DOPPLER RETINAL BLOOD FLOW MEASUREMENTS

K. Gardner & E.R. Pike
Royal Signals & Radar Establishment,
St. Andrews Road, Malvern, Worcs., England

J.A. Govan & D.W. Hill
Royal College of Surgeons of England,
35-43 Lincoln's Inn Fields, London WC2A 3PN, England

ABSTRACT

The feasibility of using laser Doppler velocimetry to measure retinal blood flow was first demonstrated in 1972 by Riva, Ross and Benedek[1]. Since that time two groups have been following independent programmes aimed at producing a clinical instrument for this purpose. This paper will describe the application of laser Doppler velocimetry to retinal blood flow measurements, paying particular attention to an instrument developed as a collaborative project between R.S.R.E. Malvern and the Royal College of Surgeons of England.

INTRODUCTION

We will begin by describing the general principle of laser Doppler velocimetry and then discuss the application of this to blood flow measurements. The instrument developed at R.S.R.E. will be described, together with a discussion of the results to date. Finally we will consider the safety aspects of such a system which are very important for a clinical instrument.

LASER DOPPLER VELOCIMETRY

Laser Doppler velocimetry is now a well established technique of flow measurement, covering the range from supersonic wind tunnel studies to molecular diffusion. However in all cases the underlying concept is that when a moving body radiates a signal, an observer will, in general, see a signal at a frequency different from that radiated. This is simply the Doppler effect by which a moving object reradiating by elastic scattering radiation incident upon it will produce a frequency shift in the scattered field. By measuring the frequency shift one can determine the velocity of the scattering target.

Referring to figure 1a, if we consider a particle moving with a velocity $\vec{v}$ being irradiated by a laser beam with incident wavevector $\vec{k}$ $(= \frac{2\pi\vec{\hat{n}}}{\lambda}$; $\vec{\hat{n}}$ is a unit vector in the direction of propagation), then an observer detecting scattered radiation with a wavevector $\vec{k}_s$ will see a Doppler shift $\Delta\omega$ given by

$$\Delta\omega \;=\; (\vec{k}_s - \vec{k}_o).\vec{v} \;=\; \vec{K}.\vec{v} \quad . \tag{1}$$

ISSN:0094-243X/81/650213-06$1.50 American Institute of Physics

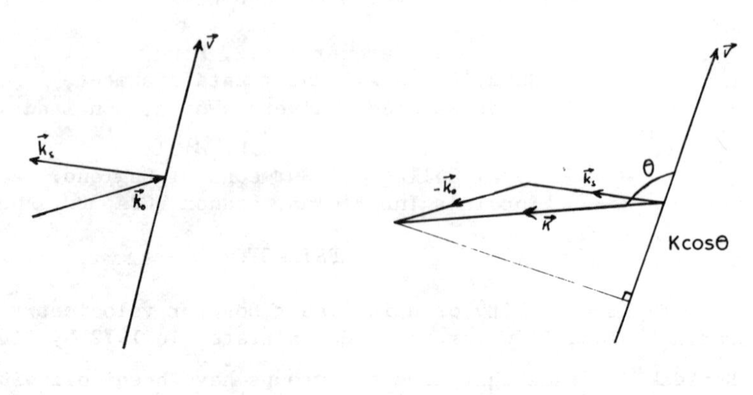

Fig. 1a Fig. 1b

This is given by the projection K cos θ of figure 1b multiplied by v.

The value of Δω could be determined by, for example, a light beating experiment on the photocathode of a photomultiplier. The scattered signal at angular frequency ω+Δω is mixed with a portion of the incident wave at frequency ω, producing a beat frequency at Δω which would modulate the rate of emission of photo-electrons from the photo-cathode at this frequency. To extract the information about Δω, the most sensitive method of signal processing is photon correlation. In this method the individual photoelectric events are amplified, discriminated and shaped to give a random train of standard height pulses whose rate is modulated at the frequency Δω. This modulated random pulse train is then fed into a real time digital photon correlator (e.g. Malvern Instrument K7027) which constructs the temporal autocorrelation function of the signal which is related to the frequency spectrum of the intensity fluctuations by a Fourier transform.

BLOOD FLOW MEASUREMENT

The measurement of blood flow in the retina presents a particularly difficult problem since the velocity of the individual cells changes over the cross section of the vessel and spatial discrimination on this scale is not feasible in the eye. The relatively large number of cells present in whole blood also produces multiple scattering of the probe laser light which again complicates interpretation.

Consider first a Newtonian fluid undergoing laminar flow in a circular pipe. The velocity profile across the pipe will be given by the usual parabolic Poiseuille formula. It has been shown[1] that for this type of flow, if the fluid is uniformly seeded with scattering particles then there will be an equal number of scattering particles in each velocity interval, up to a maximum velocity at the centre of the tube. If the whole tube is uniformly irradiated by a laser beam then the spectrum of the scattered light will be flat up to a cut off corresponding to the maximum velocity present. The correlation function recorded using photon correlation techniques would therefore be of the form sinc $(\Delta\omega_m \tau)$, where $\Delta\omega_m$ is the highest angular frequency present in the spectrum and τ is the delay time in the correlation function. One could therefore calculate the volume flow rate from the recorded correlation function and a knowledge of the tube diameter and the scattering wave vector $\vec{K}$.

The case of whole blood flowing in a vessel <u>in vivo</u> is considerably more complicated. The velocity is not constant and the fluid is certainly not Newtonian. In the case of retinal blood flow measurement the diameter of the vessel is difficult to determine and even the calculation of $\vec{K}$ is not simple due to the geometry of the eye. However the mean velocity of blood flowing in retinal vessels can still be calculated if one makes certain assumptions about the velocity profile and if one can make measurements in a time short compared to the heart cycle, or alternatively average over several heart cycles.

To avoid many of the difficulties encountered in determining $\vec{K}$ we have developed a two beam system illustrated in figure 2. In this we make two measurements simultaneously at two closely spaced points along the vessel. Since the scattered wave vector is in the opposite sense to the incident wave vector $\vec{k}_s = -\vec{k}_o$ for this configuration. If we label the two beams 1 and 2 then for the first beam

$$\Delta\omega_1 = (\vec{k}_{s1} - \vec{k}_{o1}).\vec{v} = 2\vec{k}_1.\vec{v}$$

and for the second beam

$$\Delta\omega_2 = (\vec{k}_{s2} - \vec{k}_{o2}).\vec{v} = 2\vec{k}_2.\vec{v} \quad .$$

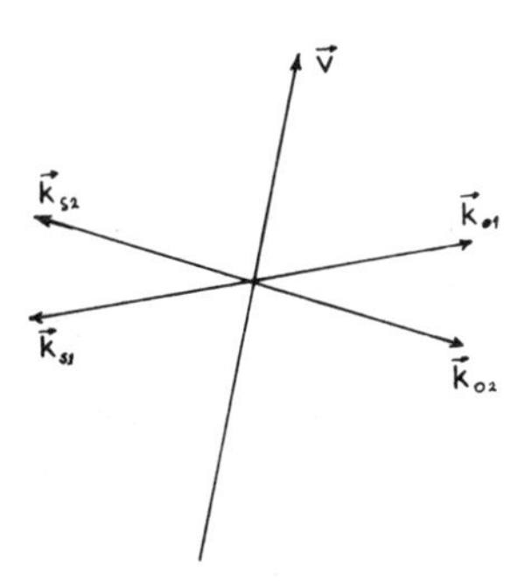

Fig. 2

The difference between the Doppler shifts is

$$\Delta\omega_1 - \Delta\omega_2 = 2(\vec{k}_1 - \vec{k}_2).\vec{v} \quad .$$

The vector $(\vec{k}_1 - \vec{k}_2)$ lies almost parallel to $\vec{v}$ and therefore the cos θ term in the scalar product is approximately unity, and the difference $(\Delta\omega_1 - \Delta\omega_2)$ can be used to measure the velocity $\vec{v}$ provided the angle between $\vec{k}_1$ and $\vec{k}_2$ is known. This angle can be calculated from a standard model of the eye and will be fairly constant when one makes measurements at various parts of the eye. By making the measurements in this way one determines the component of the velocity in a direction at right angles to the angle bisector of $\vec{k}_1$ and $\vec{k}_2$. Therefore any movement of the vessel in the direction of the angle bisector is not registered in the final result. This can be important if one uses a separate local oscillator to extract the velocity information from the signal.

EXPERIMENTAL SYSTEMS

The earlier work of Dr. Riva and his group at the Boston Retinal Foundation was based on a slit lamp and used a Goldmann contact lens to annul the corneal refraction. Difficulties in angle measurement led again to the development of a two-beam system and in his latest work in Philadelphia at the Schei clinic Dr Riva has used a Topcon fundus camera.[2] Various schemes of analogue processing of the scattered light signals have been used.

The basis for the retinal blood flow measurement system developed by the R.S.R.E./R.C.O.S. workers is a Zeiss fundus camera. This was chosen at an early stage since it is a widely used, reliable instrument which could easily be adapted to take laser optics. The modifications to the camera were minor and only entailed replacing a mirror in the camera with a smaller one. This did not impair the performance of the camera in any way. All of the optical components for the laser delivery system were attached to a side plate which was bolted to the camera body. The layout of the optical components is sketched in figure 3.

In operation, the two beams are directed at the vessel of interest by the deflection lens, and then rotated using the beam rotation device so that they lie along the vessel. The scattered light is then collected along the same direction as the input light and since it is depolarized it passes through the polarizing beam splitter and is focussed on to two optical fibres. These fibres are coupled to two high quality photomultiplier tubes (E.M.I. 9863) the outputs of which are fed to a real time digital correlator (Malvern K7027). Each experiment typically lasts for 100 milliseconds at which time the contents of the correlator is read out into the memory of a Commodore P.E.T. minicomputer and simultaneously another experiment starts. The readout time for the correlator is 12.8 milliseconds, but the next experiment can start before this operation is

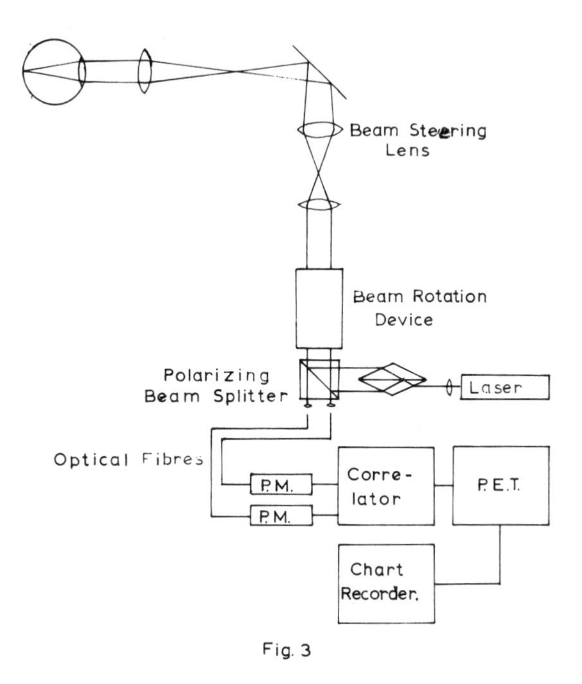

Fig. 3

completed. In the intervening period between two experiments, the data is analysed by the P.E.T. and the resultant velocity information is displayed on a chart recorder. We feel that this digital photon-correlation processing[3] maximised the information content in the signals although it is more expensive than analogue processors.

In an earlier series of experiments[4] the laser Doppler technique was directly compared with fluoroscein cineangiography. There was very good agreement between the two methods and it is anticipated that the improved two beam method will be simple to use and give reliable results, particularly if equipped with a servo-mechanical device to hold the laser spot on the retinal vessel.

LASER SAFETY

For a measurement system of this type to be ethically acceptable we must be absolutely sure that the levels of laser radiation incident on the retina will not cause damage. Many calculations and experiments on animals have been performed in recent years which give us a clear indication that the 1.5 μW we have used to date on human subjects is quite safe. We would like, however, to use up to 10 μW which should, by the same considerations, also be well within safe limits but to increase our knowledge a series of exposures are currently being performed on human subjects whose eyes are to be enucleated due to cancer or for other reasons. The power being used in these tests is 50 μW incident on the cornea. The results of histological examinations are not available at the time of writing this report, but we hope that even at this level no damage will be found.

CLINICAL EVALUATION

The current situation in this field is of an established feasibility for these rather difficult measurements. Both Dr. Riva's group and ourselves are now looking forward to a preliminary stage of clinical evaluation. We hope that a number of clinically useful applications of this type of measurement will be identified both for retinal pathology, in particular in diabetic retinopathy, and perhaps also for more general microcirculation studies.

REFERENCES

1. C. Riva, B. Ross and G.B. Benedek, Invest. Ophth. 11(11), 936 (1972).
2. C.E. Riva, G.T. Feke, B. Eberli, and V. Benary, Appl. Opt. 18(13), 2301 (1979).
3. H.Z. Cummins and E.R. Pike, (Eds.), Photon Correlation Spectroscopy and Velocimetry (Plenum Press, 1976).
4. D.W. Hill, P. Parker, E.R. Pike and S. Young, Proceedings of the "Fifth International Congress of Ophthalmology", Hamburg 1976.

ANALYSIS OF SPECKLE CORRELATION VARIATION IN HOLOGRAM INTERFEROMETRY TO DISTINGUISH RIGID BODY ROTATIONS

M. Celaya and D. Tentori
Applied Physics Dept. CICESE. Apdo. Postal 2732 Ensenada,B.C. Mexico

ABSTRACT

In the present work rigid body rotations are examined. The correlation variation of speckle patterns with the aperture stop of the viewing system is analyzed for tilt and in-plane rotation. This analysis is made by considering a mathematical relation of the correlation between the real and holographically reconstructed wavefronts, which are reflected from a rough surface before and after the rotation. This equation gives the intensity distribution on the image plane of an optical imaging system focused on the reflecting surface. From the analysis of this relation it is shown that using the same arrangement, fringe contrast variation with the aperture stop radius of the optical system is different for different types of movements. Contrast reversal of the fringe pattern is obtained for some values of the numerical aperture of the viewing system. In this work tilt and in-plane rotation are considered.

INTRODUCTION

As well known in hologram interferometry, interference fringe patterns can be similar for different kinds of movements. To be specific interference fringe patterns are formed by straight lines, for tilt, in-plane rotations, and in-plane translation, in general one cannot identify the type of movement the object has undergone by a simple observation of the fringes.[1]

In hologram interferometry, the intensity distribution on the image plane of the viewing system is given by the sum of the intensities of the wavefronts coming from the reconstructed diffuse object plus the correlation between them. Each speckle pattern on the image plane is given by the amplitude distribution on the reflecting object convolved with the impulse response of the optical system.

Under normal working conditions the observation of the interference fringes depends on the correlation between the two speckle patterns.[2] The speckle pattern produced by each wavefront depends upon the optical imaging system impulse response, which is a function of the aperture stop of the imaging system. By varying the size of the aperture stop, the average contrast of the interference fringes at the image plane can be changed. No fringes will be observed when the speckle patterns exhibit complete lack of correlation. As the correlation between the wavefronts is improved the average contrast of the interference fringes will be increased. The contrast variation

ISSN:0094-243X/81/650219-08$1.50

of the interference fringe pattern with the impulse response of the optical imaging system can be used to identify the type of movement undergone by a diffuse reflecting object.[3]

THEORY

If we consider a uniformly illuminated diffuse object located on the plane (x_0,y_0), with a complex field in the immediate vicinity of the object given by $O_1(x_0,y_0)$ and a viewing system focused on the object's surface, the complex field at the image plane (x_1,y_1) is given by

$$V_1(x_1,y_1) = \iint_{-\infty}^{\infty} h(x_1,y_1;x_0y_0)O_1(x_0,y_0)\, dx_0dy_0 \tag{1}$$

where $h\,(x_0,y_0;\, x_1,y_1)$ is the impulse response of the viewing system

Similarly for the other wavefront coming from the object, the complex amplitude on the image plane is given by

$$V_2(x_1,y_1) = \iint_{-\infty}^{\infty} h(x_1,y_1;x_0',y_0')O_2(x_0',y_0')dx_0'dy_0' \tag{2}$$

We consider the following common working conditions: the viewing system is made up of a thin lens with an aperture stop located at a distance H_0 from the object plane. The near field complex amplitudes $O_1\,(x_0,y_0)$ and $O_2\,(x_0',y_0')$ which are scattered by the surface of the rough object are assumed to be characterized by the cross-correlation function.[4]

$$\langle O_1(x_0,y_0)O_2^*(x_0',y_0')\rangle = \langle \tilde{O}_1(x_0,y_0)\tilde{O}_2^*(x_0,y_0)\rangle\delta(x_0-x_0')\delta(y_0-y_0') \tag{3}$$

where $\tilde{O}_k\,(x_0,y_0)$ is a smooth version of $O_k\,(x_0,y_0)$, $k = 1, 2$, with respect to the surface roughness. In this case the intensity distribution in the image plane is given approximately by

$$\langle(V_1+V_2)(V_1+V_2)^*\rangle \simeq I_0(1 + C\,(X_1,\, Y_1)\cos\phi) \tag{4}$$

if one retains only the first term in a Taylor series expansion of $\cos\phi$ around $(x_1/m,y_1/m)$. In eq. (4), I_0 represents the uniform intensity over the entire surface, and

$$\phi = (2\pi/\lambda)\ PD$$

where PD is the path difference measured with respect the lens vertex. In addition, the function

$$C(X_1,Y_1) = \frac{F(X_0,Y_0)\quad G(X_0,Y_0)}{\iint_{-\infty}^{\infty} F(X_0,Y_0)G(X_0,Y_0)\ dX_0dY_0} \tag{5}$$

is the normalized correlation function of both wavefronts described in terms of the new coordinates

$$(X_0,Y_0) = (x_0 - x_1/m,\ y_0 - y_1/m);$$

here m is the magnification of the viewing system and

$$(X_1,Y_1) = (a_x(x_1,y_1),\ a_y(x_1,y_1));$$

where a_x and a_y are functions of (x_1,y_1) that depend on the type of motion the object has undergone.

The functions C (X_1,Y_1) and cos ϕ, describe the fringe visibility and the shape of the fringe pattern, respectively.

In the particular case of a circular aperture we have

$$F(X_0,Y_0) = \frac{2J_1(\pi A(X_0,Y_0))}{\pi A(X_0,Y_0)} \tag{6}$$

where

$$A = A_0(X_0{}^2 + Y_0{}^2)^{1/2}$$

and

$$G\ (X_0,Y_0) = \frac{2J_1(\pi A'(X_0,Y_0))}{\pi A'(X_0,Y_0)}$$

$$A' = A_0'\left(\frac{X_0{}^2}{a} + \frac{Y_0{}^2}{b}\right)^{1/2} \tag{7}$$

where a and b are given by equations (8) and (13).
Equation (6) gives the response of a circular aperture and $C(X_1,Y_1)$ is the corresponding correlation function with $G(X_0,Y_0)$ given by Eq. (7). Apart from a phase factor, this gives the average correlation between the two speckle patterns formed by the aperture stop on the image plane (to first order in the cos ϕ expansion). Because A_0, X_1 and Y_1 are functions of the numerical aperture system NA $\simeq \rho/H_0$ (where ρ is the radius of the optical system aperture stop), wavelenght (λ) and the displacement or the object, the variation of these

parameters provides information on the motion. In what follos we analyze eq. (5) for in-plane rotation and tilt.

Let us suppose that in-plane rigid body rotation has ocurred. With the axis of rotation of the plane object located at the coordi-

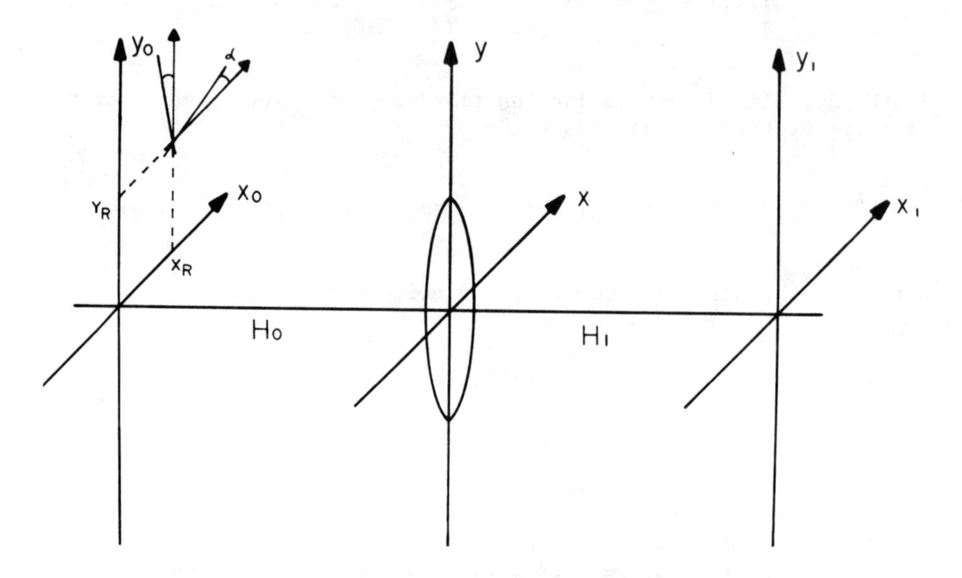

Fig.1. Coordinate systems used to analyze in-plane rotation of a plane object. The axis of rotation is located at coordinates (x_R, y_R) and the angle of rotation is given by α.

nates (x_R, y_R). A_0, A_0', a and b are given by

$$A_0 = A_0' = 2\rho/H_0\lambda, \; a = b = 1 \tag{8}$$

and

$$a_x(x_1, y_1) = - x_1/m + x_R + (x_1/m - x_R)\cos\alpha + (y_1/m - y_R)\sin\alpha$$

$$a_y(x_1, y_1) = - y_1/m + y_R + (x_1/m - x_R)\sin\alpha + (y_1/m - y_R)\cos\alpha$$

where α is the angle of rotation (Fig. 7). On substituting equation (7), (8), (9) in Eq. (5) we obtain the product of two Bessel functions in the integrand whose maxima, located at $(x_1/m, y_1/m)$ and $(x_1/m + a_x, y_1/m + a_y)$ as shown in Fig 2, are separated by a distance

$$D = (2(1 - \cos\alpha))^{1/2} \; ((x_1/m - x_R)^2 + (y_1/m - y_R)^2)^{1/2} \tag{10}$$

The first zero of both Bessel functions will be on a circle of radius.

$$r = .61\lambda H_0/\rho \tag{11}$$

centered around the respective maxima

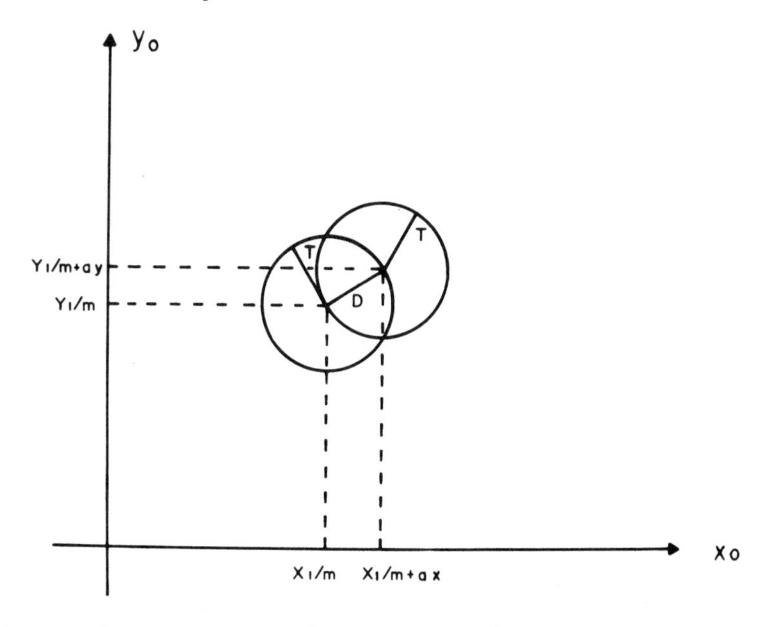

Fig. 2 For in plane rotation, the maxima of the Bessel functions corresponding to homologous points are located at coordinates $(x_1/m, y_1/m)$ *and* $(x_1/m + a_x, y_1/m + a_y)$. *Separation between these maxima is denoted by D. The circles surrounding these points, correspond to the locus of the first zero of both Bessel functions and* r *is the radius of these circles.*

For a given axis and angle of rotation, the distance between maxima is fixed, while the radius r changes as the aperture stop radius ρ is varied. This means that the normalized correlation function $C(X_1,Y_1)$ of the wavefronts will take positive values for $0 < r < D$, it will go to zero for a certain value of the radius r, and will take negative values for larger values of r. This process continues in an oscillatory manner until the correlation approaches zero. This means that for some values of the aperture stop radius a contrast reversal is obtained. In particular, for in-plane rotation the straight fringes of the interference pattern will have an annulus of contrast reversal centered on the axis of rotation. The radius of the annulus will change as the radius of the aperture stop of the viewing system is varied.

In the presence of tilt, A_0, A_0', a and b are given by

$$\begin{aligned} A_0 &= 2\rho/H_0\lambda \\ A_0' &= 1 \end{aligned} \tag{12}$$

$$\begin{aligned} a &= H_0\lambda/(2\rho \cos \beta) \\ b &= H_0\lambda/2\rho \end{aligned} \tag{13}$$

$$\begin{aligned} a_x(x_1,y_1) &= - x_1/m + x_1/(m \cos \beta) \\ a_y(x_1,y_1) &= 0 \end{aligned} \tag{14}$$

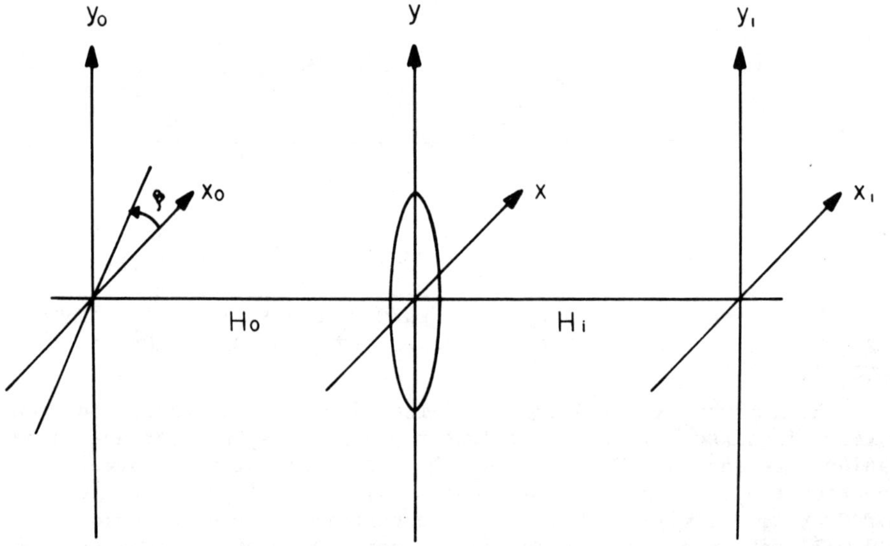

Fig. 3. Coordinate systems used in the analysis of tilt. The tilt angle is denoted by β.

where β is the tilt angle. Similarly, on substituting Eqs. (11), (12) and (13) in Eq. (5), we find that the maxima of both Bessel functions will be separated by a distance

$$D = (x_1/m)(1/\cos\beta - 1) \qquad (15)$$

and the first zero of $F(X_0,Y_0)$ will be located on a circle of radius $r = .61\lambda H_0/\rho$, centered at $(x_1/m, y_1/m)$. The first zero of $G(X_0 - X_1, Y_0 - Y_1)$ is located on an ellipse with center at $(x_1/m\cos\beta, y_1/m)$ and semiaxes a',b' directed along the axes x_0, y_0, whose values are given by

$$a' = .61\lambda H_0/(\rho\cos\beta)$$
$$b' = .61\lambda H_0/\rho \qquad (16)$$

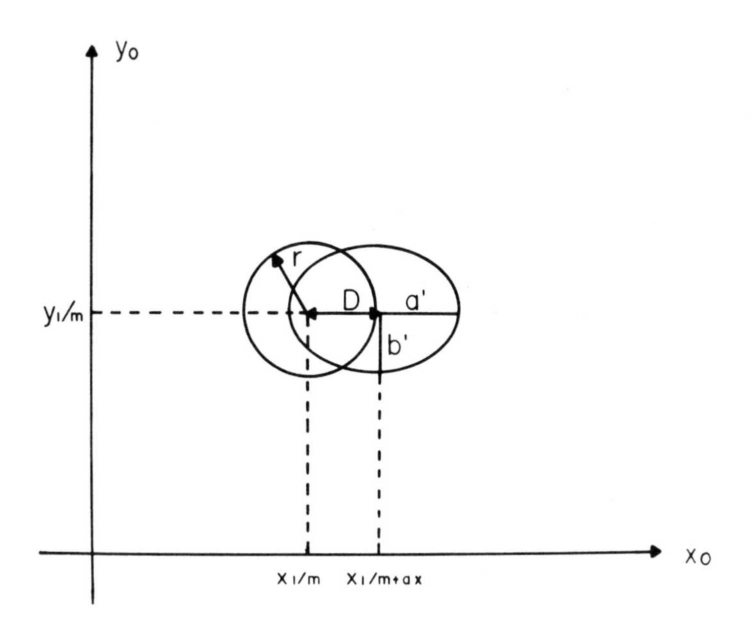

Fig.4 The maxima of the Bessel functions corresponding to homologous points are located at coordinates $(x_1/m, y_1/m)$ and $(x_1/m + a_x, y_1/m + a_y)$ The circle of radius surrounding the point $(x_1/m, y_1/m)$ represents the first zero of the Bessel Function. The ellipse of semiaxes a' and b' represents the locus of the first zero of the Bessel function of the homologous point after tilting. The separation between both maxima is given by D.

As one can see from Fig. 4, the fringe contrast should go to zero when the separation between the maxima becomes larger than the sum of the radius r of the circle and the semiaxis a' of the ellipse.

This implies that the following relation is satisfied

$$\frac{x_1}{m} = \frac{.61\lambda H_0}{\rho} \frac{1 + \cos\beta}{1 - \cos\beta} \qquad (17)$$

Because β is small, the object should be very large in order for the effect to be noticeable. Hence we can say that in this case fringe contrast does not depend on the radius of the aperture stop of the viewing system.

CONCLUSIONS

We have shown that different kind of object movements can be identified in holographic interferometry by simply controlling the system aperture stop and observing the change in the nature of the fringe contrast. For in-plane rotations, the resulting straight fringe pattern presents annular contrast variation centered on the rotation axis. When a lateral displacement takes place, a similar pattern of straight fringes is obtained, but in this case the fringe contrast varies uniformly over the whole surface of the object[3], while as shown above, the average fringe contrast of the straight fringe pattern corresponding to a tilt does not depend on the radius of the aperture stop of the viewing system.

REFERENCES

1. Charles M. Vest, Holographic Interferometry (John Wiley & Sons, N. Y. 1979) Chap. 2 and 3
2. Ch. Roychoudhuri, S. Guel, J. Opt. Soc. Am. 68, 1375 (1978)
3. M. Celaya, D. Tentori, L. Celaya, J. Opt. Soc. Am. 69, 1416(1979)
4. S. Lowental, H. Arsenault, J. Opt. Soc. Am. 60, 1478 (1970)
5. J.W. Goodman, Introduction to Fourier Optics, Mc.Graw-Hill N.Y. 1968, Chap. 2 and 5.

RECENT DEVELOPMENTS IN THREE DIMENSIONAL DISPLAYS

James F. Butterfield
Stereorama Picture Company
North Hollywood, CA 91602

ABSTRACT

Holography and stereoscopy are two optical techniques of achieving three dimensional displays. Holography can be used to produce relatively small monochromatic images within certain technical limitations in taking and displaying the picture. Stereoscopic techniques can provide full color large screen displays. Three dimensional television equipment for closed-circuit applications is now available. A stereoscopic video microscope for industrial and medical purposes has been recently developed. However, both systems require viewing aids. The Stereorama Projector now displays panoramic three dimensional color slides without glasses.

INTRODUCTION

Three dimensional displays can be achieved by means of two optical techniques, holography and stereoscopy. Holographic still images are practical and useful displays. However, they are generally limited to one color. The recording procedure restricts the size and location of the scene. Lasers are usually required for recording and viewing. True holographic moving pictures or television will probably not be developed in this century.[1]

Stereoscopic photography of still and motion pictures using special viewing means or stereo glasses has been in use for many years. The stereopticon viewer provided three dimensional entertainment to millions of homes from 1860 to 1920. Feature stereoscopic motion pictures with glasses, as shown in Figure 1, were exhibited beginning in the 1950's.[2]

FIG. 1. 3-D Motion Pictures.

FIG. 2. 3-D Home Television.

ISSN:0094-243X/81/650227-06$1.50

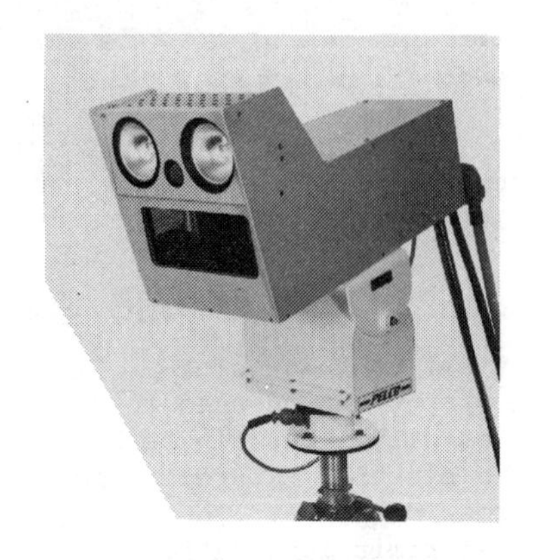

FIG. 3. 3-D TV Camera.

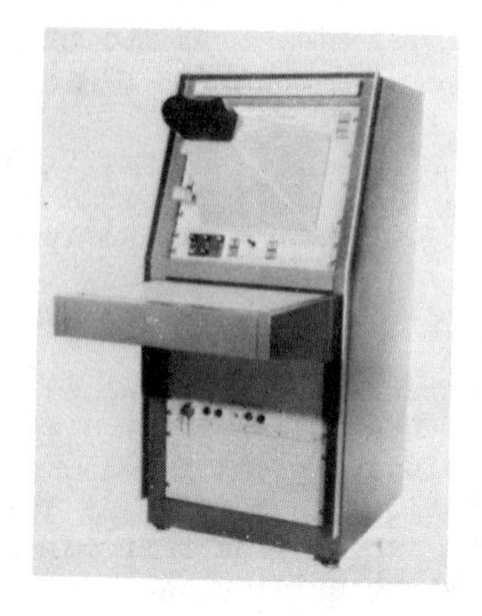

FIG. 4. 3-D TV Display Console.

THREE DIMENSIONAL TELEVISION

Three dimensional entertainment television was tested on a daily basis in 1954 and 1955 by the author in Mexico City. Figure 2 portrays the viewer at home wearing micro-mirror glasses.[3] The International Telecommunications Union (CCIR) has been developing standards for world home stereoscopic television.[4]

For some years, the author's laboratory has fabricated several configurations of three dimensional television equipment for numerous closed-circuit industrial, military and educational applications.[5,6,7]

Figure 3 illustrates a typical configuration. The 3D-TV camera housing contains: an optical beam splitter; a single lens; and a single TV camera. Included in this model are two lights and a microphone. The lens is a 10 to 1 zoom lens and the video camera is a very high resolution model with 1023 TV scan lines and 32 megahertz bandwidth.[8]

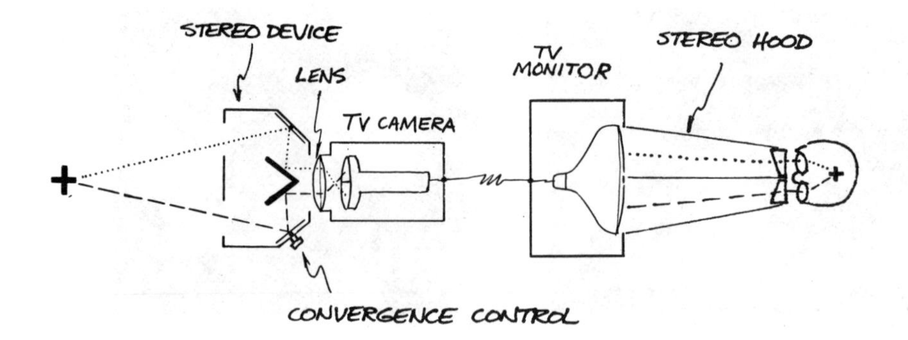

FIG. 5. Schematic of Stereoscopic TV System.

The display console in Figure 4 has a TV monitor with a stereo viewing hood for individual viewing or stereo glasses for group viewing. Also located in the console is a conventional video tape recorder to record and playback 2D or 3D pictures.

Figure 5 is a schematic showing the stereo-optical beam splitter, which picks-up two images from slightly different angles and relays them to be focused by the lens side-by-side on the television pick-up tube of a conventional black-and-white or color video camera. The video signal appears as two side-by-side images on the screen of the conventional TV monitor. A stereo-hood with a baffle down the center channels one image to each eye and prism wedges superimpose the images into one three dimensional picture for individual viewing. Group viewing of a projected 3D video picture requires the individuals to each wear quality polarized glasses.

STEREOSCOPIC VIDEO MICROSCOPE

A more recent development is the author's stereoscopic video microscope shown in Figures 6 and 7. This new instrument replaces the optical stereo microscopic with many advantages.[9,10]

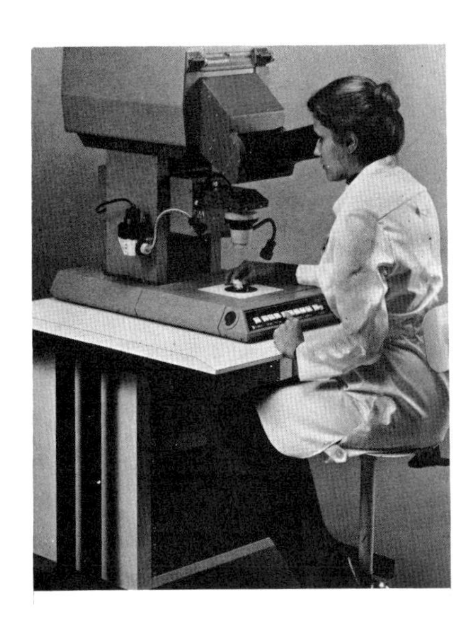

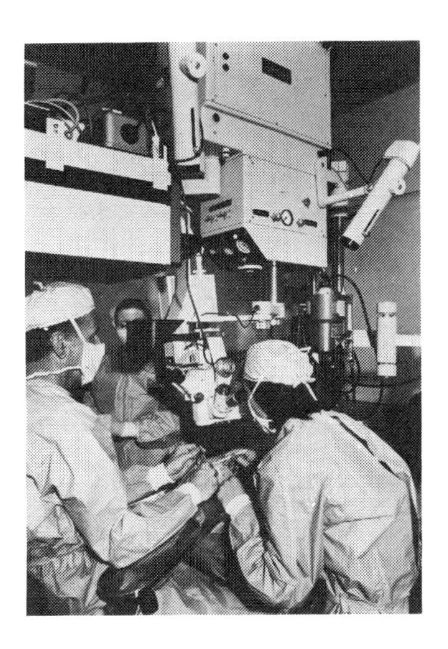

FIG. 6. Industrial Video Microscope. FIG. 7. Medical Video Microscope.

Figure 8 is a side view of the electro-optical layout of the industrial video microscope of Figure 6. The light reflected from the specimen is focused by the objective lens on to the TV tube after being reflected horizontally by a 45° mirror. The picture on the CRT of the monitor is viewed through a stereo-hood.

In the top optical schematic view of Figure 9, the two images of the specimen are picked-up from different angles by twin objective lenses. Each image is reflected by a set of mirrors to reverse them to counter a previous reversal in each of the lenses. The images are also brought in to fall side-by-side on the camera's pick-up tube. The two images appear side-by-side on the monitor's CRT. The stereo-hood has a baffle down the center to channel one image to each eye and wedge prisms to superimpose them as a single magnified three dimensional picture.

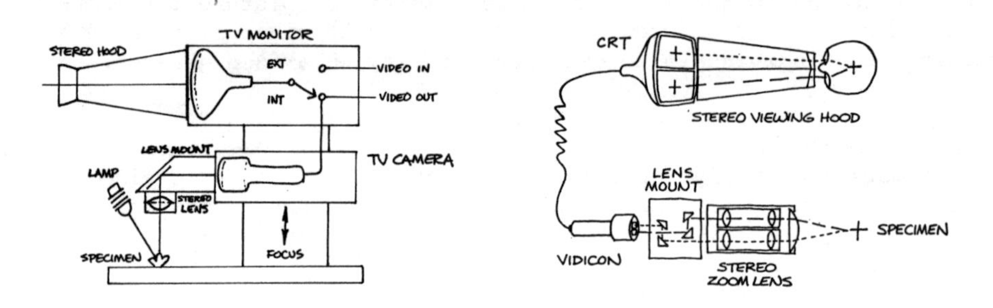

FIG. 8. Side View - Video Microscope. FIG. 9. Top View - Video Microscope.

The advantages of the video microscope are: great ease of viewing with considerable allowable head movement; group viewing by means of interconnecting several video microscopes or by use of polarized glasses with a large screen TV projection; ability to record on video tape; and capability of electronically modifying the video signal for enhancement, measurement, and comparison. The applications for the video microscope are numerous, including the fabrication and inspection of micro-electronics. Another application, illustrated in Figure 7, is for microsurgery where the doctor can perform eye, ear, or neurosurgery with great ease while viewing a 3D video picture.[11,12]

AUTOSTEREOSCOPIC DISPLAYS

Autostereoscopic (without glasses) displays have been developed and used commercially for over 40 years.[13,14,15] Lenticular pictures, such as 3D picture postcards and back-illuminated 3D transparencies are wellknown. In the 1960's, special theatres in the USSR exhibited on lenticular type screens stereo-pair motion pictures without glasses. The viewers' head movement was limited to about 65mm.[16,17] The author has recently developed the Stereorama Projector for panoramic film projection of slides and motion pictures. These are viewed autostereoscopically with almost unlimited head movement. They also provide a panoramic picture in that as the viewer moves, he can see around foreground objects.[18] The three dimensional picture can be observed from almost any position, angle or distance from the screen. Images can come out of the screen several meters and go back many meters into the screen.[19]

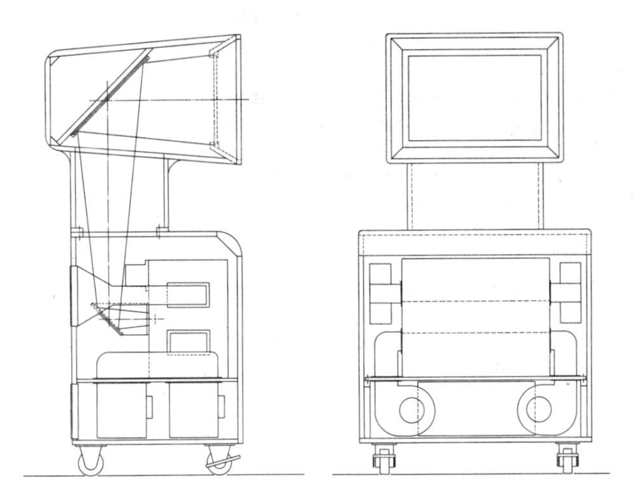

FIG. 10. Stereorama Slide Film Projector.

Figure 10 is a schematic of this self-contained projector with a capacity of 12 slides, which are presented sequentially with audio. The slides are projected from a lamp house and reflected by mirrors to back project on the lenticular-type of autostereoscopic screen. The stereoscopic picture is panoramic in that as a viewer moves his head laterally, he sees around foreground objects.

The Stereorama picture can be taken in three ways: (1) by a number of horizontally spaced cameras; (2) by a single camera with multiple horizontally spaced lenses; or (3) by a single camera moved laterally through a sequence of positions around the object of interest. The conventional 35mm film is specially mounted into Stereorama slides.

Indoor or outdoor scenes can be recorded and displayed. Scientific applications are of considerable interest. For example, a sequence of aerial or space reconnaissance photos may be mounted in a single slide and panoramically displayed autostereoscopically. Also a sequence of images representing tomographic data from X-rays, electromicrograph, seismographic, radar or sonar scans can be presented as a parallax-panoramagram. The observer views a single display which presents all of the information as an integral three dimensional picture.

CONCLUSION

Stereoscopic three dimensional displays, both with viewing aids and without, now provide a useful means of portraying information for a wide variety of applications.

REFERENCES

1. Butterfield, J.F., "A Survey of Three Dimensional Displays" Proc. Holography--a View of the Future, LA, AIAA Monograph Series Vol. 24, pp. 25-35, American Institute of Aeronautics & Astronautics, Los Angeles, January 1979.
2. Norling, J.A., "The Stereoscopic Art--A Reprint," J. of SMPTE, Vol. 60, No. 3, pp. 286-308, March 1953.
3. Report, "Fun with 3D TV Down Mexico Way," TV Guide pp. 11, Week of October 30-November 5, 1954.
4. Reports 312-2, "Constitution of a System of Stereoscopic Television" XIIIth Plenary Assembly (Geneva 1974) Vol. XI Broadcasting Services (Television) (Study Group 11) pp. 20 and 270, International Tele-Communications Union, Geneva 1975.
5. Butterfield, J.F., "Three-Dimensional Television," Proc. of 15th Annual SPIE Symp., pp. 3-9, September 1970.
6. Butterfield, J.F., "A Survey of Three-Dimensional Displays," Digest of IEEE International Convention, pp. 116-117, March 1972.
7. Butterfield, J.F., "Stereoscopic Television," OSA Meeting on Design and Visual Interface of Binocular Systems, May 1972.
8. Butterfield, J.F., "Very High Resolution Stereoscopic Television," Proc. of 21st Annual SPIE Symp., Vol. 120, Three Dimensional Imaging, pp. 208-212, August 1977.
9. Butterfield, J.F., "Video Microscope," Microscope, Vol. 26, pp. 171-181, November 1978.
10. Butterfield, J.F., "A Stereoscopic Video Microscope," Proc. 24th Annual SPIE Symp., Vol. 249, Advances in Image Transmission II, August 1980.
11. Butterfield, J.F., "A Teaching Surgical Stereo-Video Microscope," Proc. 19th Annual SPIE Symp., Vol. 66, pp. 212-214, August 1975.
12. Butterfield, J.F., "Development of a Surgical Stereo-Video Microscope," Proc. 18th Annual SPIE Symp., Vol. 54, pp. 8-11, August 1974.
13. Kaplan, S.H., "Theory of Parallax Barriers," J. of SMPTE, Vol. 59, pp. 11-21, July 1952.
14. Dudley, L.P., "Stereoptics," Macdonald & Co., Ltd., London 1951.
15. Okoshi, T., Three-Dimensional Imaging Techniques, Academic Press, 1976.
16. Valyus, N.A., Stereoscopy, Focal Press, London 1956.
17. Butterfield, J.F., "Autostereoscopic Displays Using a Stereo-Pair of Images," Proc. 22nd Annual SPIE Symp., Vol. 162, Visual Simulation and Image Realism, pp. 157-163, August 1978.
18. Butterfield, J.F., "Autosteroscopic Film Displays," Proc. of Electro-Optics/Laser '79 Conference, pp. 214-219, Oct. 25, 1979.
19. Butterfield, J.F., "Autostereoscopy Delivers What Holography Promised," Proc. 23rd Annual SPIE Symp., Vol. 199, Advances in Display Technology, pp. 42-46, August 1979.

PRODUCTION OF AGGREGATE COLOR CENTERS FOR OPTICAL MEMORIES BY ION IMPLANTATION, X-RAY AND ELECTRON IRRADIATION

C. Ortiz,* R. M. Macfarlane, R. M. Shelby and G. C. Bjorklund
IBM Research Laboratory, 5600 Cottle Road, San Jose, CA 95193

ABSTRACT

Aggregate color centers for optical memory applications have been produced by ion implantation, x-ray, γ and electron irradiation. Arbitrary spatial distributions of centers in a 6μ thin film have been obtained in e-irradiated NaF crystals.

INTRODUCTION

Photochemical hole burning provides a mechanism for substantially increasing optical storage density by using the frequency domain in addition to the spatial dimension.[1] It has recently been shown that long-lived holes can be burned at low temperatures in the inhomogeneously broadened zero phonon lines (ZPL) of color centers.[2] The important parameter is the ratio of the inhomogeneous to homogeneous linewidth. Typical numbers for this ratio are $\sim 10^3$ so that many holes can be burned at a given spatial spot.

Alkali halide crystals containing color centers have several attractive properties for these applications. Large crystals are easily obtained with good optical quality, are optically isotropic, and they can be hosts for a large number of centers covering different spectral regions. Coloring can be done by a variety of techniques with good spatial control, and we will explore this aspect in more detail here.

It has been known for some time that a systematic trend exists between the peak wavelength of the F center or aggregate center band, λ, and the lattice constant a, of the host. This is expressed as the Mollwo-Ivey relation[3]

$$\lambda = Ka^n \qquad \text{where } n \sim 1\text{-}2$$

For hole burning applications we are interested in aggregate centers because only these exhibit ZPL, which are necessary to obtain a large ratio of inhomogeneous to homogeneous linewidth. A Mollwo-Ivey plot of the wavelength of centers having prominent ZPL, that is R and N, is shown in Figure 1. The wavelengths refer to the peak of the phonon sideband not to the ZPL, which is usually no more than a few hundred Angstroms to the red. This plot was prepared from data in Reference 4 and is a useful guide in choosing suitable materials for matching to the wavelength range of various laser sources with desirable properties, e.g., GaAlAs.

*Instituto de Optica, Serrano 121, Madrid

ISSN:0094-243X/81/650233-04$1.50

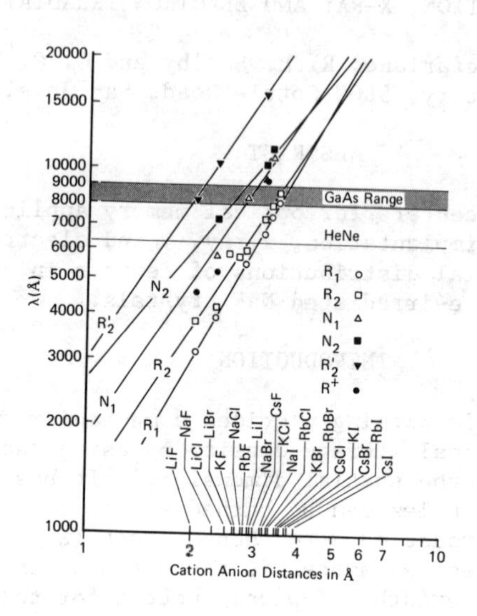

Figure 1: Mollwo-Ivey Plot for R and N Centers in Alkali Halide Crystals.[4]

Hole-burning measurements have been made on two centers in NaF: the N center (ZPL 6070Å)[5] and the F_3^+ (or R^+) center (ZPL 5456Å)[2], which are accessible to Rhodamine dye lasers. The hole-widths in these systems are ~50MHz and the inhomogeneous linewidths ~100GHz. It is know that these centers can be produced in the bulk by x-ray or γ-ray irradiation.[6] Our interest was in the production of thin films (~5μ) of colored material to enable diffraction limited focusing of the reading or writing laser beam. One approach is to fabricate films of alkali halides and use a bulk coloring agent (e.g., γ- or x-rays). In this case, control of the areal distribution by masking is possible though inconvenient. The other approach is to use a method of coloring which has a very small penetration depth. Previous work has shown that ion implantation produces F centers in alkali halides.[7] We have found that both ion implantation and electron beam irradiation produce thin layers of F_3^+ centers as well as other centers. Control of the areal distribution can be achieved by standard masking procedures in the case of ion implanted crystals or by high resolution writing with a scanning e-beam.

Ion implantation was carried out using H_2^+, D_2^+, He^+, O^+, Ne^+, B^+ and Na^+ with maximum energy of 190 KeV. The width of the zero phonon lines were somewhat broader than those in crystals colored by x-rays, γ-rays or electrons.

Electron beam coloring used a 25KeV source with a beam diameter of ~1000Å. The penetration depth P of the electrons was expected to be 6μ from the relation

$$P(\text{microns}) = \frac{0.064}{\rho\,(\text{gm/cm}^3)}\, E_0^{1.68}\ (\text{KeV})$$

A test pattern of lines was written onto a crystalline surface and observed in fluorescence following excitation with 5145Å Ar^+ laser light (see Figure 2).

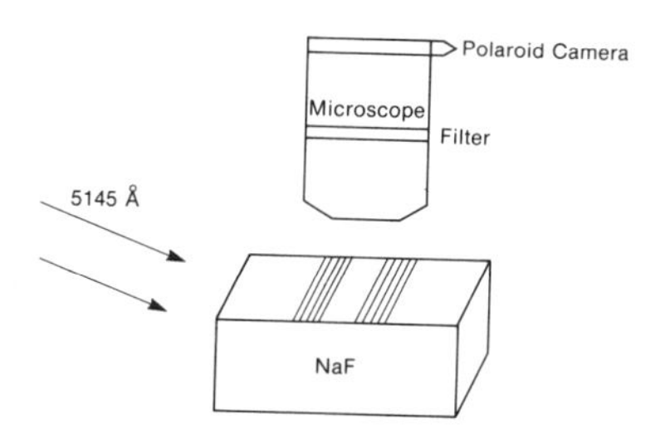

Figure 2. Room Temperature Observation of F_3^+ Pattern.

Resolution of 2μ wide lines spaced 2μ apart was clearly observed through a microscope with an objective of NA 0.25. This resolution was probably limited by the microscope. The profile of the coloring normal to the crystal surface was observed by cleaving through the test pattern and observing the fluorescence in the direction perpendicular to the cleavage plane. This showed the characteristic pear-shaped profile due to scattering of the incident electron beam.

For the thin films of color centers in NaF, the ZPL could not be observed directly in absorption. On the other hand, N centers were produced in NaCl by electron irradiation (ZPL 8373Å) and 12% absorption observed. This shows that very efficient production of these centers was being achieved and that read-out schemes based on absorption, e.g., FM spectroscopy[8] are feasible.

Figure 3. Five Line Pattern Written onto NaF Crystal (Linewidth: .3, .5, 1, 2 and 4μ).

We have shown that ion implantation and electron beam irradiation can be used to produce spatially well defined distributions of color centers. In optical memory applications the confinement of color centers to a thin layer is necessary for compatibility with the short depth of field of tightly focused reading or writing laser beams. The control of the distribution in the transverse directions is desirable for producing servo tracks or patterns to guide the location of the laser beam. Other applications are the production of distributed feedback F-center lasers and the fabrication of integrated optics elements.

REFERENCES

1. A. Szabo, U.S. Patent No. 3,896,420 (1975); G. Castro, D. Haarer, R. M. Macfarlane and H. P. Trommsdorff, U.S. Patent No. 4,101,976 (1978); D. Haarer, SPIE 177, 97 (1979).
2. R. M. Macfarlane and R. M. Shelby, Phys. Rev. Lett. 42, 788 (1979).
3. E. Mollow, Nachr. Ges. Wiss, Goettingen II Math. Physik Kl. (1931) p. 97; H. F. Ivey, Phys. Rev. 72, 341 (1947).
4. J. H. Schulman and W. D. Compton, Color Centers in Solids, Pergamon Press (Oxford, 1962); K. Amenu-Kpodo and T. J. Neubert, J. Phys. Chem. Sol. 26, 1615 (1965); C. B. Pierce, Phys. Rev. A83, 135 (1964); K. Konrad and T. J. Neubert, J. Chem. Phys. 47, 4946 (1967); I. Schneider and H. Rabin, Phys. Rev. A1983, 140 (1965); C. J. Delbecq and P. Pringsheim, J. Chem. Phys. 21, 794 (1953); F. Lanzl and W. Von der Osten, Phys. Lett. 15, 208 (1965); D. B. Fitchen, R. H. Silsbee, T. A. Fulton and E. L. Wolf, Phys. Rev. Lett. 11, 275 (1963); A. E. Hughes, Proc. Phys. Soc. 87, 535 (1966).
5. M. D. Levenson, R. M. Macfarlane and R. M. Shelby, Phys. Rev. to be published.
6. F. Seitz, Rev. Mod. Phys. 26, 7 (1954); F. Agullo-Lopez, Phys. Stat. Sol. 22, 483 (1967).
7. D. Pooley, Brit. J. Appl. Phys. 17, 855 (1966); M. Saidoh and P. D. Townsend, J. Phys. C 10, 1541 (1977).
8. G. C. Bjorklund, Opt. Lett. 5, 15 (1980).

ACOUSTO-OPTIC MODULATION AT gHz FREQUENCIES USING INTEGRATED OPTIC TECHNOLOGY*

G. B. Brandt, M. Gottlieb, R. W. Weinert, J. J. Conroy
Westinghouse R&D Center, Pittsburgh, PA 15235

ABSTRACT

Phase modulation of light guided in a thin film optical waveguide is produced by bulk acoustic waves propagating through the waveguide perpendicular to the optical wave. The principal advantages of this interaction for high frequency modulation are: (1) A wide choice of waveguide materials may be used and (2) Acoustic propagation takes place through a thin waveguide, thus attenuation does not limit performance until very high frequencies are reached. We describe the acoustic effect then show how it may be used to construct an integrated optic device which converts the phase modulation to amplitude modulation. This device integrates an interferometer structure with the thin film acoustic transducer on a silicon substrate. We have built the device and tested it.

INTRODUCTION

Modulation and detection of light at gHz frequencies are difficult problems, the solutions to which are sought as optical data communications links are operated at increasing bandwidths. One approach to modulation is the exploitation of the bulk acousto-optic effect operating on light which is guided in thin film waveguides under the acoustic transducer.[1,3] By confining the light in a thin waveguide (whose dimensions typically are on the order of one micron) passing under the transducer, acoustic losses across the optical wavefront (one limitation to the operating frequency of acoustic modulators) become negligible until very high frequencies--on the order of tens of gHz--are reached. An additional benefit of this approach results from the fact that all materials show an acoustic-optic effect to some degree, thus, no special material need be used to form the confining waveguide structure.

The bulk acoustic wave produces phase modulation of the light guided under the transducer; in order to detect this modulation some sort of interferometer is necessary. We choose to use an interferometer in which a portion of the beam is split by etched V grooves in the silicon substrate.[4] Four of these grooves form the integrated optic equivalent of a MachZehnder interferometer. When an acoustic transducer is placed in one leg of this interferometer, the output will be amplitude modulated at the drive frequency of the transducer. Figure 1 shows this arrangement.

*This work supported in part by Naval Underwater Systems Center, New London, CT 06320; Contract #N00140-79-C-6282.

ISSN:0094-243X/81/650237-07$1.50

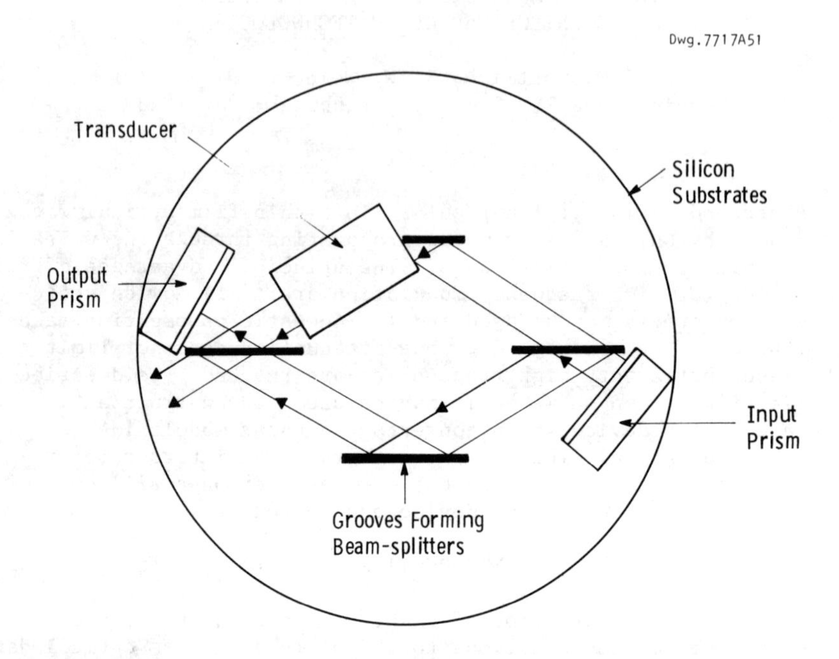

Figure 1. Schematic of integrated optic modulator using bulk transducers and V groove interferometer structure.

The acoustic transducer in this device is designed with segmented electrodes in order to increase the impedance to 50 ohms. Transducers operating at 1.3 and 2.3 gHz were fabricated using thin film deposition techniques. In the following sections we will discuss the principle of operation of such modulators, the technology needed to fabricate the interferometer structure and transducer, and the results of our optical and electrical tests.

PRINCIPLE OF OPERATION

In an earlier publication[5] we showed that bulk acoustic modulation could be detected using an interferometer fabricated from conventional optical components. Figure 1 shows our integrated optical analog to such an interferometer. Here the mirrors and beamsplitters are replaced by V grooves formed by preferential etching of [111] planes in [100] oriented silicon. Tsang and Wang[4] showed that such structures, depending upon their width, can act as partial or totally reflecting mirrors for light guided in a thin film deposited over them. An attractive feature of this approach results from the fact that the preferential etching process automatically aligns the long (1 cm), narrow (10 μm) grooves along the [110] directions thus producing a perfectly aligned interferometer independant of mask errors.

In Figure 2 is shown a schematic of the transducer structure and its relationship to the waveguide and optical buffer layers.

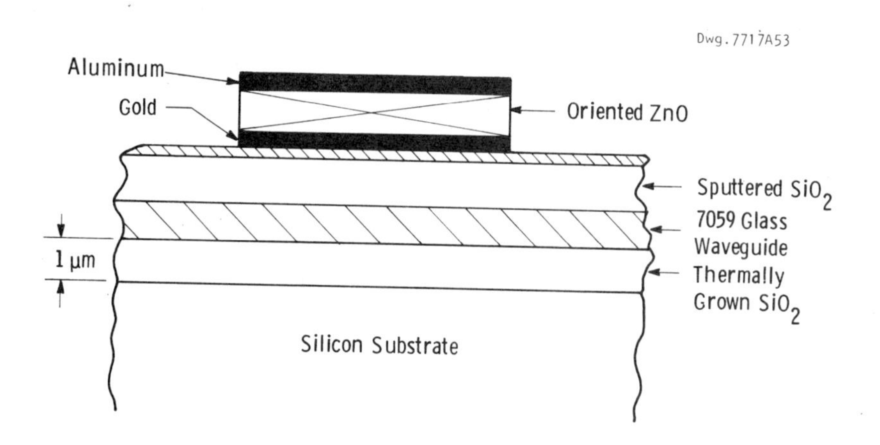

Figure 2. Cross section of transducer structure.

Sound produced by the oriented ZnO propagates through the sputtered SiO_2 layer used to optically decouple the Corning 7059 glass waveguide from the electrode structure. As the sound passes through the waveguide it induces an optical phase shift in the light trapped within the guide. This phase shift, $\Delta\phi$, is given by

$$\Delta\phi = \pi\frac{L}{\lambda}\,(2M_2\;P_A)^{1/2} \tag{1}$$

where L is the transducer length, λ is the light wavelength, P_A is the acoustic power density, and M_2 in turn is related to the material constants of the waveguide through the relation

$$M_2 = \frac{n^6p^2}{\rho v^3} \tag{2}$$

where n is the refractive index, p the photoelastic coefficient, ρ the density, and v the acoustic velocity of the waveguide. Equation 1 represents the peak phase disturbance induced in the waveguide at the acoustic frequency, f. This, in turn, gives rise to temporal side-bands on the optical beam. If the beam has frequency ν then for small excursions of $\Delta\phi$, the modulated light consists of the unshifted portion and two sidebands at $\nu \pm f$. The intensity of these sidebands is proportional to the first order Bessel function $J_1 (\Delta\phi)^2$. The analysis which leads to this result is that used to predict temporal spectra in rf phase modulation; it is formally identical to the analysis used to predict amplitudes of diffraction orders produced by a phase grating.[6]

By combining the modulated light with an unmodulated beam from the interferometer leg which does not contain the transducer, amplitude modulated light is obtained. A physical argument which supports this result is as follows. Assume the interferometer has equal length legs. In this case, since the optical paths are identical, the two beams, in the absence of the sound interfere constructively. Application of the sound field shifts the optical phase of one leg relative to the other at the rf frequency resulting in an amplitude modulation of the output.

The analysis shows that the device shown in Figure 1 can be used as a modulator. With the same arrangement, signals at frequencies other than the transducer frequency can be produced on a detector by heterodyning on the detector the AM signal from this device and another signal present on the light beam. In this way, for example, a 3 gHz signal on the light beam can be detected by mixing a 2.9 gHz modulation produced acoustically in the modulator. Thus we see that the acoustic modulator can be used both to produce modulation and to detect high frequency modulation on an optical signal.

ACOUSTIC TRANSDUCER DESIGN

Design of high frequency (1-2.5 gHz), broadband piezoelectric transducers involves several considerations, both regarding electrical response of the transducer and acoustic response of the multi-layered structures inherent in our device. For a half-wave piezoelectric plate to be resonant between 2 and 3 gHz, its thickness must be less than 1 micron, and the thickness of a well loaded broadband transducer must less than 1/3 this value. These very thin layers can only be obtained by depositing thin films of the piezoelectric onto the substrate. We use ZnO, a strong piezoelectric with hexagonal symmetry, sputtered so that the "c" axis has a high degree of orientation normal to the plane of the film. This high degree of orientation is necessary for efficient conversion of rf power to acoustic power.

Since the transducer layers are thin, on the order of thickness of the other electrode, waveguide and buffering layers, the acoustic response of the transducer is a function of the number, thickness, and acoustic impedances of these various layers. Analysis of this response can be accomplished by considering the various layers as a series of coupled acoustic delay lines. Electrical impedance is another factor in transducer design. Transducer dimensions were chosen based on optical considerations, to be 3 mm wide and 10 mm long. A single transducer of these dimensions operating at 2.5 gHz has an unreasonably low impedance, namely .005-i .035 ohms. Since this impedance is impractically low, the transducer cross sectional area must be reduced without decreasing the overall dimensions. This is accomplished by fabricating the transducer as a number of series connected segments. Each segment of this mosaic has a higher impedance by virtue of its small cross sectional area, and the overall impedance level is raised even further by the series connection. We use a computer program to balance the various acoustical and electrical characteristics of the transducer and have produced a design at 2.3 gHz for which the impedance is 125-i101 ohms. This design and a similar design for center frequency of 1.3 gHz were fabricated for further experiments.

FABRICATION

We chose silicon as a substrate for several reasons. First we have considerable experience fabricating good waveguides on thermally grown SiO_2 low index buffer layers. The second attraction is the possibility of etching the interferometer structure along crystallographic planes to insure alignment. Finally, silicon offers the promise of integrating electronic functions at a later date using existing technology.

Starting with a (100) oriented silicon wafer, standard semiconductor masking and etching techniques are used to define the openings in the SiO_2 through which the V grooves will be etched. A KOH etch is used to preferentially etch the exposed (111) planes, after which the SiO_2 mask is removed. Then an SiO_2 layer is thermally grown on the wafer and the grooves; next the waveguide layer is sputtered followed by a sputtered SiO_2 layer to optically isolate the waveguide from the transducer electrodes. Finally in a series of masked sputtering and evaporation steps, the transducer and its electrodes are applied. Final assembly involves mounting the 2 inch wafer in a holder, connecting electrodes to the transducer and attaching the input and output prisms. Figure 3 shows a photo of the completed assembly.

Figure 3. Completed modulator assembly.

TESTING

Electrical tests of the completed 1.3 gHz transducer performed both at Westinghouse and at Naval Underwater Systems Center showed the low standing wave ration expected from the measured impedance which ranged between 40 and 60 ohms. In addition, transducer operation was observed from 0.9 to 1.5 gHz bandwidth, somewhat broader than that predicted by the computer design program. The lower impedance and broader bandwidth we observed result from attenuation and acoustic coupling effects which decrease the Q of the structure over that assumed in the calculations. The 2.3 gHz transducer was also tested and, while it showed activity, the practical difficulties of making measurements at that frequency made it impossible to make the same comparisons as were made at 1.3 gHz.

During the process of fabrication various optical tests were made. Waveguide properties of low scatter were important and unfortunately in some of the fabrication steps, waveguide quality was compromised. One interesting result was the strong effect of incident angle and reflectivity observed on the V groove beamsplitters. Near grazing incidence, the groove becomes more reflecting; the effect is strongly dependant upon angle; the transmission can range from 0 to 50 percent with a 5 degree change in incidence angle. This factor presented us with problems since the groove geometry was fixed by our initial masking. As a result, our completed device showed the desired reflectivity and an angle that was too large and the guided beams recombined at a point beyond the edge of the final beamsplitter. This fact prevented us from demonstrating the modulation of the light beam with the acoustic transducer.

CONCLUSIONS

We have designed and fabricated an acousto-optic modulator combined with an integrated optic interferometer which should amplitude modulate guided light at 2.3 gHz with a bandwidth of several hundred mHz. Even though we have not yet demonstrated modulation with the integrated optical interferometer our previous experiments in which modulation was observed with a laboratory inerferometer, combined with our experiences in fabricating the integrated optical version give us no fundamental reason to believe that such an approach should not work. We have uncovered some interesting problems, in particular, we have no good theoretical explanation for the behavior of the V groove beamsplitters with incident angle. Because of the dimensions involved, the answer to this behavior lies somewhere between the domains of geometrical and diffractive optics. We have also observed that the problems of integrating a number of components into a working device are much more difficult than the problems associated with making the components work in the first place.

REFERENCES

1. G. B. Brandt, M. Gottlieb and R. W. Weinert, "Gigahertz Modulators Using Bulk Acousto-Optic Interactions in Thin Film Waveguides", Fiber and Integrated Optics 1, 417-430 (1978).
2. G. B. Brandt, M. Gottlieb and J. J. Conroy, "Bulk Acoustic Wave Interaction with Guided Optical Waves", Appl. Phys. Lett. 23, 53-54 (1973).
3. M. L. Shah, "Fast Acousto-Optic Waveguide Modulators", Appl. Phys. Lett. 23, 75-77 (1973).
4. W. T. Tsang and S. Wang, "Thin-Film Beam Splitter and Reflector for Optical Guided Waves", Appl. Phys. Lett. 588-590 (1975).
5. M. Gottlieb, G. B. Brandt and R. W. Weinert, "Integrated Optic Waveguide Modulation at gHz Frequencies", Proc. Electro-optics Conf., NY (1976).
6. H. M. Smith "Principles of Holography" Wilay, NY (1969); p 49 ff.

ACOUSTIC HOLOGRAPHIC IMAGE STORAGE AND ITS COHERENT OPTICAL RECONSTRUCTION

Hua-Kuang Liu
The University of Alabama, University, AL. 35486

ABSTRACT

A theory of a thermoplastic device for real-time in-situ acoustical holographic recording is presented. The theoretical analysis takes into account the possible existence of an in-line reference beam and its influence on the real-time coherent optical image reconstruction.

Deformable thermoplastic materials have been used with considerable success in in-situ optical holographic recordings.[1-22] Advantages of a device made of this material include the fact that the device has the capability of near real-time holographic reconstruction, it can be recycled, wet processed, and does not require dark-room handling. Resolution in excess of 4000 cycles per millimeter can be achieved. Nevertheless, it appears that the potential of this material for the recording of acoustical holograms has received very little attention following the original suggestion by Young and Wolfe.[23] In this paper, a detailed theoretical analysis of a thermoplastic acoustical holographic recording devices will be presented.

The analysis of the holographic recording is divided into two parts.

The first contains a description of the interaction of the ultrasound with the liquefied thermoplastic surface. The second part deals with the principle of image reconstruction by laser light. The analysis is similar to the one adopted by Brenden[24], but is different from his in that the existence of an auxiliary reference wave, which originated from the object beam transducer, is also considered. This additional reference beam originates in some cases as a result of the geometry of the test objects. Our assumption will make the mathematical analysis more tedious. This is necessary, on the other hand for a complete interpretation of the experimental results.

In the second part, the process of coherent optical image reconstruction of the acoustical hologram is described.

A. Interaction of Ultrasound With the Liquefied Thermoplastic Surface

In the acoustical holographic system shown in Figure 1, a reference beam of amplitude A_r, generated by the reference transducer, is incident upon the surface (z=0) of the thermoplastic layer at an angle Θ_r, while the object beam with amplitude A_{ob} is incident at an angle Φ_0 with respect to the perpendicular to the surface and carries an object-dependent phase $\phi(x,y)$. If the object is simple, such as a

ISSN:0094-243X/81/650244-14$1.50

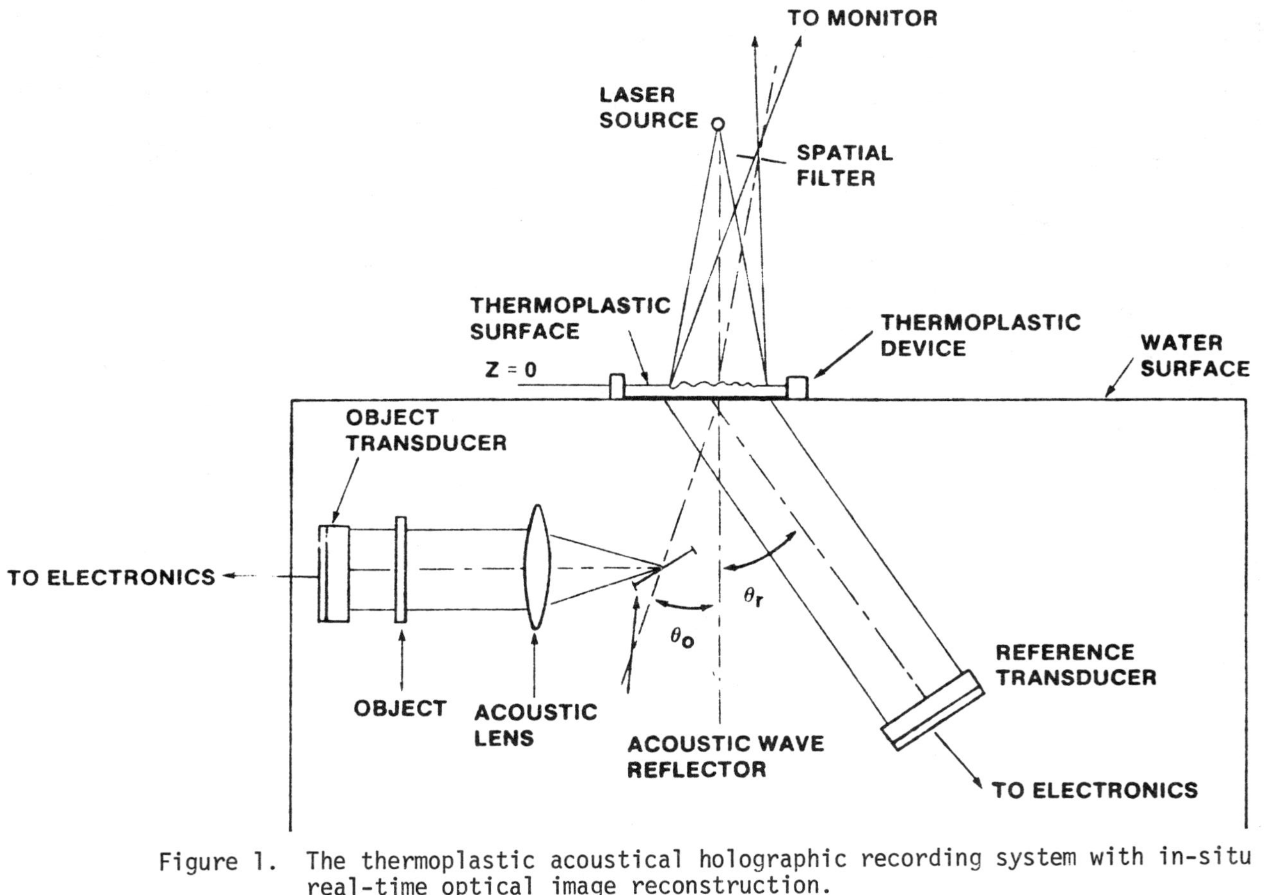

Figure 1. The thermoplastic acoustical holographic recording system with in-situ real-time optical image reconstruction.

coarse grating structure, part of the beam generated by the object transducer may propagate through the object plane without being modulated by the object. The portion of the beam with amplitude A_o, and incidence angle Θ_o, may be considered as a reference beam for the hologram. These waves are listed below:

$$\text{Reference beam No. 1:} \quad R_1 = A_r \exp(j\, k_r\, y) \quad , \tag{1}$$

$$\text{Object beam:} \quad 0 = A_{ob} \exp[-j(k_o y + \phi(x,y))] \ , \tag{2}$$

$$\text{Reference beam No. 2:} \quad R_2 = A_o \exp(-j\, k_o\, y) \ , \tag{3}$$

where

$$k_r = \frac{2\pi}{\lambda} \sin\Theta_r, \text{ and } k_o = \frac{2\pi}{\lambda} \sin\Theta_o \ , \tag{4}$$

and λ is the wavelength of the acoustic wave.

The simultaneous presence of these waves, and the change of momentum on reflection at the surface produces a radiation pressure given by

$$p(x,y) = |0 + R_1 + R_2|^2 \ / \ (\rho\, v_s^{\,2}) \ , \tag{5}$$

where ρ is the density of the thermoplastic medium and v_s is the velocity of sound.

The radiation pressure is counterbalanced by gravity, ρgZ, and surface tension

$$p_t = - c_s\left(\frac{\partial^2 z}{\partial x^2} + \frac{\partial^2 z}{\partial y^2}\right) \ , \tag{6}$$

where c_s is the surface tension coefficient. For a unit area, the balance of force equation is

$$p(x,y) = \rho g z - c_s\left(\frac{\partial^2 z}{\partial x^2} + \frac{\partial^2 z}{\partial y^2}\right) \ . \tag{7}$$

Solution of the above equation for z, yields the displacement of the thermoplastic surface.

Assuming a steady-state solution, as shown in Eq. (B1) of Appendix B, and applying the results obtained from both Appendix A and Appendix B, Eq. (7) can be rewritten as

$$\frac{1}{\rho v_s^{\,2}} \{ A_r^{\,2} + A_{ob}^{\,2} + A_o^{\,2} + 2A_r A_{ob} \cos[(k_r + k_o)y + \phi]$$

$$+ 2A_r A_o \cos[(k_r + k_o)y] + 2A_o A_{ob} \cos\phi \}$$

$$= 2\rho g \{ \alpha \cos[(k_r + k_o)y + \phi] + \beta \cos[(k_r + k_o)y] + \gamma \cos\phi + \tfrac{1}{2}\kappa \}$$

$$
\begin{aligned}
&- 2C_s \{[(\frac{\partial^2\alpha}{\partial x^2} + \frac{\partial^2\alpha}{\partial y^2}) - \alpha [(\frac{\partial\phi}{\partial x})^2 \\
&+ (\frac{\partial\phi}{\partial y} + k_r + k_o)^2]] \cos[(k_r + k_o)y + \phi] + [(\frac{\partial^2\beta}{\partial x^2} + \frac{\partial^2\beta}{\partial y^2}) \\
&- \beta(k_r + k_o)^2] \cos[(k_r + k_o)y] + [(\frac{\partial^2\gamma}{\partial x^2} + \frac{\partial^2\gamma}{\partial y^2}) - \gamma[(\frac{\partial\phi}{\partial x})^2 \\
&+ (\frac{\partial\phi}{\partial y})^2]] \cos \phi - [2([\frac{\partial\alpha}{\partial x}\frac{\partial\phi}{\partial x} + \frac{\partial\alpha}{\partial y}\frac{\partial\phi}{\partial y} + \frac{\partial\alpha}{\partial y} (k_r + k_o)] \\
&+ \alpha[(\frac{\partial^2\phi}{\partial x^2}) + \frac{\partial^2\phi}{\partial y^2}]) \sin [(k_r + k_o)y + \phi] \\
&- 4 \frac{\partial\beta}{\partial y}(k_r + k_o) \sin [(k_r + k_o)y] \\
&- 2 [2(\frac{\partial\gamma}{\partial x} \frac{\partial\phi}{\partial x} + \frac{\partial\gamma}{\partial y} \frac{\partial\phi}{\partial y}) + \gamma(\frac{\partial^2\phi}{\partial x^2} + \frac{\partial^2\phi}{\partial y^2})] \sin \phi + \frac{\partial^2\kappa}{\partial y^2}\} .
\end{aligned}
\tag{8}
$$

Equation (8) is very complicated. However, if it is assumed that the following conditions are satisfied;

$$
\begin{aligned}
&(k_r + k_o) >> \frac{\partial\phi}{\partial y} , \frac{\partial\phi}{\partial x} , \\
&\alpha(k_r + k_o)^2 >> \frac{\partial^2\alpha}{\partial x^2} + \frac{\partial^2\alpha}{\partial y^2} , \\
&\beta(k_r + k_o)^2 >> \frac{\partial^2\beta}{\partial x^2} + \frac{\partial^2\beta}{\partial y^2} , \\
&\frac{\partial\alpha}{\partial x} \simeq 0 , \\
&\frac{\partial\alpha}{\partial y} \simeq 0 , \\
&\frac{\partial\beta}{\partial y} \simeq 0 ,
\end{aligned}
\tag{9}
$$

$$\frac{\partial \gamma}{\partial x} \simeq 0 ,$$

$$\frac{\partial \gamma}{\partial y} \simeq 0 ,$$

$$\frac{\partial^2 \phi}{\partial x^2} + \frac{\partial^2 \phi}{\partial y^2} \simeq 0 ,$$

$$\rho g \kappa >> c_s \left(\frac{\partial^2 \kappa}{\partial x^2} + \frac{\partial^2 \kappa}{\partial y^2}\right) ,$$

then the right-hand side of Eq. (8) can be put into the form

$$\begin{aligned} &\frac{1}{\rho v_s^2} \{A_r^2 + A_{ob}^2 + A_o^2 + 2A_r A_{ob} \cos[(k_r + k_o)y + \phi] \\ &\quad + 2A_r A_o \cos[(k_r + k_o)y] + 2A_o A_{ob} \cos \phi \} \\ &= 2\alpha [\rho g + c_s(k_r + k_o)^2] \cos[(k_r + k_o)y + \phi] \\ &\quad + 2\beta[\rho g + C_s(k_r + k_o)^2] \cos[(k_r + k_o)y] \\ &\quad + 2\gamma \rho g \cos \phi + \rho g \kappa . \end{aligned} \tag{10}$$

From the above equation, one finds

$$\alpha = \frac{A_r A_{ob}}{\rho v_s^2 [\rho g + c_s(k_r + k_o)^2]} , \tag{11}$$

$$\beta = \frac{A_r A_o}{\rho v_s^2 [\rho g + c_s(k_r + k_o)^2]} , \tag{12}$$

$$\gamma = A_o A_{ob}/(\rho^2 g v_s^2) , \tag{13}$$

$$\kappa = [A_r^2 + A_{ob}^2 + A_o^2]/(\rho^2 g v_s^2) . \tag{14}$$

The solution in the form of Eq. (B1) is rewritten as follows:

$$z_s = 2\alpha \cos [(k_r + k_o)y + \phi]$$
$$+ 2\beta \cos [(k_r + k_o)y]$$
$$+ 2\gamma \cos \phi + \kappa \qquad (15)$$

The physical meaning of Eq. (15) is described below. There is a bulge of height $2\gamma \cos\phi(x,y) + \kappa$ in the region where the acoustical wave is applied. Impressed upon this bulge are two interference patterns with the same spatial wave number, $k_r + k_o$, but different amplitudes 2α and 2β. One of the patterns is different from the other by a phase $\phi(x,y)$. If the amplitude of the reference beam Number 1 is set equal to zero, then from Eqs. (11) and (12) we find $\alpha=\beta=0$ and $\kappa = (A_{ob}^2 + A_o^2)/(\rho^2 g v_s^2)$. The bulge is slightly reduced in height, and the interference patterns still exist. The object geometry still modifies the liquefied surface, in a similar way as in the case of an in-line hologram. On the other hand, if the amplitude of the reference beam Number 2 is zero, Eq. (15) becomes

$$z_s = 2\alpha \cos [(k_r + k_o)y + \kappa' \quad , \qquad (16)$$

where

$$\kappa' = (A_r^2 + A_{ob}^2)/(\rho^2 g v_s^2) \quad . \qquad (17)$$

Eq. (16) represents the interference pattern resulting from a typical off-axis hologram.

B. Optical Reconstruction of the Acoustical Surface Hologram

The detection of the acoustical hologram can be achieved by the technique of coherent optical image reconstruction. Assume that a laser beam plan wave of amplitude A_ℓ and wavelength Λ is normally incident (oblique incidence is also possible with additional complication in the mathematical manipulations) on the surface where the acoustical hologram is recorded. The light beam is expressed by

$$s(z) = A_\ell e^{-j\frac{2\pi}{\Lambda} z} = A_\ell e^{-jk_\ell z} \quad , \qquad (18)$$

where $k_\ell = \frac{2\pi}{\Lambda}$ is the wave number of the light wave.

After reflection from the surface hologram, the amplitude s(z) is reduced by a factor R and the phase of the beam is modified by the added term 2 $z_s(x,y)$, where $z_s(x,y)$ is in general given by Eq. (15). The reflected light beam may then be written as

$$S_r(x,y) = R\, e^{jk_\ell z}\, e^{j2k_\ell z_s}$$

$$+ R\, e^{-jk_\ell z} \exp[2jk_\ell\{2\alpha \cos[(k_r + k_o)y + \phi]$$

$$+ 2\beta \cos[(k_r + k_o)y]$$

$$+ 2\gamma \cos\phi + \kappa\}] \quad . \tag{19}$$

Since

$$\exp[j\, a \cos b] = \sum_{n=-\infty}^{\infty} j^n\, J_n(a) \exp(-jnb) \quad , \tag{20}$$

where J_n is the nth order Bessel function, Eq. (19) can be expressed in terms of the Bessel functions as

$$S_r(x,y) = R\, e^{jk_\ell(z + 2\kappa)}$$

$$\cdot\{ \sum_{n=-\infty}^{\infty} j^n\, J_n(4\alpha k_\ell) e^{-j[n(k_r + k_o)y + n\phi]}\}$$

$$\cdot\{ \sum_{n=-\infty}^{\infty} j^n\, J_n(4\beta k_\ell) e^{-j[n(k_r + k_o)y]}\}$$

$$\cdot\{ \sum_{n=-\infty}^{\infty} j^n\, J_n(4\gamma k_\ell) e^{-jn\phi}\} \quad . \tag{21}$$

The property of $J_n(x)$ indicates that when $\underline{a}$ is sufficiently small, $J_0(a)$ approaches 1, $J_{+1}(a)$ approaches $\frac{a}{2}$, $J_{-1}(a)$ approaches $-\frac{a}{2}$, and $J_n(a)$ for n not equal to -1, 0, or 1 approaches 0. Therefore if 4 αk_ℓ, 4 βk_ℓ, and 4 γk_ℓ are all sufficiently small, $S_r(x,y)$ can be approximately given by

$$S_r(x,y) = R\, e^{jk_\ell(z + 2\kappa)}$$

$$\cdot \{1 + j2\alpha k_\ell e^{-j[(k_r + k_o)y + \phi]}$$

$$+ j2\alpha k_\ell e^{j[(k_r + k_o)y + \phi]}\}$$

$$\cdot \{1 + j2\beta k_\ell e^{-j[(k_r + k_o)y + \phi]}$$

$$+ j2\beta k_\ell e^{j[(k_r + k_o)y + \phi]}\}$$

$$\cdot \{1 + j2\gamma k_\ell e^{-j\phi} + j2\gamma k_\ell e^{-j\phi}\} \quad . \tag{22}$$

Expansion of Eq. (22) yields

$$S_r(x,y) = R\, e^{jk_\ell(z + 2\kappa)}$$

$$\{1 + j\; 2[(\alpha k_\ell + \beta k_\ell)e^{-j(k_r + k_o)y} + \gamma k_\ell]e^{-j\phi}$$

$$+ j\; 2[(\alpha k_\ell + \beta k_\ell)e^{j(k_r + k_o)y} + \gamma k_\ell]e^{j\phi}$$

$$- \;[\![16\; \gamma\beta k_\ell^{\;2} \cos\phi \cos[(k_r + k_o)y + \phi]$$

$$+ 16\; \alpha\beta k_\ell^{\;2} \cos^2[(k_r + k_o)y + \phi]$$

$$+ 16\; \alpha\gamma k_\ell^{\;2} \cos\phi \cos[(k_r + k_o)y + \phi]]\!] \}$$

$$\simeq R\, e^{jk_\ell(z + 2\kappa)} \{1 + j\; 2[(\alpha k_\ell + \beta k_\ell)e^{-j(k_r + k_o)y} + \gamma k_\ell]e^{-j\phi}$$

$$+ j\; 2\,[(\alpha k_\ell + \beta k_\ell)e^{j(k_r + k_o)y} + \gamma k_\ell]e^{j\phi}\} \quad . \tag{23}$$

The reason why the terms within the $[\![\quad]\!]$ brackets can be omitted is not only because they are negligibly small as compared to unity, but also because their phases are of no importance for the reconstruction process. What is of importance for the optical image reconstruction is the positive first order diffraction term, or the term that contains $e^{-j\phi}$. The phase term is identical to that of the object beam given by Eq. (2), and the amplitude A_{ob} of the original object beam is also contained in α and γ. The following two special cases are considered:

(1)

$A_o = 0$, so that $\beta = 0$, $\gamma = 0$,

$$S'_{r_1}(x,y) = j2\,R\,e^{jk_\ell(z + 2\kappa)}\,[\alpha k_\ell\, e^{-j(k_r + k_o)y}]e^{-j\phi}$$

$$= S_{r_1}(x,y) \quad . \tag{24}$$

A spatial filter can be placed in the Fourier plane and the term $Sr_1(x,y)$ can be isolated for the holographic image reconstruction.

(2)

$A_r = 0$ but $A_o \neq 0$, this implies $\alpha = 0$ and $\beta = 0$, hence

$$S_r(x,y) = j2R\gamma k_\ell\, e^{jk_\ell(z + 2\kappa)}\, e^{-j\phi} \quad . \tag{25}$$

The amplitude and phase of the object beam is still contained in $S_r(x,y)$.

Therefore the image of the in-line acoustical hologram can also be optically reconstructed by the proper spatial filtering process.

APPENDIX A

DERIVATION OF EQUATION (10)

$$\rho(x,y) = \frac{1}{\rho v_s^2}\left| A_r \exp jk_r y + A_{ob} \exp[-j(k_o y + \phi(x,y)] + A_o \exp(-j\,k_o y)\right|^2$$

$$= \frac{1}{\rho v_s^2}\Big| [A_r \cos k_r y + A_{ob} \cos(k_o y + \phi(x,y))$$

$$+ A_o \cos k_o y]$$

$$+ j\, [A_r \sin k_r y - A_{ob} \sin[k_o y + \phi(x,y)]$$

$$- A_o \sin k_o y]\Big|^2$$

$$= \frac{1}{\rho v_s^2} \Big\{ [A_r \cos k_r y + A_{ob} \cos(k_o y + \phi(x,y))$$

$$+ A_o \cos k_o y]^2$$

$$+ [A_r \sin k_r y - A_{ob} \sin [k_o y + \phi(x,y)]$$

$$- A_o \sin k_o y\,]^2 \Big\}$$

$$= \frac{1}{\rho v_s^2} \Big\{ A_r^2 + A_{ob}^2 + A_o^2$$

$$+ 2\, A_r\, A_{ob} [\cos k_r\, y \cos[k_o y + \phi(x,y)]$$

$$- \sin k_r y \sin[k_o y + \phi(x,y)]$$

$$+ 2\, A_r\, A_o [\cos k_r y \cos k_o y - \sin k_r y \sin k_o y]$$

$$+ 2\, A_o\, A_{ob}\, [\cos k_o y \cos[k_o y + \phi(x,y)]$$

$$+ \sin k_o y \sin[k_o y + \phi(x,y)]\Big\}$$

$$= \frac{1}{\rho v_s^2} \Big\{ A_r^2 + A_{ob}^2 + A_o^2 + 2\, A_r\, A_{ob} \cos[(k_r + k_o)y + \phi(x,y)] + 2\, A_r\, A_o \cos(k_r + k_o)y + 2\, A_o\, A_{ob} \cos \phi(x,y) \Big\} \quad (10)$$

APPENDIX B

Let α, β, γ, and κ be functions of x and y. Assume a steady state solution of Eq. (7) to be of the form

$$z_s\,(x,y) = 2\, \alpha \cos[(k_r + k_o)y + \phi(x,y)] + 2\, \beta \cos[(k_r + k_o)y] + 2\, \gamma \cos \phi(x,y) + k\ , \quad (B1)$$

then,

$$\frac{\partial z_s}{\partial x} = 2\,\frac{\partial \alpha}{\partial x}\cos[(k_r + k_o)y + \phi(x,y)]$$

$$- 2\,\alpha\,\frac{\partial \phi}{\partial x}\sin[(k_r + k_o)y + \phi(x,y)]$$

$$+ 2\,\frac{\partial \beta}{\partial x}\cos[(k_r + k_o)y]$$

$$+ 2\,\frac{\partial \gamma}{\partial x}\cos\,\phi(x,y)$$

$$- 2\,\gamma\,\frac{\partial \phi}{\partial x}\sin\,\phi(x,y) + \frac{\partial k}{\partial x} \quad , \tag{B2}$$

$$\frac{\partial^2 z_s}{\partial x^2} = 2\,\frac{\partial^2 \alpha}{\partial x^2}\cos[(k_r + k_o)y + \phi(x,y)]$$

$$- 4\,\frac{\partial \alpha}{\partial x}\,\frac{\partial \phi}{\partial x}\sin[(k_r + k_o)y + \phi(x,y)]$$

$$- 2\,\alpha\left(\frac{\partial \phi}{\partial x}\right)^2\cos[(k_r + k_o)y + \phi(x,y)]$$

$$- 2\,\alpha\,\frac{\partial^2 \phi}{\partial x^2}\sin[(k_r + k_o)y + \phi(x,y)]$$

$$+ 2\,\frac{\partial^2 \beta}{\partial x^2}\cos[(k_r + k_o)y] + 2\,\frac{\partial^2 \gamma}{\partial x^2}\cos\,\phi(x,y)$$

$$- 4\,\frac{\partial \gamma}{\partial x}\,\frac{\partial \phi}{\partial x}\sin\,\phi(x,y) - 2\,\gamma\left(\frac{\partial \phi}{\partial x}\right)^2\cos\,\phi(x,y)$$

$$- 2\,\gamma\,\frac{\partial^2 \phi}{\partial x^2}\sin\,\phi(x,y) + \frac{\partial^2 k}{\partial x^2} \quad , \tag{B3}$$

$$\frac{\partial z_s}{\partial y} = 2\,\frac{\partial \alpha}{\partial y}\cos[(k_r + k_o)y + \phi(x,y)]$$

$$- 2\alpha\,[\,\frac{\partial \phi}{\partial y} + (k_r + k_o)]\,\sin[(k_r + k_o)y + \phi(x,y)]$$

$$+ 2\,\frac{\partial \beta}{\partial y}\cos[(k_r + k_o)y]$$

$$- 2\,\beta\,(k_r + k_o)\,\sin[(k_r + k_o)y]$$

$$+ 2 \frac{\partial \gamma}{\partial y} \cos \phi \ (x,y)$$

$$- 2 \gamma \frac{\partial \phi}{\partial y} \sin \phi(x,y) + \frac{\partial k}{\partial y} \quad , \tag{B4}$$

and

$$\frac{\partial^2 z_s}{\partial y^2} = 2 \frac{\partial^2 \alpha}{\partial y^2} \cos[(k_r + k_o)y + \phi]$$

$$- 4 \frac{\partial \alpha}{\partial y} \left(\frac{\partial \phi}{\partial y} + k_r + k_o\right) \sin[(k_r + k_o)y + \phi]$$

$$- 2 \alpha \frac{\partial^2 \phi}{\partial y^2} \sin[(k_r + k_o)y + \phi]$$

$$- 2 \alpha \left(\frac{\partial \phi}{\partial y} + k_r + k_o\right)^2 \cos[(k_r + k_o)y + \phi]$$

$$+ 2 \frac{\partial^2 \beta}{\partial y^2} \cos[(k_r + k_o)y] - 4 \frac{\partial \beta}{\partial y} (k_r + k_o) \sin[(k_r + k_o)y]$$

$$- 2 \beta (k_r + k_o)^2 \cos[(k_r + k_o)y]$$

$$+ 2 \frac{\partial^2 \gamma}{\partial y^2} \cos \phi - 4 \frac{\partial \gamma}{\partial y} \frac{\partial \phi}{\partial y} \sin \phi$$

$$- 2 \gamma \frac{\partial^2 \phi}{\partial y^2} \sin \phi - 2 \gamma \left(\frac{\partial \phi}{\partial y}\right)^2 \cos \phi + \frac{\partial^2 k}{\partial y^2} \quad . \tag{B5}$$

REFERENCES

1. Glenn, W.E., "Thermoplastic Recording," Applied Physics, Volume 30 (1959), p. 1870.
2. Gaynor, J. and Aftergut, S., "Photoplastic Recording," Photo. Science Engr., Volume 7 (1963), p. 209.
3. Wolff, N.E., "A Photoconductive Thermoplastic Recording System," RCA Review, Volume 25 (1964), p. 200.
4. Giaimo, E.C., "Thermoplastic Organic Photoconductive Recording Media-Electrophotographic Characteristics and Processing Techniques," RCA Review, Volume 25 (1964), p. 692.
5. Urbach, J.C. and Meier, R.W., "Thermoplastic Xerographic Holography," Applied Optics, Volume 5 (1966).

REFERENCES (Continued)

6. Lin, L.H. and Beauchamp, M.L., "Write-Read-Erase in-Situ Optical Memory Using Thermoplastic Holograms," Applied Optics, Volume 9 (1970), p. 2088.
7. Credelle, T.L. and Spong, F.W., "Thermoplastic Media for Holographic Recording," RCA Review, Volume 33 (1972), p. 206.
8. Colburn, W.S. and Tompkins, E.N., "Improved Thermoplastic Photoconductor Devices for Holographic Recording," Applied Optics, Volume 12 (1974), p. 2934.
9. Colburn, W.S. and DuBow, J.B., "Photoplastic Recording Materials," Final Technical Report AFAL-TR-73-255, August 1973.
10. Fienup, J.R., "Optical Processors Using Holographic Optical Elements," Final Technical Report, Report No. 119400-2-F, Environmental Research Institute of Michigan, Ann Arbor, April 1978.
11. Schaffert, R.M., "A New High-Sensitivity Organic Photoconductor for Electrophotography," IBM Journal of Research and Development, Volume 15 (1971), p. 75.
12. Bergen, R.F., "Characterization of a Xerographic Thermoplastic Holographic Recording Material," Photographic Science and Engineering, Volume 17 (1973), p. 473.
13. Kriz, K., "Spectral Response and Sensitometric Curves of Doped Poly-N-Vinylcarbazole Films in Electrophotography Under Positive and Negative Charging Modes," Photographic Science and Engineering, Volume 16 (1972), p. 58.
14. Anderson, H.R., Jr., Bartkus, E.A., and Reynolds, J.A., "Molecular Engineering in the Development of Materials for Thermoplastic Recording," IBM Journal of Research and Development, Volume 15 (1971), p. 140.
15. Urbach, J.C. and Meier, R.W., "Properties and Limitations of Hologram Recording Materials," Applied Optics, Volume 8 (1969), p. 2269.
16. Falconer, D.G., "Role of the Photographic Process in Holography," Photographic Science and Engineering, Volume 10 (1966), p. 133.
17. Cathey, W.T., Optical Information Processing and Holography, Wiley & Sons, New York, pp. 68-70.
18. Maloney, W.T. and Gravel, R.L., "Submillisecond Development of Thermoplastic Recordings," Applied Optics, Volume 13 (1974), p. 2471.
19. Gill, W.D., "Drift Mobilities in Amorphous Charge-Transfer Complexes of Trinitrofluorenone and Poly-n-vinylcarbazole," Journal of Applied Physics, Volume 43 (1972), p. 5033.
20. Lo, D.S., Johnson, L.H., and Honebrink, R.W., "Thermoplastic Recording," SPIE Proceedings on Practical Applications of Low Power Lasers, Volume 92, San Diego, California, August 26-27, 1976.
21. Reich, S., Rav-Noy, Z., and Friesem, A.A., "Frost Suppression in Photoconductor-Thermoplastic Holographic Recording Devices," Applied Physics Letters, Volume 31 (1977), p. 654.

REFERENCES (Concluded)

22. Saito, T., Oshima, S., Honda, T., and Teujiuchi, J., "An Improved Technique for Holographic Recording on a Thermoplastic Photoconductor," Optical Communications, Volume 16 (1976), p. 90.
23. Young, J.D. and Wolfe, J.E., "A New Recording Technique for Acoustic Holography," Applied Physics Letters, Volume 11 (1967), p. 294.
24. Brenden, B.B., Optical and Acoustical Holography, Editor: E. Cannatini (Plenum Press, N.Y.) (1972).
25. Irelan, V.G., Mullinix, B.R., and Castle, J.G., "Real-time Acoustical Holography Systems," Technical Report T-78-10, US Army Missile Research and Development Command, October 1977.

APPLICATION OF AN ELECTRO-OPTICAL PHASE MEASUREMENT TECHNIQUE TO SURFACE PROFILING AND SENSING SMALL OBJECT DISPLACEMENTS

G. Makosch, B. Solf
IBM Deutschland GmbH, 7032 Sindelfingen, Germany

ABSTRACT

Many geometrical measurement problems can be solved by measuring the phase difference of two coherent laser beams interacting with the object of measurement. In this paper a simple electro-optical technique is described that permits high speed measurement of a phase difference between two orthogonally polarized, coherent laser beams. With a response time of the order of microseconds and a measuring resolution better than $6 \cdot 10^{-3} \pi$ this method could be a key in solving numerous measurement problems encountered in semiconductor manufacturing. Of primary interest is the application of this method to measuring topographies of machined and etched or sputtered surfaces. Moreover, this method is suitable for very accurate sensing of lateral displacements and motions of grating like structures.

The principle of the phase measurement technique is described and results of different applications are presented.

INTRODUCTION

In semiconductor manufacturing a large field of measurement applications is covered by interference methods. They have the great advantage of being extremely precise and nondestructive. Commonly, these methods are incremental - the smallest measurement increment is a fixed fraction of a light wavelength.

The subject of this paper is a newly developed interference technique that measures the phase difference between two interfering light beams continuously [1,2] at considerable speed. This method opens numerous measurement applications, some of which will be described in the following.

PRINCIPLE OF THE PHASE MEASURING TECHNIQUE

STEP-HEIGHT MEASUREMENT

In our laboratory, this phase measuring technique was first applied to the measurement of step-heights on etched and sputtered silicon wafers. In this application an optical arrangement was used that is illustrated in Fig. 1. A laser beam passing through a double refracting crystal is split into two parallel beams with perpendicular directions of polarization. At reflection by a step as shown in the figure, these beams are subjected to a phase difference that is given by:

$$\varphi_M = \frac{4\pi}{\lambda} h , \qquad (1)$$

ISSN:0094-243X/81/650258-13$1.50

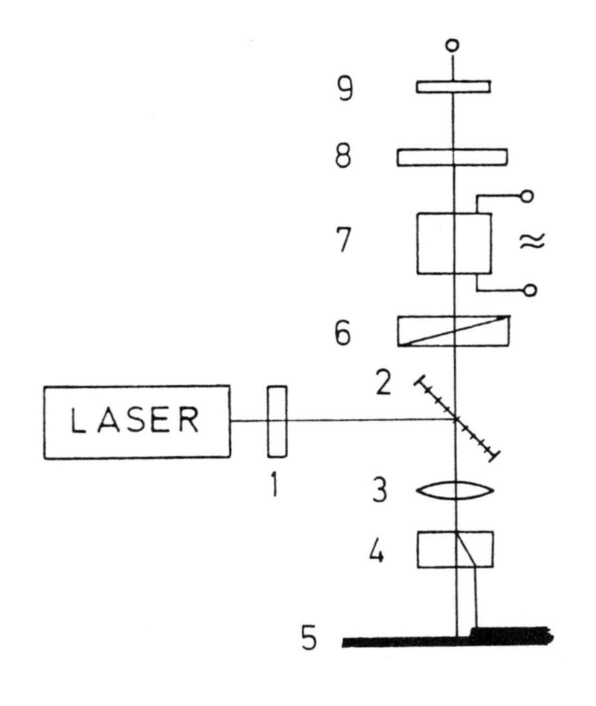

1. λ/2 Plate
2. Beam Splitter
3. Lens
4. Calcite Crystal
5. Test Object
6. Soleil-Babinet Compensator
7. Light Modulator
8. Polarizer
9. Photodetector

Fig.1 Optical arrangement for step-height measurement

where h is the step-height to be measured and λ is the laser wavelength.

On reflection by passing through the crystal again, the two beams are recombined. Now, they propagate in one common beam with perpendicular polarizations, and their phase difference is a measure of the step-height h. These two recombined beams are passed through a Babinet Compensator and an electro-optical light modulator. Finally, they are allowed to interfere after passing through a polarizer. The light intensity measured by photodetector versus the phase variation φ_M is given by:

$$J = \frac{1}{2} J_o [1 - \cos (\varphi_M + \varphi)] \quad . \tag{2}$$

J_o is the maximum intensity at constructive interference and φ is an additional phase difference introduced by the light modulator .

For simplicity, only equal intensities of the interfering beams are assumed. The phase measuring technique, however, is not restricted to this particular case.

The phase difference φ_M occurring due to the step is determined by a very fast null method. This method consists of compensating the phase difference φ_M by a counterbalancing phase difference φ introduced by the electro-optical modulator. Compensation is indicated by a minimum intensity when φ equals $-\varphi_M$.

The phase shift introduced by the electro-optical modulator is linearly related to the voltage applied to the modulator. Hence, the voltage V_M producing an intensity minimum is a measure of the phase difference φ_M.

With a sinusoidal voltage applied to the modulator as shown in Fig. 2 the phase difference φ is varied periodically between $-\pi$ and $+\pi$. The corresponding intensity signal is shown below. Two intensity minima indicate compensation during one modulation period. The phase measurement is reduced to measuring the voltage of the modulator pertaining to an intensity minimum. This is done electronically by differentiating the intensity signal and correlating the zero passage of a minimum to the corresponding voltage V_M of the modulator. At a modulation frequency of 50 kHz the measurement time is of the order of microseconds. This method has the advantage that intensity fluctuations of the light source and the two beams reflected at the test object do not affect the measurement result.

Fig. 3 shows a calibration curve of this phase measuring technique that was obtained by varying the measurement phase difference φ_M over a range of 4π which is equivalent to a step increase by one wavelength λ. This was done employing a Babinet Compensator.

The result is a saw tooth curve with a period 2π which is exactly linear over one period. It is evident that an ambiguity exists when steps exceeding a half wavelength are to be measured. However, step-heights can be measured without ambiguity if the step-height is either smaller than a half wavelength or if the step-height is measured continuously, for example, during a sputter or

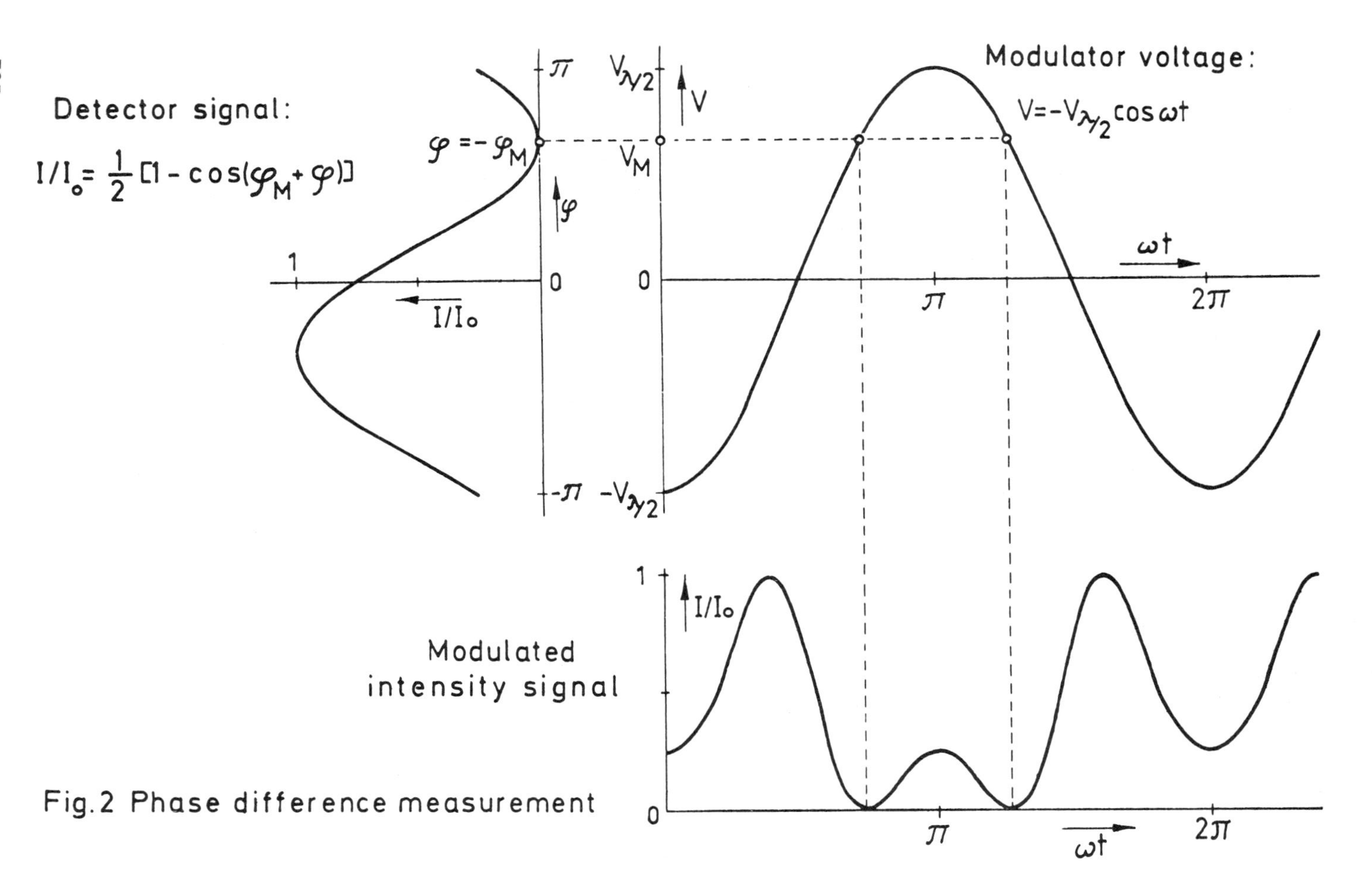

Fig. 2 Phase difference measurement

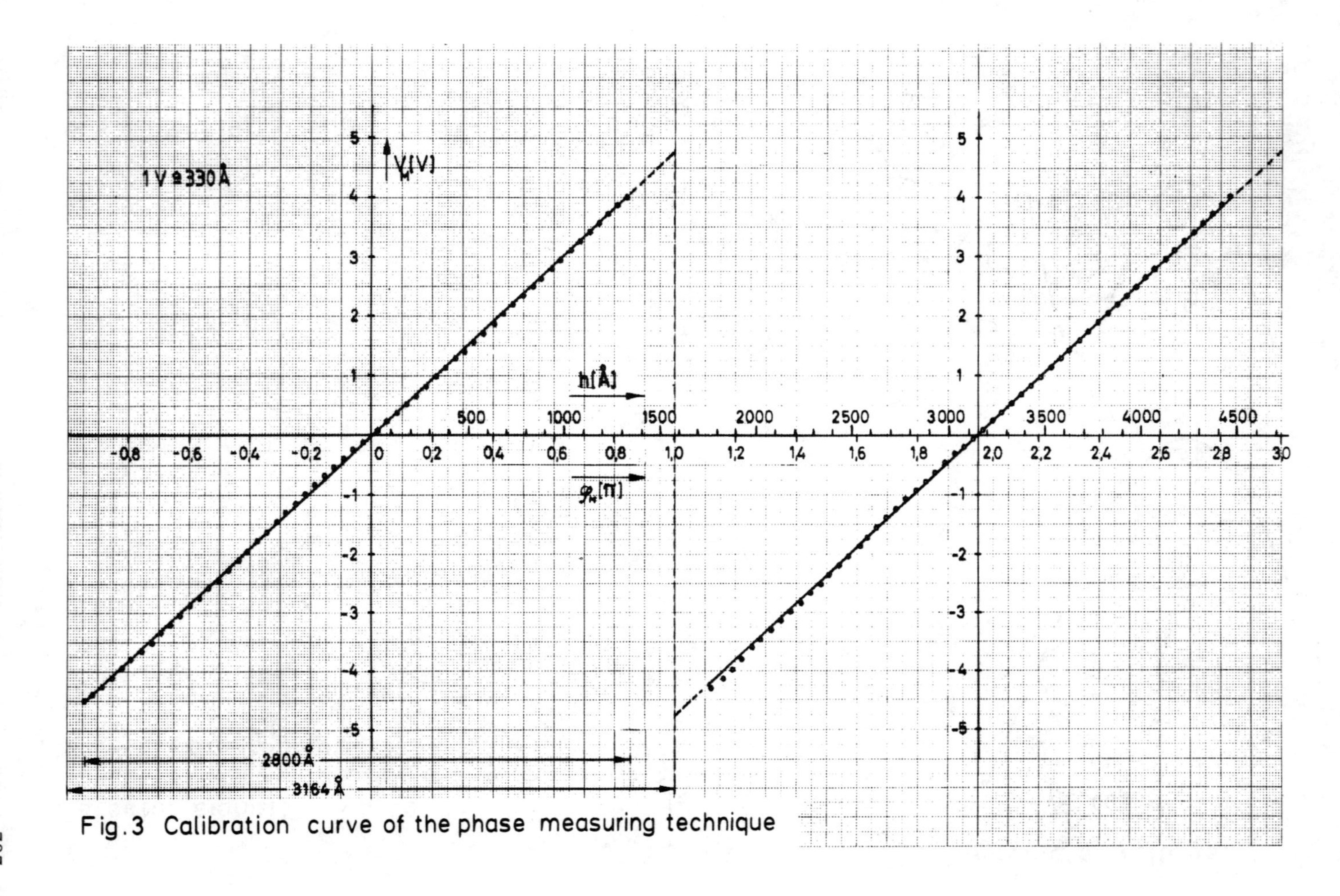

Fig.3 Calibration curve of the phase measuring technique

etching process. In our laboratory, this method was successfully applied to "in-situ" crater depth measurements during sputter etching on a SIMS Auger System. Moreover, this method was applied to contactless measurements of steps etched into silicon. For these investigations a special pattern, as shown in Fig. 4a, was etched into the silicon surface. A vertical profile of this pattern obtained with a diamond stylus is shown in Fig. 4b. The groves etched into silicon have an average depth of 500 nm. A comparison scan recorded with our optical step-height measuring method is shown below. This record was obtained by translating the object on an XY-table past the two beams, one of which serving as reference (broken line), the other as signal beam (continuous line). Since the phase measuring technique suppresses multiples of $\lambda/2$, the scan shown in this figure represents only the groves profile below a depth $\lambda/2$ (= 316 nm by using a HeNe-laser). This step-height measuring method has the great advantage of being insensitive to distance variations of the object relative to the optical measuring head. The measurement resolution is better than 1 nm.

SURFACE PROFILING

The ambiguity of the above method for steps exceeding one half wavelength can be eliminated by implementing a different sensing head as shown in Fig. 5.

The main difference compared to the former method consists in using two laser beams incident at different angles on the object surface. They are focused at one point on the surface. One beam is reflected perpendicularly and the other at an angle α. On reflection both beams are subjected to a phase shift which is approximately given by:

$$\varphi_M = \frac{4\pi}{\lambda} h (1 - \cos\alpha) , \qquad (3)$$

where h is the vertical deviation of the object surface from the focal plane. By measuring the phase difference φ_M after recombining the two beams as shown in the figure, the distance variation h can be measured. If in addition the angle α is chosen such that the resulting phase difference is smaller than 2π, h can be measured unequivocally. It can be shown that the range of unequivocal measurement is indentical with the depth of focus of the optical head. In order to optically profile a surface, this system has to be used similar to a diamond stylus. The object has to be moved laterally on an XY-table past the laser focus. However, unlike the method described before, this optical profiler records not only the profile of the surface but vertical fluctuations of the table as well. The signal plots shown in Fig. 6 are obtained from two measurement scans, one taken across the etched wafer pattern shown earlier, and the other taken across a flat wafer surface using the same table range. A comparison makes it apparent that the first scan shows a superposition of the etch profile and table errors.

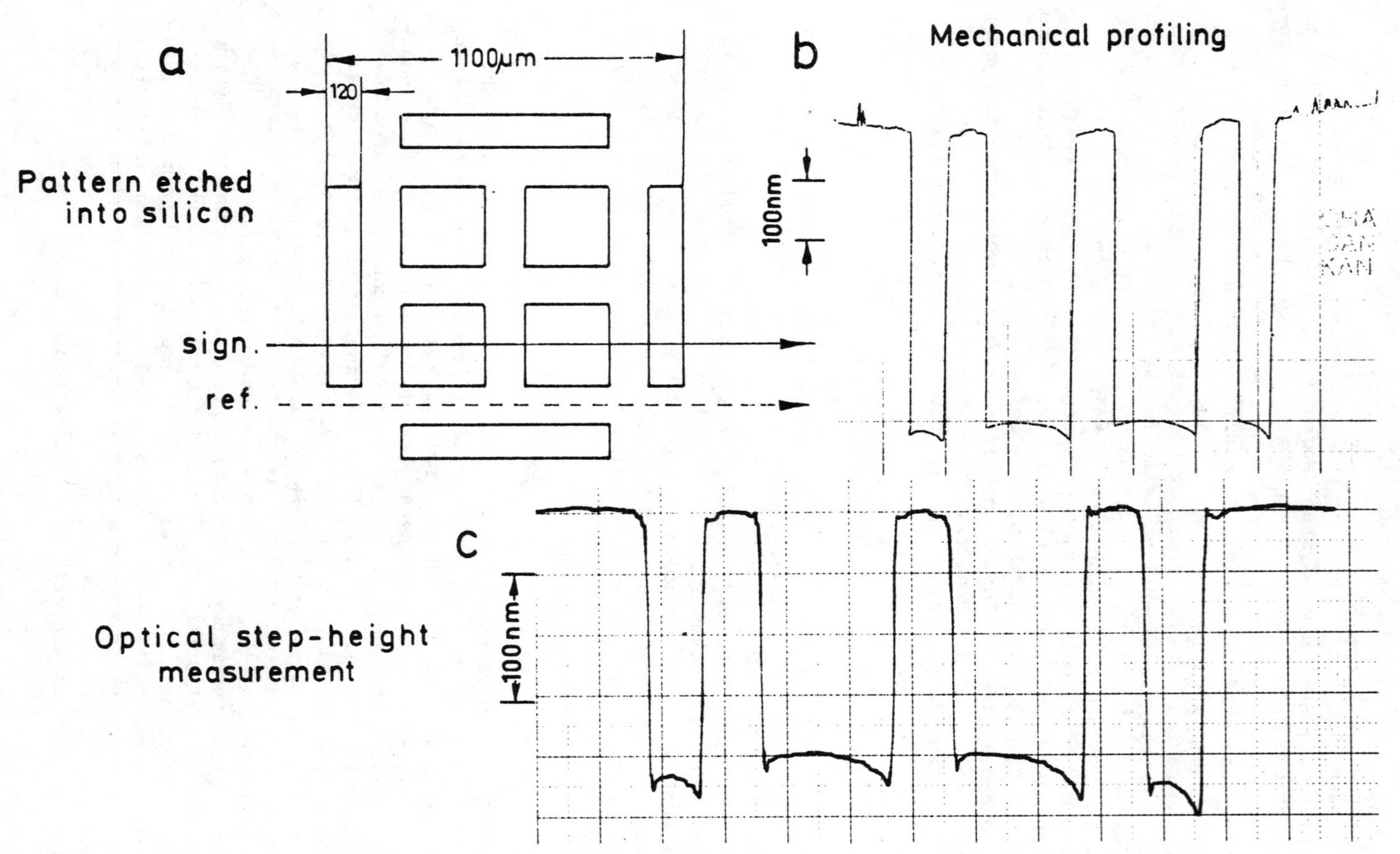

Fig.4 Comparison mechanical profiling versus optical step-height measurement

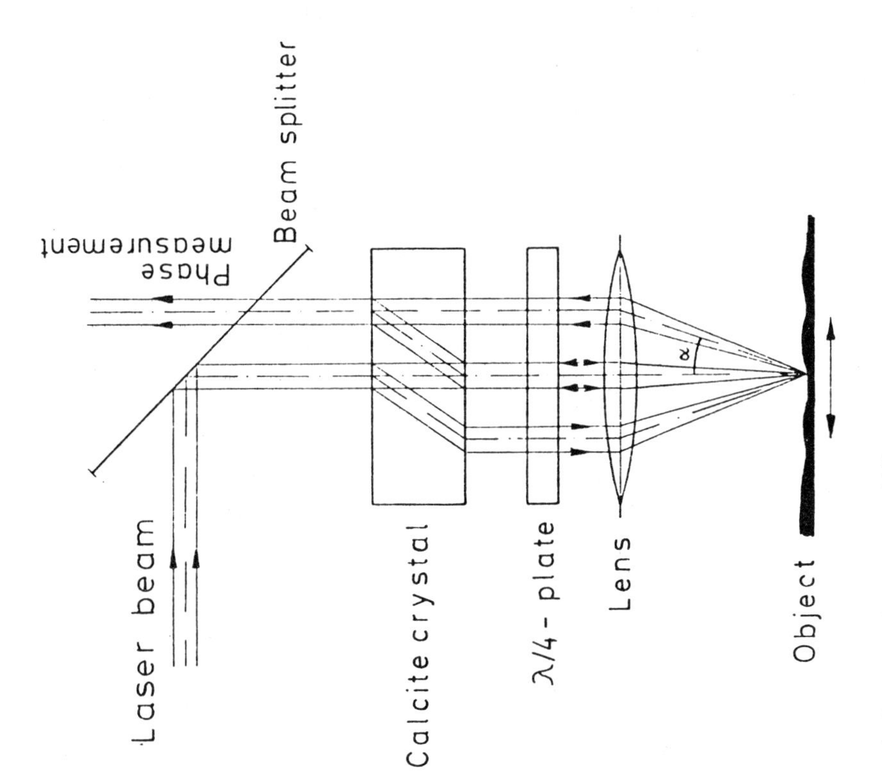

Fig.5 Optical head for surface profiling

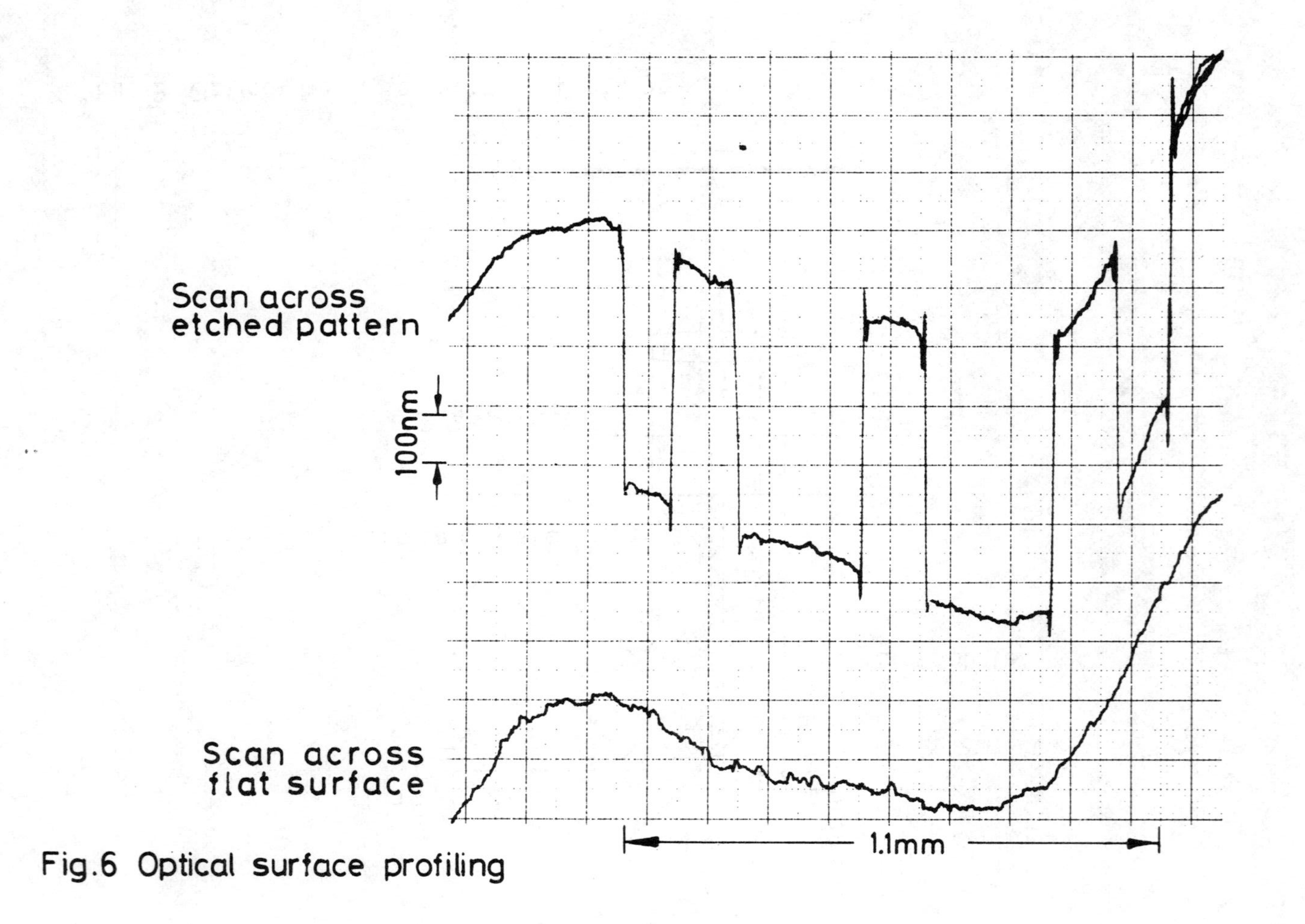

Fig.6 Optical surface profiling

MEASUREMENT OF LATERAL DISPLACEMENT

The phase measuring technique described above is not only useful for measuring vertical surface topographies. It may also be applied to measuring horizontal displacements of an object. In this application, a laser beam is diffracted at a tiny grating imprinted on the object surface. The phase difference of two diffraction orders varies in accordance with a grid displacement [4]. It can be measured by using an optical head as illustrated in Fig. 7. A laser beam is passed without deflection through a calcite crystal, a hole in a half wave plate, and then is focused by a lens on the grid surface. The grating will decompose an incident beam into several diffraction orders. According to diffraction theory, a phase difference

$$\varphi_M = 2\pi \frac{\Delta x}{g} \qquad (4)$$

occurs between the zero order diffraction beam and a first order diffraction beam if the grating is displaced laterally by a distance Δx. g is the grating constant. In the optical arrangement illustrated in Fig. 7 the zero order diffraction beam is reflected perpendicularly. The first order, however, is reflected at an angle ϑ and then collimated by the lens. By passing through the half wave plate its direction of polarization is rotated by 90°. As extraordinary beam, this beam is deflected in the crystal and recombined with the zero order diffraction beam.

By measuring the phase difference between these two orthogonally polarized beams, the amount of the grid displacement as well as its direction can be determined very accurately. Fig. 8 shows the phase difference signal obtained by translating a grid with a grid constant g = 10 µm. The period of this saw tooth curve is equivalent to a displacement by a grating constant. The measurement resolution is about one hundredth of the grid constant. An enhancement of the measurement sensitivity could be obtained by using higher diffraction orders for phase measurement. This last technique may be very useful for aligning flat surfaces as masks and wafers in lithographic processes. Moreover, it can be used for overlay control in these technologies. In conjunction with a grid ruler, this method can be applied to microposition control of highly precise XY-tables.

CONCLUSION

In summary, the phase measuring technique described above consists of a null method in which an optical sensor detects a minimum intensity occurring by introducing a phase compensation. The phase measurement is done electronically at high speed and extremely high precision. Suitability of this method for step-height measurement, surface profiling and measuring lateral displacements has been demonstrated. However, this technique is not restricted to these applications only. Numerous other applications are feasible, particularly the extension of this method to evaluating two dimensional interference and Moiré patterns.

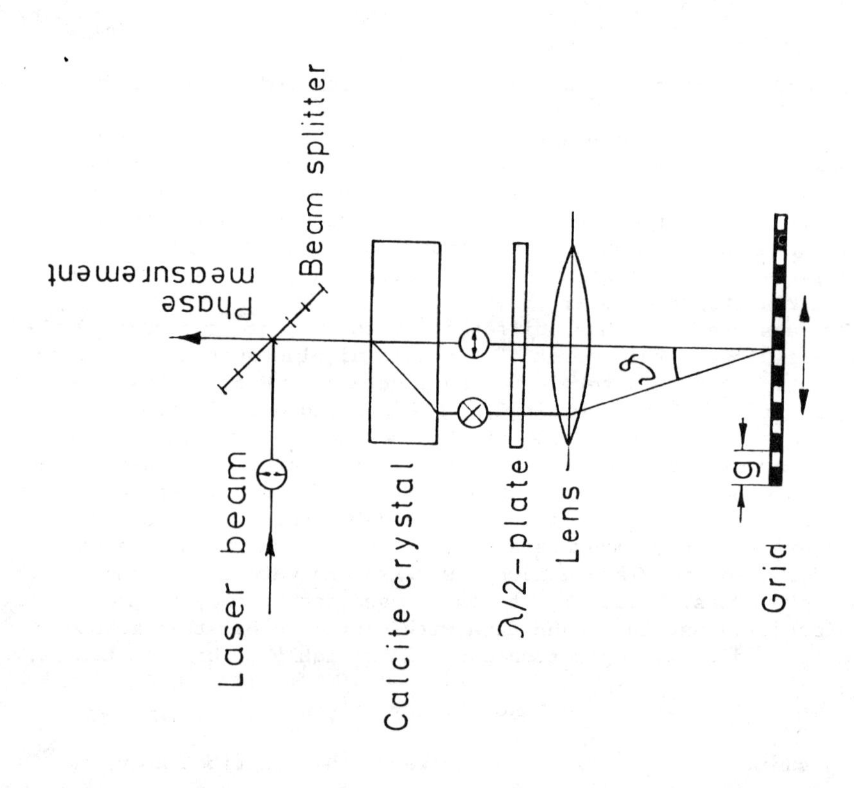

Fig.7 Optical head for measuring lateral displacements

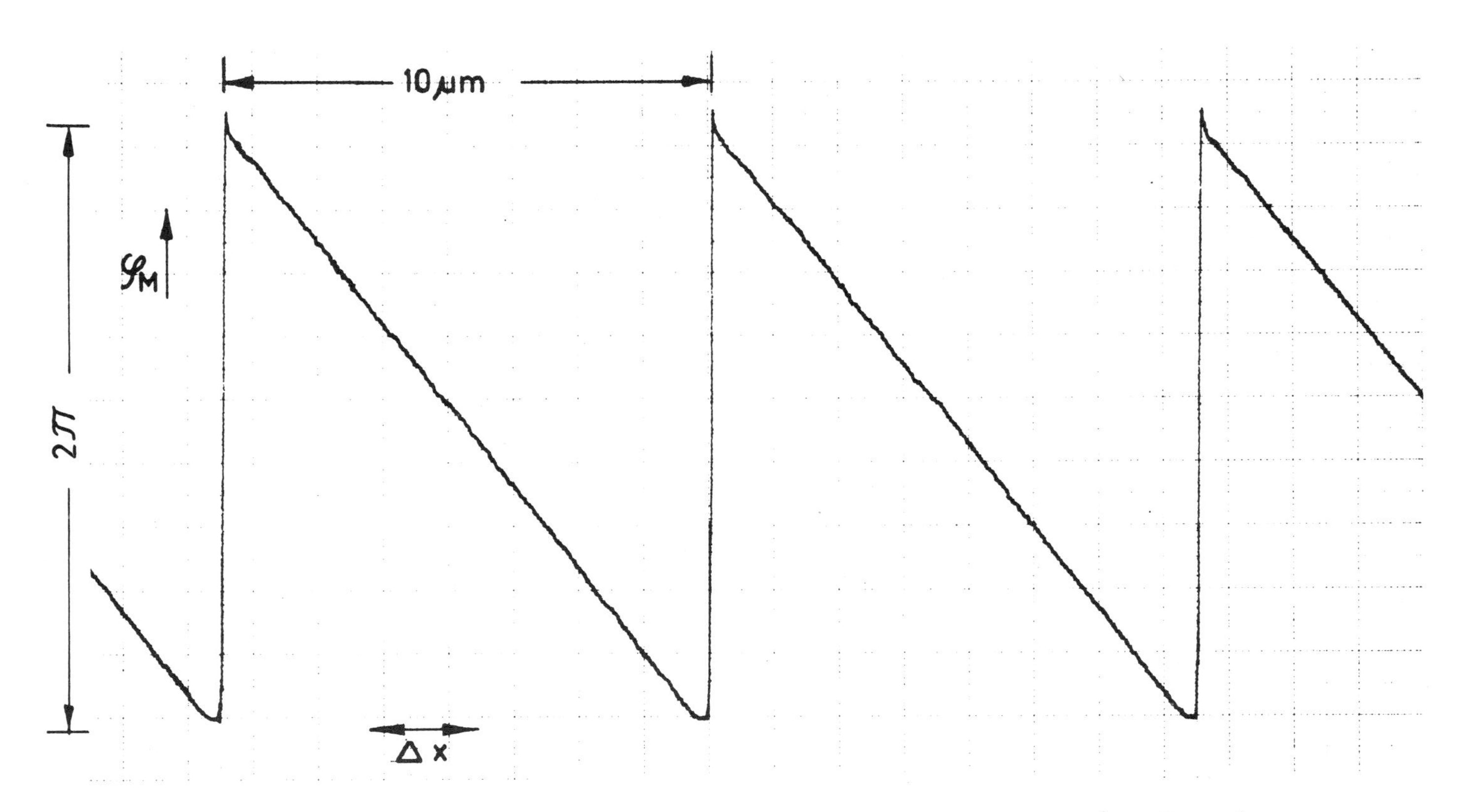

Fig.8 Phase difference measured versus grid displacement (g=10µm)

REFERENCES

1. J. Dyson, Interferometry as a measuring tool (The machinery publishing CO., LTD 1970), p. 190.
2. D.T. Moore, D.P. Ryan, J. Opt. Soc. Am. 68, (1978).
3. R.M. Pettigrew, F.J. Hancock, IPC Business Press (1979).
4. D.C. Flanders, H.J. Smith, Appl. Phys. Lett. Vol. 31, No. 7 (1977).

SOME UNSOLVED PROBLEMS IN IMAGE FORMATION

A. Walther
Worcester Polytechnic Institute, Worcester, MA 01609

During the last thirty years the field of optical image formation has seen many new developments. In addition to the invention of the laser, with all its new applications, and the formulation of coherence theory, two innovations have been of crucial importance: lens design with large computers and the use of Fourier transforms in the diffraction theory of image formation. These developments have, indeed, been so successful that some unsolved problems that cannot be attacked by these methods have faded into the background. With this paper, I would like to direct your attention towards two of these problems, an old one in geometrical optics and a more modern one in physical optics.

It is necessary to summarize some background material. In geometrical optics, lenses are used to transform rays in the object space into rays in an image space. A convenient choice of coordinate systems is shown in Fig. 1. It is assumed that a ray travelling along the z-axis emerges from the lens travelling along the z'-axis. We specify an arbitrary ray in the object space by the x- and y-coordinate of its point of intersection

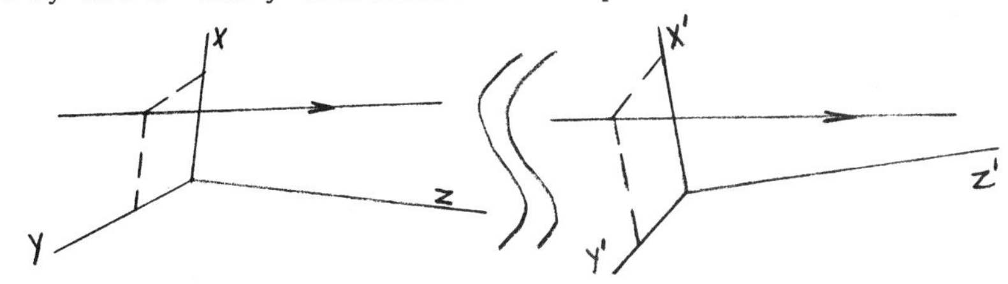

Fig. 1. Coordinate systems.

with the x-y plane, and its directions cosines L and M relative to the x and y directions. Rays in the image space are similarly specified by the four variables x', y', L', and M'. Each point in a finite region of the (x,y,L,M) space represents a ray in the object space. Rays in the image space are similarly associated with points in a (x',y',L',M') space. These mathematical spaces are often called phase spaces because multiplication of the direction cosines with h/λ turns them into components of the linear momentum of the photon. The four-dimensionality of these phase spaces is in keeping with the main thrust of this conference.

One might think that the properties of a lens need to be pinned down by specifying the four variables (x',y',L',M') in terms of the four variables (x,y,L,M), i.e., by specifying four functions of four variables. Hamilton has shown that this is not the case. Fermat's principle is a restriction which dictates interrelations between these four functions, so much so that they can all be derived from a single generating function depending

ISSN:0094-243X/81/650271-05$1.50

on four properly chosen variables. In the period between the wars these "eikonal" methods were studied extensively, notably by M. Herzberger[1] and T. Smith.[2] In this paper, we only state the basic point of departure which, foregoing some niceties, can be stated as follows: given a lens and a set of coordinates chosen in the manner shown in Fig. 1, there is always a "point eikonal function" S(x,y,x',y') such that

$$nL = -\frac{\partial S}{\partial x}, \quad nM = -\frac{\partial S}{\partial y},$$

$$n'L' = \frac{\partial S}{\partial x^1}, \quad n'M' = \frac{\partial S}{\partial y^1}.$$

Here n and n' are the refractive index in the object- and image space. Physically, the function S represents the optical distance between the point (x,y) in the plane z = 0 and the point (x',y') in the plane z' = 0, measured along the ray connecting these two points. If this ray is not uniquely determined, owing for instance to the fact that the plane z = 0 is imaged into the plane z' = 0, other eikonal functions can be used that need not be discussed here.

The existence of eikonal functions constitutes a profound limitation on the art of lens design. A set of design goals cannot be met unless each of the goals is derivable from the same eikonal function. It is easy to run afoul of this restriction. Take, for instance, an enlarging lens that must be perfectly corrected for 1:1 magnification. This requirement does not determine the eikonal function uniquely. The remaining degrees of freedom can be used to meet some other requirements, but the choice is limited. It is, for instance, possible to also image the axial point at some other magnification perfectly; but it is impossible to image an off axis point at another magnification perfectly. A numerical example may be found in[3].

We can now formulate the first unsolved problem referred to in the title of this paper. The design goals must be compatible with a single eikonal function. Suppose that this requirement has been met, and that the eikonal function needed has been determined mathematically. Can we then guarantee that the lens can, at least in principle, be designed?

References to this existence problem are scant. In Caratheodory's monograph "Geometrische Optik", published in 1937, we find[4]:

> "Die Trennung zwischen den Abbildungen dieser Art die man optisch realisieren kann, und den übrigen ist ein mathematisches Problem, das nie in Angriff genommen worden ist und ausserordenlich schwierig sein dürfte".

(The distinction between optically realizable image formations and those that are not is a mathematical problem that has never been studied and that might well be extraordinarily difficult.) Around the same time, similar questions were raised by Herzberger and by T. Smith.[5]

Since then the problem does not seem to have attracted any attention. And yet, its solution would by no means merely satisfy academic curiosity. If it can be established unequivocally that a desired eikonal function cannot be realized under a specified set of restrictions on the size of the lens, its location relative to the reference planes, and the range of refractive indices, countless hours of computer time, programming, and just plain frustration can be saved. Even more important, we would obtain valuable insights in the existence of inherent aberrations, i.e., aberrations that cannot be corrected in principle. A very modest example of such inherent aberrations is the well known inevitable presence of field curvature and astigmatism in aplanatic thin lenses.

A slightly different but closely connected problem deserves to be mentioned separately. It is well known that, as far as eikonal theory is concerned, perfect imaging of a finite three dimensional region of space is possible, provided that the angular magnification is unity. Few systems are known that realize this image formation. A common example is the plane mirror. The only other known examples are certain concentric systems of which we have sketched an example in Fig. 2.

radius +1.00 +0.75 -1.00 -1.41
index 1.33 1.88 1.33 1.00 1.33

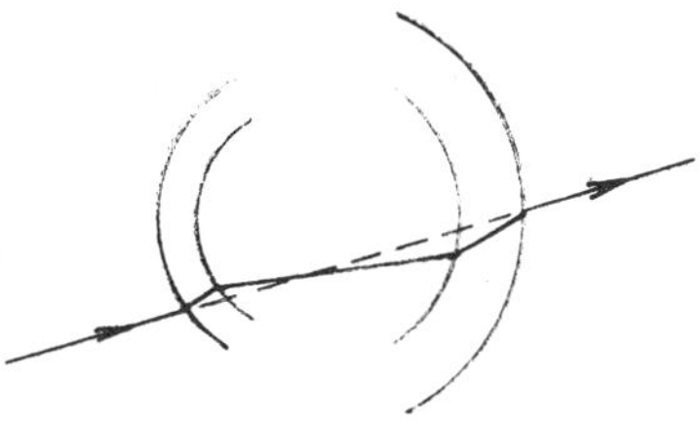

Fig. 2. Perfect afocal concentric system.

All these systems, the plane mirrors as well as the concentric systems, have a zero optical path between object and image. This raises the question whether a system exists that gives perfect three dimensional imaging, and has a non zero optical path between object and image. Again, this is not merely an intellectual exercise. If we had a design for a perfect Fourier lens, two of these lenses used back to back would give perfect three dimensional imaging with a rather large optical path length between object and image. A proof that such an imaging cannot be realized would imply that Fourier lenses have inherent aberrations that no amount of computer programming can correct.

The second broad area on which I wish to comment is Fourier optics. With suitable precautions, the amplitude distribution in an image can be calculated by convolving the amplitude distribution in the object with a point spread function, which is the Fourier transform of a pupil function characterizing the lens. This approach to the wave theory of image formation has been so successful that little time has been available to contemplate its limitations. One of these limitations is that the theory provides the aerial image only; the interaction between field and detector is ignored.

Let us choose for a detector a layer of photoresist, so that we need not worry about the inhomogeneity of silver-halide materials, or the many arbitrary design parameters of photoelectric detection systems. If the layer of photoresist is flat, the field in the photoresist close to the boundary caused by an incident aerial image can be calculated without great difficulty. In the spirit of Fourier optics, we decompose the incident field into a superposition of plane waves. The interaction of each plane wave with the air-photoresist interface can be treated in the usual manner, and the result is that the interaction can be described mathematically as a change in the apodization of the pupil function, and a usually very slight change in its phase.

However, in many practical applications, such as the fabrication of integrated circuits, the photoresist surface is not flat. Let the interface be specified by a surface profile function $\zeta(x,y)$, and let L and M again be direction cosines relative to the x and y axis. Then the amplitude of the incident aerial image evaluated on the interface is given by an integral of the type

$$u(x,y) = \iint u(L,M) \exp ik[Lx + My + \zeta(x,y) \sqrt{1-L^2-M^2}] \,.$$

If ζ is periodic, a Fourier series development of the part of the exponential function containing ζ is of some limited utility. If ζ is a random Gaussian process, various averages can be calculated. But this approach does not address the heart of the problem, because it merely provides us with the value of the <u>incident field on</u> the air-detector interface. The recorded image is determined by the <u>actual field in</u> the photoresist close to the interface. This field can only be determined by finding the solution to a boundary value problem that is usually extremely difficult to solve. We are, therefore, driven from the pleasant ambiance of linear systems theory to the severe mathematical environment of rigorous grating theory and the exact theory of scattering by rough surfaces.

The extensive literature on this boundary value problem is, of course, at our disposal, but much work remains to be done. We choose, without prejudice, one example. Waterman[6] reduces the boundary value problem to a single linear integral equation for the angular spectrum of the transmitted field. He deals with gratings; therefore, the integral equation appears as an infinite set of linear algebraic equations. These equations are solved by a suitable truncation of the infinite set. This truncation may be reasonable if one is primarily interested in the far field created by a grating. Our goal is different; we need to find the field close to the boundary, i.e., in a region where even evanescent waves are important. It is unclear whether the truncation is still appropriate in this case, especially because the kernel of the integral equation seems at times to diverge for very high spatial frequencies.

A second, more direct approach to the problem is the development of large scale computer programs, pioneered by, e.g., D. Maystre.[7] These techniques promise to become extremely useful and will undoubtedly contribute much by way of numerical results. However, in addition to numerical results there is also a need for analytical insights if the theory is to become of widespread use in the engineering community.

These are two of my favorite unsolved problems. I do not doubt that the second one will attract a great deal of attention in the near future. The future of the existence problem in geometrical optics is less clear, but the current interest in the optics of inhomogeneous media may be an impetus to have a new look at this old problem. I would be very pleased if the title of this talk were soon to be proved obsolete.

REFERENCES

1. M. Herzberger, Strahlenoptik (Springer, Berlin 1931).
2. T. Smith, e.g. Trans. Opt. Soc. 23, 311-322 (1921/22).
3. A. Walther, J. Opt. Soc. Am. 60, 918-920 (1970).
4. C. Caratheodory, Geometrische Optik (Springer, Berlin 1937).
5. T. Smith, Rep. Prog. Ph. 2, 357 (1936).
6. P.C. Waterman, J. Acoust. Soc. Am. 57, 791-802 (1973).
7. D. Maystre, J. Opt. Soc. Am. 68, 490-495 (1978).

DIGITAL HETERODYNE HOLOGRAPHIC INTERFEROMETRY OF FLOW FIELDS

Jeff Hartlove
Rockwell International, Rocketdyne Division
6633 Canoga Avenue, Canoga Park, California 91304

ABSTRACT

Flow medium inhomogeneities associated with the gain region of large scale laser systems can result in significant degradation of the outcoupled laser beam quality. To achieve beam quality goals, relative density fluctuations will have to be below 10^{-5}. An optical technique called digital heterodyne holographic interferometer (HHI) is presented which measures the wavefront distortion of a visible probe beam by the medium. HHI is similar to conventional holography in that both use very short averaging times to study the homogeneity fluctuations. HHI, however, derives an increase in sensitivity of an order of magnitude over traditional holography by the use of an electronic phase readout technique during reconstruction. This increase in measurement sensitivity is required to diagnose low level perturbations. Experimental results are presented of the flowfields from a set of supersonic nozzles.

The Heterodyne Holographic Technique

The HHI technique consists of taking a double exposure, double reference beam hologram of the flowfield.[1] This hologram is then reconstructed using two reference beams of slightly different optical frequencies. The reconstructed image is amplitude modulated at the difference frequency of the two reference beams. A photodetector placed in the beam produces an AC current at the difference frequency whose phase is equal to the optical phase recorded on the hologram. The phase difference between two detectors located at different spatial points on this reconstructed image is equal to the optical phase difference (OPD) of the object beam wavefront at these points between the two exposures. Thus by keeping one detector at a fixed location on the reconstructed image, and then by scanning a second detector over the image, a reconstruction of the OPD map can be recorded.

In our system, the double reference beam hologram is made by using a Holobeam ruby oscillator operating single mode with a temperature controlled etalon. Typically a 20 ns, 30 mj pulse is generated. The first exposure is made at one reference beam angle without gas flow; the second exposure is recorded at another reference beam angle with gas flow. High efficiency, low noise holograms were required for the heterodyne detection scheme.

ISSN:0094-243X/81/650276-06$1.50

The heterodyne reconstruction system is shown in Fig. 1. A single frequency krypton ion laser operating at 647.1 nm was beam split and differentially frequency shifted using acousto-optical modulators. These two beams were expanded and directed onto the hologram. The amplitude modulated reconstructed image was projected onto an image dissection camera (IDC). The IDC has an area photocathode which can be randomly accessed by electronic control to scan the image. A programmable desktop calculator controls the image scanning process of the IDC and stores the digitized electronic phase signal in its memory. This OPD data array can then be least square fitted to a plane to determine the residual rms or peak to valley OPD fluctuations of the gas flow. These OPD fluctuations can then be related to the density fluctuations of the flow.

Systems Testing

To determine systems performance and reproducibility, repeated double exposure double reference beam holograms of fixed objects were taken. These holograms were reconstructed using the heterodyne technique and stored as digital arrays. These arrays were compared to one another. The arrays were found to be the same to an equivalent precision of $\lambda/85$ rms at 694.3 nm, within our experimental requirements. We see no reason to limit achieving yet higher accuracies. A plot of the difference between these two arrays is shown in Fig. 2 where we plot isometrically the phase error for a 40 x 40 point data array with an arbitrary piston valve added. The vertical scale is Optical Pathlength Difference (OPD) from a fixed point on the image in microns. The OPD phase mapping of a subsonic gas nozzle is shown in Fig. 3.

Experimental Results

The test cell is shown schematically in Fig. 4. A set of supersonic nozzles is located in the object beam path. These measurements sought to examine the existence of small scale turbulence near the nozzles associated with the mixing of the flows between adjacent nozzles, as well as to measure downstream mixing. The nozzles are spaced 3 mm apart and optical quality data 3 cm downstream of the nozzles was required. The HHI technique was required for our experiment because conventional fringe holography with interpolation between fringes could neither produce the sensitivity or the spatial resolution required to resolve the anticipated small scale turbulence of the flow field.

Acoustic sound fields in our test cell were high, averaging 105 db during the operation of the nozzles due to the vacuum and water cooling systems.

A hologram was made of the nozzle flow field, developed and reconstructed. The flow field OPD mapping is shown in Fig. 5, where we plot the medium inhomogeneities of two sets of nozzles. A total of 100 points across the 3 mm nozzle set was taken. Plotted is the OPD versus the distance across as well as downstream of the nozzles in centimeters. Diffraction from the nozzles prevented accurate results closer than 1 mm to the nozzle plane.

Conclusions

Digital Heterodyne Holographic Interferometry has proven to be a very useful diagnostic tool in this characterization of flow fields in severe test environments. The technique provides a direct measure of the OPD of flowfields without the fringe approximations or manual digitization necessary with conventional holography. HHI also provides higher sensitivity than possible with any other holographic technique.

References

1. R. Dandliker, Opt. Commun. 9, 412 (1973).

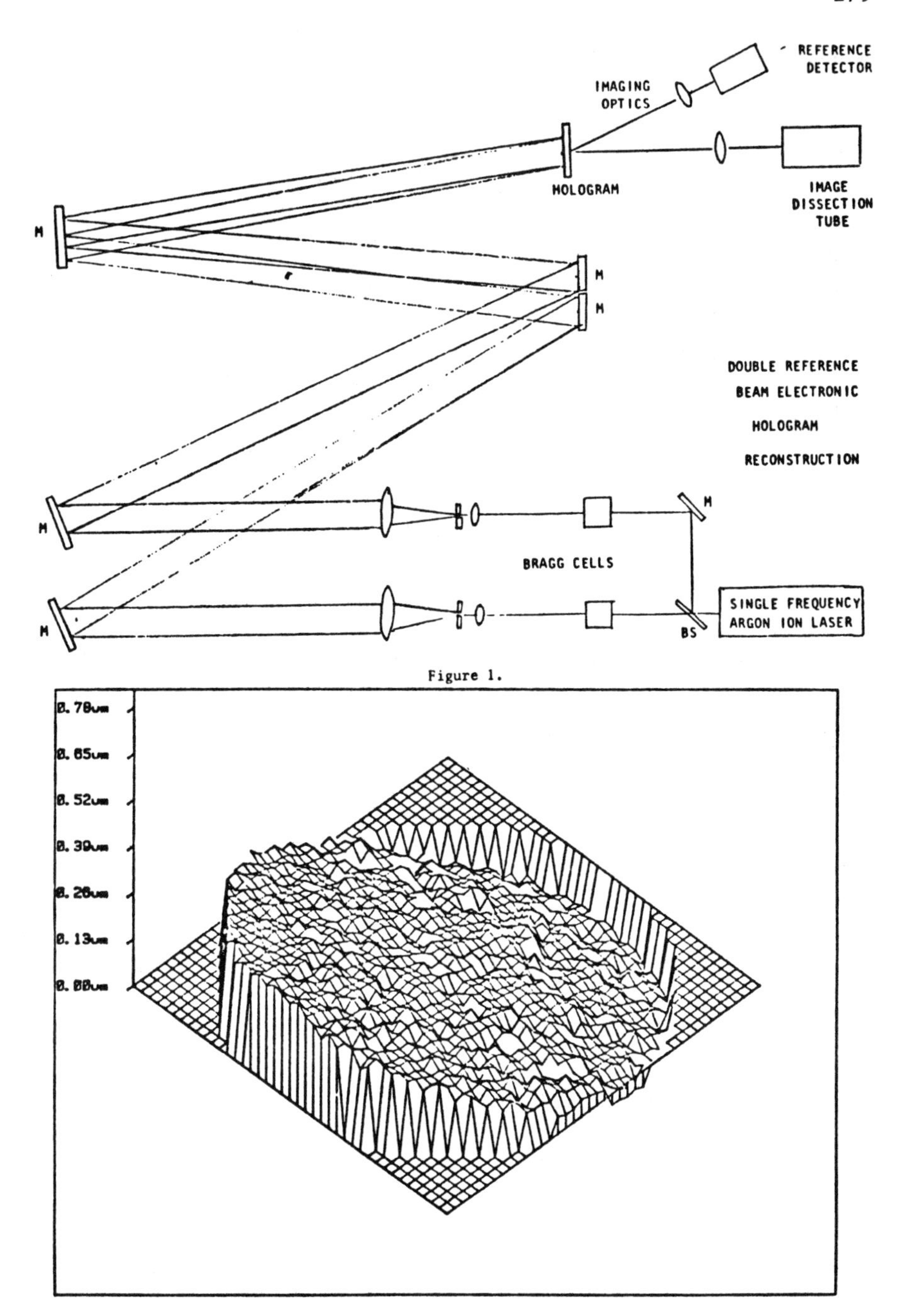

Figure 1.

Heterdyne Holography

Baseline Accuracy with tilt added

Figure 2.

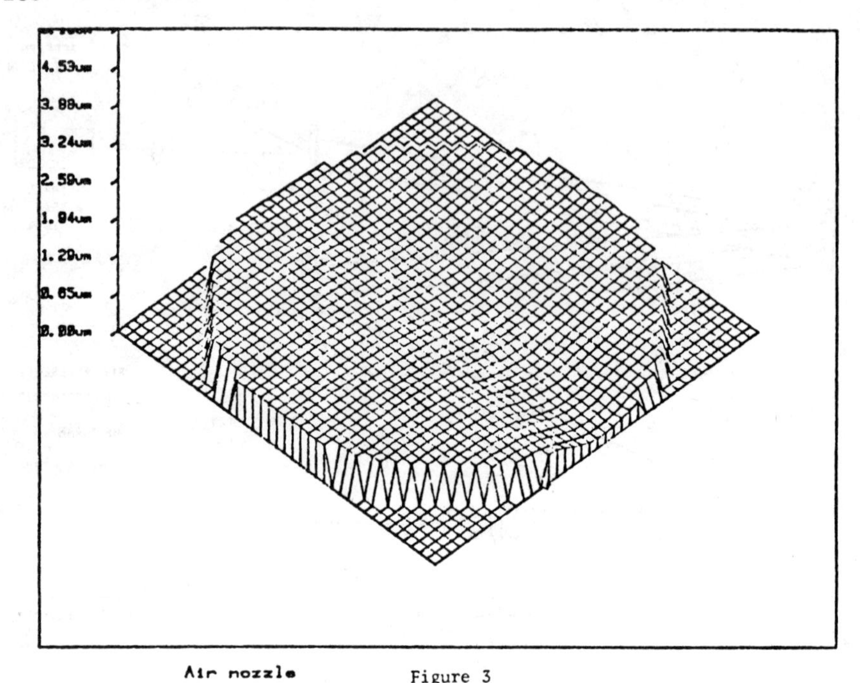

Figure 3

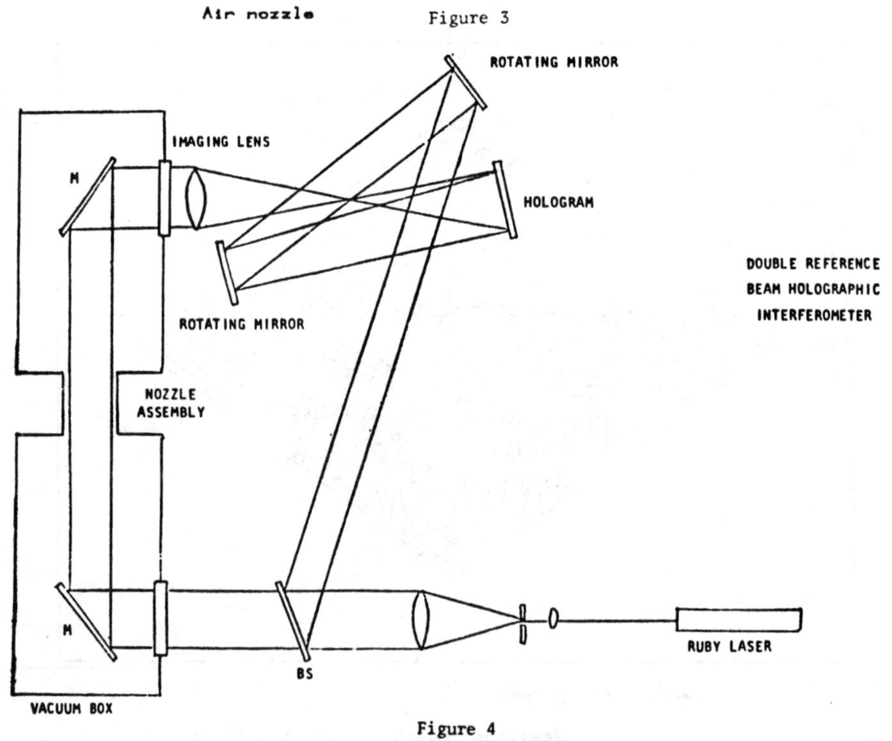

Figure 4

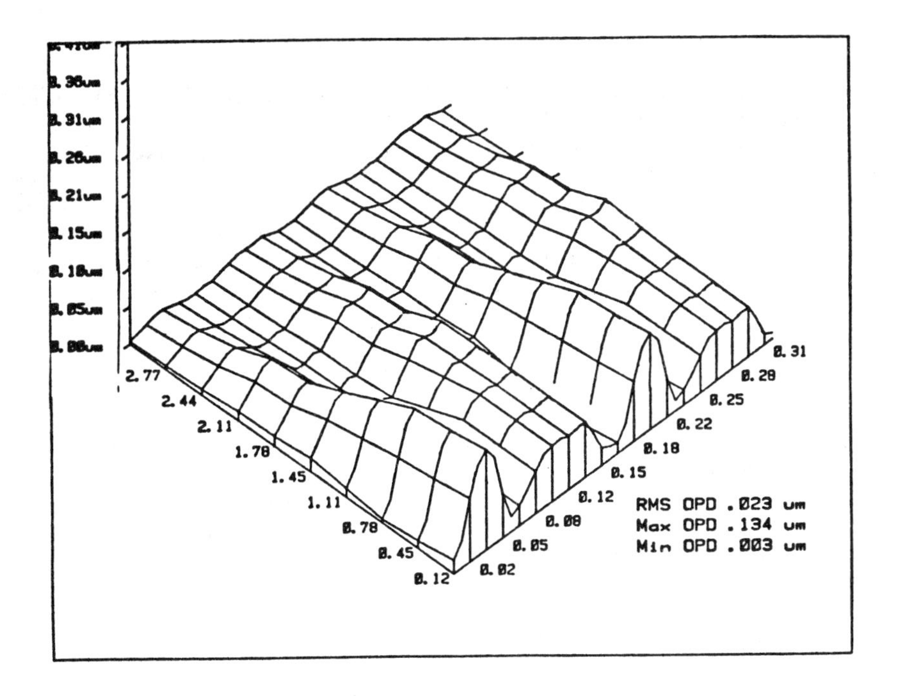
0.36um
0.31um
0.26um
0.21um
0.15um
0.10um
0.05um
0.00um
2.77
2.44
2.11
1.78
1.45
1.11
0.78
0.45
0.12
0.02
0.05
0.08
0.12
0.15
0.18
0.22
0.25
0.28
0.31
RMS OPD .023 um
Max OPD .134 um
Min OPD .003 um

Figure 5.

Research on Optical Methods for Correcting Aperture Distortion in Vidicon Tubes

V.K. Sokolov
A.F. Ioffe Physical Technical Institute
194021 Leningrad, USSR

ABSTRACT

The results of an experimental investigation of an optical scheme for correcting aperture distortions of Vidicon tubes are discussed. Our procedure is based on a spatial filtering scheme that utilizes an inverse Gaussian amplitude filter in the frequency plane. Our experiments with Vidicon tubes have shown that the optical aperture correction method may be preferable to traditional electronic schemes because it can provide full aperture correction in the desired spatial frequency band, at favorable signal to noise ratio and at high spatial frequency.

INTRODUCTION

The possibility of carrying out distortion corrections of Vidicon tube apertures by optical means has been proposed in Ref. 1. The method is based on an appropriate apodization process of the image projected into the tube target. This is accomplished by optical spatial filtering. Figure 1 illustrates the ideas behind the correction scheme in the case when the aperture response is approximated by the following equation

$$A(x) = \exp(-k^2) = \exp(-r_e{}^2\omega^2/2) \quad (1)$$

where r_e is the effective radius of the electron beam that scans the image on the target and k is a generalized spatial frequency.

For the case of interest in this paper, the amplitude transmission function of a physically realizable spatial filter that is capable of performing a full correction of the aperture distortion is given by

$$T(\omega) = \begin{cases} To\ \exp\ (a_k{}^2\omega^2/2), & \omega < \omega_o \\ 1 & , \omega > \omega_o \end{cases} \quad (2)$$

where $T_o = \sqrt{A(ko)}$ is the minimum transmission of the filter at zero spatial frequency. This parameter is to be chosen under conditions of complete correction of the Vidicon aperture distortion in a given band; ko is the boundary frequency of the correction, A(ko) is the value of the Vidicon aperture response at this boundary frequency, and a_k is a constant that characterizes the spatial filter.

As one can see from Eq. (2), a spatial filter in the form of an inverse Gaussian should be used for correction of the Vidicon aperture distortion.

ISSN:0094-243X/81/650282-04$1.50

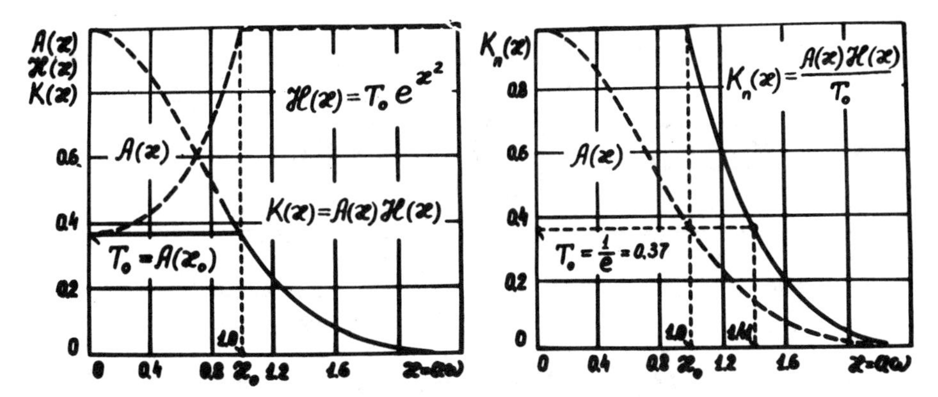

Figure 1

Theoretical amplitude and frequency response of the Vidicon TV with the optical aperture corrector for the case $T_0 = 1/e$; nonnormalized (left), normalized (right); H(x) is the response transfer function of the aperture correction.

EXPERIMENTAL RESULTS

The experiment was conducted in two stages: the preparation of the spatial filter and the measurement of the aperture response on the Vidicon TV camera with the optical aperture correction and without it.

A technique similar to the one described in Ref. (2) was used for the preparation of the inverse Gaussian filter. In this case, however, a concave lens was used instead of a convex one. Under conditions where the thickness of the lens formed by the ink was much smaller than its radius, the intensity of the transmitted light acquired a profile which could be accurately represented by the expression

$$I = I_0 \exp(-\alpha r/2R) \tag{3}$$

where α is a measure of the transmittance of the lens in the center, and R is the lens radius.

A two-step photoprocess was used to produce a linear record of the intensity distribution (3). The selection of the required diameter of the spatial filter was made in the second step.

In our experiments, the spatial filter was prepared using a LP-1 photoplate with an intensity transmittance of 0.2 at zero frequency. A standard closed-circuit TV camera with a Vidicon tube was used as shown in Fig. 2 together with the rest of the experimental set-up. The aperture response of the Vidicon was measured with a contrast factor K = 0.54 using a line selector oscilloscope.

In Figure 3 we show several oscillograms of video signals taken without correction (left column) and with correction (right column) which have been taken from corresponding lines. The experimental aperture correction (2) and without it (1) is shown in Figure 4.

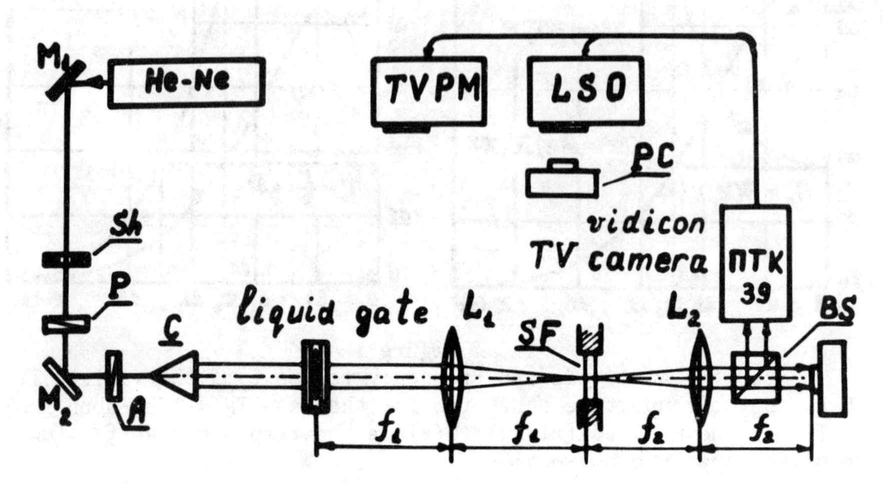

Figure 2

Schematic diagram of the experimental set-up. The main components include: He-Ne laser; mirrors M_1-M_2; shutter, Sh; polarizer, P; analyzer, A; collimator, C; lenses L_1, L_2; f_1 = 620 mm, f_2 = 400 mm; inverse Gaussian spatial filter, SF; beam splitter, BS, photographic camera, pc; closed circuit Vidicon TV camera, πTK-39; Line selector oscilloscope, LS; television picture monitor, TV PM.

As seen from Figs. 3 and 4, the optical method of correcting the distortion of a Vidicon aperture which we have realized by means of an inverse Gaussian spatial filtering is fairly effective in spite of the fact that the filter used in this work did not take into account the deformation of the electron beam cross section in the target scanning process.

REFERENCES

1. V.K. Sokolov, Proceedings of ICO-11 Congress. Madrid, Spain, 1978, p. 283-286.
2. C. Lloria, G. Mas, J. Roig, J. Opt., 1977, 8, p. 385-392.

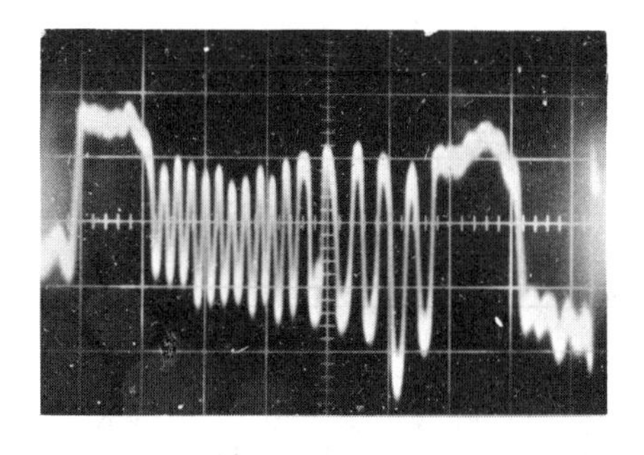

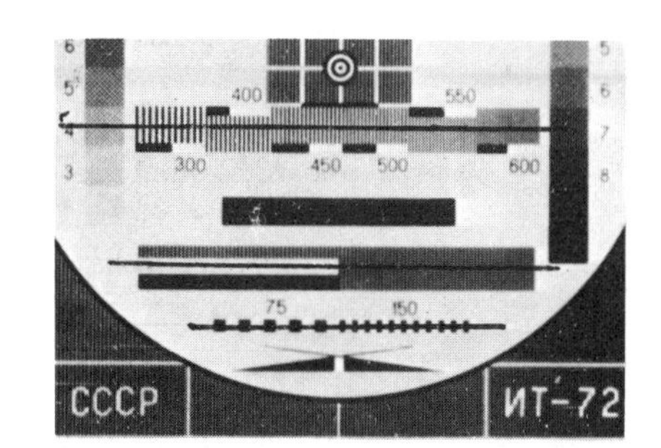

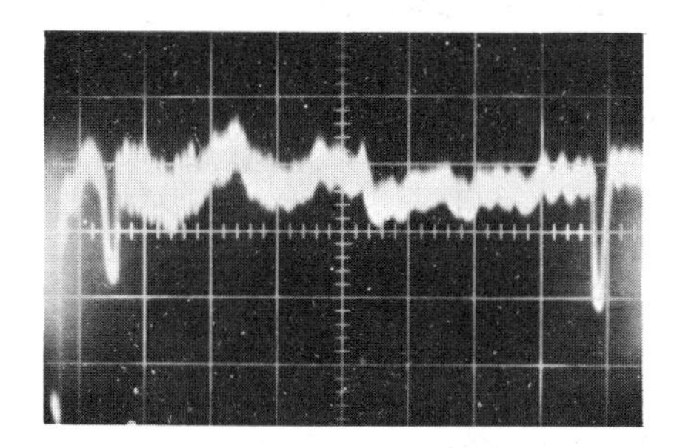

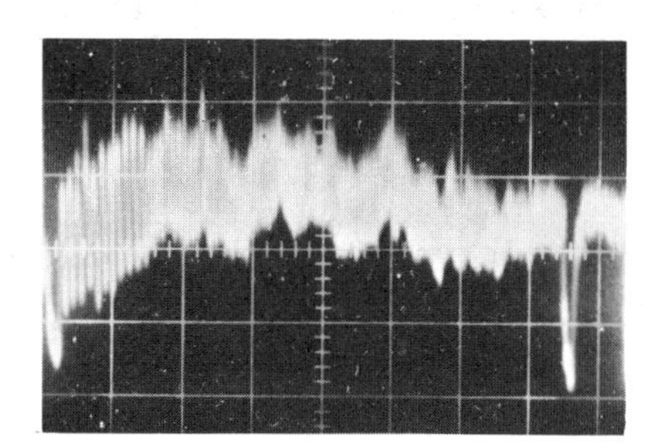

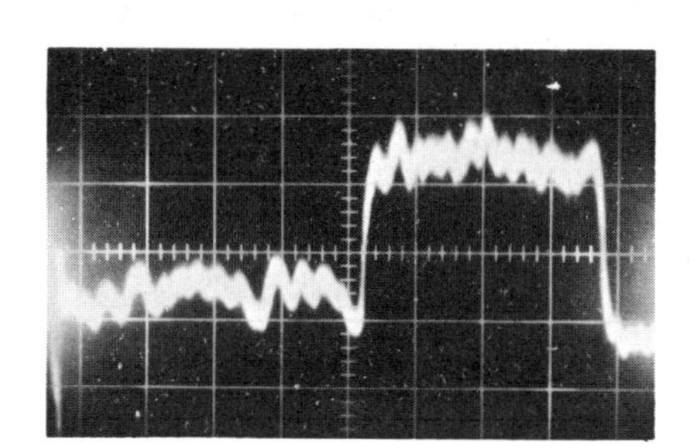

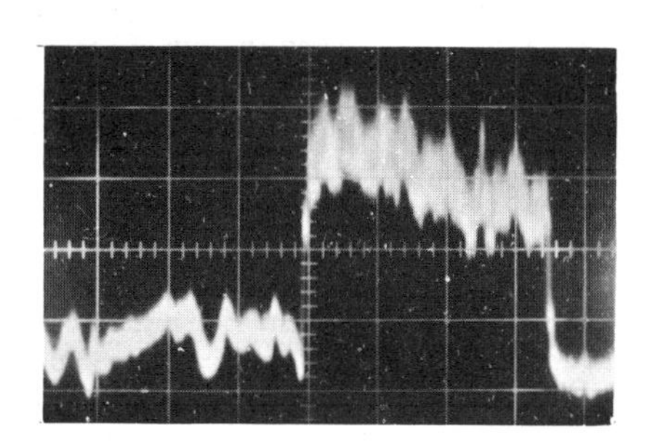

FIGURE 3

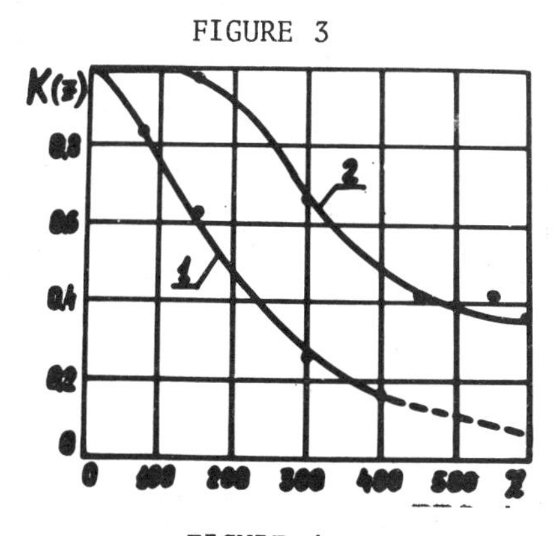

FIGURE 4

NOVEL APPLICATIONS OF IMAGE SUBTRACTION

J. Ojeda-Castañeda, E. Jara and J. Ibarra
Instituto Nacional de Astrofísica, Optica y Electrónica
A. P. 216, Puebla, Pue., MEXICO

ABSTRACT

It is shown that the optical image subtraction techniques can be used to implement two novel operations: firstly, to map the complex amplitude of the object into image irradiance, and in this way to obtain a square root operation; secondly, for pseudocoloring gray level information.

INTRODUCTION

The techniques for detecting the difference, or the subtraction between two images have been used mainly to indicate dissimilarities or for feature extraction[1]. This paper presents two novel applications of image irradiance subtraction.

In the first case, it is shown that a square root operation can be performed optically by means of the difference of two images of the same object. The first image is the irradiance distribution associated with almost any filtered image of the object (this image irradiance can be obtained in a conventional optical processor). The second image is a dark-field, or strioscopic[2], version of the first image. That is, the second image irradiance is obtained in the same way as the first one, except that a central stop is placed in the Fourier plane in order to have a dark-field image. The difference between the above images can be rephrased in the following terms. The first image can be thought of as formed by the interference between all the orders of the Fourier spectrum. In a similar manner, the second image can be thought of as formed only by the interference between the non zero orders. Then it is possible to say that the image difference is formed only by the interference between the zero order and the non zero orders. In other words, the *output image irradiance* is formed in the same manner as the *complex amplitude* of the object. Consequently, a square root operation has been implemented. This result can be employed to visualize the complex amplitude of the object as image irradiance variations, in particular to render phase structures visible. The mathematical treatment of the method is presented. As a second application of the difference of two images, we discuss a technique for pseudocoloring the gray levels of a black and white object. The method proposed exploits the following results of the technique of image difference.

i) The difference between a uniform background and a black and white object gives a negative or a reversed contrast version of the object provided that the irradiance transmittance of the object is smaller than the uniform background.

ISSN:0094-243X/81/650286-06$1.50

ii) It is possible to obtain simultaneously the sum of the two images and the difference. Now, the sum of a uniform background plus the object gives still a positive version of the object.

These results can be used to obtain a positive and a negative version of the object; each of them coded in a different color. Thus a pseudocolor method through contrast reversal[3] can be implemented.

In principle, both new applications (square root operation and pseudocolor of gray levels) can be implemented with any technique for detecting the difference between two incoherent images. The choice of a particular method would depend on the application. In this work, for the sake of simplicity, the mathematical model is based on the method of subtraction by carrier-frequency photography[4].

THEORY

The subtraction between two image irradiances F(x) and G(x) is obtained, using the method of carrier-frequency photography, as follows. A double exposure photograph is made over the same photographic plate. In the first exposure the irradiance recorded is given by

$$I_1(x) = F(x)\left[1 + \cos 2\pi s_o x\right], \tag{1}$$

where s_o is the spatial frequency of the cosinusoidal irradiance grating. In the second exposure, the irradiance recorded is

$$I_2(x) = G(x)\left[1 - \cos 2\pi s_o x\right]. \tag{2}$$

Thus, the total image irradiance is given by the sum of Eqs. (1) and (2), i. e.

$$I(x) = F(x) + G(x) + \left[F(x) - G(x)\right] \cos 2\pi s_o x. \tag{3}$$

The difference between the two images can be obtained by selecting the Fourier spectrum being carried with frequency $\pm s_o$.

1. *Square Root Operation.*

Let us consider the case in which the signal F(x) is the image irradiance of an object with complex amplitude transmittance A(x) which is formed in a coherent optical processor with amplitude PSF P(x), that is

$$F(x) = |P(x) \otimes A(x)|^2 . \tag{4}$$

Then, the second signal G(x) should be given by

$$G(x) = |P(x) \otimes A(x) - K|^2, \tag{5}$$

where K is the complex background over the image.

The difference between these two signals is then given by the difference between Eqs. (4) and (5), that is

$$F(x) - G(x) = 2\,\mathrm{Re}\{K^* P(x) \otimes A(x)\} - |K|^2 . \tag{6}$$

In the particular case of a real object being imaged by an ideal optical system, Eq. (6) reduces to

$$F(x) - G(x) = 2K[A(x) - K] . \tag{7}$$

From Eq. (7) it is easy to appreciate that a square root operation is implemented . In the general case, $A(x) = |A(x)|\exp(j\Phi(x))$ and $P(x) \neq \delta(x)$, Eq. (6) can be written explicitly as

$$\begin{aligned} F(x) - G(x) &= 2\,\mathrm{Re}\{K^* P(x)\} \otimes |A(x)| \cos\Phi(x) \\ &- 2\,\mathrm{Im}\{K^* P(x)\} \otimes |A(x)| \sin\Phi(x) \\ &- |K|^2 . \end{aligned} \tag{8}$$

Equation (8) indicates that the image irradiance variations are, in general, related to both the real and the imaginary parts of the object. This result can be used for phase retrieval in similar manners as suggested previously [5,6].

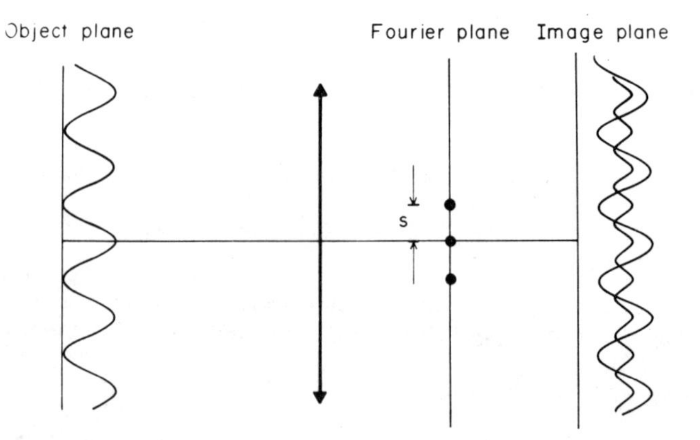

Fig. 1. Double frequency term produced by the quadratic detector.

Another very suggestive proof of the square root operation can be made in the following terms. Let us consider an object in the form of a cosinusoidal grating,

$$A(x) = 1 + \cos 2\pi s\, x, \tag{9}$$

as shown in Fig. (1). Assuming an ideal optical system, with unit magnification, the image irradiance, Eq. (4), is given by

$$F(x) = |A(x)|^2 = 1.5 + 2 \cos 2\pi s\, x + 0.5 \cos 2\pi(2s)\, x; \tag{10}$$

and as is indicated in Fig. (1), the image irradiance shows a spurious term with twice the spatial frequency of the object. This is due to the fact that the detector is a quadratic device. Then, the squared root operation consists in eliminating this double frequency grating. This is achieved by subtracting the dark-field image, which is given by

$$G(x) = 0.5 \cos 2\pi(2s)\, x. \tag{11}$$

Then, from Eqs. (10) and (11) it follows that the image irradiance difference shows the amplitude of the object as irradiance variations, namely

$$F(x) - G(x) = 1.5 + 2 \cos 2\pi s\, x. \tag{12}$$

It is to be noted that the contrast in the image is not the same as that of the object. This analysis can be used to decompose a general object in terms of cosinusoidal gratings; and in this way it is possible to obtain a transfer function, which can be associated with the square root operation.

2. *Pseudocolor Equidensitometry.*

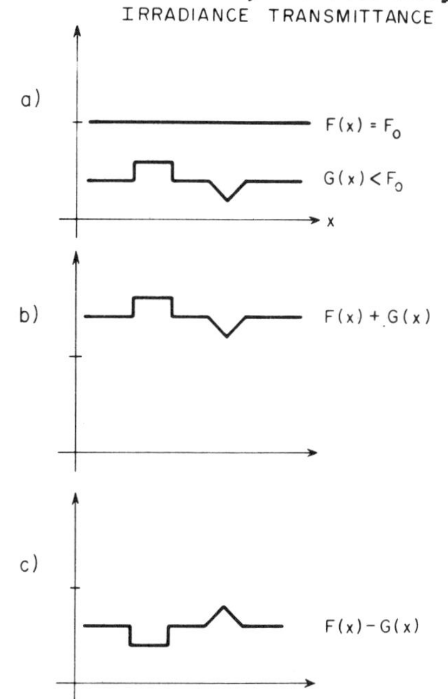

Fig. 2. Addition and subtraction.

For this application, the first signal is a constant, that is

$$F(x) = F_o. \tag{13}$$

The second signal is the gray-level object to be coded, in color (Fig. (2a)). While the addition of the object with a uniform background still gives a positive version of the object —Fig. (2b)—, the subtraction gives a negative one[7] —Fig. (2c)—.

Now, the frequency spectrum of the negative and positive versions (or equivalently, the subtraction and the addition) can be selected and coded in different colors, over the Fourier plane, see Fig. (3). In this way, a pseudocolor technique by contrast reversal can be implemented[3].

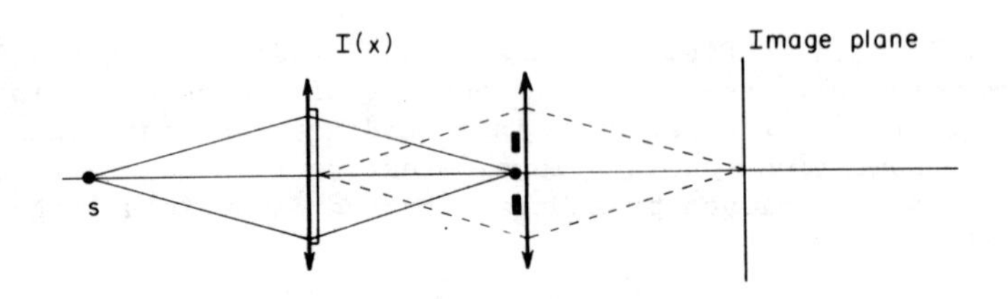

Fig. 3. Fourier spectrum of the positive and negative versions of a gray-level object.

CONCLUSIONS

Two novel applications of the image subtraction techniques were presented. In the first application, it was shown that the complex amplitude of the object can be displayed as image irradiance variations. In the second application, it was indicated that positive and negative versions of a gray-level object can be obtained from the addition and subtraction of the object with a uniform background; provided that the background is greater than the signal. Negative and positive versions can be used to codify in colors gray-level information, by the method of contrast reversal.

We are indebted to Prof. H. A. Fewerda for calling our attention to Ref. 5. Financial support from CONACYT for attending this meeting is acknowledged.

REFERENCES

1. J. F. Ebersole, Opt. Engineer. 14, 436 (1975).

2. L. C. Martin, Theory of the Microscope (Blackie, London, 1967) 350.

3. J. Santamaría, M. Gea, and J. Bescós, J. Optics 10, 151 (1979).

4. K. S. Pennington, P. M. Will, and G. L. Shelton, Opt. Commun. 2, 113 (1970).

5. J. Frank, Optik (Stuttgart) 38, 582 (1973).

6. J. Santamaría, J. Ojeda-Castañeda, and L. R. Berriel-Valdós (This proceeding, session M-4).

7. C. Roychoudhuri and D. Malacara, Appl. Optics 14, 1683 (1975).

THE WIGNER DISTRIBUTION FUNCTION AND ITS APPLICATIONS TO OPTICS

Martin J. Bastiaans
Technische Hogeschool Eindhoven, Afdeling der Elektrotechniek,
Postbus 513, 5600 MB Eindhoven, The Netherlands

ABSTRACT

The paper presents a review of some applications of the Wigner distribution function to optics. The Wigner distribution function $F(x,u)$ of a signal describes the signal in space x and spatial frequency u, simultaneously, and can be considered as a local spatial frequency spectrum of the signal. Although derived in terms of Fourier optics, the description of a signal by means of its Wigner distribution function closely resembles the ray concept in geometrical optics. Some examples are given to show this resemblance.

Properties of the Wigner distribution function are discussed, showing, for instance, its relation to Heisenberg's uncertainty principle and to some well-known radiometric quantities.

The concept of the Wigner distribution function is not restricted to deterministic signals; it can easily be extended to stochastic signals. Some examples of stochastic signals are considered.

The propagation of signals through linear systems can readily be expressed in terms of Wigner distribution functions. Again, the description of systems by Wigner distribution functions can be interpreted directly in terms of geometrical optics; for Luneburg's first-order system, for instance, such a description immediately yields the ray transformation matrix of the system, and the relation between the input and the output Wigner distribution function reads $F_o(x,u) = F_i(Ax+Bu,Cx+Du)$.

Transport equations for the Wigner distribution function in a general linear medium are derived. In the Liouville approximation they take the form of a first-order partial differential equation, which allows, once more, a geometric-optical interpretation: along the path of a geometric-optical light ray the Wigner distribution function has a constant value.

1. INTRODUCTION

Let us describe a scalar optical signal by $\tilde{\phi}(x,y,z,t)$, where x,y,z denote space variables and t represents the time variable; a tilde on top of a symbol will mean throughout that we are dealing with a function in the time domain. The *temporal Fourier transform* $\phi(x,y,z,\omega)$ of a signal will be defined as

$$\phi(x,y,z,\omega) = \int\tilde{\phi}(x,y,z,t)\exp[i\omega t]dt \quad . \tag{1.1}$$

(Unless otherwise stated, all integrations and summations in this paper extend from $-\infty$ to $+\infty$). For the sake of convenience we shall omit the temporal frequency variable ω in the formulas, since in the present discussion the temporal frequency dependence is of no importance. In other words, we can restrict ourselves to a time-harmonic

ISSN:0094-243X/81/650292-21$1.50

signal, which is described by its *complex amplitude* $\phi(x,y,z)$. For the time being we shall consider signals in a plane $z=$ constant, and therefore we can omit the space variable z in the formulas. Furthermore, we shall restrict ourselves to the one-dimensional case, where the signals are functions of the space variable x only; the extension to two dimensions is straightforward. The signals we are dealing with are thus described by a function $\phi(x)$.

It is sometimes convenient to describe a signal $\phi(x)$ not in the space domain, but in the (spatial) frequency domain by means of its *(spatial) frequency spectrum*, i.e., the *(spatial) Fourier transform* $\bar{\phi}(u)$ of the function $\phi(x)$, which is defined by

$$\bar{\phi}(u) = \int \phi(x)\exp[-iux]dx \; ; \tag{1.2}$$

a bar on top of a symbol will mean throughout that we are dealing with a function in the frequency domain. The frequency spectrum shows us the *global* distribution of the energy of the signal as a function of frequency. However, one is often more interested in the *local* distribution of the energy as a function of frequency. Geometrical optics, for instance, is usually treated in terms of rays, and the signal is described by giving the directions (i.e., frequencies) of the rays that should be present at a certain point. Hence, we look for a description of the signal that might be called the *local frequency spectrum* of the signal.

The need for a local frequency spectrum arises in other disciplines, too. It is the case in music, for instance, where a signal is usually described not by a time function nor by the temporal Fourier transform of that function, but by its *musical score*; indeed, when a composer writes a score, he prescribes the frequencies of the tones that should be present at a certain time. It arises also in mechanics, where the position and the momentum of a particle are given simultaneously, leading to a description of mechanical phenomena in a *phase space*. It was in mechanics that Wigner introduced a distribution function[1] which provided a description of mechanical phenomena in this phase space.

In section 2 we shall introduce the *Wigner distribution function* $F(x,u)$ in optics. The Wigner distribution function describes the signal in space x and frequency u, simultaneously. It is thus a function of two variables, derived, however, from a function of one variable. Therefore, it must satisfy certain restrictions, or, to put it another way: not any function of two variables is a Wigner distribution function. The restrictions that a Wigner distribution function must satisfy correspond to *Heisenberg's uncertainty principle* in mechanics, which states the impossibility of a too accurate determination of both the position and the momentum of a particle. These and other properties of the Wigner distribution function are considered in section 3. In section 4 the concept of the Wigner distribution function will be extended to partially coherent light, while the propagation of the Wigner distribution function will be studied in sections 5 and 6.

2. THE WIGNER DISTRIBUTION FUNCTION

We shall now define the Wigner distribution function[2,3,4] $F(x,u)$ of the signal $\phi(x)$ by

$$F(x,u) = \int \phi(x+\tfrac{1}{2}x')\phi^*(x-\tfrac{1}{2}x')\exp[-iux']dx' \; ; \tag{2.1a}$$

the asterisk in this definition denotes complex conjugation. The Wigner distribution function is a function that may act as a local frequency spectrum of the signal; indeed, with x as a parameter, the integral in definition (2.1a) represents a Fourier transformation (with frequency variable u) of the product $\phi(x+\frac{1}{2}x')\phi^*(x-\frac{1}{2}x')$. Instead of the definition in the space domain, there exists an equivalent definition in the frequency domain, reading

$$F(x,u) = \frac{1}{2\pi}\int \bar{\phi}(u+\tfrac{1}{2}u')\bar{\phi}^*(u-\tfrac{1}{2}u')\exp[iu'x]du' \; . \tag{2.1b}$$

The Wigner distribution function $F(x,u)$ represents the signal in space and frequency, simultaneously. It thus forms an intermediate signal description between the pure space representation $\phi(x)$ and the pure frequency representation $\bar{\phi}(u)$. Furthermore, this simultaneous space-frequency description closely resembles the ray concept in geometrical optics, where the position and direction of a ray are also given simultaneously. In a way, $F(x,u)$ is the amplitude of a ray passing through the point (=position) x and having a frequency (=direction) u.

The Wigner distribution function $F(x,u)$ is related to the *ambiguity function* $A(x',u')$, which is defined by the equivalent definitions

$$A(x',u') = \int \phi(x+\tfrac{1}{2}x')\phi^*(x-\tfrac{1}{2}x')\exp[iu'x]dx \tag{2.2a}$$

and

$$A(x',u') = \frac{1}{2\pi}\int \bar{\phi}(u+\tfrac{1}{2}u')\bar{\phi}^*(u-\tfrac{1}{2}u')\exp[-iux']du \; , \tag{2.2b}$$

and which is an intermediate signal description, as well[5,6,7]. It is easy to show that the Wigner distribution function and the ambiguity function are related through a double Fourier transformation.

We shall give some examples to illustrate the concept of the Wigner distribution function.

2.1. POINT SOURCE

The Wigner distribution function of the point source

$$\phi(x) = \delta(x-x_o) \tag{2.3}$$

takes the form

$$F(x,u) = \delta(x-x_o) \; . \tag{2.4}$$

We remark that at only one point $x=x_0$, all frequencies are present, while there is no contribution from other points. In fact, this is exactly what we expect as the local frequency spectrum of a point source.

2.2. PLANE WAVE

As a second example we consider a plane wave, described in the space domain by

$$\phi(x) = \exp[iu_0 x] , \tag{2.5a}$$

or, equivalently, in the frequency domain by

$$\bar{\phi}(u) = 2\pi\delta(u-u_0) . \tag{2.5b}$$

A plane wave and a point source are *dual* to each other, i.e., the Fourier transform of one function has the same form as the other function. Due to this duality, the Wigner distribution function of a plane wave will be the same as the one of the point source, but rotated in the space-frequency plane over $90°$. Indeed, the Wigner distribution of the plane wave (2.5) takes the form

$$F(x,u) = 2\pi\delta(u-u_0) . \tag{2.6}$$

We remark that at all points, only one frequency $u=u_0$ manifests itself, which is exactly what we expect as the local frequency spectrum of a plane wave.

2.3. QUADRATIC-PHASE SIGNAL

The quadratic-phase signal

$$\phi(x) = \exp[i\tfrac{1}{2}\alpha x^2] \tag{2.7}$$

represents, at least for small x, a *spherical wave* whose *curvature* is equal to α. The Wigner distribution function for such a signal reads

$$F(x,u) = 2\pi\delta(u-\alpha x) , \tag{2.8}$$

and we conclude that at any point x, only one frequency $u=\alpha x$ manifests itself. This corresponds exactly to the "ray picture" of a spherical wave.

2.4. SMOOTH-PHASE SIGNAL

The Wigner distribution function of the smooth-phase signal

$$\phi(x) = \exp[i\psi(x)] , \tag{2.9}$$

where $\psi(x)$ is a sufficiently smooth function of x, takes the form

$$F(x,u) \simeq 2\pi\delta\left(u-\frac{d\psi}{dx}\right) . \tag{2.10}$$

We remark that at any point x, only one frequency $u=d\psi/dx$ manifests itself. Note that relation (2.10) is exact when $\psi(x)$ varies only linearly or quadratically in x (see Examples 2.2 and 2.3). The concept of a smooth-phase signal and its Wigner distribution function may be useful, for instance, in the geometrical theory of aberrations[8] and in describing Bryngdahl's geometrical transformations[9].

2.5. GAUSSIAN SIGNAL

As a final example we shall consider the Gaussian signal

$$\phi(x) \overset{\Delta}{=} g(x;x_o,u_o) = \left(\frac{2}{\sigma}\right)^{\frac{1}{4}} \exp[-\frac{\pi}{\sigma}(x-x_o)^2+iu_o x] \; ; \tag{2.11}$$

in relation (2.11) σ is a positive quantity. The Wigner distribution function of this Gaussian signal reads

$$F(x,u) \overset{\Delta}{=} F_g(x-x_o,u-u_o) = 2\exp[-\frac{2\pi}{\sigma}(x-x_o)^2-\frac{\sigma}{2\pi}(u-u_o)^2] \; ; \tag{2.12}$$

note that it is a function that is Gaussian in both x and u. Let us consider this function a little further. The *center of gravity* of this function follows from the first-order moments

$$\langle x\rangle = \frac{\iint xF(x,u)\,dxdu}{\iint F(x,u)\,dxdu} = \frac{\int x|\phi(x)|^2dx}{\int|\phi(x)|^2dx} \tag{2.13a}$$

and

$$\langle u\rangle = \frac{\iint uF(x,u)\,dxdu}{\iint F(x,u)\,dxdu} = \frac{\int u|\bar{\phi}(u)|^2du}{\int|\bar{\phi}(u)|^2du} \; , \tag{2.13b}$$

from which we conclude that the function is located around the space-frequency point (x_o,u_o). The *effective widths* follow from the second-order central moments

$$\langle(x-x_o)^2\rangle = \frac{\iint (x-x_o)^2F(x,u)\,dxdu}{\iint F(x,u)\,dxdu} = \frac{\int (x-x_o)^2|\phi(x)|^2dx}{\int|\phi(x)|^2dx} \tag{2.14a}$$

and

$$\langle(u-u_o)^2\rangle = \frac{\iint (u-u_o)^2F(x,u)\,dxdu}{\iint F(x,u)\,dxdu} = \frac{\int (u-u_o)^2|\bar{\phi}(u)|^2du}{\int|\bar{\phi}(u)|^2du} \, . \tag{2.14b}$$

We find

$$\langle(x-x_o)^2\rangle = \frac{1}{2}\left(\frac{\sigma}{2\pi}\right) \tag{2.15a}$$

and

$$\langle(u-u_o)^2\rangle = \frac{1}{2}\left(\frac{2\pi}{\sigma}\right) \, . \tag{2.15b}$$

From the effective widths (2.15) we recognize Heisenberg's inequality[6,10]

$$<(x-x_o)^2>\cdot<(u-u_o)^2> \geq \frac{1}{4} , \tag{2.16}$$

where, in the present case of a Gaussian signal, the equality sign holds.

When we consider *Gaussian beams*, we have to deal with a Gaussian signal that is multiplied by a quadratic-phase signal:

$$\phi(x) = \exp[-\frac{\pi}{\sigma}x^2 + i\tfrac{1}{2}\alpha x^2] . \tag{2.17}$$

The Wigner distribution function of such a signal takes the form

$$F(x,u) = 2\exp[-\frac{2\pi}{\sigma}x^2 - \frac{\sigma}{2\pi}(u-\alpha x)^2] . \tag{2.18}$$

3. PROPERTIES OF THE WIGNER DISTRIBUTION FUNCTION

In this section we list some properties of the Wigner distribution function. Other properties can be found elsewhere[2,3,11].

3.1. INVERSION FORMULAS

The inverse relations of relations (2.1) read

$$\phi(x_1)\phi^*(x_2) = \frac{1}{2\pi}\int F\left(\tfrac{1}{2}(x_1+x_2),u\right)\exp[iu(x_1-x_2)]du \tag{3.1a}$$

and

$$\bar{\phi}(u_1)\bar{\phi}^*(u_2) = \int F\left(x,\tfrac{1}{2}(u_1+u_2)\right)\exp[-i(u_1-u_2)x]dx , \tag{3.1b}$$

respectively. In fact, these inverse relations formulate the conditions that a function of two variables must satisfy in order to be a Wigner distribution function: a function of x and u is a Wigner distribution function if and only if the pertinent right-hand integral in relation (3.1a) or (3.1b) is separable in the form of the left-hand side of that relation. From relations (3.1) we conclude that the signal $\phi(x)$ and its frequency spectrum $\bar{\phi}(u)$ can be reconstructed from the Wigner distribution function apart from a constant phase-factor.

3.2. REALITY

It follows immediately from the definitions (2.1) that the Wigner distribution function is real. Unfortunately, the Wigner distribution function is not necessarily non-negative, which prohibits a direct interpretation of this function as an energy density function.

3.3. SPACE AND FREQUENCY LIMITATION

It follows directly from the definitions that, if the signal $\phi(x)$ is limited to a certain space interval, the Wigner distribution function is limited to the same interval. Similarly, if the frequency spectrum $\bar{\phi}(u)$ is limited to a certain frequency interval, the Wigner distribution function is limited to the same interval.

3.4. SPACE AND FREQUENCY SHIFT

It follows immediately from the definitions that a space shift of the signal $\phi(x)$ yields the same shift for the Wigner distribution function. Similarly, a frequency shift of the frequency spectrum $\bar{\phi}(u)$, which corresponds to a modulation of the signal $\phi(x)$, yields the same shift for the Wigner distribution function.

3.5. SOME EQUALITIES AND INEQUALITIES

From the relation

$$\frac{1}{2\pi}\iint F(x,u)\,dx\,du = \int|\phi(x)|^2dx = \frac{1}{2\pi}\int|\bar{\phi}(u)|^2du\ , \qquad (3.2)$$

which follows directly from the definitions (2.1), we conclude that the integral of the Wigner distribution function over the entire space-frequency plane represents the *total energy* of the signal. It is not difficult to recognize Parseval's theorem in relation (3.2).

An important relationship between the Wigner distribution functions of two signals and the signals themselves has been formulated by Moyal[12]; it reads

$$\frac{1}{2\pi}\iint F_1(x,u)F_2(x,u)\,dx\,du =$$

$$= \left|\int\phi_1(x)\phi_2^*(x)\,dx\right|^2 = \left|\frac{1}{2\pi}\int\bar{\phi}_1(u)\bar{\phi}_2^*(u)\,du\right|^2 . \qquad (3.3)$$

This relationship has an application in averaging one Wigner distribution function with another Wigner distribution function, which, unlike the Wigner distribution function itself, always results in a non-negative function. Averaging the Wigner distribution function with the Gaussian Wigner distribution function (2.12) is of some practical importance[11,13,14,15]; in that case relation (3.3) takes the special form

$$\frac{1}{2\pi}\iint F(x,u)F_g(x-x_o,u-u_o)\,dx\,du = \left|\int\phi(x)g^*(x;x_o,u_o)\,dx\right|^2 . \qquad (3.4)$$

There appears to be a relationship with the *complex spectrogram*[13,16,17]

$$G\{\phi(x);x_o,u_o\} = \int\phi(x)g^*(x;x_o,u_o)\,dx$$

of the signal, and with *Gabor's expansion*[16,17,18,19]

$$\phi(x) = \sum_{mn}\sum a_{mn}\,g\left(x;mX,n\frac{2\pi}{X}\right)$$

of the signal into a discrete set of shifted and modulated Gaussian signals.

Relations (3.2) and (3.3), together with Schwartz' inequality, yield the relationship

$$\frac{1}{2\pi}\iint F_1(x,u)F_2(x,u)\,dxdu \le \left(\frac{1}{2\pi}\iint F_1(x,u)\,dxdu\right)\left(\frac{1}{2\pi}\iint F_2(x,u)\,dxdu\right), \quad (3.5)$$

which can be considered as Schwartz' inequality for Wigner distribution functions.

Another important inequality, which has been formulated by De Bruijn[11], reads

$$\iint\left(\frac{2\pi}{\sigma}x^2+\frac{\sigma}{2\pi}u^2\right)^n F(x,u)\,dxdu \ge n!\iint F(x,u)\,dxdu \; ; \quad (3.6)$$

in relation (3.6) n is a non-negative integer. For the special case n=1, this inequality reduces to

$$\frac{2\pi}{\sigma}\langle x^2\rangle + \frac{\sigma}{2\pi}\langle u^2\rangle \ge 1 \; , \quad (3.7)$$

from which relation we can directly derive Heisenberg's inequality [6,10]

$$\langle x^2\rangle\cdot\langle u^2\rangle \ge \frac{1}{4} \, . \quad (3.8)$$

One might think that relation (3.6) expresses the fact that the Wigner distribution function cannot be concentrated upon a small neighbourhood of a single point. Properly speaking, however, inequalities of the type (3.6) do not prove the impossibility of such a strong concentration. For example, if 99% of the energy of the signal lies very close to the origin, and the remaining 1% lies in a part of the space-frequency plane far away from the origin, then that 1% can make the left-hand side of relation (3.6) large. This objection does not hold against the inequality[11]

$$\iint\left\{1-\exp\left[-\frac{1}{2}\left(\frac{x^2}{X^2}+\frac{u^2}{U^2}\right)\right]\right\} F(x,u)\,dxdu \ge \frac{1}{1+2XU}\iint F(x,u)\,dxdu \, . \quad (3.9)$$

3.6. RADIOMETRIC QUANTITIES

Several integrals of the Wigner distribution functions can be interpreted as radiometric quantities. The integral over the frequency variable u,

$$\frac{1}{2\pi}\int F(x,u)\,du = |\phi(x)|^2 \; , \quad (3.10)$$

is proportional to the *intensity* of the signal, while the integral over the space variable x,

$$\int F(x,u)\,dx = |\bar{\phi}(u)|^2 \; , \quad (3.11)$$

is, apart from a factor $\cos^2\theta$ (where θ is the angle of observation), proportional to the *radiant intensity*[20,21]. The integral

$$J_z = \frac{1}{2\pi}\int\sqrt{k^2-u^2}\,F(x,u)\,du \quad (3.12a)$$

(where $k = 2\pi/\lambda = \omega/c$ is the usual wave number) is proportional to the *radiant emittance*[20,21]; when we combine the latter integral with the integral

$$J_x = \frac{1}{2\pi} \int uF(x,u)\,du \ , \tag{3.12b}$$

we can construct the two-dimensional vector

$$\vec{J} = (J_x, J_z) \ , \tag{3.12c}$$

which is proportional to the *geometrical vector flux*[22].

Although not strictly a radiometric quantity, we shall define the *local frequency* U(x) of the signal by

$$U(x) = \frac{\int uF(x,u)\,du}{\int F(x,u)\,du} = \text{Im}\left\{\frac{d}{dx}\ln\phi(x)\right\} \ ; \tag{3.13}$$

it is merely a normalized version of J_x. Note that when we represent the signal $\phi(x)$ by its absolute value $|\phi(x)|$ and its phase arg $\phi(x)$, the local frequency can be expressed as

$$U(x) = \frac{d}{dx}\arg\phi(x) \ . \tag{3.14}$$

4. THE WIGNER DISTRIBUTION FUNCTION FOR STOCHASTIC SIGNALS

The concept of the Wigner distribution function is not restricted to deterministic signals, but can be applied to stochastic signals, as well. Hence, partially coherent light can be described by a Wigner distribution function, too.

Let the signal $\tilde{\phi}(x,t)$ be a temporally stationary stochastic signal; the ensemble average of the product $\tilde{\phi}(x_1,t_1)\tilde{\phi}^*(x_2,t_2)$ is then a function of the time difference t_1-t_2 only:

$$E\,\tilde{\phi}(x_1,t_1)\tilde{\phi}^*(x_2,t_2) = \tilde{\Gamma}(x_1,x_2,t_1-t_2) \ . \tag{4.1}$$

The function $\tilde{\Gamma}(x_1,x_2,\tau)$ is known as the *mutual coherence function*[10,23] of the signal $\tilde{\phi}(x,t)$. The *mutual power spectrum*[10,23] $\Gamma(x_1,x_2,\omega)$ is defined as the temporal Fourier transform of the mutual coherence function:

$$\Gamma(x_1,x_2,\omega) = \int \tilde{\Gamma}(x_1,x_2,\tau)\exp[i\omega\tau]\,d\tau \ . \tag{4.2}$$

We shall again restrict ourselves to mono-chromatic signals, and shall omit the temporal frequency variable ω in the formulas.

We can now define the Wigner distribution function of a stochastic signal as

$$F(x,u) = \int \Gamma(x+\tfrac{1}{2}x', x-\tfrac{1}{2}x')\exp[-iux']\,dx' \ , \tag{4.3}$$

which definition is equivalent to definition (2.1a) for a deterministic signal, but with the product $\phi(x+\frac{1}{2}x')\phi^*(x-\frac{1}{2}x')$ replaced by the mutual power spectrum. Note that this definition is similar to the definition of Walther's *generalized radiance*[21,24,25].

We shall give some examples of Wigner distribution functions of stochastic signals.

4.1. INCOHERENT LIGHT

Incoherent light can be represented by the mutual power spectrum

$$\Gamma(x+\tfrac{1}{2}x',x-\tfrac{1}{2}x') = p(x)\delta(x') \ . \qquad (4.4)$$

The corresponding Wigner distribution function takes the form

$$F(x,u) = p(x) \ . \qquad (4.5)$$

4.2. SPATIALLY STATIONARY LIGHT

The mutual power spectrum of spatially stationary light reads

$$\Gamma(x+\tfrac{1}{2}x',x-\tfrac{1}{2}x') = s(x') \ . \qquad (4.6)$$

We remark that spatially stationary light and incoherent light are each other's duals, which is, in fact, the Van Cittert-Zernike theorem. The Wigner distribution function of spatially stationary light takes the form

$$F(x,u) = \bar{s}(u) \ , \qquad (4.7)$$

and has indeed the same form as the one for incoherent light, but rotated in the space-frequency plane over 90°.

4.3. QUASI-HOMOGENEOUS LIGHT

Quasi-homogeneous light[20,21] can locally be considered as spatially stationary, having, however, a slowly varying intensity. Its mutual power spectrum reads

$$\Gamma(x+\tfrac{1}{2}x',x-\tfrac{1}{2}x') \simeq p(x)s(x') \ , \qquad (4.8)$$

where $p(\cdot)$ is a slowly varying function compared to $s(\cdot)$. The Wigner distribution function of such quasi-homogeneous light takes the form

$$F(x,u) \simeq p(x)\bar{s}(u) \ . \qquad (4.9)$$

Note that incoherent light and spatially stationary light are special cases of quasi-homogeneous light, and note also that for quasi-homogeneous light the Wigner distribution function is non-negative.

4.4. COHERENT LIGHT

For coherent light, whose mutual power spectrum can be represented in the form

$$\Gamma(x_1,x_2) = q(x_1)q^*(x_2) \ , \qquad (4.10)$$

the Wigner distribution function reads

$$F(x,u) = \int q(x+\tfrac{1}{2}x')q^*(x-\tfrac{1}{2}x')\exp[-iux']dx' \ . \qquad (4.11)$$

Note that the Wigner distribution function (4.11) is similar to definition (2.1a) of the Wigner distribution function for deterministic signals.

5. PROPAGATION OF THE WIGNER DISTRIBUTION FUNCTION THROUGH LINEAR SYSTEMS

In this section we shall consider the propagation of the Wigner distribution function through linear, time-invariant optical systems. A linear, time-invariant optical system that transforms a signal ϕ_i in the input plane into a signal ϕ_o in the output plane, can be represented in four different ways, depending on whether we describe the input and the output signal in the space or in the frequency domain. We thus have four equivalent input-output relations:

$$\phi_o(x_o) = \int g(x_o,x_i)\phi_i(x_i)dx_i \ , \qquad (5.1a)$$

$$\bar{\phi}_o(u_o) = \int h(u_o,x_i)\phi_i(x_i)dx_i \ , \qquad (5.1b)$$

$$\phi_o(x_o) = \frac{1}{2\pi}\int \hat{h}(x_o,u_i)\bar{\phi}_i(u_i)du_i \ , \qquad (5.1c)$$

$$\bar{\phi}_o(u_o) = \frac{1}{2\pi}\int f(u_o,u_i)\bar{\phi}_i(u_i)du_i \ . \qquad (5.1d)$$

Relation (5.1a) is the usual system representation in the space domain by means of the *point spread function* $g(x_o,x_i)$; we remark that the function $g(x,x_i)$ is the response of the system in the space domain to the input signal $\phi_i(x) = \delta(x-x_i)$. Relation (5.1d) is a similar system representation in the frequency domain, where the function $f(u,u_i)$ is the response of the system in the frequency domain to the input signal $\bar{\phi}_i(u) = 2\pi\delta(u-u_i)$. Since the latter signal represents a plane wave, we shall call $f(u_o,u_i)$ the *wave spread function* of the system. Relations (5.1b) and (5.1c) are hybrid system representations, since the input and the output signal are described in different domains. We shall call the functions $h(u_o,x_i)$ and $\hat{h}(x_o,u_i)$ *hybrid spread functions*.

There is a similarity between the four *system functions* f,g,h and $\hat{h}$ and the four *Hamilton characteristics*[8] that can be used to describe geometric-optical systems. Indeed, for a geometric-optical system the *point characteristic* is nothing but the phase of the point spread function[26]; a similar relation holds between the *angle characteristic* and the wave spread function, and between the *mixed characteristics* and the hybrid spread function.

Unlike the *four* system representations (5.1), there is only *one* system representation when we describe the input and the output signal by their Wigner distribution functions. Indeed, combining the system representations (5.1) with the definitions of the Wigner distribution function (2.1) results in the relationship

$$F_o(x_o,u_o) = \frac{1}{2\pi}\iint K(x_o,u_o,x_i,u_i)F_i(x_i,u_i)dx_i du_i \ , \qquad (5.2)$$

in which the Wigner distribution functions in the input and the output plane are related through a superposition integral. The function $K(x_o,u_o,x_i,u_i)$ is completely determined by the system, and can be expressed in terms of the four system functions. We find

$$K(x_o,u_o,x_i,u_i) =$$

$$= \iint g(x_o+\tfrac{1}{2}x_o',x_i+\tfrac{1}{2}x_i')g^*(x_o-\tfrac{1}{2}x_o',x_i-\tfrac{1}{2}x_i')\exp[-i(u_o x_o'-u_i x_i')]dx_o'dx_i' \ , \qquad (5.3)$$

and similar expressions for the other system functions[26]. Relation (5.3) can be considered as the definition of a *double* Wigner distribution function; hence, the function $K(x_o,u_o,x_i,u_i)$ has all the properties of a Wigner distribution function, for example the property of reality.

In a formal way, the function $K(x,u,x_i,u_i)$ is the response of the system in the space-frequency domain to the input signal $F_i(x,u) = 2\pi\delta(x-x_i)\delta(u-u_i)$; in a formal way only, since there does not exist an actual signal yielding the Wigner distribution function $2\pi\delta(x-x_i)\delta(u-u_i)$. Nevertheless, such an input signal could be considered as a *single ray*, entering the system at the point x_i with a frequency u_i. Hence, we might call the function $K(x_o,u_o,x_i,u_i)$ the *ray spread function* of the system.

It is not difficult to express the ray spread function of a *cascade* of two systems in terms of the respective ray spread functions $K_1(x_o,u_o,x_i,u_i)$ and $K_2(x_o,u_o,x_i,u_i)$. The ray spread function of the overall system reads

$$K(x_o,u_o,x_i,u_i) = \frac{1}{2\pi}\iint K_2(x_o,u_o,x,u)K_1(x,u,x_i,u_i)dxdu \ . \qquad (5.4)$$

Some examples of ray spread functions of elementary optical systems[27] might elucidate the concept of the ray spread function.

5.1. THIN LENS

A thin lens having a focal distance f can be described by the point spread function

$$g(x_o,x_i) = \exp[-i\frac{k}{2f}x_o^2]\delta(x_o-x_i) \ . \qquad (5.5)$$

The corresponding ray spread function takes the form

$$K(x_o,u_o,x_i,u_i) = 2\pi\delta(x_i-x_o)\delta(u_i-u_o-\frac{k}{f}x_o) \ , \qquad (5.6)$$

and the input-output relation (5.2) for a thin lens reduces to

$$F_o(x,u) = F_i(x,u+\frac{k}{f}x) \ . \qquad (5.7)$$

Relation (5.6) represents exactly the geometric-optical behaviour of a thin lens: if a single ray is incident on a thin lens, it will leave the lens from the same position, but its direction will be changed according to the actual position.

5.2. FREE SPACE IN THE FRESNEL APPROXIMATION

The point spread function of a section of free space having a length z reads, at least in the Fresnel approximation,

$$g(x_o,x_i) = \left(\frac{k}{2\pi iz}\right)^{\frac{1}{2}}\exp[i\frac{k}{2z}(x_o-x_i)^2] \quad , \tag{5.8a}$$

while the wave spread function reads

$$f(u_o,u_i) = \exp[-i\frac{z}{2k}u_o^2]2\pi\delta(u_i-u_o) \quad . \tag{5.8b}$$

The similarity between the wave spread function of free space and the point spread function of a lens shows that these two systems are each other's duals. This becomes apparent also from their ray spread functions, which for free space takes the form

$$K(x_o,u_o,x_i,u_i) = 2\pi\delta(x_i-x_o+\frac{z}{k}u_o)\delta(u_i-u_o) \quad : \tag{5.9}$$

we conclude that the frequency behaviour of one system is similar to the space behaviour of the other system. The input-output relation (5.2) for a section of free space reduces to

$$F_o(x,u) = F_i(x-\frac{z}{k}u,u) \quad . \tag{5.10}$$

Relation (5.9) again represents exactly the geometric-optical behaviour of a section of free space: if a single ray propagates through free space, its direction will remain the same, but its position will change according to the actual direction.

5.3. FOURIER TRANSFORMER

For a Fourier transformer whose point spread function reads

$$g(x_o,x_i) = \left(\frac{\beta}{2\pi i}\right)^{\frac{1}{2}}\exp[-i\beta x_o x_i] \quad , \tag{5.11}$$

the ray spread function takes the form

$$K(x_o,u_o,x_i,u_i) = 2\pi\delta(x_i+\frac{u_o}{\beta})\delta(u_i-\beta x_o) \quad , \tag{5.12}$$

and the input-output relation (5.2) reduces to

$$F_o(x,u) = F_i(-\frac{u}{\beta},\beta x) \quad . \tag{5.13}$$

We conclude that the space and frequency domains are interchanged as can be expected for a Fourier transformer.

5.4. MAGNIFIER

Let a magnifier be represented by a point spread function

$$g(x_o,x_i) = \sqrt{m}\delta(mx_o-x_i) \quad . \tag{5.14}$$

Then its ray spread function will read

$$K(x_o,u_o,x_i,u_i) = 2\pi\delta(x_i-mx_o)\delta(u_i-\frac{u_o}{m}) \ , \tag{5.15}$$

and the input-output relation (5.2) reduces to

$$F_o(x,u) = F_i(mx,\frac{u}{m}) \ . \tag{5.16}$$

We note that the space and frequency domains are merely scaled, as can be expected for a magnifier.

5.5. FIRST-ORDER OPTICAL SYSTEM

Examples 5.1, 5.2, 5.3 and 5.4 are special cases of Luneburg's first-order optical systems[28]. A first-order optical system can, of course, be characterized by its system functions f,g,h and $\hat{h}$: they all have a constant absolute value, and their phases vary quadratically in the pertinent variables. (Note that a Dirac function can be considered as a limiting case of such a quadratically varying function). A system representation in terms of Wigner distribution functions, however, is far more elegant: the ray spread function of a first-order optical system has the form[29]

$$K(x_o,u_o,x_i,u_i) = 2\pi\gamma\delta(x_i-Ax_o-Bu_o)\delta(u_i-Cx_o-Du_o) \ , \tag{5.17}$$

and its input-output relation (5.2) reduces to

$$F_o(x,u) = \gamma F_i(Ax+Bu,Cx+Du) \ . \tag{5.18}$$

The constant γ is non-negative; it equals unity if the system is lossless[27,29], i.e., if for any input signal the total energy in the input plane equals that in the output plane. The four real constants A,B,C and D constitute a matrix

$$\begin{pmatrix} A & B \\ C & D \end{pmatrix} \tag{5.19}$$

which is *symplectic*[28,29,30]; for a 2x2 matrix, symplecticity can be expressed by the condition $AD-BC=1$.

From relation (5.17) we conclude that a *single* input ray entering a first-order system at the point x_i with a frequency u_i, will yield a *single* output ray leaving the system at the point x_o with a frequency u_o, where x_i,u_i,x_o and u_o are related by

$$\begin{pmatrix} x_i \\ u_i \end{pmatrix} = \begin{pmatrix} A & B \\ C & D \end{pmatrix} \begin{pmatrix} x_o \\ u_o \end{pmatrix} \ . \tag{5.20}$$

Relation (5.20) is a well-known geometric-optical matrix description of a first-order optical system[28]; the ABCD-matrix (5.19) is known as the *ray transformation matrix*[30].

Quadratic-phase signals (see Example 2.3) fit very well to a first-order system, since their general character remains unchanged when they are propagating through such a system. We recall that, through relation (2.8), a quadratic-phase signal is completely described by its curvature α. Let the input curvature be α_i; then the output curvature α_o is related to α_i by the bilinear relation[29]

$$\alpha_i = \frac{C+D\alpha_o}{A+B\alpha_o} , \tag{5.22}$$

which follows immediately from relations (2.8) and (5.18). In fact, the bilinear relation (5.22) also applies to Gaussian beams, if we describe such a beam formally by a *complex* curvature α[29].

6. TRANSPORT EQUATIONS FOR THE WIGNER DISTRIBUTION FUNCTION

In the previous section we studied, in Example 5.2, the propagation of the Wigner distribution function through free space, by considering a section of free space as an optical system. It is possible, however, to find the propagation of the Wigner distribution function through free space directly from the differential equation that the signal must satisfy. To show this, we let the space variable z enter into the formulas again, and we thus express the signal by $\phi(x,z)$ and its Wigner distribution function by $F(x,u;z)$; for convenience, we recall the definition of the Wigner distribution function

$$F(x,u;z) = \int \phi(x+\tfrac{1}{2}x',z)\phi^*(x-\tfrac{1}{2}x',z)\exp[-iux']dx' . \tag{6.1}$$

In the Fresnel approximation of free space, the signal $\phi(x,z)$ satisfies a differential equation which has the form of a diffusion equation of the parabolic type[10]:

$$-i\frac{\partial\phi}{\partial z} = \left(k + \frac{1}{2k}\frac{\partial^2}{\partial x^2}\right)\phi . \tag{6.2}$$

The propagation of the Wigner distribution function is described by a transport equation[31,32,33,34,35,36], which in this case takes the form

$$\frac{u}{k}\frac{\partial F}{\partial x} + \frac{\partial F}{\partial z} = 0 . \tag{6.3}$$

(Relation (6.3) is a special case of the transport equation (6.6) that corresponds to the more general differential equation (6.5), which will be studied in the next paragraph). The transport equation (6.3) has the solution

$$F(x,u;z) = F(x-\frac{u}{k}z,u;0) , \tag{6.4}$$

which is equivalent to the previously derived relation (5.10).

The differential equation (6.2) is a special case of the more general equation

$$-i\frac{\partial\phi}{\partial z} = L(x,-i\frac{\partial}{\partial x};z)\phi , \tag{6.5}$$

where L is some explicit function of the space variables x and z, and of the partial derivative of ϕ contained in the operator $\partial/\partial x$. The transport equation that corresponds to this differential equation reads

$$-\frac{\partial F}{\partial z} = 2\,\mathrm{Im}\left\{L\left(x+\frac{i}{2}\frac{\partial}{\partial u}, u-\frac{i}{2}\frac{\partial}{\partial x}; z\right)\right\}F \quad ; \tag{6.6a}$$

a derivation of this formula can be found in the Appendix. In the elegant, symbolic notation of Besieris and Tappert[36], the transport equation (6.6a) takes the form

$$-\frac{\partial F}{\partial z} = 2\,\mathrm{Im}\left\{L(x,u;z)\exp\left[\frac{i}{2}\left(\frac{\overleftarrow{\partial}}{\partial x}\frac{\overrightarrow{\partial}}{\partial u}-\frac{\overleftarrow{\partial}}{\partial u}\frac{\overrightarrow{\partial}}{\partial x}\right)\right]\right\}F \quad , \tag{6.6b}$$

where, depending on the directions of the arrows, the differential operators on the right-hand side operate on L(x,u;z) or F(x,u;z). In the *Liouville approximation* (or *geometric-optical approximation*) the transport equation (6.6b) reduces to

$$-\frac{\partial F}{\partial z} = 2\,\mathrm{Im}\left\{L(x,u;z)\left[1+\frac{i}{2}\left(\frac{\overleftarrow{\partial}}{\partial x}\frac{\overrightarrow{\partial}}{\partial u}-\frac{\overleftarrow{\partial}}{\partial u}\frac{\overrightarrow{\partial}}{\partial x}\right)\right]\right\}F \quad , \tag{6.7a}$$

which is a *linearized* version of relation (6.6b). In the usual notation this linearized transport equation reads

$$-\frac{\partial F}{\partial z} = 2(\mathrm{Im}L)F + \frac{\partial \mathrm{Re}L}{\partial x}\frac{\partial F}{\partial u} - \frac{\partial \mathrm{Re}L}{\partial u}\frac{\partial F}{\partial x} \quad . \tag{6.7b}$$

Relation (6.7b) is a first-order partial differential equation, which can be solved by the method of characteristics[37]: along a path described in a parameter notation by $x=x(s)$, $z=z(s)$, $u=u(s)$, and defined by the differential equations

$$\frac{dx}{ds} = -\frac{\partial \mathrm{Re}L}{\partial u} \quad , \quad \frac{dz}{ds} = 1 \quad , \quad \frac{du}{ds} = \frac{\partial \mathrm{Re}L}{\partial x} \quad , \tag{6.8}$$

the *partial* differential equation (6.7b) reduces to the *ordinary* differential equation

$$\frac{dF}{ds} + 2\,(\mathrm{Im}L)\,F = 0 \quad . \tag{6.9}$$

In the special case that L(x,u;z) is a *real* function of x,u and z, equation (6.9) implies that along the path defined by relations (6.8) the Wigner distribution function has a *constant* value.

Let us consider some examples.

6.1. FREE SPACE IN THE FRESNEL APPROXIMATION

In free space in the Fresnel approximation the signal is governed by equation (6.2), and the function L(x,u;z) reads

$$L(x,u;z) = k - \frac{u^2}{2k} \quad . \tag{6.10}$$

The corresponding transport equation (6.3) and its solution (6.4) have already been mentioned in the first paragraph of this section.

6.2. FREE SPACE

In free space (but not necessarily in the Fresnel approximation) the signal $\phi(x,z)$ must satisfy the Helmholtz equation, which we write in the form

$$-i\frac{\partial\phi}{\partial z} = \left(k^2+\frac{\partial^2}{\partial x^2}\right)^{\frac{1}{2}}\phi \ . \tag{6.11}$$

The function L(x,u;z) reads

$$L(x,u;z) = \sqrt{k^2-u^2} \ , \tag{6.12}$$

and the linearized transport equation takes the form

$$\frac{u}{k}\frac{\partial F}{\partial x}+\frac{\sqrt{k^2-u^2}}{k}\frac{\partial F}{\partial z} = 0 \ . \tag{6.13}$$

This linearized transport equation can again be solved explicitly; the solution reads

$$F(x,u;z) = F(x-\frac{u}{\sqrt{k^2-u^2}}z \ ,u;0) \ . \tag{6.14}$$

Note that in the Fresnel approximation the relations (6.11), (6.12), (6.13) and (6.14) reduce to (6.2), (6.10), (6.3) and (6.4), respectively.

When we integrate the linearized transport equation (6.13) over the frequency variable u and we use the definitions (3.12), we find

$$\frac{\partial J_x}{\partial x}+\frac{\partial J_z}{\partial z} = 0 \ , \tag{6.15}$$

which shows that the geometrical vector flux $\vec{J}$ has zero divergence[22].

6.3. WEAKLY INHOMOGENEOUS MEDIUM

In a weakly inhomogeneous medium the differential equation that the signal must satisfy, and the function L(x,u;z), are described by relations (6.11) and (6.12), respectively, but now with $k = k(x,z)$. The linearized transport equation now takes the form

$$\frac{u}{k}\frac{\partial F}{\partial x}+\frac{\sqrt{k^2-u^2}}{k}\frac{\partial F}{\partial z}+\frac{\partial k}{\partial x}\frac{\partial F}{\partial u} = 0 \ , \tag{6.16}$$

which, in general, cannot be solved explicitly. With the method of characteristics we conclude that along the path defined by

$$\frac{dx}{ds} = \frac{u}{k} \ , \ \frac{dz}{ds} = \frac{\sqrt{k^2-u^2}}{k} \ , \ \frac{du}{ds} = \frac{\partial k}{\partial x} \ , \tag{6.17}$$

the Wigner distribution function has a constant value. When we eliminate the frequency variable u from equations (6.17), we are immediately led to

$$\frac{d}{ds}\left(k\frac{dx}{ds}\right) = \frac{\partial k}{\partial x} \ , \ \frac{d}{ds}\left(k\frac{dz}{ds}\right) = \frac{\partial k}{\partial z} \ , \tag{6.18}$$

which are the equations for an optical light ray in geometrical optics[8]. Note that in a homogeneous medium (i.e., $\partial k/\partial x \equiv \partial k/\partial z \equiv 0$) the linearized transport equation (6.16) reduces to (6.13), and that the ray paths become straight lines.

NOTE

Until now the space variables x and z in the Wigner distribution function were treated differently: a frequency variable u was associated with the space variable x, but there was not such a frequency variable associated with the space variable z. This different treatment of the space variables corresponds to the fact that the signal $\phi(x,z)$ may be chosen arbitrarily, for instance, in a plane $z = \text{constant}$; the z-dependence then follows from the properties of the medium through which the signal is propagating. We can, however, define a *higher-dimensional* Wigner distribution function, treating the space variables x and z in like manner, by

$$\mathcal{F}(x,u,z,w) =$$

$$= \iint \phi(x+\tfrac{1}{2}x',z+\tfrac{1}{2}z')\,\phi^*(x-\tfrac{1}{2}x',z-\tfrac{1}{2}z')\exp[-i(ux'+wz')]dx'dz' \ . \qquad (6.19)$$

We can always regain the original Wigner distribution function $F(x,u;z)$ from the higher-dimensional one through the relation

$$F(x,u;z) = \frac{1}{2\pi}\int \mathcal{F}(x,u,z,w)dw \ . \qquad (6.20)$$

The use of the higher-dimensional Wigner distribution function may lead to some mathematical elegance, as we shall show in the next paragraph.

The differential equation (6.5) is a special case of the more general equation

$$L(x,-i\frac{\partial}{\partial x},z,-i\frac{\partial}{\partial z})\,\phi = 0 \ . \qquad (6.21)$$

The corresponding equation for the Wigner distribution function $\mathcal{F}(x,u,z,w)$ can easily be derived to read

$$2\,\mathrm{Im}\left\{L(x,u,z,w)\exp\left[\frac{i}{2}\left(\frac{\overleftarrow{\partial}}{\partial x}\frac{\overrightarrow{\partial}}{\partial u}-\frac{\overleftarrow{\partial}}{\partial u}\frac{\overrightarrow{\partial}}{\partial x}+\frac{\overleftarrow{\partial}}{\partial z}\frac{\overrightarrow{\partial}}{\partial w}-\frac{\overleftarrow{\partial}}{\partial w}\frac{\overrightarrow{\partial}}{\partial z}\right)\right]\right\}\mathcal{F} = 0 \ . \qquad (6.22)$$

As an example we shall again study the propagation in free space, which is governed by the Helmholtz equation

$$\frac{1}{k}\left(\frac{\partial^2}{\partial x^2}+\frac{\partial^2}{\partial z^2}+k^2\right)\phi = 0 \ . \qquad (6.23)$$

The function $L(x,u,z,w)$ reads

$$L(x,u,z,w) = k - \frac{u^2+w^2}{k} \ , \qquad (6.24)$$

and the higher-order Wigner distribution function satisfies the partial differential equation

$$\frac{u}{k}\frac{\partial\mathcal{F}}{\partial x}+\frac{w}{k}\frac{\partial\mathcal{F}}{\partial z} = 0 \ , \qquad (6.25)$$

which resembles relation (6.13), but, unlike the latter one, is exact.

APPENDIX: DERIVATION OF THE TRANSPORT EQUATION

Starting from the differential equation

$$\left\{i\frac{\partial}{\partial z}+L(x,-i\frac{\partial}{\partial x};z)\right\}\phi = 0 , \tag{A.1}$$

we can formulate the relation

$$\left\{i\frac{\partial}{\partial z}+L(x_1,-i\frac{\partial}{\partial x_1};z) - L^*(x_2,-i\frac{\partial}{\partial x_2};z)\right\}\phi(x_1,z)\phi^*(x_2,z) = 0 \tag{A.2}$$

for the product $\phi(x_1,z)\phi^*(x_2,z)$. Expressing this product in terms of the Wigner distribution function $F(x,u;z)$ through the inversion formula (3.1a) yields

$$\frac{1}{2\pi}\int\left\{i\frac{\partial}{\partial z}+L(x_1,-i\frac{\partial}{\partial x_1};z) - L^*(x_2,-i\frac{\partial}{\partial x_2};z)\right\}\times$$

$$\times F\left(\tfrac{1}{2}(x_1+x_2),u_o;z\right)\exp[iu_o(x_1-x_2)]du_o = 0 , \tag{A.3}$$

which can be expressed as

$$\frac{1}{2\pi}\int\left\{i\frac{\partial}{\partial z}+L(x+\tfrac{1}{2}x',u_o-\frac{i}{2}\frac{\partial}{\partial x'};z) - L^*(x-\tfrac{1}{2}x',u_o-\frac{i}{2}\frac{\partial}{\partial x'};z)\right\}\times$$

$$\times F(x,u_o;z)\exp[iu_o x']du_o = 0 . \tag{A.4}$$

Multiplication of relation (A.4) by $\exp[-iux']$, and writing the integral over x' yields

$$\frac{1}{2\pi}\iint\left\{i\frac{\partial}{\partial z}+L(x+\tfrac{1}{2}x',u_o-\frac{i}{2}\frac{\partial}{\partial x'};z) - L^*(x-\tfrac{1}{2}x',u_o-\frac{i}{2}\frac{\partial}{\partial x'};z)\right\}\times$$

$$\times F(x,u_o;z)\exp[i(u_o-u)x']du_o dx' = 0 , \tag{A.5}$$

which can be expressed as

$$\frac{1}{2\pi}\iint\left\{i\frac{\partial}{\partial z}+L(x+\frac{i}{2}\frac{\partial}{\partial u},u_o-\frac{i}{2}\frac{\partial}{\partial x};z) - L^*(x+\frac{i}{2}\frac{\partial}{\partial u},u-\frac{i}{2}\frac{\partial}{\partial x};z)\right\}\times$$

$$\times F(x,u_o;z)\exp[i(u_o-u)x']du_o dx' = 0 . \tag{A.6}$$

Carrying out the integrations in relation (A.6) leads to

$$\left\{i\frac{\partial}{\partial z}+L(x+\frac{i}{2}\frac{\partial}{\partial u},u-\frac{i}{2}\frac{\partial}{\partial x};z) - L^*(x+\frac{i}{2}\frac{\partial}{\partial u},u-\frac{i}{2}\frac{\partial}{\partial x};z)\right\}F = 0, \tag{A.7}$$

which is equivalent to the desired result

$$-\frac{\partial F}{\partial z} = 2\,\mathrm{Im}\left\{L(x+\frac{i}{2}\frac{\partial}{\partial u},u-\frac{i}{2}\frac{\partial}{\partial x};z)\right\}F = 0 . \tag{A.7}$$

REFERENCES

1 E. Wigner, "On the quantum correction for thermodynamic equilibrium", Phys.Rev. 40, 749-759 (1932).

2 H. Mori, I. Oppenheim, and J. Ross, "Some topics in quantum statistics: The Wigner function and transport theory", in *Studies in Statistical Mechanics*, edited by J. de Boer and G.E. Uhlenbeck (North-Holland, Amsterdam, 1962), Vol. 1, 213-298.

3 N.G. de Bruijn, "A theory of generalized functions, with applications to Wigner distribution an Weyl correspondence", Nieuw Archief voor Wiskunde (3) 21, 205-280 (1973).

4 M.J. Bastiaans, "The Wigner distribution function applied to optical signals and systems", Opt.Commun. 25, 26-30 (1978).

5 A. Papoulis, "Ambiguity function in Fourier optics", J.Opt.Soc. Am. 64, 779-788 (1974).

6 A. Papoulis, *Signal Analysis* (McGraw-Hill, New York, 1977).

7 J.-P. Guigay, "The ambiguity function in diffraction and isoplanatic imaging by partially coherent beams", Opt.Commun. 26, 136-138 (1978).

8 M. Born and E. Wolf, *Principles of Optics* (Pergamon, Oxford, 1975).

9 O. Bryngdahl, "Geometrical transformations in optics", J.Opt. Soc.Am. 64, 1092-1099 (1974).

10 A. Papoulis, *Systems and Transforms with Applications in Optics* (McGraw-Hill, New York, 1968).

11 N.G. de Bruijn, "Uncertainty principles in Fourier analysis", in *Inequalities*, edited by O. Shisha (Academic, New York, 1967), 57-71.

12 J.E. Moyal, "Quantum mechanics as a statistical theory", Proc. Cambridge Philos.Soc. 45, 99-132 (1949).

13 W.D. Mark, "Spectral analysis of the convolution and filtering of non-stationary stochastic processes", J.Sound Vib. 11, 19-63 (1970).

14 H.O. Bartelt, K.-H. Brenner, and A.W. Lohmann, "The Wigner distribution function and its optical production", Opt.Commun. 32, 32-38 (1980).

15 A.J.E.M. Janssen, "Weighted Wigner distributions vanishing on lattices", to be published in J.Math.Anal.Appl.

16 C.W. Helstrom, "An expansion of a signal in Gaussian elementary signals", IEEE Trans.Inform.Theory IT-12, 81-82 (1966).

17 M.J. Bastiaans, "A sampling theorem for the complex spectrogram, and Gabor's expansion of a signal in Gaussian elementary signals", to be published in the Proceedings of the International Optical Computing Conference, Washington, D.C., 1980; to be published also in Opt.Eng.

18 D. Gabor, "Theory of communication", J.Inst.Elec.Eng. 93 (III), 429-457 (1946).

19 M.J. Bastiaans, "Gabor's expansion of a signal into Gaussian elementary signals", Proc.IEEE 68, 538-539 (1980).

20 W.H. Carter, and E. Wolf, "Coherence and radiometry with quasi-homogeneous planar sources", J.Opt.Soc.Am. 67, 785-796 (1977).

21 E. Wolf, "Coherence and radiometry", J.Opt.Soc.Am. 68, 6-17 (1978).

22 R. Winston, and W.T. Welford, "Geometrical vector flux and some new nonimaging concentrators", J.Opt.Soc.Am. 69, 532-536 (1979).

23 M.J. Bastiaans, "A frequency-domain treatment of partial coherence", Opt.Acta 24, 261-274 (1977).

24 A. Walther, "Radiometry and coherence", J.Opt.Soc.Am. 58, 1256-1259 (1968).

25 A.T. Friberg, "On the existence of a radiance function for finite planar sources of arbitrary states of coherence", J.Opt. Soc.Am. 69, 192-198 (1979).

26 M.J. Bastiaans, "The Wigner distribution function and Hamilton's characteristics of a geometric-optical system", Opt.Commun. 30, 321-326 (1979).

27 H.J. Butterweck, "General theory of linear, coherent optical data-processing systems", J.Opt.Soc.Am. 67, 60-70 (1977).

28 R.K. Luneburg, *Mathematical Theory of Optics* (University of California, Berkeley and Los Angeles, 1966).

29 M.J. Bastiaans, "Wigner distribution function and its application to first-order optics", J.Opt.Soc.Am. 69, 1710-1716 (1979).

30 G.A. Deschamps, "Ray techniques in electromagnetics", Proc.IEEE 60, 1022-1035 (1972).

31 H. Bremmer, "General remarks concerning theories dealing with scattering and diffraction in random media", Radio Sci. 8, 511-534 (1973).

32 J.J. McCoy, and M.J. Beran, "Propagation of beamed signals through inhomogeneous media: A diffraction theory", J.Acoust. Soc.Am. 59, 1142-1149 (1976).

33 H. Bremmer, "The Wigner distribution and transport equations in radiation problems", J.Appl.Sci.Eng. A 3, 251-260 (1979).

34 M.J. Bastiaans, "Transport equations for the Wigner distribution function", Opt.Acta 26, 1265-1272 (1979).

35 M.J. Bastiaans, "Transport equations for the Wigner distribution function in an inhomogeneous and dispersive medium", Opt.Acta 26, 1333-1344 (1979).

36 I,M, Besieris, and F.D. Tappert, "Stochastic wave-kinetic theory in the Liouville approximation", J.Math.Phys. 17, 734-743 (1976).

37 R. Courant and D. Hilbert, *Methods of Mathematical Physics* (Interscience, New York, 1960), Vol. 2, Chap. 2.

PHASE-SPACE METHODS FOR PARTIALLY COHERENT WAVEFIELDS*

Ari T. Friberg
The Institute of Optics, The University of Rochester
Rochester, NY, 14627, USA

ABSTRACT

As is well known, the expectation value of a quantum mechanical operator may be expressed in a form of a classical c-number average with respect to the so-called generalized phase-space distribution functions. The observable quantities in the theory of radiometry with partially coherent light may be represented as appropriate integral transforms of the so-called generalized radiance functions that bear some formal resemblance to such distribution functions. In the present paper, we consider analogies between various quantum mechanical phase-space distribution functions and generalized radiance functions associated with an optical wavefield of any state of coherence. Some misconceptions surrounding the implications of the possible negativeness of the generalized radiance are discussed. Several unsolved problems in the area of generalized radiometry with partially coherent wavefields are also mentioned.

1. INTRODUCTION

In 1932 Wigner[1] introduced a mathematical technique which made it possible to calculate the expectation values of quantum mechanical operators in a manner similar to that of classical statistical mechanics.[2] At the root of this method is a quantity now commonly known as the Wigner distribution function (WDF), which may be regarded as a generalized phase-space distribution function.[3] In recent years some analogs of the WDF have been employed in a number of investigations dealing with Fourier optics[4-6] and with radiometry of partially coherent light.[7,8] An analog appearing in the Fourier optical studies, often called the Wigner function,[9] has been interpreted to represent either a generalized geometrical light ray or a local spatial frequency spectrum, whereas an analog appearing in the radiometric studies has acquired a different physical meaning. However, the analogy between the Wigner function and the WDF, as presented in many of these investigations, appears to a large degree to be a rather formal one, based mostly on the apparent similarity of their mathematical structure.

Since its introduction, it has become gradually realized that the WDF is merely one of many possible distribution functions that may be used to calculate quantum mechanical averages in a generalized phase-space. In the present paper we show that, when properly interpreted, there exists, indeed, a deep analogy between the Wigner function and the WDF. We also consider, in a broader context, some analogies between other phase-space distribution functions and some quantities that have been employed in conventional optics in recent years. Particular emphasis will be placed on quantities arising in the analysis of radiative energy transfer in partially coherent wavefields. Our main object of

ISSN:0094-243X/81/650313-19$1,50

of interest will be the so-called generalized radiance (or generalized specific intensity) associated with an optical field of any state of coherence.

We begin with a brief review of the basic results of the phase-space formulation of quantum mechanics and we will discuss a number of commonly used phase-space distribution functions. We will recall that the phase-space distribution functions cannot be considered to be true probability densities, because they may, in general, become negative or violate some other basic postulates of probability theory. For this reason, they are sometimes referred to as quasi-probabilities.

After briefly summarizing the fundamental concepts of conventional radiometry and of optical coherence theory, we will consider expressions for the generalized radiance associated with a planar source of any state of coherence. We note that no generalized radiance function can be introduced that would possess, for all possible states of coherence of the source, all the properties that one normally attributes to the radiance in conventional radiometry.

Some misconceptions regarding an often suggested relationship between the possible negativeness of the generalized radiance and the reversal of the direction of the energy flow will also be briefly discussed.

We conclude this paper by mentioning some unsolved problems relating to the foundations of radiometry with partially coherent light.

2. REVIEW OF THE PHASE-SPACE FORMULATION OF QUANTUM MECHANICS

The generalized phase-space methods, originated in the work of Wigner[1] and later developed by Moyal[10] and others,[11] have proven to be very powerful mathematical techniques for treatments of various quantum mechanical problems. Such methods have played a particularly important role in several investigations in quantum optics, especially in studies concerning the coherence properties of light[12] and in the formulation of laser theory.[13] A systematic approach to the calculus of functions of non-commuting operators and to the general phase-space methods in quantum optics was developed by Agarwal and Wolf.[14]

In the present review, we consider only systems with one degree of freedom, such as a particle constrained to move in one dimension. The basic dynamical variables are then the position q and the momentum p, which in the classical theory commute with each other. In the quantized theory these quantities are replaced by Hermitian operators[15] $\hat{q}$ and $\hat{p}$ that obey the commutation relation

$$[\hat{q},\hat{p}] \equiv \hat{q}\hat{p}-\hat{p}\hat{q} = i\hbar, \tag{1}$$

where $\hbar$ is Planck's constant divided by 2π. When one considers physical quantities that depend on products of position and momentum, the order in which these variables appear is then of importance. This raises the question of a rule of association between classical functions of the commuting variables q and p and operator functions of the non-commuting operators $\hat{q}$ and $\hat{p}$.

As an example, let us consider the so-called standard rule of association. According to this rule the operator corresponding to the classical product $p^m q^n$, where m and n are non-negative integers, is given by the expression $\hat{q}^n\hat{p}^m$, i.e., the operator is constructed in such a way that position operators $\hat{q}$ precede momentum operators $\hat{p}$. It will be convenient to introduce a linear mapping operator Ω which transforms the classical product $p^m q^n$ onto the operator $\hat{q}^n\hat{p}^m$:

$$\Omega\{p^m q^n\} = \hat{q}^n\hat{p}^m. \tag{2}$$

With the help of the commutation relation (1), the operator $\hat{q}^n\hat{p}^m$ may, of course, be written in a number of different ways. To map an arbitrary function $F(p,q)$ of p and q onto an operator function $\hat{G}(\hat{p},\hat{q})$ of the operators $\hat{p}$ and $\hat{q}$, the function $F(p,q)$ must first be expanded in a power series and the operator Ω is then applied to each term.

It is evident that one may also define the corresponding inverse mapping operator Ω^{-1} which, in the case of the standard rule of association, has the property that

$$\Omega^{-1}\{\hat{q}^n\hat{p}^m\} = p^m q^n. \tag{3}$$

If one deals with an arbitrary operator $\hat{G}(\hat{p},\hat{q})$, the commutation relation (1) must first be used to arrange $\hat{G}(\hat{p},\hat{q})$ in standard order and then Ω^{-1} may be applied separately to each term of the standard-ordered form.

In Table I we summarize some of the better known rules of association between elementary classical functions and operators. The symbol $(\hat{p}^m\hat{q}^n)_W$ denotes the Weyl-symmetrized form of the operator product $\hat{p}^m\hat{q}^n$. It is defined as the sum of all products containing m momentum operators $\hat{p}$ and n position operators $\hat{q}$, divided by the number of terms, e.g.,

$$(\hat{p}^2\hat{q})_W = \frac{1}{3}(\hat{p}^2\hat{q} + \hat{p}\hat{q}\hat{p} + \hat{q}\hat{p}^2). \tag{4}$$

Rule of Association	Correspondence
Standard	$p^m q^n \longleftrightarrow \hat{q}^n\hat{p}^m$
Anti-Standard	$p^m q^n \longleftrightarrow \hat{p}^m\hat{q}^n$
Weyl's Symmetrization	$p^m q^n \longleftrightarrow (\hat{p}^m\hat{q}^n)_W$
Rivier's Symmetrization	$p^m q^n \longleftrightarrow \frac{1}{2}\left\{\hat{p}^m\hat{q}^n + \hat{q}^n\hat{p}^m\right\}$
Born-Jordan	$p^m q^n \longleftrightarrow \frac{1}{m+1}\sum_{\ell=0}^{m} \hat{p}^{m-\ell}\hat{q}^n\hat{p}^\ell$

TABLE I: Some of the commonly used rules of association between products of the commuting variables q and p and functions of the non-commuting operators $\hat{q}$ and $\hat{p}$. Here m and n are non-negative integers.

Sometimes, especially in quantum optics, it is convenient to employ a complex representation. The basic variables then are:

$$z = \sqrt{\frac{1}{2\hbar}}\,(q+ip); \qquad z^* = \sqrt{\frac{1}{2\hbar}}\,(q-ip), \tag{5}$$

$$\hat{a} = \sqrt{\frac{1}{2\hbar}}\,(\hat{q}+i\hat{p}); \qquad \hat{a}^\dagger = \sqrt{\frac{1}{2\hbar}}\,(\hat{q}-i\hat{p}). \tag{6}$$

In Eq. (5), z^* is the complex conjugate of z. In Eq. (6), $\hat{a}$ is the boson annihilation operator and its Hermitian conjugate $\hat{a}^\dagger$ is the boson creation operator. They obey the commutation relation

$$[\hat{a},\hat{a}^\dagger] = 1. \tag{7}$$

One may then consider rules of association between classical functions of the complex variables z and z^* and operator functions of the non-commuting operators $\hat{a}$ and $\hat{a}^\dagger$. In Table II, we list some of the better known rules of association involving there variables. The symbol $(\hat{a}^{\dagger m}\hat{a}^n)_W$, corresponding to the Weyl symmetrization, is defined in a similar way as $(\hat{p}^m\hat{q}^n)_W$.

Rule of Association	Correspondence
Normal	$z^{*m}z^n \longleftrightarrow \hat{a}^{\dagger m}\hat{a}^n$
Anti-Normal	$z^{*m}z^n \longleftrightarrow \hat{a}^n\hat{a}^{\dagger m}$
Weyl's Symmetrication	$z^{*m}z^n \longleftrightarrow (\hat{a}^{\dagger m}\hat{a}^n)_W$
Moment	$z^{*m}z^m \longleftrightarrow (\hat{a}^\dagger\hat{a})^m$

TABLE II: Some of the commonly used rules of association between products of the complex numbers z and z^* and functions of the boson annihilation and creation operators $\hat{a}$ and $\hat{a}^\dagger$. Here m and n are non-negative integers.

To each rule of association listed in Tables I and II, and more generally, to each rule of a wider class introduced by Agarwal and Wolf,[14] there corresponds a linear mapping operator Ω and its inverse Ω^{-1} that may be defined in a strictly similar way as discussed in connection with the standard rule of association [cf. Eqs. (2) and (3)]. The operator Ω maps an arbitrary function $F(p,q)$ [or $F(z,z^*)$] onto an operator function $\hat{G}(\hat{p},\hat{q})$ [or $\hat{G}(\hat{a},\hat{a}^\dagger)$] of the non-commuting operators $\hat{p}$ and $\hat{q}$ [or of $\hat{a}$ and $\hat{a}^\dagger$]. Similarly, the operator Ω^{-1} maps the operator function $\hat{G}(\hat{p},\hat{q})$ [or $\hat{G}(\hat{a},\hat{a}^\dagger)$] onto a classical function $F(p,q)$ [$F(z,z^*)$] according to the rule that determines how the non-commuting operators $\hat{p}$ and $\hat{q}$ [or $\hat{a}$ and $\hat{a}^\dagger$] are ordered in expressions involving their products. Hence, associated with each rule of association, specified by the mapping operator Ω, there exists a unique, one-to-one correspondence between the phase-space (c-number) functions F and the operator (q-number) functions G. This correspondence is illustrated schematically in Fig. 1.

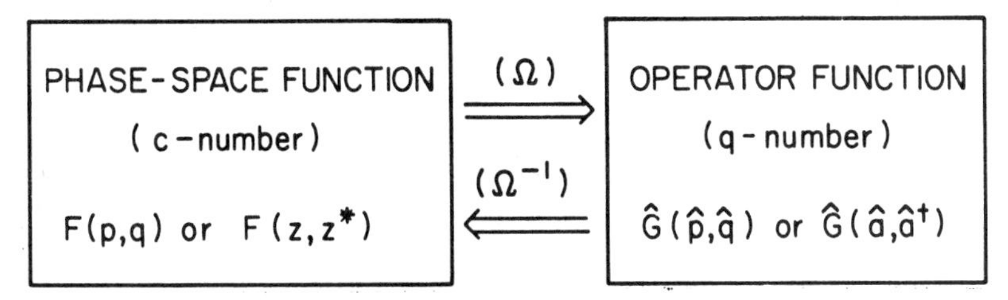

FIG. I: A schematic illustration of the correspondence, established via a linear mapping operator Ω, between arbitrary functions of the c-numbers p and q (or z and z*) and functions of the operators $\hat{p}$ and $\hat{q}$ (or $\hat{a}$ and $\hat{a}^{\dagger}$) that obey the commutation relation $[\hat{q},\hat{p}] = i\hbar$ (or $[\hat{a},\hat{a}^{\dagger}] = 1$).

From the physical point of view, the important quantities are, of course, the expectation values of observables. Suppose that the operator $\hat{G}(\hat{p},\hat{q})$ represents an observable and that $\hat{\rho}(\hat{p},\hat{q})$ is the density operator that characterizes the statistical state of the system. The quantum mechanical expectation value of $\hat{G}(\hat{p},\hat{q})$ in the state specified by $\hat{\rho}(\hat{p},\hat{q})$ is then given by the formula

$$\langle\hat{G}(\hat{p},\hat{q})\rangle_{QM} = \mathrm{Tr}\{\hat{\rho}(\hat{p},\hat{q})\hat{G}(\hat{p},\hat{q})\}, \tag{8}$$

where Tr denotes the trace. The procedure of calculating the quantum mechanical expectation value of an observable in this way is, of course, quite different from the corresponding procedure used in classical statistical mechanics. In classical mechanics the expectation value of a function F(p,q) of position q and momentum p is calculated according to the rules of ordinary probability theory:

$$\langle F(p,q)\rangle_{CL} = \iint \Phi(p,q)F(p,q)\,dpdq. \tag{9}$$

The c-number function $\Phi(p,q)$, known as the phase-space distribution function, is a joint probability density of position q and momentum p.

The central question of the phase-space formulation of quantum mechanics is then the following: It is possible to map all the relevant operators onto c-number functions,

$$\hat{q} \rightarrow q, \tag{10}$$

$$\hat{p} \rightarrow p, \tag{11}$$

$$\hat{G}(\hat{p},\hat{q}) \rightarrow F(p,q), \tag{12}$$

$$\hat{\rho}(\hat{p},\hat{q}) \rightarrow \Phi(p,q), \tag{13}$$

in such a way that the quantum mechanical expectation value (8) of any observable $\hat{G}(\hat{p},\hat{q})$ may be expressed in a form of a classical c-number average with respect to the "phase-space" distribution function $\Phi(p,q)$, i.e., as

$$\langle\hat{G}(\hat{p},\hat{q})\rangle_{QM} = \iint F(p,q)\Phi(p,q)dpdq? \tag{14}$$

It is known that, in general, this may be done. In fact, there exists an infinite number of possible ways of expressing $\langle\hat{G}(\hat{p},\hat{q})\rangle_{QM}$ as a classical c-number average, one associated with each of the Ω-mappings discussed earlier. If the correspondence between $\hat{\rho}(\hat{p},\hat{q})$ and $\Phi(p,q)$ [cf. Eq.(13)] is established via some given Ω-mapping, the correspondence between $\hat{G}(\hat{p},\hat{q})$ and $F(p,q)$ [cf. Eq. (12)] is expressible[16] in terms of a mapping operator $\tilde{\Omega}$ which in a certain sense represents a mapping reciprocal to Ω. For example, the mappings corresponding to standard and anti-standard ordering are reciprocal to each other in this sense. Similarly, the mappings corresponding to normal and anti-normal ordering are reciprocal. The Weyl-ordering, on the other hand, is self-reciprocal.

For the purposes of later discussion, it will be convenient to present some explicit expressions of the c-number functions $\Phi(p,q)$ representing the density operator $\hat{\rho}(\hat{p},\hat{q})$ via the Ω-mappings corresponding to some common rules of association [cf. Eqs. (2) and (3) and Tables I and II].

The Wigner distribution function (WDF) is obtained via the Weyl symmetrization rule and may be expressed in the following two equivalent forms:

$$\Phi(p,q) = \hbar\int \langle q+\tfrac{1}{2}\hbar v|\hat{\rho}(\hat{p},\hat{q})|q-\tfrac{1}{2}\hbar v\rangle\, e^{-ivp}dv \tag{15a}$$

$$= \hbar\int \langle p+\tfrac{1}{2}\hbar u|\hat{\rho}(\hat{p},\hat{q})|q-\tfrac{1}{2}\hbar u\rangle\, e^{+iuq}du\ . \tag{15b}$$

In Eq. (15a) and (15b), $|q\rangle$ is the position state vector and $|p\rangle$ is the momentum state vector. It is seen from these expressions that the WDF has essentially the same mathematical structure both in the coordinate representation [Eq. (15a)] and in the momentum representation [Eq. (15b)]. This fact is a consequence of the symmetry of the Weyl ordering in $\hat{q}$ and $\hat{p}$ [cf. Table I].

The phase-space (c-number) representative of $\hat{\rho}(\hat{p},\hat{q})$ that corresponds to standard ordering can likewise be expressed in two equivalent forms:

$$\Phi(p,q) = \hbar\int \langle q|\hat{\rho}(\hat{p},\hat{q})|q-\hbar v\rangle\, e^{-ivp}dv \tag{16a}$$

$$= \hbar\int \langle p+\hbar u|\hat{\rho}(\hat{p},\hat{q})|p\rangle\, e^{+iuq}du. \tag{16b}$$

Eq. (16a) is the coordinate representation and Eq. (16b) is the momentum representation of the phase-space distribution function $\Phi(p,q)$.

The c-number representative of the density operator in anti-normal ordering [cf. Table II] may be expressed in the form

$$\Phi(z,z^*) = \frac{1}{\pi} e^{|z|^2} \int \langle -\alpha|\hat{\rho}(\hat{a},\hat{a}^\dagger)|\alpha\rangle\, e^{|z|^2} e^{-(\alpha z^*-\alpha^* z)} d^2\alpha. \quad (17)$$

Here α is a complex number, $|\alpha\rangle$ denotes the coherent state[17] labeled by α and the integration extends throughout the complex α-plane. The mapping corresponding to the anti-normal rule of association is closely related to the so-called Sudarshan-Glauber diagonal coherent state representation of an operator.[18,19] In that representation the operator is expressed as an appropriate weighted superposition of coherent state projection operators. The formula (17) represents the weighting function for an arbitrary density operator $\hat{\rho}(\hat{a},\hat{a}^\dagger)$.

The c-number representation of the density operator in anti-normal ordering plays an important role in the study of the statistical properties of light by means of photoelectric correlation and coincidence techniques.[20] This is a consequence of the fact that in the correlation experiments the operators representing physical observables turn out to be naturally expressed in terms of normally ordered products of the photon creation and annihilation operators. The possibility of expressing the quantum mechanical expectation value of any normally ordered operator in a form of a classical statistical average with respect to the c-number representative of the density operator that is anti-normally ordered is the essence of the so-called optical equivalence theorem.[21]

As mentioned earlier, one may always express the quantum mechanical expectation value of any observable $\hat{G}(\hat{p},\hat{q})$ in a form of statistical average in phase-space [cf. Eq. (14)]. However, unlike in classical statistical mechanics, the quantum mechanical phase-space weighting functions $\Phi(p,q)$ do not possess all the properties of a true probability density. In fact, there is a fundamental theorm due to Wigner[22] that asserts, in essence, the following:[23] The four requirements

$$\Phi(p,q) = \Omega\{\hat{\rho}(\hat{p},\hat{q})\}, \quad (18)$$

$$\Phi(p,q) \geqslant 0, \quad (19)$$

$$\int \Phi(p,q)\,dp = P(q), \quad (20)$$

$$\int \Phi(p,q)\,dq = P(p), \quad (21)$$

cannot be simultaneously satisfied for all possible statistical states of a quantum mechanical system characterized by the density operator[24] $\hat{\rho}(\hat{p},\hat{q})$. In Eq. (18), Ω denotes a linear mapping operator and in Eqs. (20) and (21), $P(q)$ and $P(p)$ are the probabilities of position and of momentum, respectively. In particular, the phase-space weighting functions $\Phi(p,q)$ may occasionally take on negative values, they may become singular for some states of the system or they may violate some other basic postulate of probability theory. For this reason, the phase-space weighting functions $\Phi(p,q)$ are sometimes referred to as quasi-probabilities.

We mention two facts concerning the implications of Wigner's theorem: First, for some statistical states of the quantum mechanical system it is, nonetheless, entirely possible to construct phase-space distribution functions that meet all the conditions (18) - (21). Generally, quantum mechanical systems in states with well-defined classical analogs (such as an optical field in thermal equilibrium) admit such phase-space distribution functions. However, for states with no classical analog (such as the Fock states[25]) the phase-space distribution functions frequently violate the postulates of ordinary probability theory. And second, by strengthening some of the conditions of Wigner's theorem and by weakening or omitting others, one may formulate other sets of requirements that the phase-space distribution functions should satisfy. Such sets of requirements may or may not admit solutions.[26]

To conclude this review of the phase-space formulation of quantum mechanics, we wish to emphasize the fact that even though any one of the infinite number of generalized phase-space distribution functions may be used when calculating the expectation values of quantum mechanical observables, none of them - including the WDF - is a true joint probability density of position and momentum.

3. BASIC CONCEPTS OF CONVENTIONAL RADIOMETRY AND OF THE OPTICAL COHERENCE THEORY

Let us now turn our attention to some recent developments in classical optics where several analogs of the WDF and of other generalized phase-space distribution functions have appeared. In the present paper we will be mostly concerned with quantities that play a role in the theory of radiometry with partially coherent light.

Conventional radiometry describes the transfer of radiant energy on a phenomenological basis involving intuitive notions such as tubes of light rays. The central quantity in that theory is the radiance (also known as the brightness or the specific intensity). It is defined in the following way[27] (Fig. 2): Let dP represent the energy flux

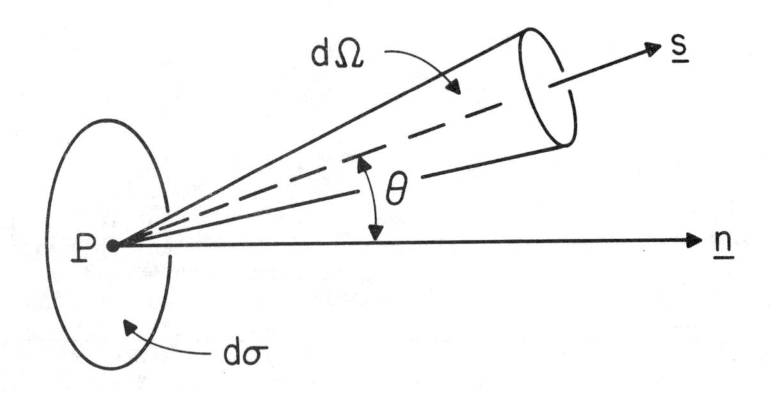

FIG. 2: Illustration of the notation relating to the traditional definition of the radiance.

(at frequency ω) radiated by surface element $d\sigma$ (possibly fictitious) surrounding a point P into an element $d\Omega$ of solid angle around a direction specified by a unit vector $\underline{s}$. Then the formula

$$dP = B(\underline{r},\underline{s})\cos\Theta d\sigma d\Omega, \tag{22}$$

where $\underline{r}$ denotes the position vector of the point P and Θ is the angle betweeen the $\underline{s}$ direction and the unit normal $\underline{n}$ to $d\sigma$, defines the radiance $B(\underline{r},\underline{s})$ at the point $P(\underline{r})$, in the $\underline{s}$ direction.

In terms of the radiance, the other radiometric quantities, most notably the radiant emittance and the radiant intensity, are then obtained as appropriate integral transforms.[28] The radiant emittance $E(\underline{r})$ may be defined by the formula

$$E(\underline{r}) = \int_{(2\pi)} B(\underline{r},\underline{s})\cos\Theta d\Omega, \tag{23}$$

where the integration extends over the 2π solid angle formed by all the possible $\underline{s}$ directions. The radiant emittance $E(\underline{r})$ represents the energy flux radiated by the source per unit area around the point $P(\underline{r})$. The radiant intensity $J(\underline{s})$, on the other hand, may be defined by the formula

$$J(\underline{s}) = \cos\Theta \int_{\sigma} B(\underline{r},\underline{s})d\sigma, \tag{24}$$

with the integration extending over the source area σ (possibly infinite). The radiant intensity $J(\underline{s})$ represents the energy flux radiated by the source per unit solid angle around the $\underline{s}$ direction.

We note that all these radiometric quantities are defined at a single temporal frequency ω of the light and that they are regarded as measurable, at least in principle. Furthermore, a simple additive superposition of energy from the various parts of the source is assumed to hold and all effects due to diffraction and interference are neglected.

In order to be able to incorporate the effects caused by the fluctuating nature of light in a convenient way, we must first recall a few quantities employed in optical coherence theory. For the present purposes we only need to discuss the so-called second-order theory of partial coherence.

We consider a statistically stationary, scalar optical field. It is represented by a complex analytic signal[29] $V(\underline{r},t)$, where $\underline{r}$ is the position vector of a point in space and t denotes the time. The mutual coherence function is defined by the formula

$$\Gamma(\underline{r}_1,\underline{r}_2;\tau) = \langle V(\underline{r}_1,t)V^*(\underline{r}_2,t+\tau)\rangle , \tag{25}$$

where the brackets denote the average over the statistical ensemble. In terms of $\Gamma(\underline{r}_1,\underline{r}_2;\tau)$ one may then define the cross-spectral density (or the cross-power spectrum) as

$$W(\underline{r}_1,\underline{r}_2;\omega) = \frac{1}{2\pi}\int_{\infty}^{\infty} \Gamma(\underline{r}_1,\underline{r}_2;\tau)e^{i\omega\tau}d\tau. \tag{26}$$

With the appropriate physical interpretation of $W(\underline{r}_1,\underline{r}_2;\omega)$, the integral relationship (26) is an optical analog of the generalized Wiener-Khintchine theorem for stationary random processes.

As the theory of radiative energy transfer in fluctuating optical fields is more naturally formulated in the space-frequency domain, rather than in the space-time domain, the fundamental quantity representing the (second-order) coherence properties of the field is the cross-spectral density function $W(\underline{r}_1,\underline{r}_2;\omega)$. It characterizes the correlation that exists between the values of the optical field at points specified by $\underline{r}_1$ and $\underline{r}_2$, at frequency ω. The normalized form of the cross-spectral density function $W(\underline{r}_1,\underline{r}_2;\omega)$, defined by the formula[30,31]

$$\mu(\underline{r}_1,\underline{r}_2;\omega) = \frac{W(\underline{r}_1,\underline{r}_2;\omega)}{[W(\underline{r}_1,\underline{r}_1;\omega)W(\underline{r}_2,\underline{r}_2;\omega)]^{\frac{1}{2}}}, \tag{27}$$

is called the complex degree of spatial coherence. The function $\mu(\underline{r}_1,\underline{r}_2;\omega)$ satisfies, for all values of $\underline{r}_1$, $\underline{r}_2$ and ω, the condition

$$0 \leqslant |\mu(\underline{r}_1,\underline{r}_2;\omega)| \leqslant 1. \tag{28}$$

The limiting cases 0 and 1 in Eq. (28) indicate, respectively, complete incoherence and complete coherence of the light at frequency ω. The relationship of the complex degree of spatial coherence to the usual complex degree of coherence and its role in the description of the spectral structure of the classic two-beam interference pattern has been recently studied.[30]

In order to fully appreciate the physical meaning of the directional variable $\underline{s}$ in the definition (22) of the radiance $B(\underline{r},\underline{s})$ we take guidance from the so-called angular spectrum representation[32] of an optical field. Under fairly general conditions[33] a monochromatic (at frequency ω) wavefield $U(\underline{r})$ may be expressed throughout a free half-space $z > 0$ in the form

$$U(x,y,z) = \iint_{-\infty}^{\infty} a(u,v)e^{ik(ux+vy+wz)}du dv, \tag{29}$$

where

$$w = \sqrt{1-u^2-v^2}, \quad \text{when } u^2+v^2 \leqslant 1, \tag{30a}$$

$$w = i\sqrt{u^2+v^2-1}, \quad \text{when } u^2+v^2 > 1, \tag{30b}$$

$\underline{r} = (x,y,z)$ and $k = \omega/c$ with c being the vacuum speed of light. The formula (29) expresses the field $U(\underline{r})$ as a superposition of plane waves, all of the same frequency ω. The waves associated with values (u,v) for which $u^2+v^2 < 1$ are homogeneous plane waves, propagated into the half-space $z > 0$, in directions specified by the unit vector $\underline{s} = (u,v,w)$. The waves for which $u^2+v^2 > 1$ are evanescent plane waves, propagated in directions parallel to the plane $z = 0$ and decaying exponentially in amplitude as the distance from that plane increases. Unlike the usual Fourier decomposition, the angular spectrum representation (29) is a genuine mode-decomposition of the field $U(\underline{r})$.

On setting $z = 0$ in Eq. (29) we find that the spectral amplitude $a(u,v)$ that characterizes the "strength" of the plane wave propagating in the direction $\underline{s} = (u,v,w)$ is simply related to the two-dimensional spatial Fourier transform of the field across the plane $z = 0$:

$$a(u,v) = k^2\tilde{U}(ku,kv,0), \tag{31}$$

where

$$\tilde{U}(f_x,f_y,0) = \left(\frac{1}{2\pi}\right)^2 \iint u(x,y,0)e^{-i(f_x x+f_y y)}dxdy. \tag{32}$$

The formula (31) establishes a correspondence between the spatial frequencies (f_x,f_y) of the field $u(x,y,0)$ and the variables (u,v) labeling the plane waves in the angular spectrum representation (29), via the relationship

$$f_x = ku \; ; \quad f_y = kv , \tag{33}$$

illustrated schematically in Fig. 3. Hence the spatial frequency variable $\underline{f} = (f_x,f_y)$ and also the directional variable $\underline{s} = (u,v,w)$ are disguised momentum variables, associated with a plane wave of momentum $\underline{p} = \hbar\underline{k}$ where $\underline{k} = k\underline{s}$.

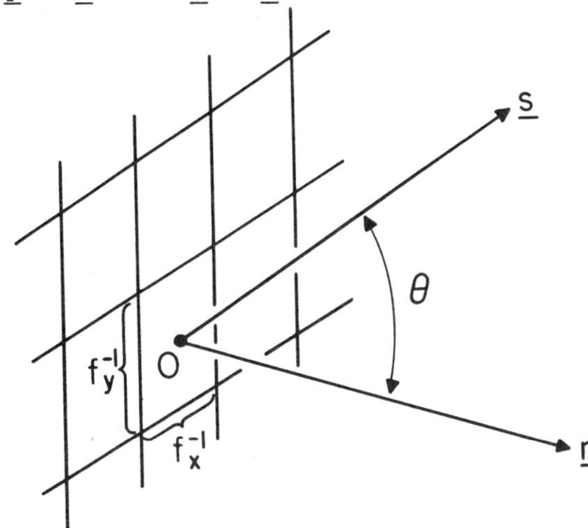

FIG. 3: A schematic illustration of the relationship between the spatial frequency variable $\underline{f} = (f_x,f_y)$ and the directional variable $\underline{s} = (u,v,w)$. The plane wave propagating in the direction $\underline{s}$ for which $u^2+v^2 < 1$ carries a momentum $\underline{p} = \hbar\underline{k} = \hbar k\underline{s}$.

4. ANALOGIES BETWEEN EXPRESSIONS OF THE RADIANCE AND PHASE-SPACE REPRESENTATIONS OF THE DENSITY OPERATOR

We begin by considering some explicit expressions that have been proposed in recent years for the generalized radiance associated with a planar source of any state of coherence. These expressions show several analogies to the quantum mechanical phase-space distribution functions discussed earlier in Section 2. We conclude this section by presenting a radiometric result that is to some extent analogous to Wigner's theorem mentioned earlier and by discussing some of its consequences.

In 1968 Walther[7] considered a partially coherent planar source located in the plane $z = 0$, and defined a generalized radiance of the source by the expression

$$B(\underline{r},\underline{s}) = \left(\frac{k}{2\pi}\right)^2 \cos\Theta \iint W^{(0)}(r + \tfrac{1}{2}r', r - \tfrac{1}{2}r') e^{-ik\underline{s}_{\perp}\cdot\underline{r}'} d^2r'. \quad (34)$$

In this formula $W^{(0)}(\underline{r}_1,\underline{r}_2)$ denotes the cross-spectral density of the light across the source (dependence on the frequency ω is not shown explicitly), and $s_{\perp}$ denotes the vectorial projection of the unit vector $\underline{s}$ onto the plane $z = 0$. The appearance of $\underline{s}_{\perp}$ rather than $\underline{s}$ in the expression (34) is a consequence of the fact that we are considering a two-dimensional source. In view of the momentum relationshio $\underline{p} = \hbar k\underline{s}$, the variables $\underline{r}$ and $\underline{s}_{\perp}$ are thus essentially conjugate variables. The expression (34) for the generalized radiance $B(\underline{r},\underline{s})$ may in some instances become negative and it may be non-zero outside the source area in the source plane.[8] Clearly, the generalized radiance (34) is related to the Wigner function $\chi(\underline{r},\underline{f})$ [cf. footnote 9] of the light across the plane $z = 0$ by the formula

$$B(\underline{r},\underline{s}) = \left(\frac{k}{2\pi}\right)^2 \cos\Theta\, \chi(\underline{r},k\underline{s}_{\perp}). \quad (35)$$

The generalized radiance (34) should be compared to the Wigner distribution function (WDF), given in a one-dimensional form by Eqs. (15a) and (15b). Expressed in a slightly different form the WDF is given by

$$\Phi(p,q) = \int \rho(q + \tfrac{1}{2}q', q - \tfrac{1}{2}q') e^{-i\beta pq'} dq' , \quad (36)$$

where $\beta = 1/\hbar$ and $\rho(q_1,q_2) \equiv \langle q_1|\hat{\rho}|q_2\rangle$ denotes the coordinate representation of the density operator $\hat{\rho}(\hat{p},\hat{q})$. As noted earlier, the WDF is the phase-space representative corresponding to the Weyl symmetrization of $\hat{\rho}(\hat{p},\hat{q})$.

Quite apart from the apparent mathematical similarity of the expressions (34) and (36), we note two facts concerning these formulae: First, both functions are proportional to the Fourier transform of a spatial (coordinate) representation of a quantity that characterizes the statistical state of the system. And second, in both cases the Fourier transform variable is a momentum variable.

Another expression proposed for the generalized radiance associated with a partially coherent planar source may be expressed in the form[34]

$$B(\underline{r},\underline{s}) = \left(\frac{k}{2\pi}\right)^2 \cos\Theta \,\mathrm{Re} \iint W^{(0)}(\underline{r}',\underline{r}) e^{-iks_{\perp}\cdot(\underline{r}'-\underline{r})} d^2r' , \quad (37)$$

where Re denotes the real part. It has been shown that the expression (37) for the generalized radiance $B(\underline{r},\underline{s})$ may occasionally take on negative values.[35]

The generalized radiance (37) should be compared to another widely used quantum mechanical phase-space distribution function introduced by Margenau and Hill[36] and given in one-dimensional form by Eqs. (16a) and (16b). It may be shown to be expressible in the form[37]

$$\Phi(p,q) = \mathrm{Re} \int \rho(q',q) e^{-i\beta p(q'-q)} dq' , \tag{38}$$

where $\beta = 1/\hbar$ and $\rho(q',q)$ denotes the coordinate representation of the density operator $\hat{\rho}(\hat{p},\hat{q})$. The phase-space distribution function (38) corresponds to the standard ordering of $\hat{\rho}(\hat{p},\hat{q})$.

Observations similar to those presented in connection with Eqs. (34) and (36) pertain also to the generalized radiance (37) and the generalized phase-space distribution function (38).

Just as there are an infinite number of generalized phase-space distribution functions $\Phi(p,q)$ that may be used in calculations of the expectation values of observables, one may also introduce an infinite number of possible generalized radiance functions $B(\underline{r},\underline{s})$ that have some of the properties normally postulated for the radiance in conventional radiometry. However, it has been shown[38] that no generalized radiance function $B(\underline{r},\underline{s})$ satisfies, for all possible states of coherence of the source,[39] the following requirements:[40]

$$B(\underline{r},\underline{s}) = L\,\{W^{(0)}(\underline{r}_1,\underline{r}_2)\} , \tag{39}$$

$$B(\underline{r},\underline{s}) \geqslant 0 , \tag{40}$$

$$B(\underline{r},\underline{s}) = 0 , \quad \text{when } \underline{r} \notin \sigma , \tag{41}$$

$$\cos\Theta \int_\sigma B(\underline{r},\underline{s})\, d\sigma = [J(\underline{s})]_{\text{phys. opt.}} . \tag{42}$$

In Eq. (39), L denotes a linear operator. Eqs. (40) and (41) express the requirements that the radiance $B(\underline{r},\underline{s})$ be non-negative and vanish identically in the source plane outside the source area σ. Eq. (42) expresses the condition that the radiometric expression on the left-hand side [cf. Eq. (24)] correctly represents the radiant intensity $[J(\underline{s})]_{\text{phys. opt.}}$ produced by the source when the radiant intensity is calculated directly on the basis of its definition using the notions of physical optics.[41]

The above result relating to generalized radiometry with partially coherent wavefields bears a strong formal resemblance to Wigner's theorem discussed earlier in Section 2. In particular, the requirement (41), ensuring the condition that the radiant emittance $E(\underline{r})$ [cf. Eq. (23)] be zero in the source plane outside the source area σ, is to some extent analogous to the condition (20) on the marginal probability $P(q)$ of position. Similarly, the requirement (42) is analogous to the condition (21) on the marginal probability $P(p)$ of momentum.

As mentioned earlier, the requirements (39) - (42) on the generalized radiance $B(\underline{r},\underline{s})$ cannot, in general, be all simultaneously satisfied. However, by weakening or omitting some of these conditions one obtains other sets of requirements that may admit solutions.[42] On the other hand, if one delimits the possible states of coherence of the source, expressions may be given for the generalized radiance $B(\underline{r},\underline{s})$ that satisfy the requirements (39) - (42) for such sources.

An important class of sources of this type are the so-called quasi-homogeneous sources.[43] A quasi-homogeneous planar source is characterized by a cross-spectral density function [cf. Eq. (26); the frequency ω is not shown explicitly below] of the form[44]

$$W^{(0)}(\underline{r}_1,\underline{r}_2) = I^{(0)}\left(\frac{\underline{r}_1+\underline{r}_2}{2}\right)\mu^{(0)}(\underline{r}_1-\underline{r}_2), \tag{43}$$

where $I^{(0)}(\underline{r}) \equiv W^{(0)}(\underline{r},\underline{r})$ is a slowly varying intensity distribution while $\mu^{(0)}(\underline{r}_1-\underline{r}_2)$, assumed to be a fast function of its argument $\underline{r}' = \underline{r}_1-\underline{r}_2$, denotes the complex degree of spatial coherence [cf. Eq. (27)]. The behavior of these functions is illustrated schematically in Fig. 4. In a global sense a quasi-homogeneous source is always spatially rather incoherent whereas locally it may or may not be coherent, depending on whether the effective width of $\mu^{(0)}(\underline{r}')$ is or is not large compared to the wavelength $\lambda = 2\pi c/\omega$. For a quasi-homogeneous planar source the two formulae (34) and (37) reduce to the same expression that has been shown[45] to satisfy simultaneously all the four requirements (39) - (42).

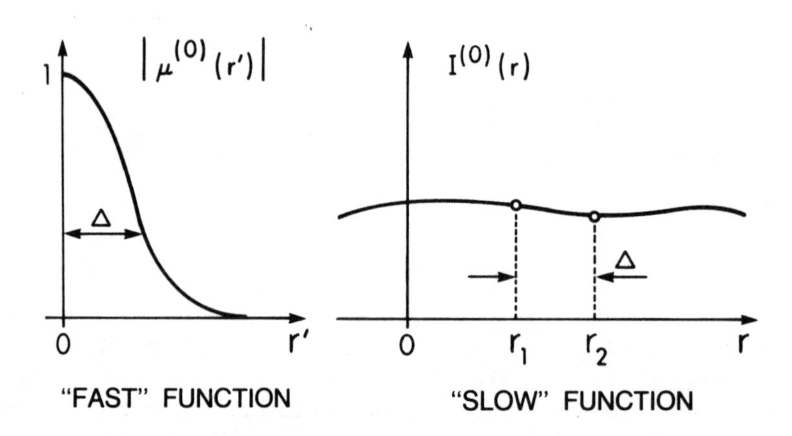

FIG. 4: A schematic (one-dimensional) illustration of the behavior of the degree of spatial coherence and of the optical intensity of light across a quasi-homogeneous source.

5. NEGATIVENESS OF THE GENERALIZED RADIANCE FUNCTIONS

There appears to be some confusion about the implications of the possible negativeness of the generalized radiance functions. It has been suggested[46] that a negative value for the generalized radiance $B(\underline{r},\underline{s})$ for some values of its arguments $\underline{r}$ and $\underline{s}$ implies that the direction of the energy flow at the point specified by $\underline{r}$ has been reversed. We put forward below two different arguments which show that this view is incorrect.

The first argument is based on the fact that there are an infinite number of possible generalized radiance functions $B(\underline{r},\underline{s})$. Two commonly used examples of such functions are given by the expressions (34) and (37). In general, the different generalized radiance functions $B(\underline{r},\underline{s})$ may become negative at different points. Consequently, any conclusion based on the negativeness of one radiance function would be contradicted by other radiance functions.

The second argument relies on the physical significance associated with the radiance $B(\underline{r},\underline{s})$. Consider the energy flow at some point specified by $\underline{r}_0$. The strength of the flow in the direction $\underline{s}_0$ is given by $B(\underline{r}_0,\underline{s}_0)$, while the strength in the opposite direction is given by $B(\underline{r}_0,-\underline{s}_0)$. Now, if a negative value for the radiance $B(\underline{r},\underline{s})$ implied a reversal of the direction of the energy flow, then a negative value for $B(\underline{r}_0,\underline{s}_0)$ would necessarily mean that $B(\underline{r}_0,-\underline{s}_0)$ must be positive. However, the quantities $B(\underline{r}_0,\underline{s}_0)$ and $B(\underline{r}_0,-\underline{s}_0)$ may or may not be related, depending on the coherence properties of the wavefield. In particular, these two quantities need not have opposite signs, as the argument would require. An example of this is provided by an electromagnetic radiation field at thermal equilibrium. It has been shown[47] that a generalized radiance $B(\underline{r},\underline{s})$ associated with such a field may be defined by the formula

$$B(\underline{r},\underline{s}) \equiv B_0 = \frac{2\hbar\omega^3}{(2\pi c)^2} \frac{1}{e^{\hbar\omega/KT}-1} , \tag{44}$$

where K is Boltzman's constant and ω denotes the frequency and T denotes the temperature of the field. Since in this case $B(\underline{r},\underline{s})$ is independent of $\underline{s}$, no reversal of sign appears as the directional argument $\underline{s}$ is changed into $-\underline{s}$.

6. SOME UNSOLVED PROBLEMS

Despite much effort during the last decade or so, several unsolved problems still remain in the field of generalized radiometry with partially coherent wavefields. We will conclude this paper by pointing out some areas that would benefit from further study.

A quantity known as the ambiguity function, originally introduced in the radar theory[48] but recently also applied in optical investigations[49,50], seems to share some of the properties of the Wigner function. For a fluctuating optical field, the ambiguity function in the plane z = 0 may be defined by the formula

$$A(\underline{r}',\underline{f}) = \iint W^{(0)}(\underline{r} + \tfrac{1}{2}\underline{r}',\underline{r} - \tfrac{1}{2}\underline{r}')e^{-i\underline{f}\cdot\underline{r}}d^2r , \tag{45}$$

where $W^{(0)}(\underline{r}_1,\underline{r}_2)$ is the cross-spectral density of the field across the plane $z = 0$ (the frequency ω is not shown explicity). The formula (45) shows that the ambiguity function $A(\underline{r}',\underline{f})$ is proportional to the two-dimensional spatial Fourier transform of the function $\overline{W}^{(0)}(\underline{r},\underline{r}') \equiv W^{(0)}(\underline{r} + \frac{1}{2}\underline{r}',\underline{r} - \frac{1}{2}\underline{r}')$ with respect to the variable $\underline{r}$. The ambiguity function $A(\underline{r}',\underline{f})$ is related to the Wigner function $\chi(\underline{r},\underline{f})$ [cf. footnote 9] by a double Fourier transform.[6] If $W^{(0)}(\underline{r}_1,\underline{r}_2)$ is represented in a form of a suitable ensemble average $\langle V(\underline{r}_1)V^*(\underline{r}_2)\rangle$, then for

$\underline{f}$ = 0 the ambiguity function $A(\underline{r}',\underline{f})$ can be shown to reduce to the averaged correlation function of $V(\underline{r})$ and $V^*(\underline{r})$, whereas for $\underline{f}$ = 0 the Wigner function $\chi(\underline{r},\underline{f})$ was seen to be the averaged convolution of $V(\underline{r})$ and $V^*(\underline{r})$ [cf. footnote 9]. The mathematical structure of the ambiguity function $A(\underline{r}',\underline{f})$ suggests that it is a phase-space representation of a quantity that represents the statistical state of the field. However, its physical meaning and its relation to the various functions $B(\underline{r},\underline{s})$ obtained as linear transformations of $W^{(0)}(\underline{r}_1,\underline{r}_2)$ are not known at the present time.

Another area with several unsolved problems concerns the properties of the transformations, characterized by the operator L, of the source cross-spectral density function $W^{(0)}(\underline{r}_1,\underline{r}_2)$ onto the functions $B(\underline{r},\underline{s})$. It would be desirable, for example, to specify the class of all permissible operators L, i.e., transformations leading to functions $B(\underline{r},\underline{s})$ that can be employed in calculations of measurable radiometric quantities via the usual expressions.[51] It seems that this problem is deeply rooted in the understanding of the underlying reasons for the possibility of associating several different generalized radiance functions $B(\underline{r},\underline{s})$ with a given optical field.

Perhaps solutions to these problems would give new insight into the true significance of the concept of generalized radiance and also make it possible to develop further the phase-space methods for partially coherent wavefields

ACKNOWLEDGEMENTS

The author wishes to thank Professor E. Wolf for several stimulating discussions concerning the subject matter of this manuscript.

REFERENCES AND FOOTNOTES

*Supported by the Academy of Finland and by the U.S. Army Research Office.

1. E. Wigner, Phys. Rev. 40, 749 (1932).
2. See, for example, a) E.C.G. Sudarshan, Lectures in Theoretical Physics (Benjamin, New York, 1961), Vol. II; b) T.F. Jordan and E.C.G. Sudarshan, Rev. Mod. Phys. 33, 515 (1961).
3. A generalized phase-space distribution function is a function that can be used to calculate expectation values according to the formulae of ordinary probability theory but that does not have all the properties of a true probability distribution function.
4. M.J. Bastiaans, Opt. Commun. 25, 26 (1978).
5. M.J. Bastiaans, J. Opt. Soc. Am. 69, 1710 (1979).
6. H.O. Bartelt, K.-H. Brenner and A.W. Lohmann, Opt. Commun. 32, 32 (1980).
7. A. Walther, J. Opt. Soc. Am. 58, 1256 (1968).
8. E.W. Marchand and E. Wolf, J. Opt. Soc. Am. 64, 1219 (1974).
9. The Wigner function, sometimes also called the Wigner distribution function or the Wolf function, must be distinguished from the actual WDF employed in the phase-space representation of quantum mechanics. Using the notation of Sections 3 and 4, the Wigner function associated with a fluctuating optical field in the plane z = 0 may be defined by the formula

9. (cont'd)

$$\chi(\underline{r},\underline{f}) = \iint W^{(0)}(\underline{r} + \tfrac{1}{2}\underline{r}', \underline{r} - \tfrac{1}{2}\underline{r}') e^{-i\underline{f}\cdot\underline{r}'} d^2r' ,$$

i.e., it is proportional to the two-dimensional spatial Fourier transform of the function $W^{(0)}(\underline{r},\underline{r}') \equiv W^{(0)}(\underline{r} + \frac{1}{2}\underline{r}', \underline{r} - \frac{1}{2}\underline{r}')$ with respect to the variable $\underline{r}'$. If $W^{(0)}(\underline{r}_1,\underline{r}_2)$ is represented as a suitable ensemble average $\langle V(\underline{r}_1)V^*(\underline{r}_2)\rangle$, then for $\underline{f} = 0$ the Wigner function $\chi(\underline{r},\underline{f})$ reduces to the average value of the convolution of $V(\underline{r})$ with respect to its complex conjugate $V^*(\underline{r})$.

10. J.E. Moyal, Proc. Cambridge Phil. Soc. 45, 99 (1949).
11. For a review that describes the use of the WDF in quantum statistical mechanics, see a) K. Imre, E. Ozizmir, M. Rosenbaum and P.F. Zweifel, J. Math. Phys. 8, 1097 (1967). Other phase-space distribution functions are discussed, for example, in b) J.R. Shewell, Am. J. Phys. 27, 16 (1959); c) C.L. Mehta, J. Math. Phys. 5, 677 (1964); and d) L. Cohen, J. Math. Phys. 7, 781 (1966). The papers cited in Ref. 14 below contain an extensive bibliography on the subject of the general phase-space methods in quantum mechanics.
12. See, for example, L. Mandel and E. Wolf, Rev. Mod. Phys. 37, 231 (1965).
13. See, for example. M. Lax and W. H. Louisell, IEEE J. Quant. Electron. QE-3, 47 (1967), and the references therein.
14. G.S. Agarwal and E. Wolf, a) Phys. Rev. D2, 2161 (1970); b) Phys. Rev. D2, 2187 (1970); and c) Phys. Rev. D2, 2206 (1970).
15. Quantum mechanical operators are denoted by carets.
16. For a fuller discussion of this subject and for a proper definition of the reciprocal mapping $\tilde{\Omega}$, see the papers cited in Ref. 14.
17. For a discussion of the properties of the coherent states, see, for example, J.R. Klauder and E.C.G. Sudarshan, *Fundamentals of Quantum Optics* (Benjamin, New York, 1968), Chapter 7.
18. E.C.G. Sudarshan, Phys. Rev. Lett. 10, 277 (1963).
19. R.J. Glauber, Phys. Rev. 131, 2766 (1963).
20. Fuller accounts of such techniques and references to original contributions are given, for example, in Ref. 12 and in Ref. 17, Chapter 8.
21. This theorem, relating to the equivalence of the semi-classical and the quantum theory of optical coherence, was first discovered by Sudarshan in Ref. 18 and also noted, in a somewhat more restricted form, by Glauber in Ref. 19. It was further elaborated by J.R. Klauder, Phys. Rev. Lett. 16, 534 (1966). A more detailed discussion of the optical equivalence theorem is given in Ref. 17, Sec. 8-4.
22. E.P. Wigner, "Quantum Mechanical Distribution Functions Revisited", in *Perspectives in Quantum Theorey*, ed. by W. Yourgrau and M. van der Merwe (M.I.T. Press, Cambridge, Mass, 1971), pp. 25-36.
23. For a precise statement of Wigner's theorem, see M.D. Srinivas and E. Wolf, Phys. Rev. D11, 1477 (1975).
24. Actually, in his original work, Wigner considered only a pure state $|\psi\rangle$ of the quantum mechanical system. However, any statistical state (pure or mixed) may be characterized by a density operator

24. (cont'd)

$$\hat{\rho} = \sum_i a_i |\psi_i\rangle\langle\psi_i| ,$$

where the $\{|\psi_i\rangle\}$ denote a complete set of orthonormal states and the a_i's are non-negative numbers representing the probability of the states $|\psi_i\rangle$ $(\Sigma a_i = 1)$.

25. The Fock states are the eigenstates of the number operator $\hat{n} = \hat{a}^\dagger\hat{a}$. The relationship between the Fock states and the coherent states is discussed in Ref. 17, Chapter 7.
26. Some such sets of requirements and other related researches are discussed in Ref. 23.
27. S. Chandrasekhar, Radiative Transfer (Dover, New York, 1960), Sec. 2.1.
28. See, for example, a) M. Born and E. Wolf, Principles of Optics, fifth edition (Pergamon Press, Oxford and New York, 1975), Sec. 4.8.1; b) G.C. Pomraning, The Equations of Radiation Hydrodynamics (Pergamon, Oxford and New York, 1973), Sec. I.2.
29. The precise definition of the complex analytic signal is given in Ref. 12 and in Ref. 28a, Sec. 10.2.
30. L. Mandel and E. Wolf, J. Opt. Soc. Am. 66, 529 (1976).
31. M.J. Bastiaans, Optica Acta 24, 261 (1977).
32. For a discussion of the angular spectrum representation see, for example, a) J.W. Goodman, Introduction to Fourier Optics (McGraw-Hill, New York, 1968), Sec. 3-7; b) J.R. Shewell and E. Wolf, J. Opt. Soc. Am. 58, 1596 (1968); c) G.C. Sherman and H.J. Bremermann, J. Opt. Soc. Am. 59, 146 (1969).
33. Conditions for the validity of the angular spectrum representation are discussed in a) E. Lalor, J. Opt. Soc. Am. 58, 1235 (1968); b) G.C. Sherman, J. Opt. Soc. Am. 59, 697 (1969).
34. A. Walther, J. Opt. Soc. Am. 63, 1622 (1973).
35. E.W. Marchand and E. Wolf, J. Opt. Soc. Am. 64, 1973 (1974).
36. H. Margenau and R.N. Hill, Prog. Theoret. Phys. (Kyoto) 26, 722 (1961).
37. The phase-space distribution function $\Phi(p,q)$ given in Eqs. (16a) and (16b) is not, in general, real. However, as the expectation values of observables are necessarily real, there are no contributions for the imaginary part of $\Phi(p,q)$, and we may thus consider only its real part.
38. A.T. Friberg, a) "On the existence of a radiance function for a partially coherent planar source", in Coherence and Quantum Optics IV, ed. by L. Mandel and E. Wolf (Plenum, New York, 1978), pp. 449-455; b) J. Opt. Soc. Am. 69, 192 (1979).
39. Actually, it would be sufficient to consider here only spatially completely coherent sources. The cross-spectral density function of any fluctuating source may be represented as a linear combination of contributions from spatially completely coherent elementary sources [cf. E. Wolf, "A new description of second-order coherence phenomena in the space-frequency domain", in these proceedings; cf. also footnote 24].
40. See also E. Wolf, J. Opt. Soc. Am. 68, 6 (1978).
41. The radiant intensity produced by a planar source of any state of coherence is discussed in E. Wolf, J. Opt. Soc. Am. 68, 1597 (1978).

42. See, for example, A. Walther, Opt. Lett. 3, 785 (1978).
43. W.H. Carter and E. Wolf, J. Opt. Soc. Am. 67, 785 (1977).
44. It should not be assumed (as has been sometimes incorrectly implied in recent publications) that a cross-spectral density function of the form of Eq. (43) necessarily characterizes a quasi-homogeneous source. For a quasi-homogeneous source not only must the cross-spectral density function $W^{(0)}(\underline{r}_1,\underline{r}_2)$ be of the form given by Eq. (43), but in addition the optical intensity $I^{(0)}(\underline{r})$ must be a "slow" function of $\underline{r}$ and the complex degree of spatial coherence $\mu^{(0)}(\underline{r}')$ must be a "fast" function of $\underline{r}'$.
45. A.T. Friberg, Optica Acta (to be published).
46. See, for example, A. Walther, J. Opt. Soc. Am. 64, 1275 (1974).
47. E. Wolf, Phys. Rev. D13, 865 (1976).
48. P.M. Woodward, Probability and Information Theory with Applications to Radar (Pergamon Press, Oxford and New York, 1953), Chapter 7.
49. A. Papoulis, J. Opt. Soc. Am. 64, 779 (1974).
50. J.-P. Guigay, Opt. Commun. 26, 136 (1978).
51. The corresponding problem in the phase-space formulation of quantum mechanics has been solved: Any linear operator Ω of a well-defined class acting on the density operator $\hat{\rho}(\hat{p},\hat{q})$ yields a permissible c-number function that may be used to calculate the expectation value of any observable when the system is in the state characterized by $\hat{\rho}(\hat{p},\hat{q})$.

THE WIGNER DISTRIBUTION FUNCTION - EXPERIMENTS AND APPLICATIONS

H. Bartelt
K.-H. Brenner
Physikalisches Institut der Universität Erlangen
Erwin-Rommel-Straße 1, D-8520 Erlangen
Fed. Rep. of Germany

ABSTRACT

The Wigner Distribution Function (WDF), defined for a phase space representation in quantum mechanics, offers an alternate form for representing signals simultaneously in Fourier reciprocal variables such as time and frequency. A purely nonnegative function that is closely related to the WDF, is the so-called "spectrogram". In this paper we describe three types of optical setups for the production of the spectrogram. They use coherent, partially coherent or temporally incoherent light as illumination. As application example speech signals are processed for a speaker verification.

MOTIVATION

The Wigner Distribution Function (WDF) is a useful tool for the description of a signal simultaneously in Fourier reciprocal variables. In acoustics it could be used to describe a signal simultaneously in time and frequency. In optics it would correspond to a representation in space and spatial frequency /1/. Indeed such descriptions have been applied before, e.g. in spatial filtering /2/ or in pseudo coloring /3/ without mentioning the WDF. But the WDF for all these methods provides a mathematical basis and thus helps to understand the process better from a system theoretical point of view. The WDF, introduced into optics by Bastiaans /4/, had been defined by Wigner /5/ as phase space representation in quantum mechanics and was further investigated by de Bruijn /6/. We will first recall the basic properties of the WDF. Then we will present three setups for an optical implementation using coherent, partially coherent and temporally incoherent illumination. They will be applied to represent one-dimensional signals simultaneously in space and spatial frequency. The spatial variables x and ν_x will be represented in two different ways by two space coordinates x and y or by one space coordinate and the wavelength λ. Using the wavelength λ as a third parameter we also can describe two-dimensional struc-

ISSN:0094-243X/81/560332-09$1.50

tures in three variables such as x,y and one frequency component e.g. in radial direction ν_r.

As an application for one dimensional signals we will show that the optically produced spectrogram of spoken words can be useful for speaker verification.

We would like to mention also that Wiersma /7/ proposed a setup for producing a spectrogram with completely incoherent light.

BASIC PROPERTIES OF THE WDF

The WDF for one-dimensional signals can be defined in two equivalent forms:

$$W(x,\nu) = \int u(x+\frac{x'}{2})u^*(x-\frac{x'}{2})\exp(-2\pi i\nu x')\,dx', \quad (1)$$

$$W(x,\nu) = \int \tilde{u}(\nu+\frac{\nu'}{2})\tilde{u}^*(\nu-\frac{\nu'}{2})\exp(+2\pi i\nu' x)\,d\nu'. \quad (2)$$

It produces a two-dimensional function of x and ν out of a one-dimensional function u. The complete symmetry between x and ν indicates that x and ν have equal weight in this description.

From the WDF the intensity, the power spectrum and the total energy of the signal can be retrieved by simple projections:

$$|u(x)|^2 = \int W(x,\nu)d\nu \ , \quad (3)$$

$$|\tilde{u}(\nu)|^2 = \int W(x,\nu)dx \ , \quad (4)$$

$$E_{total} = \int\int W(x,\nu)dxd\nu \ . \quad (5)$$

The complex amplitude u(x) can also be retrieved from the WDF apart from a constant phase factor:

$$u(x) = e^{i\phi(0)} \frac{\int W(x/2,\nu)\exp(2\pi i\nu x)}{(\int W(0,\nu)d\nu)^{1/2}} \quad (6)$$

$$e^{i\phi(0)} = \frac{u(0)}{|u(0)|} \ .$$

As the WDF is the Fouriertransform of a hermitian function, apparently $W(x,\nu)$ is real:

$$W(x,\nu) = W^*(x,\nu) \ . \quad (7)$$

Further properties of the WDF are reported i.e. by N.G. de Bruijn /4/. He also showed, that the WDF, which may be negative, will become positive everywhere, if we convolve it with a two-dimensional Gaussian:

$$G(x;\delta x)G(\nu;\delta\nu) = \frac{1}{\sqrt{\delta x \delta\nu}} \exp(-2\pi[(\frac{x}{\delta x})^2+(\frac{\nu}{\delta\nu})^2]) , \quad (8)$$

$$\text{with } \delta x \ \delta\nu \geqslant 1 . \quad (9)$$

If we choose the minimum uncertainty $\delta x \ \delta\nu = 1$ then the following equations hold:

$$\hat{W}(x,\nu;\delta x)=\int\int W(x',\nu')G(x-x';\delta x)G(\nu-\nu';1/\delta x)dx'd\nu' \quad (10)$$

$$=|\int u(x')G(x-x';\sqrt{2}\delta x)\exp(-2\pi i\nu x')dx'|^2 . \quad (11)$$

It says, the convolved WDF is equal to something we would call momentary spectrum or spectrogram. This equality can be generalized to all kinds of window functions, but with a Gaussian the uncertainty product is minimal in the mean square sense.

OPTICAL IMPLEMENTATION

Here we will describe three methods for the optical implementation of the convolved WDF. The first setup works with completely coherent light (fig. 1). It just converts the definition of a spectrogram (11) into an optical arrangement. The one-dimensional signal u(x) is multiplied by a tilted gaussian slit G:

$$u(x) \rightarrow u(x) \ G(x-y;\sqrt{2}\delta x) \quad (12)$$

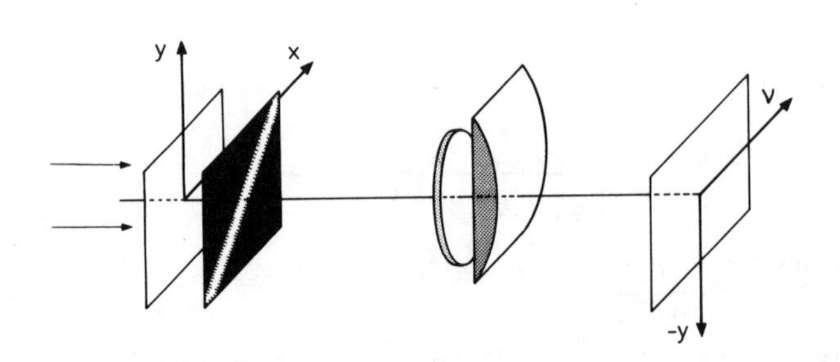

Fig. 1 Coherent optical setup for producing a spectrogram

A one-dimensional Fourier transformation results in the spectrogram:

$$|\int u(x)G(x-y;\sqrt{2}\delta x)\ \exp(-2\pi i\nu x)dx|^2=\hat{W}(y,\nu;\delta x). \quad (13)$$

An example with a test pattern is shown in fig.2. Of course the spectrogram images suffer from speckle noise due to the coherent processing. The speckle noise can be suppressed either by a serial processing as shown in fig. 3 or by illumination with a partially coherent light source (fig. 4). To this end the light source consists of a line with mutually incoherent light sources. Therefore in ν_x-direction, intensities are added giving a smoother image (fig. 5).

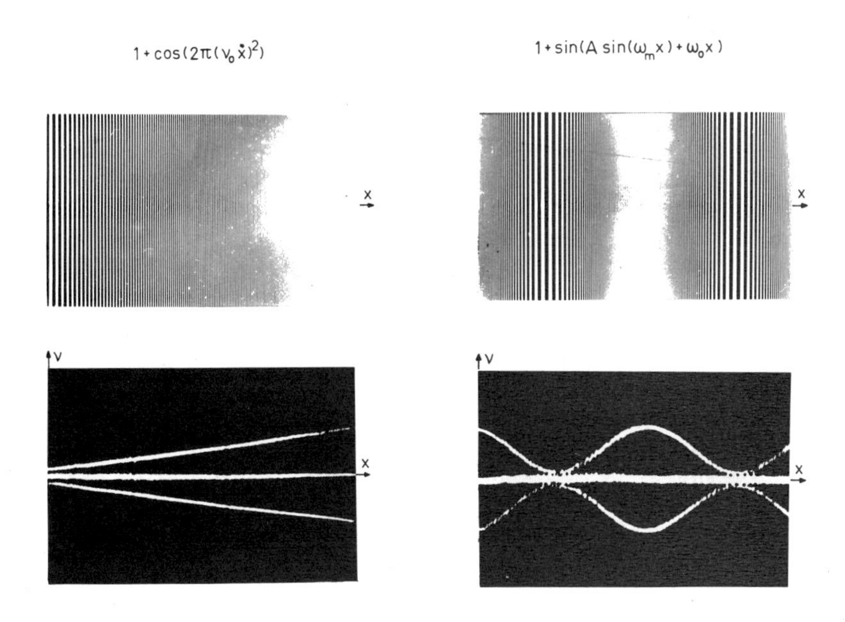

Fig. 2 Test signals and their spectrograms, linearly increasing frequency and sinusoidally variing frequency

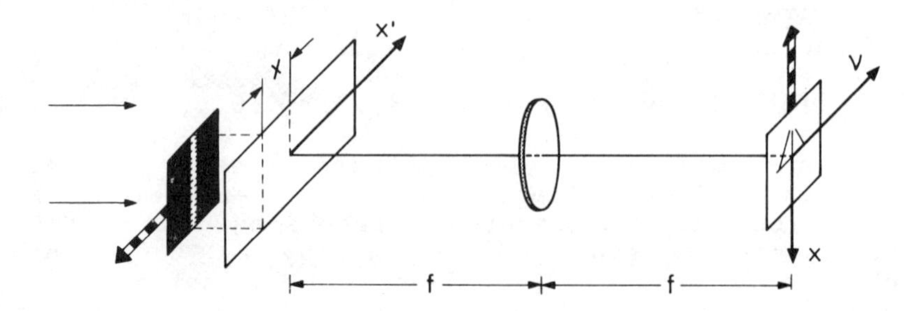

Fig. 3 Serial optical setup for producing a spectrogram

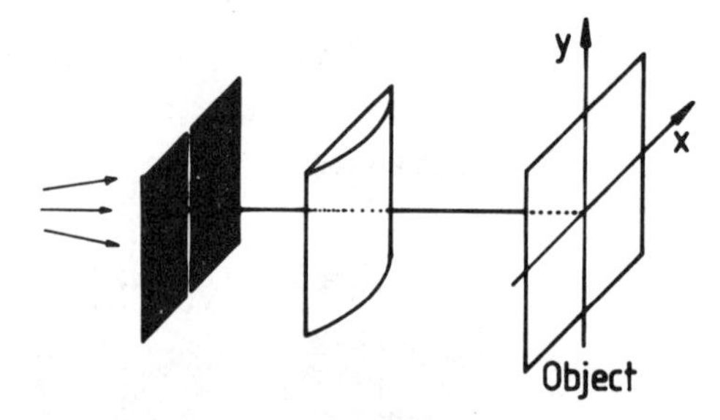

Partially coherent light source

Fig.4
A setup producing partially coherent illumination

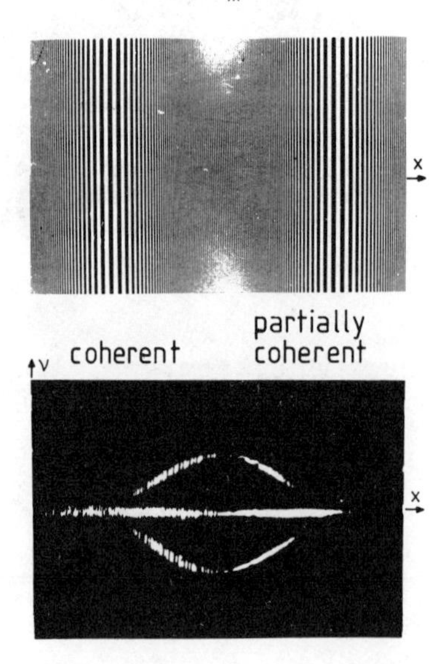

Fig. 5 Comparison of spectrograms from a test signal (sinusoidally variing frequency) produced with coherent light (left half of the spectrogram) and partially coherent light (right half of the spectrogram)

A different approach to the WDF representation could use the wavelength as an additional parameter. Then we are working with temporally incoherent light. Pseudo coloring methods belong to this type of description. White light allows to represent three different variables simultaneously. The radial frequency component ν_r or one of the frequency components ν_x or ν_y could be incorporated, in addition to the x- and y-dependance of the signal. Then the four-dimensional spectrogram of a special class of two-dimensional functions can be displayed by two three-dimensional representations in x,y,ν_x and x,y,ν_y.

For pseudo coloring we used the setup of fig. 6. A slit in the Fourier plane at x_o selects the spatial frequencies $\nu_x=x_o/\lambda f$. In the one-dimensional case it results again in an image of the spectrogram type (11):

$$\hat{W}(x,\lambda^{-1},\delta x)=\left|\int u(x')G(x-x';\delta x)\exp(-2\pi i\frac{x_o}{\lambda f}x')dx'\right|^2. \quad (14)$$

In fig. 7 black and white copies of a color representation for a two-dimensional function are shown. A radial filter would allow a translation of radial frequency components into wavelength.

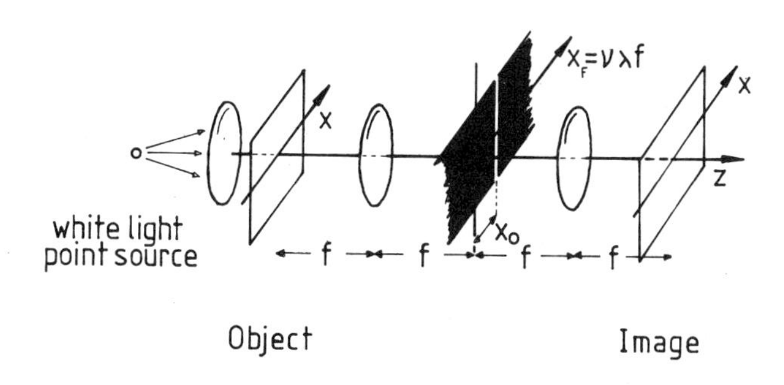

Fig. 6 Optical setup for encoding spatial frequency by wavenumber

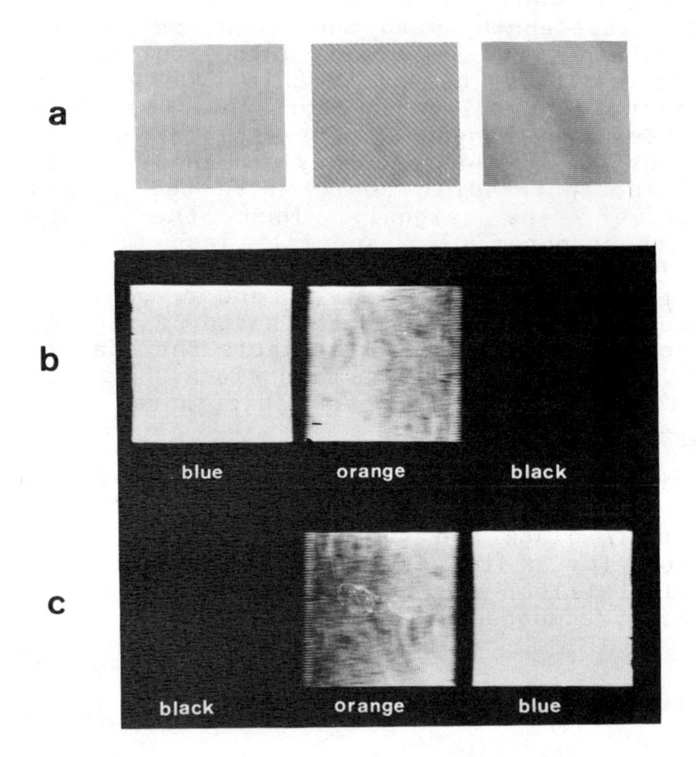

Fig. 7 Wavenumber encoding
(a) test gratings with vertical, 45° rotated and horizontal orientation
(b) black and white picture of the x,y,ν_x representation
(c) black and white picture of the x,y,ν_y representation

APPLICATION OF SPECTROGRAMS

For an optical WDF representation one-dimensional functions such as acoustical signals are well suited. Especially the processing of speech or spoken words is of practical importance. Three different types of such an application exist: word recognition, speaker identification and speaker verification. As an example we used optically produced spectrograms for speaker verification. Obviously it is the simplest application because a cooperative speaker is assumed. We recorded the spoken word "sesame" of three different persons onto film. Results are shown in fig. 8. The film transparencies were processed in the coherent setup of fig. 1 and compared to the spectrogram of the same word spoken earlier by one of these speakers. By visual comparison we can already decide which was the corresponding speaker. For a quantitative decision a crosscorrelation with the reference spectrogram was measured. Spectrograms of the same speaker at different times show a high correlation peak while the crosscorrelation with other speakers results in considerably lower values.

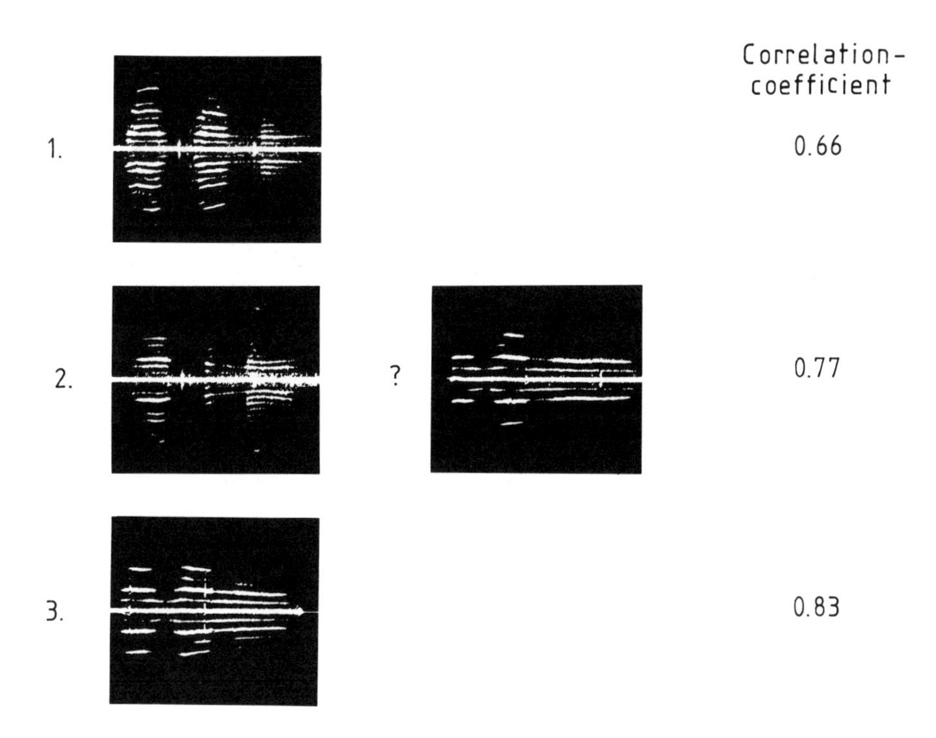

Fig.8 Speaker verification with spectrograms of the spoken word "sesame" by three different speakers; the spectrogram of the unidentified speaker (middle), recorded earlier, corresponds to speaker3.

ACKNOWLEDGMENT

The authors would like to thank Prof. A. W. Lohmann for fruitful discussions.

REFERENCES

1. H. O. Bartelt, K.-H. Brenner, A. W. Lohmann, Opt. Comm. 32 ,32 (1980)
   H. Bartelt, K.-H. Brenner, to be published in Israel Journal of Technology
2. H. Platzer, H. Glünder. Conference of the DGaO 1978 in Berlin
   J. D. Armitage, A. W. Lohmann, Appl. Opt. 4,399 (1965)
3. J. Bescos, T. C. Strand, Appl. Opt. 17,2524 (1978)
   I. Glaser, to be published in J. Optics
   H. Bartelt, to be published
4. M. J. Bastiaans, Opt. Comm. 25, 26 (1978)
   M. J. Bastiaans, Optica Acta 26, 1265 (1979)
   M. J. Bastiaans, Optica Acta 26, 1333 (1979)
   M. J. Bastiaans, Opt. Comm. 30, 321 (1979)
   M. J. Bastiaans, J. Opt. Soc. Am. 69,1710 (1979)
5. E. Wigner, Phys. Rev. 40, 749 (1932)
6. N.G. de Bruijn, Nieuw Archief voor Wiskunde 21, 205 (1973)
7. P. Wiersma, Opt. Comm. 33,237 (1980)

THE VIRTUAL FOURIER TRANSFORM

AND ITS APPLICATION IN THREE DIMENSIONAL DISPLAY

N.H. Farhat and C.Y. Ho
University of Pennsylvania
The Moore School Graduate Research Center
Philadelphia, PA 19104

ABSTRACT

In contrast to the well known and widely used instantaneous Fourier transforming property of the convergent lens in coherent (laser) light, the "Virtual Fourier Transform" (VFT) capability of the divergent lens is less widely known or used despite many advantages. We will review the principle of the VFT and discuss its advantages in certain applications. In particular a method for viewing the virtual Fourier transform of a two dimensional function with the naked eye using an ordinary point source will be presented. A scheme for three-dimensional image display based on a "Fourier domain projection theorem" utilizing varifocal VFT is described and a discussion of the properties of the displayed image given.

INTRODUCTION

Several sophisticated three dimensional (3-D) imaging techniques such as x-ray tomography[1], electron microscopy[2], crystallography[2], wave-vector diversity imaging and inverse scattering[3], involve measurements that give access to a finite volume in the 3-D Fourier space of a 3-D object function. A 3-D image of the original object can then be reconstructed by computing the inverse 3-D Fourier transform. The retrieved image normally represents the spatial distribution of a relevant parameter of the object such as absorption, reflectivity, scattering potential, etc.

Obviously, the required inverse transform can be performed digitally. Digital techniques however often preclude real-time operation particularly when the object being imaged is not simple but contains considerable resolvable intricate detail. More importantly, because of the inherent two dimensionality of CRT computer displays, direct true 3-D image display is not possible. Present day computer graphic displays are capable of displaying 3-D image detail either in separate cross-sections or slices, or in a computed perspective (isometric) view of the object, or in some instances stereoscopically where an illusion of a 3-D scene is created in the mind of the observer who is required usually to use special viewing glasses[4,5].

ISSN:0094-243X/81/650341-14$1.50

Hybrid (opto-digital) computing techniques offer an alternate approach to 3-D image retrieval from 3-D Fourier space data. They furnish as shown in this paper the ability to display true 3-D image detail. The approach is based on "Fourier Domain Projection Theorems"[2,3] that are dual to "Spatial or Object Domain Projection Theorems" used in radio-astronomy[6,7] and tomography[1]. These theorems permit the reconstruction of 3-D image detail tomographically* i.e. in slices from 2-D projections of the 3-D Fourier space data[2,3]. Although the required 2-D Fourier transform can be carried out digitally, the emphasis in this paper is on coherent optical techniques for performing the 2-D Fourier transform with particular attention to implementations that permit the execution of the necessary 2-D optical transforms of the various projection hologram sequentially in real-time. Specific attention is given to a technique that utilizes the virtual Fourier transform which permits the viewing of a virtual 3-D image in real-time.

FOURIER DOMAIN PROJECTION THEOREMS

There are two Fourier domain projection theorems. One leads to tomographic object reconstruction in parallel slices and is called the "weighted Fourier domain projection theorem", the other leads to tomographic object reconstruction in meridional or central slices and can therefore be called the "meridional or central slice Fourier domain projection theorem".

We begin by considering a 3-D object function $f(\bar{r})$ with $\bar{r} = x\bar{1}_x + y\bar{1}_y + z\bar{1}_z$ being a position vector in object space. Let $F(\bar{w})$ be the 3-D Fourier transform of $f(\bar{r})$ defined by,

$$F(\bar{w}) = \int f(\bar{r})\, e^{-j\bar{w}.\bar{r}}\, d\bar{r} \tag{1}$$

where $d\bar{r} = dx\,dy\,dz$ and $\bar{w} = w_x\bar{1}_x + w_y\bar{1}_y + w_z\bar{1}_z$ is a position vector in the Fourier or spatial frequency domain.

Consider next the projection of $F(\bar{w})$ on w_x,w_y plane defined by,

$$F_p(w_x,w_y) = \int_{w_z} F(\bar{w})\, dw_z \tag{2}$$

and combining eq. (1) and (2),

$$F_p(w_x,w_y) = \int_{w_z} \{\iiint_{xyz} f(x,y,z)\, e^{-j(w_x x + w_y y + w_z z)}\, dxdydz\} dw_z \tag{3}$$

*From the Greek work Tomos meaning slice.

Integrating with respect to w_z first, and assuming that the volume in $\bar{w}$ space occupied by $F(\bar{w})$ is sufficiently large, we obtain

$$F_p(w_x,w_y) = \int_x\int_y\int_z f(x,y,z)\,\delta(z)e^{-j(w_x x + w_y y)}\,dxdydz \quad (4)$$

$$= \int_x\int_y f(x,y,o)\,e^{-j(w_x x + w_y y)}\,dxdy. \quad (5)$$

The 2-D Fourier domain projection $F_p(w_x,w_y)$ and the central slice $f(x,y,o)$ through the object form thus a Fourier transform pair. This may be symbolically expressed as,

$$F_p(w_x,w_y) \leftrightarrow f(x,y,o). \quad (6)$$

Other parallel slices through the object at $z = z_n$, z_n being a constant describing the z coordinate of the n-th parallel slice, can be related in a similar manner to "weighted" Fourier domain projections of $F(\bar{w})$ defined by,

$$F_{p,n}(w_x,w_y) = \int_{w_z} F(w)e^{jz_n w_z}\,dw_z\,. \quad (7)$$

Making use of eq. (1), and again performing the integration with respect to w_z first, we obtain

$$F_{p,n}(w_x,w_y) \leftrightarrow f(x,y,z_n) \quad (8)$$

which indicates that the weighted projection $F_{p,n}(w_x,w_y)$ and the n-th parallel object slice $f(x,y,z_n)$ form a Fourier transform pair. Equation (6) is seen to be a special case of eq. (8) when $z_n = o$.

Given the 3-D Fourier space data manifold $F(\bar{w})$ one can digitally compute and display a set of "weighted projection holograms" $F_{p,n}(w_x,w_y)$. A corresponding set of images of parallel slices or cross-sectional outlines of the 3-D object can then be retrieved via 2-D Fourier transform operations which can most conveniently be carried out optically from photographic transparency records of the weighted projection holograms displayed by the computer.

Returning to eqs. (1) and (2) one can also show that projections of $F(\bar{w})$ on arbitrarily oriented planes other than the w_x, w_y plane chosen for eq. (2), yields "meridional projection holograms" that are 2-D Fourier transforms of corresponding merdional (central) slices of the object. This is the "meridional Fourier domain projection theorem. It furnishes the basis for angular multiplexing of the resulting meridional projection holograms into a single composite hologram which can be used to form a 3-D image of the object in a manner similar to that in integral holography[8] which is increasingly being referred to as Cross holography*.

THE VIRTUAL FOURIER TRANSFORM

In contrast to the well known spatial Fourier transforming property[9] of the convergent lens widely used in coherent optical computing, the complementary virtual Fourier transform capability of a divergent lens[10] is less widely known or used despite many attractive features. This is surprising since the power spectrum associated with the VFT is a phenomenon that is frequently observed in daily life when one happens to look at a distant point source such as a street light through a fine mesh screen or the fine fabric of transparent curtain material. The spectrum of the screen transmittance appears then as a virtual image in the plane of the point source.

The VFT concept of the divergent lens is easily derived from the Fourier transform expression of the convergent lens. Figure 1 illustrates the well known process of forming a real Fourier transform with a convergent lens. The object transparency, with complex transmittance $t(x_o, y_o)$, is placed at a distance d in front of a convergent lens of focal length F and illuminated with a normally incident collimated laser beam. The complex field amplitude of the wavefield in the back focal plane, the transform plane, is given by the well known formula

$$T(x,y) = \frac{j}{\lambda F} e^{-j\frac{k}{2F}\left[\left(1 - \frac{d}{F}\right)(x^2 + y^2)\right]} \times \iint_{-\infty}^{\infty} t(x_o, y_o)\, e^{j\frac{k}{F}(x\, x_o + y\, y_o)}\, dx_o dy_o \qquad (1)$$

in which the integral is recognized as the two dimensional Fourier transform of the object transmittance. $T(x,y)$ becomes the exact Fourier transform of $t(x_o, y_o)$ when $d = F$, that is when the object transparency is placed in the front focal plane of the lens. The power spectrum associated with the transform is real and can be projected on a screen placed in the back focal plane. It is also well known that a scaled version of the transform can be obtained in the back focal plane by placing the object tranparency in the converging laser beam to the right of the lens[9].

---

*Named after Lloyd Cross the originator of integral holography.

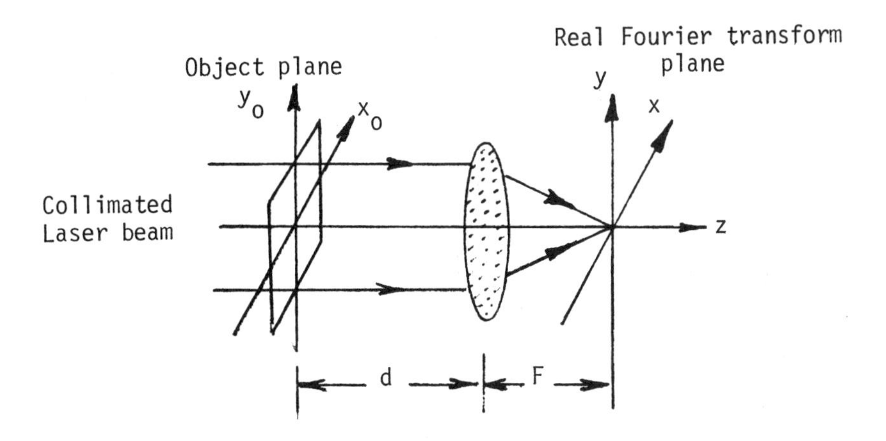

Fig. 1. Real Fourier transform formed with a convergent lens

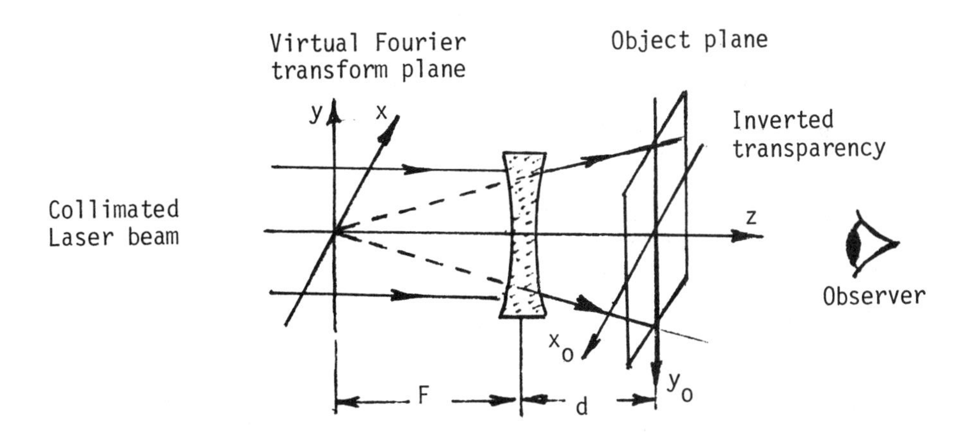

Fig. 2. Virtual Fourier transform formed with a divergent lens

Noting that eq. (1) does not change when we replace d by -d, F by -F, x_0 and y_0 by $-x_0$ and $-y_0$ respectively, we can arrive at the complementary VFT arrangement illustrated in Fig. 2. An inverted transparency $t(x_0,y_0)$ is placed now in the divergent coherent beam to the right of the divergent lens (of focal length -F) and a VFT given by eq. (1) is observed in the virtual focal plane of the lens. The same VFT can be seen by removing the divergent lens and replacing the laser beam with a point source placed at the origin of the VFT plane as depicted in Fig. 3. Thus a simple way of viewing the power spectrum associated with the VFT of a given diffracting screen (which is usually a Fourier transform hologram or a projection hologram of the type described above) is to hold the screen close to the eye and look through it at a distant bright point source. The point source used need not be derived from a laser. In fact it is preferable for safety purposes to use an LED or a spectrally filtered minute white light source such as a "grain-of-wheat" subminiature incandescent lamp or a miniature Christmas tree decorating lamp covered by a color or interference filter. This has the added advantage of furnishing a measure of control over the coherence properties of the wavefield impinging on the screen providing thereby a means for reducing coherent noise in the observed VFT and also, as will be discussed below, a means for coherent or noncoherent superposition of VFT's. As the distance of the point source from the diffracting screen is decreased in order to make it compatible with typical laboratory or optical bench dimensions, the size of the observed VFT decreased because of the change in the curvature of the wavefield illuminating the diffracting screen. To compensate for this effect it is necessary to reduce the size of the diffracting screen or transparency often to such a scale where viewing the VFT throught the small available aperture becomes difficult. To overcome this limitation the displacement property of the Fourier transform can be utilized. A composite transparency containing an ordered or random array of reduced replicas of the transmittance function $t(x_0,y_0)$ arranged side by side as illustrated in Fig. 4 is prepared. When such a composite transparency is viewed with the point source, the VFT's formed by the individual elements will overlap in the virtual Fourier plane. The VFT's are identical except for linear phase dependence on x,y which depends in each VFT on the central position of each element in the composite transparency. This leads to a desirable noise averaging effect and the appearance of fine checkered texture in the image detail. All this leads to an enhancement of the quality of the observed power spectrum. Both coherent and noncoherent superposition of the overlapping VFT's is possible using this scheme by varying the coherence area of the wavefield illuminating the composite transparency. When the coherence area is roughly equal to the size of the individual elements of the composite transparency, noncoherent superposition results, while a coherence area equal or greater than the size of the composite transparency would yield coherent superposition.

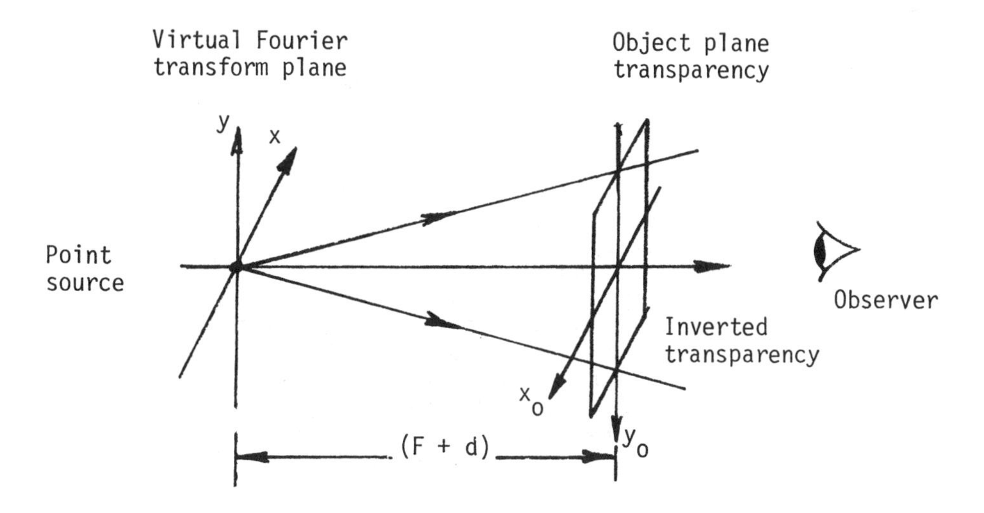

Fig. 3. Arrangement for viewing a virtual Fourier transform with a point source

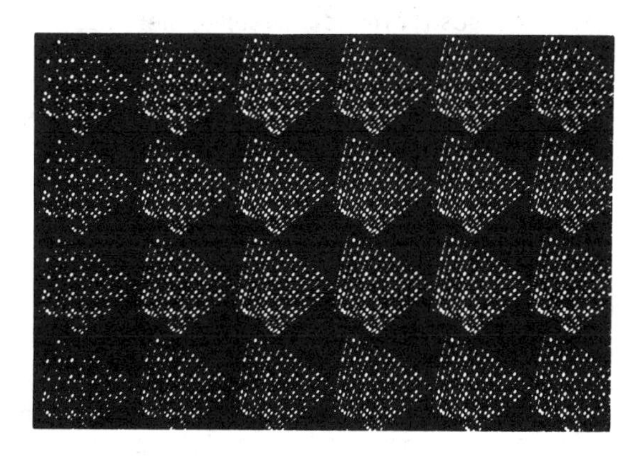

Fig. 4. A composite screen consisting of an ordered array of identical Fourier transform projection holograms.

THREE DIMENSIONAL DISPLAY

The VFT concept and the "weighted Fourier domain projection theorem" discussed above can be combined in an attractive scheme for the reconstruction and display of a 3-D image from a series of weighted projection holograms corresponding to different parallel slices through the object. The scheme is based on viewing a series of weighted projection holograms sequentially in the proper order of the occurrence of their corresponding slices in the original object while displacing the point source axially from one hologram to the next by an axial increment proportional to the spacings between adjacent object slices. In this fashion the reconstructed virtual images of the various slices are seen in depth at different VFT planes that are determined by the positions of the axially incremented point source. Repeated rapid execution of this procedure by displacing the point source back and forth leads the observer to see a virtual 3-D image tomographically in parallel slices or sections as he looks through the series of projection holograms passed rapidly, as in a motion picture film, infront of his eyes.

More specifically the scheme is based on preparing a series of N weighted Fourier domain projection holograms from the 3-D Fourier domain data $F(\bar{w})$ of a given object $f(\bar{r})$ as described in the preceeding sections. Each of the projection holograms would correspond to a different parallel slice through the object. A composite transparency similar to that shown in Fig. 4 is formed for each projection hologram. In fact Fig. 4 is an example of a computer generated composite hologram containing an array of identical weighted projection holograms corresponding to one slice of the test object shown in Fig. 5. The chosen test object consisted of eight point scatterers arranged as shown. The 3-D Fourier space of this test object was accessed in a computer simulation of wavelength diversity imaging as described in a companion paper in this volume*. The resulting computer generated Fourier space data manifold $F(\bar{w})$ was used to compute three weighted projection holograms corresponding to the three planes R_z'=1m,0,1m of Fig. 5 containing the three different distributions of point scatterers. A composite array such as that of Fig. 4 was formed and displayed by the computer for each of the three projection holograms, each was photographed yielding a set of three projection hologram composite transparencies. Copies of these were then mounted on a rotating wheel as shown in Fig. 6 (a) and viewed with an axially scanned point source. Four sets of transparency copies of these three composite projection holograms were mounted in the order 1,2,3,2,1,2 ... on the periphery of a rotating wheel as shown in Fig. 6 (a). The wheel is driven by a computer controlled stepper motor. The axially scanned point source was produced by scanning a focused laser beam back and forth on a length of fine nylon thread with the aid of a deflecting mirror mounted on the shaft of a second computer controlled stepper motor as shown in Fig. 6 (b). The laser and optical bench arrangement for forming the scanned focused beam appear in the background of Fig. 6 (a). The computer controlled steppers enable precise positioning of the secondary point source on the scattering thread in synchronism with the hologram

*See paper entitled "Holography, Wavelength Diversity Inverse Scattering" in this volume.

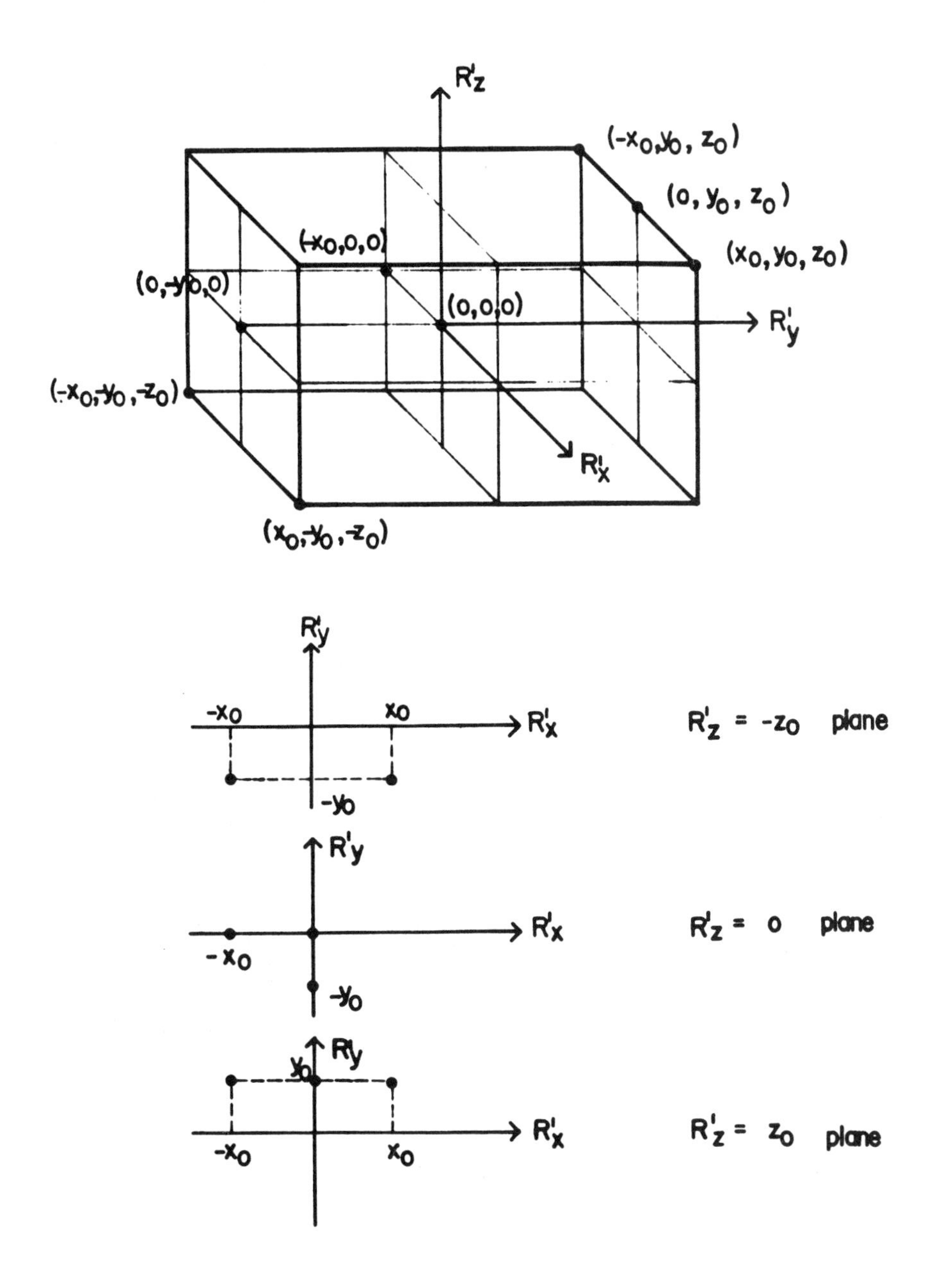

Fig. 5. A three-dimensional test object consisting of a set of eight point scatterers shown in isometric and R_x'-R_x' plane views at $R_z'=-z_o,0,z_o$. $x_o=y_o=z_o=100$ cm.

(a)

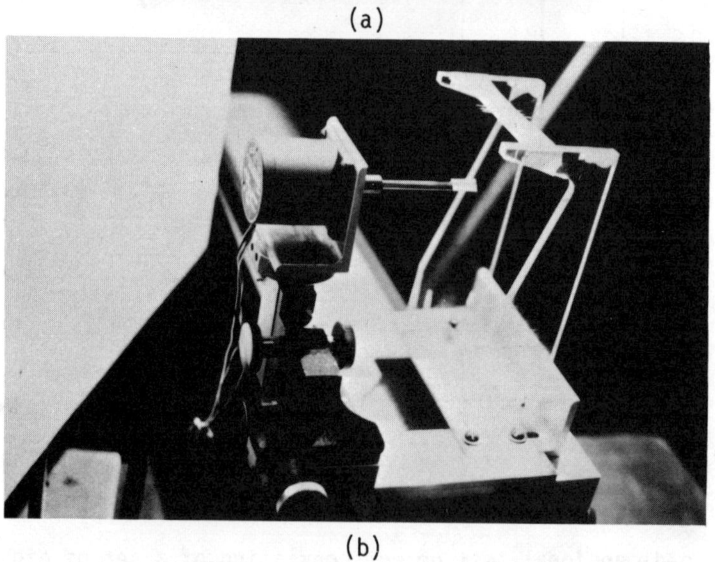

(b)

Fig. 6. Quasi real-time three-dimensional image reconstruction and tomographic display in successive slices from a series of projection holograms mounted on rotating wheel seen in forefront of (a); Detail of laser scanner used to produce linearly scanned point source is shown in (b).

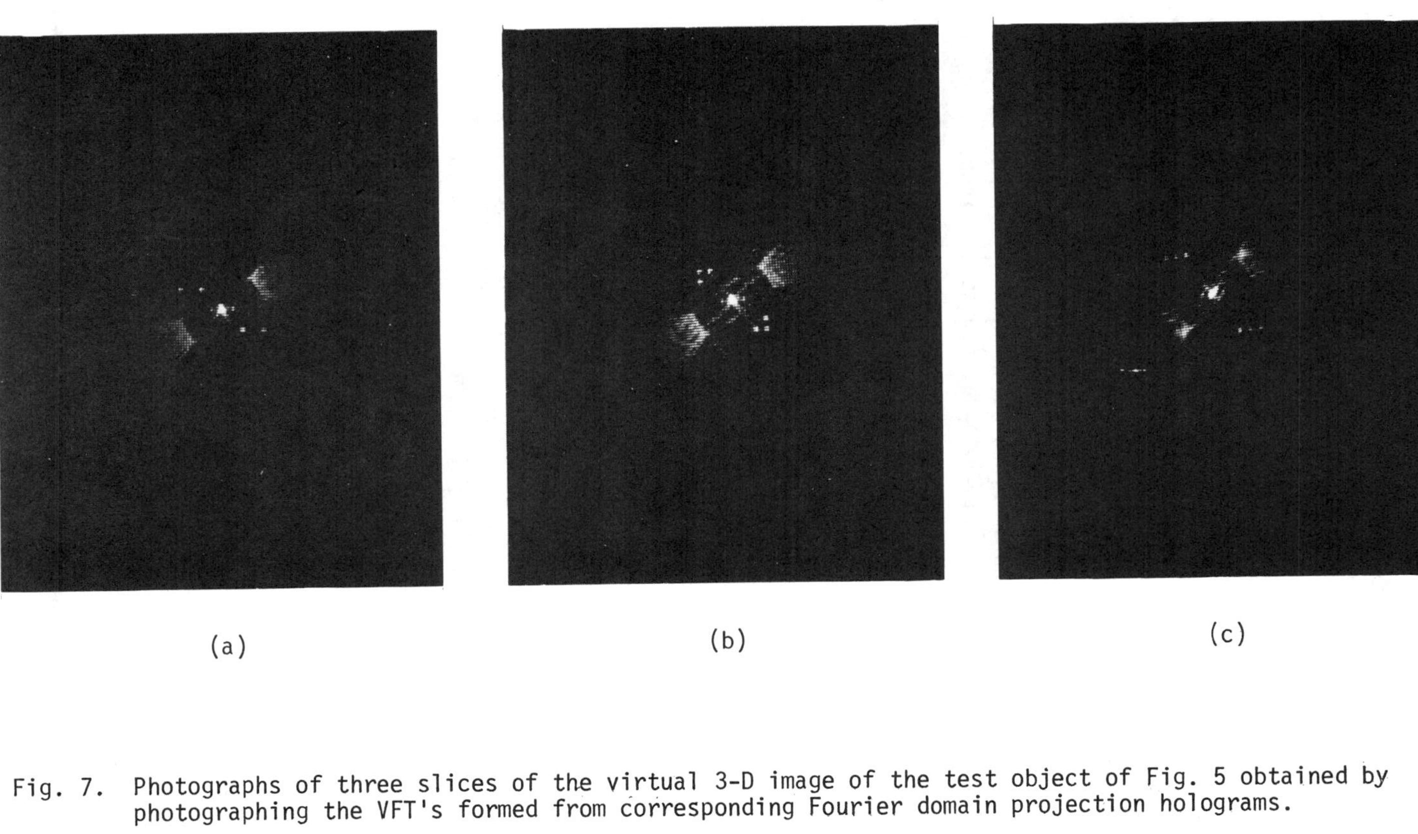

(a) (b) (c)

Fig. 7. Photographs of three slices of the virtual 3-D image of the test object of Fig. 5 obtained by photographing the VFT's formed from corresponding Fourier domain projection holograms.

being viewed so that the VFTs are formed in their proper planes. A viewer looking at the axially displaced point source through each transparency mounted on the wheel as it passes infront of his eye will see a 3-D virtual image. Photographs of the three virtual images seen by an observer in this fashion are shown in Fig. 7. An opto-digital scheme for rapid real-time implementation of the procedure realized above is shown in Fig. 8. This scheme, presently under study, utilizes a rapid recyclable spatial light modulator (SLM) such as the Itek PROM in order to form VFT's of the projection holograms displayed by the computer in real-time.

CONCLUSIONS

We have presented the basic principles of tomographic 3-D image display based on Fourier domain projection theorems. One possible implementation of the principle using the virtual Fourier transform and a series Fourier domain projection holograms has been described. There are several advantages for using the VFT rather than the real Fourier transform (RFT), the most important of which is the ease with which the position of the VFT plane can be moved axially by simply moving the position of the reconstruction point source. The VFT approach was adopted in the present study because it is much easier to move a point source rapidly than to move the display screen needed in the RFT approach. Furthermore focusing in the VFT approach is carried out by the observer while in the RFT approach it must be performed by the system. Other attractive features of the VFT are:

(a) Simplicity - enables direct viewing of the power spectrum of a transparency or a hologram with a variety of simple point sources.

(b) The scale of the observed VFT can be easily altered by changing the distance between the projection hologram transparency and the reconstruction point source.

(c) Lower speckle noise and therefore higher reconstructed image quality can be attained by using nonlaser point sources in the reconstruction such as LED or miniature spectrally filtered incandescent lamps. Further reduction in speckle noise occurs when an array of the projection hologram rather than a single hologram is used and when the hologram is slightly vibrated or is in motion because of a noise averaging effect.

(d) Coherent and noncoherent superposition of VFT's is possible by altering the coherence area of the reconstruction wavefield.

(e) Because of the Fourier transform nature of the projection holograms utilized, the resolution requirements from the storage medium (photographic film or the CRT/SLM system of Fig. 8) are much lower than would be needed in the recording of a Fresnel hologram of the object as a means of 3-D image display. The 3-D image detail contained in the single Fresnel hologram is now distributed over a series of lower resolution projection holograms which are used to form the 3-D image sequentially in time in individual slices.

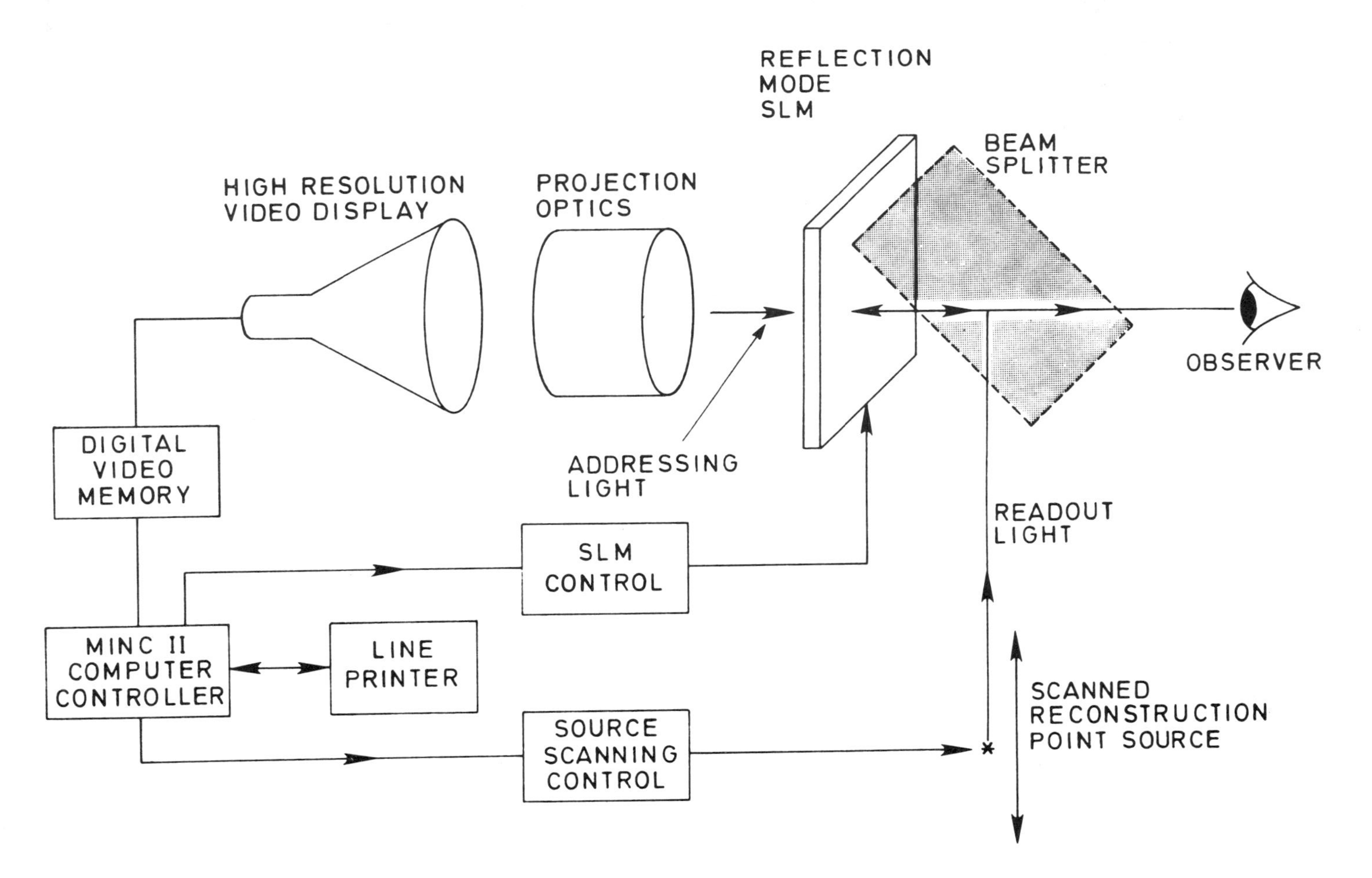

Fig. 8. Opto-digital scheme for the reconstruction and display of 3-D images using a recyclable spatial light modulator and a point source to view the VFT in real-time.

(f) Because 3-D image reconstruction is tomographic (in separate slices) there is no interference between the wavefields forming the various slices.

(g) Permits other forms of 3-D image display involving spatial or angular multiplexing in a fashion similar to integral holography.

ACKNOWLEDGEMENT

Research work reported here was supported by the U.S. Army Research Office under Grant No. DAAG29-80-K-0024 and by the U.S. Air Force Office of Scientific Research, Air Force Systems Command, under Grant No. AFOSR-77-3256-D.

REFERENCES

1. H.H. Barret and M.Y. Chiu, Optica Hoy Y Manãna, J. Bescos et.al., (eds.) 136, (Proc. of ICO-11, Sociedad Española De Optica, 1978).

2. N.H. Farhat and C.K. Chan, Acoustical Imaging, 8, 499, A. Metherel (ed.), (Plenum Press, New York, 1980).

3. G.W. Stroke and M. Halioua, Trans. Amer. Crystallographic Assoc., 12, 27, (1976).

4. T. Okoshi, Three-Dimensional Imaging Techniques, (Academic Press, New York, 1976).

5. T. Okoshi, Proc. IEEE, 68, 548, (1980).

6. R.N. Bracewell, Australian Journal of Physics, 9, 148, (1956).

7. R.N. Bracewell and S.J. Wernecke, J. Opt. Soc. Am., 65, 1342,(1975).

8. D.L. Vickers, Lawerence Livermore Laboratory Report, No. UCID-17035, (February 1976).

9. J.W. Goodman, Introduction to Fourier Optics, 83, (McGraw Hill, New York, 1968).

10. J. Knapp and M.F. Becker, App. Optics, 17, 1669, (1976).

WIGNER DISTRIBUTION AND AMBIGUITY FUNCTION

H. Harold Szu and J. A. Blodgett
Applied Optics Branch, Naval Research Laboratory
Washington, D. C. 20375

ABSTRACT

The Wigner phase space distribution and the Woodward radar ambiguity function have each been optically realized in idealized situations. The realistic performance will be assessed under simulated noise and clutter. A general cross Wigner distribution and cross ambiguity function are also compared as to their advantage in signal analysis. The cross Wigner distribution is a general convolution of non-zero frequency while the cross ambiguity function is a general correlation of non-zero frequency. The former is commutative while the latter is not. The significance of this difference is discussed. While the ambiguity function is generally complex, the Wigner distribution is real but not always positive. Both are known to be intermediate signals and are related through a double Fourier transform. However, optical equivalence of one from the other has not been demonstrated. Bartelt et al[1] have only produced the even part of the signal whose Wigner distribution is positive. Marks et al[2] have presented in the noise-free case the modulus of the ambiguity function, which is known as the ambiguity surface in the radar community. The present extension with noise and clutter provides a better understanding of the optimum condition for signal to noise ratio. Such a comparison study, which takes into account both knowledge of quantum mechanical symmetry and practical experience of radar clutter and noise, turns out to be fruitful.

1. INTRODUCTION

The Wigner distribution function used in quantum mechanics

$$W_{\psi}(x,\nu) = \int \psi(x + \frac{x'}{2})\ \psi^*(x - \frac{x'}{2})\ \exp(-i2\pi\nu x')dx', \quad (1)$$

and the Woodward ambiguity function used in radar/sonar echoes

$$A_u(\mu,y) = \int u(x + \frac{y}{2})\ u^*(x - \frac{y}{2})\ \exp(-i2\pi\mu x)dx, \quad (2)$$

are equivalent, in principle, in describing the wavefunction $\psi(x)$ and the signal envelope $u(x)$ that combines the wave localization with its frequency spread. However, different practical advantages will be discussed in Section 3 for radar/sonar echoes, whose time delay, y, and Doppler shift, μ, are related respectively to the target range and the target radial velocity, and in Section 4 for a moving quantum mechanical wave packet used as Gabor's elementary signal.

ISSN:0094-243X/81/650355-27$1.50

A cross Wigner distribution function and a cross Woodward ambiguity function are compared as to their utility for a phase space covert communication optical system since both are connected with the zero, first and second order integral transforms. Such a novel scheme of coding and decoding in the Wigner and Woodward phase spaces can be optically implemented for real time covert communications. This possibility may become evident after the basic properties of the cross Woodward ambiguity function (exemplified in Section 3) have been systematically compared with those of the cross Wigner distribution function (illucidated in Section 4). Some new formulae relating the two will be developed in Section 5 and optically demonstrated with additive noise in Section 6. The actual demonstration of such a phase space covert communication optical system is, however, not reported due to its preliminary nature.

2. A PHASE-SPACE COVERT COMMUNICATION OPTICAL SYSTEM[3]

One possibility for a phase space covert communication optical system would propagate (eg. video link) the cross Wigner distribution function

$$W_{uK}(x,\nu) = \int u(x + \frac{x'}{2})\, K^*(x - \frac{x'}{2})\, \exp(-i2\pi\nu x')dx' \qquad (3a)$$

$$= \int \tilde{u}(\nu + \frac{\nu'}{2})\, \tilde{K}^*(\nu - \frac{\nu'}{2})\, \exp(i2\pi\nu' x)d\nu' \qquad (3b)$$

and perhaps use the cross Woodward ambiguity function

$$A_{uK}(\mu,y) = \int u(x + \frac{y}{2})\, K^*(x - \frac{y}{2})\, \exp(-i2\pi\mu x)dx \qquad (4a)$$

$$= \int \tilde{u}(\mu' + \frac{\mu}{2})\, \tilde{K}^*(\mu' - \frac{\mu}{2})\, \exp(i2\pi\mu' y)d\mu' \qquad (4b)$$

for optical processing. The overhead tilde always denotes the Fourier transform:

$$\tilde{u}(\nu) = \int u(x)\, \exp(-i2\pi\nu x)dx \qquad (5)$$

An asterisk between two functions denotes the convolution at zero frequency, while two asterisks indicate the correlation at zero frequency; the conventional superscript asterisk still denotes the complex conjugate; the integration domain without boundaries implies integration over the real axis.

Communication of a coded signal in phase space is secret because the key function K(x) is known only to the communicating parties. Knowing the key function, the unknown signal u(t) can be deciphered as follows

$$u(t) = \iint K(2x' - t)\, W_{uK}(x',\nu')\, \exp(-i4\pi\nu'(x' - t))\, dx'd\nu'$$

$$= \iint u(x' + \frac{x''}{2})\, K(2x'-t)K^*(x' - \frac{x''}{2})\, \delta(2x'-2t+x'')dx'dx''$$

$$= u(t) \int K(t - x'')\, K^*(t - x'')dx'' , \qquad (6)$$

$$u(t) = \iint K(t - y') \, A_{uK}(\mu',y') \exp(i2\pi\mu'(t - \frac{y'}{2}))d\mu'dy', \quad (7)$$

provided that the key function is square integrable and normalized to unity.

$$(K,K) = \int K^*(x) \, K(x)dx = \int \tilde{K}^*(\nu) \, \tilde{K}(\nu)d\nu = (\tilde{K},\tilde{K}) = 1. \quad (8)$$

Sometimes it is convenient to display the Fourier transform $\tilde{u}(\nu)$ directly from

$$\tilde{u}(\nu) = \iint \tilde{K}(2\nu' - \nu) \, W_{uK}(x',\nu') \exp(i4\pi x'(\nu' - \nu))dx'd\nu', \quad (9)$$

$$\tilde{u}(\nu) = \iint \tilde{K}(\nu - \mu') \, A_{uK}(\mu',y') \exp(-i2\pi y' \, (\nu - \frac{\mu'}{2})) \, d\mu'dy'. \quad (10)$$

The key function is not to be mistaken with the so called N-bit Barker code[6] used[4,5], for the ratio (=N) of the central peak and (N-1) sidelobes in the radar ambiguity function. The key function is chosen with the following consideration

$$W_{uK}(x,o) = \int u(x + \frac{x'}{2}) \, K^*(x - \frac{x'}{2}) \, dx' = W^*_{Ku}(x,o), \quad (11)$$

$$A_{uK}(o,y) = \int u(x + \frac{y}{2}) \, K^*(x - \frac{y}{2})dx = A^*_{Ku}(o, -y). \quad (12)$$

The time-order and the causality condition must be maintained in the cross correlation function. The linear, associative, and commutative properties of the convolution operation give a smoothing operation. The width of u*K is the sum of the width of u and the width of K, while the area under u*K is the product of areas under u and K. The slowly varying components of the signal are reproduced with little degradation, while any structure significantly finer than the width of the key function is severely attenuated. Thus, the key functions should be chosen with narrow widths. They should also provide the phase space redundancy sufficient to restore propagation, degradation, and distortion. Therefore a good choice for the key function must be periodic, have narrow width

$$\tilde{K}(\nu) = \Sigma_n \, k_n \, \delta(\nu - n) \quad (13)$$

$$K(x) = \Sigma_n \, k_n \, \exp(i2\pi nx) \quad (14)$$

and be such that the Parseval formula insures the normalization

$$(\tilde{K},\tilde{K}) = (K,K) = \Sigma_n \, |k_n|^2 = 1 \, . \quad (15)$$

The key function is related to the Woodward sampling function, comb(x), if all expansion coefficients are unity. However, the uniform key function is no longer square integrable. The fabrication of the key function is possible by various means. For example, a computer controlled laser beam scanner, used successfully to fabricate circuit chips, can be programed to produce dots of different transmission in a computer-generated hologram.

3. CROSS WOODWARD AMBIGUITY FUNCTIONS

A matched filter is a linear time-invariant filter with an impulse response h(t) defined by

$$h(t) = S(y - t); \; S(t) = 0, \; t < 0, \tag{16}$$

where S(t) is the real signal to which the filter is matched, and y is the value of time delay. The transfer function of a matched filter has the form

$$\tilde{h}(\mu) = \tilde{S}^*(\mu) \exp(- i2\pi\mu y) \tag{17}$$

where $\tilde{S}(\mu)$ is the Fourier transform of the signal S(t). The range R of a target is measured using the delay

$$y = 2R/C, \tag{18a}$$

while, simultaneously, the radial velocity V of a target is measured using the Doppler frequency shift

$$\mu = 2V/\lambda. \tag{18b}$$

C is the speed of light and λ is the transmitted wavelength. The complex envelope u(t) of radar/sonar echoes when Fourier transformed is correlated with the delayed but matched filter in order to suppress the background noise and clutter

$$\chi(y,\mu) \equiv \int \tilde{u}(\mu') \, \tilde{u}^*(\mu'-\mu) \, e^{-i2\pi\mu' y} d\mu' = e^{-i\pi\mu y} A_u(\mu, - y). \tag{19}$$

The phase factor preceeding the integral is physically insignificant. This function considered first by Y. W. Lee et al (1950)[7] and Woodward et al[8] is a measure of the degree of similarity between a signal envelope and a replica that is shifted in time y and frequency μ. In order to reduce clutter effects over a specific control area, the Woodward ambiguity function[8] is naturally extended to the cross ambiguity function,

$$\chi_{uv}(y,\mu) = \int u(x) \, v^*(x - y) \, e^{-i2\pi\mu x} dx = e^{-i\pi\mu y} A_{uv}(\mu,y), \tag{20}$$

where the best processing wave form v has been determined by Stutt et al (1968)[9] using the calculus of variations. The basic integral of all cross ambiguity functions is S. M. Sussman's (1962)[10] product formula

$$\iint A_{uv}(\mu,y) \, A^*_{KZ}(\mu,y) \, e^{i2\pi(\mu x - \nu y)} d\mu dy = A_{uK}(\nu,x) \, A^*_{vZ} (\nu,x), \tag{21}$$

which shows that the Fourier transform of the complex product of cross ambiguity functions equals the product itself except that v and K must be interchanged. Since the result seems to contradict the common sense expectation of a correlation, a simple proof of Sussman's product formula is included as follows:

$$\text{LHS} = \iiint u(x' + \tfrac{y}{2})\, v^*(x' - \tfrac{y}{2})\, e^{-i2\pi\mu x'} dx' \cdot \int K^*(x'' + \tfrac{y}{2})\, Z(x'' - \tfrac{y}{2})\, e^{i2\pi\mu x''} dx''\, e^{i2\pi(\mu x - \nu y)} d\mu dy.$$

Since

$$\int \exp(i2\pi\mu(x - x' + x''))d\mu = \delta(x + x'' - x')$$

then integrating dx' and changing variables from (x'', y) to (t', t'')

$$x + x'' + \frac{y}{2} = t'' + \frac{x}{2}\ , \quad x + x'' - \frac{y}{2} = t' + \frac{x}{2}, \quad y = t'' - t'$$

the proof is thus completed:

$$\text{LHS} = \int u(t'' + \tfrac{x}{2})\, K^*(t'' - \tfrac{x}{2})\, e^{-i2\pi\nu t''} dt'' \cdot \int v^*(t' + \tfrac{x}{2})\, Z(t' - \tfrac{x}{2})\, e^{i2\pi\nu t'} dt'$$

$$= \text{RHS}.$$

Reconciliation with what is expected from common sense will lead, in Section 4, to a new formula useful for covert phase space communication as a general smoothing operation. If u=K and v=Z, then C. A. Stutt's (1961)[11] formula for converting cross to auto ambiguity functions follows from Eq. (21),

$$\iint |A_{uv}(\mu,y)|^2\, e^{i2\pi(\mu x-\nu y)}\, d\mu dy = A_u(\nu,x)\, A_v^*(\nu,x). \tag{22}$$

If u=K=v=Z then W. M. Siebert's (1958)[12] self-reciprocal property follows from Eq. (22)

$$\iint |A_u(\mu,y)|^2\, e^{i2\pi(\mu x - \nu y)}\, d\mu dy = |A_u(\nu,x)|^2\ . \tag{23}$$

In other words the ambiguity surface, $|A_u|^2$, is its own 2D Fourier transform. In particular, when ν=x=0, Woodward's (1953) theorem follows

$$\iint |A_u(\mu,y)|^2\, d\mu dy = |A_u(0,0)|^2 = 1\ , \tag{24}$$

which states that the volume under the ambiguity surface is independent of the signal wave form. The Woodward theorem, in conjunction with the central peak theorem proved using Schwarz inequality for Eq. (2)

$$|A_u(\mu,y)| \leqq |A_u(0,0)|^2 \tag{25}$$

provides the basic constraint on the ambiguity surface. Synthesization of the Woodward ambiguity surface is similar to the jelly-squeezing phenomenon in that a sharp central peak of small volume implies that somewhere else there must be broad humps to accommodate the remaining volume. This phenomenon gives rise to self-clutter which may mask targets relatively far removed in range and range rate. More details of volume-redistribution have been given by E. C. Westerfield et al (1963)[13] and D. Richman (1966)[14] as follows. Setting $\nu=0$ or $x=0$ in the Siebert integral, Eq. (23), and Fourier transforming respectively on the x variable or ν variable gives

$$\frac{\delta \text{ Volume}}{\delta \text{ Doppler}} = \int |A_u(\mu,y)|^2 \, dy = \int |A_u(0,x)|^2 \, e^{-2\pi\mu x} dx, \tag{26}$$

$$\frac{\delta \text{ Volume}}{\delta \text{ Delay}} = \int |A_u(\mu,y)|^2 d\mu = \int |A_u(\nu,0)|^2 \, e^{i2\pi\nu y} d\nu \, . \tag{27}$$

If the central peak of a thumbtag shape[15] is squeezed toward the origin of the delay axis y (sharp and accurate delay measurement), then the volume must be spread out along the Doppler axis (broad and inaccurate Doppler measurement) and vice versa. Such a radar uncertainty principle embodied in the Woodward ambiguity function is consistent with the quantum mechanical uncertainty principle embodied in the Wigner distribution function.

If a signal has a chirp factor, $\exp(-i\pi p\mu^2)$, then the associated Woodward ambiguity function

$$\begin{aligned} A_{u\exp(-i\pi p\mu^2)}(\mu,y) &= e^{-i\pi\mu y} \int \tilde{u}(\mu')e^{-i\pi p\mu'^2} \, \tilde{u}^*(\mu'-\mu) \\ &\quad \cdot e^{i\pi p(\mu'-\mu)^2} \, e^{i2\pi\mu' y} \, d\mu' \\ &= e^{-i\pi(y-p\mu)\mu} \int \tilde{u}(\mu') \, \tilde{u}^*(\mu'-\mu) \, e^{i2\pi\mu'(y-p\mu)} d\mu' \\ &= A_u(\mu, y' = y - p\mu), \end{aligned} \tag{28}$$

shows the shearing in Fig. 1. This was implemented electronically by a dispersive filter for the quadratic phase or optically by a lens. An approaching target that has a positive Doppler shift has an echo response attenuated in amplitude, and shifted to an earlier time ($y<0$). This is shown in Fig. 1 where $y'=y-p\mu$ together with a representative ambiguity contour for a long wave train of constant frequency and one for a short wave train also of a constant frequency. A uniform amplitude chirp signal has a sidelobe diminishing in inverse proportional to y' in the Woodward phase space, while the nonuniform amplitude Hermite wave form of J. R. Klauder (1960)[17] makes the sidelobes of the ambiguity function, which is highly peaked at the center and is rotationally symmetric, fall off slower than the chirp signal.

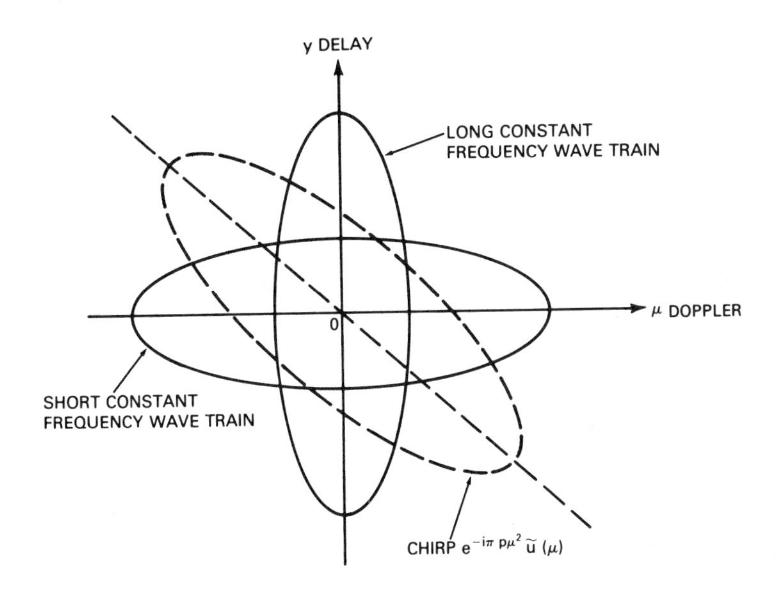

Fig. 1. Typical contours for the ambiguity surface are plotted for three cases. The short constant frequency wave train signal gives a more accurate delay time, y, of its echo used for the target range measurement, Eq. (18a), while the long constant frequency wave train signal gives a more accurate doppler frequency shift μ used for the target velocity measurement, Eq. (18b).

The chirp signal, Eq. (28), gives a compromise between the two by a rotation in the phase space, $y'=y-p\mu$. Also note that they are all centered at the origin according to Eqs. (29) and (25).

4. CROSS WIGNER DISTRIBUTION FUNCTIONS

The Wigner distribution is sometimes preferred to the Woodward ambiguity function because of the inner product formula of J. E. Moyal (1949)[16].

$$\iint W_{uv}(x,\nu)\, W^*_{KZ}(x,\nu)\, dx d\nu = (K,u)\,(Z,v)^*. \tag{29}$$

$$\text{LHS} = \iiint u(x + \frac{x'}{2})\, v^*(x - \frac{x'}{2})\, e^{-i2\pi\nu x'}\, dx' \int K^*(x + \frac{x''}{2})\, Z(x - \frac{x''}{2}) \cdot e^{i2\pi\nu x''}\, dx'' dx d\nu = \text{RHS} \tag{30}$$

The proof by inspection is simple: Changing integration variables x and x' to y and y', $y=x+(x'/2)$, $y'=x-(x'/2)$, after dx'' having been integrated using the Dirac delta $\delta(x''-x')$ obtained first by integrating $\int d\nu$, one obtains the Moyal inner product formula. Two important modulus formulae follow from the Moyal formula Eq. (29) in two special cases. If u=v and v=K=Z then,

$$\iint W_u(x,\nu)\, W_v(x,\nu)\, dx d\nu = |(u,v)|^2 = |(\tilde{u},\tilde{v})|^2 \tag{31}$$

and furthermore if u=v, then

$$\iint W_v(x,\nu)^2\, dx d\nu = |(v,v)|^2 = |(\tilde{v},\tilde{v})|^2 \tag{32}$$

N.G. de Bruijn (1973)[18] used the second modulus formula to prove that the moments of W are the moments of the signal energy density $|v(x)|^2$ and the moments of the power spectrum density $|\tilde{v}(\nu)|^2$ respectively. The first modulus formula also used by Bruijn follows. While a small moving particle can not have a very precise quantum mechanical momentum, similarly a short sound wave can not have a very precise frequency either. Such a parallelism led D. Gabor (1946)[19] to apply to signal communication the quantum mechanical minimum uncertainty wave packet which is the gaussian wave of spread δx associated with the moving particle of momentum $p=h\mu=\mu$ (the unit of h=1 is used) and coordinate x.

$$G_\mu(x-y|\delta x) = (2/\delta x)^{1/4} \exp\left[-\frac{\pi}{\delta x}(x-y)^2 + i2\pi\mu(x - \frac{y}{2})\right] \tag{33}$$

$$\tilde{G}_\mu(\nu-\mu|\delta\nu) = G_{(-y)}(\nu-\mu|\frac{1}{\delta x}) \tag{34}$$

Both the moving signal and the standing signal are square integrable and identically normalized to unity, because the phase difference

$$G_o(x-y|\delta x) \equiv (\frac{2}{\delta x})^{1/4} \exp(-\frac{\pi}{\delta x}(x-y)^2) = G_\mu(x-y|\delta x)\, e^{-i2\pi\mu(x-\frac{y}{2})}, \quad (35)$$

does not change the conservation of the Born probability. The Wigner distribution of the Gabor elementary signal can be easily calculated and written in a compact form

$$W_{G_\mu(x-y|\delta x)}(x,\nu) = \int G_\mu(x+\frac{x'}{2}-y\,|\,\delta x)\; G_\mu^*(x-\frac{x'}{2}-y|\,\delta x)\; e^{i2\pi\nu x'}\,dx'$$

$$= G_o(x-y|\delta x)^2\; G_o(\nu-\mu|\frac{1}{\delta x})^2\;, \quad (36)$$

where use is made of the identity

$$(2\pi)^{-1/2}\int \exp(-ikx-\frac{1}{2}x^2)\,dx = \exp\,(-\frac{1}{2}k^2). \quad (37)$$

Note that Bartelt et al (1980)[1] and M. J. Bastiaans (1978)[26] used different normalization constants; the present choice is convenient to put Eq. (36) and the optical equivalence theorem (Eq. 41) in compact form.

Substituting v for G_μ in the first and second modulus formulae, Bruijn (1973)[18] obtained the following version of the smoothed Wigner distribution function together with its normalization

$$\langle W_u(x,\nu)\rangle_{G_\mu(x-y|\delta x)} \equiv \iint W_u(x,\nu)\; W_{G_\mu(x-y|\delta x)}(x,\nu)\,dx\,d\nu$$

$$= |(u,G_\mu)|^2\;, \quad (38)$$

$$||W_{G_\mu}(x,\nu)||^2 = ||G_\mu(x-y|\delta x)||^4 = 1. \quad (39)$$

Bartelt et al (1980)[1] have optically implemented such a smoothed version of the Wigner distribution. Because of the importance of their technique, the following basic new analysis is given. The optical setup for the technique is familiar[2]. This is not surprising because from our optical equivalence theorem the inner product is reduced precisely to the cross ambiguity function

$$(u(x),\; G_\mu(x-y|\delta x)) = A_{uG_o(x|\delta x)}(\mu,y)$$

$$= \int u(x+\frac{y}{2})\; G_o(x-\frac{y}{2}|\,\delta x)\, e^{-i2\pi\mu x}dx. \quad (40)$$

Thus, the optical implementation of the Bruijn-Wigner formula is reduced to the optical implementation of the modulus of the cross ambiguity function

$$\langle W_u(x,\nu) \rangle_{G_\mu(x-y|\delta x)} = |A_{uG_o(x|\delta x)}(\mu,y)|^2. \quad (41)$$

To show that this coincidence is general, and not just limited to Gabor's elementary signal, a second order transform formula analogous but not equivalent to Sussman's product formula Eq. (21) can be directly derived as follows,

$$\iint W_{uv}(x,\nu)\, W^*_{KZ}(x,\nu)\, e^{-i2\pi(\mu x-\nu y)} dx d\nu = A_{uv}(\mu,y) ** A^*_{KZ}(\mu,y) \quad (42)$$

$$LHS = \iiiint A_{uv}(\mu' y')\, e^{i2\pi(\mu' x-\nu y')} d\mu' y' \iint A^*_{KZ}(\mu'', y'')\, e^{-i2\pi(\mu'' x-y''\nu)} d\mu'' dy''\, e^{-i2\pi(\mu x-\nu y)} dx d\nu$$

where the first order Klauder transform[17] is used for simplicity. The integrations of $d\nu$ and dx give $\delta(y-y'+y'')$ and $\delta(\mu'-\mu-\mu'')$ and the integrations of dy' and $d\mu$ give the correlation of cross ambiguity functions. Now setting $\mu=y=0$ in Eq. (42) and setting $x=\nu=0$ in Sussmans product formula Eq. (21), one obtains the interplay between Eqs. (42) and (21), a general smoothing formula.

$$\langle W_{uv}(x,\nu)\rangle_{KZ} \equiv \iint W_{uv}(x,\nu) W^*_{KZ}(x,\nu)\, dx d\nu = A_{uK}(0,0)\; A^*_{vZ}(0,0). \quad (43)$$

In particular if $u=v$ and $K=Z=G_\mu(x-y|\delta)$, then the phase difference equation (35) for the standing ($\mu=0$) and the moving ($\mu\neq 0$) Gabor elementary signals can be used to prove

$$A_{uG_\mu(x-y|\delta x)}(0,0) = A_{uG_o(x|\delta x)}(\mu,y). \quad (44)$$

Substituting Eq. (44) into the general smoothing formula Eq. (43), our optical equivalence theorem Eq. (41) is again verified. The formulae analogous to Stutt's and Siebert's follows from the second order transform formula Eq. (42) when $u=K$ and $v=Z$ and when $u=K=v=Z$.

$$\iint |W_{uv}(x,\nu)|^2\, e^{-i2\pi(\mu x-\nu y)} dx d\nu = A_{uv}(\mu,y) ** A^*_{uv}(\mu,y). \quad (45)$$

This verifies, at $\mu=y=0$ together with Stutt's convertion formula Eq. (22), the Parseval formula for cross Wigner and cross Woodward functions

$$\iint |W_{uv}(x,\nu)|^2 dx d\nu = \iint |A_{uv}(\mu,y)|^2 d\mu dy = A_u(0,0) A^*_v(0,0) \quad (46)$$

and a Woodward like theorm for cross Wigner distribution that states the volume under $|W(x,\nu)|^2_{uv}$ is independent of the wave forms u and v.

The geometrical meaning of the inner product $(u(x), G_\mu(x-y|\delta x))$ in Hilbert space is simple, namely the projection of the signal vector u(x) onto the Gabor elementary signal $G_\mu(x-y|\delta x)$. Therefore the optical equivalent cross ambiguity function A_{uG_o} should be similarly interpreted. In order to take full advantage of quantum mechanics operator calculus, the Dirac bracket notation for c (complex) number inner product is adopted

$$\langle x|y,\ \mu,\delta x\rangle \equiv G_\mu(x-y\ |\delta x)\cdot \tag{47}$$

It is a complete set for a given δx

$$\iint dyd\mu\ \ |y,\mu,\delta x\rangle\langle\delta x,\mu,y| \ = \ \hat{1}_{\delta x} \tag{48}$$

$$\int dx \qquad |x\rangle\langle x| \ = \ \hat{1} \tag{49}$$

where a caret (∧) denotes an operator.

$$\langle x|\hat{1}|x'\rangle = \delta(x-x') \tag{50}$$

$$\langle\delta x,\ \mu,\ y'|y,\ \mu,\delta x\rangle = \delta(y-y') \tag{51}$$

It is not orthogonal[20] except when $\mu=\mu'$. Using this technique, C. W. Helstrom's (1966)[21] expansion of a signal u(t) in Gabor elementary signals easily rederived using these completeness relationships, Eqs. (48) and (49), is

$$\begin{aligned} u(x) &= \langle x|u\rangle = \langle x|\hat{1}_{\delta x}\ \hat{1}|u\rangle \\ &= \iint dyd\mu \langle x|y,\mu,\delta x\rangle \int \langle\delta x,\mu,y|x'\rangle\langle x'|u\rangle\, dx' \\ &= \iint dyd\mu\ G_\mu(x-y|\delta x)\ (u(x),\ G_\mu(x-y|\delta x)) \end{aligned} \tag{52}$$

This is consistent with our formulae for covert decoding, Eq. (7), if the specific property, Eq. (40), of Gabor's elementary signals, Eq. (35), is used. The geometrical meaning of the Wigner distribution is simple in Hilbert space. A. Royer (1977)[23] has shown that the Wigner distribution is the expectation value of the parity operator at a phase space point and he has rederived the S. R. de Groot and L. G. Suttorp (1972)[24] bounds on the Wigner distribution function

$$W_\psi(x,\nu) \ = \ 2\ \langle\psi|\hat{\pi}_{x\nu}|\psi\rangle \tag{53}$$

$$-\ 2 \leqq W_\psi \leqq 2; \qquad \langle\psi|\psi\rangle = 1 \tag{54}$$

This operator technique can be extended to $W_{uK}(x,\nu)$ and $A_{uK}(\mu,y)$ as follows. Since the Glauber (1963) phase space displacement operator[25] can be used to shift the origin in x and ν, then, without lose of generality, we can consider $x=\nu=0$ as the coordinate origin of the phase space

$$W_{uv}(0,0)= \int v^*(-\tfrac{x'}{2})\, u(\tfrac{x'}{2})\, dx' = 2\int v^*(-x)\, u(x)\, dx$$

$$= 2\int (v^{(+)*}(-x) + v^{(-)*}(-x))(u^{(+)}(x) + u^{(-)}(x))\, dx$$

$$= 2(v^{(+)}, u^{(+)}) - 2\,(v^{(-)}, u^{(-)}) \tag{55a}$$

$$v(x) = \tfrac{1}{2}\,(v(x) + v(-x)) + \tfrac{1}{2}\,(v(x) - v(-x)) = v^{(+)}(x) + v^{(-)}(x) \tag{55b}$$

Let $\hat{\pi}_{oo}$ be the parity operator at the origin $x=\nu=0$,

$$\hat{\pi}_{oo} \equiv \int |-x\rangle\langle x|\, dx, \tag{56}$$

then Eq. (55a) can be rewritten as

$$W_{uv}(0,0) = 2\int \langle v|-x\rangle\langle x|u\rangle\, dx = 2\,\langle v|\hat{\pi}_{oo}|u\rangle \tag{57}$$

The results obtained in Eq. (55a) and (55b) can be rederived in operator notation and this will be found useful for subsequent developments. It will be recalled that

$$\hat{1} = \hat{\pi}_{oo}^{\,2} = \int dx'' \int dx' |-x'\rangle\, \langle x'|-x''\rangle\, \langle x''|$$

$$= \int dx'' \int dx' |-x'\rangle\;\; \delta(x' + x'')\langle x''| \tag{58a}$$

and the sign of the ket vector changes

$$\hat{\pi}_{oo}|x\rangle = \int dx' |-x'\rangle\, \langle x'|x\rangle = \int dx' |-x'\rangle\, \delta(x'-x) = |-x\rangle, \tag{58b}$$

it has $\pm$ 1 eigenvalues corresponding to the even and odd eigenkets

$$\hat{\pi}_{oo}|\phi^{\pm}_{oo}\rangle = \pm 1|\phi^{\pm}_{oo}\rangle. \tag{59}$$

Such an identification is useful because the parity operator can be conveniently decomposed into the projection operators

$$\hat{P}^{\pm}_{oo} \equiv \tfrac{1}{2}\,(\hat{1} \pm \hat{\pi}_{oo}) \tag{60}$$

which can give useful equations such as

$$\hat{P}^{+}_{oo} - \hat{P}^{-}_{oo} = \hat{\pi}_{oo},\quad \hat{P}^{+}_{oo} + \hat{P}^{-}_{oo} = \hat{1} \tag{61)(62}$$

$$(\hat{P}^{+}_{oo})^2 = \hat{P}^{+}_{oo} \tag{63}$$

A single projection operation does the projection completely that the second projection will have no effect on the ket vector.

$$(\hat{P}^{+}_{oo})^2 |\phi\rangle = \hat{P}^{+}_{oo} |\phi^+\rangle = |\phi^+\rangle \tag{64}$$

Using Eqs. (57) (61) and (63) we prove

$$\begin{aligned} W_{uv}(0,0) &= 2\langle v|\hat{\pi}_{oo}|u\rangle - 2\langle v|\hat{P}^{+}_{oo} - \hat{P}^{-}_{oo}|u\rangle \\ &= 2\langle v|(P^{+}_{oo})^2|u\rangle - 2\langle v|(P^{-}_{oo})^2|u\rangle \\ &= 2\langle v^+|u^+\rangle - 2\langle v^-|u^-\rangle \end{aligned} \tag{65}$$

In general

$$u(x) = (\,u(x) + u(-x) \;+\; u(x) - u(-x)\,)/2 = u^{(+)}(x) + u^{(-)}(\,x)$$

and similarly

$$v(x) = v^{(+)}(x) + v^{(-)}(\,x).$$

Therefore, if both u and v are even and real functions then

$$W_{u^{(+)}\,v^{(+)}} = 2\langle v^+|u^+\rangle$$

is real and positive. Since

$$||u^+||^2 + ||u^-||^2 = 1$$

then

$$||u^+||^2 < 1$$

and therefore

$$-2 \leq W(0,0) \leq 2\ . \tag{66}$$

Thus W_{uv} is proportional to the overlap of u with the mirror image of v about the phase space point, which is clearly a measure of the symmetry of u with respect to the mirror image of v at the phase space point.

The Glauber phase space displacement operator[25] is unitary

$$\hat{D}(x,\nu) = \exp(i2\pi(\nu\hat{R} - x\hat{P})),\ [\hat{R},\hat{P}] = 1 \tag{67}$$

The similarity transformation can be used to shift the origin in x,ν, and one finds

$$\hat{\pi}_{x\nu} = \hat{D}(x,\nu)\hat{\pi}_{oo}\ \hat{D}(x,\nu)^{-1} = \int dx'\ e^{-i4\pi\nu x'}\ |\ x - x'\rangle\ \langle x' + x|. \quad (68)$$

We can show that the cross ambiguity function can also be written as

$$A_{uv}(\mu,y) = 2\ \langle\ v\ |\ \tilde{\pi}_{\mu y}\ |u\ \rangle\ , \quad (69)$$

where

$$2\tilde{\pi}_{\mu y} \equiv \int dx\ e^{-i2\pi\mu x}|x - \frac{y}{2}\rangle\langle\frac{y}{2} + x|. \quad (70)$$

Note that

$$2\tilde{\pi}_{\mu y} = \iint dx d\nu\ e^{-i2\pi(\mu x - \nu y)} 2\hat{\pi}_{x\nu} \quad (71)$$

and

$$2\hat{\pi}_{x\nu} = \iint d\mu dy\ e^{i2\pi(\mu x - \nu y)}\ 2\tilde{\pi}_{\mu y} \quad (72)$$

Since from Eqs. (70) (70) and (49)

$$2\tilde{\pi}_{oo} = \iint dx d\nu\ 2\hat{\pi}_{x\nu} = \hat{1} \quad (73)$$

$$2\hat{\pi}_{oo} = \iint d\mu dy\ 2\tilde{\pi}_{\mu y} \quad (74)$$

then one verifies Fourier's zeroth order theorem

$$A_{uv}(0,0) = \iint dx d\nu\ W_{uv}(x,\nu) = \langle\ v|u\ \rangle \quad (75)$$

$$W_{uv}(0,0) = \iint d\mu dy\ A_{uv}(\mu,y) = 2\langle v^{+}|u^{+}\rangle - 2\langle v^{-}|u^{-}\rangle \quad (76)$$

where the RHS is a special property of the function.

Besides the example of signal expansion, Eq. (52), a simple application of the Dirac bracket and operator notation is given as follows. M. J. Bastiaans (1978)[26] has elaborated[27] the close resemblance[28] between the Wigner distribution function and the ray concept in geometrical optics and introduced the ray spread function as a double Wigner distribution function based on two arguments of the point spread function $S(x,x')$,

$$\psi(x) = \int S(x,\ x')\ \phi(x')\ dx' \quad (77)$$

$$W_{\psi}(x_o,\nu_o) = \iint D(x_o,\nu_o,\ x_i,\nu_i)\ W_{\phi}(x_i,\ \nu_i)\ dx_2\ d\nu_2. \quad (78)$$

It is of interest to express the phase space propagation in the Dirac operator notation:

$$|\psi\ \rangle = \hat{S}|\phi\ \rangle \quad (79)$$

$$W_\psi(x_o,\nu_o) = 2\langle\psi|\hat{\pi}_{x_o\nu_o}|\psi\rangle = 2\langle\psi|\hat{\pi}_{x_o\nu_o}\hat{S}\hat{1}|\phi\rangle$$

$$= 2\langle\psi|\hat{\pi}_{x_o\nu_o}\hat{S}\hat{\pi}^2_{x_i\nu_i}|\phi\rangle. \tag{80}$$

The special case of space invariance and the usefulness in treating ray propagation and transport problems will not be presented here.

The linearly increasing frequency,

$$b(x) = b_o + ax\frac{1}{2},$$

gives for the chirp signal a simple Wigner distribution function

$$u(x) = \exp(i2\pi b(x)x) \tag{81a}$$

$$W_u(x,\nu) = \int u(x+\frac{x'}{2})\, u^*(x-\frac{x'}{2})e^{-i2\pi\nu x'}dx'$$

$$= \int dx'\, \exp(i2\pi x'(ax+b_o-\nu)) = \delta(ax+b_o-\nu). \tag{81b}$$

The corresponding Woodward ambiguity function is plotted in Fig. 2 for comparison

$$A_u(\mu,y) = \iint dxd\nu\, e^{-i2\pi(\mu x-\nu y)}\, W_u(x,\nu)$$

$$= \iint dxd\nu\, e^{-i2\pi(\mu x-\nu y)} \int dx'\, e^{i2\pi x'(ax+b_o-\nu)}$$

$$= \frac{1}{2}\,\delta(y-\frac{\mu}{2})\, e^{i2\pi\frac{b_o}{a}\mu}. \tag{82}$$

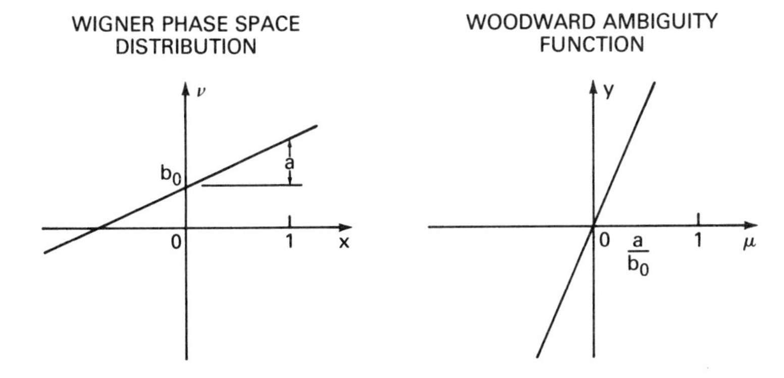

Fig. 2. Comparison of the Wigner distribution function, Eq. (81b), with the ambiguity function, Eq. (82).

5. ANALYTIC RELATIONSHIP BETWEEN TWO CROSS FUNCTIONS

Various integral transform relationships between the cross Wigner distribution function and the cross Woodward ambiguity function will be placed into zero, first, and second order groups as follows.

The zeroth order has no integral sign. If

$$K(t)^* = K(-t)^*$$

then an equivalent equation follows for arbitrary u

$$W_{uK}(x,\nu) = 2\, A_{uK}(2\nu, 2x). \tag{83}$$

The symmetry property is also listed here for comparison:

$$W_{uK}(x,\nu) = W^*_{Ku}(x,\nu) \tag{84}$$

$$A_{uK}(\mu,y) = A^*_{Ku}(-\mu,-y)\ . \tag{85}$$

The first order transform between a single W and a single A was first discovered by Klauder (1960)[17] who gave the 2D-transform

$$W_{uv}(x,\nu) = \iint A_{uv}(\mu,y)\ e^{i2\pi(\mu x-\nu y)}\ d\mu dy \tag{86}$$

$$A_{uv}(\mu,y) = \iint W_{uv}(x,\nu)\ e^{-i2\pi(\mu x-\nu y)}\ dx d\nu. \tag{87}$$

The second order transform between a double W and a double A is new.

$$\iint W_{uv}(x,\nu)\ W^*_{KZ}\ (x,\nu)\ e^{-i2\pi(\mu x-\nu y)} dx d\nu = A_{uv}(\mu,y)**A^*_{KZ}(\mu,y). \tag{88}$$

$$\iint A_{uv}(\mu,y)\ A^*_{KZ}\ (\mu,y)\ e^{i2\pi(\mu x-\nu y)} d\mu dy = W_{uv}(x,\nu)**W^*_{KZ}(x,\nu). \tag{89}$$

A simple proof by inspection is possible for Eq. (89)

$$\text{RHS} = \iiint u(x+x'+\tfrac{y}{2})\ v^*(x+x'-\tfrac{y}{2})\ e^{-i2\pi(\nu+\nu')y} dy$$

$$\int K^*(x'+\tfrac{y'}{2})\ Z(x'-\tfrac{y'}{2})\ e^{i2\pi\nu' y'} dy' dx' d\nu'$$

$$= \int u(t'+\tfrac{x}{2})\ K^*(t'-\tfrac{x}{2})\ e^{-i2\pi\nu t'} dt'$$

$$\int v^*(t''+\tfrac{x}{2})\ Z(t''-\tfrac{x}{2})\ e^{i2\pi\nu t''} dt''$$

$$= \quad A_{uK}(\nu,x)\ A^*_{vZ}(\nu,x) = \text{LHS by Eq. (21)}$$

where the ν' integral gives $\delta(y'-y)$ which sets $y'=y$ and the change of variables

$$x + x' + \frac{y}{2} = t' - \frac{x}{2}, \quad x + x' - \frac{y}{2} = t'' + \frac{x}{2}$$

gives $y=t'-t''$ in the final result. Some limiting cases are given as follows. If $u=K$ and $v=Z$ then the auto-correlation of the cross Wigner distribution gives

$$\iint W_{uv}(x+x', \nu+\nu')\, W^*_{uv}(x',\nu')\, dx'd\nu' = A_u(\nu,x)\, A^*_v(\nu,x) \tag{90}$$

which implies at the origin, $x=\nu=0$, the smoothing of a self-matching filter

$$\iint W_{uv}(x,\nu)\, W^*_{uv}(x,\nu)\, dxd\nu = A_u(0,0)\, A^*_v(0,0) \tag{91}$$

If $u=K=v=Z$, then the auto correlation of the Wigner distribution function is the surface of the ambiguity function

$$\iint W_u(x+x', \nu+\nu')\, W_u(x',\nu')\, dx'd\nu' = |A_u(\nu,x)|^2 \tag{92}$$

which implies at $x=\nu=0$

$$\iint W_u(x,\nu)^2\, dxd\nu = |A_u(0,0)|^2 \tag{93}$$

i.e. the Parseval relationship, using Eq. (24),

$$\iint W_u(x,\nu)^2\, dxd\nu = \iint |A_u(\mu,y)|^2\, d\mu dy. \tag{94}$$

All of these special case relationships can be verified on an optical bench, using Fourier transform (FT) lens, and films,

$$FT^{-1}\, [FT(W_u)\, FT(W_u)^*] = |A_u(\nu,x)|^2\ . \tag{95}$$

A generalization of the Bruijn local smoothing, Eq. (38) i.e. substituting the Gabor elementary signal into the Moyal's inner product formula, is possible using Eq. (21) and correlation smoothing Eq. (89)

$$\iint W_{G_\mu(x-y|\delta x)}(x+x', \nu+\nu')\, W_u(x',\nu')\, dx'd\nu' = |A_{G_\mu(x-y|\delta x)u}(\nu,x)|^2 \tag{96}$$

It may be of interest to verify the consistency of Eqs. (83) and (89) by substituting the zero order relationship into the second order relationship. If $u(t) = u(-t)$, an even function, then

$$W_u(x,\nu) = 2A_u(2\nu, 2x)$$

according to the zeroth order result, Eq. (83). Substituting the zeroth order result, Eq. (83), into the RHS of the second order relationship, Eq. (89), one finds using Eq. (21) in the LHS

$$4\iint A_u(2(\nu+\nu'),\ 2(x+x'))\ A_u(2\nu',\ 2x')\ dx'd\nu' = |A_u(\nu,x)|^2.$$

This interesting result shows that the auto correlation of A_u is the ambiguity surface itself

$$\iint A_u(\nu+\nu',\ x+x')\ A_u(\nu',\ x')\ dx'd\nu' = |A_u(\tfrac{\nu}{2},\ \tfrac{x}{2})|^2 \qquad (97)$$

(if u(t) is even). After $\int d\nu'$ integration giving $\delta(y+y')$, one uses the following algebra to derive Eq. (97) directly. One uses the symmetry

$$u(-y + \frac{x'}{2}) = u(y- \frac{x'}{2}),$$

$$u^*(-y- \frac{x'}{2}) = u^*(y+ \frac{x'}{2})$$

and changes of variable

$$y+ \frac{x}{2} + \frac{x'}{2} = y'' + \frac{x}{4},$$

$$y- \frac{x}{2} - \frac{x'}{2} = x'' - \frac{x}{4}$$

such that $2y=y''+x''$ and

$$y- \frac{x'}{2} = x'' + \frac{x}{4}$$

and

$$y+ \frac{x'}{2} = y'' - \frac{x}{4}.$$

This completes the derivation.

The differential relationship is known but given for completeness. The zeroth order moment follows, using Klauder's formula,

$$E_{total} = \iint W_u(x,\nu)\ dxd\nu = (u,u) = (\tilde{u},\ \tilde{u}) = A_u(0,0) = 1 \qquad (98)$$

The signal intensity and power spectrum are

$$|u(x)|^2 = \int W_u(x,\nu)\ d\nu = \int A_u(\mu,0)\ e^{i2\pi\mu x} d\mu \qquad (99)$$

$$|\tilde{u}(\nu)|^2 = \int W_u(x,\nu)\ dx = \int A_u(0,y)\ e^{-i2\pi\nu y} dy\ . \qquad (100)$$

The first order moments are

$$\overline{\begin{pmatrix} x^n \\ \nu^n \end{pmatrix}} \equiv \iint \begin{pmatrix} x^n \\ \nu^n \end{pmatrix} W_u(x,\nu)\ dxd\nu = \left(\frac{-1}{2\pi i}\right)^n \begin{pmatrix} \dfrac{\partial^n A_\mu}{\partial \mu^n} \\ \dfrac{\partial^n A_\mu}{\partial(-y)^n} \end{pmatrix}_{\mu=y=0} \quad (101)\ (102)$$

The second order moments
can be computed using the second modulus formula of Moyal's equation (32)

$$\overline{\overline{x}} \equiv \iint x\ W_u^2(x,\nu)\ dxd\nu = \int x|u(x)|^2 dx \quad |\tilde{u}(\nu)|^2 d\nu \qquad (103)$$

$$= (u,xu)$$

$$\overline{\overline{\nu}} \equiv \iint \nu\ W_u^2(x,\nu)\ dxd\nu = (\tilde{u},\nu\tilde{u}). \qquad (104)$$

6. PERFORMANCE ASSESSMENT UNDER ADVERSE CONDITIONS

The probabilistic treatment of radar/sonar echoes is not new, nor is the stochastic treatment of quantum mechanics. There is a famous Wigner theorem[35,36] on the impossibility of real, positive and 'correct' probability description of stochastic quantum mechanics. The cross over implication through the zero, first, and second order transforms between the Wigner distribution function and the Woodward ambiguity function has not been fully explored. Since this is necessary for a thorough probabilistic assessment of the performance under noise and clutter, then without it we are for the time being content (i) to reproduce the literature results and then use the setup to verify formulae, (ii) to add simulated noise to verify for the Wigner distribution the Woodward-like volume theorem Eq. (46). We expect the volume under the quadratic Wigner function to be independent of the signal waveform rect(x) perturbed by 5% and 10% noisy black dots if we keep a constant area of transmission.

Extensive studies of coherent optical processors have been given under noisy conditions. One of the recent reviews[37] on optical frequency plane correlators gives the performance of the input, $(s(x)+n(x))B(x)$ from a signal $s(x)$ under additive noise $n(x)$ and a blur function $B(x)$, as a function of the signal spatial width (L) and the blur function frequency bandwidth (D_ν). The best performance is at the space blur-bandwidth product $LD_\nu=1$ independent of the input, e.g. the Gaussian blur of frequency function $\tilde{B}(\nu)$ of halfwidth D_ν equals the inverse of the signal spatial width L. Such an optimum condition, $LD_\nu=1$, has been demonstrated on an A-scope for IR radar threshold signals in internal noise and reported by J. L. Lawson and G. E. Uhlenbeck (1950)[38]. Because the Fourier transform pairs $B(x)$ and $\tilde{B}(\nu)$ have reciprocal widths, the halfwidth of $B(x)$, D, equals the reciprocal of that $\tilde{B}(\nu)$, $D=D_\nu^{-1}$. Thus the optimum condition $LD_\nu = 1$ implies equal input widths

$$L = D. \tag{105}$$

(i) An ideal signal is represented by the rect(x) function for clear light transmission.

$$\text{rect}\left(\frac{x}{\Delta x}\right) = \begin{cases} 1 & |x| \leq \frac{1}{2}\Delta x \\ 0 & \text{otherwise} \end{cases} \tag{106}$$

The product of two crossed rect functions, rect(x+y) and rect(x-y), gives a transparent diamond shape of 4 hard edges. This was 1-D Fourier transformed on the x variable by Marks et al to exemplify the Woodward ambiguity function $A(\mu,y)$. Five studies for five purposes are given as follows.

(A) Optical Setup:[39,40] The astigmatic coherent processor of 3 alternating cylindrical lenses of proper focal lengths (f,2f,f) (f=75mm) which gives no spatially varying quadratic phase terms in its output was used. Three alternating and orthogonal cylindrical lenses, that are separated by equal distances from the input plane to the output plane for a total distance of 4f, can Fourier transform along the two common cylindrical axes while imaging along the single cylindrical axis. The set up can reproduce the result of Marks et al

$$A_{rect}(\mu,y) = e^{i\pi\mu y}\chi(\mu,y) = \begin{cases} \frac{1}{\pi\mu} \sin(\pi\mu(\Delta x-|y|)), & |y| \leq \Delta x \\ 0 & \text{otherwise} \end{cases} \tag{107}$$

The sine function vanishes at the piecewise hyperbola $\mu(y)$

$$\pi\mu\cdot(\Delta x-|y|) = n\pi,\ \mu = n/(\Delta x-|y|),\ n=0,1,2,\ \ldots \tag{108}$$

The characteristic symmetric hyperbolas are seen in Fig. 3

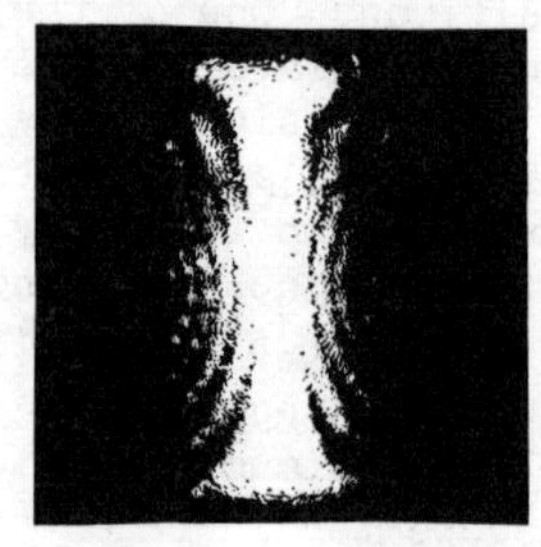

Fig. 3. Ambiguity function for an ideal rect signal (no artificial noise added to input).

(B) Optical equivalence theorem Eq. (41). Bartelt et al[1] have optically implemented the Bruijn version of the smoothed Wigner distribution. It is clear from our analysis, Eq. (41), that such a smoothed Wigner distribution equals a cross ambiguity function. The implementation of Bartelt et al for the ideal signal amounts to replacing two hard edges of the transparent diamond with two soft Gaussian edges (Fig. 4a,b).

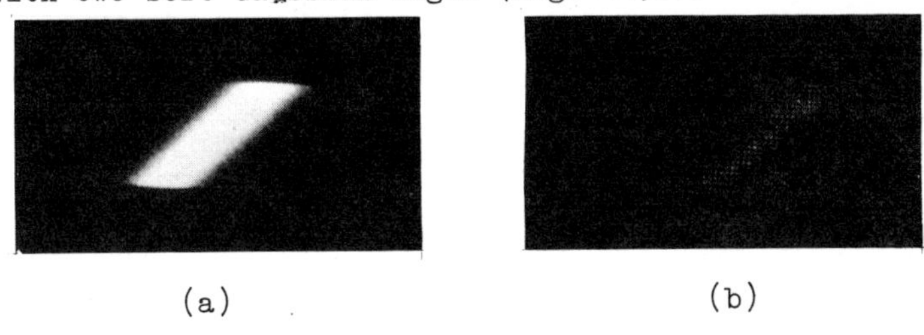

(a) (b)

Fig. 4. Input aperture for ambiguity function generation.

(a) Product of rect function and gaussian function at 45°

(b) Product of rect function with a square wave and gaussian function at 45°

For a finite monotonic signal

$$u(x) = [1+ \cos(2\pi\mu_o x)]\ \mathrm{rect}(\frac{x}{\Delta x})$$

$$= \left[1 + \frac{1}{2}\, e^{i2\pi\mu_o x} + \frac{1}{2}\, e^{-i2\pi\mu_o x}\right] \mathrm{rect}(\frac{x}{\Delta x}) \quad (109)$$

the Fourier transform is

$$\tilde{u}(\mu) = \Delta x[\mathrm{sinc}(\mu\Delta x) + \frac{1}{2}\,\mathrm{sinc}((\mu+\mu_o)\Delta x) + \frac{1}{2}\,\mathrm{sinc}((\mu-\mu_o)\Delta x)] \quad (110)$$

where

$$\mathrm{sinc}(\mu) \equiv \sin(\pi\mu)/\pi\mu .$$

The optical equivalent theorem Eq. (41) says

$$\langle W_u(x,\nu)\rangle_{G_\mu(x-y|\delta x)} = \left| A_{uG_o(x|\delta x)}(\mu,y) \right|^2 \quad (111)$$

$$= \left|\int u(x+\frac{y}{2})\, G_o(x-\frac{y}{2}|\,\delta x)e^{-i2\pi\mu x}dx\right|^2$$

$$= \left|e^{i\pi\mu y}\int u(x)G_o(y-x|\,\delta x)e^{-2\pi i\mu x}dx\right|^2 .$$

In the limit $\delta x \to \infty$ $G_o = 1$ and Eq. (111) is reduced to

$$\langle W_u(x,\nu)\rangle_{G_\mu(x-y|\infty)} = |\tilde{u}(\mu)|^2 \tag{112}$$

which is the signal spectral density. In the limit $\delta x \to 0$, $G_o \to \delta(x-y)$ and Eq. (111) is reduced to

$$\langle W_u(x,\nu)\rangle_{G_\mu(x-y|0)} = |u(y)|^2 \tag{113}$$

which is the signal energy density. For the present case the limit of a flat top gaussian ($\delta x \to \infty$) yields by substituting Eq. (110) into Eq. (112) a combination of squared sinc functions centered at $\mu=0, \pm\mu_o$ of period Δx^{-1}, while the limit of a sharp gaussian gives the rect-width Δx. Therefore for a finite δx, $\langle W\rangle$ is equivalent to the Fourier transform of the signal u smeared along y by a biased gaussian blur function. Because of the rapidly diminishing intensity of the sinc function squared, we expect the following result on the film. (Fig. 5).

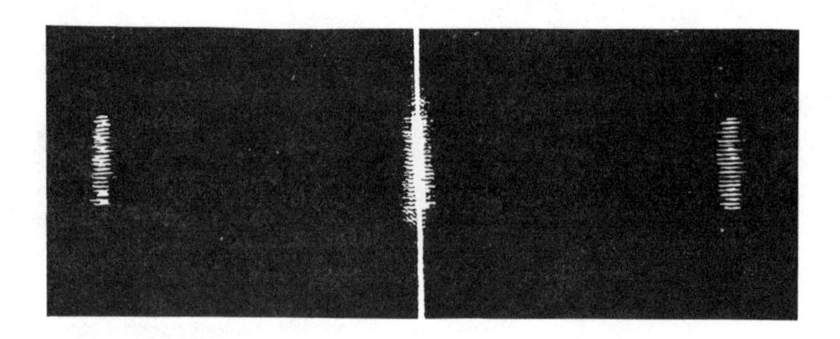

Fig. 5. Ambiguity function produced from input 4b. The horizontal bars seen are produced by imaging the "checkerboard" pattern, which was the discrete approximation to a cosine function at 45°, in the y direction.

The limiting case $\mu_o=0$ is clear from Fig. 6.

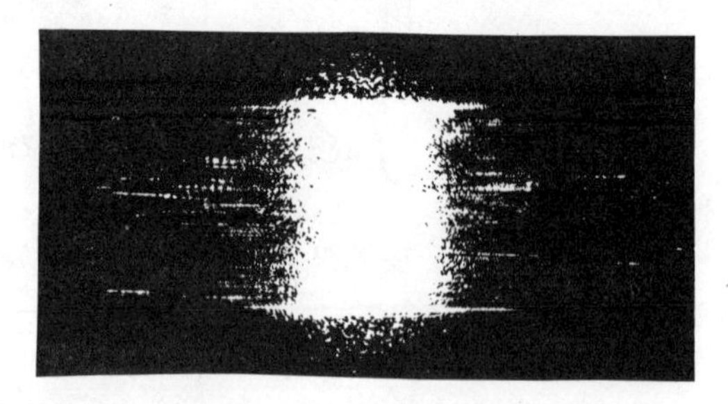

Fig. 6. Ambiguity function produced from input 4a.

We expect the $\Delta x=\delta x$ to be the optimum condition under additive random noise (see Eq. (105)).

(C) Wigner distribution function as expectation of parity operator, cf.Eq. (53). Bastiaans (1980)[42] has pointed out that the mirror technique that Marks and Hall (1979)[41] used to form the product rect(x+y) rect(x-y) by reflecting a single 1-D input rect(x+y) through a mirror and a pellicle, can also be Fourier transformed along the y axis to give the scaled version of the Wigner distribution function. The mirror reflection brings out the parity operator whose expectation value is the Wigner distribution Eq. (53). It is quite clear that the identical setup of (A) when the 3 cylindrical lenses are rotated 90° can give

$$\frac{1}{2} W_{rect}(x, \frac{\nu}{2}) = \int \text{rect}\ (x+y)\ \text{rect}\ (x-y)\ e^{-i2\pi\nu y} dy \qquad (113)$$

Due to the symmetry of the transparent diamond shape, the Wigner distribution becomes formally identical to the Woodward ambiguity function.

(D) Klauder's first order transform[17] Eq. (86). If a 2D-Fourier transform lens is added along the optical axis at one focal length behind the output of the setup (a), the Woodward ambiguity function should be transformed into the Wigner distribution function.

(E) The zeroth order relationship Eq. (83). Because rect(-x) = rect(x), Eq. (83) predicts

$$W_{rect}(x,\nu) = 2\ A_{rect}(2\nu,2x) \qquad (114)$$

and they have equal volume under their modulus surfaces although $W_{rect}(0,0) = 2A_{rect}(0,0)$ is brighter.

(ii) A noise input film was obtained using the computer filmwriter. The noise was generated by a lens-capped TV camera which gave rise to electronically amplified dark current. Two types of noise are considered. One type is the regulated $1/f^n$ noise, i.e. the high intensity noise occurs less frequently. This is produced by thresholding the histogram at its peak and shifting the upper portion such that the threshold level is at zero intensity.

$$I(N) = \Lambda(N) \equiv \text{rect}(N)*\text{rect}(N),$$

$$\tilde{I}(f) = \text{sinc}(f)^2 \simeq (1 + f^2/3)^{-2}.$$

The second type is obtained by a nonlinear threshold of the intensity to make binary dots scattered on the film. The first type is of general interest while the second type can be more easily used to control the total transmission energy. The input with the second type is presented in Fig. 7.

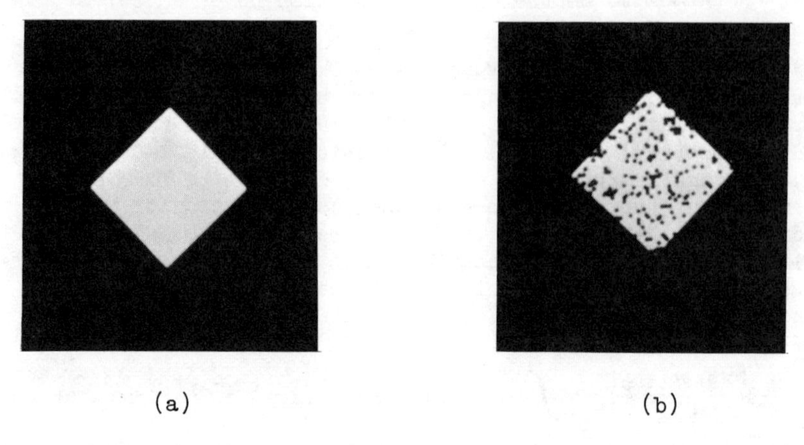

(a) (b)

Fig. 7. Input aperture for ambiguity function generation

(a) Product of two rect functions at 90°

(b) Product of two rect functions at 90° with 10% additive binary noise. Diamond area has been expanded to keep the clear area equal to that in 7a.

The ambiguity function for a rect signal is presented for three separate cases of additive noise in Fig. 8. The result for a "clear" rect function but with residual film grain noise is seen in Fig. 8a. The results for a rect function with 5% and 10% additive binary noise are seen in Figs. 8b and 8c respectively.

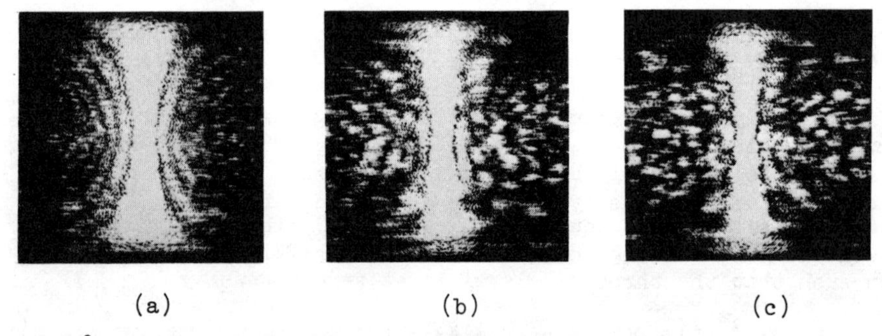

(a) (b) (c)

Fig. 8. Ambiguity function produced from input as in Fig. 7.

(a) residual film grain noise - no binary noise

(b) 5% binary noise

(c) 10% binary noise

The Woodward-like theorem Eq. (46) agrees well with the experiment. Given a constant input width Δx the larger blur width $\delta x > \Delta x$ means a flat top gaussian window for more noise energy transmission while for narrower blur width $\delta x<\Delta x$ means a sharp top of a gaussian window for less signal energy transmission. To sum up the analytical relationships, we mention several early works. A. Walther (1968)[32] and E. Wolf (1978)[31] used the similar Wigner distribution function for generalized radiance while K. Dutta and J. W. Goodman (1977)[33] used the ambiguity function for partial coherent imaging. Bastiaans (1978)[26] introduced the Wigner distribution to optics while Papoulis (1974)[34] introduced the ambiguity function to optics and matrix transform and propagation. Because Marks et al (1977)[2] and Casasent et al (1979)[4] optically implemented the ambiguity functions and because Bartelt et al (1980)[1] optically produced the smoothed Wigner distribution function, it becomes natural to examine, both more deeply and in slightly more general forms, what is needed to implement a phase space covert communication optical system.

ACKNOWLEDGEMENT

We thank R. L. Easton, Jr. for discussions and helping us to generate the input films. We also wish to thank Professor Lorenzo Narducci for many helpful comments and suggestions. Finally we wish to acknowledge the skill and patience Mrs. Barbara Eckert exhibited in the preparation of this rather difficult manuscript.

REFERENCES

1. H. O. Bartelt, K. H. Brenner and A. W. Lohmann, "The Wigner distribution function and its optical production," Opt. Comm. 32, 32-38 (1980).
2. R. J. Marks II, J. F. Walkup and T. F. Krile, "Ambiguity function display: an improved coherent processor," Appl. Opt. 16, 746-750 (1977); addendum 16, 1777-1777 (1977).
3. H. H. Szu, patent filed through U.S. Naval Research Laboratory.
4. D. Casasent and B.V.K. Vijaya Kumar, "Optical Image plane correlator for ambiguity surface computation," Appl. Opt. 18, 1673-1678 (1979).
5. D. Psaltis and D. Casasent, "Space-variant ambiguity function processor," Appl. Opt. 18, 1869-1874 (1979).
6. M. Skolnik, "Radar Handbook" (McGraw-Hill, New York, 1970), Ch. 3.
7. Y. W. Lee, T. P. Cheatham, Jr., and J. B. Wiesner, "Application of correlation analysis to the detection of periodic signals in noise," Proc. IRE 38, 1165-1172 (Oct. 1950).
8. P. M. Woodward and I. L. Davies, "A theory of radar information", Phil. Mag. 41, 1001-1017 (Oct. 1950).
9. C. A. Stutt, "Some results on real part/imaginary part and magnitude-phase relations in ambiguity functions," IEEE Trans. Inf. Theo. IT-10 321-327 (1964).
   C. A. Stutt and L. J. Spafford, "A 'best' mismatched filter response for radar clutter discrimination," IEEE Trans. Inf. Theo. IT-14, 280-287 (1968).

10. S. M. Sussman, "Least-square synthesis of radar ambiguity functions," IRE Trans. Inf. Theo. IT-8 246-254 (1962).
11. C. A. Stutt, "The application time/frequency correlation functions to the continuous-waveform encoding of message symbols," General Elec. Res. Lab., Schenectady, N. Y., Rept. No. 61-RL-2761E (July, 1961).
12. W. M. Siebert, "Studies of Woodward's uncertainty function," MIT Res. Lab. Elec. Quart. Progr. Rept. (April, 1958).
13. E. C. Westerfield, R. H. Prager and J. L. Stewart, "Processing gains against reverberation (clutter) using matched filters," IRE Trans. Inf. Theo. IT-6, 342-348 (1960).
14. D. Richman, "Resolution of multiple targets in clutter," Inst. Defense Anal., Res. Paper 158 (June 1966).
15. A. W. Rihaczek, "Principles of high-resolution radar," McGraw-Hill, N. Y., (1969), Ch. 5.
S. S. Haykin, editor, "Detection and estimation with application to radar," IEEE Benchmark papers (1976).
16. J. E. Moyal, "Quantum mechanics as a statistical theory," Proc. Cambridge Phil. Soc. 45, 99-132 (1949).
17. J. R. Klauder, Bell Syst. Tech. J., 39, 809-820 (1960).
18. N. G. de Bruijn, "A theory of generalized functions with application to Wigner distribution and Weyl correspondence," Nieuw Archief voor Wiskunde (3) 21, 205-280 (1973), cf. Eq. (27.12.1.3).
19. D. Gabor, "Theory of communication," J. IEE (London) 93 (III), 429-457 (1946).
20. G. Patsakos, "Classical particles in quantum mechanics," Am. J. Phys. 44, 158-166 (1976).
21. C. W. Helstrom, "An expansion of a signal in Gaussian elementary signals," Trans. IEEE Inf. Theo. IT-12, 81-82 (1966).
22. M. J. Bastiaans, "Gabor's expansion of a signal into Gaussian elementary signals," Proc. IEEE, 68, (1980).
23. A. Royer, "Wigner function as the expectation value of a parity operator," Phys. Rev. A15, 449-450 (1977).
24. S. R. de Grout and L. G. Suttorp, "Foundations of electrodynamics" (North-Holland, Amsterdam, 1972).
25. R. J. Glauber, "Coherent and incoherent states of the radiation filed," Phy. Rev. 131, 2766-2788 (1963).
26. M. J. Bastiaans, "The Wigner distribution function applied to optical signal and systems," Opt. Comm. 25, 26-30 (1978).
27. M. J. Bastiaans, "The Wigner distribution function and Hamiltonian's characteristics of a geometric optical system," Opt. Comm. 30, 321-326 (1979).
28. M. J. Bastiaans, "Wigner distribution functions and its application to first order optics," J. Opt. Soc. Am. 69, 1710-1716 (1979).
29. M. J. Bastiaans, "Transport equations for the Wigner distribution function in an inhomogenous and dispersive median," Opt. Acta 26, 1333-1344 (1979).

30. M. J. Bastiaans, "Transport equations for the Wigner distribution function," Opt. Acta 26, 1256-1272 (1979).
31. E. Wolf, "Coherence and radiometry," J. Opt. Soc. Am. 68, 6-17 (1878).
32. A. Walther, "Radiometry and coherence," J. Opt. Soc. Am. 58, 1256-1259 (1968).
33. K. Dutta, J. W. Goodman, "Reconstution of images of partially coherent objects from samples of mutual intensity," J. Opt. Soc. Am. 67, 796-803 (1977).
34. A. Papoulis, "Ambiguity function in Fourier optics," J. Opt. Soc. Am. 64, 779-788 (1974).
35. E. P. Wigner, "On the quantum correction for thermo-dynamic equilibrium," Phys. Rev. 40, 749-759 (1932).
36. M. D. Srinivas and E. Wolf, "Some non-classical features of phase space representations of quantum mechanics," Phys. Rev. D11, 1477-1485 (1975).
37. B. V. K. Vijaya Kumar and D. Casasent, "Space-blur bandwidth product in correlator performance evaluation," J. Opt. Soc. Am. 70, 103-110 (1980).
38. J. L. Lawson and G. E. Uhlenbeck, "Threshold signal," MIT Radiation Lab. Series Vol. 24 (McGraw-Hill, New York, 1950) (Dover, New York, 1965) Sect. 8.6 Fig. 8.9.
39. P. N. Tamura and J. C. Wyant, "Matrix multiplication using coherent optical techniques," SPIE Proc. 83, 13 (San Diago, 1976).
40. R. J. Marks II and S. V. Bell, "Astigmatic coherent processor analysis," Opt. Engn. 17, 167-169 (1978).
41. R. J. Marks II and M. W. Hall, "Ambiguity function display using a single 1-D," Appl. Opt. 18, 2539-2540 (1979).
42. M. J. Bastiaans, "Wigner distribution function display: a supplement to ambiguity function display using a single 1-D input," Appl. Opt. 19, 192 (1980).

NEW METHODS AND RESULTS IN FOCUSING

Jakob J. Stamnes
Central Institute for Industrial Research
Blindern, Oslo 3, Norway

ABSTRACT

In an earlier paper[1] methods have been presented for efficient computation of two-dimensional wave fields in focal regions. In this paper the methods of Ref. 1 are extended to cover three-dimensional waves. The computational techniques are based on asymptotic approximations by means of the method of stationary phase for double integrals, two-dimensional Fast Fourier Transform algorithms, and two-dimensional phase and amplitude linearization procedures. Explicit analytical results pertaining to a perfect three-dimensional wave are given. Also, numerical results are presented for comparison of the integrated energy of perfect and nonperfect two-dimensional waves.

1 INTRODUCTION

Most methods for computing wave fields in focal regions that have been discussed in the literature in the past suffer from at least one of two major drawbacks, one being the reliance on the paraxial approximation and the other lack of computational speed.

The reliance on the paraxial approximation limits the validity of the method so that the transverse extent of the aperture and the observation area must be small compared to the distance from the aperture to the observation point[2].

Even though the paraxial approximation is <u>not</u> imposed in some earlier works[3], the computational scheme, which is based on direct numerical integration, is not sufficiently fast to make the method valuable as a practical tool.

In our efforts to design low f-number lenses for focusing electromagnetic, acoustical, or gravity waves, the existing methods for evaluating diffraction integrals are inadequate for both of the two reasons just quoted. We therefore set out to develop methods that remain valid for arbitrarily large relative apertures and that are sufficiently fast at short as well as at long wave lengths.

In an earlier paper[1] we have presented methods for rapid computation of two-dimensional wave fields in focal regions. The primary objective of the present paper is therefore to generalize the methods in Ref. 1 to cover three-dimensional waves.

In Sec. 2 we generalize the computational method of Ref. 1, based on the angular-spectrum representation, to cover three-dimensional waves. The method of stationary phase for double integrals is used to obtain asymptotic approximations to the angular spectrum of the field.

ISSN:0094-243X/81/650382-16$1.50

The technique of phase linearization, presented in Ref. 1 for single integrals, is generalized in Sec. 3 to double integrals so that the computational method based on the impulse-response integral may be extended to cover three-dimensional waves. In contrast to the procedure given in Ref. 1 it is now found necessary to linearize also the amplitude function. Besides a better choice than in Ref. 1 is given for the coefficients of the linearized functions.

In Sec. 4 a perfect incident wave in three dimensions is defined. Explicit formulas are given for the perfect field distribution in the aperture and for the focal plane field distribution it produces in the Debye approximation.

Sec. 5 is devoted to the focusing of perfect and nonperfect waves in two dimensions. These waves are treated in detail in Ref. 1. However, we present here numerical results which show that, as far as integrated energy is concerned, the perfect wave always gives a better focus than the nonperfect waves considered.

In Sec. 6 we comment on the validity of scalar theory and on the applicability of our methods to vectorial waves. Also we present some ideas for applications of the methods and comment on how the methods of this paper compare to Keller's geometrical theory of diffraction[4] and Mittra's spectral methods[5].

2 ANGULAR-SPECTRUM METHOD FOR THREE-DIMENSIONAL WAVES

In this section we generalize the procedure of Sec. 2 in Ref. 1 to three-dimensional waves. The temporal behavior of the field may be handled as in Eqs. (1)-(2) in Ref. 1. Thus, we restrict our attention to the spatial part $u(x,y,z)$ of wave fields that may be expressed as angular spectra of plane waves, so that we may write

$$u(x,y,z) = \left(\frac{1}{2\pi}\right)^2 \iint_{-\infty}^{\infty} U(k_x,k_y)\exp\{i[k_x x+k_y y+(k^2-k_x^2-k_y^2)^{\frac{1}{2}}z]\}dk_x dk_y, \tag{2.1}$$

where $k = 2\pi/\lambda$, λ being the wave length at frequency ω, and the angular spectrum $U(k_x,k_y)$ is the Fourier transform of the aperture field $u(x,y,0)$, i.e.

$$U(k_x,k_y) = \iint_{-\infty}^{\infty} u(x,y,0)\ \exp[-i(k_x x+k_y y)]\ \ dxdy\ . \tag{2.2}$$

Since we expect the field to be focused at a point $x=x_o$, $y=y_o$, $z=z_o$, we write the aperture field in the form

$$u(x,y,0) = \begin{cases} u_o(x,y)\exp[ik(\phi_o(x,y)-r_o)] & \text{for } x,y\in A \\ 0 & \text{for } x,y\notin A \end{cases}\ , \tag{2.3}$$

where $u_o(x,y)$ and $\phi_o(x,y)$ are real functions and

$$r_o = [(x-x_o)^2 + (y-y_o)^2 + z_o^{\ 2}]^{\frac{1}{2}}. \tag{2.4}$$

The function $\phi_o(x,y)$ represents the deviation of the incident wave front from a perfect spherical wave front converging towards the focal point at (x_o,y_o,z_o). Thus, if $\phi_o(x,y) = 0$, the incoming wave front has a perfect spherical shape. In (2.3), A is some bounded region in the aperture plane z=0 outside of which the field vanishes. If f.ex. the Kirchhoff boundary conditions are employed, then A is the aperture opening and u(x,y,0) is equal to the incident field inside the aperture and is zero outside the aperture.

2.1 Asymptotic Evaluation of the Angular Spectrum

In preparation for an asymptotic evaluation of the angular spectrum we introduce dimensionless variables

$$\xi = x/z_o,\ \eta = y/z_o;\ \alpha = k_x/k,\ \beta = k_y/k, \tag{2.5}$$

and substitute (2.3) in (2.2) to obtain

$$U(\alpha k,\beta k) = J^*(\Omega), \tag{2.6}$$

where the asterisk denotes complex conjugation and

$$J(\Omega) = \iint_{A'} g(\xi,\eta)\exp[i\Omega f(\xi,\eta)]d\xi d\eta, \tag{2.7}$$

with

$$g(\xi,\eta) = z_o^2\, u_o(z_o\xi,z_o\eta), \tag{2.8}$$

$$f(\xi,\eta) = \alpha\xi+\beta\eta+[1+(\xi-\xi_o)^2+(\eta-\eta_o)^2]^{\frac{1}{2}} - \phi_o(z_o\xi,z_o\eta)/z_o\ , \tag{2.9}$$

$$\xi_o = x_o/z_o;\ \eta_o = y_o/z_o;\ \Omega = kz_o\ . \tag{2.10}$$

A' is the domain in ξ,η space corresponding to the domain A in x,y space.

We now assume that $\Omega = kz_o \gg 1$ so that the integral $J(\Omega)$ in (2.7) may be evaluated asymptotically by the method of stationary phase for double integrals[6]. According to this method, only neighborhoods of critical points of various kind contribute to the integral $J(\Omega)$ in (2.7) in the limit as $\Omega \to \infty$. Critical points of the first kind are points (ξ_1,η_1) inside A' or at its boundary at which the phase is stationary, i.e. $f_\xi(\xi_1,\eta_1) = f_\eta(\xi_1,\eta_1) = 0$. Critical points of the second kind are points (ξ_2,η_2) at the boundary of A' at which the tangential derivative of the phase function vanishes, i.e. if $\partial/\partial s$ denotes differentiation along the boundary curve, then $[\partial f/\partial s]_{\xi = \xi_2,\ \eta=\eta_2} = 0$. We will assume for simplicity that the curve bounding the aperture is smooth so that there are no contributions from critical points of the third kind, i.e. corner points. This is not a very severe restriction since such points, if present, give contributions of a higher order than those due to critical points of the first or second kind[7].

The contribution due to an interior stationary point is simply given by[7]

$$J_1(\Omega) = 2\pi\sigma\Omega^{-1}\left\{\left|f_{\xi\xi}f_{\eta\eta}-f_{\xi\eta}^{\;2}\right|^{-\frac{1}{2}} g(\xi,\eta)\exp[i\Omega f(\xi,\eta)]\right\}_{\xi=\xi_1,\eta=\eta_1}, \tag{2.11}$$

where

$$\sigma = \begin{cases} 1 & \text{if} \quad H < 0 \\ +i & \text{if} \quad H > 0; \quad f_{\eta\eta}(\xi_1,\eta_1) < 0 \\ -i & \text{if} \quad H > 0; \quad f_{\eta\eta}(\xi_1,\eta_1) < 0 \end{cases}, \tag{2.12}$$

with

$$H = \left[f_{\xi\xi}f_{\eta\eta} - f_{\xi\eta}^{\;2}\right]_{\xi=\xi_1,\eta=\eta_1}. \tag{2.12a}$$

The contribution J_2 due to a critical point of the second kind also is of a simple nature as long as there is no stationary point (i.e. critical point of the first kind) in its vicinity. To obtain the formula for the contribution J_2, we first make a transformation to integration variables X,Y so that X = 0 becomes the equation for the boundary in the neighborhood of the point (ξ_2,η_2) corresponding to X = Y = 0, and so that the positive X-axis points into the integration demain[8]. In terms of the new variables the phase function possesses an expansion of the form

$$f(\xi,\eta) = F(X,Y) = F_{0,0}+F_{1,0}X+F_{2,0}X^2+F_{1,1}XY+F_{0,2}Y^2+\dots \quad . \tag{2.13}$$

If $F_{1,1} \neq 0$, we make the additional transformation

$$X = u \;;\; Y = v + cu, \tag{2.14}$$

to obtain

$$\tilde{f}(\xi,\eta) = \tilde{f}(u,v) = \tilde{f}_{0,0}+\tilde{f}_{1,0}u+\tilde{f}_{1,1}uv+\tilde{f}_{2,0}u^2+\tilde{f}_{0,2}v^2+\dots \quad , \tag{2.15}$$

where

$$\tilde{f}_{0,0} = F_{0,0} \;;\; \tilde{f}_{1,0} = F_{1,0} \;;\; \tilde{f}_{0,2} = F_{0,2}, \tag{2.16}$$

$$\tilde{f}_{2,0} = F_{2,0} + cF_{1,1} + c^2F_{0,2}, \tag{2.17}$$

$$\tilde{f}_{1,1} = F_{1,1} + 2cF_{0,2}, \tag{2.18}$$

and choose

$$c = - F_{1,1}/2F_{0,2} \quad , \tag{2.19}$$

so that $\tilde{f}_{1,1}$ vanishes. With this choice the phase function in (2.15) is in a form suitable for application of asymptotic methods.

The integral in (2.7) now takes the form

$$J(\Omega) = \int_{-\infty}^{\infty} \int_{0}^{\infty} \tilde{g}(u,v)e^{i\Omega\tilde{f}(u,v)}dudv, \tag{2.20}$$

where the Jacobian of the transformation from (ξ,η) variables to (u,v) variables is included in $\tilde{g}(u,v)$, and where infinite limits have been put on the v integral to indicate that it gives no end point contribution.

The v integration in (2.20) may be performed by the method of stationary phase for single integrals. Since $\tilde{f}(u,v)$ has a power series expansion of the form

$$\tilde{f}(u,v) = \tilde{f}(u,0) + \frac{1}{2}\left.\frac{\partial^2\tilde{f}(u,v)}{\partial v^2}\right|_{v=0} v^2 + \ldots.$$

$$= \tilde{f}(u,0) + \tilde{f}_{0,2}(u)v^2 + \ldots. \;,$$

where $\tilde{f}_{0,2}(u)$ is bounded away from zero, it is clear that $\tilde{f}(u\text{=const.},v)$ satisfies all the necessary requirements for this method to be applicable[9]. To lowest order we have

$$\int_{-\infty}^{\infty} \tilde{g}(u,v)e^{i\Omega\tilde{f}(u,v)}dv = \{\pi/\Omega|\tilde{f}_{0,2}(u)|\}^{\frac{1}{2}}\,\tilde{g}(u,0)\exp[i\Omega\tilde{f}(u,0) + i\;\mathrm{sgn}(\tilde{f}_{0,2}(u))\pi/4] \;. \tag{2.20a}$$

Substitution of this result in (2.20) yields to lowest order

$$J(\Omega) = (\pi/\Omega)^{\frac{1}{2}}\; e^{i\beta_2\pi/4}\; I(\Omega,0) \quad , \tag{2.21}$$

where

$$\beta_2 = \mathrm{sgn}(\tilde{f}_{0,2}(u)) \;;\; \tilde{f}_{0,2}(u) = \left.\frac{\partial^2}{\partial v^2}\tilde{f}(u,v)\right|_{v=0} \;, \tag{2.22}$$

and

$$I(\Omega,u_a) = \int_{u_a}^{\infty} h(u)e^{i\Omega q(u)}du \;, \tag{2.23}$$

with

$$h(u) = \tilde{g}(u,0)|\tilde{f}_{0,2}(u)|^{-\frac{1}{2}} , \tag{2.24}$$

$$q(u) = \tilde{f}(u,0) . \tag{2.25}$$

A uniform asymptotic approximation $I_U(\Omega,u_a)$ to the integral in (2.23) that remains valid as the stationary point u_s (at which $q'(u_s) = 0$) moves out of the integration region, is given by Eq. (16) of Ref. 1.

If the stationary point u_s is far away from the end point $u_a=0$, then (2.23) yields - via integration by parts - the following contribution $I_E(\Omega,0)$ due to the isolated end point

$$I_E(\Omega,0) = -\frac{e^{i\Omega q(0)}}{i\Omega q'(0)} h(0) = i\,\frac{e^{i\Omega\tilde{f}_{0,0}}}{\Omega\,\tilde{f}_{1,0}}\,\frac{\tilde{g}_{0,0}}{|\tilde{f}_{0,2}|^{\frac{1}{2}}} , \tag{2.26}$$

Substitution of (2.26) in (2.21) and the use of (2.16) produce the well-known result[10]

$$J_2(\Omega) = \pi^{\frac{1}{2}}\Omega^{-3/2}\, G_{0,0}\exp[i\Omega F_{0,0}+i\alpha\tfrac{\pi}{2}]/|F_{0,2}|^{\frac{1}{2}}F_{1,0} , \tag{2.27}$$

where

$$\alpha = \begin{cases} 3/2 & \text{if} \quad F_{0,2} > 0 \\ \frac{1}{2} & \text{if} \quad F_{0,2} < 0 \end{cases} . \tag{2.28}$$

If the stationary point is at the boundary of the integration domain, [i.e. $u_s = u_a = 0$], so that $\tilde{f}_{1,0}$ in (2.15) vanishes, then we obtain by application of the method of stationary phase for single integrals [cf. (2.20a)] to (2.23) the contribution $I_{SB}(\Omega,0)$ given by

$$I_{SB}(\Omega,0) = \frac{1}{2}\,\{\pi/\Omega|\tfrac{1}{2}q''(0)|\}^{\frac{1}{2}}\, h(0)\exp[i\Omega q(0) + i\,\mathrm{sgn}(q''(0))\pi/4] , \tag{2.29}$$

which upon substitution in (2.21) and the use of (2.16), (2.17) and (2.19) yield the familiar result which is just one half of the contribution due to an interior stationary point [cf. (2.11)].

To obtain the contribution due to an isolated interior stationary point a similar development as in (2.15) is made about that point, and the lower limit of the u integration in (2.20) is extended to $-\infty$. The transformation from ξ,η to X,Y variables [cf. (2.13)] is superfluous in this case, and the coefficient $\tilde{f}_{1,0}$ in (2.15) vanishes. Repeated application of the method of stationary phase for single integrals to (2.20) then produces the well-known result (2.11).

The single integral in (2.23) gives a contribution due to an isolated interior stationary point that may be obtained by letting the lower limit of integration extend to $-\infty$ and applying the method of stationary phase for single integrals. When the contribution thus obtained is substituted in (2.21) we arrive at a result that is identical in form to the expression for $J_1(\Omega)$ in (2.11). However, the point $v = 0$, $u = u_s$ generally does not coincide with the stationary point (ξ_1, η_1) when $u_s \gg u_a$. Therefore, in general (2.21) is not uniformly valid as the stationary point moves far away from the boundary into the integration domain.

Thus, the use of a uniform asymptotic approximation to the single integral in (2.23) generally does not provide a uniform asymptotic approximation to our original double integral in (2.7), the reason simply being that (2.21) is not a uniformly valid asymptotic representation of (2.7). It is reasonable to expect, however, that the region of validity of (2.21) overlaps with that of (2.11) and (2.27).

The formula (2.27) is seen to break down, not only in the case as $F_{1,0}$ vanishes, as discussed above, but also in the case as $F_{0,2}$ approaches zero. This case corresponds to the coalescence of two critical points of the second kind. To obtain a formula that covers this case we first do the u integration in (2.20) by integration by parts to get

$$J(\Omega) = \int_{-\infty}^{\infty} \bar{g}(v) e^{i\Omega\bar{f}(v)} dv \ , \tag{2.30}$$

where

$$\bar{g}(v) = i\tilde{g}(0,v)/\Omega\tilde{f}'(0,v) \ ; \ \bar{f}(v) = \tilde{f}(0,v) \ . \tag{2.31}$$

Next, $\bar{f}(v)$ is expanded in a Taylor series about $v = 0$ up to third order and $\bar{g}(v)$ is put equal to $\bar{g}(0)$. The integral (2.30) then yields to lowest order[11]

$$J(\Omega) = 2\pi\bar{g}(0)\exp\{i\Omega\bar{f}(0)+i(\Omega/3)[\bar{f}''(0)]^3[\bar{f}'''(0)]^{-2}\} \times \left[\frac{2}{\Omega|\bar{f}'''(0)|}\right]^{1/3} \mathrm{Ai}(\xi), \tag{2.32}$$

where

$$\xi = \mp (\tfrac{1}{2}\Omega)^{2/3} \ |\bar{f}''(0)|^2 |\bar{f}'''(0)|^{-4/3} \ , \ f'''(0) \gtrless 0, \tag{2.33}$$

and Ai(ξ) is the Airy function defined as

$$\mathrm{Ai}(\xi) = \frac{1}{2\pi} \int_{-\infty}^{\infty} \exp\{i(\xi t+t^3/3)\} \ dt \ . \tag{2.34}$$

We point out that the formula (2.32) is not uniformly valid as the two critical points of the second kind moves far apart from each other. However, the region of validity of (2.32) is expected to overlap with that of (2.27).

For cases in which $F_{1,0}$ and $F_{0,2}$ go to zero simultaneously, none of the formulas given above remain valid. In such cases the technique of phase and amplitude linearization, given in the next section, is recommended for evaluation of the integral in (2.7).

2.2 Computation of the Field in the Focal Region

Substitution of the angular spectrum $U(\alpha k, \beta k)$, computed as explained in the previous subsection, in (2.1) yields

$$u(x,y,z) = \left(\frac{k}{2\pi}\right)^2 \int_{\alpha_1}^{\alpha_2}\int_{\beta_1}^{\beta_2} \left\{U(\alpha k, \beta k)\exp[ikz(1-\alpha^2-\beta^2)^{\frac{1}{2}}]\right\} \cdot \exp[ik(\alpha x+\beta y)]d\alpha d\beta \quad , \tag{2.35}$$

where (α_1,β_1) and (α_2,β_2) are respectively the smallest and largest values of (α,β) for which $U(\alpha k,\beta k)$ gives a nonnegligible contribution. To get an idea about the values of these parameters the reader is referred to the explanation given in Sec. 2 of Ref. 1 in the case of two-dimensional waves.

The integral in (2.35) may be computed in planes z = const. parallel to the aperture plane with the help of a two-dimensional Fast Fourier Transform algorithm. The angular-spectrum method for computing the three-dimensional field distribution around focus is now complete.

For an explanation as to why asymptotic techniques cannot be used to evaluate the angular spectrum integral in (2.35) for observation points close the focal point, we refer to Sec. 2 of Ref. 1. Also, we mention that Sec. 4 of Ref. 1 contains a comparison between the Debye and the Kirchhoff approximations. In particular it is shown that in the Kirchhoff approximation one takes into account diffraction at the aperture edge when calculating the angular spectrum of the field, whereas edge diffraction is neglected in the Debye approximation to the angular spectrum. Thus, in the Debye approximation the angular spectrum is simply given by the contribution in (2.11) (or the sum of several such contributions if there are more than one stationary point). The difference between the two approximations is a consequence of the additional assumption that the aperture is infinitely far away from the focal region in the latter case.

3 IMPULSE-RESPONSE METHOD FOR THREE-DIMENSIONAL WAVES

Applying the convolution theorem to (2.1) it may be cast into the form of an impulse-response integral, which is nothing but the familiar Rayleigh-Sommerfeld's diffraction integral[12], i.e.

$$u(x,y,z) = \frac{-1}{2\pi} \iint_{-\infty}^{\infty} u(x',y',0) \frac{d}{dz}\left(\frac{e^{ikr}}{r}\right) dx'dy' \quad , \tag{3.1}$$

where

$$r = [(x-x')^2 + (y-y')^2 + z^2]^{\frac{1}{2}} \ . \tag{3.2}$$

Substitution of the aperture field (2.3) in (3.1) yields

$$u(x,y,z) = \iint_A g(x',y') \exp[if(x',y')] \, dx'dy' \ , \tag{3.3}$$

where

$$g(x',y') = -ikz \, u_o(x',y')(1+i/kr)/2\pi r^2 \ , \tag{3.4}$$

$$f(x',y') = k[\Delta r+\phi_o(x',y')] \ , \tag{3.5}$$

with

$$\Delta r = r-r_o = \{(x-x_o)(x+x_o-2x') + (y-y_o)(y+y_o-2y') + (z-z_o)(z+z_o)\}/(r+r_o) \quad , \tag{3.6}$$

$$r_o = [(x'-x_o)^2 + (y'-y_o)^2 + z_o^2]^{\frac{1}{2}} \ . \tag{3.7}$$

Thus, to base the computation of the field around the focal point on the impulse-response integral, one must look for an expedient way of computing the double integral in (3.3).

To simplify the discussion we assume that the integration domain A in (3.3) is a rectangle with sides parallel to the x',y' axes. If in reality the aperture is of elliptical shape, a rectangular integration domain in (3.3) may be obtained by a simple transformation of coordinates: First one rotates the coordinate system to align the new axes with those of the ellipse; then one changes to polar coordinates.

In preparation to implement the technique of phase and amplitude linearization we divide the integration domain in MxN rectangular subdomains, so that we may write (3.3) in the form

$$u(x,y,z) = \sum_{n=1}^{N} \sum_{m=1}^{M} \int_{y_n^L}^{y_n^U} \int_{x_m^L}^{x_m^U} g(x',y')\exp[if(x',y')]dx'dy' \ . \tag{3.8}$$

For notational convenience we introduce the quantities [with n=1,2,... N; m=1,2,... M]

$$\Delta x_m = \frac{1}{2}(x_m^U - x_m^L); \quad y_n = \frac{1}{2}(y_n^U - y_n^L) \;, \tag{3.9a}$$

$$x_m^A = \frac{1}{2}(x_m^U + x_m^L); \; y_m^A = \frac{1}{2}(y_n^U + y_n^L) \;. \tag{3.9b}$$

First we perform the x' integration in (3.8). To that end, we linearize the phase function, i.e. we write

$$f(x',y') = f_{LIN}(x',y') = a_m(y') + b_m(y')x' \;;\; x_m^L \leq x' \leq x_m^U, \tag{3.10}$$

so that we obtain from (3.8)

$$u(x,y,z) = 2 \sum_{n=1}^{N} \sum_{m=1}^{M} g(x_m^A, y_m^A)\Delta x_m \int_{y_n^L}^{y_n^U} \text{sinc}(b_m(y')\Delta x_m) \;.$$

$$. \exp[if_{LIN}(x_m^A, y')]dy' \;, \tag{3.11}$$

where $\text{sinc}(x) = \sin x/x$, and where we have assumed that the amplitude function $g(x',y')$ varies so slowly that it may be replaced by its value at the midpoint of the subdomain in question. We will return in a moment to the question of how to determine the coefficients $a_m(y')$ and $b_m(y')$ in (3.10).

Suppose now that $b_m(y')$ is so small that the amplitude function $\text{sinc}\,(b_m(y')\Delta x_m)$ in (3.11) varies slowly compared to $\exp[if_{LIN}\,(x_m^A, y')]$. In that case we linearize both the amplitude function and the phase function $f_{LIN}(x_m^A, y')$, whereafter the integration in (3.11) is readily performed by means of the formula

$$\int_{x-\Delta/2}^{x+\Delta/2} (\alpha+\beta t)\exp[i(\gamma+\delta t)]dt = \Delta[\alpha\,\text{sinc}(\delta)\cos\gamma + \beta h(\delta)\sin\gamma]$$

$$+ i\Delta[\alpha\,\text{sinc}(\delta)\sin\gamma - \beta h(\delta)\cos\gamma] \;, \tag{3.12a}$$

where

$$h(\delta) = (\delta\cos\delta - \sin\delta)/\delta^2. \tag{3.12b}$$

If the sinc-function in (3.11) does not vary slowly compared to the exponential function, we split up the former function in its two exponential parts. As a result the integral in (3.11) becomes a sum of two integrals, both with the same amplitude function $[2ib_m(y')\Delta x_m]^{-1}$, and with phase functions $f_{LIN}(x_m^A, y') \pm b_m(y')\Delta x_m$. In each integral we again linearize both the amplitude function and the phase function, and perform the integration by means of the formula (3.12).

We now return to the question of how to determine the coefficients in the linearization procedure. In Ref. 1 the coefficients were determined by requiring that the error at the end points be of the same magnitude but of oppposite sign compared to the error at the midpoint. However, it can be shown[13] that if we approximate the integral

$$I = \int_{x_o-\Delta/2}^{x_o+\Delta/2} a(x)\exp[i\phi(x)]dx \quad \text{by } I_A = \int_{x_o-\Delta/2}^{x_o+\Delta/2} (\alpha+\beta x)\exp[i(\gamma+\delta x)]dx,$$

then the coefficients α,β,γ and δ should be chosen as

$$\begin{aligned}
\beta &= [a(x_o+\Delta/2) - a(x_o-\Delta/2)]/\Delta \\
\alpha &= \frac{2}{3}\, a(x_o) + \frac{1}{6}[a(x_o+\Delta/2) + a(x_o-\Delta/2)]-x_o\beta \\
\delta &= [\phi(x_o+\Delta/2) - \phi(x_o-\Delta/2)]/\Delta \\
\gamma &= \frac{2}{3}\, \phi(x_o) + \frac{1}{6}[\phi(x_o+\Delta/2) + \phi(x_o-\Delta/2)]-x_o\delta
\end{aligned} \tag{3.13}$$

to minimize the error $I-I_A$.

With regard to the computational speed of the method in this section compared to that of the previous section, we mention that arguments are given at the end of Sec. 3 of Ref. 1 to show that the angular-spectrum method is substantially faster than the impulse-response method when the aperture covers a large number of wave lengths.

4 FOCUSING OF A PERFECT THREE-DIMENSIONAL WAVE

In Ref. 1 a perfect two-dimensional wave has been defined. We now extend that definition to three-dimensional waves and present some pertinent results.

A perfect incident wave is defined as one that produces a δ-function field distribution in the focal plane if the aperture is extended to infinity and account is taken also of evanescent waves. Clearly, the angular spectrum $U_P(k_x,k_y)$ of a perfect wave is given by

$$U_P(k_x,k_y) = \exp[-i(k^2-k_x^2-k_y^2)^{\frac{1}{2}}z_o] \; ; \tag{4.1}$$

for if this angular spectrum is substituted in (2.1) we obtain

$$u_P(x,y,z_o) = k^2\delta(kx,ky) \; . \tag{4.2}$$

However, as explained in Ref. 1, this definition of a perfect wave is rather unphysical, since the angular spectrum in (4.1) grows exponentially with k_x,k_y for $k_x^2+k_y^2>k^2$. Therefore, as in Ref. 1, we redefine our perfect wave so that it has the same angular spectrum as in (4.1) for the homogeneous plane waves $(k_x^2+k_y^2<k^2)$ and has an angular spectrum $\exp[-(k_x^2+k_y^2-k^2)z_o]$ for the inhomogeneous plane waves $(k_x^2+k_y^2>k^2)$.

It can be shown[14] that (2.1), with the above definition of the angular spectrum, yields an aperture field distribution given by

$$u_P(x,y,o) = \frac{-1}{2\pi}\frac{d}{dx_o}\left(\frac{e^{-ikr_o}}{r_o}\right) \quad ; \quad r_o = (x^2+y^2+z_o^2)^{\frac{1}{2}} \quad . \tag{4.3}$$

From (4.3) it follows that the aperture field distribution of a perfect wave is that of a dipole wave which converges towards the focal point.

If the Debye assumption is used to obtain an approximate representation for the aperture field, it can be shown[14] that the field distribution in the focal plane produced by a perfect incident wave, is given by

$$u_P(x,y,z_o) = \frac{k^2}{2\pi}\sin^2\Theta_o \frac{J_1(kr\sin\Theta_o)}{kr\sin\Theta_o} \quad , \tag{4.4}$$

where $r=(x^2+y^2)^{\frac{1}{2}}$, and $2\Theta_o$ is the angular aperture. Thus, we may conclude that in the Debye approximation a perfect incident wave produces the well-known Airy pattern intensity distribution in the focal plane, irrespective of the size of the relative aperture.

5 INTEGRATED ENERGY OF PERFECT AND NONPERFECT TWO-DIMENSIONAL WAVES

The focusing properties of perfect and nonperfect two-dimensional waves are treated in Ref.1. The perfect two-dimensional wave has an aperture field distribution equal to that of a dipole wave which converges towards the focal point. The nonperfect wave considered in Ref. 1 has the same phase distribution over the aperture as the perfect wave. However, the amplitude distribution over the aperture of the nonperfect wave is different from that of the perfect wave: It is equal to the amplitude distribution that a point source placed one focal distance z_o from the aperture would produce.

As in Ref.1, we restrict our attention here to nonperfect waves that have a perfect phase distribution over the aperture. However, the equivalent point source generating the amplitude distribution over the aperture may be situated at any distance z_1 from it, as indicated in Fig. 1. In particular, we are interested in the limiting case $z_1 \to \infty$, in which case we obtain a uniform amplitude distribution over the aperture corresponding to that of an incident plane wave.

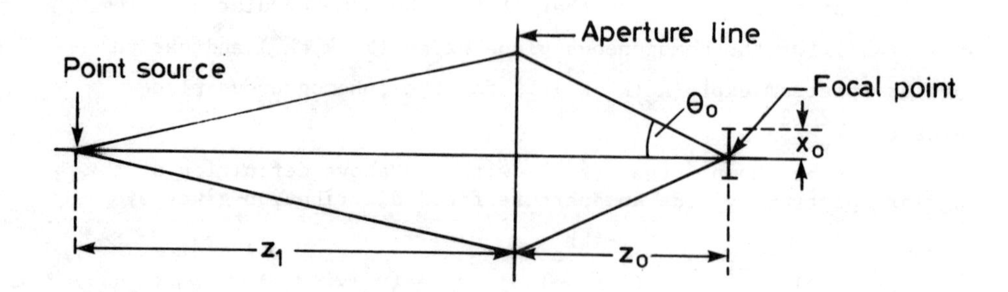

Fig. 1. The two-dimensional focusing geometry used in the integrated energy calculations shown in Fig. 2.

We assume that the energy is proportional to the intensity of the field, and consider the symmetric geometry in Fig. 1. Then the integrated energy in the aperture plane is given by

$$L(x_o) = \int_{-x_o}^{x_o} |u(x,z_o)|^2 \, dx \Big/ \int_{-a}^{a} |u(x,0)|^2 \, dx = N/D , \qquad (5.1)$$

where 2a is the aperture width.

We have used the impulse-response method of Ref.1 to compute $u(x,z_o)$ in (5.1) and Simpsons's integration formula to compute the integral in the numerator. The integrals in the denominator of (5.1) may be evaluated analytically: We have

$$|u(x,0)|^2 = \begin{cases} z_o^2 / (x^2+z_o^2)^{3/2} & \text{for a perfect wave} \\ z_1 / (x^2+z_1^2)^{\frac{1}{2}} & \text{for a nonperfect wave} \end{cases} \qquad (5.2)$$

Substituting these expressions into the integral in the denominator of (5.1), we find

$$D = \begin{cases} 2z_o \sin\Theta_o & \text{for a perfect wave} \\ 2a \ln[a/z_1 + (1+(a/z_1)^2)^{\frac{1}{2}}] / (a/z_1) & \text{for a nonperfect wave} \end{cases}$$

where $2\Theta_o$ is the angular aperture [cf. Fig. 1].

If we apply the Debye approximation, so that the focal plane field distribution in the case of a perfect wave is a sinc-function [cf. Eq. (73) of Ref. 1], then (5.1) reduces to the simple expression

$$L(x_o) = \frac{2}{\pi} \int_0^{kx_o \sin\Theta_o} (\sin(t)/t)^2 \, dt . \qquad (5.4)$$

We have evaluated also this integral by Simpson's method as a check on our other results.

Numerical results for $L(x_o)$ are presented in Figs. 2a) and 2b) for the geometry shown in Fig. 1. The angular aperture $2\Theta_o$ is 90° in the case of Fig. 2a) and 120° in the case of Fig. 2b), and the focal distance z_o is 25λ in Fig. 2a) and 14.4λ in Fig. 2b). In each figure the upper curve corresponds to a perfect wave; the curve in the middle corresponds to a nonperfect wave with $z_1 = z_o$; and the lower curve corresponds to a plane incident wave ($z_1 = \infty$).

The figures show that as far as integrated energy is concerned, the perfect wave always gives a better focus than the nonperfect waves considered. Also, as is to be expected, the difference in integrated energy between the two wave types increases as the angular aperture increases and as the distance z_1 increases.

6 DISCUSSION

In this paper, as well as in Ref. 1, the developments have been based on scalar theory. Therefore, the results are strictly speaking valid only when they are applied to truly scalar wave phenomena, such as f.ex. to focusing of acoustical waves in liquids or gases.

The developments are, however, easily extended to cover vector waves. To that end one merely applies the methods of this paper or Ref. 1 to each scalar component of the vector field. Thus, the theory may be applied to study such diverse phenomena as focusing of electromagnetic waves, acoustical waves in solids, and gravity waves. As an example in the case of gravity waves, we mention that we have made extensive use of the theory in Ref. 1 to study focusing of ocean swells, and in particular the movements of water particles in focal regions[15].

As another example, the angular spectrum methods, developed in this paper and in Ref. 1, combined with the work in Ref. 16, are ideally suited to study focusing of electromagnetic energy into material media through plane interfaces.

Our asymptotic methods bear close resemblance to the geometrical theory of diffraction (GTD) of Keller[4], the main difference being that we apply our methods to the angular spectrum of the field rather than to the field itself. One important advantage of the present methods over the GTD is that they remain valid in cases in which GTD breaks down, that is at shadow boundaries and at caustics.

There is an even closer resemblance between our methods and the spectral theory of diffraction developed by Mittra[5]. Aside from the fact that we study different physical problems than he does, the main difference between our approach and his seems to be that we base our asymptotic developments on the method of stationary phase while he uses the method of steepest descents.

a)

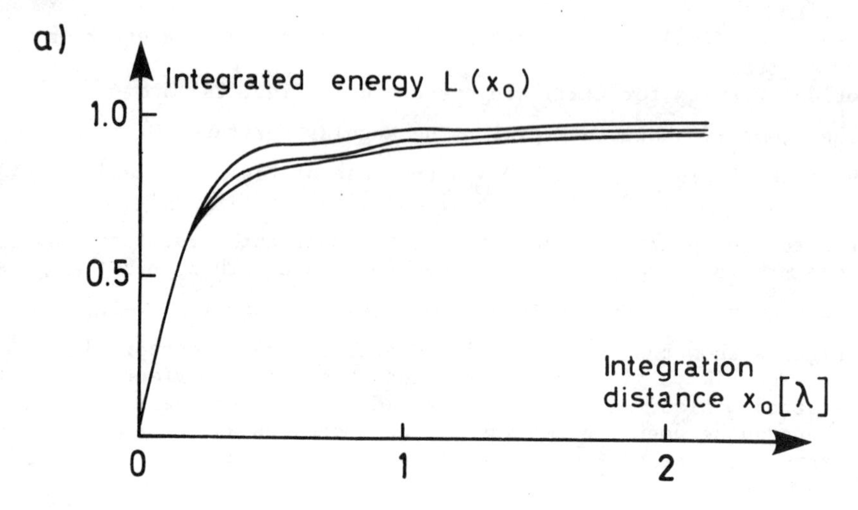

b)

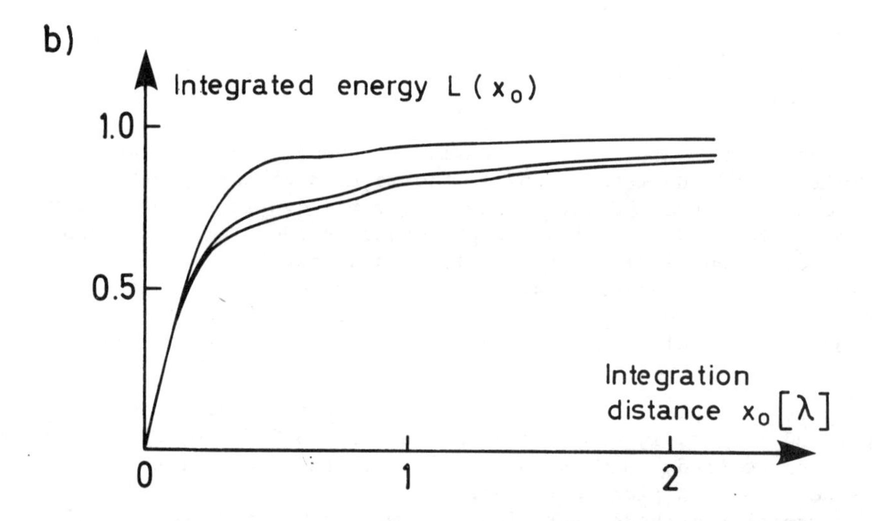

Fig. 2. Integrated energy on the focal line for two-dimensional waves. The focusing geometry is shown in Fig. 1. The angular aperture $2\Theta_o$ [cf. Fig. 1] is 90° in a) and 120° in b), and the focal distance z_o is 25λ in a) and 14.4λ in b). In all cases the phase of the incident wave is assumed to be perfect. The upper curve in each figure corresponds to an incident wave with perfect phase and amplitude; the curve in the middle corresponds to a wave with nonperfect amplitude, generated by a point source at $z_1 = z_o$ [cf. Fig. 1]; and the lower curve corresponds to a nonperfect wave with uniform amplitude, i.e. a plane incident wave ($z_1 = \infty$).

ACKNOWLEDGEMENT

This work was supported by the Royal Norwegian Council for Scientific and Industrial Research. The computer programs needed to produce Fig. 2 were written by F.S. Ljunggren and B. Spjelkavik.

REFERENCES

1. J.J. Stamnes, Focusing of two-dimensional waves, J. Opt. Soc. Am. (in press).
2. M. Born and E. Wolf, Principles of Optics (Pergamon, N.Y., 1970), 4th ed., Sec. 8.8.
3. B. Richards and E. Wolf, Proc. Roy. Soc. A 253,358 (1959); A. Boivin and E. Wolf, Phys. Rev. B 138, 1561 (1965); A. Boivin and E. Wolf, J. Opt. Soc. Am. 57, 1171 (1967); A. Boivin, N. Brousseau and S.C. Biswas, Optica Acta 25, 415 (1978).
4. J. Keller, J. Opt. Soc. Am. 52, 116 (1962).
5. R. Mittra, Spectral Theory of Diffraction, in Modern topics in electromagnetics and antennas, (Peter Peregrinus Ltd., 1977).
6. J. Focke, Ber. Verh. Saechs. Akad. Wiss. Leipzig, 101, 1 (1954); G. Braun, Acta Phys. Austriaca 10, 8 (1956); D.S. Jones and M. Kline, J. Math. Phys. 37, 1 (1958); N. Chako, J. Inst. Maths Applics 1, 372 (1965).
7. Ref. 2, Appendix III.
8. D.S. Jones and M. Kline, Ref. 6, p. 10; J. Focke, Ref. 6, p. 38.
9. A. Erdélyi, Asymptotic Expansions (Dover, N.Y., 1956).
10. D.S. Jones and M. Kline, Ref. 6, Eq. (24).
11. G.L. James, Geometrical theory of diffraction for electromagnetic waves (Peter Peregrinus Ltd., 1976), p. 30.
12. G.C. Sherman, J. Opt. Soc. Am. 57, 546 (1967).
13. H.M. Pedersen and J.J. Stamnes (in preparation).
14. J.J. Stamnes (in preparation).
15. J.J. Stamnes and F.S. Ljunggren, Propagation and Focusing of Linear Waves in Water of Constant Depth (in preparation).
16. J. Gasper, G.C. Sherman and J.J. Stamnes, J. Opt. Soc. Am. 66, 955 (1976); J.J. Stamnes and G.C. Sherman, J. Opt. Soc. Am. 67, 683 (1977).

GABOR'S EXPANSION OF A SIGNAL INTO A DISCRETE SET OF GAUSSIAN ELEMENTARY SIGNALS

Martin J. Bastiaans
Technische Hogeschool Eindhoven, Afdeling der Elektrotechniek,
Postbus 513, 5600 MB Eindhoven, The Netherlands

ABSTRACT

It is shown how a signal can be expanded in Gaussian elementary signals. The expansion is exactly the one suggested by Gabor in 1946, when he proposed to expand a signal into a discrete set of properly shifted and modulated Gaussian elementary signals; determining the expansion coefficients, however, seemed difficult, since the set of Gaussian elementary signals is not orthogonal. A set of functions is described, which is bi-orthonormal to the set of Gaussian elementary signals; this bi-orthonormality property allows an easy determination of the expansion coefficients.

SUMMARY

Since the complete paper can be found elsewhere[1], we shall give here a short summary of it, only.

In 1946 Dennis Gabor suggested the expansion of a signal into a discrete set of properly shifted and modulated Gaussian elementary signals[2]. As an introduction to the present paper, a quotation from Gabor's original paper might be useful. Gabor writes in the summary: "Hitherto communication theory was based on two alternative methods of signal analysis. One is the description of the signal as a function of time; the other is Fourier analysis. ... But our everyday experiences ... insist on a description in terms of *both* time and frequency. ... Signals are represented in two dimensions, with time and frequency as co-ordinates. Such two-dimensional representations can be called 'information diagrams', as areas in them are proportional to the number of independent data which they can convey. ... There are certain 'elementary signals' which occupy the smallest possible area in the information diagram. They are harmonic oscillations modulated by a probability pulse. Each elementary signal can be considered as conveying exactly one datum, or one 'quantum of information'. Any signal can be expanded in terms of these by a process which includes time analysis and Fourier analysis as extreme cases."

In the case of Fourier optics, where the signals are functions of two space variables, Gabor's suggestion results in the expansion of the signal into a discrete set of Gaussian beams. For the sake of convenience, the entire analysis will be presented for one-dimensional time signals; the extension to two-dimensional optical signals is straightforward. For time signals Gabor's expansion can be considered as a signal representation in terms of musical notes that appear at discrete moments and that have discrete pitches. Such a representation thus resembles the musical score of the time signal.

ISSN:0094-243X/81/650398-04$1.50

We shall define the Gaussian elementary signal by

$$g(t) = \left(\frac{\sqrt{2}}{T}\right)^{\frac{1}{2}} \exp[-\pi \left(\frac{t}{T}\right)^2] . \tag{1}$$

With this Gaussian elementary signal Gabor's expansion of the signal $\phi(t)$ reads

$$\phi(t) = \sum_{m}\sum_{n} a_{mn} g(t-mT) \exp[in\Omega t] , \tag{2}$$

where Ω and T satisfy the relation

$$\Omega T = 2\pi . \tag{3}$$

(Unless otherwise stated, all integrations and summations in this paper extend from $-\infty$ to $+\infty$). Equation (2) represents the signal as a superposition of a set of Gaussian elementary signals that are shifted over discrete distances mT and that are modulated with discrete angular frequencies $n\Omega$.

There seems to be a problem with this expansion: the expansion coefficients a_{mn} cannot be determined in the usual way, since the Gaussian elementary signals are not orthonormal, i.e.,

$$\int g(t) g^*(t-mT) \exp[-in\Omega t] dt \neq \delta_m \delta_n , \tag{4}$$

where δ_k is the Kronecker delta ($\delta_0=1$, $\delta_k=0$ for $k \neq 0$) and the asterisk denotes complex conjugation. Nevertheless, there does exist an easy way to find the expansion coefficients.

With the help of the Fourier transforms

$$\tilde{\phi}(\tau,\omega) = \sum_{m} \phi(\tau+mT) \exp[-im\omega T] , \tag{5a}$$

$$\tilde{g}(\tau,\omega) = \sum_{m} g(\tau+mT) \exp[-im\omega T] \tag{5b}$$

and

$$\tilde{a}(\tau,\omega) = \sum_{m}\sum_{n} a_{mn} \exp[-i(m\omega T - n\Omega\tau)] , \tag{6}$$

Gabor's signal expansion (2) can be transformed into

$$\tilde{\phi}(\tau,\omega) = \tilde{a}(\tau,\omega) \tilde{g}(\tau,\omega) . \tag{7}$$

Note that the function $\tilde{a}(\tau,\omega)$ is periodic in ω and τ (with periods Ω and T, respectively), and that the functions $\tilde{\phi}(\tau,\omega)$ and $\tilde{g}(\tau,\omega)$ are periodic in ω (with period Ω) and quasi-periodic in τ (with quasi-period T).

We can now determine the expansion coefficients a_{mn} as follows. Under the assumption that division by $\tilde{g}(\tau,\omega)$ is allowed, the function $\tilde{a}(\tau,\omega)$ can be found via Eq. (7). The expansion coefficients can then be determined through the relation

$$a_{mn} = \frac{1}{2\pi} \iint_{T\Omega} \tilde{a}(\tau,\omega)\exp[i(m\omega T - n\Omega\tau)]d\tau d\omega \ , \tag{8}$$

which is, in fact, the inverse of the Fourier transformation (6); the integrations in Eq. (8) extend over one period T and one period Ω, respectively.

In the previous paragraphs we showed how the coefficients a_{mn} of the Gabor expansion (2) could be determined when the signal $\phi(t)$ and the Gaussian elementary signal $g(t)$ are known. There is, however, a simpler way to find these expansion coefficients. Under the assumption, again, that division by $\tilde{g}(\tau,\omega)$ is allowed, we define the function $\tilde{\gamma}(\tau,\omega)$ by

$$\tilde{\gamma}(\tau,\omega)\tilde{g}^*(\tau,\omega) = \frac{1}{T} \ . \tag{9}$$

Substitution of this relation into Eq. (7) yields

$$\frac{1}{T}\tilde{a}(\tau,\omega) = \tilde{\phi}(\tau,\omega)\tilde{\gamma}^*(\tau,\omega) \ . \tag{10}$$

With the relationship

$$\gamma(\tau+mT) = \frac{1}{\Omega}\int_{\Omega}\tilde{\gamma}(\tau,\omega)\exp[im\omega T]d\omega \ , \tag{11}$$

which is, in fact, the inverse of a Fourier transformation similar to Eq. (5), Eq. (10) can be transformed into

$$a_{mn} = \int \phi(t)\gamma^*(t-mT)\exp[-in\Omega t]dt \ . \tag{12}$$

We conclude that, when the signal $\phi(t)$ and the function $\gamma(t)$ are known, the expansion coefficients can be determined immediately by means of Eq. (12).

We see that the coefficients a_{mn} can be determined as follows: multiply the signal $\phi(t)$ by a function γ that is shifted over the distance mT and that is modulated with the frequency $n\Omega$, and carry out the integration. Why does this procedure work? It works because the sets of shifted and modulated functions g and γ are bi-orthonormal, although they are not orthonormal themselves! Indeed, from Eq. (10) we can derive, by means of an inverse Fourier transformation, again, the relation

$$\int \gamma(t)g^*(t-mT)\exp[-in\Omega t]dt = \delta_m\delta_n \ . \tag{13}$$

Until now we did not really use the fact that g(t) is a Gaussian function, and it is true that the ideas of this paper are not restricted to the case of a Gaussian elementary signal. The entire analysis holds, in principle, when g is an arbitrary function of time[3]: for any function g(t) we can find a bi-orthonormal function $\gamma(t)$.

Let us now confine ourselves to the Gaussian case. For a Gaussian elementary signal g(t), the functions $\tilde{g}(\tau,\omega)$ and $\tilde{\gamma}(\tau,\omega)$ can be expressed in terms of theta-functions, and the function $\gamma(t)$ takes the form

$$\gamma(t) = \left(\frac{1}{T\sqrt{2}}\right)^{\frac{1}{2}} \left(\frac{K_0}{\pi}\right)^{-\frac{3}{2}} \exp[\pi (\tfrac{t}{T})^2] \sum_{n+\frac{1}{2}\geq\frac{t}{T}} (-1)^n \exp[-\pi(n+\tfrac{1}{2})^2] \ , \qquad (14)$$

where $K_0 = 1.85407468$ is the complete elliptic integral for the modulus $\frac{1}{2}\sqrt{2}$. The function $\gamma(t)$ is plotted in Figure 1.

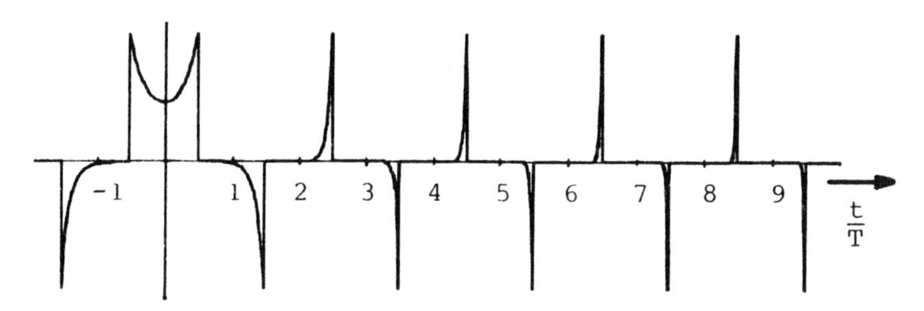

Figure 1. The function $\gamma(t)$.

REFERENCES

1 M.J. Bastiaans, "The expansion of an optical signal into a discrete set of Gaussian beams", in *Erzeugung und Analyse von Bildern und Strukturen* (Informatik-Fachberichte, Band 29), edited by S.J. Pöppl and H. Platzer (Springer, Berlin, 1980), 23-32; to be published also in Optik.

2 D. Gabor, "Theory of communication", J.Inst.Elec.Eng. 93 (III), 429-457 (1946).

3 M.J. Bastiaans, "A sampling theorem for the complex spectrogram, and Gabor's expansion of a signal in Gaussian elementary signals", to be published in the Proceedings of the International Optical Computing Conference, Washington, D.C., 1980; to be published also in Opt.Eng.

FUNDAMENTAL ASPECTS OF THE PHASE RETRIEVAL PROBLEM

H.A. Ferwerda

University of Groningen, Department of Applied Physics,
Nijenborgh 18, 9747 AG Groningen, The Netherlands.

ABSTRACT

A review is given of the fundamental aspects of the phase retrieval problem in optical imaging for one dimension. The phase problem is treated using the fact that the wavefunction in the image-plane is a band-limited entire function of order 1. The ambiguity of the phase reconstruction is formulated in terms of the complex zeros of entire functions. Procedures are given how the relevant zeros might be determined. When the zeros are known one can derive dispersion relations which relate the phase of the wavefunction to the intensity distribution. The phase problem of coherence theory is similar to the previously discussed problem and is briefly touched upon. The extension of the phase problem to two dimensions is <u>not</u> straightforward and still remains to be solved.

INTRODUCTION

Phase problems arise in many different areas of physics such as X-ray crystallography, electron microscopy, stellar interferometry, nuclear & high-energy physics and also appear in instrumental applications such as adaptive optics [27] and the construction of (phase)gratings with a prescribed efficiency [30]. The different branches of physics have developed their own techniques for handling their phase problems. In this paper I will discuss the optical phase problem. This problem was noticed by Wolf [24] when one has to extract the spectral lineshape from the visibility of the interference fringes formed in an interferometer. Almost simultaneously Walther [2] noticed that in optical image formation the phase of the wavefunction may turn out to be retrievable, because the wavefunction belongs to a very special class of functions, namely the class of entire functions of order 1 and finite type. In the sixties and seventies the phase problem received more and more attention, in particular in electron microscopy where a quantitative evaluation is a prerequisite [1,10]. In this paper we shall concentrate on the phase problem of image formation and coherence theory.

THE OPTICAL PHASE PROBLEM

Most optical imaging systems can be schematically represented as in Fig.1. We consider an object which is illuminated by spatially coherent quasi-monochromatic radiation (light or electrons). We assume the radiation to be completely polarized so that the imaging process can be described in terms of a <u>scalar</u> wave function (wf). After interaction of the beam with the object the wf in a plane immediately behind the object is $U_o(\vec{x}_o)$, which will be called object

ISSN:0094-243X/81/650402-10 $1.50

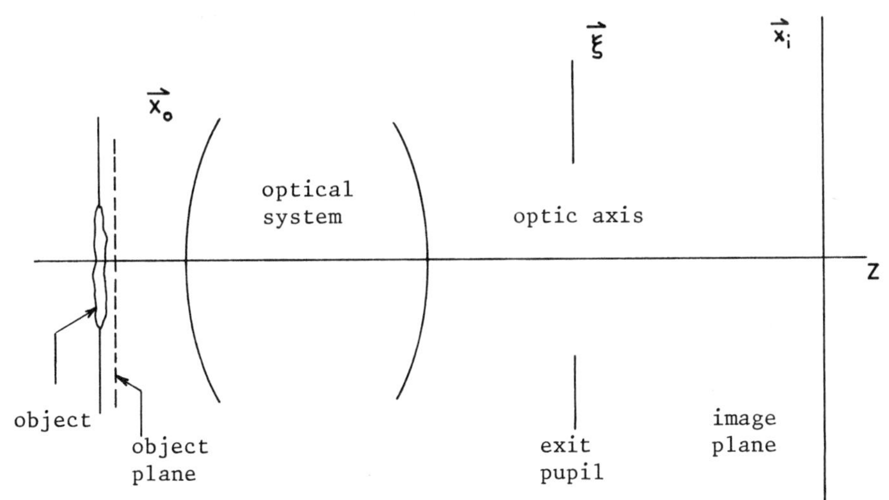

Fig.1 Scheme of imaging process

wf. The wf propagates through the microscope which we do not specify in further detail. The parameters of the microscope are contained in the expression for the wf in the exit pupil which will be denoted by $P(\vec{\xi})$. The wf in the plane where the image is recorded (the image plane) is denoted by $U_i(\vec{x}_1)$. In most imaging processes the detector only records the intensity which is proportional to $|U_i(\vec{x}_1)|^2$, i.e. apparently the phase of $U_i(\vec{x}_1)$ gets lost. In the early 1960's Walther [2] has already pointed out that this conclusion may be premature, because $U_i(\vec{x}_1)$ belongs to a special class of functions, namely the class of band-limited entire functions of order unity as is proved as follows: we assume the object wf $U_o(\vec{x}_o)$ to have a finite support D_o in the object plane. From standard Fourier Optics [3] we know that the wf in the exit pupil is given by

$$\tilde{P}(\vec{\xi}) = \iint_{D_o} U_o(\vec{x}_o)\exp(2\pi i\vec{\xi}.\vec{x}_o)d\vec{x}_o, \tag{1}$$

where $\vec{x}_o$ is measured in units of the wavelength λ while $\vec{\xi}$ will be measured in units of the focal length f of the optical system. $\tilde{P}(\vec{\xi})$ is modified by the wave-aberration function of the system, $\exp[i\phi(\vec{\xi})]$ (assuming isoplanatic imaging), leading to the wf

$$P(\vec{\xi}) = \tilde{P}(\vec{\xi})\exp[i\phi(\vec{\xi})] \tag{2}$$

from which the wf in the image plane follows by Fourier transformation:

$$U_i(\vec{x}_1) = \iint_{D_e} P(\vec{\xi})\exp(-2\pi i\ \vec{\xi}.\vec{x}_1)d\vec{\xi}, \tag{3}$$

where D_e is the aperture in the exit pupil and $\vec{x}_1$ is measured in units $M\lambda$ (M=magnification). From standard complex analysis [4] it follows that $\tilde{P}(\vec{\xi})$, $P(\vec{\xi})$ and $U_i(\vec{x}_1)$ are entire functions. From (3) we immediately infer the band-limitation of $U_i(\vec{x}_1)$. As $U_i(\vec{x}_1)$ belongs to a special class of functions we may expect a relationship between $|U_i(\vec{x}_1)|$ and arg $U_i(\vec{x}_1)$, because log $U_i(\vec{x}_1)$ is an analytic function of the components of $\vec{x}_1$. $\ell n|U_i(\vec{x}_1)|$ and arg $U_i(\vec{x}_1)$ are its real and imaginary part, respectively. In the case of analytic functions of one complex variable we know that its real and imaginary part are connected by the Cauchy-Riemann equations [4]. The phase problem consists of the determination of the phase of the wavefunction from intensity data only. The following questions have to be answered:

1) does the problem have a solution?
2) is the solution unique?

In order to answer this question we have still to make a very restrictive assumption:
$U_o(\vec{x}_o)$ etc. depends only on one component of the position vector $\vec{x}_o$. In that case we can draw upon the well-established theory of entire functions of one (complex) variable. The solution of the phase problem for two transverse directions is still in its infancy and is certainly not an obvious extension of the results to be presented here.

UNIQUENESS OF THE PHASE PROBLEM

We try to deduce the complex image wf U(x) (we drop the suffices "i" and "1") from its modulus $|U(x)|$, a quantity which follows from experiment: $|U(x)| = \{I(x)\}^{\frac{1}{2}}$ where I(x) is the intensity. As U(x) is an entire function of order unity U(z) can be written as [5]

$$U(z) = c_1 \exp(c_2 z)\ \Pi(1-z/a_n)\exp(z/a_n), \qquad (4)$$

where $c_1=U(0)$ and $c_2=U'(0)/U(0)$ and z is the complex variable $z=x+iy$. So, apart from the constants c_1 and c_2 U(z) is determined by its zeroes. The intensity I(x) can be analytically continued from the real axis according to

$$I(z) = U(z)U^*(z^*). \qquad (5)$$

In principle the zeroes of I(z) can be determined by the following device [6]: consider the Fourier transform i(t) of I(x):

$$i(t) = \int I(x)\exp(2\pi itx)dt,$$

$$I(x) = \int i(t)\exp(-2\pi itx)dt.$$

Next replace x by z=x+iy, yielding

$$I(z) = \int i(t)\exp(2\pi itx-2\pi ty)dt$$

from which the zeros can be deduced. It should be stressed that this is a very formal procedure which becomes unreliable "far" from the real axis. In practice we have to deal with only a restricted number of zeros around z=o, because our measurement of the intensity distribution only extends along a finite interval on the real axis. Zeros far away from z=o give contributions that are comparable with or less than the uncertainties in U(z) due to noise, as can be seen from (4). Let these "relevant" zeros be denoted by $a_1, a_2, \ldots, a_N$. The corresponding relevant zeros of I(z) are then given by $a_1, a_1^*, \ldots; a_N, a_N^*$. When determining the zeros of I(z), e.g. by using the procedure outlined above, we do not know whether a particular zero belongs to U(z) or U*(z*). This means that these are 2^{2N} wavefunctions U(z) consistent with I(z). Moreover, it can be straightforwardly verified that all these functions are also band-limited with the same bandwidth [7]. So we need additional information. Such additional information may consist of prior knowledge about the object. Such a situation arises when we know that the wavefunction satisfies certain symmetry properties (e.g. definite parity, positivity). Such prior knowledge has been advantageously exploited by Fienup [8] and Greenaway [9] who uses deliberately placed aperture stops. If there is no sufficient prior information available we need additional measurements to reduce the degree of ambiguity. In the next sections we shall review the following proposals:

a. phase determination using diffraction and image pattern,
b. phase determination using two defocused images,
c. methods based on dispersion relations (half-plane apertures),
d. combination of bright-field and dark field-images.

PHASE DETERMINATION FROM DIFFRACTION AND IMAGE PATTERN

The diffraction pattern is the intensity distribution in the exit pupil. As the wf in the exit pupil and image planes are Fourier conjugate to each other we have the problem to construct a complex wf from its modulus and the modulus of its Fourier transform. An algorithm, also applicable to two transverse spatial dimensions has been proposed by Gerchberg and Saxton [6,10]. The algorithm does not always converge in both diffraction and image plane as has been observed by Van Toorn and Ferwerda [11]. Even in those cases where the iteration converges we have to establish the uniqueness of the solution. In a series of papers [12] we have shown the following results: if we exploit the prior knowledge that the wfs P(ξ) and U(x) are entire functions we find uniqueness of the solution and at most a two-fold ambiguity if the object possesses certain symmetries. Unfortunately, this result cannot be implemented in practical situations, because we do not know how to enforce the requirement that the solution has to be analytic. This means that we have to investigate the existence of solutions belonging to a larger class of functions for which we chose functions with a continuous first derivative. In that case the solution does not have to be unique. The trouble is that the question of uniqueness or non-uniqueness cannot be settled beforehand because the quantity which governs the uniqueness depends upon the solution still to be obtained. Choosing the smoothest possible solution sometimes leads to a unique solution but lacks a satisfactory physical basis. The theoretical predictions were confirmed in numerical simulations. Schiske[13] constructed different solutions for the same data.

PHASE DETERMINATION FROM TWO DEFOCUSED IMAGES

This procedure is strongly related to the preceding proposal and has been put forward by Misell [14]. For an appraisal of the advantages of this method the reader is referred to ref.10, Chap.7. Misell has formulated an algorithm, similar to the one of Gerchberg and Saxton, exploiting the relation between the wfs in different image planes. The uniqueness has been indepedently established by Schiske [15] and the workers of the Groningen group [16]. An alternative, one-dimensional algorithm has been given by Van Toorn. The problem to be solved in the latter case is the solution of two coupled non-linear integral equations:

$$f_1(\xi) = \int_{\xi}^{\beta} P(\xi')P^*(\xi'-\xi-\beta)d\xi',$$

$$f_2(\xi) = \int_{\xi}^{\beta} P(\xi')P^*(\xi'-\xi-\beta)\exp[i\Delta\{\xi'^2-(\xi'-\xi-\beta)^2\}]\,d\xi',$$

where the exit pupil is characterized by $-\beta\leq\xi\leq\beta$, Δ is a measure for the amount of defocus between the two images.
Simulations have revealed that the algorithm is very sensitive to noise which is particularly embarassing in electron microscopy where the incident electrons give rise to specimen damage. In that field of application we cannot build up statistics without changing the specimen in an uncontrollable way. As the number of electrons which can participate in the imaging process is very limited (only a few electrons per Å^2) we have to study the effect of quantum noise (shot noise) on the quantities to be calculated systematically. A first approach has been made by Van Toorn et al. [18]. The crux of one of the proposals in this paper is to use two defocused images together with the intensity in the diffraction plane. The latter quantity has a stabilizing influence on the sensitivity of the final results to noise in the data. A more fundamental approach to the noise problem is under construction by C.H. Slump using techniques of statistical estimation theory and stochastic integral equations.

PHASE RETRIEVAL WITH THE LOGARITHMIC HILBERT TRANSFORM (DISPERSION RELATIONS)

An interesting proposal was put forward by Misell and Greenaway [19]: they advocated the use of an opaque stop in the exit pupil such that $P(\xi)=0$ for $\xi<0$. Because of Titchmarsh's theorem (quoted in [20]) the real and imaginary part of the wf in the image plane are Hilbert transforms of each other:

$$\text{Re } U(x) = \frac{1}{\pi} P \int_{-\infty}^{\infty} \frac{\text{Im } U(x')}{x'-x}\, dx',$$

$$\text{Im } U(x) = -\frac{1}{\pi} P \int_{-\infty}^{\infty} \frac{\text{Re } U(x')}{x'-x} dx'. \tag{6}$$

The phase retrieval problem for weak objects can now be solved. In this case the image wf can be decomposed into

$$u(x) = 1 + u_s(x) \ ; \quad |u_s(x)| \ll 1,$$

where 1 denotes the unscattered wf and $u_s(x)$ the scattered wf. The intensity in the image plane can be calculated according to

$$I(x) = |U(x)|^2 \simeq 1 + 2 \text{ Re } u_s(x).$$

It can easily be checked that also Re $u_s(x)$ and Im $u_s(x)$ are Hilbert related. Experiment yields Re $u_s(x)$, its Hilbert transform gives Im $u_s(x)$, which supplies us with a first approximation of the complex scattered wf, $u_s^{(1)}(x)$. A second approximation can be obtained as follows:

$$I(x) = 1 + 2 \text{ Re } u_s^{(2)}(x) + |u_s^{(1)}(x)|^2, \text{ etc.}$$

which leads us beyond the traditional weak object approximation (provided the iterative procedure converges). The phase retrieval procedure only yields $P(\xi)$ for $\xi>0$. We need a second exposure with the complementary half plane to obtain $P(\xi)$ for $\xi<0$.

The method may also work for general objects if we add a coherent reference wave to the object wavefunction [21]. For a half-plane aperture we have $U(x) = \int_0^{\beta} P(\xi)\exp(-2\pi i\xi x)d\xi$ which is an entire function bounded in Im $x \leq 0$. We now add a coherent reference wave A (for which we choose a positive constant) such that the resulting wave is $w(x) = A + U(x)$. A is chosen such that $|U(x)|<A$ on Im $x=0$. According to Rouché's theorem [22] $w(x)$ then has no zeros in Im $x \leq 0$, log $w(x)$ then is a causal transform as may be shown by

$$\log w = \log A + \log(1+U/A) = \log A + U/A - \tfrac{1}{2}(U/A)^2 + \ldots\ldots,$$

where $U(x)$ is a causal transform. So the real and imaginary part of log w are related by Hilbert transforms:

$$\arg w(x) = -\frac{1}{\pi} P \int_{-\infty}^{\infty} \frac{\ell n|w(x')|}{x'-x} dx'. \tag{7}$$

From elementary geometrical considerations we deduce arg $U(x)$ from arg $w(x)$, which solves the phase problem in this case.

In cases where it is impossible to find a reference wave with the desired properties we may proceed as follows : we start from the relation

$$U(x) = \int_{-\beta}^{\beta} P(\xi)\exp(-2\pi i\xi x)d\xi. \tag{8}$$

Partial integration yields its asymptotic behavior

$$U(z) \sim -(2\pi i z)^{-1}[P(\beta)\exp(-2\pi i\beta z) - P(-\beta)\exp(2\pi i\beta z)] + \dots, \tag{9}$$

The asymptotic distribution of the zeros follows from (9):

$$z_n = (4\pi i\beta)^{-1}\,\ell n\{P(\beta)/P(-\beta)\} + \frac{n}{2\beta}, n \text{ integer}, \tag{10}$$

which shows that depending upon whether $|P(\beta)/P(-\beta)|$ is greater than or smaller than unity the zeros lie in the lower or upper half plane, except for a finite number. Let us assume the case $|P(\beta)/P(-\beta)|>1$. Then the number of zeros of $U(x)$ in the u.h.p. is finite: $a_1,\dots,a_N$. We now construct the function $\tilde{U}(z)$ which is obtained from $U(z)$ by fipping its zeros $a_1,\dots,a_N$:

$$\tilde{U}(z) = U(z)\prod_{n=1}^{N}\frac{z-a_n^*}{z-a_n}. \tag{11}$$

$\tilde{U}(z)$ has no zeros in the u.h.p. consequently $\log\tilde{U}(z)$ is regular there. From (9) and (11) we straightforwardly conclude that asymptotically $\log[\tilde{U}(z)\exp(+2\pi i\beta z)]$ remains finite for $z\to\infty$ in the u.h.p. We now apply the calculus of residues to

$$\frac{1}{2\pi i}\int_C \frac{\log[\tilde{U}(z')\exp(+2\pi i\beta z')]}{z'(z'-x)}\,dz' = 0,$$

where the contour C has been sketched in Fig.2. The large semicircle with radius R gives a vanishing contribution in the limit $R\to\infty$ while the contributions from the semicircles around the simple poles $z'=0$ and $z'=x$ can be calculated in a standard fashion making the inessential assumption that $U(0)\neq 1$. A straightforward calculation yields:

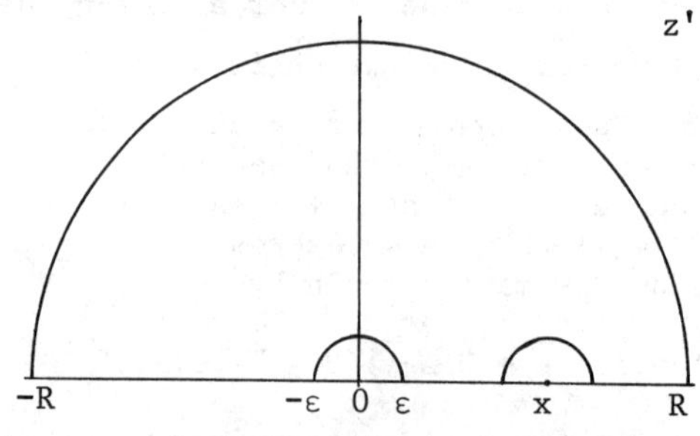

Fig.2.

$$\phi(x)-\phi(0) = -\frac{x}{\pi}\,P\int_{-\infty}^{\infty}\frac{\ell n|U(x')|}{x'(x'-x)}\,dx + 2\sum_{n=1}^{N}\arg(x-a_n)$$

$$-\,2\sum_{n=1}^{N}\arg a_n + 2\pi\beta x. \tag{12}$$

where $\phi(x)=\arg U(x)$. So $\phi(x)$ can be calculated from the intensity if we know the value of $|P(\beta)/P(-\beta)|$ and the finite number of zeros in one of the halfplanes (u.h.p. or l.h.p.), a result obtained by Hoenders [23]. Obviously, this result may be hard to be put into practice, because the locations of the zeros is very difficult to establish (see also [21] and the ingeneous proposal by Fiddy and Greenaway [28]).

PHASE RETRIEVAL FROM COMBINED BRIGHT-FIELD AND DARK-FIELD IMAGES

A very simple proposal has been suggested by Frank[29]. The bright-field image reads

$$I_b = 1 + 2\,\mathrm{Re}\,u_s(x) + |u_s(x)|^2.$$

The corresponding dark-field image is obtained by intercepting the unscattered beam in the exit pupil. The dark-field image reads

$$I_d = |u_s(x)|^2.$$

From I_b-I_d we can calculate Re $u_s(x)$. I_d then yields Im $u_s(x)$, up to a sign.

THE PHASE PROBLEM OF COHERENCE THEORY

In the theory of temporal coherence (ref.3, §7.5.8) the visibility of the interference fringes is proportional to the modulus square of the Fourier transform of the spectral distribution $g(\omega)$ of the object. In principle, the position of the interference fringes determines the phase of the Fourier transform, but in practice it turns out to be virtually impossible to measure the shift of the interference fringes. The phase problem is closely related to the one discussed in the previous sections. Here we have the constraint that $g(\omega)\geq 0$. This case has been studied by Wolf and Dialetis [24].

A similar problem arises in the case of spatial coherence. According to the Van Cittert - Zernike theorem (ref.3, §10.4.2) the complex degree of spatial coherence in the far field of an incoherent source is the Fourier transform of the intensity distribution across the source. The visibility of the interference fringes only yields the modulus of the Fourier transform. Proposals for the solution of the phase problem have been reviewed by Kohler and Mandel [25] while a suggestion by Ferwerda [26] has still to be tested.

PHASE RETRIEVAL WITH PARTIALLY COHERENT ILLUMINATION

In the preceding treatment we assumed perfectly spatially coherent, quasi-monochromatic illumination. The question arises how phase retrieval can be achieved with partially coherent quasi-monochromatic radiation. In a recent paper Huiser [31] has shown that it is possible in most cases to calculate the phase from two images of the object (image plane and exit pupil or two defocused images). Of course, the coherence function has to be known from a separate measurement for which ref. 26 might prove to be useful.

OPEN PROBLEMS

The one-dimensional phase problem is fairly well understood by now. The application to practical cases still poses problems [10]. A systematic study of the sensitivity of the results to noise in the experimental data should be performed. Such a project is under way in the Groningen group (by C.H. Slump). Another problem which should be investigated is the phase problem in two dimensions. There are indications that the two-dimensional phase problem is less ambiguous than its one-dimensional counterpart [8].

REFERENCES

1. H.A. Ferwerda in Chapter 3 of "Inverse Source Problems in Optics", H.P. Baltes (Ed.), Springer, New York(1979)where several references can be found.
2. A. Walther, Optica Acta 10, 41 (1963).
3. M. Born and E. Wolf, Principles of Optics, Pergamon, Oxford (1975),Sec. 10.5 or J.W. Goodman, Introduction to Fourier Optics, McGraw-Hill, New York(1968).
4. E.C. Titchmarsh, Theory of Functions, Oxford University Press, Oxford (1968), pp.99
5. Reference 4, Ch.8.
6. W.O. Saxton in Chap.2 of Computer Processing of Electron Microscope Images, P.W. Hawkes (Ed.), Springer, New York (1980), formula (2.44).
7. See e.g. B.J. Hoenders, J. Math. Phys. 16, 1719 (1975).
8. J.R. Fienup, Opt. Lett. 3, 27 (1978).
9. A.H. Greenaway, Opt. Lett. 1, 10 (1977).
10. W.O. Saxton, Computer Techniques for Image Processing in Electron Microscopy, Academic Press, New York (1978), Chapt. 4,5,6 [Advances in Electronics and Electron Physics, Supplement 10],
11. P. van Toorn and H.A. Ferwerda, Optik 47, 123 (1977).
12. A.M.J. Huiser, A.J.J. Drenth and H.A. Ferwerda, Optik 45, 303 (1976); A.M.J. Huiser and H.A. Ferwerda, Optik 46, 407 (1976); A.M.J. Huiser, P. van Toorn and H.A. Ferwerda, Optik 47, 1(1977).
13. P. Schiske, Optik 40, 261 (1974).
14. D.L. Misell, J. Phys. D6, L6-9 (1973), ibid. D6, 2200, 2217 (1973).
15. P. Schiske, J. Phys. D8, 1372 (1975).
16. A.J.J. Drenth, A.M.J. Huiser and H.A. Ferwerda, Optica Acta 22, 615 (1975); A.M.J. Huiser and H.A. Ferwerda, Optica Acta 23, 445 (1976).
17. P. van Toorn and H.A. Ferwerda, Optica Acta 23, 457, 469 (1976).
18. P. van Toorn, A.M.J. Huiser and H.A. Ferwerda, Optik 51, 309 (1978).
19. D.L. Misell and A.H. Greenaway, J. Phys. D7, 832 (1974).
20. J. Hilgevoord, Causal Description and Dispersion Relations, North-Holland, Amsterdam (1960), Ch.3.

21. R.E. Burge, M.A. Fiddy, A.H. Greenaway and G. Ross, Proc. R. Soc. London A 350, 191 (1976).
22. Reference 4, §3.42.
23. B.J. Hoenders, J. Math. Phys. 16, 1719 (1975).
24. E. Wolf, Proc. Phys. Soc. London 80, 1269 (1962); D. Dialetis and E. Wolf, Nuovo Cimento 47, 113 (1967).
25. D. Kohler and L. Mandel, J. Opt. Soc. Am. 60. 280 (1970); 63, 126 (1973).
26. H.A. Ferwerda, Optics Comm. 19, 54 (1976).
27. R.A. Gonsalves, J. Opt. Soc. Am. 66, 961 (1976); A.J. Devaney and R. Chidlaw, J. Opt. Soc. Am. 68, 1352 (1978).
28. M.A. Fiddy and A.H. Greenaway, Optics Comm. 29, 270 (1979).
29. J. Frank, Optik 38, 582 (1973).
30. A. Roger and D. Maystre, Optica Acta 26, 447 (1980); A. Roger, Optics Comm. 32, 11 (1980).
31. A.M.J. Huiser, Optik 55, 241 (1980).

High Quality Focused-Image-Hologram for Color Image Preservation

C. S. Ih
University of Delaware
Department of Electrical Engineering
Newark, Delaware 19711

ABSTRACT

Very high quality FIH's have been made using an all-spherical-mirror system. High quality reconstruction from these FIH's can be made with a Random-Spatial-Phase modulated laser light or white light without the usual coherent noise.

INTRODUCTION

Preserving color images using Fourier-Color-Hologram (FCH) has recently been proposed and demonstrated [1,2]. One of the most difficult obstacles associated with using coherent light for serious imaging applications is the well known coherent noise problem. Any imperfections in the optical system, such as scratches, dents, inhomogeneity or air bubbles in the optical material and even dusts, can produce very distracting diffraction patterns and seriously affect the final image quality. When holograms are involved, then the holograms themseves and their recording and reconstruction optics become part of the optical system. Distracting diffraction patterns can be reduced or eliminated by using diffused laser lights. However, diffused laser light produces speckles which in many cases are far worse than the diffraction patterns. The speckle noise is particularly serious for an optical system with a limited aperture.

In order to minimize color cross-talk, the reconstruction optics for the FCH must inherently have a small effective aperture [1]. It is therefore inadvisable to use diffused laser light for the proposed FCH system. High quality images must be obtained from good and "clean" optics. We have constructed a very good and simple imaging system using only spherical mirrors [3]. These mirrors were made precisely and were polished to have minimum scattering. For this simple optical system, we have been able to keep the mirror surfaces clean. The holograms were processed under controlled environment. The developing solutions were under constant circulation and filtration. We have found that under these conditions, the noise introduced by the optical system including the imaging mirrors is minimal and that the major noise source is the hologram emulsion. Since they

ISSN:0094-243X/81/650412-06$1.50

must be panchromatic, only very limited types of films are availabe for our experiments. All the films we have tried were not satisfactory because of either excess scattering or imperfections in the emulsions themselves. These imperfections include ridges, bumps (apparently large dust particles inbedded in the emulsion during the coating process), scratches and pinholes. These imperfections are not normally found in ordinary photographic films. Among all the films we have tried, the most frequently used Kodak 649-F film appears to be the worst in term of emulsion imperfections. Most of these imperfections excepting the pinholes can be made less objectionable if Focused-Image-Holograms (FIH) instead of Frensel holograms are used. Very high quality FIH's have been made using the all-spherical-mirror system. When FIH's are used, the RSPM (Random-Spatial-Phase-Modulation) modulated laser light [4] can be used for the reconstruction thus alleviating the coherent noise problems. The reconstruction can also be made with white light, even though the reconstruction may suffer from very low intensity or reduced color saturation, or both.

DESCRIPTIONS AND RESULTS

The FIH system is schematically shown in Fig. 1. The three-color laser beam, which can be derived either from a single laser or multiple lasers, is first split into two beams, the object and the reference beam. The two beams are then redirected toward the hologram by two mirrors (M1 and M2). In the path of each of the two beams, there is a SPF (Spatial Pinhole Filter) which expands and "cleans" up the laser beam. M3 and M4 are included in the reference beam in order to compensate the increase in optical path introduced by the mirror imaging system. During the recording process, the original color slide is imaged onto the hologram by the mirror imaging system. The point light source of the object beam is focused at the Fourier plane where the Fourier spatial filter is located. After exposure and processing, the hologram is rotated 180 degrees (about an axis perpendicular to the paper) and placed at the position of the original slide. M6 is now inserted in the reference beam and together with a SPF, forms the reconstruction beam. During the reconstruction, the object beam is blocked (not shown). The reconstructed true color image will be formed at the original hologram position. When the reconstruction beam is properly placed, the Fourier transform of the reconstructed image will also be in the original Fourier plane. A Fourier spatial filter of appropriate size (in the plane of the paper only) is used to stop undesired color cross-talk [1].

Since, for the FIH, the image is exactly on the hologram, we do not need a coherent source for the reconstruction. If a tungsten lamp is used, the filament should be placed at the position of the SPF and oriented perpendicular to the plane of the paper. The Fourier spatial filter which now assumes the form of a slit is also in the same direction. In order to have good color saturation, a fine filament and narrow slit should be used. However, this combination will produce a very dim reconstruction. Since all the laboratory incandescent light sources have very limited operating temperatures, these impose a fundamental limitation for the white light reconstruction. This limitation can be overcome by using a spatially randomized laser light. The arrangement described in Ref. 4 can effectively randomize the laser light to a point that pictures taken with RSPM light is virtually indistinguishable with those taken with white ligtht [4]. We have adapted the RSPM for the three-color laser beam for FIH reconstruction. A black and white print of a full color reconstruction is shown in Fig. 2. A b&w print of a color duplicate by the conventional process is shown in Fig. 3. The difference is minimal. Most of the spots on Fig. 2 are due to pinholes on the hologram film.

DISCUSSION AND CONCLUSION

The most important single item for making high quality FIH's is a high quality and cosmetically clean imaging system. Ordinary lens systems require multiple elements to correct aberrations. Not only all the surfaces must be polished to have low scattering, but also they must essentially contain no air bubbles. The cost for making such a lens is prohibitive Also more important is that even if all the surfaces are coated with multi-layer coatings to reduce the surface reflection to 0.1%, the reflected waves can concentrate to a small area on the hologram, thus creating 'hot' spots. If the linear concentration is 100 to 1 (say 50 mm to .5 mm), the intensity will be increased by 4 orders. Therefore the intensity at the 'hot' spot is ten times higher than the average. The all-spherical-mirror system used in this experiment uses only three reflecting surfaces. These surfaces can be made to have low scattering and be maintained clean. The reflection system is free from 'hot' spots. Computer ray tracings have shown that this system is diffraction-limited (for f/8) for the entire image area of a standard 35 mm film [3].

In conclusion, we have shown that high quality FIH's can be made using a mirror imaging system. High quality reconstructions can be made from these FIH' using either the RSPM laser light or white light without the usual coherent noise. Even at this stage special precautions and facilities must be exercised and are necessary to make the FIH's, the reconstruction is less critical and therefore can be carried out in ordinary photo laboratories.

This research is supported in part jointly by the American Film Institute (AFI) and The Library of Congress through a grant from AFI, grant reference number #7931300006.

1. C. S. Ih, Multicolor Imagery from Holograms by Spatial Filtering, Applied Optics, 14, 438 (Feb. 1975).
2. C. S. Ih, Fourier Color Hologram and Color Image Preservation, Proceedings of ICO-11 Conference, Madrid, Spain, 1978.
3. C. S. Ih and K. Yen, The Imaging and Fourier Transform Properties of A Spherical-Mirror System, to be published.
4. C. S. Ih and L. A. Baxter, Improved Random Spatial Phase Modulation (RSPM) for Speckle Elimination, Applied Optics, 17, 1447 (May 1978).

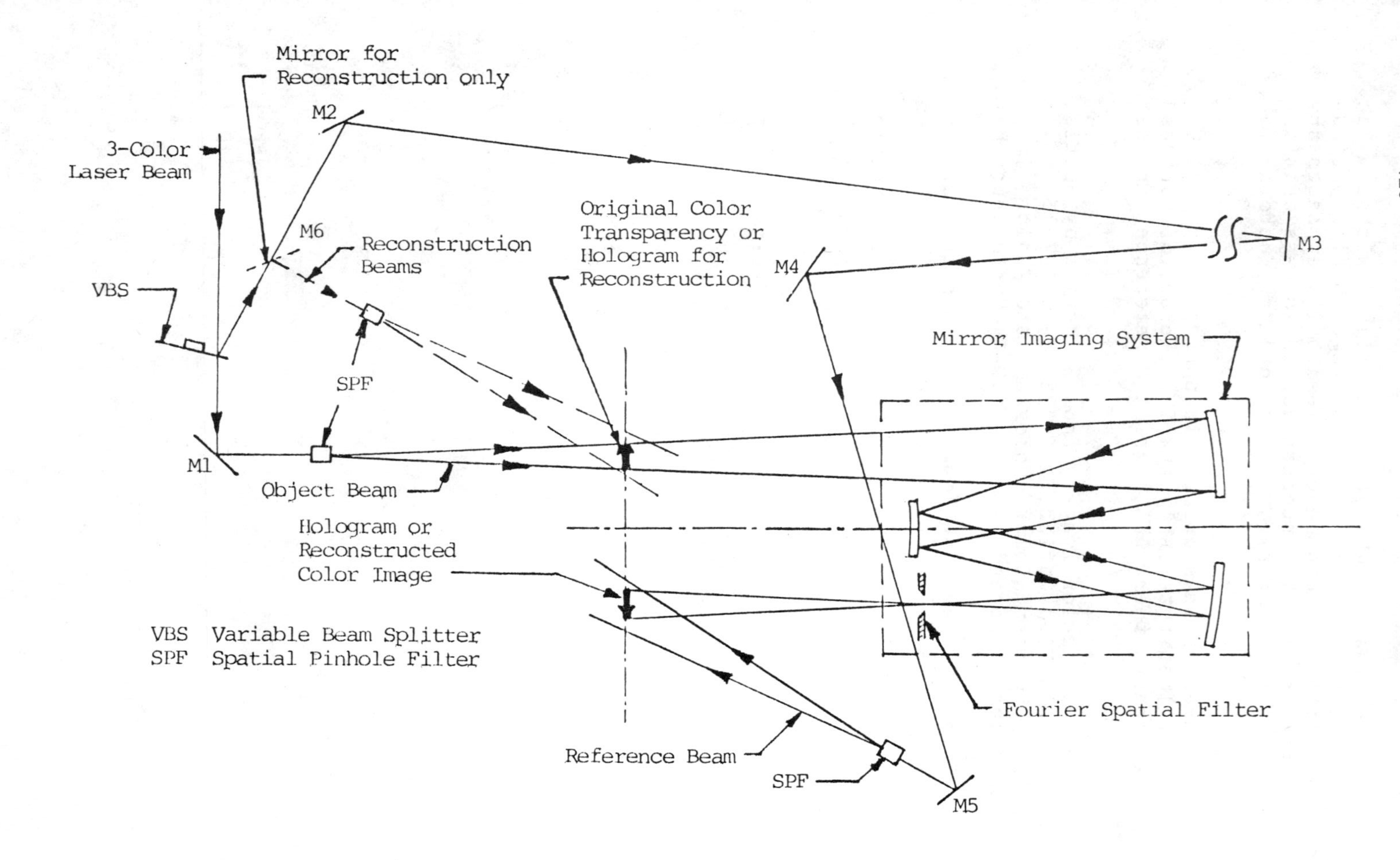

Fig. 1 Schematics of the Foucused-Image-Hologram System

Fig. 2 Reconstruction from Fourier Color Hologram

Fig. 3 Conventional Duplicate

COMMUNICATION VIA ELECTRO-OPTIC MODULATION OF WHITE LIGHT

J.P. Goedgebuer, J. Salcedo (•), R. Ferrière
J.Ch. Viénot

Lab. d'Optique, Ass. CNRS, U. de Franche-Comté, 25030 Besançon, France

ABSTRACT

It is well-known that the introduction of a path-difference between two beams of white light likely to interfere corresponds to a sinusoidal filtering of the temporal frequencies of the light observable as a channelled spectrum. This property can be used in information transmission, the power spectrum of white light being then filtered. The device proposed here is based on the Pockels effect induced by an electric field in an electro-optic crystal behaving as a two-wave interferometer whose path-difference is proportional to the electric field. As a result, a white light beam travelling through the crystal undergoes a chromatic filtering that encodes the electric signal. The decoding is carried out by spectral analysis of the light using either grating- or prism- spectroscope, or birefringent quartz plate acting as a Fourier Transform Spectroscope. The electro-optic bandwidth is expected to range up to 100 Mhz. Preliminary experiments show that an optical fiber used as transmission line fits the whole range of the light frequencies, independently of any dispersion effects.

PRINCIPLE OF OPERATION

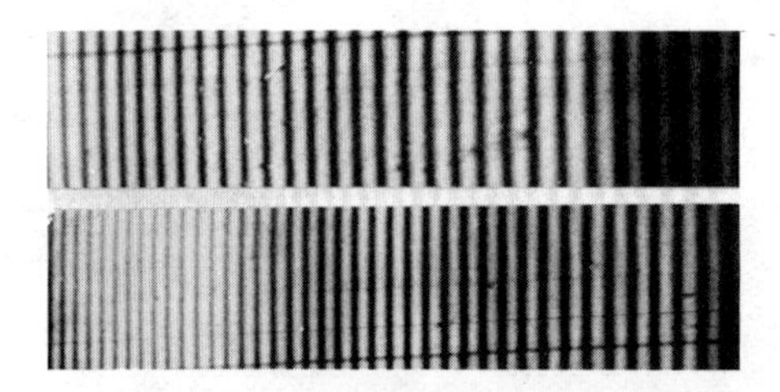

Fig. 1. The difference between the two channelled spectra shown here is due to a slight variation $m(t)$ of the optical delay in the interferometric part set in front of the spectrometer.

It is well-known that the introduction of a path-difference between two beams of white light results in a sinusoidal filtering of the power-spectrum of white light and leads to the channelled spectrum phenomena [1]. Fig.1.1 is a photograph of a channelled spectrum generated by an original path-difference D_o. Bright fringes occur along λ-*axis* (λ = wavelength) whenever

$$D_o = k\lambda \ (k : \text{integer}) \quad (1)$$

Any slight modification of the initial path-difference, that will be called the message $m = dD_o$, can vary with time. It induces a chromatic displacement $d\lambda$ (Fig. 1.2) of the fringes proportional to the

(•) On leave of absence from Depto Fisica, Universidad Industrial de Santander, BUCARAMANGA, Colombia.

ISSN;0094-243X/81/650418-07$1.50

message *m(t)* :

$$m = dD_o = k\ d\lambda \text{ (with } m \ll D_o \text{)} \qquad (2)$$

In turn, it induces a time-varying modulation of the power-spectrum of white light. This statement is the basic operation of the transmission system presented in the paper, the modulation *m(t)* of a path-difference D_O being considered as the information to be transmitted and detected.

Fig. 2 summarizes details of the process. At left, phase-modulated

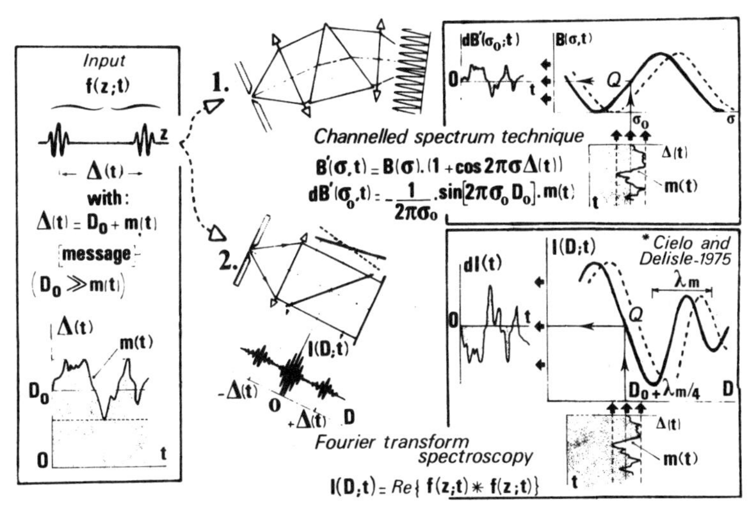

Fig. 2. Transmission by spectral modulation of white light.

white light is represented as two wave-groups separated by a path-difference *Δ(t)* varying with time by a quantity *m(t)* - the message - around a mean value D_O. The right side of the figure is related to the temporal domain as the spectral analysis of the two wave-groups is performed either with a spectroscope (Fig.2.1) or a 2-beam interferometer (Fig.2.2). In Fig. 2.1, the power-spectrum $B'(\sigma,t)$ displayed at the output of the spectroscope yields the channelled spectrum [1, 2] discussed so far :

$$B'(\sigma,t) = B(\sigma).\ \{1 + \cos(2\pi\sigma\Delta(t))\} \qquad (3)$$

where $B(\sigma)$: power-spectrum of the white light
$\sigma = 1/\lambda$: wave-number

The displacement of the fringes presented above takes place along a *σ-axis* (the dots represent another position of the fringes), i.e. a time-varying spectral modulation of white light power-spectrum. The decoding process is straightforward. Around the wave-number $\sigma = \sigma_o$

corresponding to the inflexion point Q of the fringe function, the time variation of energy $dB'(\sigma, t)$ is proportional to the message $m(t)$ as long as one works in a region where the slope is quasi-linear. Practically, the linear region of the fringe function can be approximated to a quarter of a fringe ; this implies that the peak-to-peak amplitude of $m(t)$ is smaller than $\lambda_0/4 = 0.1 \mu m$.

An alternative way of decoding the message $m(t)$ is shown in Fig.2.2. In place of a conventional spectroscope, a 2-beam interferometer (variable path-difference D generated by an air-wedge) performs the spectral analysis of the phase-modulated wave-groups of white light $f(z;t)$. This is Fourier Transform Spectroscopy [2,3,4]. At the output of the interferometer, the intensity $I(D;t)$ is described by the real part of the autocorrelation function of the input phase-modulated signal $f(z;t)$:

$$I(D;t) = Re\{f(z;t) \quad f(z;t)\} \tag{5}$$

Moreover, one observes (Fig.3) three achromatic fringe patterns centered at $D = 0$, and $D = \pm\Delta(t)$ respectively. The fringe spacing along D-axis is λm (λm : mean wavelength of white light). The side band fringes are of interest since their position directly depends on the value of $\Delta(t)$. In other words, any modulation $m(t)$ of the path-difference $\Delta(t)$ gives rise to a geometrical displacement of the side bands along D-axis, proportional to $m(t)$ (the dots represent another position of the fringe pattern). So the decoding process is quite similar to the previous one. From Fig.2.2, it can be seen that a point of operation Q corresponding to the inflexion of the side band fringes insures a linear variation of energy $dI(D;t)$ proportional to the message $m(t)$.

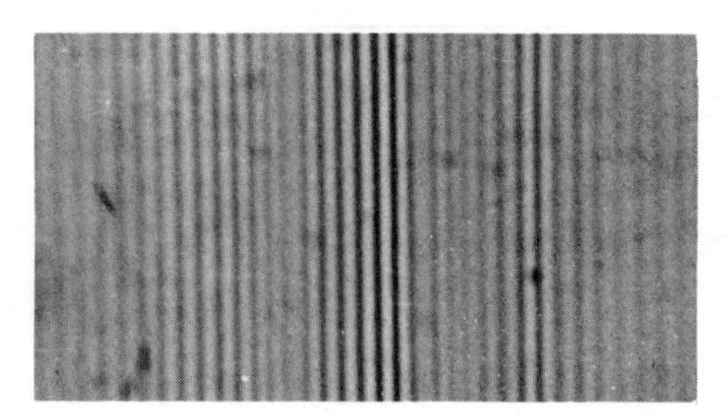

Fig. 3. Fringe patterns as result of autocorrelation performed by a 2-beam interferometer.

ELECTROOPTICAL PHASE MODULATION OF WHITE LIGHT

Emitter : The arrangement (Fig.4) reminds us those used in monochromatic transmission systems. The set-up essentially consists of a birefringent quartz plate and a transverse electrooptical modulator (E O M) set in cascade between two parallel polarizers. The latters are oriented so that their common direction of polarization is at 45° to either x- or y- crystal axes. The slow and fast axes x and y of the birefringent plate and E O M are parallel with respect to each other, yielding an overall path-difference $\Delta(t)=D_o + m(t)$ by dedoubling any incident wave-group of white light. The path-difference D_o introduced by the quartz plate (typically $D_o = 10\ \mu m$) plays the role of an "offset" path-difference. The EOM is a 45° X-cut ADP crystal that induces a time varying path-difference *m(t)* proportional to the voltage *V(t)*. The 45° X-cut crystal configuration permits a high degree of temperature stability with low voltage operation and is free from piezoelectric resonances. The intensity of the output light beam is constant if $\Delta(t) \gg L_c$, coherence length of the white light (L_c=2µm). The phase-modulated signal is :

$$f(z,t) = f(z) \otimes \left[\delta\{z-\Delta(t)/2\}+\delta\{z+\Delta(t)/2\}\right]$$

the time-averaged intensity $\langle|f(z,t)|^2\rangle$ being constant.

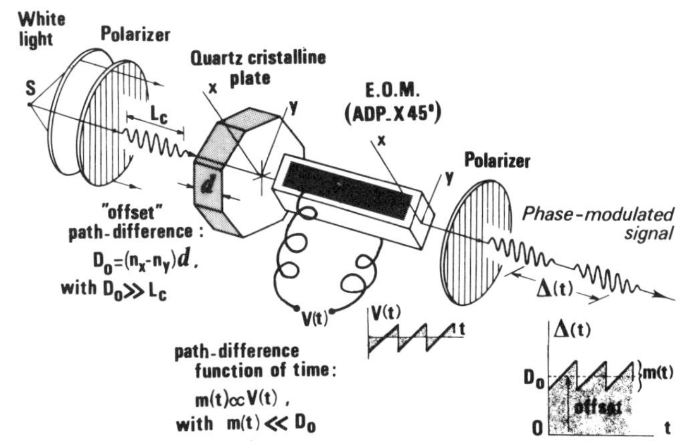

Fig. 4. Electrooptical phase modulation of white light.

Detection.(a) channelled spectrum technique : assessing the message is carried out either with a conventional spectroscope (Fig. 5.1.), or, better, with a double spectroscopic device (Fig. 5.2.).

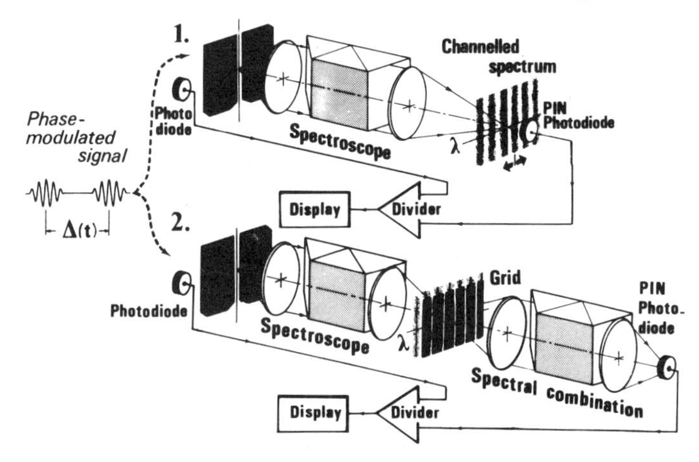

Fig. 5. Phase decoding by conventional spectroscopy.

In this case, a grid performs chromatic filtering of the channelled spectrum, and a second dispersor combinates the chromatic components. The advantage is in wavelength multiplexing, corresponding to an increase of energy at the photodetector.

One may wonder why a divider is required in the electronic detection circuit. We must recall here that, unlike the case of a monochromatic transmission system, we are not dealing with a phase modulation of the information carrier, but instead with a phase modulation. This means that any intensity fluctuation that may occur in the present case is due to external causes, e.g., instability of the white light source, or time-varying inhomogeneities in the propagation medium itself. In any event, intensity fluctuations at the input of the spectroscope cause fluctuations in the output. Therefore, as pointed out in Ref. 3, measuring the input/output intensity ratio allows the elimination of the external fluctuations at the detection stage. Compared to intensity modulation systems, that is a major advantage since the signal detected at the end of the transmission line is free from external fluctuations introduced by the light source or the propagation medium (dusts, diffusion, absorption, etc...).

(b) Fourier Transform Spectroscopy : in the arrangement used here (Fig.6) the quartz cristalline plate still works as an interferometer.

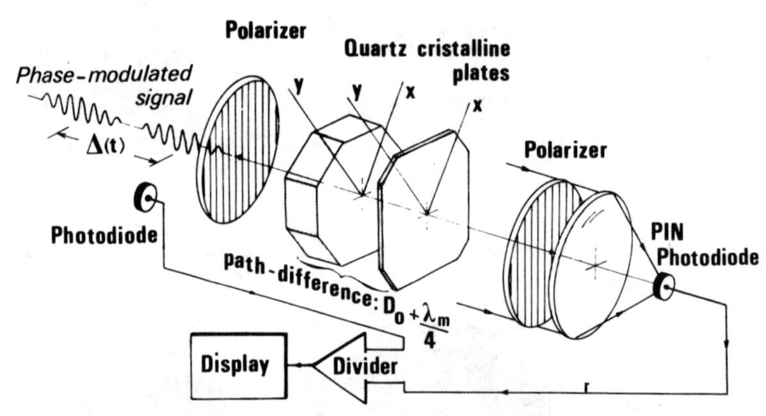

Fig. 6. Phase decoding by Fourier Transform Spectroscopy.

It introduces a path-difference $D_o+\lambda m/4$ between two polarized beams of white light when placed between two parallel polarizers oriented at 45° to the crystal axes. The additional path-difference $\lambda m/4$ sets the operation point Q in the inflexion region of the autocorrelation function $I(D,t)$ of the input phase-modulated signal $f(z,t)$; this insures a linear detection. Compared to the channelled spectrum technique, the present one offers a complete multiplexing, all the wavelengths occuring in the process as in usual Fourier Transform Spectroscopy. Here also, measuring the ratio between the input and output energies allows retrieval of the message without any stray signal, and regardless of the intensity fluctuations of the light source and of the perturbations induced by the propagation medium.

PERFORMANCES AND CONCLUSION

The results in Fig. 7 deal with various types of signals. First experiments have shown the capacity of transmitting signals in a bandwidth ranging from 1 Hz to 1 Mhz, and over distances of several tens of meters in free space. Multimode fibers are also suitable as medium. The fairly good quality of such transmitted signals could be surprising since one expects that dispersion effects combined with multimode propagation could spoil the information. Indeed, Fig. 8 illustrates the actual situation. The dispersion effects result in a spreading of the wave-groups of white light when propagating in the fiber, but do not degrade the information, i.e. $\Delta(t)$. The various distances of propagation attached to each mode in the fiber correspond to various propagation times of the wave groups. Finally, the overall signal at the output of the fiber is a sequence of couples of phase-modulated wave-groups, these couples being randomly time distributed. However, for each couple of wave-groups corresponding to a particular mode, the path-difference $\Delta(t)$ remains stationary in the sense that $\Delta(t)$ is not altered by propagation and dispersion effects. Therefore, such a stationary information remains accessible through a spectral analysis of the stochastic signal transmitted by the fiber. Experiments have shown the possibility of transmitting messages with 12 meter-long multimode fibers (core diameter 100 μm). In fact, such a distance does not seem to be the upper limit of transmission since only considerations about losses determine the transmission distance.

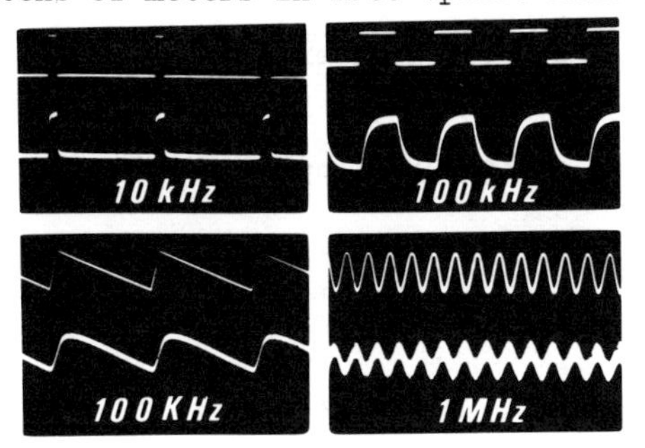

Fig. 7. Transmission of various types of signals.

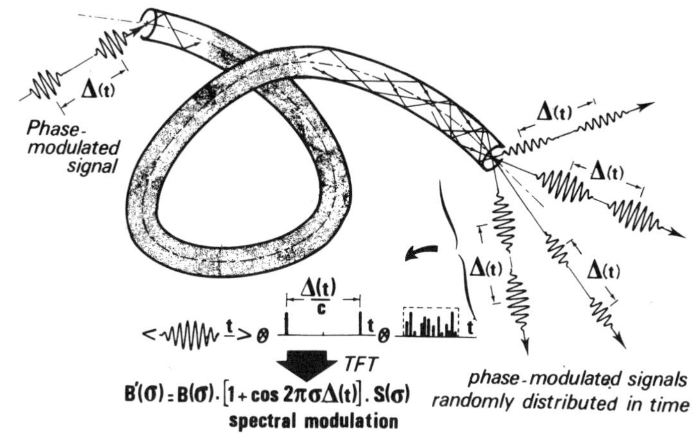

Fig. 8. Propagation in a multimode fiber.

REFERENCES

1 - See for example H. Bouasse, Z. Carrière, Interférences (Delagrave Ed. 1923 Paris).

2 - J.Ch. Viénot, J.P. Goedgebuer, A. Lacourt, Applied Optics, 16, 2 (1977), 454-461.

3 - C. Delisle, P. Cielo, Canadian J. of Physics, 53, 11 (1975), 1047-1053.

4 - P. Cielo, C. Delisle, Canadian J. of Physics, 54, 23 (1976), 2322-2331.

HOLOGRAPHIC PSEUDOCOLOURING OF GRAY LEVEL INFORMATION

S. Guel-Sandoval, J. Santamaría* and J. H. Altamirano**
Instituto National de Astrofísica, Optica y Electrónica
Apartado Postal 216. Puebla, Pue. México

ABSTRACT

A new method for the pseudocolouring of transparencies using holographic techniques is presented. An image hologram of a transparency is first registered. When this hologram is illuminated by two waves of different wavelengths, one of the two waves produces a positive holographic image while the other produces a negative transmitted image. The positive and negative images are added at the final image plane resulting in a colour coded image. Conventional white light lamps can be used for the reconstruction and the method can be very easily implemented. A detailed description of the method and some experimental results are shown.

INTRODUCTION

The concept of pseudocolouring and its interest as a technique for introducing colours in a black and white object, have been extensively put forward in many cases. Pseudocolouring was first implemented by photographic and digital techniques, but recently several optical methods have been proposed to encode the information content of the gray levels or frequency distributions of a black and white image. Halftone techniques, optical filtering and rainbow holography have been successfully applied with different degrees of sophistication.

In this paper a method is presented to encode gray level information in colours, based on the use of conventional holographic techniques. The encoding is carried out through the reconstruction of an image hologram so that the temporal and spatial coherence requirements are strongly reduced and therefore, no lasers are needed for the reconstruction step. Only two primary colours are allowed by this encoding but the method can be very easily implemented, thus avoiding the sophistication of halftone techniques and the low frequency noise problems of the optical filtering methods.

DESCRIPTION OF THE METHOD

As in the case of pseudocolouring through contrast filtering[1],

---

* Instituto de Optica, Serrano 121. Madrid-6, España

** On study leave from Escuela Superior de Física y Matemáticas. Instituto Politécnico Nacional. Mexico 14, D.F.

ISSN:0094-243X/81/650425-06$1.50

the method is based upon the addition of a positive and a negative image of an object transparency, each image formed with a different colour. In our case, both the positive and the negative images are encoded together by making an image hologram of the object transparency. A conventional two-lens imaging system is used to form an image on the holographic plate and a coherent reference plane wave is added to the image as it is represented in figure 1.

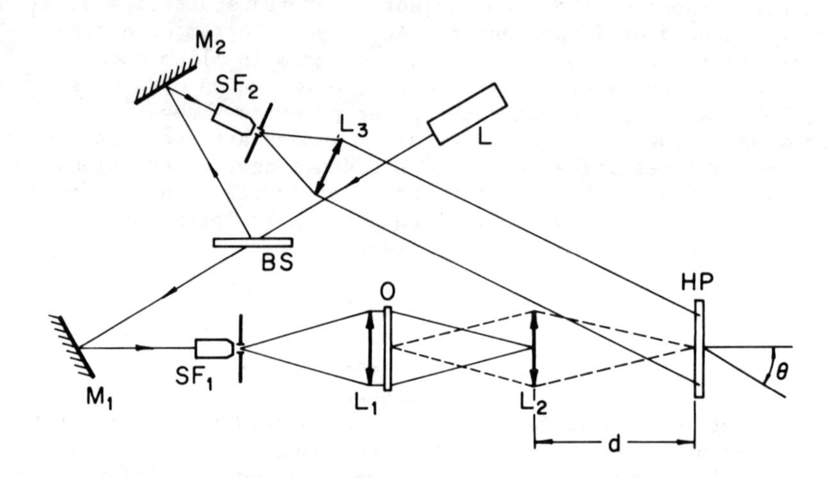

Fig. 1. Recording set-up. L:He-Ne laser; BS: beam splitter; SF_1 and SF_2 : spatial filters; L_1, L_2 : coherent two-lens imaging system; HP: holographic plate; M_1, M_2: mirrors; O: object transparency.

Let S(x, y) represent the amplitude distribution at the holographic plate corresponding to the image and $R\exp(2\pi ix\sin\theta/\lambda)$ the reference wave travelling at an angle θ with the axis. The amplitude transmittance of the holographic plate after developing in linear condition with respect to the amplitude-exposure curve is

$$t = t_0 - \beta(|S(x, y)|^2 + R^2 + S(x, y)\,Re^{2\pi iux} + S^*(x, y)\,Re^{-2\pi iux}) \qquad (1)$$

where $u = \sin\theta/\lambda$, t_0 is the bias transmittance and β is the modulus of the slope of the t-E curve.

When this hologram is reconstructed with the original reference, a diffracted wave (t') containing the image propagates along the optical axis:

$$t' = -\beta R^2\, S(x, y) \qquad (2)$$

The minus sign of this wave disappears at the detection step and a positive image is always produced.

On the contrary, if the hologram is illuminated by an on axis plane wave of amplitude K the transmitted wave is:

$$t'' = \left[t_0 - \beta(|S(x, y)|^2 + R^2)\right]K \qquad (3)$$

which is in fact a contrast reversed image.

By illuminating the hologram simultaneously with the two waves (reference and on axis) a positive and a negative object waves are travelling together on axis. If each wave has a different colour, they produce a pseudocoloured image, after imaging with a lens, or just by viewing.

It must be pointed out that when registering the image hologram, the amplitude corresponding to the image is always affected by a quadratic phase factor, which represents a diverging spherical wave, and that unless corrected by a lens on the holographic plate, it also appears at the reconstruction step. The result of this phase factor is that the two waves emerging on axis from the hologram do not have the same curvature, so that the observation or registration of both images simultaneously without vignetting becomes difficult; for this reason the curvatures of both waves must be equalized.

If the hologram is reconstructed with the reference conjugated wave, the diffracted beam is also the conjugated of the object amplitude, $S^*(x, y)$. This results in a converging wave instead of the

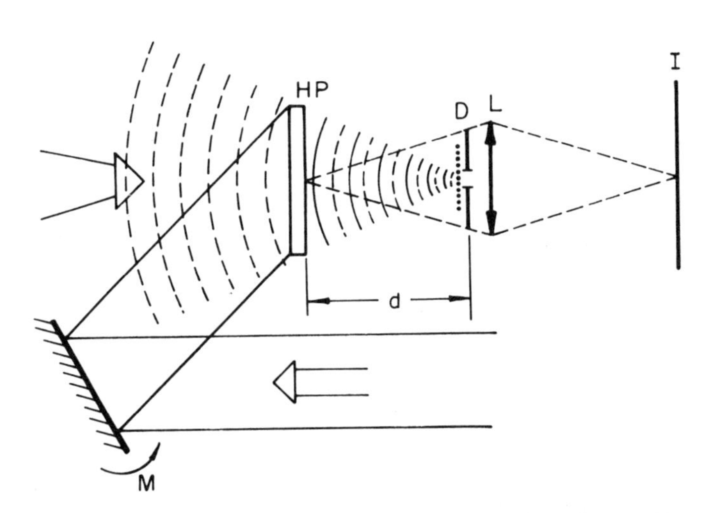

Fig. 2. Reconstruction set-up; HP: holographic plate; M: mirror; D: diaphragm; L: imaging lens; I: image.

diverging wave that would be obtained if the original reference were used. This converging wave forms the spectrum of the original object transparency at a distance d away from the hologram. The transmitted wave (Eq. 2) can also be made to converge to the same point by illuminating the hologram with an adequate converging wave (Figure 2). If the eye or a lens are placed at the center of the spherical waves, two incoherently superposed bright images are formed at the corresponding image plane producing the pseudocoloured image.

EXPERIMENTAL TECHNIQUES

As the spatial and temporal coherence requirements are very limited in image hologram reconstruction, extended white light sources can be used. The colour for the transmitted beam is selected by a conventional colour filter placed anywhere between the source and the hologram. For the diffracted beam, and due to the wavelength dependent character of the process, the images of the source for the different wavelengths spread on the source image plane. The desired colour is then selected by a diaphragm properly placed in that plane. The reconstructing beam is reflected by a mirror, which changes the incidence angle on the hologram; then different wavelengths pass through the diaphragm for different angles of the mirror. In this way, the colour of the diffracted beam can be continuously selected just by rotating the mirror. It must be pointed out that the size of the reconstructing sources is limited by the fact that the mentioned diaphragm for the colour selection of the diffracted beam allows transmission of a spectral interval of finite width. This width obviously increases with the source size and if it becomes too large, colour variation will appear in the image of a constant transmittance object. Nevertheless, sources of the order of 3 mm have been used in practice allowing a confortable manipulation without noticeable colour variation.

On the other hand, there is a non-linearity inherent to the method that must be mentioned. It is due to the fact that, after reconstruction, the holographic image has an amplitude distribution proportional to the amplitude transmittance of the transparency (Eq. 2) while the transmitted image is a contrast reversed image from the square of the object transmittance (Eq. 3). However, taking into account the non-euclidean character of the colour space, the mentioned non-linearity is not relevant for qualitative pseudocolouring.

With respect to the operating condition, photographic parameters for the recording process have to be determined first. For a good holographic image, the linear region of the amplitude transmittance versus exposure curve should be used. However, for a good transmitted image, the holographic plate behaves as a conventional incoherent photographic plate and so the linear region of the optical density versus logarithm of the exposure curve should be used. The linear region of the amplitude-exposure curve corresponds to a region (in the toe) of the H-D curve, that, although non-linear, is less pernicious to the transmitted image than it would be for the holographic image to shift the working point to higher exposure regions. For this reason, exposure conditions for the holographic requirements have been chosen in spite of the fact that the transmitted image will present a decreased contrast without appreciable distorsion. On the other hand, this contrast is also affected by the coherent reference, that behaves as an incoherent background for the transmitted image. Fortunately, image holography allows the use of 1:1 object to reference beam ration[2], so that the contrast, although reduced, is still satisfactory for this applications. The only effect on the pseudocolouring process is a barely appreciable loss of encoding efficiency.

RESULTS

Figures 3 and 4 show in black and white the holographic positive and the transmitted negative images obtained from an image hologram

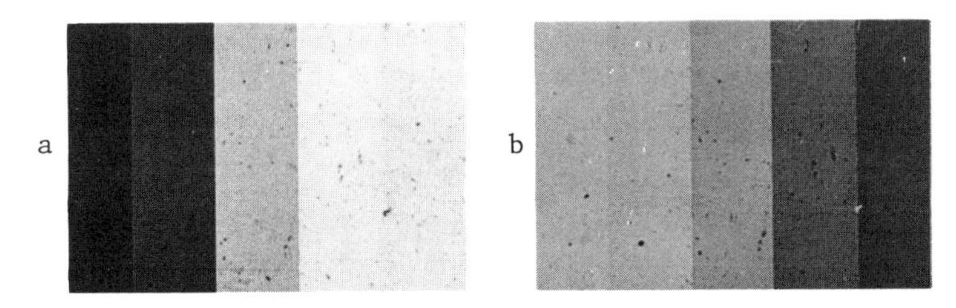

Fig. 3. Gray tone scale; a: holographic image; b: transmitted image.

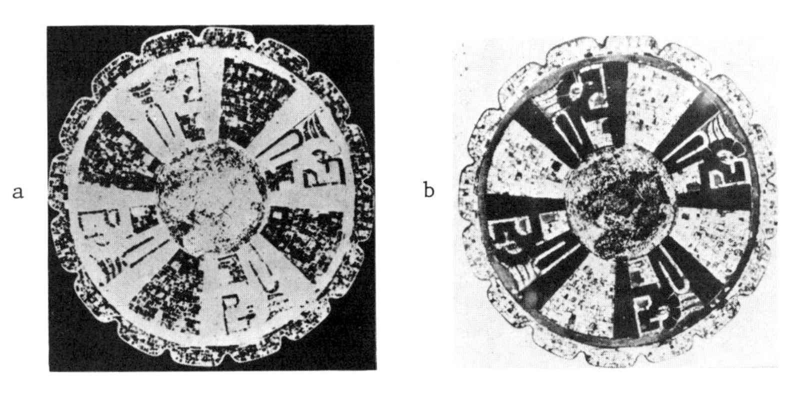

Fig. 4. Ancient Mexican God; a: holographic image; b: transmitted image.

registered with the set-up represented in figure 1. The first image shows a gray tone scale and the second one a picture of an ancient Mexican God. The reduction of contrast of the transmitted image with respect to the holographic image can be clearly appreciated in the figures. The pseudocoloured image is obtained by simultaneous recording of the positive and negative images, each one with a different colour*.

CONCLUSIONS

A new method for pseudocolouring gray level information of transparencies is presented. As in the case of pseudocolouring through contrast reversal filtering, the method is based upon the addition of a positive and a negative image, each having a different colour. The positive and the negative images are encoded together in an image hologram and therefore the reconstruction does not require the use of a laser. The method can then be very easily implemented, avoiding the sophistication of halftone techniques and the low frequency noise introduced by optical filtering techniques.

ACKNOWLEDGMENTS

This work was partly supported by Secretaría de Educación Pública of Mexico, code 37/08/01/78.

One of the authors (S.G.S.) also acknowledges CONACYT for the partial financial support given to attend this meeting.

REFERENCES

1. G. B. Brandt, Appl. Opt. 8, 1421 (1969).
2. J. Santamaría, M. Gea and J. Bescós, J. Optics 10, 151 (1979).

---

*The corresponding colour pictures are not included in the text but will be projected in the session.

PSEUDOCOLOR ENCODING OF DEPTH INFORMATION USING TALBOT EFFECT

P. Chavel* and T.C. Strand

Image Processing Institute
University of Southern California
Los Angeles, California 90007

ABSTRACT

When an object is placed in the Fresnel diffraction field of a grating, the Talbot or self images of the grating delineate constant range contours on the object. If the image of the object with the superimposed Talbot images is bandpass filtered and normalized, an image is generated with bright regions corresponding to the range contours. If several such images are made with different illumination wavelengths so as to produce several contours, each filtered image can be assigned a different color to produce a pseudocolor display of depth information.

INTRODUCTION

The problem of extracting depth information from a scene is an issue that has prompted diverse solutions depending upon the application and the associated constraints. One must consider such factors as range, sensitivity, accuracy, speed and ease of implementation. Medium to long range ($1-10^8$ m) applications can be handled by time-of-flight measurement devices such as laser range finders or ultrasonic systems. However, these systems only measure distance at one given point. They must be scanned to measure distances over an entire scene. Triangulation techniques such as those found in rangefinder cameras have long been applied to measuring medium range ($10^{-1}-10^4$ m) distances. The above methods measure absolute range with relatively coarse sensitivity or resolution. Other techniques have been developed for measuring relative distances with a very fine sensitivity. Interferometric systems which measure distance variations down to a fraction of the wavelength of illumination exemplify such techniques. Also in this category are the moiré methods which have been developed recently for metrological applications. The main operational distinction between moiré techniques and interferometric approaches is one of sensitivity or resolution, the moiré techniques are typically applied to relative depth variations over a range of $10^{-4}-10$ m whereas interferometry at visible wavelengths is useful for distances in the $10^{-10}-1$ m range. Both of these techniques can be

*Permanent Address: Institut d'Optique, Orsay, France.

ISSN:0094-243X/81/650431-06$1.50

used to process a three-dimensional volume in parallel. However, they share the problem that the depth information is generally not displayed directly so that the output images must be scanned and processed to extract the desired information. Obtaining the sign of depth differences with these techniques is an added complication.

Here we describe a new technique for extracting depth information based on the self-imaging property of gratings as discovered by Talbot [1]. This method provides absolute distance information over its range of operation which lies typically between 10^{-4} and 10 m. The depth information for all points in a scene is obtained directly and can be easily decoded and displayed.

METHOD

The Talbot effect consists of the fact that in the parabolic approximation to the Fresnel-Kirchhoff diffraction equation, the Fresnel diffraction pattern of a periodic object is also a periodic function along the direction of propagation. This implies that "images" of the original object appear at periodic intervals. The amplitude of the diffraction pattern is equivalent to the original periodic object in equally spaced planes separated by a distance given by

$$d_T = 2d^2/\lambda \qquad (1)$$

where d is the grating period and λ is the wavelength of illumination. The images thus formed are referred to as Talbot, Fresnel, of self-images [1-3]. Talbot images which are shifted by one-half period from the original object are also formed; these are also spaced a distance d_T apart and lie midway between the in-phase Talbot images.

Since these images occur at periodic, known intervals, their locations can be used as depth or range markers. If an object is placed in the Fresnel diffraction zone behind a grating illuminated by a coherent source, images of the grating will form on the object in the regions where the Talbot image planes intersect the object as indicated in Fig. 1. If one views the object along the direction of propagation by means of a beam splitter, the grating structure will be visible at the regions of intersection. The contrast of the grating fundamental frequency at any point on the object will be a function of the exact range at that point. Thus, by measuring the contrast on the image of the object, range information is obtained. Furthermore, since the period of the Talbot images depends upon λ, different depth planes can be selected by changing λ.

Once the depth information is encoded, the problem of decoding and displaying the information remains. The system for encoding, decoding and displaying is indicated schematically in Fig. 2. A bandpass filter tuned to the fundamental grating frequency is used to generate a signal proportional to the amplitude of the grating fundamental. This filtering can be accomplished by partially coherent optical filtering, electronic analog filtering of a video signal, or digital electronic filtering of a digitized image. If the object has a varying reflectance or illumination the output of the filter should be normalized to get a measure of grating contrast rather than simply grating amplitude. The coding and decoding steps must be done separately for each wavelength used. A convenient means of displaying the depth information in a two-dimensional image is to encode it as color in the image. Since each wavelength channel in the coding process corresponds to a different depth plane, the image from each wavelength channel can be simply displayed as a different color. Such a pseudocolor display permits easy observer interpretation.

EXPERIMENT

The experimental arrangement used for recording the projected Talbot images is shown in Fig. 1. The illumination source was an argon-ion laser. Recordings were made at three separate wavelengths, 515, 488 and 458 nm, to encode three separate depth planes. The grating was an absorptive element with a square-wave transmittance having a 10% duty-cycle. The fundamental grating frequency was 8 mm^{-1}. The object was placed in the region of the first Talbot image. The Talbot fringes on the object were photographed by viewing along the direction of propagation via a beamsplitter. This produced constant frequency fringes independent of the object's surface orientation.

For these experiments, the coded images were recorded on Kodak SO-115, 35 mm film, then scanned on an Optronix C4100 drum scanner with a 25x25 μm pixel spacing and an aperture of the same size. The resulting 512x512 digitized images were band-pass filtered using 1-D discrete Fourier transformation and a Gaussian-shaped filter centered at the fundamental grating frequency. The filtered images, in turn, were normalized to a low-pass filtered version of the initial image to compensate for changes in reflectance or illumination. Finally, the normalized results were output on film.

Figure 3 shows three coded images corresponding to the three different wavelengths, and Fig. 4 shows the resulting processed images. The object was a cylinder whose axis was inclined with respect to the optical axis. While this shape is not obvious from Fig. 3, it is more easily understood from the three intersections of the cylinder with three parallel planes shown in Fig. 4. In the present case, the distance between planes was close to 4 mm. A more precise depth information could be derived by using the exact value of the fringe contrast at each point rather than merely looking at the different images.

By displaying the different filtered images with different colors, a pseudocolor image can be formed where the color indicates the depth. For the results shown above, a pseudocolor image could be generated by assigning Figs. 4 (a), (b), and (c) to the red, green, and blue channels, respectively.

CONCLUSION

We have presented a simple technique for extracting and displaying depth information using Talbot images to indicate specific depth planes. For a limited range of applicability this technique offers the advantage of providing absolute range information that is easily decoded by a simple filtering operation. It can be used to process a three-dimensional volume without scanning. Filtering techniques and fundamental characteristics of the system are being studied more thoroughly.

This work has received partial support from the Joint Services Electronics Program and Batelle Pacific Northwest Laboratories which is gratefully acknowledged by the authors.

REFERENCES

1. F.H. Talbot, Philos, Mag. (London), 9, 401 (1836).

2. J.M. Cowley and A.F. Moodie, Proc. Phys. Soc. London, Ser. B: 70, 486 (1957).

3. John T. Winthrop and C.R. Worthington, J. Opt. Soc. Am., 55, 373, (1965).

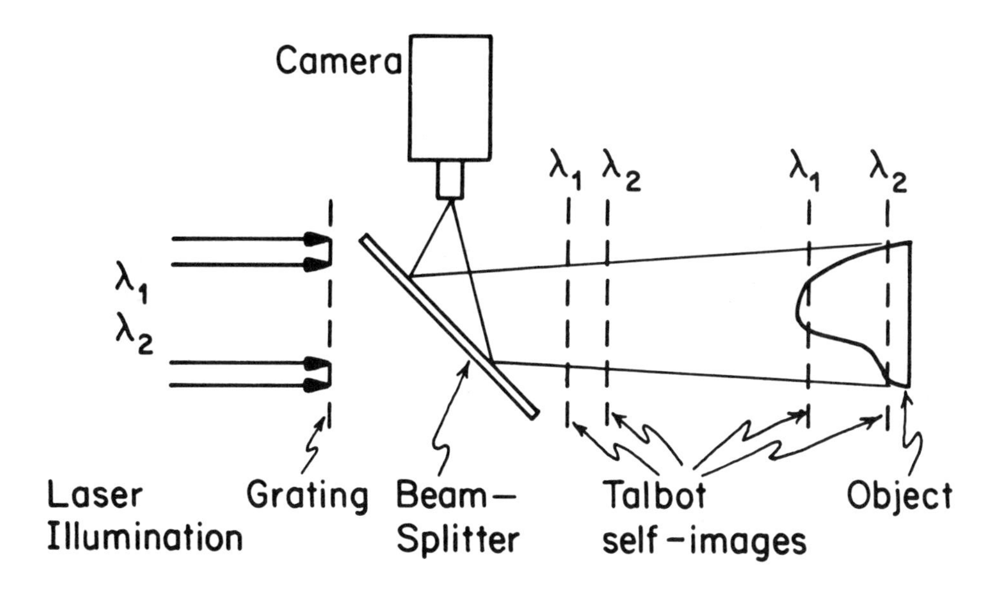

Figure 1. Projecting Talbot self-images on a 3-D object.

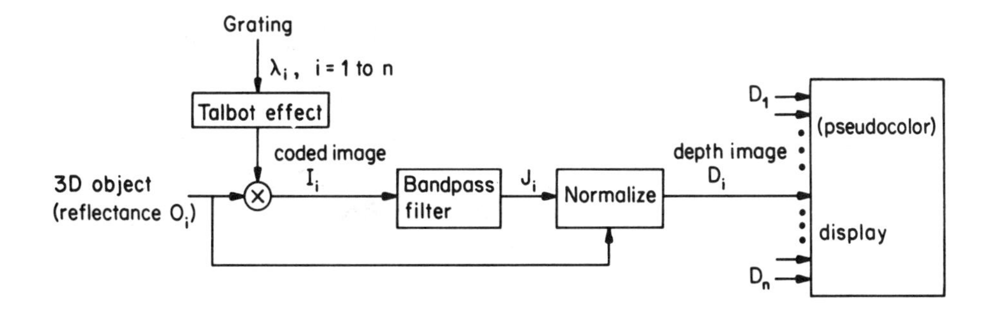

Figure 2. Block diagram of the method.

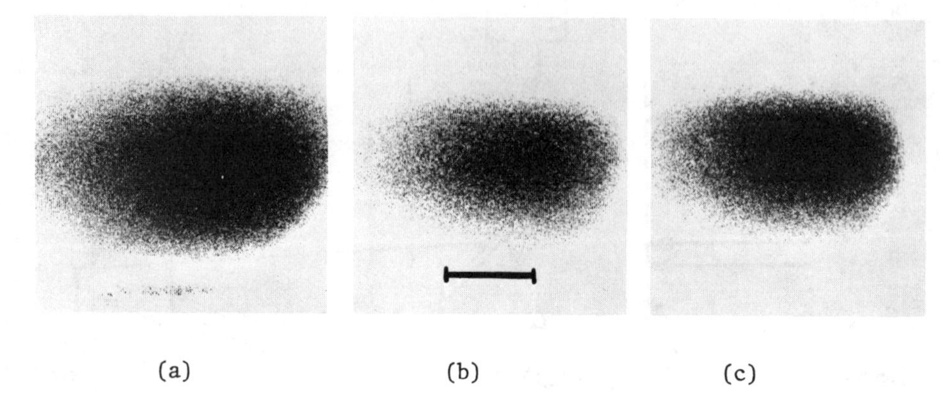

(a) (b) (c)

Figure 3. Coded images of a cylinder inclined with respect to the optical axis. a) λ = 515 nm, b) λ = 488 nm, c) λ = 458 nm. The length of the segment below b) is equal to 50 Talbot fringe spacings.

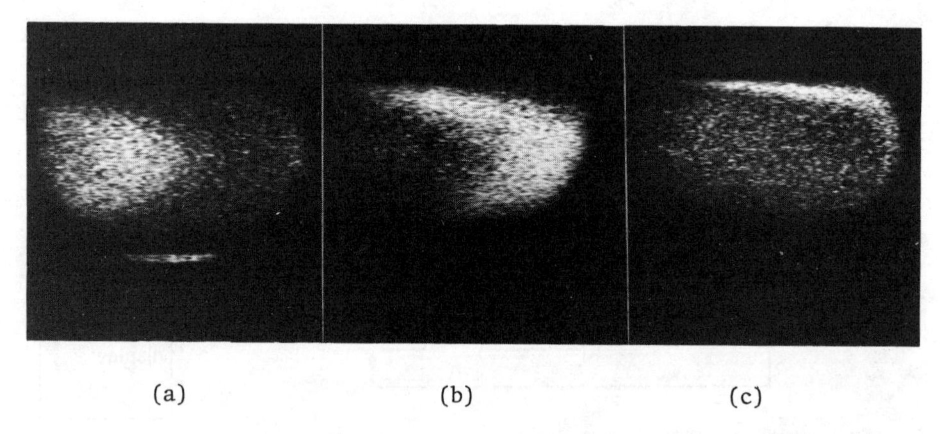

(a) (b) (c)

Figure 4. Processed images obtained after band-pass filtering and normalization of the coded images shown in Fig. 3.

HOLOGRAPHIC METHODS IN OPTICAL TESTING

D. Malacara
Centro de Investigaciones en Optica, A.C.
Apartado Postal 948
León, Gto; México

ABSTRACT

Holographic methods are very useful for testing optical systems, specially those producing aspherical wavefronts. Some testing procedures using holograms with this purpose, are presented.

A holographic common path as well as radial shear interferometers are described.

Also, a lateral shear interferometer and an improvement of the normal Ronchi test using holograms, called side band Ronchi test, are shown.

INTRODUCTION

Holographic methods can be used with advantage to test optical systems (Wyant 1978).

The wavefront under test is stored in a hologram for future reconstruction when testing the wavefront produced by an optical system.

There are two basic types of holograms. A real hologram, when an actual wavefront is used to record a photographic plate, and a synthetic hologram obtained by computer interference fringes.

Wyant (1978) has described a holographic Twyman-Green interferometer, as shown in Fig 1. The imaging lens and the spatial filter shown are used in the reconstruction process.

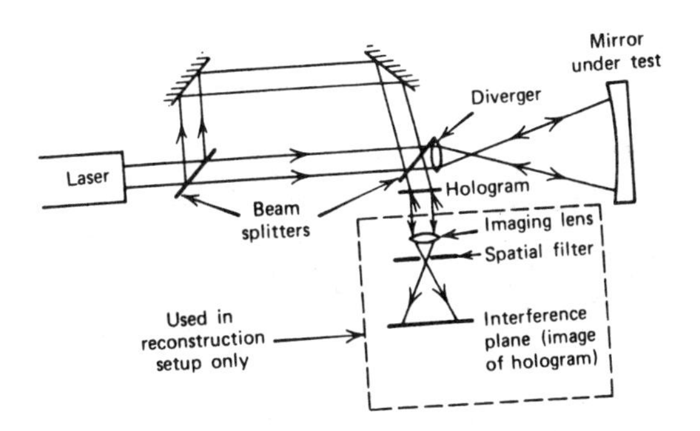

Figure 1. Modified Twyman-Green interferometer used for holographic testing.(From Wyant 1978)

ISSN:0094-243X/81/650437-13$1.50

In a general manner, any wavefront could be stored and later analysed without having to use the optical system, as in Fig 2.

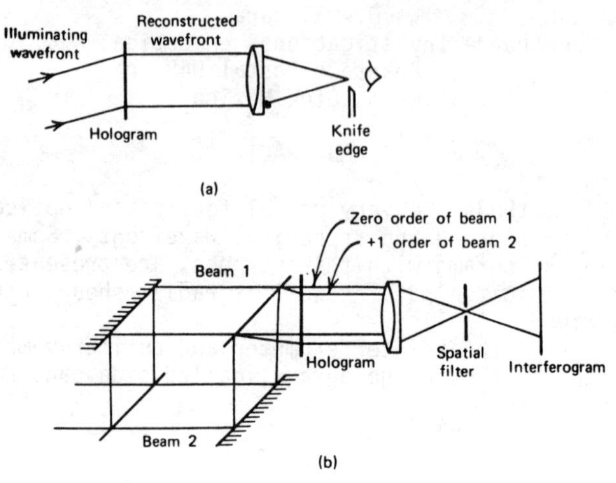

Figure 2. Hologram "reconstruction setup" for analysis of stored wavefront. (a)Foucault knife-edge test. (b) Interferometric test.(From Wyant 1978)

Fig 3 shows some applications of holographic interferometry.

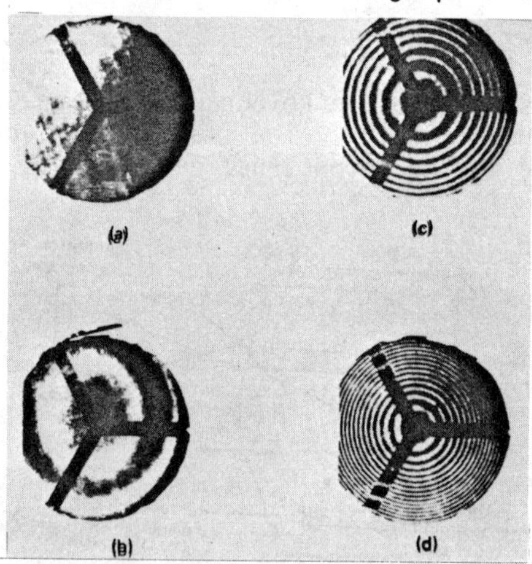

Figure 3. Interferograms from the thermal test of a 70 cm fused silica egg-crate mirror blank.(a)Taken just before initiation of heating.(b)Taken 6 min after start of test. (c)Taken 10 min after start of test.(d)Taken 20 min after start of test. (From Van Deelen and Nisenson 1969.)

COMMON PATH HOLOGRAPHIC INTERFEROMETER

A common path type of holographic interferometer was described by Broder and Malacara (1975). The optical set up used is shown in Fig 4. Let us now consider a photographic plate at the Fresnel zone plate is formed. If the mirror is perfect, the image is another Fresnel zone plate, but this is not true if the mirror is defective. Then the photographic plate may be considered as a hologram.

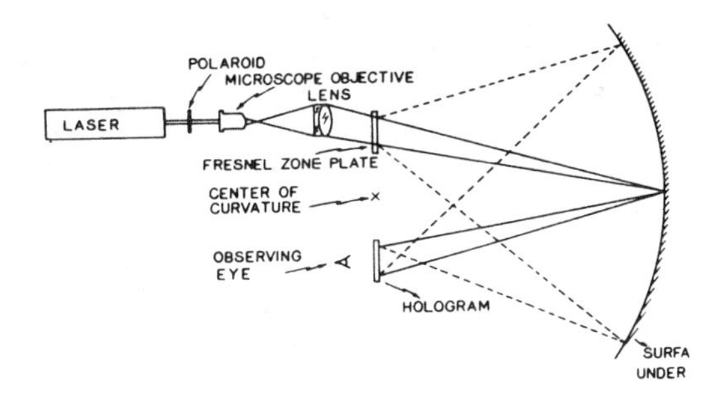

Figure 4. Interferometer layout. (From Broder and Malacara 1975)

Several diffracted orders are produced by a Fresnel zone plate, as shown in Fig 5

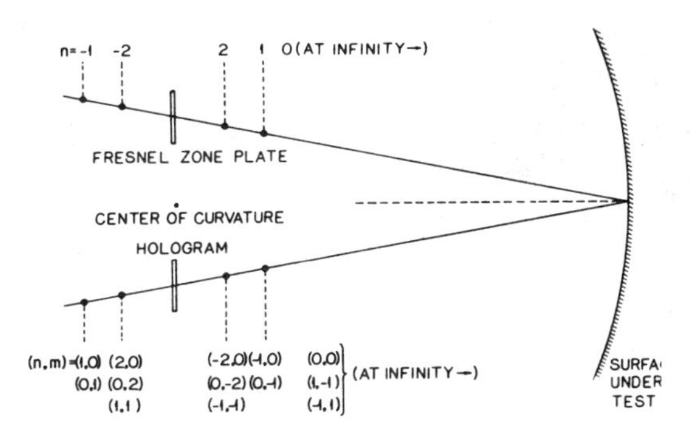

Figure 5. Model schematics for the interferometer. (From Broder and Malacara 1975)

In turn, they produce several interference patterns, as shown in Table 1

|n| + |m|	n + m	n	m	ASPHERICITY PRODUCED BY	EFFECT PRODUCED
0	0	0	0		CENTRAL SPOT
1	1	1	0	MIRROR	NULL INTERFERENCE PATTERN
		0	1	HOLOGRAM	
	-1	-1	0	MIRROR	NON-NULL INTERFERENCE PATTERN
		0	-1	HOLOGRAM	
2	0	1	-1		CENTRAL SPOT
		-1	1		
	2	2	0		FOG REDUCING CONTRAST
	-2	-2	0		
	2	1	1		EXTERNAL HALO WITH TWICE THE MIRROR DIAMETER
		0	2		
	-2	-1	-1		
		0	-2		

Table 1. Wavefronts in the Interferometer.(From Broder and Malacara 1975)

Fig 6 shows the interferogram of an aspheric wavefront in a normal common path interferometer.

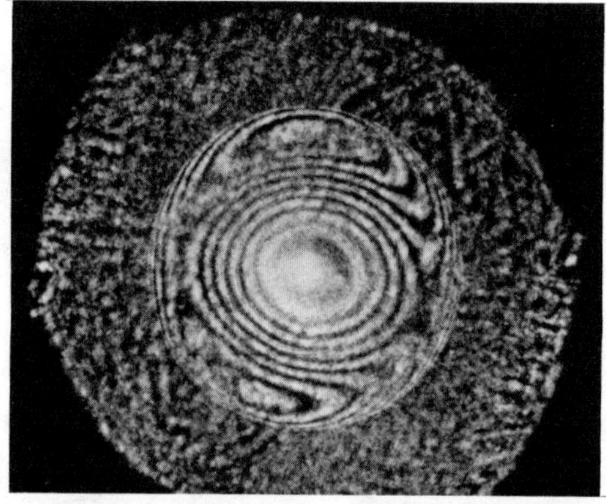

Figure 6. Interferogram of aspherical surface in Murty's Fresnel zone plate interferom eter.

Fig 7 shows an interferogram of the same wavefront in the holographic interferometer.

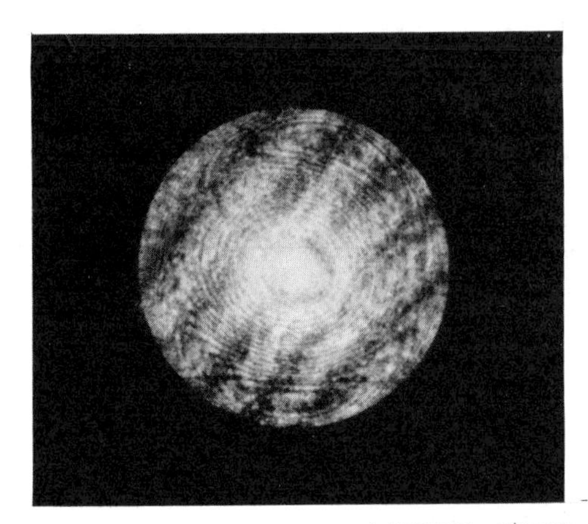

Fig. 7 interferogram in holographic interferometer. (from Broder and Malacara 1975).

LATERAL SHEAR INTERFEROMETER

Another interesting holographic interferometer is the lateral shear interferometer described by Malacara and Mallick (1976). The setup used to make the hologram is shown in fig. 8.

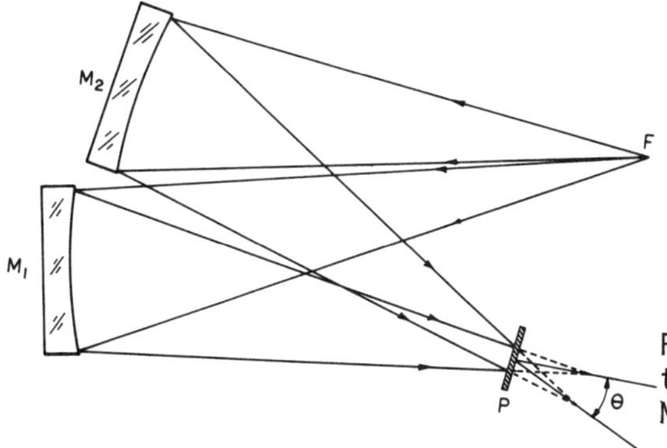

Fig. 8 Setup for making the hologram, (From Mallick and Malacara).

If the reconstruction process is performed as in Fig. 9.

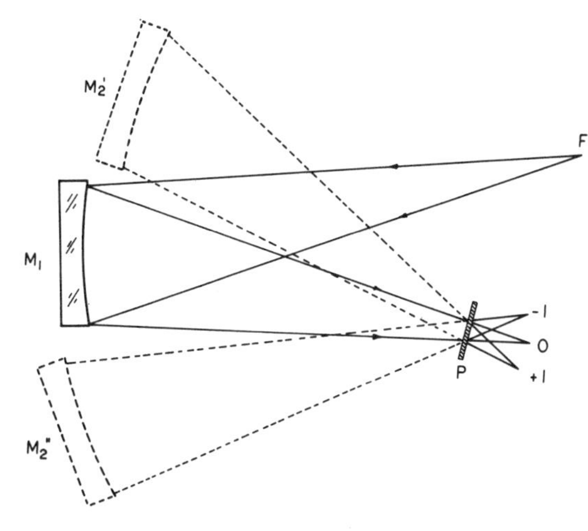

Fig. 9. Hologram reconstruction (From Mallick and Malacara).

two lateral sheared interferograms are obtained as shown in Fig. 10.

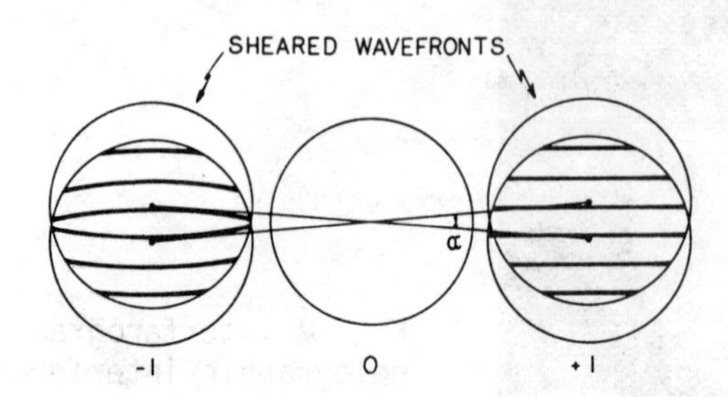

Figure 10 Observed interferometer patterns(From Malacara and Mallick)

The interferometer pattern in the diffracted order + 1 gives straight lines if the aspheric wavefront is the same wavefront used in the reconstruction. The interfering wavefronts travel in the directions explained in Fig. 11

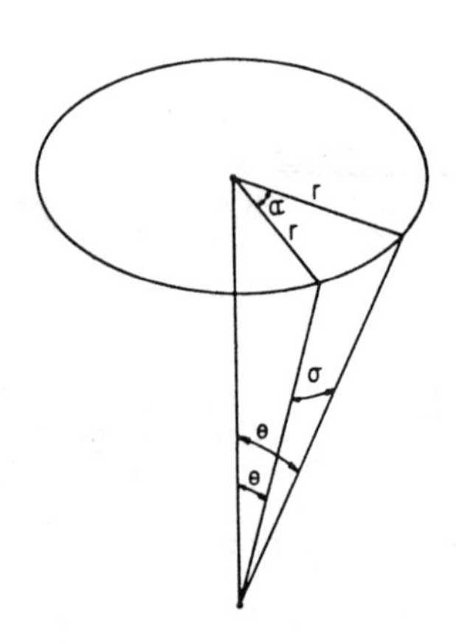

Figure 11 Geometry of interferometer beams (From Malacara and Mallick)

Lateral shear and tilt of the wavefronts are produced by small displacements of the hologram, as shown in Fig 12

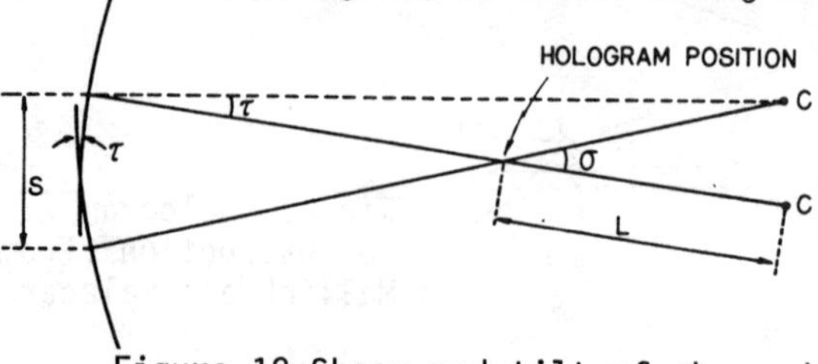

Figure 12 Shear and tilt of sheared wavefronts(From Malacara and Mallick)

The inteferogram obtained with the same aspheric wavefront in the recording and the reconstruction steps are shown in Figs. 13 and 14.

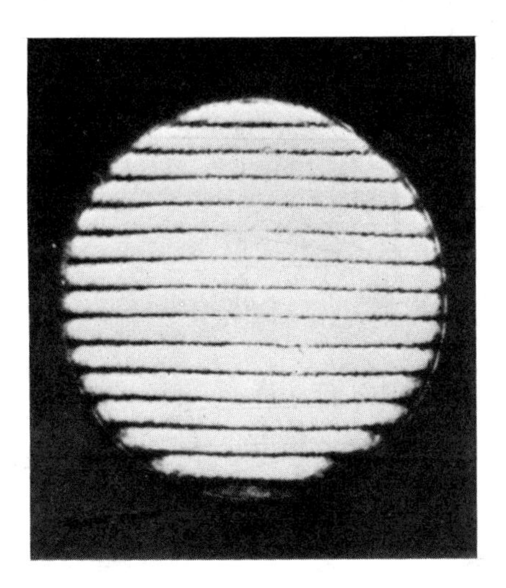

Figure 13. Null pattern for aspherical surface in order + 1 (From Malacara and Mallick)

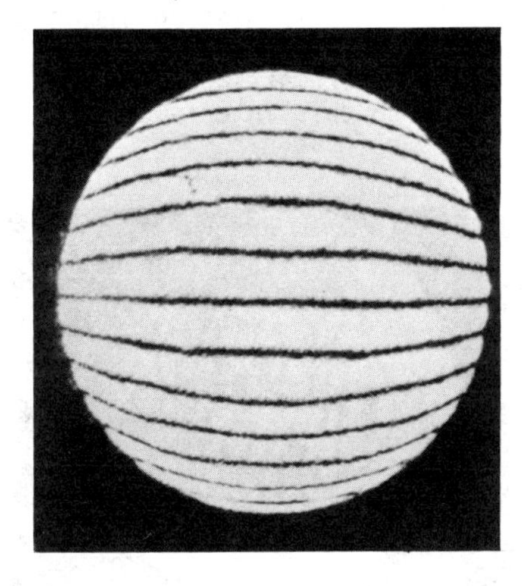

Figure 14. Nonnull pattern for aspherical surface in order -1 (From Malacara and Mallick)

RADIAL SHEAR INTERFEROMETER

A holographic radial shear interferometer was described by Fouéré and Malacara (1974).

The first step is to make a Gabor zone plate as shown in Fig. 15 (A)

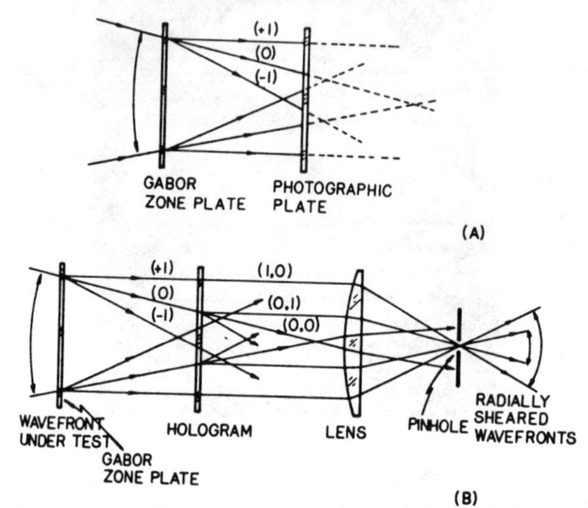

Figure 15 Holographic radial shear interferometer

Then, the Gabor plate is illuminated with the convergent wavefront under test, as shown in Fig 15(B), which clearly explains why two radially sheared wavefronts are obtained. Fig 16 shows an interferogram obtained with this interferometer.

Figure 16 Interferogram obtained in holographic interferometer (From Fouéré and Malacara 1974)

SIDE BAND RONCHI TEST

A holographic Ronchi test, also called side band Ronchi test was devised by Malacara and Cornejo (1976).

The first step is to fabricate a special Ronchi ruling by the procedure illustrated in Fig. 17.

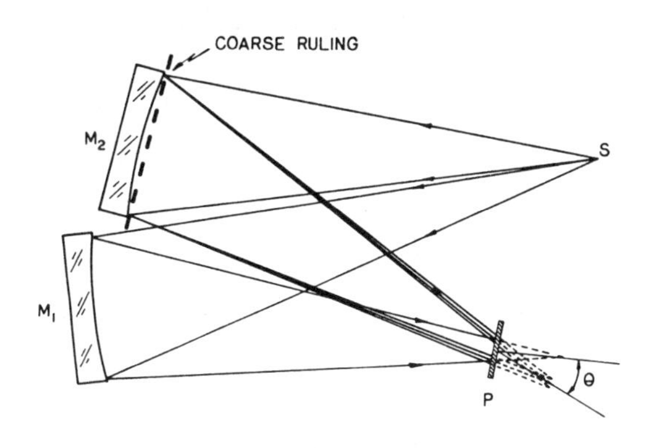

Figure 17 Fabrication of the band Ronchi ruling. (From Malacara and Cornejo, 1976)

The shape of the mirror could be spherical or aspherical.

The method essentially consists of forming a hologram to reconstruct the image of the mirror M_2, including the coarse ruling in front of it, using the beam from mirror M_1 as a reference.

During the reconstruction step (Fig 18) two images M_2' and M''_2 are obtained.

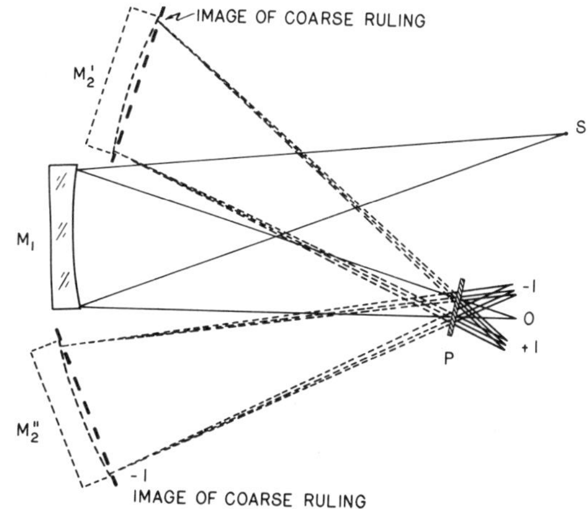

Figure 18 Observing the patterns with a side band Ronchi ruling (From Malacara and Cornejo 1976)

If the mirror M_2 is identical to the one used in the recording step, the fringes in the image M'_2 (-1 order of diffraction) will be straight. (see fig 19)

Figure 19 Null Ronchi pattern with an aspherical surface with first positive order. (From Malacara and Cornejo, 1976)

If the mirror M_2 was aspherical, however, the fringes in the reconstructed image M''_2 will not be straight (see fig 20)

Figure 20 Ronchi pattern for aspherical surface used for Fig 3 with first negative order (From Malacara and Cornejo , 1976)

When the mirror M_2 is aspherical in the recording phase but is spherical in the reconstruction step, the fringes in the image M'_2 are curved. (see fig 21)

Figure 21 Ronchi pattern for spherical surface using null side band Ronchi ruling prepared for the aspherical surface of fig. 3 (first positive order) (From Malacara and Mallick)

INTERFEROMETERS USING SYNTHETIC HOLOGRAMS

When a masteroptics is not available for making a real hologram, a computer generated synthetic hologram can be made (Mac Govern and Wyant, 1971). The resulting hologram, as shown in Fig. 22 is a representation of the interferogram that would be obtained if the ideal aspheric waverfront interfered with a tilted plane wavefront.

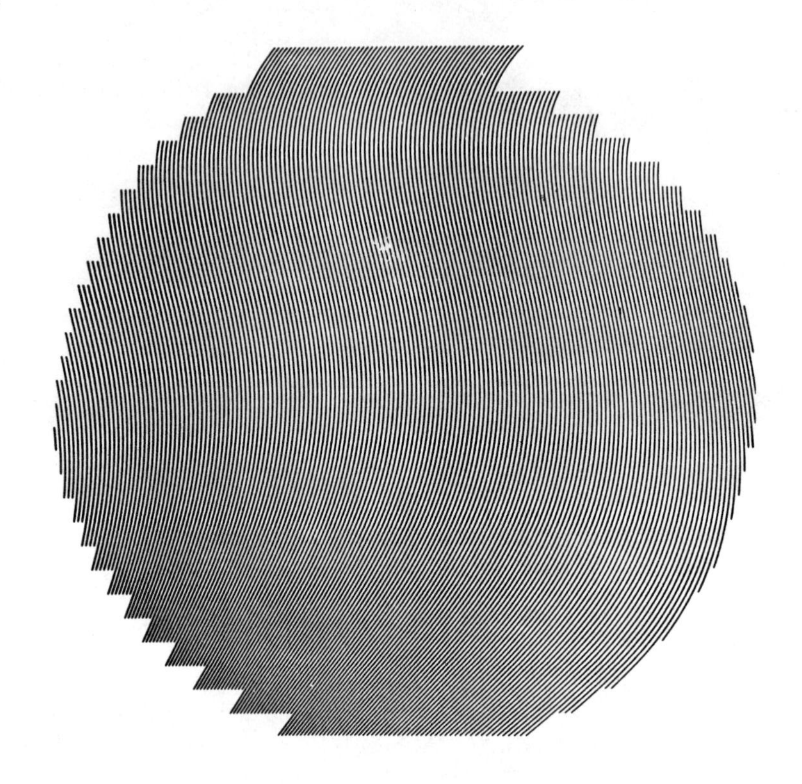

Figure 22 Master CGH plotted by a laser beam recorder

Figure 23 illustrates the result of testing an aspheric wavefront using a synthetic hologram.

Figure 23 CGH test of aspheric wavefront having a maximun slope of 126 waves per radius and 64 waves departure. (From Wyant and Bennett 1972)

REFERENCES

Broder-Bursztyn. F., and D. Malacara Hernández, "Holographic Interferometer to Test Optical Surfaces", Appl. Opt., 14 2280 (1975)

Fouéré, J.C., and D. Malacara "Holographic Radial Shear Interfero meter," Appl. Opt., 13, 2035 (1974)

Mac Govern, A.J., and J.C., Wyant. "Computer Generated Holograms For Tes-ing Optical Elements," Appl. Opt., 10, 619 (1971)

Malacara, D., and M. Josse. "Testing Of Aspherical Lenses Using Side Band Ronchi Test." Appl . Opt. 17, 17 (1978)

Malacara, D., and S. Mallick. "Holographic Lateral Shear Interferometer". Appl. Opt., 15 2695 (1976).

Malacara, D., and M. Josse. "Testing Of Aspherical Lenses Using Side Band Ronchi Test." Appl. Opt. 17, 17 (1978).

Wyant, J.C., "Holographic and moire techniques." Chap. 12 In Optical Shop Testing, Ed. D. Malacara, Wiley Interscience. (1978)

Wyant, J. C., and V. P. Bennett "Using Computer Generated Holograms To Test Aspheric Wavefronts," Appl. Opt. 11,2833 (1972).

HOLOGRAPHY ON THE NASA SPACE SHUTTLE

R.F.Wuerker, L.O.Heflinger, J.V.Flannery, A.Kassel, & A.M.Rollauer
TRW/DSSG Redondo Beach, CA 90278

ABSTRACT

The SL-3 flight on the NASA Space Shuttle will carry a 25 mW He-Ne laser holographic recorder for recording the solution growth of Triglycine Sulfate (TGS) crystals under low-zero gravity conditions. Three hundred holograms (two orthogonal views) will be taken (on SO-253 film) of each (∿3) growth experiment. Processing and analysis (i.e., reconstructed imagery, holographic Schlieren, reverse reference beam microscopy, and stored beam interferometry) of the holographic records will be done at NASA/MSFC. Other uses of the recorder on the Shuttle have been proposed.

INTRODUCTION

A wide variety of near-zero gravity material processing experiments will be conducted on the NASA Space Shuttle, shown in Figure 1. One experiment will be the growth of Triglycine Sulfate $(NH_2CH_2COOH)_3 \cdot H_2SO_4$ or TGS ferroelectric crystals from solution. In orbit a nonbuoyant environment will be realized, resulting in an extended symmetrical diffusion boundary around the crystal in contrast to the usual gravitational convection pattern around crystals grown on earth. Holography was chosen to record this unique crystal growth phenomenon. One hundred and fifty holograms of two different orthogonal views of each growth attempt period (∿24 hours in duration) will be recorded. The exposed recordings will be brought back to earth for developing and detailed reconstruction and analysis. The holographic recorder will include, in addition, a conventional Schlieren system for monitoring in real time the crystal and its diffusion boundary. The Schlieren image will be displayed on television monitors both in the spacecraft and on the ground (via data link), for use by the on board attendants (payload specialist) and the experiment's principal investigators, R. B. Lal, Alabama A&M, and R. L. Kroes, MSFC).

Figure 1. The NASA Space Shuttle

Work performed under NASA Contract NAS-8-33300 for NASA/MSFC

ISSN:0094-243X/81/650450-10$1.50

The recorder must survive launch, must be easy to operate and reliable, not require tedious alignments, not be dangerous, require minimal power and cooling, and must not subject the crystal growth environment to forces in excess of 10^{-5}g. These requirements lead to the choice of a single He-Ne laser as illuminator of both the holographic recorders and the Schlieren monitor, to the choice of 70mm wide SO-253 ESTAR base roll film as the holographic recording media, and no chemical processing of the exposed rolls during flight.

THE LASER ILLUMINATOR

A He-Ne laser was chosen as illuminator for reasons of reliability, low electrical requirement, and ready availability as precision (no change in wavelength) holographic reconstructor. The largest He-Ne laser that could fit in the allocated space (1.28 m high, 0.96 m wide, 0.48 m deep) was selected namely, a Spectra Physics Model 107A with nominal output at 0.6328μ of $\gtrsim$ 25 mW (TEM_{00}), 13 centimeters coherence length, and 250 VA electrical input. Tests of a factory model on a shaker table showed that with only minor modification this commercial laser will survive launch.

THE CRYSTAL GROWTH CELL

Each TGS crystal will be grown in separate 10 x 10 x 10 cm (inside dimensions) sealed temperature controlled (45°C) cells with front, rear, and side electrically heated double pane insulated windows which act as thermal guards.* Figure 2 shows a top view of a growth cell. Three or four of these cells will be stored on the spacecraft. Once in orbit, a cell will be transferred by the payload specialist to a preheating oven where it will be warmed to 70°C. Afterwards it will be placed in the optical recorder. The temperature of the solution will be reduced (via a process computer), the seed crystal will be exposed to the solution, and the growth recording and monitoring cycles will begin.

The large (10cm clear aperture) rear window on the cell is for illumination of the interior by the laser. The opposing front window views the crystal in silhouette. It is used by one of the holographic recorders and the Schlieren monitor. The side window is for the second holographic recorder, which takes only reflected or scattered light holograms of the growing crystal.

THE OPTICAL APPARATUS

The optical design is shown schematically in Figure 3. The components of the two holographic recorders and Schlieren monitor and the crystal growth cell are all mounted on one side of an invar honeycomb plate. The laser illuminator is mounted on the other side.

---

*The crystal seed is attached to a sting which is temperature programmed.

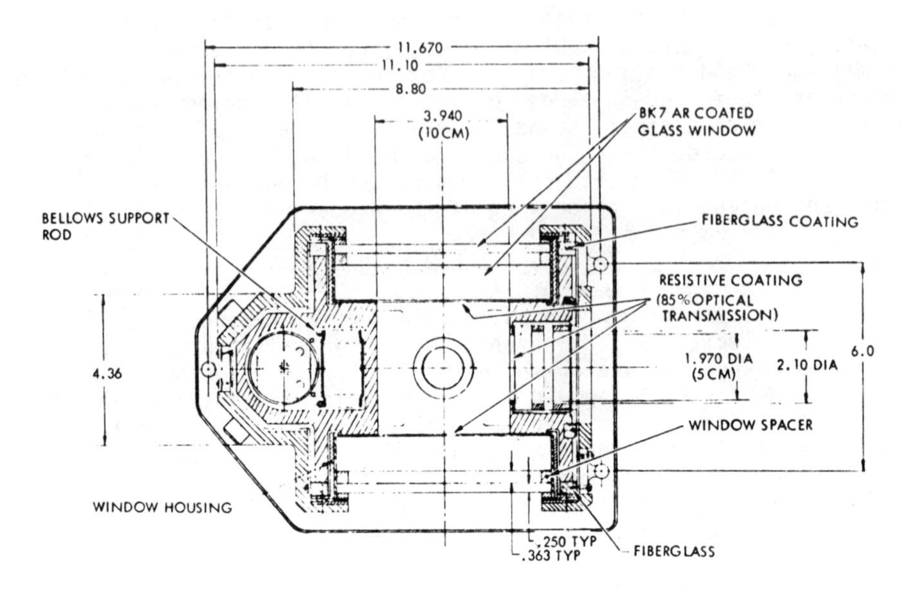

Figure 2. Top View of the Crystal Growth Cell

The whole assembly fits in a standard Space Lab double rack along with storage enclosures, the preheating oven enclosure, and control electronics.

Referring to Figure 3 it can be seen that the output beam from the laser is turned by a pair of small mirrors, passes through a microscope objective (L1), is reflected by another pair of periscoping mirrors, and is collimated to 15cm diameter by lens L2. Mirror MA directs the expanded beam onto a beam splitter BS1 (35%R) which generates the holographic reference beam for the transmission recorder. This beam is reflected by mirrors M1 and M3 through BS3, which is specially coated to be 90% transmitting at normal incidence. Laser light transmitted through BS1 is partially reflected BS2 (86%R) and is incident upon the cell's rear window. Its intensity is reduced by the two (15%R each) electrical heating coatings on the inner windows. Laser light transmitted through the cell is reflected by BS3 (86%T at 50° angle of incidence) into the Schlieren monitor.* The remaining light is incident on the film in the cassette as the holographic scene beam. At the film, the scene reference intensity ratio is 1/5, optimum for SO-253 emulsion. Exposure time is approximately 1/30 second.

---

*Light reflected from BS3 is focused by collimator L4 onto a variable density glass plate which converts a 100:1 refractive gradient dynamic range into a corresponding intensity change. Light refracted by the crystal's diffusion boundary passes around the variable optical density plate and is imaged by a camera lens onto a 525 line vidicon, which relays the image to television monitors in the craft and on the ground.

Laser light scattered by the immersed crystal passes through the cell's side window (after attenuation the 15%R heating coating) and is incident upon the second holographic film as its scene beam. The reference beam for this hologram comes from light transmitted through BS2 and reflected by partial mirror M1A (a piece of optically polished black glass of 10% surface reflectivity) and mirror M1B. This reference beam intensity is purposefully reduced to compensate for the inherently weak scattering of laser light by the immersed crystal (∿ 0.17% facet reflectivity), resulting in scene-reference ratios at the film of 1/2 at best, but more typically 1/50. This holographic recording mode depends on the seldom appreciated sensitivity (and linearity) of holography to record weak scene beams. Exposures of 1/2 — 1 second will be needed with the chosen SO-253 film, meaning that the whole apparatus had to be carefully designed against any thermal and vibrational changes in the optical positions

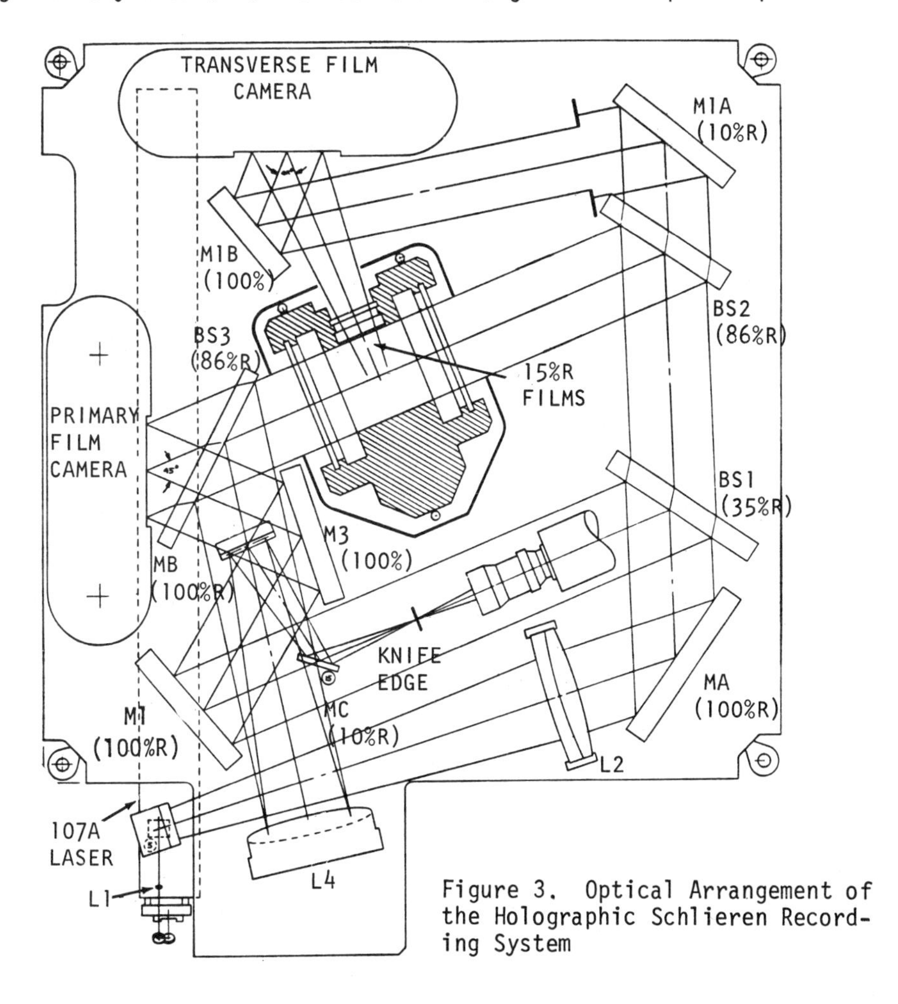

Figure 3. Optical Arrangement of the Holographic Schlieren Recording System

of the mirrors and beam splitters.* Scattered light holograms of water immersed crystal have been successfully recorded under conditions that simulate the flight system.

Both holographic recording modes are temporally matched, that is to say, from their respective beam splitters the scene and reference beam optical path lengths are equal. As a result all requirements on laser temporal coherence are minimized.

The transmitted holographic recording mode is additionally spatially matched, which means that a ray from the initial collimated illuminating beam recombines with its scene reference beam component at the hologram. As a result, a spatial filter or pinhole, at the focus of the microscope objective (L1) is not essential, eliminating a delicate (and possibly traumatic) in orbit alignment. Use of a pinhole is optional. Without it, the holograms will not be as clear. Tests have shown that removal of the pinhole in a spatially matched system, does not degrade the resolution of the reconstruction ($\lesssim$ 50%).

A ground glass diffuser can be slid into the illuminating laser beam (before the cell's entrance window) for recording diffuse rear illumination transmission holograms with the primary recorder. Such holograms will give three dimensional silhouette parallax images of the growing crystal, while double exposure recordings will give differential interferograms of the diffusion boundary which three dimensionally localize the regions of constant optical path change (another one of holography's unique features).

Both holographic recording modes have scene reference beam angles of 45° consistent with the resolution of SO-253 film.

The two film planes are symmetrically oriented relative to their scene reference beam directions. This optimizes the resolution of the reconstructed images.[1] Reference beam light is, however, reflected from the emulsion back along the scene beam, and is returned to the hologram by the 15%R thermal window coatings. Finite coherence of the laser light, however, reduces the presence of this back reflect light in the reconstructions. The Schlieren beam splitter BS3 acts to further suppress the reference beam light reflected from the emulsion in the primary camera.

Both holographic films are placed as close to their respective viewing windows as possible to maximize the viewing angles and resolutions of their reconstructions.

THE RECORDING FILM

From the onset, the program had a requirement that the reconstructed holographic images should give (two bar) resolutions of 20μ for the scattered or reflected light recording mode. In

---

*Precision control of the interior of the growth cell requires window cooling. Air will be drawn from the recorder's environment through ports opposite the viewing windows. Tests have shown that such flows do not introduce any fluctuation in optical path over time periods of several seconds.

addition, it should be possible to interferometrically compare the collimated light reconstructions against a collimated comparison beam for purposes of interferometrically measuring the crystal's diffusion boundary, as well as analyzing the reconstructed wavefront by Schlieren techniques. Later, flexible SO-253 film was chosen as the holographic recording medium for reasons of automatic transport, storage, and safety. Thus, one of the program challenges was to make flexible holograms perform like glass plate holograms. Such performance has essentially been realized by sucking the film onto an optically flat surface oriented symmetrically to the scene reference beam direction. After exposure the film can be removed for chemical processing, stored and reflattened onto another transparent optical flat whenever it is reconstructed. Flat suction substrates were chosen because they are easiest to reproduce. Initially, optical flats with only a peripheral (0.4mm wide) vacuum groove were used. These gave satisfactory results, namely, collimated light holograms with reconstructed image resolutions (two bar) of 8μ at 24cm object hologram distances and reflected scattered light holograms with 25 micron resolutions.[2] It was observed, however, that plattens with a single peripheral vacuum channel did not give reliable adhesion in the center due to mm sized trapped air pockets. An improved configuration has been adopted. The optically flat platten is now inscribed with a grid pattern of 100 micron grooves spaced 1 cm apart. These micro channels are connected to the wider peripheral vacuum channel. The small grooves hardly bother the reconstructions, and are not even seen when the holographic image is magnified. Flattening of the recording film on the substrate by suction is still a statistical phenomenon. Small differences in film contour between the recording and reconstruction steps (due to dust and incipient air pockets) act to degrade the fidelity of the reconstructed wavefronts and resolution of the images. To minimize these contour effects the platten is oriented symmetrically to the scene reference beam direction. With this orientation, departures from film flatness have a minimal effect on image quality.[1] Experiments have verified these predictions, showing a reduction in resolution by a factor of two when vacuum platten SO-253 holograms are misaligned from the symmetrical position by only $\pm$ 4°.

The exposed holographic rolls will not be processed until after the end of the mission. Flight durations of several weeks are anticipated. Preliminary tests with exposed films stored in air at ambient temperatures have shown that exposures must be at least doubled (16ergs/cm^2) to maintain desired short development times (2 minutes in stock D-19). Storage at reduced temperatures ($\sim$5°C) has not improved the sensitivity. Use of inert atmospheres or chemical treatments are presently being investigated. Other important considerations will be protection of films on the spacecraft against fogging by solar x-radiation.

RECONSTRUCTION AND ANALYSIS

With the return of the Shuttle to earth, the holograms (after chemical development) will be available for viewing and analysis. This will be done at MSFC. Figure 4 shows a reconstruction apparatus. It has been included as a means of describing how the holograms can be reconstructed and used. The beam from another He-Ne laser (essentially the same wavelength as the one in the Shuttle) would be split by a variable beam splitter into 2 beams, which after passing through individual shutters, would be collimated by separate telescopes. One collimated beam would be directed at the hologram holder along the original reference beam direction. This would be the reconstructing beam. The other collimated beam is directed along the original collimated scene beam direction for the primary camera. This would be the comparison beam which would be used only for interferometric measurements of the collimated light holograms. The hologram holder would be a transparent vacuum platten essentially identical to one used in the Shuttle recorder (ideally the original should be used in the reconstructor). Like the original, it would be precisely oriented to bisect the reconstruction and comparison beam directions. Both types of Shuttle holograms can be reconstructed in the Figure 4 arrangement since both have the same scene reference beam angles and directions.

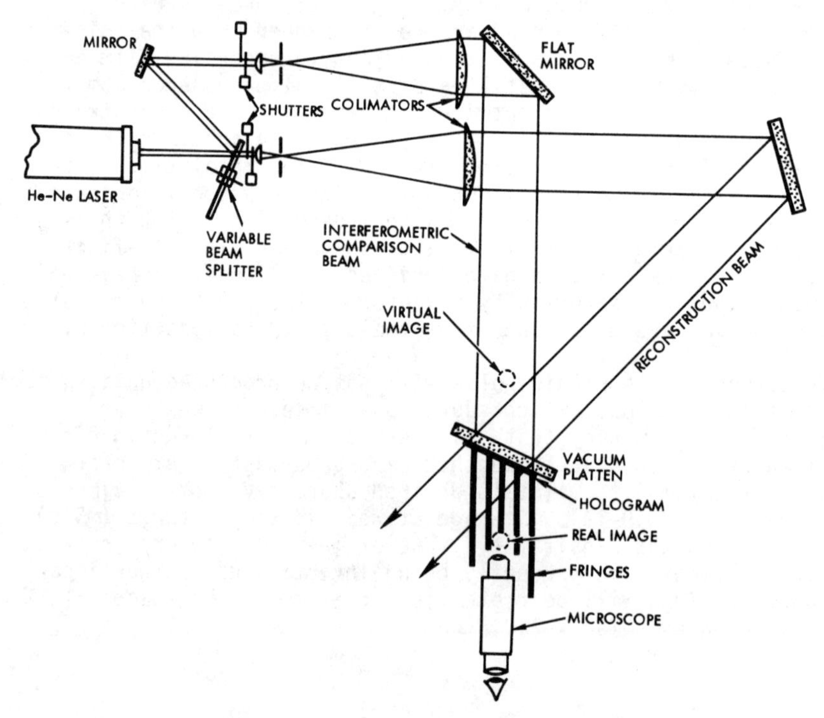

Figure 4. A Proposed Reconstructor

When a hologram is sucked against the platten, with the emulsion out and facing the incident light as it did during recording, the reference beam reconstructs an image which appears as a virtual image behind the hologram in exactly the same position as the original crystal image. The comparison beam is blocked. The reconstruction intensity is maximized to produce the brightest reconstruction (using the variable beam splitter in the Figure 4 apparatus). The virtual image can be viewed with the unaided eyes (in the case of the scattered light and rear diffuse transmission holograms) as a normal three dimensional image. Both recording modes have a 30° field of view. The images can also be examined with simple magnifiers, telescopes, or recorded with a conventional photographic camera. When the holograms are reconstructed in this manner, the wavefront which fell on the hologram at the time it was recorded is recreated. The collimated transmission holograms can be analyzed by Schlieren methods or compared interferometrically against the collimated beam provided by the reconstructing apparatus. For Schlieren viewing of the diffusion boundary, a collimator is set up behind the vacuum platten with a variable optical density glass plate at its focus, to stop the non-refracted portion of the reconstructed wavefront. The holographic Schlieren images should be similar to those recorded during flight with the on board monitor. Interferometry will be used to quantitatively measure the diffusion boundary. The reconstructed wavefront is compared in the Figure 4 apparatus against the comparison beam which is adjusted (by the variable beam splitter) to maximize fringe contrast. The interference fringes are due to differences in the two wavefronts, with neighboring fringes indicating a change in optical path length (i.e., the integral of the product of change in refractive index and physical path through the cell) of one wavelength of laser light. One counts fringes from chamber wall to the image of the surface of the crystal, to measure the diffusion boundary. In the terminology of holography, one would be doing a version of "stored beam holographic interferometry;" comparing a reconstructed wavefront against a plane wavefront. Such analysis is routine for holograms recorded on glass plates. Under this program we have learned that it can be done with flexible films when they are sucked onto a vacuum platten.

More interesting scientifically will be observations on the holographic images at high magnifications. One does not use the virtual image due to the fact that long (24 cm) working distance microscopes cannot be made. Instead, the hologram itself becomes the determiner of image resolution and a conventional short working distance microscope is used to explore the reconstructed image. The hologram is reconstructed by the reverse reference beam technique. In practice, the whole platten is rotated (in the plane of Figure 4) through 180°, from its position during recording, to the position shown in Figure 4. The emulsion now faces away from the reconstruction beam, which passes through the hologram counter to its original recording direction. The hologram now reconstructs the "true" image which is formed by the original wavefront propagating backwards. In contrast to the usual virtual image, this wavefront reconstructs a real aerial image which is accessible for viewing with a conventional microscope.* This operation requires a precisely collimated wavefront reference in the recording step, otherwise the reconstructed wavefront is distorted leading to poorer resolution and extraneous fringes when the reconstructed wave is compared interferometrically. When interferometric comparisons are done, fringes of 8μ spacing can be resolved.

Double exposed holograms made with the ground glass diffuser will show interference fringes localized in space. These will be seen in both the virtual and real image reconstructions. For the latter, the fringes can be viewed with a microscope, using the microscopes narrow focal plane to locate their position in space.

We have even found that for the real image reconstruction, the whole platten does not have to be reversed. One merely needs to flip the film hologram around, laying the emulsion against the vacuum surface. The platten does not have to be realigned. However one runs the risk of scratching the emulsion which may be too risky for our one-of-a-kind Shuttle holograms. Moreover, the reconstruct image is in principle aberrated by the platten.

Once the platten is tuned to maximum hologram resolution, the holograms can (after fine tuning) be copied. In copying a hologram, one records a hologram of a hologram. Several methods have been developed using either virtual or reverse reference beam methods. The copies could be glass plate holograms, which could be disseminated and reconstructed without the need for a vacuum platten (using an apparatus similar to Figure 4).

Inspection and analysis of the 1000 or so hologram recorded on a single Shuttle crystal growth flight will at best be an arduous task, which could require several man years of work per flight. Holography always seems to complicate the problems of analysis since more information is recorded than by any other optical schemes. One can view the images of the growing crystal superficially or in great detail and sophistification. For these reasons holography was chosen to make the archival records of crystal growth under zero gravity conditions.

OTHER APPLICATIONS

Though the holographic recorder was designed specifically to record the TGS crystal growth experiment on the Space Shuttle, it is also being considered for recording of fluid convection phenomena. In the latter experiment the trajectories of neutrally buoyant microballoons in a water filled cell, with hot and cold side walls, would be recorded by multiple exposure holograms.[2] Fluid velocity throughout the cell is determined by measurement in a reconstruction of different particle displacements, dividing the measured distances by the known exposure intervals. Convecting

---

*Spherical aberrations due to beam splitter BS3 and the near cell windows and fluid can be removed by passing the conjugate wave back through glass blocks in the original beam splitter-window positions.

particles in water can be easily recorded with the designed recorder as verified by tests. Such an application capitalizes on the ability of holography to record small distributed unpredictable phenomena.

SUMMARY

A holographic recorder using flexible 70 mm wide SO-253 film has been designed for the NASA Space Shuttle for recording the growth of Triglycine Sulfate crystals under near zero gravity conditions. The apparatus uses a reliable 25 mW helium neon laser, has a single beam expanding collimator, and uses only flat mirrors and beam splitters. It records two separate orthogonal holograms. Spatial filtering is not essential due to the fact that the collimated transmission mode is spatially matched. Without the spatial filter, precision collimation of the illuminating beam is the only in-orbit adjustment anticipated. The reconstructions can be viewed microscopically to $\lesssim$ 20 micron resolutions. The collimated light transmission holograms can be interferometrically compared against an earth generated plane wavefront.

ACKNOWLEDGEMENTS

B. P. Hildebrand and J. D. Trolinger of Spectron Development Laboratories contributed significantly to the perfection of the recorder's design.[1] Other contributors were W. J. Skinner and R. A. Briones of TRW.

REFERENCES

1. B. P. Hildebrand and J. D. Trolinger, these proceedings.
2. R. F. Wuerker, et al, SPIE, Recent Advances in Holography, 215, 76-84, 1980.

A STATISTICAL ANALYSIS OF A HOLOGRAPHIC SYSTEM INTENDED FOR THE SPACE SHUTTLE*

B. P. Hildebrand and J. D. Trolinger
Spectron Development Laboratories, Inc.
3303 Harbor Blvd., Suite G-3
Costa Mesa, California 92626

ABSTRACT

This paper describes an analytical study to determine the stability requirements for a holographic system intended for the NASA space shuttle. The primary emphasis is on optimization of the holographic geometry to minimize the effect of film variations between hologram exposure in space, and wavefront reconstruction on the ground. It is shown that it is extremely important to use a balanced geometry (equal reference and object beam angles) and if possible, a specific object beam angle.

INTRODUCTION

One of the experiments to be performed on the space shuttle is crystal growth in a zero gravity environment. Part of the diagnostics package will consist of two holographic cameras looking at the crystal in orthogonal directions. The cameras will make about one exposure per minute for the duration of the experiment on SO253 film drawn across a vacuum platen. The film will be developed and analyzed on the ground after completion of the mission. The data that is to be extracted from the holograms is the index of refraction gradient in the liquid surrounding the crystal, and the light scattered by the crystal. The gradient is to be obtained from the hologram taken in transmission through the cell, and the crystal image from the hologram of light scattered at right angles to the illumination beam.

Because there will be no opportunity to examine the holograms during the experiment, and because flexible film will be used, it is imperative that the system be designed to tolerate minute changes and operate properly without any on-board adjustments. TRW commissioned Spectron Development Laboratories, Inc., to perform a series of analyses to ascertain the optimum configuration of the system and to establish the tolerances on the stability of the components[(1)]. This paper which is based on a part of the analysis, describes a theoretical treatment based on random process theory, which established that the optimum geometry requires the reference and object beams to be directed at the film plane at equal angles from the normal. This solution is completely independent of the usual physical explanation, of fringe orientation in the emulsion, used to justify this arrangement. A second result specifies that the object beam should impinge on the film plane at a specific angle

*This work was sponsored by TRW Defense and Space Systems.

ISSN:0094-243X/81/650460-11$1.50

determined by the F-number of the hologram. This turns out to be a rather small angle, so is usually not practical. However, if both conditions are fulfilled, the error tolerance on the system is remarkably large. A simple physical interpretation justifying these results is also presented.

STATISTICAL ANALYSIS

Since flexible film is to be used in the shuttle holocamera, it is not possible to guarantee that its position and shape will remain unchanged between recording and image extraction. This analysis attempts to evaluate the effect of these parameters on the final image.

Consider Figure 1, a general arrangement for recording the hologram. The light from a point object at position (x_o, z_o) is interfered with a collimated reference beam striking the film at the an-

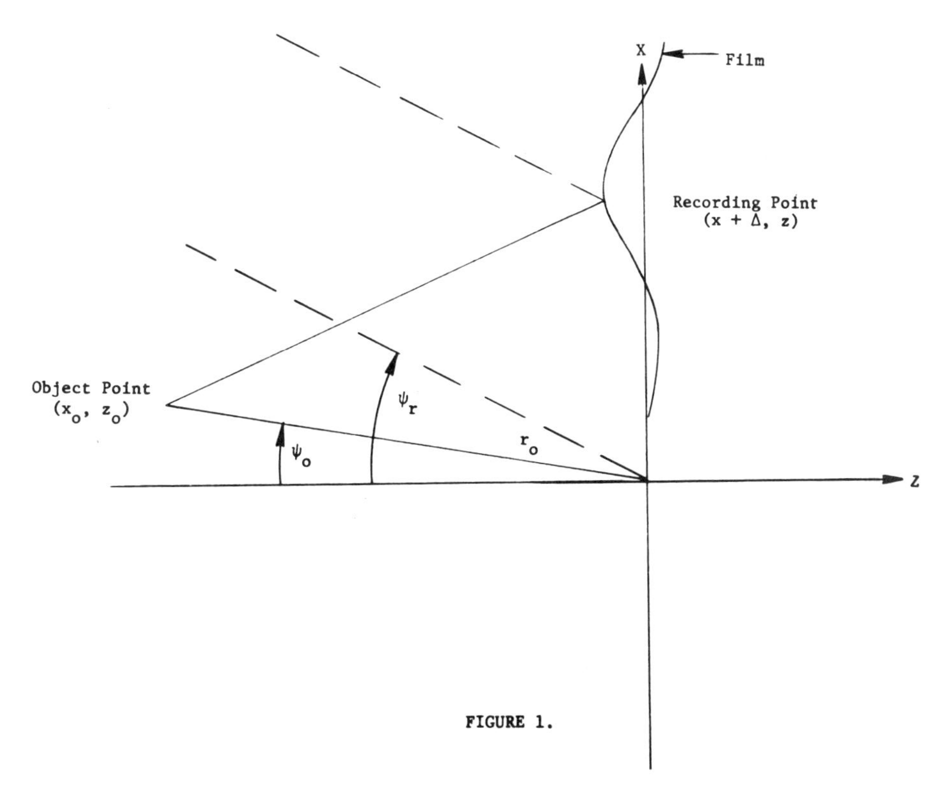

FIGURE 1.

gle ψ_r with respect to the film plane. The exact position of the film is unknown because of various parameters that can change between the recording and image extraction states. Therefore, the recording point is designated by $(x + \Delta, z)$ where Δ and z are random deviations from the z = o plane. Although, the statistics of the random process are not known, we assume them to be Gaussian. That is, the joint probability density is

$$p(\Delta,z) = \left(\frac{1}{2\pi\sigma_\Delta\sigma_z(1-\rho^2)^{1/2}}\right) \exp\left\{-\frac{\sigma_z^2\Delta^2-2\sigma_z\sigma_\Delta\rho\Delta z+\sigma_\Delta^2 z^2}{2\sigma_\Delta^2\sigma_z^2(1-\rho^2)}\right\} \quad (1)$$

where we have assumed zero means and σ_Δ, σ_z are standard deviations of the random variables Δ, z, and ρ is the correlation coefficient.

The problem is to find the mean and standard deviation of the reconstructed wavefront. If we record the hologram and reconstruct it with a beam duplicating the reference beam exactly, we have a wavefront emanating from the hologram with phase distribution

$$\phi(x,\Delta,z) = \left\{\frac{2\pi}{\lambda}(x+\Delta)\sin\psi_r + z\cos\psi_r + [(x+\Delta-x_o)^2 + (z-z_o)^2]^{1/2}\right\}. \quad (2)$$

After the usual Fresnel approximations are applied to the binomial expansion of the square root term, the equation becomes

$$\phi(x,\Delta,z) = \frac{2\pi}{\lambda}\left\{x(\sin\psi_r - \sin\psi_o) + z(\cos\psi_r - \cos\psi_o) + \Delta(\sin\psi_r - \sin\psi_o) + \frac{x^2+\Delta^2+z^2}{2r_o} + \frac{x\Delta}{r_o} + r_o\right\} \quad (3)$$

where $r_o = (x_o^2 + z_o^2)^{1/2}$. The mean of this function becomes

$$m_\phi = E(\phi) = \frac{2\pi}{\lambda}\left\{x(\sin\psi_r - \sin\psi_o) + \frac{x^2}{2r_o} + r_o + \frac{\sigma_\Delta^2+\sigma_z^2}{2r_o}\right\}. \quad (4)$$

We recognize this as being the phase for a wavefront diverging from the point (x_o,z_o). The first term represents the spatial carrier and the second the quadratic focusing term. The random variables have contributed only a constant phase term. Of course, this is the average behavior, when many holograms are made. Individual members of the ensemble of holograms would behave differently. On the average, however, the holograms will behave like perfect ones. The deviation from perfection can be obtained by computing the standard deviation of the phase front. The variance becomes $V(\phi) = E(\phi^2) - E^2(\phi)$. After much algebra we obtain

$$V(\phi) = \frac{4\pi^2}{\lambda^2} \left\{ \frac{\sigma_\Delta^4 + 2\sigma_\Delta^2\sigma_z^2\rho + \sigma_z^4}{2r_o^2} + \sigma_z^2 (\cos\psi_r - \cos\psi_o)^2 + \sigma_\Delta^2 \left(\frac{x}{r_o} + \sin\psi_r - \sin\psi_o\right)^2 + 2\sigma_\Delta\sigma_z\rho \left(\frac{x}{r_o} + \sin\psi_r - \sin\psi_o\right) \cdot (\cos\psi_r - \cos\psi_o) \right\}. \quad (5)$$

An observation that can be made immediately is that the variance can be minimized by arranging the geometry so that

$$\psi_r = -\psi_o = -\sin^{-1}\left(\frac{x}{r_o}\right), \quad (6)$$

yielding the minimum

$$V(\phi)/_{MIN} = \frac{4\pi^2}{\lambda^2} \left(\frac{\sigma_\Delta^4 + 2\sigma_\Delta^2\sigma_z^2\rho + \sigma_z^4}{2r_o^2} \right). \quad (7)$$

Since x is a variable, we can not satisfy Equation 6 for all values. However, by examining Equation 5, we can see that for fixed values of ψ_r and ψ_o, V is largest at the maximum value of x. Setting the hologram size to A, we can set x to A/2 in Equation 5, with the result that Equation 6 becomes

$$\psi_r = -\psi_o = -\sin^{-1}\left(\frac{A}{4r_o}\right) \quad (8)$$

The correlation coefficient ρ, has an important bearing. To see this, examine Equation 5 at the extremes, $\rho = \pm 1$ (complete correlation) and $\rho = 0$. Then we have

$$V(\phi) = \frac{4\pi^2}{\lambda^2} \left\{ \frac{(\sigma_\Delta^2 \pm \sigma_z^2)^2}{2r_o^2} + [\sigma_z (\cos\psi_r - \cos\psi_o) \cdot \pm \sigma_\Delta \left(\frac{A}{2r_o} + \sin\psi_r - \sin\psi_o\right)]^2 \right\} \text{ for } \rho = \pm 1 \text{ and} \quad (9)$$

$$V(\phi) = \frac{4\pi^2}{\lambda^2} \left\{ \frac{\sigma_\Delta^4 + \sigma_z^4}{2r_o^2} + [\sigma_z^2 (\cos\psi_r - \cos\psi_o)^2 + \sigma_\Delta^2 \left(\frac{A}{2r_o} + \sin\psi_r - \sin\psi_o\right)^2] \right\} \text{ for } \rho = 0. \quad (10)$$

These results can be related to resolution by considering the hologram as a lens. The perfect hologram produces a wavefront converging to an image point. Thus, the wavefront is a sphere, and the resolution will be determined by the aperture alone. One convenient measure of resolution is the Rayleigh criterion. Another is the Strehl definition.

Rayleigh Criterion - the distance between two point objects that can just be resolved in the image. For circular apertures this occurs when the Airy diffraction spots in the image are separated so that the peak of one falls on the first minimum of the other. This separation is

$$\delta_R = 1.22\ \lambda F$$

$$\text{where } F = \frac{r_o}{A \cos \psi_o}$$

where A = diameter of the hologram.

The other part of Rayleigh's criterion is not often stated, but is as follows. The image resolution will not deteriorate significantly if the maximum wavefront deformation from the spherical shape does not exceed $\lambda/4$. In other words, we cannot allow phase errors greater than $\pi/2$ over the extent of the aperture.

Strehl Ratio - ratio of the light intensity at the maximum of the Airy disk to that of a perfect diffraction limited system. This definition was investigated by Marechal, who determined that there was a direct relation between the rms wavefront deformation and the peak intensity of the Airy disk. Demanding that the Strehl ratio exceeds 0.8 requires that the rms wavefront deviation be held to $\lambda/13.5$ or less.

Assuming that all the errors have been included in the recording step, and that the reconstruction beam perfectly reproduces the reference beam, we need concern ourselves only with the phase factor evaluated earlier.

Since the variance of the wavefront is expressed by Equation 5, we may proceed from there. If we use Rayleigh's criterion we may set the standard deviation of phase according to

$$3\sigma_\phi = 3\sqrt{V(\phi)} \leq \pi/2. \tag{11}$$

The reason for setting $3\sigma_\phi \leq \pi/2$ rather than $\sigma_\phi \leq \pi/2$, is that this will assure that the maximum phase error will not exceed $\pi/2$ with 99.7% probability. Thus, the inequality we have to work with is

$$\sigma_\phi \leq \pi/6 \tag{12}$$

Interestingly, when the Strehl definition is used we get nearly the same result. That is,

$$\sigma_\phi \leq \frac{2\pi}{13.5} = \frac{\pi}{6.75} \tag{13}$$

To provide a feeling for the magnitudes involved, consider the geometrically balanced case. Then, applying Equation 5 to Equation 12, we can state that

$$\left[\frac{\sigma_\Delta^{\ 4} + 2\sigma_\Delta^{\ 2}\sigma_z^{\ 2}\rho + \sigma_z^{\ 4}}{2r_o^{\ 2}} + \sigma_\Delta^{\ 2}\left(\frac{A}{2r_o} - 2\sin\psi_o\right)^2\right]^{1/2} \leq \frac{\lambda}{12} \tag{14}$$

We can provide a few special cases to get representative numbers. Let r_o = 25 cm, A = 7 cm and $\lambda = 6.328 \times 10^{-5}$ cm and $\psi_r = -\psi_o = 10°$.

(1) $\sigma_\Delta = 0$, $\sigma_z \leq 0.0136$ cm = 216 λ

(2) $\sigma_z = 0$, $\sigma_\Delta \leq 3.6 \times 10^{-5}$ cm = 0.57 λ

Thus, the balanced system is much more sensitive to in-plane errors than out-of-plane, except when $\psi_r = -\psi_o = -\sin^{-1}(A/4r_o)$, in which case the sensitivity is the same in both dimensions. If the holographic system is geometrically balanced, it is seen to be quite tolerant to film deformations. In order to illustrate the strong effect of geometrical unbalance, we consider the two cases $\sigma_\Delta = 0$ and $\sigma_z = 0$. For the first case Equation 5 becomes

$$V(\phi) = \frac{4\pi^2}{\lambda^2}\left\{\frac{\sigma_z^{\ 4}}{2r_o^{\ 2}} + \sigma_z^{\ 2}(\cos\psi_r - \cos\psi_o)^2\right\}. \tag{15}$$

Then, we have the inequality

$$\left(\frac{\sigma_z^{\ 4}}{2r_o^{\ 2}} + \sigma_z^{\ 2}(\cos\psi_r - \cos\psi_o)^2\right)^{1/2} \leq \frac{\lambda}{6\sqrt{2}} \tag{16}$$

Figure 2 contains plots of the solution σ_z, to the upper bound of Equation 16 as a function of ψ_r for values of ψ_o = o°, 5°, 10° and 15°. These curves present a dramatic illustration of the great dependence on geometric balance.

Furthermore, the angular range over which the large tolerance prevails is very small, especially at larger values of ψ_o. Thus it behooves us to design the system as near to geometrical

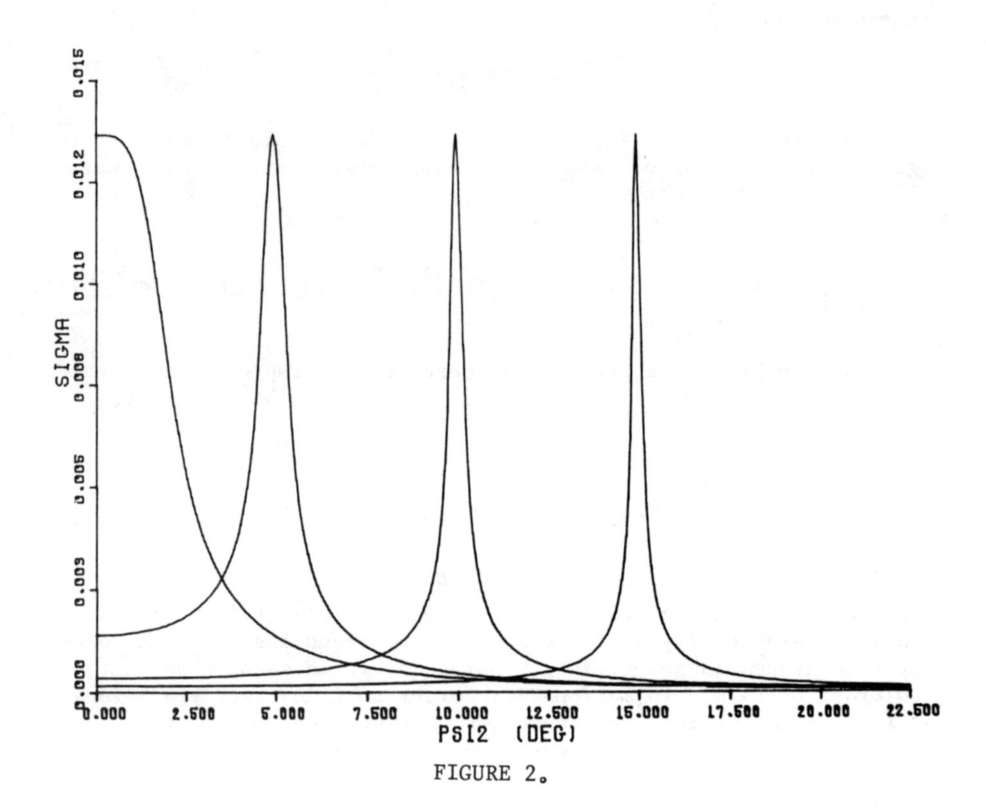

FIGURE 2.

balance as possible ($\psi_r = -\psi_o$) but also to keep ψ_o as small as possible. This, of course, will occur if Equation 8 is obeyed.

For example, if we use 70 mm film, (A = 7.0 cm) and r_o = 25 cm, the optimum arrangement yields $\psi_r = -\psi_o = -4°$. This, of course, is not practically possible.

If only in-plane motions occur ($\sigma_z = 0$) the variance becomes

$$V(\phi) = \frac{4\pi^2}{\lambda^2}\left\{\frac{\sigma_\Delta^{\ 4}}{2r_o^{\ 2}} + \sigma_\Delta^{\ 2}\ (\frac{A}{2r_o} - 2\sin\psi_o)^2\right\} \tag{17}$$

and

$$\left[\frac{\sigma_\Delta^{\ 4}}{2r_o^{\ 2}} + \sigma_\Delta^{\ 2}\ (\frac{A}{2r_o} - 2\sin\psi_o)\right]^{1/2} \leq \frac{\lambda}{6\sqrt{2}} \tag{18}$$

Figure 3 is a plot of σ_Δ versus ψ_o for A = 7 cm and r_o = 25 cm. Note that the optimum operating point occurs at $\psi_o = \sin^{-1}(A/4r_o)$ as predicted earlier. Note also, that tolerance on random in-plane errors rapidly decreases when the optimum object beam angle is not used.

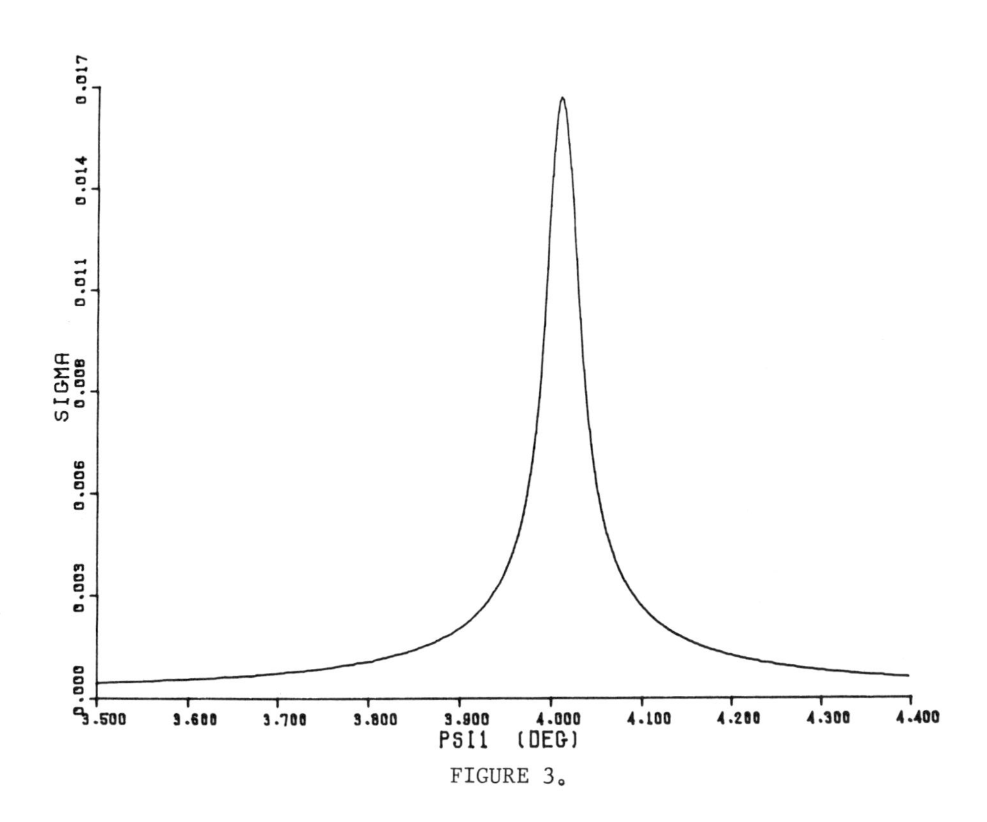

FIGURE 3.

The conclusions that can be inferred from this analysis are that

(1) Object and reference beams should enter the hologram at equal and opposite angles to the normal.

(2) For a balanced geometry the out-of-plane tolerance is much greater than in-plane.

(3) The in-plane tolerance can be made large for the specific balanced geometry, $\psi_r = -\psi_o = -\sin^{-1} A/2r_o$.

Therefore, for practical object angles, most of the tolerance lies in the z-dimension. Very little random in-plane error can be accommodated. Fortunately, this is also the most likely situation to occur in practice. That is, dust particles and air bubbles trapped under the film will cause out-of-plane deformations that are much larger than corresponding in-plane variations.

DIFFRACTION ANALYSIS

A simple deterministic analysis can be used to arrive at similar but less general results. In this analysis the reference and object beams are assumed to be collimated and incident on the hologram plane at ψ_r and ψ_o respectively. The expression for the recorded phase becomes

$$\phi(x) = -\frac{2\pi x}{\lambda}(\sin\psi_r - \sin\psi_o), \tag{19}$$

which represents a linear structure capable of diffracting light. If another collimated beam impinges on the hologram at angle ψ_a, the resultant phase and direction of the propagated wave are

$$\begin{aligned} \phi_i(x) &= \frac{2\pi}{\lambda}\, x\,(\sin\psi_a - \sin\psi_r + \sin\psi_o), \text{ and} \\ \sin\psi_i(x) &= \sin\psi_a - \sin\psi_r + \sin\psi_o. \end{aligned} \tag{20}$$

Hence, if $\psi_a = \psi_r$, the image beam exactly duplicates the object beam.

Suppose that the hologram becomes deformed in such a way that the angle ψ_a changes by σ. We wish to determine the effect on the image wave. Part of the effect will be simple image beam rotation by σ. Part will be an error term ε. That is, let

$$\begin{aligned} \psi_a &= \psi_r + \theta \text{ and} \\ \psi_i &= \psi_o + \theta + \varepsilon. \end{aligned} \tag{21}$$

Algebraic manipulation of Equation 20, and the small angle approximations $\sin\varepsilon \cong \varepsilon$ and $\cos\varepsilon = 1$, yield

$$\varepsilon = \frac{\sin\theta\,(\cos\psi_r - \cos\psi_o) + 1 - \cos\theta)\quad(\sin\psi_o - \sin\psi_r)}{\cos\psi_o\cos\theta - \sin\psi_o\sin\theta} \tag{22}$$

Once again, a balanced geometry, $\psi_r = -\psi_o$, minimizes the error to

$$\varepsilon = \frac{2\,(1 - \cos\theta)\sin\psi_o}{\cos(\psi_o + \theta)} \tag{23}$$

A further approximation,

$$\cos\theta \cong 1 - \theta^2/2$$

reduces Equation 23 to (24)

$$\varepsilon \cong \theta^2 \frac{\sin\psi_o}{\cos(\psi_o + \theta)}$$

Note that, once again, the error is sensitive to the object beam position, being zero for $\psi_o = o$. This differs from our previous result (Equation 6) because here we are considering a plane object wave.

This analysis can be related to a resolution criterion by considering the angle, ε, which would move a focused image point one Rayleigh resolution element. That is,

$$\varepsilon r_o \leq \frac{1.22\lambda r_o}{A}$$

or

$$\varepsilon \leq \frac{1.22\lambda}{A} \qquad (25)$$

Figure 4 is a plot of Equation 22 for several combinations of ψ_o, ψ_r as noted. Note how rapidly the error increases with unbalance. This is entirely consistant with our statistical analysis.

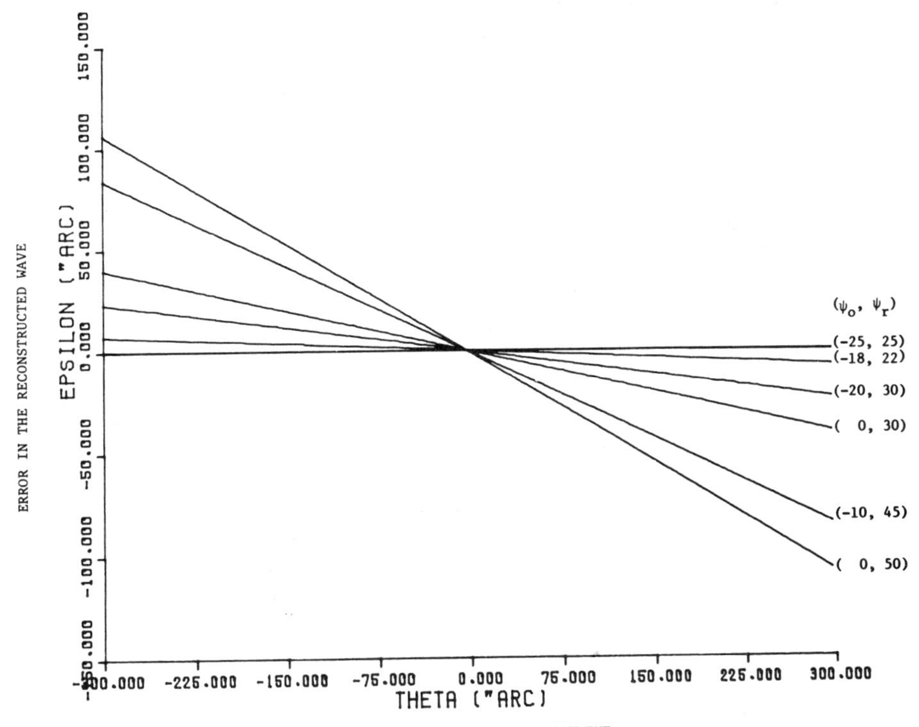

FIGURE 4.

For in-plane errors we can consider the film to stretch or shrink by a factor ρ. Then the image position may be written

$$\begin{aligned} \sin \psi_i &= (1 + \rho)\,(\sin \psi_a - \sin \psi_r + \sin \psi_o) \\ &= (1 + \rho) \sin \psi_o \quad \text{for } \psi_a = \psi_r \end{aligned} \tag{26}$$

If we set $\psi_i = \psi_o + \varepsilon$, and assume $\cos \varepsilon \cong 1$ and $\sin \varepsilon \cong \varepsilon$ the directional error ε, of the reconstructed beam becomes

$$\varepsilon \cong \rho \tan \psi_o . \tag{27}$$

From Equation 25 we find that for $\lambda = 6.328 \times 10^{-5}$ cm and A = 7 cm, $\varepsilon \leq 1.1 \times 10^{-5}$ rad. From Equations 24 and 27 with $\psi_o = 10°$, this allows a rotation of $\theta \cong 0.45°$ and $\rho \cong 6.3 \times 10^{-5}$. Thus, the maximum out-of-plane deformation allowable over the aperture is about 900λ whereas the maximum in-plane deformation is about 3.6λ. These numbers are larger than arrived at by the statistical analysis but the ratios are of the same order. The magnitudes from the statistical analysis are smaller because they represent average estimates which are more conservative.

CONCLUSION

It is concluded that, with proper design, a holographic camera will provide high resolution imaging even though film is used with different construction and reconstruction platens. The amount of allowable out-of-plane distortion is surprisingly high, but in-plane variation must be minimized. These theoretical results have been experimentally verified by the authors of the preceding paper.

REFERENCES

(1) R. Wuerker, L. O. Heflinger, J. V. Flannery, A. Kassel, A. M. Rollauer, "Holography on the NASA Space Shuttle", Paper TH1-1 CICESE, Ensenada, Mexico, August 4-8, 1980.

LARGE FACTOR WAVELENGTH SCALING FOR A HIGH EFFICIENCY VOLUME HOLOGRAPHIC COLLECTOR

M. P. Owen and L. Solymar
Department of Engineering Science, University of Oxford,
Parks Road, Oxford, OX1 3PJ, England

ABSTRACT

A regime for making volume holographic components capable of collecting radiation at wavelengths perhaps twenty times longer than that at recording is investigated.

A method for determining the recording geometry by a K-vector closure technique is shown and the properties of the resulting element are investigated by means of ray-tracing. A localised form of Kogelnik's coupled wave theory is used to show efficiencies.

Approximate analytic solutions and computer evaluated results are presented showing efficiencies and aberrations for the design and nearby wavelengths.

It is found that high efficiencies are possible if the modulation and thickness of the recording material are chosen with care but there will be large spherical aberration unless the numerical aperture is small.

As there is a notable lack of suitable infra-red volume recording materials the applicability of this technique to form a high efficiency collector of far infra-red radiation using visible light at recording is discussed.

INTRODUCTION

Latta and Pole[1] have developed a ray-tracing design technique for reconstruction of a volume holographic lens with red light after recording in the blue region. A similar ray-tracing approach has been used to determine the best regime to make the element, and then using ray-tracing again with a localised form of Kogelnik's coupled wave theory[2] to determine the aberrations and the efficiency of the collector.

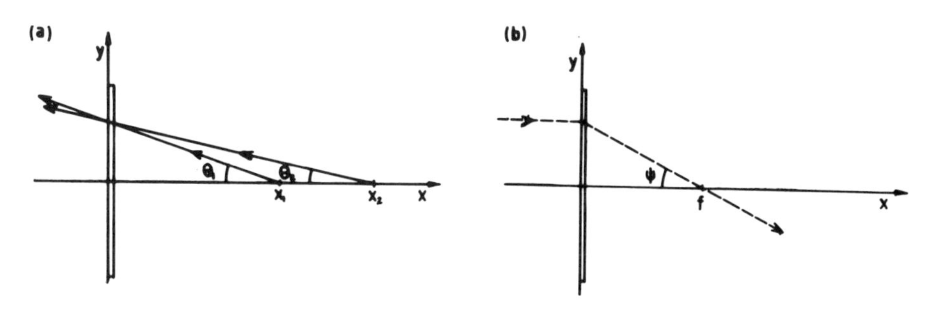

Fig.1. (a) Recording the hologram; $\lambda = \lambda_o$
(b) Diffraction of a paraxial ray; $\lambda = \lambda_1$

ISSN:0094-243X/81/650471-07$1.50

The element is recorded at wavelength λ_o with two point sources located on the axis at distances x_1 and x_2 from the hologram (Fig.1). For a certain choice of the positions of the sources one ray at wavelength λ_1 approaching the hologram from the other side parallel to the axis, will pass through the design focus. Other paraxial rays will still be in the vicinity of the focus. Hence the holographic element may concentrate the radiation within a narrow area in the focal region.

COUPLED WAVE THEORY

If we consider a small region in the recording material the two spherical waves emanating from the point sources may be regarded as plane waves provided the thickness of the material is small compared to the distance to the sources. Neglecting the vectorial nature of the waves and assuming that ε_{ro}, the dielectric constant of the recording material, is the same as that of the surrounding medium (which does not affect the ray-tracing analysis although it will introduce a small inaccuracy into our results on efficiency) the time independent complex amplitudes may be written in the form

$$E_{io} = A_{io} \exp(-j\gamma_{io}) \quad i = 1,2 \tag{1}$$

where A_{io} are constants

$$\gamma_{io} = \beta_o(-x\cos\theta_i + y\sin\theta_i), \quad \beta_o = \frac{2\pi}{\lambda_o} \tag{2}$$

and θ_i is defined in Fig.1. Assuming that the development process will produce a change in the local value of the dielectric constant proportional to the field intensity then the resulting dielectric constant will be of the form

$$\varepsilon = \varepsilon_{ro} + \varepsilon_1 \cos(\gamma_{10}-\gamma_{20}) \tag{3}$$

The grating vector $\underline{K}$ is given by

$$\underline{K} = \nabla(\gamma_{10}-\gamma_{20}) = \hat{\underline{i}}_x\beta_o(\cos\theta_2-\cos\theta_1) + \hat{\underline{i}}_y\beta_o(\sin\theta_1-\sin\theta_2) \tag{4}$$

where $\hat{\underline{i}}_x$ and $\hat{\underline{i}}_y$ are unit vectors in the x and y directions respectively.

Illuminating the same region with a paraxial beam of wavelength

$$\lambda_1 = \mu\lambda_o \tag{5}$$

(where μ is the scaling factor) the time independent complex amplitude in the hologram will be

$$E_1 = A_1(x) \exp(-j\gamma_1) \tag{6}$$

$$\gamma_1 = \beta_1 x, \quad \beta_1 = \frac{2\pi}{\lambda_1}.$$

If the Bragg conditions are close to being satisfied a diffracted wave

$$E_2 = A_2(x)\exp(-j\gamma_2) \tag{7}$$

will appear where, by Floquet's theorem

$$\nabla\gamma_2 = \nabla\gamma_1 - \underline{K} \tag{8}$$

By substituting $E = E_1 + E_2$ into the wave equation

$$\nabla^2 E + \beta_1{}^2 \frac{\varepsilon E}{\varepsilon_{ro}} = 0 \tag{9}$$

and by making suitable approximations, the amplitude of the diffracted wave at the exit boundary, and the efficiency η, can be found

$$\eta = \sin^2(\upsilon^2+\xi^2)^{\frac{1}{2}}/(1+\xi^2/\nu^2) \tag{10}$$

$$\nu = \frac{\beta_1\varepsilon_1}{4\varepsilon_{ro}}\frac{d}{(\cos\psi)^{\frac{1}{2}}} \tag{11}$$

$$\xi = \delta\beta d\sin\psi \tag{12}$$

where δ is the angular deviation from the Bragg condition and ψ is the angle subtended by the ray direction and the x axis.

RELATIONSHIP BETWEEN DESIGN FACTORS, SCALING FACTOR AND RECORDING GEOMETRY

For the Bragg relationship to hold for the illuminating wave and to bring the diffracted wave through the design focus f, the grating vector must equal

$$\underline{K} = \hat{\underline{i}}_x\beta_1(1-\cos\psi) + \hat{\underline{i}}_y\beta_1\sin\psi \tag{13}$$

Equating this to the recording geometry by considering that the expressions for the grating vector in equations (4) and (13) must be identical and by making small angle approximations, that is

$$\sin\psi = \frac{y}{f}; \quad \sin\theta_i = \frac{y}{x_i}; \quad i = 1,2 \tag{14}$$

it can be found that for small numerical apertures the Bragg relationship will hold if

$$x_1 = \frac{2f}{1+\frac{1}{\mu}}, \quad x_2 = \frac{2f}{1-\frac{1}{\mu}} \tag{15}$$

APPROXIMATE SOLUTIONS FOR EFFICIENCY AND CONVERGING PROPERTIES

Although the collector has been designed to satisfy the Bragg condition for rays near and parallel to the axis, a ray further away will not in general fulfil the condition.

Fig.2 shows an off-Bragg situation where the diffracted wave has a propagation coefficient of $\nabla\gamma_2$ not equal in magnitude to β_1, and

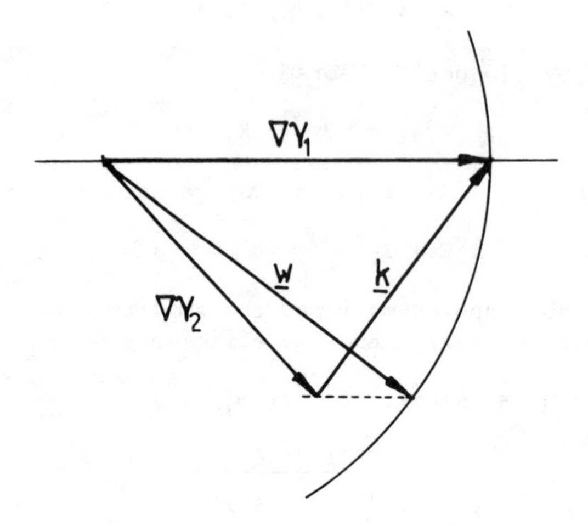

Fig.2. K-vector diagram

hence this wave cannot propagate outside the modulated region. At the exit boundary the matching conditions require that the emerging wave has the same tangential component of wave vector (represented outside the hologram by $\underline{w}$ in Fig.2).

If the incident beam is rotated by a small angle δ to allow the Bragg condition to be satisfied then the diffracted beam will have magnitude β_1, that is

$$\nabla\gamma_1 \simeq \hat{\underline{i}}_x\beta_1 - \hat{\underline{i}}_y\beta_1\delta \tag{16}$$

$$|\nabla\gamma_2| = \beta_1 \tag{17}$$

Hence by equation (8)

$$|\hat{\underline{i}}_x\beta_1 - \hat{\underline{i}}_y\beta_1\delta - \underline{K}| = \beta_1 \tag{18}$$

Substituting the value of $\underline{K}$ from equation (4), using equations (5) and (15) and making the approximations

$$\cos\theta_i = 1 - \tfrac{1}{2}\left(\frac{y}{x_i}\right)^2 + \frac{1}{24}\left(\frac{y}{x_i}\right)^4 \tag{19}$$

$$\sin\theta_i = \frac{y}{x_i} - \frac{1}{6}\left(\frac{y}{x_i}\right)^3 ; \quad i = 1,2 \tag{20}$$

we obtain, for sufficiently large scaling factor μ, the following expression

$$\delta = \frac{1}{16}\left(\frac{y}{f}\right)^3 + 0\left(\left(\frac{y}{f}\right)^5\right) \tag{21}$$

From Kogelnik, to give efficiencies greater than 80% the

restriction

$$|\xi| < \tfrac{1}{2} \tag{22}$$

applies, which by equations (12) and (21) gives the condition

$$\frac{y}{f} < \sqrt[4]{\frac{8\lambda_1}{\pi d}} \tag{23}$$

Inspection of equations (10) and (11) shows that the efficiency η can be 100% if $\nu = \pi/2$ which leads to the requirement

$$\frac{\varepsilon_1}{\varepsilon_{ro}} = \frac{\lambda_1}{d} \tag{24}$$

From equations (23) and (24) a material which can produce a 1% dielectric constant modulation may have efficiencies of 80% up to $y/f = 0.5$. However aberrations could be expected to be large. The

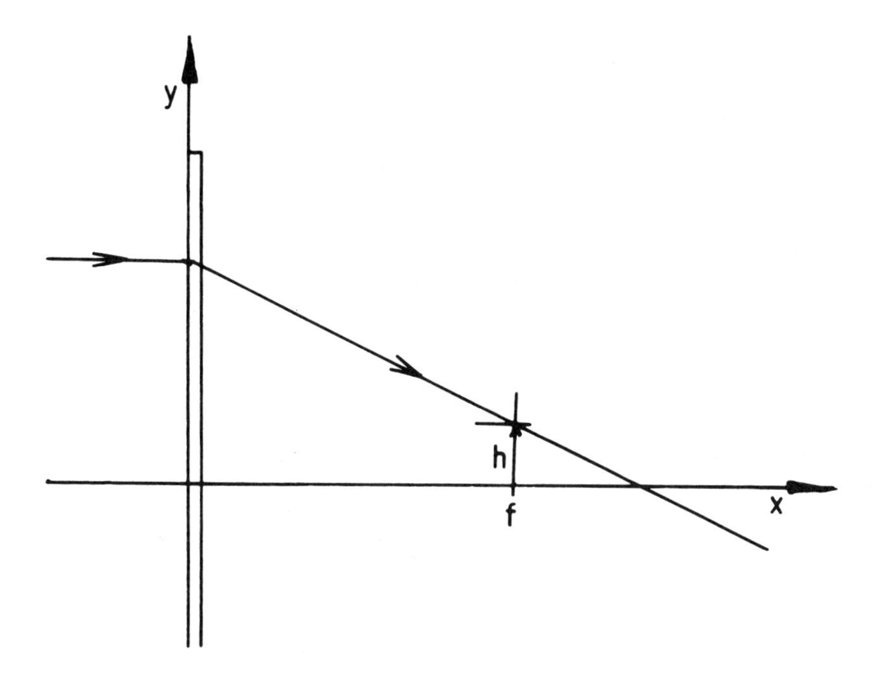

Fig.3. Definition of h

distance a ray misses the design focus in the y direction is defined in Fig.3 as h which, with our previous approximations, may be calculated as

$$\frac{h}{f} = -\frac{1}{8}\left(\frac{y}{f}\right)^3 + 0\left(\left(\frac{y}{f}\right)^5\right) \tag{25}$$

Thus a reduction of the radiation onto a disc 1% of the collector's diameter may be achieved by an element of numerical aperture 0.5. However this result applies only to the design frequency.

Surprisingly, neither equations (21) nor (25) contain the scaling factor μ; apparently for μ large the properties of the element become independent of the actual value.

NUMERICAL RESULTS

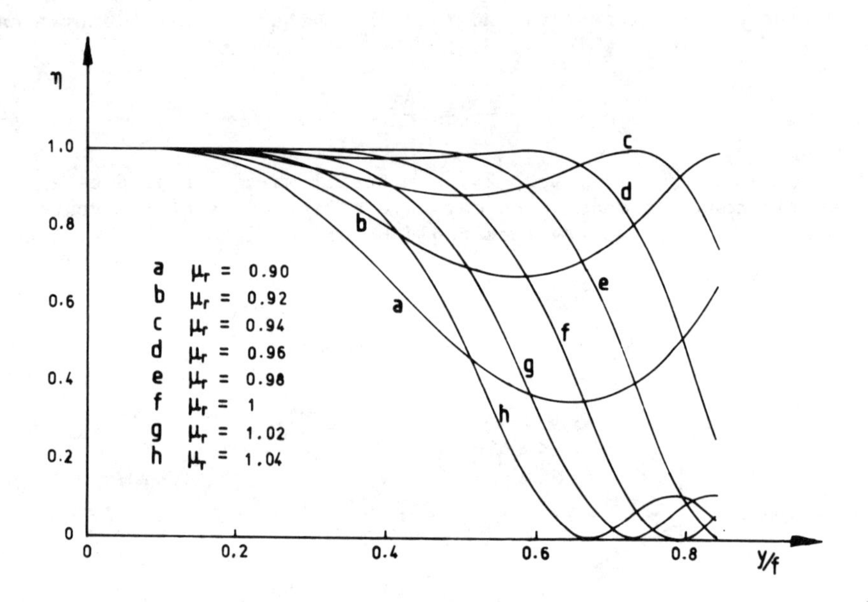

Fig.4. Efficiency vs. y/f for $\varepsilon_1/\varepsilon_{ro} = 0.01$ and $d/\lambda_1 = 100$

Computer evaluated curves are presented showing efficiency against y/f in Fig.4 and h/f against y/f in Fig.5 for the design and nearby frequencies. A new scaling factor μ_r is introduced which relates the actual wavelength of the radiation λ_2 to the design wavelength λ_1

$$\mu_r = \frac{\lambda_1}{\lambda_2} \tag{26}$$

It may be seen in Fig.4 that the efficiency remains high for larger values of y/f as the wavelength exceeds the design wavelength. Unfortunately the collecting properties, under the same conditions, decay substantially. With the aid of Figs.4 and 5 it is now possible to arrive at a design, trading bandwidth and collected power against each other. For example, for an element with unity numerical aperture a 10% frequency bandwidth could be efficiently concentrated within a region with a diameter 0.08 of the diameter of the hologram.

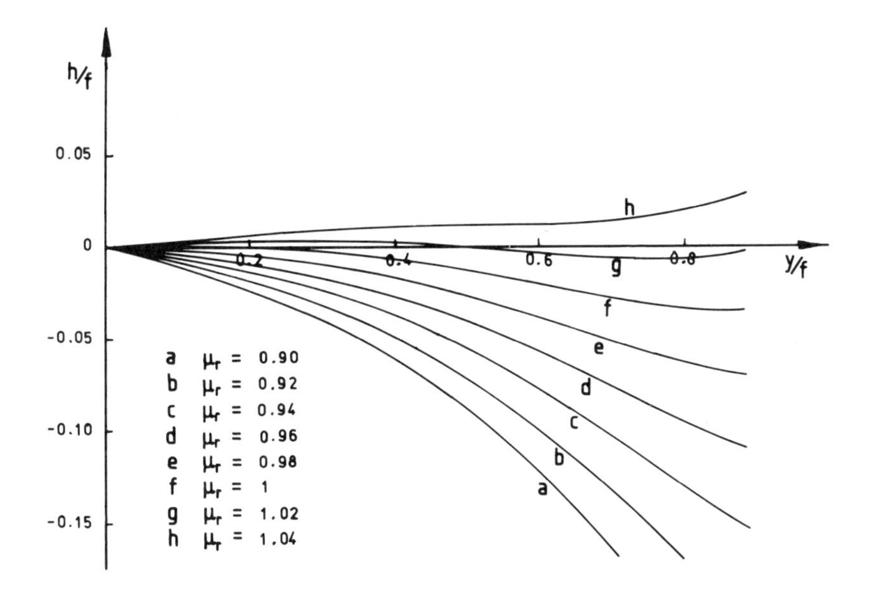

Fig.5. h/f vs. y/f

CONCLUSIONS

It is possible to produce a volume holographic element that will concentrate a small spectral bandwidth of radiation that has been produced at much shorter wavelengths. An application of this technique could be in the recording of a hologram with visible light in a volume material in order to use the element as an efficient collector of infra-red radiation.

ACKNOWLEDGEMENTS

The authors would like to thank the British Science Research Council for its financial support.

REFERENCES

1. M. R. Latta and R. V. Pole, Applied Optics, **18**, No.14, p.2418, July, 1979.
2. H. Kogelnik, Bell Syst.Tech.J., **48**, No.9, p.2909, November, 1969.

COLOR STORAGE AND IMAGE PROCESSING THROUGH YOUNG'S FRINGES MODULATED SPECKLE

E.E. Sicre, N. Bolognini, H.J. Rabal and M. Garavaglia

Centro de Investigaciones Opticas (CONICET - UNLP - CIC)
Casilla de Correo 124, 1900 La Plata, Argentina

ABSTRACT

A simple application of the Young's fringes modulated speckle for color storage in black and white film is presented. As the speckle grains in the image are coded by Young's fringes, information processing operations, like subtraction, spatial derivative, contrast-enhancement, etc., can be performed.

I- INTRODUCTION

Young's fringes modulated speckle (YFMS) can be used for storing several pictures in a single photographic plate, as suggested by Kopf[1], changing fringes orientation by in-plane rotation of the double aperture between exposures.

In this communication we present a simple application of this technique for color storage in black and white film. Actually, pictures stored on it can be different color images of the same object. Obviously, changes in double aperture orientation must be accompanied by changes in the illuminating wavelength.

Color images can be sequentially or simultaneously obtained by using a set of adequate color and spatial filters.

As the images are YFMS coded, some information processing operations can be separately performed on each of them.

II- COLOR STORAGE AND IMAGE PROCESSING

This section describes a modification of Kopf's method for storing color images in black and white photographic plates and to perform optical processing of the images.

COLOR STORAGE

Young's fringes modulated speckles appear when the image of a laser illuminated object is formed by an optical system whose pupil consists of two identical holes[2]. If the photographic record of that image is Fourier transformed in a conventional way[3], Young's fringes act as a carrier frequency, and two diffracted orders appear in the Fourier plane symmetrically located to the zero order. Using spatial filtering techniques in order to select only one diffracted order, and this one being again Fourier transformed, an image of the object can be reconstructed.

If more than one recording is made in the photosensitive mate-

ISSN:0094-243X/81/650478-05$1.50

rial, but changing the orientation of the Young's fringes by in-plane rotating the pupil with the two identical apertures, several pictures can be stored in a single photosensitive plate. Then, after processing, the pictures of the different objects can be re-obtained by properly locating the spatial filter to select in each operation the corresponding diffracted order[1].

By modifying Kopf's method, different pictures can be stored in the photographic plate which actually are the different color images of the same object. This can be done by recording the images of the object in a sequential way by changing the wavelength and pupil orientation between exposures. Figure 1 shows the corresponding experimental set-up.

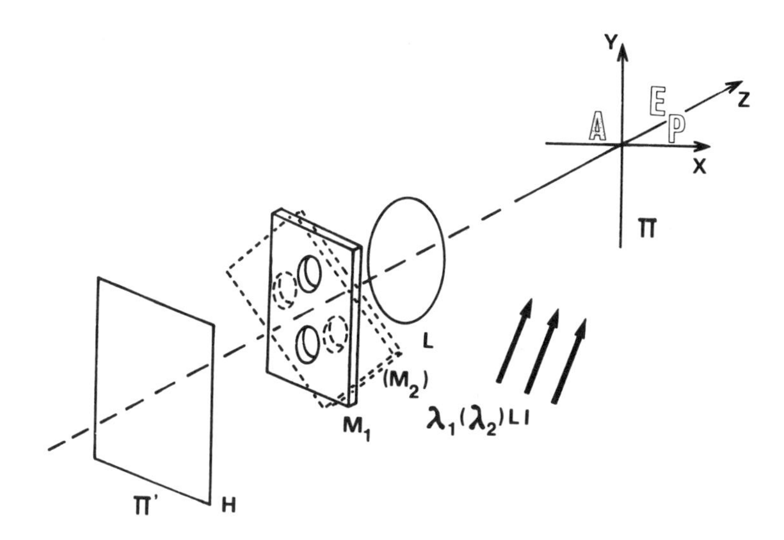

Figure 1.- Experimental set-up used to store different colors of an object in the same black and white photographic plate. LI , laser illumination at two wavelengths; λ_1 = 514.5 nm is the green argon laser and λ_2 = 632.8 nm is the red helium-neon laser wavelengths; M , two hole mask placed at two different angular positions; L , achromatic lens; H , photographic plate. π and π' are conjugate planes of the lens L . The three dimensional object consisted of a green letter A and two red letters, E and P .

After processing, the color images can be obtained in the following way. If the recording wavelengths were very different reconstruction can be done with all of them at the same time. In this case, in order to avoid that an image constructed with a wavelength be reconstructed with another wavelength, adequate color or interference filters should be placed behind the spatial filters in the Fourier plane. On the other hand, monochromatic images of each color image, stored in this way, can be sequentially reconstructed by illuminating the photographic plate with the corresponding wavelength, and placing the spatial filter in the adequate diffracted order. Figure 2 shows the employed arrangement.

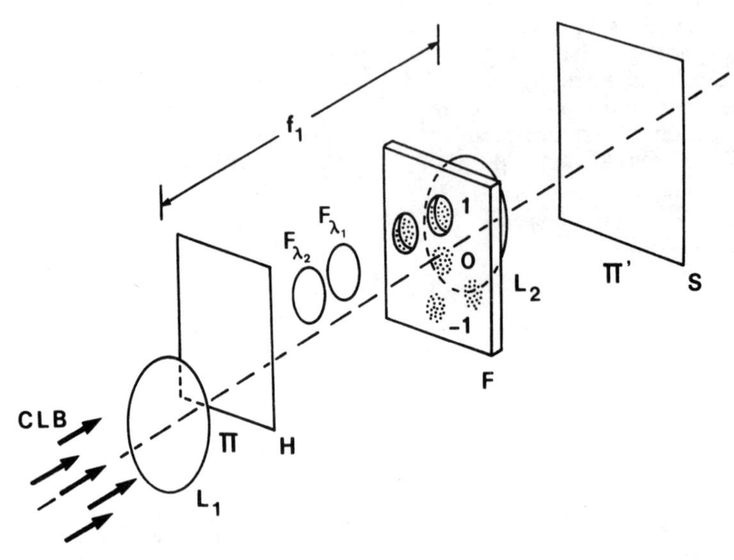

Figure 2.- Reconstruction of a color image stored in black and white film. Experimental set-up. CLB , collimated λ_1 and λ_2 laser beams; L_1 and L_2 , achromatic transforming lenses; H , developed photographic plate; F , spatial filter with a set of two holes to select the appropriate diffracted orders; F_λ , wavelength filters and S , observing screen. π and π' are conjugate planes of the L_2 transforming lens.

IMAGE PROCESSING

As the color images stored by using the method described previously are YFMS coded, then some information processing operations can separately be performed on each of them, as subtraction, spatial derivatives, contrast-enhancement, etc.

To perform image subtraction, which is the basic operation of all those above mentioned, a two-step procedure was employed. In the first step the YFMS image of a three dimensional object or of a transparency, is recorded on a photographic plate. These speckles are obtained by directly illuminating the object with a laser, or the transparency through a diffuser. The Young's fringes modulation of speckles is produced in the forementioned way. Then, Young's fringes are shifted in half a period, and the second image is recorded. If the two exposures are equal, the Young's fringes of speckles of identical parts of both images are added to a constant background, while the contrast of the Young's fringes of speckles corresponding to non-identical parts increases in accordance with their mismatch. When the developed plate is Fourier transformed as before, in the second step of the procedure, the non-cancelled Young's fringes give raise to two diffracted orders in the Fourier plane. If a spatial filter is placed in this plane, in order to observe only one diffracted order, and this order is again Fourier transformed, an image of the differences between the images will be obtained.

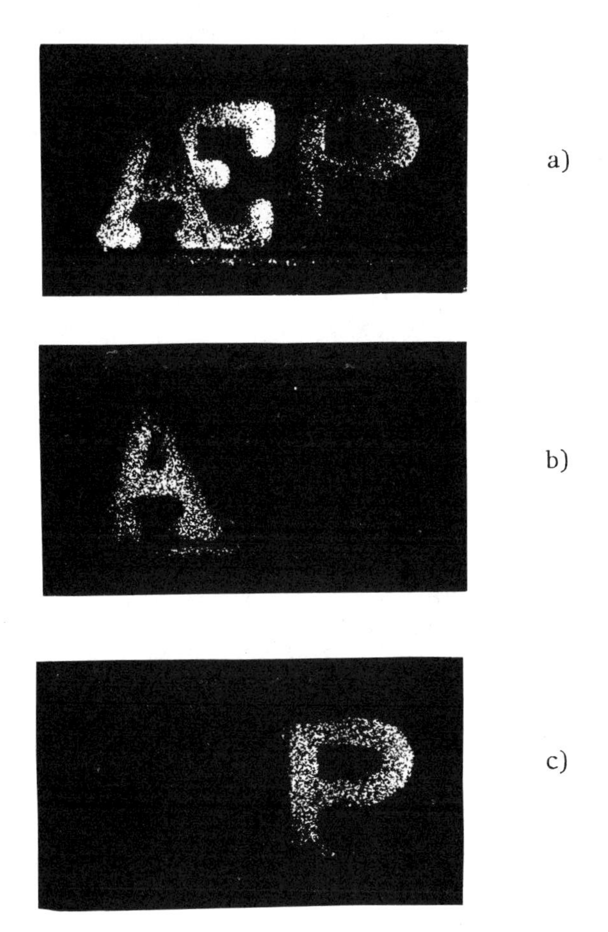

Figure 3.- a) Positive reproduction of unfiltered image recorded on Ilford FP4 common film.
b) Filtered "green image" registered with argon laser illumination.
c) Filtered "red image" registered with He-Ne laser illumination. The letter E has been subtracted.

Young's fringe shifting can be performed by in-plane moving the photographic plate in the direction perpendicular to the fringes. In this case, the cancellation of Young's fringes is exact in the whole plane. However, in some cases it is more practical to produce the π-phase-shift by rotating a plane-parallel plate placed behind one of the holes of the M mask in Figure 1, in spite of the fact that cancellation of Young's fringes is not complete all over the plane.

The influence of experimental parameters on the non-exact π-phase-shift for common image points far enough from the optical axis was previously analyzed in some detail[4].

Figure 3 shows preliminary results of two color images of an object having only two colors, each one obtained by illuminating the recording plate with only one color.

In order to diminish the speckle noise, an optic fiber was used in the reconstruction procedure. The laser beams were focused in one of the fixed ends of the fiber, so the other fixed end acted as a point-like source. Between ends, the fiber forms a loose loop which was randomly moved by a turbulent air flux that reduced the spatial coherence of the light.

III- ACKNOWLEDGMENTS

Partial financial support from the Secretaría de Estado de Ciencia y Tecnología (SECYT), Argentina, and the Organization of American States (OAS), is gratefully acknowledged.

IV- REFERENCES

1.- V. Kopf, International Optical Computing Conference, Zurich (1974). IEEE Catalog N° 74, p. 862-3C.
2.- E.D. Duffy, Appl. Opt., 11, 1778 (1972).
3.- J.W. Goodman, Introduction to Fourier Optics (McGraw-Hill, New York, 1968).
4.- H.J. Rabal, N. Bolognini, E.E. Sicre and M. Garavaglia. To be published. Opt. Comm. (1980).

SUBRESOLUTION IN MICROWAVE HOLOGRAPHY

G. Tricoles, E.L. Rope, R.A. Hayward
General Dynamics Electronics Division
P.O. Box 81127, San Diego, CA 92138, USA

ABSTRACT

Microwave holography is a useful experimental technique for imaging remote or inaccessible objects and for diagnostics of antennas, radomes, and scatterers; however, diffraction restricts image resolution. This paper describes a method for improving resolution in microwave holography. The holograms are either spherical or circular. Porter's scalar theory of curved holograms is extended to vector fields by using rectangular components to treat the effects of wave polarization. The mathematical formulation is a Helmholtz diffraction integral. This integral is written as a convolution for currents on line segments. The convolution is applied to the spatial frequency spectra of images. The spectra of dipole antennas are analytically continued, and the current distributions are exactly reconstructed. An experimental example is described; it is diffraction of a half-wavelength wide slit in a conducting screen. The analytic continuation of the holographically reconstructed nearfield produced images with resolution approximately 1/4 wavelength. Before continuation, resolution was 0.6 wavelength. In addition, the boundary condition of vanishing tangential electric field over the metal screen was better satisfied in the image produced by continuation.

INTRODUCTION

Holography is an imaging technique that is new compared to imaging with lenses, but it produces three-dimensional images and has applications in optical data processing and metrology.[1-3] Holography is also being done with microwaves and radio waves for diagnostics of radiating structures and to study radio wave propagation.

The holographic method has two steps. The first, called hologram formation, is the measurement of the

ISSN:0094-243X/81/650483-10$1.50

field diffracted by an object. In optics, the detector is photographic film, but for microwaves phase and intensity can be explicitly measured with antennas and receivers. The second holographic step is called wavefront reconstruction. A wave illuminates a hologram, the record of the measured field. In optics, a hologram is a photographic transparency, but for microwaves digital reconstruction is preferable to avoid the image distortions that arise when the reconstruction is done with visible light.

In long wavelength (microwave or acoustic) holography, image resolution is limited by aperture size. Although apertures can be synthesized in the laboratory by a moving probe or an antenna array, costs become significant.[4] Of course airborne synthetic apertures are well known, but these systems are for particular applications such as mapping terrain.

This paper describes a procedure for increasing resolution in microwave holography. The intended applications are diagnostics of antennas, radomes, and scatterers. We start with a scalar theory of curved holograms.[5] It is extended to vector fields to describe wave polarization. Image enhancement is a main topic. The diffraction integral for reconstructing images of sources inside a spherical surface is written as a convolution for linear current elements. Spatial frequency filtering is utilized. Analytic continuation is employed with a basis in diffraction theory. A theoretical example, the current on a half-wave dipole is reconstructed exactly in closed form. An experimental example, diffraction by a half-wavelength slit in a metal screen is presented. Analytic continuation improved resolution from 0.6 wavelength to 1/4 wavelength, and it better approximated the boundary condition of vanishing tangential field.

THEORY

Holography

In hologram formation, an object emits or reflects, and a receiving antenna scans over a surface such as a sphere. Figure 1 illustrates this step. The measured quantities are intensity S and phase Φ. Field amplitudes are assumed proportional to $|S|^{1/2}$. A rectangular field component satisfies the scalar Helmholtz equation; that is $(\nabla^2 + k^2)u = -\rho$, where ρ is the source distribution.

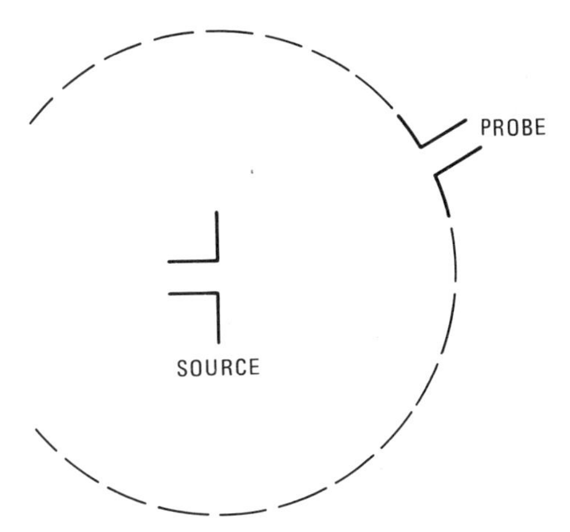

Fig. 1. Formation step. A probe scans a spherical surface.

The reconstruction step produces images by evaluating the Helmholtz diffraction integral over a surface coincident with the measurement surface. The measured data determine surface sources. The surface sources launch a wave converging toward the image region which had been occupied by the object. Since the object is absent, the field satisfies the homogeneous Helmholtz equation.

To describe the reconstruction for arbitrary surfaces, Porter[5] defined an ideal, basic imaging system for scalar fields. This system is ideal in the sense that it images a point source at $(\underset{\sim}{r} - \underset{\sim}{r}')$, where $\underset{\sim}{r}'$ gives the source location. We summarize Porter's theory. The point source produces an expanding wave $g(\underset{\sim}{r} - \underset{\sim}{r}') = (4\pi|r - r'|)^{-1} \exp(ik|r - r'|)$. The reconstruction sources launch a converging wave g^* toward the image region, were $*$ denotes complex conjugation. A wave x' diverges from the image region. The image of a point source is $g^* + x'$. Because $g^* + x'$ satisfies the homogeneous scalar equation, x' is $-g$. The field inside the measurement surface is $g^* - g$. Outside, x' expands, so the field is -g. The surface sources are defined by field discontinuities.

The sources are $\sigma_o = (g^* - g) - (-g) = g^*$, and $\partial\sigma_o/\partial n$. These sources are integrands in the Helmholtz integral, $I = \int (\sigma_o\ \partial g/\partial n - g\ \partial\sigma_o/\partial n)\, ds$, where the integration is over a sphere.

For an extended object, the basic imaging system generalizes. If u represents a rectangular field component, the reconstructed field is

$$u_I^* = \int \left(u_h^* \frac{\partial g}{\partial n} - g \frac{\partial}{\partial n} u_h^*\right) ds \qquad (1)$$

where u_h is measured. The use of the complex conjugate of the field is familiar in holography; it leads to the conjugate or real image.

RECONSTRUCTION AS CONVOLUTION

The integral in Equation 1 can be written as a convolution for linear currents of finite length. For a linear antenna at the origin of a rectangular coordinate system, with length 2L parallel to the z-axis, let the current be F(z) on the antenna and zero elsewhere. The electric field at farfield distances is $E_\theta = -ik(\eta/4\pi)\, g(r) \sin\theta I_\theta$, where r is the distance between integration and observation points, and η is the intrinsic impedance, and I_θ is $\int F(z') \exp(-ikz' \sin\theta)\, dz'$. When the values of E_θ are utilized in Equation 1, the reconstruction is

$$E_{zi}^* (z_i) = 4\eta\, k^2 \int_{-\infty}^{\infty} F^* (z')\, \text{rect}\, (z'/L) \times h(z_i - z')\, dz' \qquad (2)$$

where $h(z_i - z') = [\text{sinc}\ k(z_i - z') - \cos k\,(z_i - z')] \times [k\,(z_i - z')]^{-2}$, with sinc $z = z^{-1} \sin z$, and rect(z'/L) has value 1 for $|z'/L| < 1$ and zero elsewhere. Thus the reconstructed field is a convolution, and $h(z_i - z')$ is the impulse reponse of the spherical imaging system.

Fourier transformation of Equation 2 gives

$$\mathrm{FT}\ [E_i^*\ (z_i)] = 4\eta\ k^2\ \mathrm{FT}\ [F^*(z')\ \mathrm{rect}\ (z'/L)]$$

$$\mathrm{FT}\ [h(z_i - z')], \tag{3}$$

where FT denotes the transform operation. When division by FT [h] is valid, an inverse transform gives

$$F^*\ (z')\ \mathrm{rect}\ (z'/L) = (4\eta\ k^2)^{-1} \Big\{ \mathrm{FT}^{-1}$$

$$\mathrm{FT}\ [E_i^*\ (z_i)]/\mathrm{FT}\ [h] \Big\} \tag{4}$$

EXAMPLE: HALF-WAVE DIPOLE

For a half-wave dipole parallel to the z-axis we assume the current has a cosine distribution. The electric field at large distances has a z component $g(r)\ (\sin\theta)^{-1}$ $\cos(\pi v/2)$, where v is $\cos\theta$, so the z component is $E_z = -g(r) \cos(\pi v/2)$. From Equation 1, the field on the z-axis is given by

$$E_{iz}^*\ (z_i) = -4\pi ik\ [\mathrm{sinc}\ (\frac{\pi}{2} - k_{zi}) +$$

$$\mathrm{sinc}\ (\frac{\pi}{2} + k_{zi})] \tag{5}$$

Image enhancement requires the spectrum of the field in Equation 5. The spectrum is $\mathrm{FT}\ [E_{iz}^*\ (z_i)] = -4\pi ik$ $\cos(\pi\lambda/2)\ \mathrm{rect}\ \lambda$. The continued spectrum is FT_c $[E_{iz}^*\ (z_i)] = 4\pi ik \cos(\pi\lambda/2)$. By performing the operations in Equation 7, we obtain

$$F^*\ (z')\ \mathrm{rect}\ (z'/L) = 4\pi i\eta^{-1}\ (I_+ + I_-) \tag{6}$$

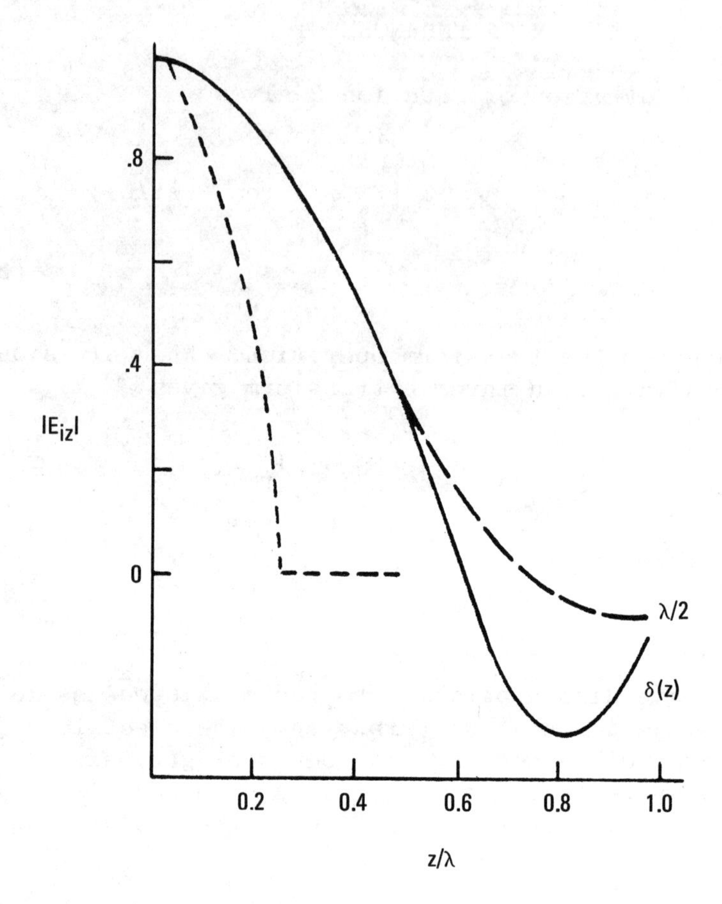

Fig. 2. Theoretical reconstructions. Image of infinitesimal dipole (—). Image of half-wave dipole (— —), Eq. 5. Enhanced image of half-wave dipole (---) Eq. 7.

The reconstructed current function is explicitly and exactly the assumed current distribution. The result is graphed in Figure 2.

FILTERING THEORY FOR A SLIT

By considering the Helmholtz diffraction formula for a slit we are led to a procedure for enhancing reconstructed fields. Assume a component E_y in the plane of the slit and orthogonal to the slit length. The steps are as follows:

1. Reconstruct the nearfield by evaluating Equation 1 at the slit.
2. Fourier transform the reconstructed field
3. Divide by $1 + (1 - \nu^2)^{1/2}$
4. Inverse transform

To reconstruct H_z, the divisor in step 3 is $1 + (1 - \nu^2)^{-1/2}$.

ANALYTIC CONTINUATION FOR A SLIT

The spectra decrease abruptly for $|\nu|>\lambda^{-1}$. To continue them we utilize the filtering functions defined above. For $|\nu|\le 1$, we use the filtered spectra, those from the reconstruction following division by $1 + (1 - \nu^2)^{1/2}$. For $|\nu|>1$, we utilize the filtering functions, with unit value at $\nu = 1$.

MEASUREMENT

Wavefront reconstruction, filtering, and continuation are applied to diffraction by a slit.

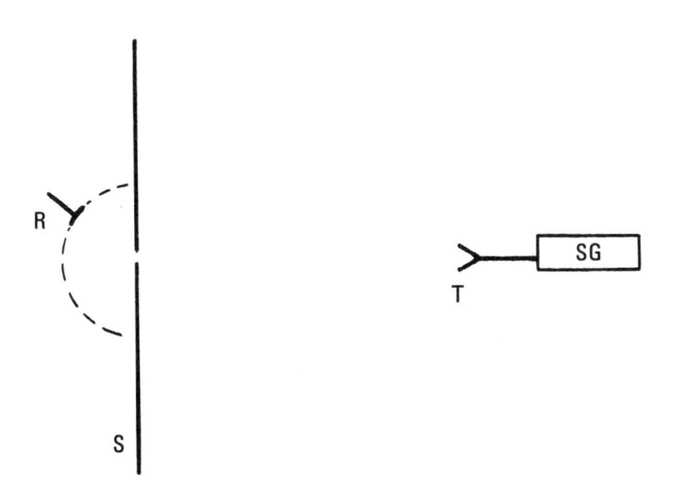

Fig. 3. Hologram formation for a slit. SG is a signal generator. T is the transmitting antenna, R is the receiving half-wave dipole, and S is the aluminum sheet with half-wavelength wide slit.

Figure 3 shows the measurement setup. A signal generator produced 6.000 GHz waves. The transmitting antenna, a horn with aperture 26.7 cm by 31.8 cm, was 411 cm from an aluminum sheet, width 244 cm and height (normal to the figure) 122 cm. Slit width was 2.5 cm or a half wavelength. The diffracted field was a dipole antenna that scanned an arc, with radius 46 cm and centered at the slit. The receiver was a network analyzer. The measurements were made in an anechoic chamber, with absorber about the edges of the diffracting screen. Figure 4 shows the measured field on an arc.

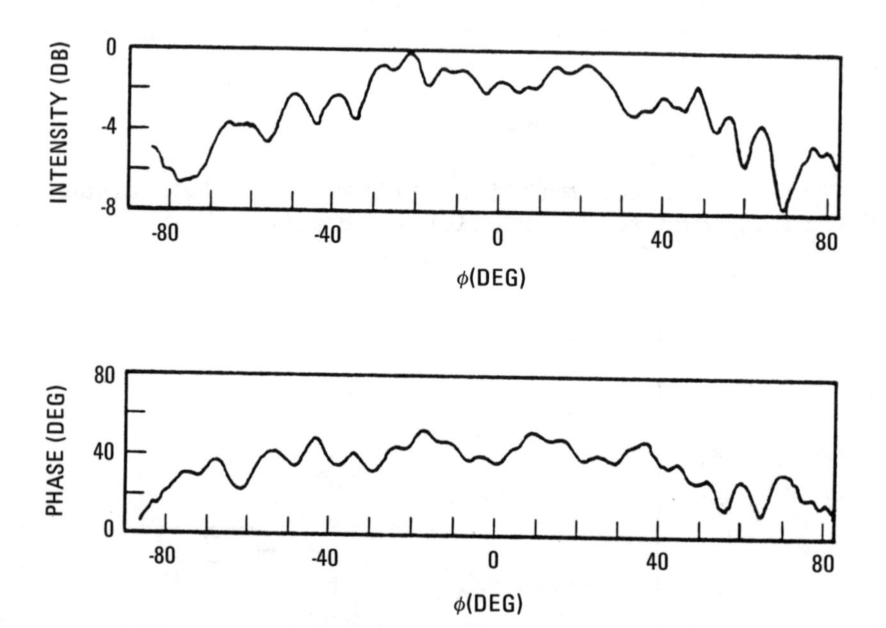

Fig. 4. Measured phase and intensity for slit diffraction.

The field in the plane of the slit was computed by evaluating Equation 1. A Dirac delta function is understood because the measurements were on an arc. Figure 4 shows the electric field component perpendicular to the slit and in the plane of the screen. The field was

Fourier transformed, and the spectrum was multiplied by the filtering function $(1 + \sqrt{1 - \nu^2})^{-1}$ and the spectrum was continued according to the procedure described in the preceding section. Inverse transformation produced the narrower image in Figure 5. Notice that the boundary condition of vanishing total field is better satisfied by the reconstruction produced with the filtered and continued spectrum.

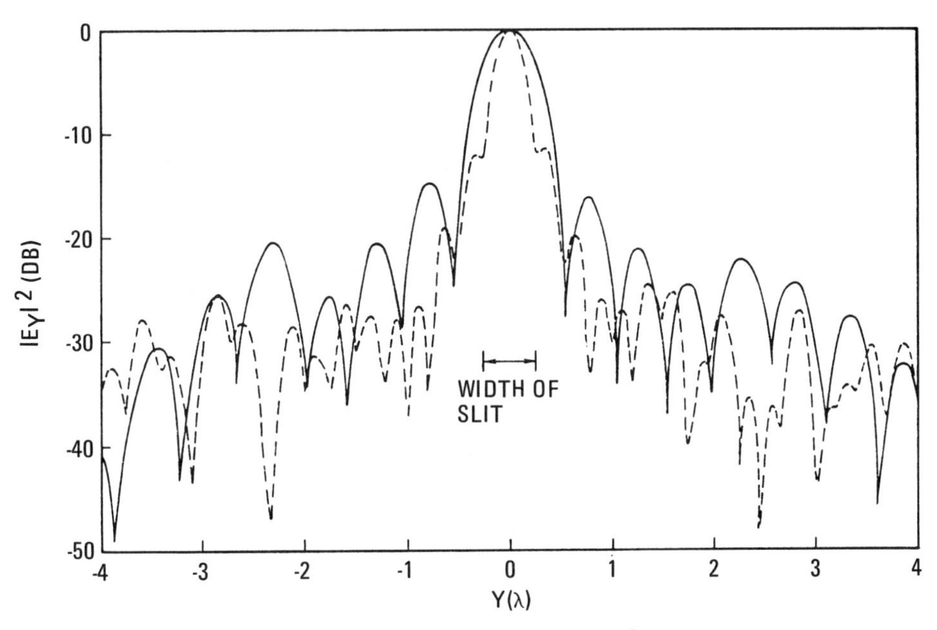

Fig. 5. Reconstruction values of $|E_y|^2$ in plane of slit aperture. Without filtering or continuation (—), with both (---) for | |<2.

Farfield patterns were computed from the reconstructed fields. The agreement with measured values was better for the farfield computed from the nearfield reconstructed from the continued spectrum.

REFERENCES

1. D. Gabor, "A New Microscopic Principle," Nature, Vol. 161, No. 4098, pp 777-778 (1948).
2. E.N. Leith and J. Upatnieks, "Wavefront Reconstruction with Continuous Tone Objects," Jour. Opt. Soc. Am., Vol. 53, pp 1377-1381 (1963).
3. J.N. Goodman, "Operations Achievable with Coherent Optical Information Processing Systems," Proc. IEEE, Vol. 65, pp 29-38 (1977).
4. G. Tricoles and N.H. Farhat, "Microwave Holography: Applications and Techniques," Proc. IEEE, Vol. 65, pp 108-121 (1977).
5. R.B. Porter, "Diffractions-Limited, Scalar Image Formation with Holograms of Arbitrary Shape," Jour. Opt. Soc. Am., Vol. 60, pp 1051-1059 (1970).

DIFFRACTION & DYNAMIC PROPERTIES OF PHOTOSENSITIVE ELECTROOPTIC MEDIA

M.P. Petrov

A.F. Ioffe Physico-Technical Institute
of the Academy of Sciences of the USSR, Leningrad

ABSTRACT

A short review of diffraction effects on volume holograms in photorefractive media is presented. Special attention is given to anisotropic diffraction, and diffraction control by an electric field.

The main properties of a new space-time light modulator PRIZ⁺, developed on the basis of cubic photorefractive crystals, are described. The PRIZ provides image contouring and suppression of diffraction in given directions. One of the PRIZ modifications allows selection of nonstationary parts of an image (dynamic image selection). A phenomenological theory is developed to describe the operation and properties of the modulator in the linear approximation.

In this paper we discuss two major problems concerning photorefractive electrooptic media: (1) Diffraction of light from volume phase holograms in anisotropic crystals, and (2) Space-time modulation of light by means of a new PRIZ⁺ type modulator using crystals with cubic symmetry. [Because some results related to the first problem have recently been published, our main concern will be with the second problem.]

Photorefractive media are substances in which the index of refraction undergoes fairly pronounced changes under the action of incident light.[1 2] Sometimes the effect of photorefraction is also referred to as optical damage. Schematically, the physical process called optical writing in photorefractive media can be described as follows: at first, the incident light produces excitation of charge carriers in the crystal; the carriers then diffuse or drift into non-illuminated sites; finally, a charge volume distribution becomes localized by the traps which are present in the crystal. As a result, an electric field pattern and a corresponding spatial distribution of refractive index are formed which reflect the spatial intensity profile of the incident light. Holographic records in photorefractive media are phase holograms which can be thick because of the high transparency of the crystals. The most important feature of the diffraction process occurring in anisotropic photorefractive media is the strong dependence of the Bragg condition on the state of polarization of the incident light and on the external conditions (e.g., the external electric field, pressure, etc.).

⁺PRIZ is the abbreviated form of "preobrasovatel izobrazheniy" (i.e., image transformer) in Russian.

ISSN:0094-243X/81/650493-15$1.50

In the usual case of a simple volume grating with a wave vector $\vec{K}_g$, the Bragg condition takes the form

$$\vec{K}_1 = \vec{K}_2 \pm \vec{K}_g \tag{1}$$

where $|\vec{K}_1| = |\vec{K}_2| = 2\pi n/\lambda$, n is the refractive index of the medium, λ is the wavelength of light in vacuum, $\vec{K}_1$ and $\vec{K}_2$ are wave vectors of the diffracted and incident beams.

In photorefractive birefringent crystals the index of refraction is different for the ordinary and extraordinary beams. Moreover, in this case, diffraction is not only possible with a fixed plane of polarization, but also with an accompanying rotation of the plane of polarization (for example, with the transformation of an ordinary into an extraordinary beam, and viceversa). The latter process (anisotropic diffraction) can arise because in a photorefractive material the diffraction grating originates from the sinusoidal modulation of the dielectric permeability tensor $\delta\hat{\varepsilon}$, while in the traditional isotropic case, the grating is produced by the modulation of the scalar refractive index or transparency. Anisotropic diffraction can occur for special orientations of the grating in the crystal when the electric field vectors of the incident and diffracted waves are connected to one another through nondiagonal elements of the tensor $\delta\hat{\varepsilon}$.

The diffraction efficiency is described by the formula[3,4]

$$\eta = \sin^2 \left[\frac{\pi}{\lambda} \frac{(\vec{E}_1|\delta\hat{\varepsilon}|\vec{E}_2)\ d}{\sqrt{K_{1Z}\cdot K_{2Z}}}\right] \tag{2}$$

where d is the thickness of the hologram, $\vec{E}_{1,2}$ are the normalized ($\vec{E}_{1,2}\cdot\vec{E}_{1,2} = 1$) electric field vectors of the readout and reconstructed waves, and K_{1Z}, K_{2Z} are the components of the fields' wave vectors normal to the crystal surface. In the isotropic case, Eq. (2) reduces to the well known Kogelnik's result.[5]

The anisotropic diffraction produces surprising changes in both angular and spectral selectivity of the holograms, as well as in the geometry of the experiments. For example, the direction of the diffracted beams can differ considerably from the directions of the recording beams. This can be seen in Figure 1. The positions of the recording beams are shown by the brighter spots in the lower right part of the figure. The diffracted beam, however, travels in quite another direction in the case of anisotropic diffraction. The reason for such a phenomenon is also shown in Figure 1, where one can see that the wave vector of the grating connects two different wave surfaces, and that this construction defines the directions of the reconstructed beam.

Another example is the reconstruction of a complex volume hologram at a different wavelength from that used in the writing process. Usually, with the only exception of the Gabor scheme, it is not possible to reconstruct a complex volume hologram at a different wavelength because of its high spectral selectivity.
Anisotropic diffraction on the other hand, allows such reconstruction. For this purpose, it is necessary to select the crystal and readout beam orientations, so that the beams at λ_2 which are reconstructed through anisotropic diffraction travel in the same direction as those

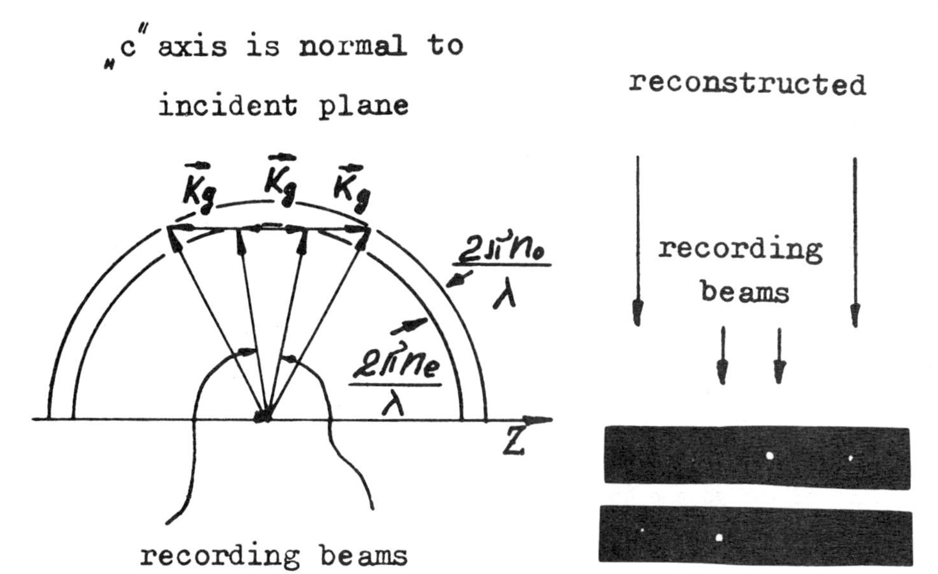

Figure 1

Wave vectors diagram explaining the behavior of anisotropic diffraction. The vector $\vec{K}_g$ can connect both the points lying on one wave vector surface and the points lying on different wave surfaces. In the latter case, anisotropic diffraction occurs.

To the right are shown the experimental points produced by the recording beams and the beams reconstructed by anisotropic diffraction.

at λ_1, which are reconstructed through the usual diffraction process. The theory of this effect is described elsewhere.[6] In our experiments we succeeded in reconstructing a good quality image with a spatial frequency range of 100-150 cycles/mm. The scheme based on anisotropic diffraction has the advantages of both the Gabor (wide band) and the Leith-Upatnieks scheme because the direction of the zero-order diffraction does not coincide with that of the reconstructed image. In practice, this method is of interest for the reconstruction of holograms recorded in photorefractive media, especially because it allows readout at a wavelength at which the recorded information is not erased.

As already mentioned, the high sensitivity of the Bragg condition to external influences allows one to control the hologram recording and retrieval processes with the help of external electric fields which affect the index of refraction of the crystal and consequently the wavelength in the medium.[7,8]

In $LiNbO_3$ crystal, several holograms were recorded[7], each with a different electric field applied to the crystal. The retrieval of a corresponding hologram was then carried out upon application of the appropriate electric field. Thus, it is possible to retrieve a hologram not only by mechanical rotation of the sample, or by changing the direction of the readout beam by means of a deflector, but also and more directly, by varying the magnitude of an external electric field.

The second part of this paper is devoted to the space-time modulation of light using the PRIZ modulator. To explain the principle of operation of the PRIZ modulator, it is convenient to compare this device with the well known PROM modulator[9] (Fig. 2).

The system consists of a working crystal slice and a current-conducting medium comprising transparent electrodes and thin dielectric layers. Cubic crystals belonging to point groups $\bar{4}3m$ and 23m are the most suitable as working material. Further on in this paper we shall discuss experimental results obtained with a $Bi_{12}SiO_{20}$ crystal (group 23). The recording light from the blue-green region of the spectrum excites charge carriers in the crystal that drift along the electric field applied to the crystalline plate. As a result, a non-uniform volume electric charge is produced in the crystal. The presence of the volume charge produces an inhomogeneous electric field having both longitudinal and transverse components, with respect to the field created by the applied voltage. Because the crystal possesses a linear electrooptic effect, spatial variations of birefringence arise, in relation to the varying magnitude and direction of the local electric field. During the readout process with a polarized beam of light at a longer wavelength, a modulation is impressed on the state of polarization of the readout beam.

One of the distinctive features of the PRIZ modulator[10] is that it employs the transverse electrooptic effect. On the contrary, the operation of the PROM is based on the longitudinal effect. This is obtained simply by orienting the crystal in a different configuration, for example, along the [111] or [110] directions, instead of the [100] direction used in the PROM.

In Figure 2, some peculiarities of the PRIZ can be seen by inspection. For instance, the zero spatial frequencies will not be read out, because the magnitude of the transverse component of the electric field decreases as the spatial frequency gets smaller. In contrast, the diffraction efficiency should rise at high spatial frequencies. The experimental results confirm these qualitative considerations.

Figure 3 demonstrates the results of measurements of diffraction efficiency versus spatial frequency using the PRIZ modulator.[11] One can see from this figure that the PRIZ performance deteriorates at low spatial frequencies. For the sake of comparison, the PROM performance under similar conditions is also given.

Our next Figure (Fig. 4) shows an example of image reconstruction using the PRIZ modulation. The most obvious feature of this figure is the absence of the zero spatial component, or background. The PRIZ operates by enhancing the contours of the image.

PROM — longitudinal
PRIZ — transverse
linear electrooptic effect

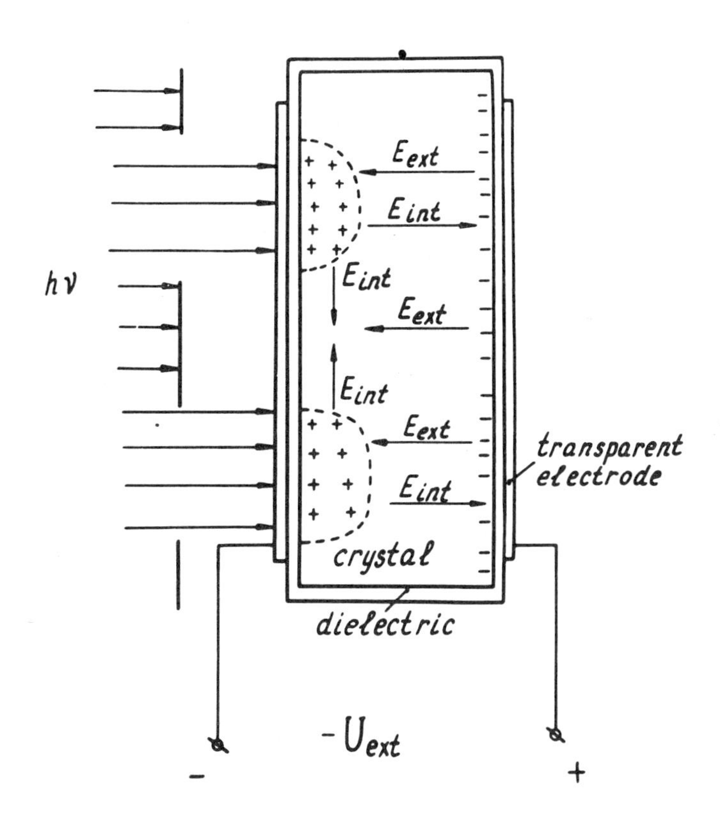

Figure 2

Cross section of a symmetric structure of the PROM and one of the modifications of the PRIZ modulator. Transparent electrodes are on both surfaces.

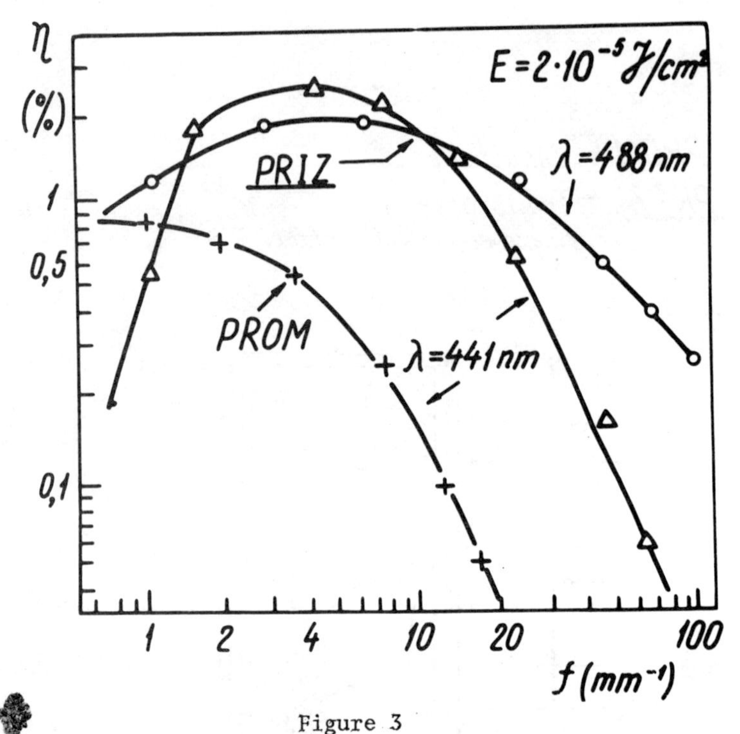

Figure 3
Diffraction efficiency of the PRIZ (0,Δ) and PROM (+) as a function of spatial frequency.

The diffraction characteristics of the PRIZ can display a sharp anisotropy as a result of the anisotropic properties of the crystal. Figure 5 shows the diffraction efficiency plotted as a function of the angle between the wave vector of the grating and the crystallographic axis for a [111]-cut crystal. The solid lines show the calculated dependence of η for linear and circularly polarized readout light. The spatial frequency of the grating was 5 cycles/mm. The linear polarization plane was oriented along the [112] axis. For the case of linear polarization, the theoretical curve is seen to be in good agreement with the experimental data.[12]

The behavior of the diffraction efficiency shown in Figure 5 implies that in the readout process with linearly polarized light, the modulator does not reconstruct all the spatial frequencies, but only those whose wave vector lies in a certain angular range, i.e., the device accomplishes a selective filtering of the images. When reading out with circularly polarized light and the crystal plate oriented along [111], no selective filtering exists. The theoretical description of these effects will be published in Reference (12).

Figure 6 shows an example where the PRIZ modulator has been used in an optical matched filter for the operation of a dual on-line image correlator.

Figure 4

Image contouring by the PRIZ (a [110] cut).
Some directions are seen in which the image
is not reconstructed due to sector filtering.

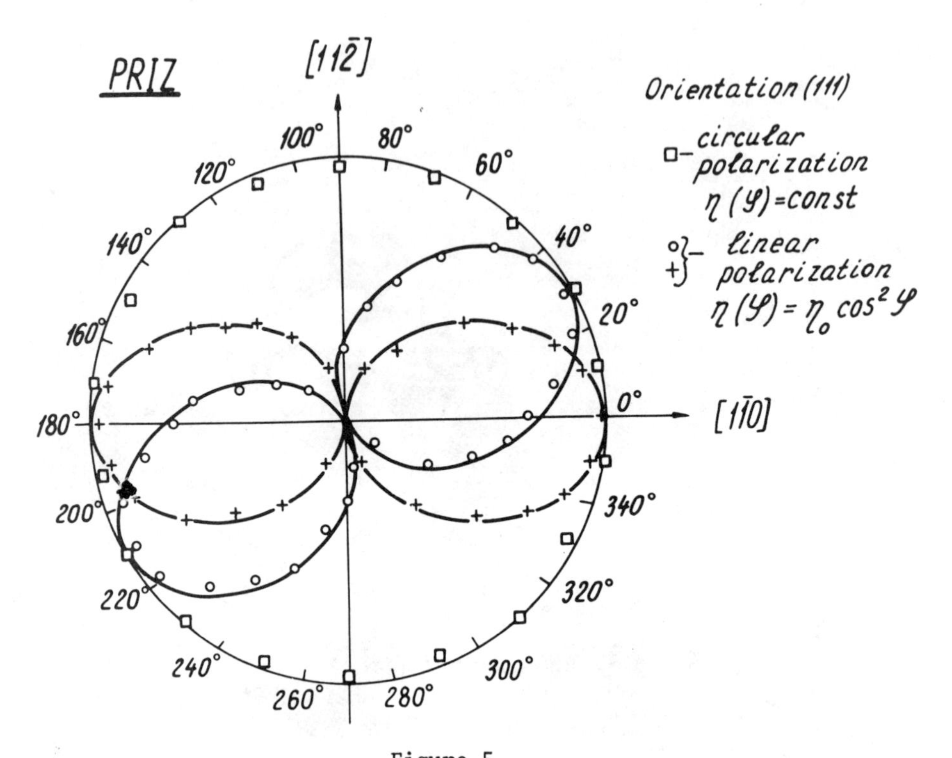

Figure 5

Angular dependence of diffraction efficiency of the PRIZ, a [111] cut. Two different dependences at linear polarization (+,0) are obtained for different polarities of the external applied field.

The PRIZ can also be used for electric signal processing, and for the compression of a LFM signal, in particular. In this case, a signal can be written as a LFM grating on the modulator, while the readout can be performed with a coherent light beam. At some distance from the modulator, one of the diffracted beams is compressed into a narrow strip.

Figure 7 illustrates the result of the signal compression obtained with the PRIZ, (deviation 8 cycles/mm, ℓ = 15 mm). The signal was recorded on the modulator from a slide illuminated by the He-Cd laser light. The experimental value of the compression coefficient was about 100, while the theoretical compression for such a signal is K_c = 120.

While investigating the image recording and readout processes with a slightly modified PRIZ modulator, the dynamic image selection effect was discovered. This effect consists of the selection of the non-stationary part of a pattern by the modulator.[13] As one can see from Fig. 8, an image appears in the modulator after switching on the recording light; the image begins to fade away (in this case, in approximately 0.5 sec.), only to reappear after the recording beam is switched off.

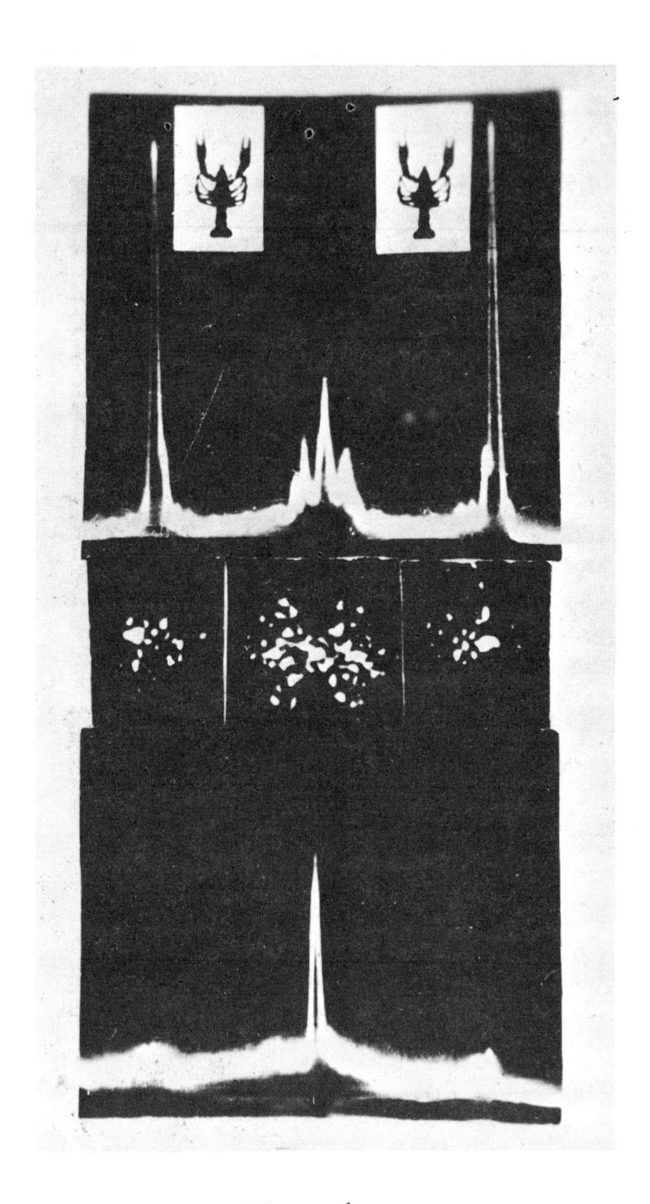

Figure 6

PRIZ application in an optical correlator. At the top, lobster image and cross-sectional scan of intensity of the correlation peak. In the middle, photograph of the light field intensity distribution in the correlation plane. At the bottom, cross-sectional scan of the correlation picture when the image of one of the lobsters is removed.

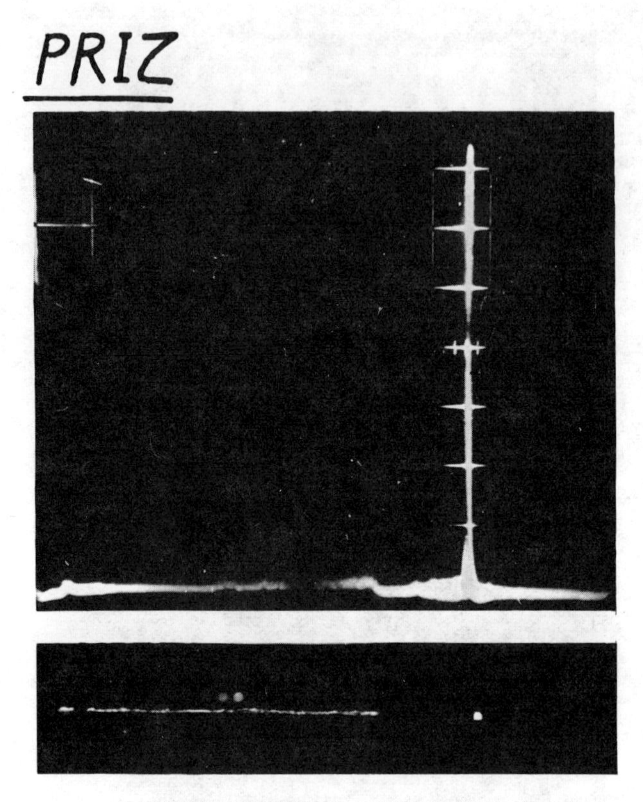

COMPRESSED PULSE

LFM PULSE

Figure 7

Compression of a frequency-modulated signal. At the top, cross-sectional scan of the output image. In the middle, a photograph of the output image.

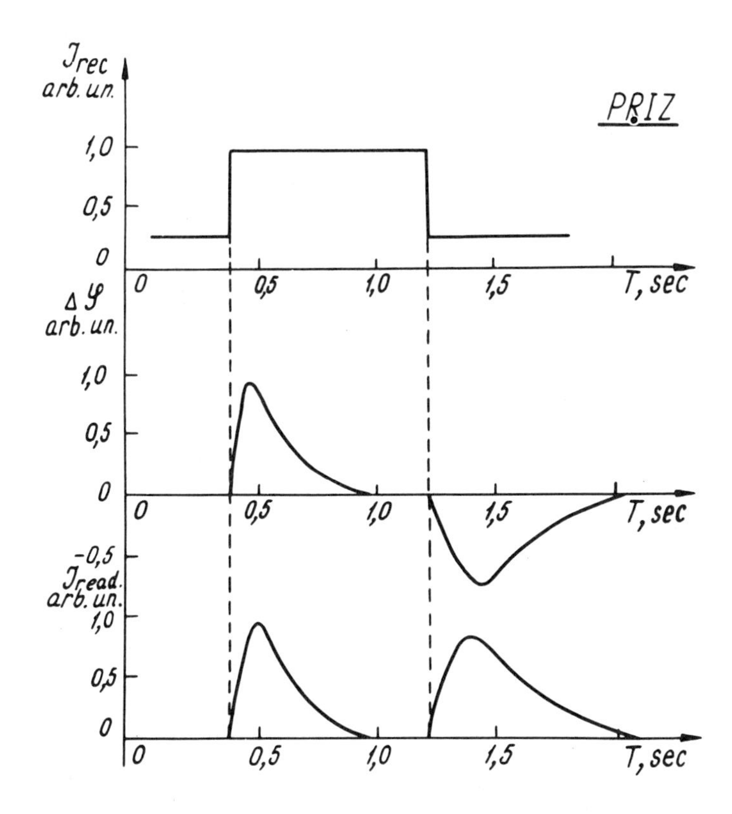

Figure 8
Time dependence of a recorded image (I_{rec}), phase modulation ($\Delta\phi$), and reconstructed image (I_{read}) in the PRIZ

The next figure (Fig. 9) shows the strip image. If the strip is stationary, its image is absent; if the image is set in motion, the leading and trailing edges become visible. The graph shows the intensity of the leading and trailing edges of the image as a function of the velocity of the strip. Thus, the time varying parts of the image are retrieved by the modulator. These effects are due to the redistribution of the volume charge that results from the image changes and are responsible for the operation of the device.

The available experimental data on the PRIZ response to nonstationary images imply that this modulator is really dynamic and that it ensures a continuous image processing, while in most known cases, discrete frame-to-frame processing is carried out. As a result, it becomes important to describe the modulator characteristics in such a way as to adequately reflect their nonstationary nature.

In this work, a phenomenological approach to the description of a dynamic modulator is developed in terms of linear systems theory.

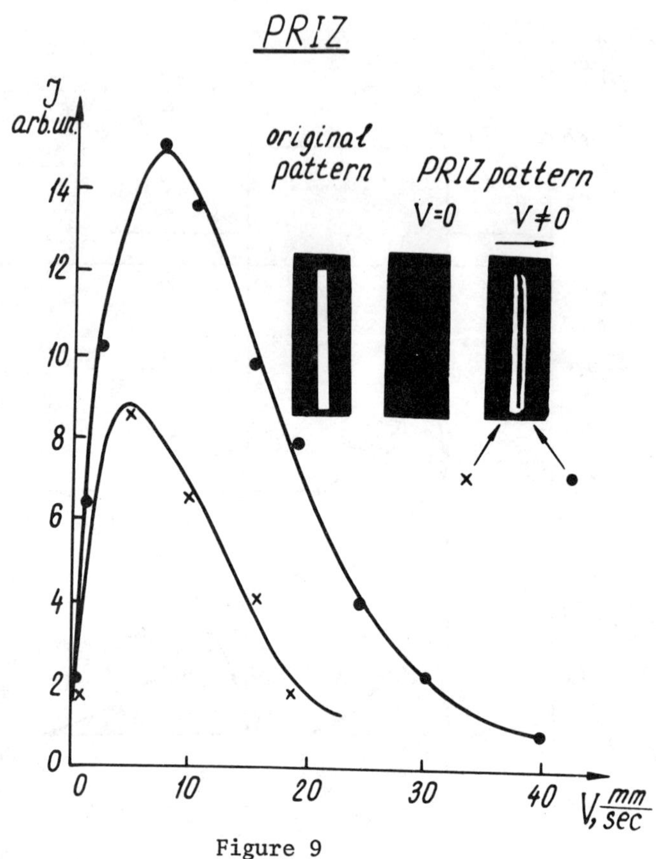

Figure 9

Image intensity of the leading and trailing edges of a strip as a function of its velocity in the plane of the modulator PRIZ aperture.

The main difference between the present analysis and the conventional approach is the fact that we analyze nonstationary images. So the Fourier analysis is performed not in terms of standing nonpropagating waves (with corresponding spatial frequencies), but in terms of running waves, each of them being characterized by a spatial and a temporal frequency.

Let us consider the response of a dynamic optically controlled modulator to a writing light having an intensity distribution $F(x,y,t)$ in the modulator plane. We shall assume the modulator to be sufficiently thin in order to be able to neglect the volume character of the image recording process. In the general case, a modulator or a photosensitive medium can be described by a second rank tensor $\hat{P}(x,y,t)$. Under the influence of the recording light, $\hat{P}(x,y,t)$ changes in such a way that

$$\hat{P}(x,y,t) = \hat{P}_0(x,y) + \Delta\hat{P}(x,y,t) \qquad (3)$$

where $\hat{P}_0(x,y)$ is a Jones matrix. In the following discussion we shall consider only $\Delta\hat{P}(x,y,t)$.

According to linear systems theory, $\Delta\hat{P}(x,y,t)$ may be represented by

$$\Delta\hat{P}(x,y,t) = \iiint \hat{K}(K_x,K_y,\omega)\cdot S(K_x,K_y,\omega)\, e^{-i(K_xX+K_yY+\omega t)}\, dK_x dK_y d\omega \quad (4)$$

where

$$S(K_x,K_y,\omega) = \iiint F(x,y,t)\, e^{i(K_xX+K_yY+\omega t)}\, dX\, dY\, dt$$

and K_x, K_y and ω are spatial and temporal frequencies, respectively. $\hat{K}(K_x, K_y, \omega)$ is the transfer function of a photosensitive modulator which is acted on by a beam of writing light. This function characterizes the variations in the diffraction properties of the modulator.

The transfer function of the modulator with respect to a readout coherent light beam, instead, is given by

$$\hat{H}(K_x,K_y,t) = \int \hat{K}(K_x,K_y,\omega)\, S(K_x,K_y,\omega)\, e^{-i\omega t}\, d\omega, \quad (5)$$

and the amplitude distribution of a coherent laser beam in the Fourier plane of the modulator is

$$\vec{E}(K_x,K_y,t) = \left(\iint \Delta P(x,y,t)\, e^{i(K_xX,K_yY)}\, dxdy\right) \vec{E}_r \quad (6)$$

Here, $\vec{E}_r$ is the amplitude of the readout plane wave.

Thus, a signal in the Fourier plane is not a simple Fourier transform of the input image, but a convolution of the temporal variables from the corresponding space-time Fourier components. In addition, the diffraction efficiency of a modulator can be determined as follows:

$$\eta(K_x,K_y,\omega) = |\hat{K}(K_x,K_y,\omega)\, S(K_x,K_y,\omega)|^2 \quad (7)$$

In the case of a traditional (non-dynamic) space-time modulator and isotropic media, the parameter $\hat{K}(K_x,K_y,\omega)$ is proportional to the well-known value of the unnormalized modulation amplitude $M_A(U')$.[14]

In order to determine experimentally the characteristic transfer function $\hat{K}(K_x,K_y,\omega)$ of the PRIZ, a sinusoidal grating with a spatial frequency K_y was written on the modulator using blue light, and the amplitude of the grating was time-modulated with a frequency ω. The intensity of the readout red light in the Fourier plane was then measured.[12] Figure 10 shows the behavior of $\hat{K}(K_x,0,\omega)$ as a function of K_x for fixed ω, and as a function of ω for fixed K_x for a modulator with a [111] cut crystal and K_x parallel to the [$\bar{1}$10] plane. It is seen from the figure that the frequency characteristic $\hat{H}(K_x,0,\omega)$ falls to zero at $\omega = 0$, and grows linearly for small values of ω. Such a behavior of the frequency characteristic indicates that in this mode of operation the PRIZ does not transmit zero temporal frequencies, (i.e., stationary images) but provides a continuous selection of the nonstationary components of the image. For small values of $\omega/2\pi$ in the interval 0-3 Hz, the dynamic selection is equivalent to a differentiation of the images with respect to time.

It can be noted that removal of the stationary part of the image by the PRIZ and selection of only the nonstationary part is reminiscent of the response of the frog's eye, which is known to react mainly to moving objects.

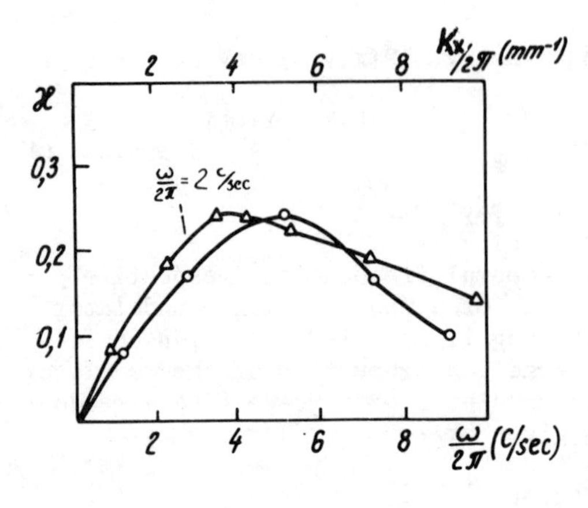

Figure 10
PRIZ transfer function as a function of spatial frequency $K_x/2\pi$ (Δ) at a temporal frequency $\omega/2\pi$ = 2 cycles/sec and as a function of temporal frequency $\omega/2\pi$ at $K_x/2\pi$ = 5 cycles/mm (o).

The second curve in Fig. 10 shows the transfer function plotted against spatial frequencies. As seen from Fig. 10, a constant spatial component (a constant background) is also removed by the PRIZ.

The development of the dynamic modulator PRIZ allows us to expand the field of application of optical methods and to perform continuous processing of nonstationary two-dimensional information.

For example, if a coherent light amplitude varies in time as $\vec{E}(X_1,Y_1,t)$ in the input plane, and a time dependent filter $\Delta\hat{P}(X_2,Y_2,t)$ is placed in the Fourier plane of a coherent optical system, then the light amplitude distribution in the correlation plane is given by

$$\vec{E}(X_3,Y_3,t) = \iint \Delta\hat{P}(X_2,Y_2,t)\cdot\vec{E}(X_2,Y_2,t)\,e^{-i(X_2X_3+Y_2Y_3)}\,dX_2dY_2 \quad (7)$$

where

$$\vec{E}(X_2,X_2,t) = \iint \vec{E}(X_1,Y_1,t)\,e^{i(X_2X_1+Y_2Y_1)}\,dX_1dY_1$$

An analysis of Eq. (7) indicates that the application of a continuously controlled filter can allow recognition of nonstationary patterns, multichannel phase detection of signals, analysis of image scale variations and a number of other operations, connected with a space-time filtering of two-dimensional images. The utilization of a dynamic filter in systems with feedback is also of interest.

In conclusion, it is necessary to make one more remark concerning the analysis given above. The analysis was performed in the framework of the linear theory. However, linearity conditions are satisfied in the modulator PRIZ over a limited dynamic range. Thus, in practice, the parameters given above are not enough to evaluate completely the range of application of the device. From a scientific point of view, on the other hand, this approach may be useful as a starting point for additional studies of the problem.

ACKNOWLEDGEMENT

The author wishes to express his sincere gratitude to his colleagues S.I. Stepanov, A.V. Khomenko, M.V. Krasinkova, A.A. Kamshilin, V.I. Marakhonov, M.G. Shlidgin for helpful discussions.

REFERENCES

(01) F.S. Chen, T.T. La Macchia, D.B. Fraser, Appl. Phys. Lett., 13, 223 (1968).

(02) D.L. Staebler, "Ferroelectric Crystals" in "Holographic Recording Materials", ed. by H.M. Smith, Springer-Verlag, Berlin, Heidelberg, NY (1977).

(03) S.I. Stepanov, M.P. Petrov, A.A. Kamshilin, Pis'ma J.T.Ph. (in Russian), 3, 849 (1977); Ferroelectrics, 21, 631 (1978). S.I. Stepanov, A.A. Kamshilin, M.P. Petrov, "Pecularities of the Holographic Recording in Birefringent Electrooptic Crystals", Proc. of IX All-Union School on Holography (1977), Leningrad (in Russian); Avtometria, 1, 16 (1978).

(04) T.G. Pencheva, M.P. Petrov, S.I. Stepanov, Avtometria, 1, 122 (1980) (in Russian).

(05) H. Kogelnik, Bell Syst. Tech. J., 48, 2909 (1969).

(06) M.P. Petrov, S.I. Stepanov, A.A. Kamshilin, Optics and Laser Technology, 11, 149 (1979).

(07) M.P. Petrov, S.I. Stepanov, A.A. Kamshilin, Opt. Comm., 29, N1, 44 (1979).

(08) J.B. Thaxter, Appl. Phys. Lett., 15, 210 (1969); O. Milami, Opt. Comm., 19, 42 (1969); T. Yasuhira, et al. Appl. Opt., 16, 2532 (1977).

(09) S.L. Hou, D.S. Oliver, Appl. Phys. Lett., 18, 325 (1971). J. Feinleib, D.S. Oliver, Appl. Opt., 11, 2752 (1972).

(10) M.P. Petrov, A.V. Khomenko, Ferroelectrics, 22, 651 (1978). M.P. Petrov, A.V. Khomenko, V.I. Berezkin, M.V. Krasinkova, Microelectronika, 8, N1, 20 (1979) (in Russian); Pis'ma J.T.Ph., 5, 334 (1979) (in Russian).

(11) M.P. Petrov, V.I. Marakhonov, M.G. Shliagin, A.V. Khomenko, M.V. Krasinkova, Pis'ma J.T.Ph., 6, 385 (1980) (in Russian).

(12) M.P. Petrov, A.V. Khomenko, V.I. Marakhonov, M.G. Shliagin, M.V. Krasinkova, J. of T. Phys., to be published (in Russian).

(13) M.P. Petrov, A.V. Khomenko, V.I. Marakhonov, M.G. Shliagin, Pis'ma J.T. Ph., 6 , 386 (1980); "Investigation of physical processes of recording information and light diffraction in complex layer structures on the base of electrooptic crystals", Proceedings of 2nd School on Optical Processing and Information, 1979, Leningrad.

(14) D. Casasent, Appl. Opt., 18, 2445 (1979).

GENERALIZED DESIGN CONSIDERATION OF HOLOGRAPHIC SCANNERS WITH ABERRATION CORRECTIONS

Yuzo Ono, Nobuo Nishida and Mitsuhito Sakaguchi
Central Research Laboratories, Nippon Electric Co., Ltd.
1-1, Miyazaki 4-chome, Takatsu-ku, Kawasaki, 213 Japan

ABSTRACT

Hologram generating method for laser scanners with aberration corrections has been proposed and demonstrated. Aberration correction consideration has led to the concept of the generalized holographic zone plate. To generate the holographic zone plate, the holographic technique for optical product operations, between a plurality of coherent spherical waves, has been applied. When spherical waves, N in number, are applied, aberration can be corrected in a scan length magnification equal to or less than N^2. Experimentally, remarkable aberration correction effect of this method holograms has been verified for scan lengths over 50 cm.

INTRODUCTION

Holographic laser scanners have been expected for bar code symbol reader applications in POS systems[1] and for document reader applications[2], because of their simple optical arrangement and easy manufacture with duplication.

It is known that desired spatial frequency characteristics of holograms for use as laser scanners are those of off-axis Fresnel zone plates[3,4], called geometric zone plates (GZP)[5]. A phase variation on the hologram is[4]

$$\phi_G(r) = \frac{\pi r^2}{\lambda F}, \tag{1}$$

where F is hologram focal length, λ is scanning beam wavelength and r is radial coordinate on the hologram. The radius of the n'th interference fringe is obtained by setting $\phi_G(r)$ equal to $2\pi n$ as

$$r_G(n) = \sqrt{2n\lambda F}\,. \tag{2}$$

Holograms generated by the interference between a divergent spherical wave and a collimated reference beam, called interferometric zone plates (IZP)[5], however, are conventionally used for GZP. In recording geometry shown in Fig. 1, a phase variation on a hologram plane is given by

$$\phi_I(r) = \frac{2\pi}{\lambda}\left(\sqrt{r^2+F^2}-F\right). \tag{3}$$

The n'th interference fringe radius is

$$r_I(n) = \sqrt{2n\lambda F + (n\lambda)^2}\,. \tag{4}$$

Radii $r_I(n)$ can be approximated to the radii $r_G(n)$, only when

ISSN:0094-243X/81/650508-07 $1.50

$$F \gg n\lambda/2. \tag{5}$$

In many cases, the scanning plane is set farther away than the focal plane to obtain the scan length required for actual applications. The hologram is illuminated with a spherical divergent wave to converge the diffracted laser beam on the scanning plane. Such an illumination wave, however, is not the same as the original recording beam. Therefore, when the IZP holograms are used, aberration is caused in large diffraction, or large scanning angle. The most serious aberration is astigmatism.

Lee has demonstrated a method for generating holograms with aberration corrections by using computer generated holograms[4]. Ikeda has presented a method to shift an aberration-free point of the IZP to a certain diffraction angle[1].

In order to correct the aberration, it is essential to generate a holographic zone plate that has the phase variation described by Eq.(1). This paper discusses aberration properties of holograms generated by using a plurality of coherent spherical waves. A procedure for generating the holograms is described. The feasibility of using the holograms as laser scanners is demonstrated.

ABERRATION PROPERTIES OF HOLOGRAMS GENERATED BY USING PLURAL SPHERICAL WAVES

Since the holographic zone plate is equivalent to a usual optical lens except that it is a diffractive element, the IZP corresponds to a plano-convex lens. In optical lens design, aberration is corrected by introducing a constitution of plural spherical lenses combination. It is considered that the analogical constitution can be introduced in the case of holographic zone plate. This consideration leads to the general phase variation in the holographic zone plate being

$$\phi_N(r) = \frac{2\pi}{\lambda}\sum_{k=1}^{N}\left(\sqrt{r^2+f_k^2}-f_k\right). \tag{6}$$

Equation (6) is interpreted as a transfer function of a lens system consisting of N spherical lenses, whose principal planes are arranged identically. Focal lengths are f_k's and the combined focal length F is

$$\frac{1}{F} = \sum_{k=1}^{N}\frac{1}{f_k}. \tag{7}$$

When f_k=NF, the n'th interference fringe radius for the generalized holographic zone plate described in Eq.(6), is

$$r_N(n) = \sqrt{2n\lambda F + (n\lambda/N)^2}. \tag{8}$$

The conventional IZP and the GZP correspond to N=1 and N=∞, respectively. The deviation from GZP is reduced to $1/N^2$, compared with the conventional IZP. It can be approximated to GZP when

$$F \gg n\lambda/(2N^2). \tag{9}$$

The holographic method for generating the phase variation described in Eq.(6) will be discussed in a following session.

Scan length magnification M, an important parameter in applications, is defined as

$$M \equiv b(0)/F, \tag{10}$$

where b(0) is the distance from the hologram to the scanning plane.

The aberration-free scan length magnification of the conventional IZP is only one, because a part of the wavefront used during recording is reconstructed with a collimated beam illumination. The above results give a useful design guide for aberration corrections. Equation (9) indicates that the same aberration properties are obtainable for the hologram, whose focal length is $1/N^2$ of that of the conventional IZP. Therefore, the scan length magnification of N^2 is expected for the generalized holographic zone plate.

Aberration properties of holograms deduced above is analyzed in the following.

Figure 2 shows a reconstruction geometry for laser beam scanner applications. A spherical wave divergent from a point source S, which has distance a from the hologram, illuminates the hologram in normal incidence. The wave is diffracted in the radial direction. Astigmatism is caused in that direction. The ray-tracing method for one dimension, in the radial direction, is applied in analyzing the aberration properties. Convergence point P for the diffracted beam is defined as the cross point of the two rays diffracted at points $A(r + \Delta r)$ and $B(r - \Delta r)$. Image distance b(r) is the distance between point P and the hologram. The hologram is equivalent to a usual convex lens. The image relation on the axis is,

$$\frac{1}{a} + \frac{1}{b(0)} = \frac{1}{F}. \tag{11}$$

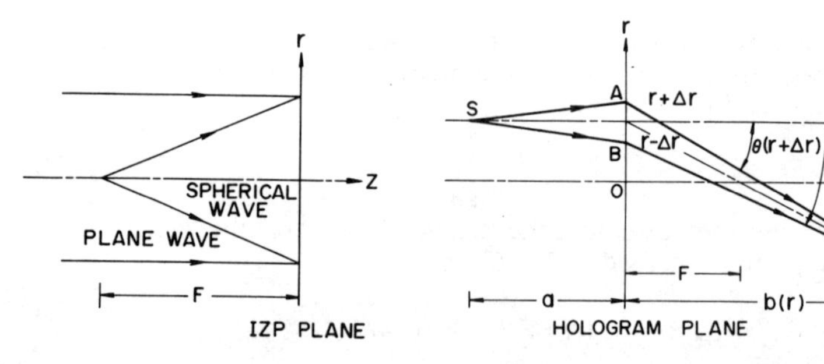

Fig. 1. Conventional IZP hologram recording geometry. Divergent spherical wave interferes with a plane wave on the IZP plane.

Fig. 2. Generalized holographic zone plate reconstruction geometry. Spherical wave divergent from point S is diffracted and converges at point P.

The radial component of the hologram grating vector K(r) is

$$K(r)=\frac{2\pi}{\lambda}\sum_{k=1}^{N}\frac{r}{\sqrt{r^2+f_k^2}}. \tag{12}$$

Diffraction angles $\theta(r\pm\Delta r)$ at points A and B are given as

$$\theta(r\pm\Delta r)=\mathrm{Sin}^{-1}\left[\frac{\lambda}{2\pi}\left\{\frac{2\pi}{\lambda}\mathrm{Sin}\left(\mathrm{Tan}^{-1}\frac{\pm\Delta r}{a}\right)-K(r\pm\Delta r)\right\}\right]. \tag{13}$$

Image distance b(r) is expressed as

$$b(r)=\frac{2\cdot\Delta r}{\tan\theta(r+\Delta r)-\tan\theta(r-\Delta r)}. \tag{14}$$

From Eqs.(12) to (14), the variation in image distance with respect to the diffraction angle of the beam central ray is calculated. Figures 3 and 4 show the results in the magnification M=4 and 9, respectively, where image distance is normalized with focal length F. These results indicate that the variation in image distance can be corrected by selecting a proper combination of f_k's within a range $M\leq N^2$, as previously theoretically predicted.

PRACTICAL PROCEDURE FOR GENERATING HOLOGRAMS

The optical product operation between coherent waves, a well known feature of holograms, can be applied in generating the phase variation required for a generalized holographic zone plate. When a hologram is illuminated with a plane wave, two wavefronts conjugate each other are reconstructed. They are conjugate product waves between the two waves recorded, that is, they have difference phases between the two waves. When this technique is repeatedly applied for N number of spherical waves, the phase variation expressed by Eq.(6) can be generated.

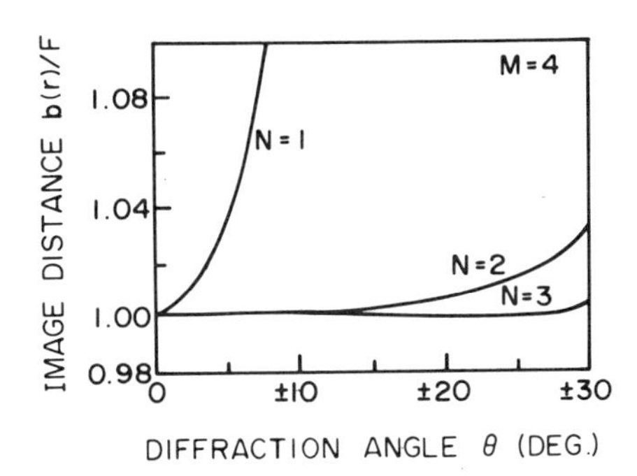

Fig. 3. Variation in image distance for magnification M=4. Spherical waves combinations $f_1=f_2=2F$ for N=2 hologram and $f_1=f_2=5.2F$, $f_3=1.625F$ for N=3 hologram are used.

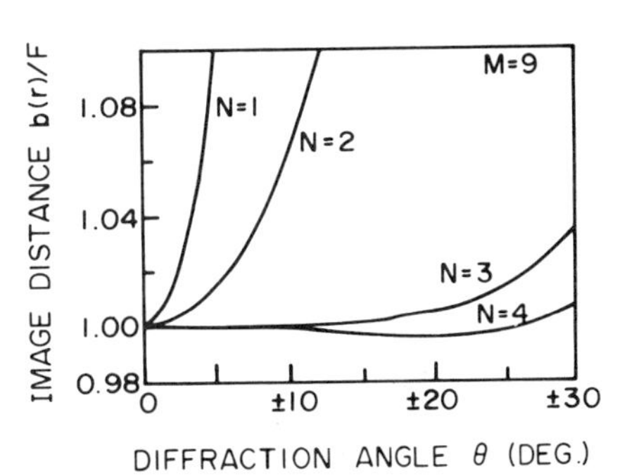

Fig. 4. Variation in image distance for magnification M=9. Spherical waves combinations $f_1=f_2=2F$ for N=2 hologram, $f_1=f_2=f_3=3F$ for N=3 hologram and $f_1=f_2=7.8F$, $f_3=f_4=2.69F$ for N=4 hologram are used.

The practical process is shown in Fig. 5. In the first step, interference fringe between divergent and convergent spherical waves is recored on photographic plate H_1, as shown in Fig. 5(a). The hologram is that of N=2 and can be used for laser scanner within M≦4. When hologram H_1 is illuminated with a plane wave P_1, as shown in Fig. 5(b), the wavefront which has the phase variation described by Eq.(6) with N=2 is reconstructed on the hologram plane.

In the second step, in order to separate the reconstructed wave from other diffracted waves, hologram H_2 is recorded by making the reconstructed wave interfere with a plane wave P_2 on a plane where the reconstructed wave does not overlap with other diffracted waves. When the hologram H_2 is illuminated with a plane wave P_3, which propagates in a direction opposite to P_2, on the opposite side of the recording plane as shown in Fig. 5(c), the same wave as the reconstructed wave from the hologram H_1, that propagates in the opposite direction, is reconstructed.

In the third step, this reconstructed wave interferes with the third spherical wave on the virtually same plane as plane H_1. The resultant hologram H_3 is the N=3 hologram. Spherical waves used in this process are shown in Fig. 5(d). The divergent and convergent points of the spherical waves are arranged in an identical axis normal to the hologram plane.

When procedures for the second and the third steps are repeated by using resultant holograms of the third step instead of hologram H_1, a hologram which has the phase variation expressed by Eq.(6) with arbitrary number N can be generated. In the above, the procedure is explained by using the first-order diffraction wave, but the other-order diffraction wave also can be used.

EXPERIMENTAL

In order to demonstrate the feasibility of this method, holograms for

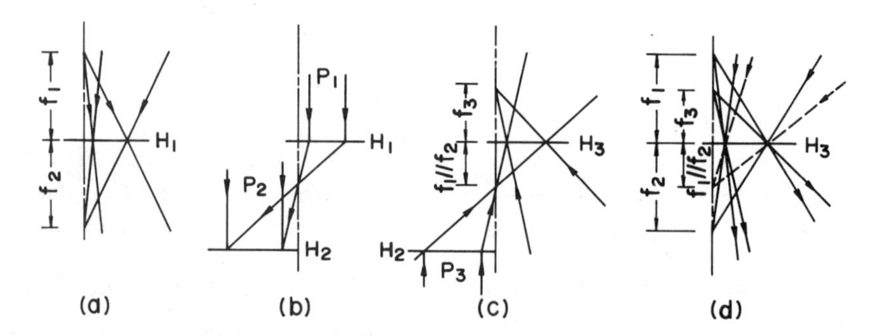

Fig. 5. Phase generating process for N=3 hologram. First step (a): Two spherical waves interfere on H_1 plane. Second step (b): Reconstructed wave from H_1 interferes with a plane wave P_2 on H_2 plane. Third step (c): Reconstructed wave from H_2 interferes with the third spherical wave on H_3 plane. Spherical waves used are shown in (d) together. Dashed line means the product wave between the first and second spherical waves.

N=1, 2 and 3 were made. He-Ne lasers (wavelength λ=6328 Å) were used for both recording and reconstruction. The hologram recording material was Kodak 649F plate.

The hologram for N=1 was made by using a divergent spherical wave where F=100 mm. The same focal length hologram for N=2 was made in accordance with the geometry shown in Fig. 5(a) with $f_1=f_2$=200 mm. Photographs of laser beam spots on the scanning plane in the magnification M=5 are shown in Fig. 6.

The hologram for N=3 was also made in accordance with the above mentioned procedure, with $f_1=f_2=f_3$=200 mm and focal length was F=66.7 mm. The laser spot photographs on the scanning plane in the magnification M=9 are shown in Fig. 7. In these experiments, the off-axis part of the holograms was made and used.

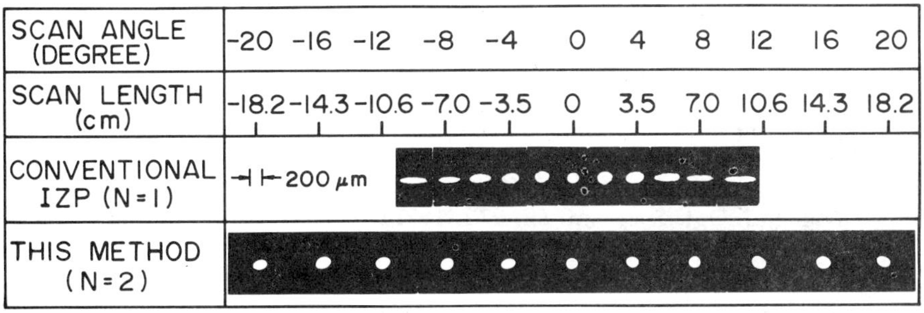

SCAN ANGLE (DEGREE)	-20	-16	-12	-8	-4	0	4	8	12	16	20
SCAN LENGTH (cm)	-18.2	-14.3	-10.6	-7.0	-3.5	0	3.5	7.0	10.6	14.3	18.2
CONVENTIONAL IZP (N=1)	⊣⊢200 µm										
THIS METHOD (N=2)											

Fig. 6. Beam spot photographs on the scanning plane (M=5 and b(0)=500 mm), reconstructed from a conventional IZP and a generalized holographic zone plate (N=2 and $f_1=f_2$=200 mm). Beam radius in the scan center is 100 µm.

SCAN ANGLE (DEGREE)	-22.6	-18.4	-14.0	-9.5	-4.8	0	4.8	9.5	14.0	18.4	22.6
SCAN LENGTH (cm)	-25	-20	-15	-10	-5	0	5	10	15	20	25
THIS METHOD (N=3)											

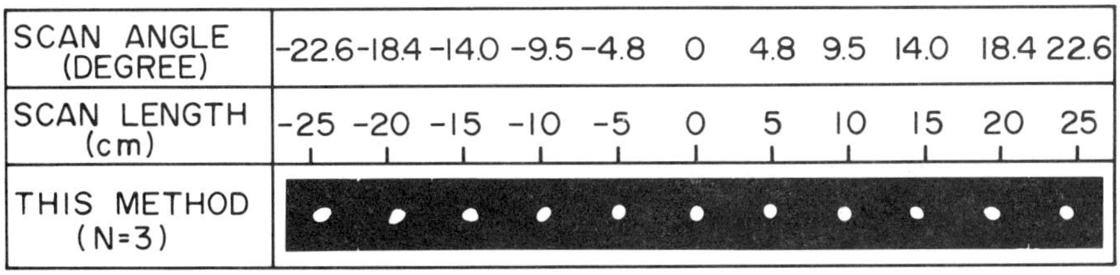

Fig. 7. Beam spots photographs on the scanning plane (M=9 and b(0)=600 mm), reconstructed from a generalized holographic zone plate (N=3 and $f_1=f_2=f_3$=200 mm). Beam radius in the scan center is 100 µm.

The scanning spots reconstructed from the conventional hologram (N=1) are enlarged in a scan direction with increasing in scan angle, as shown in Fig. 6. In contrast with this, the scanning spots reconstructed from holograms of N=2 and 3 are fixed as shown in Figs. 6 and 7.

These experimental results indicate the feasibility of aberration corrections for generalized holographic zone plates.

SUMMARY

A method for generating holograms for use as laser scanners with aberration corrections has been proposed and demonstrated. Aberration correction consideration has led to the concept of the generalized holographic zone plate. The holographic technique for optical product operations, between a plurality of coherent spherical waves, has been applied to generate holograms. When spherical waves, N in number, are applied, aberration can be corrected in a scan length magnification equal to or less than N^2. Experimentally, remarkable aberration correction effect of this method holograms has been verified for scan lengths over 50 cm.

REFERENCES

1. H. Ikeda, et al., Appl. Opt. 18, 2166 (1979).
2. R.V. Pole, et al., Appl. Opt. 17, 3294 (1978).
3. O. Bryngdahl and W.H. Lee, Appl. Opt. 15, 183 (1976).
4. W.H. Lee, Appl. Opt. 16, 1392 (1977).
5. M. Young, J. Opt. Soc. Am. 62, 972 (1972).

SURFACE PLASMON HOLOGRAPHY

J. J. Cowan

Research Laboratories, Polaroid Corporation, 750 Main St.
Cambridge, Massachusetts 02139

ABSTRACT

A new method of hologram formation is considered, whereby the reference beam is a surface plasmon. When one face of an equilateral prism is coated with a thin (30 nm) layer of Ag which in turn is overcoated with a 400 nm layer of positive photoresist, the surface plasmon can be stimulated in the photoresist layer by prism coupling using a collimated beam of light polarized parallel to the plane of incidence (p-polarized). If object light is simultaneously incident on the photoresist layer, a hologram will be formed as a result of interference between this beam and the surface plasmon. After development, reconstruction can be carried out by re-exciting the surface plasmon. This type of hologram has the characteristics of being polarization dependent, using integrated optics components,and having the reference beam always out of the field of view.

INTRODUCTION

We consider a new type of hologram formation, whereby the reference beam, instead of being an ordinary plane wave, is a surface plasmon. The surface plasmon, sometimes referred to as a surface plasma wave, may be regarded as a collective electronic oscillation at the surface of a free-electron-like metal. It has the characteristics of being a bound surface wave whose amplitude decays exponentially away from the interface into either the metal or vacuum; it always propagates at a velocity less than the free-space speed of light, c, and it can only be excited by light polarized parallel to the plane of incidence (p-polarized). The frequency-wave number relation is dispersive and is given by the equation $k = (\omega/c)\left[\epsilon_1/(1 + \epsilon_1)\right]^{1/2}$, where k is the wave number, ω is the frequency, and ϵ_1 is the real part of the dielectric constant of the metal.

ISSN:0094-243X/81/650515-04$1.50

Because the surface plasmon wave number is larger than that for free space light at any frequency, it can only be excited by light using special coupling techniques; specifically, by prism coupling or grating coupling. In earlier work we had considered hologram formation by grating coupling;[1] in the present work we consider prism coupling. The method consists of coating one face of a prism with a thin metal layer, in this case 30 nm of Ag, and allowing a collimated beam of p-polarized light to be incident from the back of the prism onto the metal layer. Because the wave number of light within the prism is larger than that in free space, it is possible to exactly match the plasmon wave number by choosing the proper angle of incidence. When this occurs the surface plasmon is resonantly excited.

SURFACE PLASMON GUIDED MODES

When the metal film is overcoated with a dielectric layer of sufficient thickness, and the same prism coupling scheme as described above is used, several plasma modes may be possible. For example, with a 400 nm thickness of photoresist, three modes at optical frequencies can be excited at three separate angles of incidence. In this case the surface wave loses some of its bound character and becomes spread out through the dielectric layer. The modes are of the TM type and correspond to standing plasma waves in a direction transverse to the plane of the film; TM_o is the fundamental plasmon mode, TM_1 is the first harmonic, and TM_2 the second harmonic.

SURFACE PLASMON HOLOGRAPHY

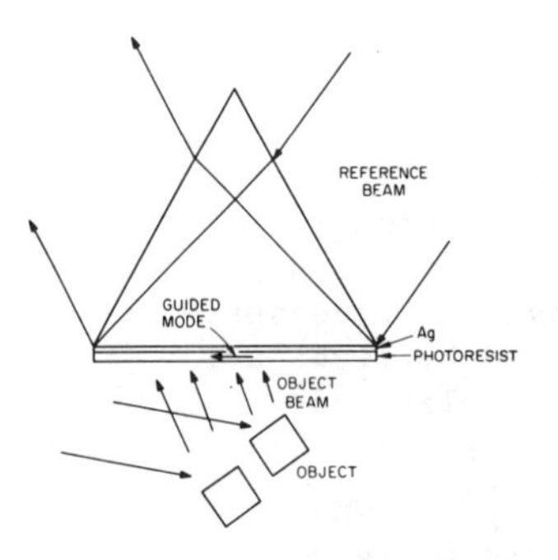

Fig. 1. Formation of hologram using surface plasmon guided mode as the reference beam

In the present configuration the overcoating dielectric layer and the recording layer are identical; a 400 nm layer of photoresist (Shipley AZ 1300). Figure 1 shows how the hologram is formed. A 488 nm beam of p-polarized light from an argon ion laser is split so that one part illuminates the object, and the other is expanded, collimated, and incident on the coupling prism from the back side so that one of the harmonic plasmon guided modes is excited.

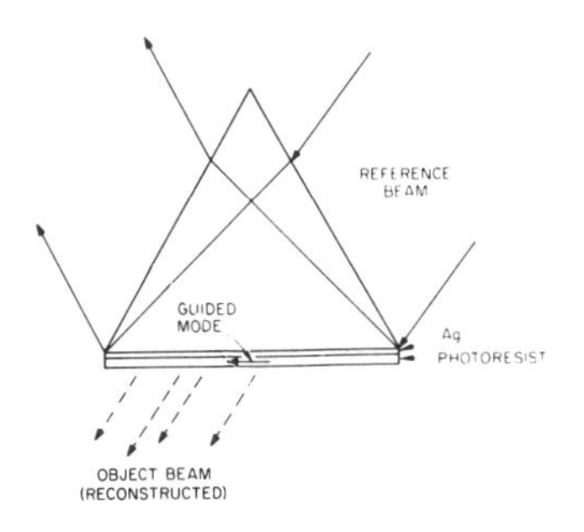

Fig. 2. Method of reconstruction of surface plasmon hologram

Light reflected from the object is incident on the photoresist from the front side and the hologram is formed by interference of the object light with the guided plasmon mode. Upon development the photoresist will etch to a depth proportional to the incident interference intensity. If the etch depth is small compared to the total dielectric thickness, the mode characteristics will be only slightly perturbed, and thus the reconstruction angle will be nearly the same as the angle of construction. Fig. 2 shows how the hologram is reconstructed, this being done by reexciting the surface plasmon. The reconstructed object light is reflected from the metal layer and can be viewed from a position in front of the prism, as shown. Fig. 3 is a picture of the reconstructed image, as seen from viewing the face of the prism. The reconstructing light is reflected completely out of the field of view, so it is never visible, although some light that strikes the edges of the prism can be seen, due to scattering.

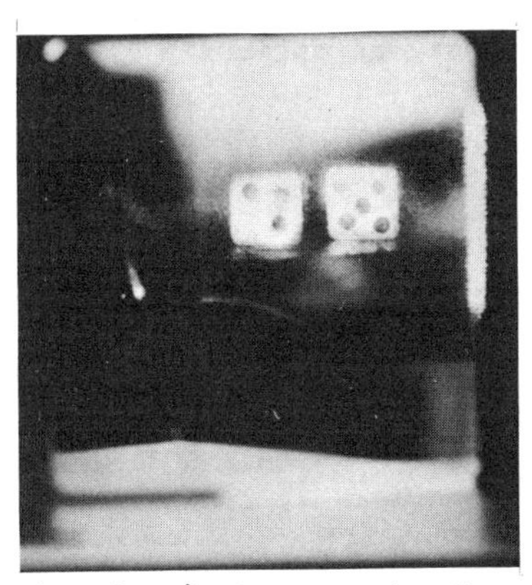

Fig. 3. a) Photograph of reconstructed image looking into prism face

b) Enlarged view of reconstructed image

SURFACE PLASMON HOLOGRAPHY CHARACTERISTICS

Holograms made with the surface plasmon as the reference beam have some interesting properties. One is that the reconstructing light, acting as it does by being totally reflected from the back side of a prism, is never in the field of view. A second characteristic is that the reconstructing process is polarization dependent. If the polarization is changed from p to s, for example, the surface plasmon is no longer excited and thus the reconstructed image disappears. Also, because the surface plasmon excitation is a resonance process that depends critically on the angle of incidence, a slight change in angle from the resonance condition will cause the reconstructed image to disappear.

A further characteristic has to do with the energy density of the surface plasmon. When excitation is achieved and the resonance condition is met, the exciting light is nearly totally absorbed in the dielectric layer. Since the layer thickness is less than a half micron, the field strength can rise to unusually high values. In the construction process this means that the amount of light that is split off to excite the plasmon can be made less than usual, and in the reconstruction process the brightness and efficiency of the reconstructed image become greater than usual.

The use of integrated optics components, including thin films and prism couplers, is important in this type of holography because of the increased significance in recent years of such components in communications and optics in general.

References

1. J. J. Cowan, "The Surface Plasmon Resonance Effect in Holography", Optics Communications, 5, 69 (1972); J. J. Cowan, "Holography with Standing Surface Plasma Waves", Optics Communications, 12, 373 (1974).

METHOD FOR INTERFEROMETRIC VIBRATION MEASUREMENT IN PRESENCE OF RANDOM PATH LENGTH FLUCTUATIONS

Hans M. Pedersen
Central Institute for Industrial Research
P.O. Box 350 Blindern, Oslo 3, Norway

ABSTRACT

It is shown theoretically and demonstrated experimentally that interferometric vibration measurements can be performed in presence of fairly strong temporal phase fluctuations. The method, which applies to any two-beam laser interferometer with phase fluctuations exceeding 2π, implies photoelectric recording and low-pass filtering of the output intensity with subsequent mean square detection of the fluctuating part of the filtered signal. For moderate phase noise the detector output exhibits the well known J_0^2-dependence on the vibration amplitude, and even for strong phase noise we have a marked amplitude dependence. Potential applications are discussed and the close analogy with time-average holographic interferometry and electronic speckle pattern interferometry is emphasized.

INTRODUCTION

Undesired path length fluctuations (phase noise) give problems in most applications of interferometry. The traditional methods for overcoming such problems have been to use passive isolation against external noise and active compensation (see, e.g., Ref.1) of object instabilities. In many cases, however, such methods are not sufficient either because the noise cannot be eliminated by isolation or because active stabilization cannot be maintained for sufficiently long measurement times. Examples of such cases are interferometric measurements through strong turbulence and measurements on living biological objects.

In this paper a new method is presented to solve the phase noise problem for interferometric vibration measurements. Rather than attempting to eliminate the path length fluctuations, we propose to utilize them to provide a random carrier wave for the interferometric fringe function which contains information about the vibration amplitude. After the output intensity of the interferometer is recorded photoelectrically, the recorded signal is low-pass filtered and the fluctuating part of the filtered signal is mean square detected.

A closely related method has been reported by Helfenstein[2]. He applied low-pass filtering of the interference signal, but instead of mean square detection he used peak-to-peak detection of the envelope signal, which is applicable only for very low frequency phase noise.

The time-domain signal processing of the present method is a counterpart to the spatial signal processing in holographic inter-

ISSN:0094-243X/81/650519-12$1.50

ferometry and ESPI (electronic speckle pattern interferometry). By these methods random spatial phase fluctuations caused by the object roughness is used to provide a speckle carrier wave for the interferometric fringe function. Indeed, the direct impetus to the present work was the good phase noise immunity obtained by combining ESPI with reference wave phase modulation and exposure modulation [3-6].

The paper is organized as follows: In section I a general description of the basic principles is given. Section II contains a more elaborate theory which is compared to experimental results in section III. Finally, the practical utilization of the method is discussed in section IV.

I. BASIC PRINCIPLES

The method applies to any two-beam laser interferometer with photoelectric recording of the output intensity. In order to illustrate the basic principles , we consider the simple Michelson interferometer shown in figure 1.

The output intensity I(t) is recorded by a PM-tube. The two mirrors M_1 and M_2 exhibit, respectively, a sinusoidal vibration with amplitude d and angular frequency ω_o, and a random noise vibration. The total phase difference between the two beams may be written,

$$\phi(t) = a \sin(\omega_o t) - n(t), \tag{1}$$

where $a=4\pi d/\lambda$ (λ=laser wavelength) and n(t) is a random process. If we assume equal intensities in the two beams the interference signal recorded by the PM-tube may be expressed,

$$I(t) = I_o(1+\cos\phi(t)) = I_o(1+\cos\{a \sin(\omega_o t)-n(t)\}), \tag{2}$$

where I_o is the mean output intensity. Equation (2) may be rewritten as,

$$I(t) = I_o(1+ \sum_{k=-\infty}^{\infty} J_k(a)\cos\{k\omega_o t-n(t)\}), \tag{3}$$

where we have made use of the relation [7],

$$\exp(ia\sin\alpha) = \sum_{k=-\infty}^{\infty} J_k(a)\exp(ik\alpha), \tag{4}$$

and J_k is the k-th order Bessel function of the first kind.

In the frequency domain the different terms in (3) represent different frequency bands centred around the harmonics $k\omega_o$ of the vibration frequency, and these terms may, to some extent, be separated by means of filtering. One easily finds the contributions to the lowest harmonics to be:

DC-band: $I_o\{1+J_0(a)\cos(n(t))\}$,

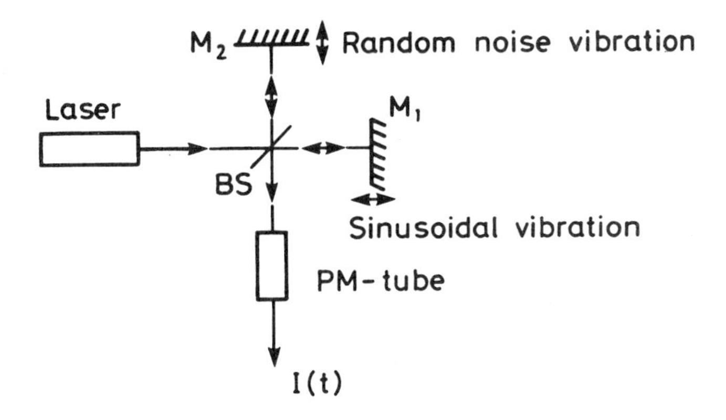

Fig. 1. Michelson interferometer set-up (M_1,M_2: mirrors; BS: beam splitter).

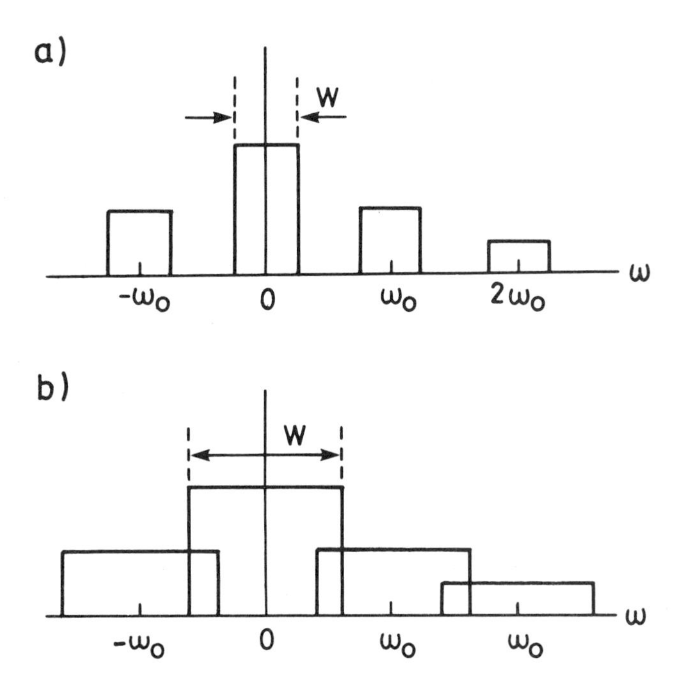

Fig. 2. Illustration of the frequency bands of I(t): (a) disjoint and (b) overlapping frequency bands (W: bandwidth of exp(in(t)), ω_o: vibration frequency).

1. harmonic band: $2I_oJ_1(a)\sin\omega_o t \sin(n(t))$,

2. harmonic band: $2I_oJ_2(a)\cos(2\omega_o t)\cos(n(t))$,

etc.

In the frequency domain the situation is as illustrated in figure 2. Figure 2(a) shows the simplest case in which the frequency bands do not overlap, i.e., the bandwidth W of exp(in(t)) is less than ω_o. In the more general case, as shown in figure 2(b), we must allow for overlapping frequency bands.

We see from figure 2 that provided W is less than $2\omega_o$, it is possible to retain contributions from the DC-band only by means of a suitably designed low-pass filter which completely removes the contributions from the higher order bands.

In the limiting case of very small W, we may remove the higher order bands without distorting the DC-band. In that case the filtered signal may be written as,

$$x(t) = I_o(1+J_0(a)\cos\{n(t)\}). \tag{5}$$

This equation has been used by Helfenstein[2] who made use of a peak-to-peak detection method to recover $J_0(a)$. By the present method we apply instead mean square demodulation of the fluctuating part of eq.(5). For the output signal, or interferometer response, we get,

$$y = \langle(\Delta x)^2\rangle = I_oJ_0^2(a)\langle\cos^2 n\rangle = \tfrac{1}{2}I_o^2J_0^2(a), \tag{6}$$

where $\langle\ldots\rangle$ denotes averaging, and where we have assumed $\langle\cos^2 n\rangle = \frac{1}{2}$, i.e., that the variation of n(t) exceeds 2π. Equation (6) exhibits the J_0^2-dependence on the vibration amplitude well known from holographic interferometry and ESPI.

In addition to the limiting case of very small W, we have an intermediate case in which the contributions from the higher order bands still may be completely removed, but the filter will distort the DC-band so that the peak-to-peak detection proposed by Helfenstein[2] cannot be applied. While the limiting case mentioned above, applies only if the bandwidth W is much less than ω_o, the intermediate case applies if this bandwidth is less than $2\omega_o$. By mean square demodulation we obtain,

$$y = I_o^2\langle(\Delta z)^2\rangle J_0^2(a), \tag{7}$$

where z(t) is a random carrier wave caused by low-pass filtering of the phase noise signal cos(n(t)). It appears that in this intermediate case the temporal phase fluctuations in the present method play a role completely analogous to that played by spatial phase fluctuations (object roughness) in holographic interferometry and ESPI. It also follows that most measurement methods developed for the latter techniques, e.g., the use of reference wave phase modulation for measurement of small vibrations[3,4], will apply also to the present technique.

In the general case we must allow for possible contributions from higher order bands in the filtered signal. A detailed evaluation is given in section II. Here, we only notice that a similar situation occurs in holographic interferometry and ESPI when the exposure time is reduced [6].

II. THEORY

Assuming the low-pass filter to be represented by an impulse response function h(t), we may write the filtered signal as

$$x(t) = \int_{-\infty}^{\infty} I(t')h(t-t')dt'. \tag{8}$$

The time averaged mean square fluctuations of x(t) may now be expressed as

$$y = \lim_{T\to\infty} \frac{1}{T} \int_0^T \sigma_x^2(t)dt, \tag{9}$$

where $\sigma_x^2(t)$ is the time-dependent variance of x(t). The variance is time-dependent because I(t) and x(t) have non-stationary statistics due to the coherent beat between the different terms in equation (3). We will see, however, that the use of a time-average estimate (equation (9)), removes the time-dependence so that the final result appears as if the different terms in (3) had been statistically independent or, equivalently, as if I(t) and x(t) had been stationary processes.

From equation (8) we find,

$$\sigma_x^2(t) = \iint_{-\infty}^{\infty} C_I(t-t_1,t-t_2)h(t_1)h(t_2)dt_1dt_2, \tag{10}$$

where,

$$C_I(t_1,t_2) = \langle\Delta I(t_1)\Delta I(t_2)\rangle = \langle I(t_1)I(t_2)\rangle - \langle I(t_1)\rangle\langle I(t_2)\rangle, \tag{11}$$

is the non-stationary auto-correlation of I(t).

By inserting equation (10) into (9) and interchanging the order of the integration we find for the output signal,

$$y = \iint_{-\infty}^{\infty} R_I(t_1,t_2)h(t_1)h(t_2)dt_1dt_2, \tag{12}$$

where,

$$R_I(t_1,t_2) = \lim_{T\to\infty} \frac{1}{T} \int_0^T C_I(t-t_1,t-t_2)dt, \tag{13}$$

is the time-averaged auto-correlation function of I(t).

From equation (2) we see that the fluctuating part of I(t) may be written as,

$$\Delta I(t) = I_o \ \mathrm{Re}\{\exp(ia \sin\omega_o t) \ \Delta z(t)\}, \tag{14}$$

where 'Re' denotes 'the real part of', and where

$$\Delta z(t) = z(t) - \langle z(t)\rangle,$$

with,

$$z(t) = \exp(-in(t)). \tag{15}$$

By substitution of equation (14) into (11) it follows that C_I may be expressed by,

$$C_I(t_1,t_2) = \tfrac{1}{2}I_o^2 \mathrm{Re}\{C_z(t_2-t_1)\exp(ia(\sin\omega_o t_1 - \sin\omega_o t_2)) + \bar{C}_z(t_2-t_1)\exp(ia(\sin\omega_o t_1 + \sin\omega_o t_2))\}, \tag{16}$$

where,

$$C_z(\tau) = \langle\Delta z(t)\Delta z^*(t+\tau)\rangle \text{ and } \bar{C}_z(\tau) = \langle\Delta z(t)\Delta z(t+\tau)\rangle. \tag{17}$$

In (17) the star denotes complex conjugation and we have assumed that n(t), and hence z(t), have stationary statistics. Substituting from equation (4) into (16), and from (16) into (13), we obtain upon integration a stationary time-averaged auto-correlation function,

$$R_I(t_1,t_2) = R_I(t_2-t_1),$$

where,

$$R_I(\tau) = \tfrac{1}{2}I_o^2\{J_0^2(a)\mathrm{Re}(C_z(\tau)+\bar{C}_z(\tau)) + 2\sum_{k=1}^{\infty} J_k^2(a)\mathrm{Re}(C_z(\tau)+(-1)^k\bar{C}_z(\tau))\cos(k\omega_o\tau)\}. \tag{18}$$

In order to compute the response function y from (12) and (18) we must consider the phase noise statistics.

Assuming n(t) to be stationary gaussian with mean value n_o, standard deviation σ_n, and normalized auto-correlation function $c_n(\tau)$, we have[8],

$$C_z(\tau) = \exp(-\sigma_n^2)\{\exp(\sigma_n^2 c_n(\tau))-1\}, \tag{19}$$

$$\bar{C}_z(\tau) = \exp(-2in_o-\sigma_n^2)\{\exp(-\sigma_n^2 c_n(\tau))-1\}. \tag{20}$$

Assuming further that $\sigma_n \gg 1$ we get the approximate relations

$$C_z(\tau) \simeq \exp(-\sigma_n^2(1-c_n(\tau))) \text{ and } \bar{C}_z(\tau) \simeq 0. \tag{21}$$

From (12), (18), and (21) we finally find for the response function:

$$y \simeq A_0 J_0^{\,2}(a) + 2\sum_{k=1}^{\infty} A_k J_k^{\,2}(a), \tag{22}$$

where,

$$A_k = I_o^{\,2}\iint h(t_1)h(t_2)\tfrac{1}{2}\{C_z(t_2-t_1)+C_z(t_1-t_2)\}\cos\{k\omega_o(t_2-t_1)\}dt_1dt_2, \tag{23}$$

and C_z is given by (21). Alternatively, we may express A_k in terms of the filter transfer function,

$$H(\omega) = \int_{-\infty}^{\infty} h(t)\exp(-i\omega t)dt, \tag{24}$$

and the power spectrum of z(t),

$$W_z(\omega) = \int_{-\infty}^{\infty} C_z(\tau)\exp(-i\omega\tau)d\tau, \tag{25}$$

which gives,

$$A_k = I_o^{\,2}\int_{-\infty}^{\infty}|H(\omega)|^2\tfrac{1}{2}\{W_z(\omega-k\omega_o)+W_z(\omega+k\omega_o)\}\frac{d\omega}{2\pi}. \tag{26}$$

Equation (26) clearly shows that A_k gives the degree of overlap between the filter frequency response and the k-th harmonic frequency band of I(t). From (22) we see that the simple cases considered in section I occur if $A_k=0$ for $k>0$.

III. COMPARISON WITH EXPERIMENTS

To test the theoretical results a simple feasibility experiment has been made. The set-up used was a simple Michelson interferometer as shown in figure 1. The two mirrors were mounted on loudspeakers which were excited by a sine wave generator and a random noise generator, respectively. The power spectrum of the random phase noise was controlled by a variable filter on the noise generator output. The interference signal was recorded by a PM-tube, low-pass filtered by a variable filter, and the mean square fluctuations were measured by a NORMA U-FUNCTIONMETER. The measurement output was finally recorded as a function of the sinusoidal excitation voltage by means of an x-y plotter.

Figure 3 shows typical experimental results for the intermediate case considered in section I. We clearly see the $J_0^{\,2}$-response function, and we also see that the signal level is reduced with increasing phase-noise bandwidth as the low-pass filter removes an increasing part of the signal energy contained in the DC-band.

For a detailed comparison between the experimental results and the theory developed in section II, the theoretical response functions were evaluated from equation (22) and (23). The filters used in the experiments were first-order low-pass filters, and therefore the evaluations were made with

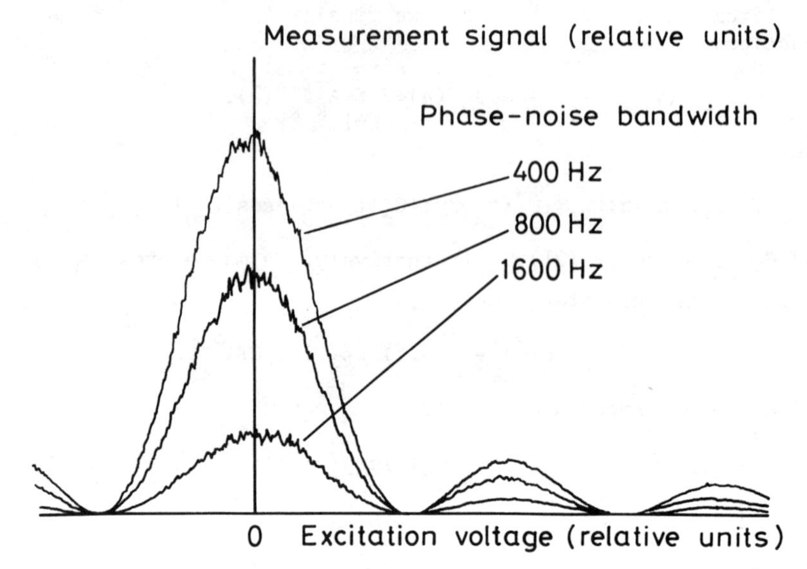

Fig. 3. Experimental results recorded for different phase-noise bandwidths, with a vibration frequency of 12500 Hz and a low-pass filter cut-off frequency of 1000 Hz.

$$h(t) = \begin{cases} \Omega_f \exp(-\Omega_f t) \ ; \ t \geq 0, \\ 0 \ ; \ t<0, \end{cases} \tag{27}$$

and,

$$c_n(\tau) = \exp(-\Omega_n |\tau|), \tag{28}$$

where Ω_f and Ω_n are the cut-off frequencies of the low-pass filter and the phase noise power spectrum, respectively. The corresponding bandwidths are $2\Omega_f$ and $2\Omega_n$. When (27) and (28) are substituted into (21) and (23), we find for A_k,

$$A_k = I_o^2 \Omega_f \ \mathrm{Re} \int_0^\infty \exp\{-(\Omega_f - ik\omega_o)\tau - \sigma_n^2(1-\exp(-\Omega_n\tau))\} d\tau$$

$$= I_o^2 \frac{\Omega_f}{\Omega_n} \ \mathrm{Re} \int_0^1 \exp(-\sigma_n^2 x) \ (1-x)^{\{(\Omega_f/\Omega_n)-1-ik(\omega_o/\Omega_n)\}} dx$$

$$= I_o^2 \ \mathrm{Re}\{M(1,(\Omega_f/\Omega_n)+1-ik(\omega_o/\Omega_n),-\sigma_n^2)/(1-ik(\omega_o/\Omega_f))\}, \tag{29}$$

where $M(a,b,z)$ is Kummer's confluent hypergeometric function for which rapidly converging series expansions exist[9]. Using (22) and (29) we have evaluated the response function y numerically.

Figure 4 shows (a) theoretical and (b) experimental results for the general case considered in section II. The results are for a vibration frequency equal to the low-pass filter cut-off frequency, and for increasing phase-noise bandwidth. In the experiments we maintained a constant output from the noise generator, so that a reduction in the phase-noise bandwidth also gave a reduction in the phase-noise variance. Therefore, the calculations were done with a phase-noise variance proportional to the phase-noise bandwidth. The results shown represent a best fit and we made no attempt to correct for the loudspeaker response function. The agreement between theory and experiment is good, and we clearly see the gradual transition from the J_0^2- response function to the case in which we have considerable contributions from higher order frequency bands. However, even when the phase noise bandwidth exceeds the vibration frequency there is a marked amplitude dependence in the response function.

IV. POTENTIAL PRACTICAL APPLICATIONS

Before we discuss the potential applications of the proposed method it should be remembered that the experiments done so far have rather limited practical validity. We have only considered an interferometer with perfect mirrors, and we have applied phase noise with known, stationary statistics. However, the theoretical model applies also for a diffusely reflecting object if the detector aperture does not exceed the average speckle size. Also, if the object is unstable so that, in addition to the phase fluctuations, we get a time-dependent rearrangement of the speckle pattern, the intensity fluctuations may be incorporated in the description by letting z(t) (equation (15)) have a fluctuating amplitude. Similarly, a moderate object translation may be incorporated by allowing for a Doppler-shift of the frequency bands. If the Doppler-shift is well below the filter bandwidth the measurements will not be seriously affected. Thus, it seems that we have a method applicable for vibration measurements on almost any object in quite noisy and unstable situations.

In a practical measurement situation the calculated response functions have little use because we generally lack control over the phase noise statistics and must assume it to be slowly changing (unless, of course, the phase noise is so weak that we may introduce additional phase noise with known statistics). We therefore need a calibration procedure.

Such a procedure is offered by the reference wave phase modulation technique introduced by Høgmoen and Løkberg[3,4] for measurement of small vibrations using ESPI. Here, a known reference vibration, with a frequency slightly offset from the object vibration frequency, is imposed on the other mirror of the interferometer. By square-law detection of the fluctuating part of the interference signal and by use of a lock-in amplifier to recover the beat at the difference frequency between the two vibrations, one obtains, for small vibrations, a linear recording of the object vibration. The

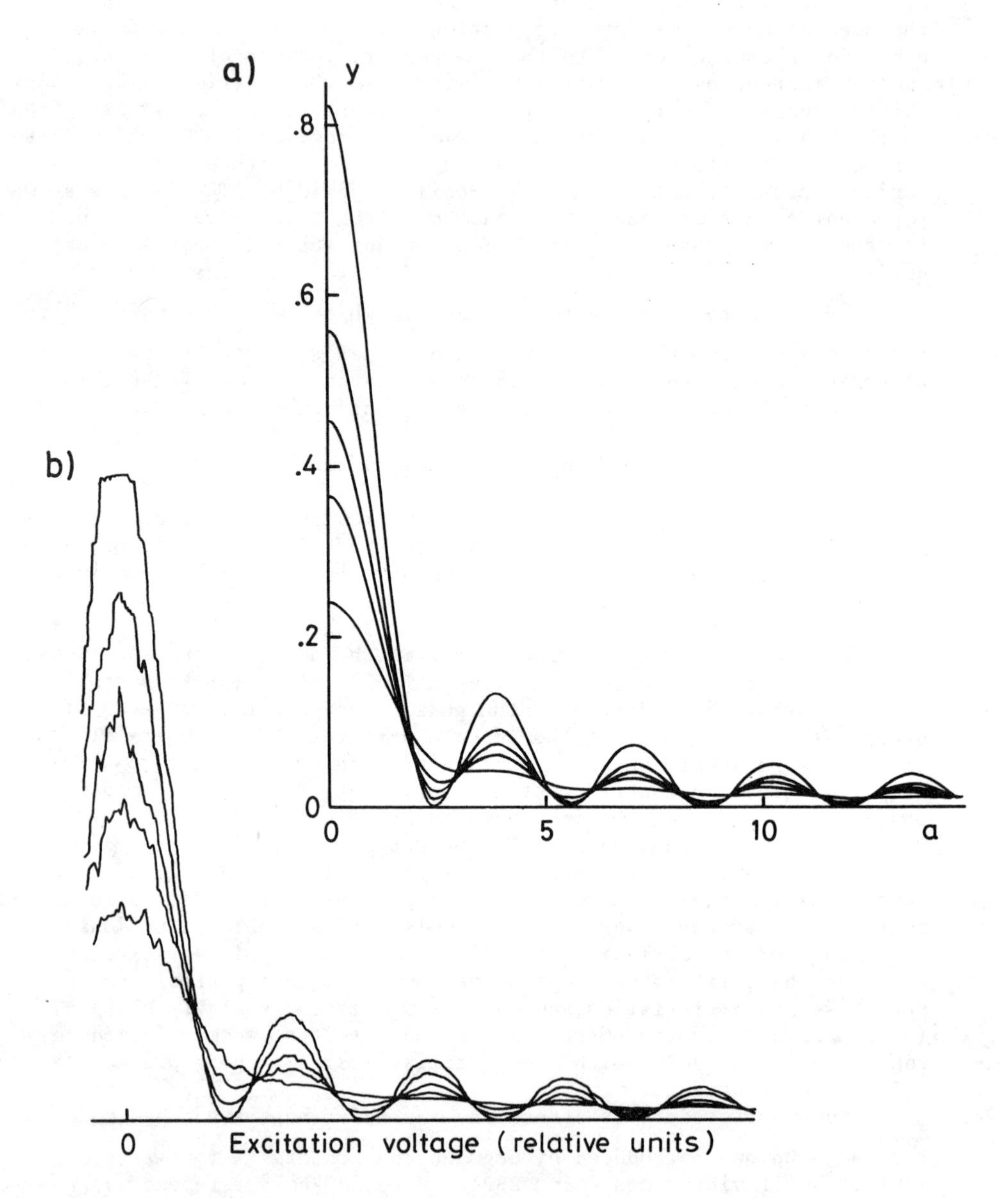

Fig. 4. Comparison of (a) theoretical and (b) experimental results for different phase-noise bandwidths, with a vibration frequency and a filter cut-off frequency of 1000 Hz. The phase-noise bandwidths are 400, 800, 1000, 1200, and 1600 Hz in the order of decreasing curve heights. (One of the experimental curves has a flat top because it exceeded the y-range of the plotter.)

latter appears as a small-signal perturbation of the response determined by the reference vibration.

It can be shown, by a straightforward, but lengthy, Bessel-series exercise similar to the theory in section II, that the method described in Ref.2 and 3 is directly applicable to the case described here. Instead of direct time-averaging of the squared fluctuations we must consider lock-in detection of the component at the beat frequency. Instead of equation (9) the theoretical response now becomes,

$$y_{lock-in} = \lim_{T\to\infty} \frac{1}{T} \int_0^T \sigma_x^{\;2}(t)\cos\{(\omega_o-\omega_r)t\}dt,$$

where ω_r is the reference vibration frequency and σ_x has to be properly corrected for the composite vibration.[10]

In our method we have full control over the low-pass filter response. In contrast to ESPI where the response function is generally unknown, we therefore may obtain continuous calibration of the system by imposing a known calibration signal at a third frequency on the reference mirror. Thus, a practical system would consist of two measurement circuits, one for measurement and one for calibration.

With such a practical version of the system in mind, it is not difficult to foresee a number of potential applications of the method. These include: (a) measurements on living biological objects, (b) measurements through turbulence (such as measurements on very hot objects or objects in liquids with turbulence and cavitation), (c) measurements in surroundings with severe mechanical noise, and (d) measurements on rotating objects by means of a derotating beam scanning optical system.

In conclusion, the present method seems to have promising potential applications for vibration measurements in noisy and unstable situations. In view of the striking analogy with holographic interferometry and ESPI, the method may be considered as an extension of these techniques to 'the fourth dimension'.

ACKNOWLEDGEMENTS

The experiments were done while I was with the University of Trondheim, The Norwegian Institute of Technology, Physics Department. I would like to thank O.M. Holje and O.J. Løkberg of that department for assistance with the experiments. I am also grateful to E. Mehlum for helpful assistance with the mathematics, and to O. Ersoy and J.J. Stamnes for useful comments on the manuscript.

REFERENCES

1. P.R. Dragsten, W.W. Webb, J.A. Paton, and R.R. Capranica, J. Acoust. Soc. Am. 60, 665 (1976).
2. W.M. Helfenstein, Neue Technik (Zürich), Nr.5/1973, p.207.
3. K. Høgmoen and O.J. Løkberg, Appl. Opt. 16, 1869 (1977).
4. K. Høgmoen and H.M. Pedersen, J. Opt. Soc. Am. 67, 1578 (1977).

5. O.J. Løkberg, K. Høgmoen, and O.M. Holje, Appl. Opt. 18, 763 (1979).
6. O.J. Løkberg, Appl. Opt. 18, 2377 (1979).
7. See, e.g., R. Courant und D. Hilbert, Methoden der Mathematischen Physik I (Springer-Verlag, Berlin, 1968) p.412.
8. See, e.g., H.M. Pedersen, J. Opt. Soc. Am. 66, 1205 (1976).
9. M. Abramowitz and I.A. Stegun, Handbook of Mathematical Functions (Dover, N.Y., 1965).
10. H.M. Pedersen, in preparation.

AUTOSTEREOSCOPIC 3-D TELEVISION EXPERIMENTS

J. Hamasaki
Institute of Industrial Science, University of Tokyo
7-22-1, Roppongi, Minatoku, Tokyo, 106, Japan.

ABSTRACT

This paper reviews theoretical and experimental backgrounds for autostereoscopic 3-D televisions, and describes, in some detail, the volume scanning method and the parallax panoramagram method, which are applied to autostereoscopic 3-D televisions. It also briefly describes some of recent experiments related to recording and reconstruction of 3-D images of various kinds.

INTRODUCTION

Thirty years ago the 2-D television had begun to prevail in the world, and after a decade, the optical holography was vividly demonstrated by Leith and Upatnieks. Since the holography can record and reconstruct a precise autostereoscopic 3-D image in a way never realized before, many workers became interested in whether it could be used for a 3-D television. Unfortunately, no one has succeeded in demonstrating a 3-D TV by this method.

To realize a 3-D TV, the amount of information at recording, reconstruction and transmission must be small to be handled in real time. This paper describes general discussions on a 3-D TV, two methods which were recently demonstrated, and other experiments related to 3-D images of various kinds.

PRELIMINARIES

1. Perception of 3-D images

Figure 1 shows the geometry of perception of a 3-D point by two eyes. When a man observes a 3-D object, two 2-D images are formed on the retinal surfaces in his eyes. Since the resolvable angle δ at the yellow point on the retina is diffraction-limited by the aperture of the eye lens, a 3-D sensation comes mainly from the parallax between the two images on the retinas. To achieve the most sensitive recognition of the parallax, and to obtain auxiliary information for the position of the 3-D object, the directions of sight of the two eyes and the focal lengths of the eye lenses are automatically adjusted.

When the interpupillary distance and the distance between the eyes and the object point are denoted by P and z, respectively, the resolvable distances in the lateral and longitudinal directions are given, respectively, by

$$\Delta x = z\delta, \quad \Delta z = 2z^2 \delta P^{-1}. \tag{1}$$

When δ=20 arcsec., P=63mm and z=1m are assumed, Δz=3.2mm is obtained.

ISSN:0094-243X/81/650531-26$1,50

(Experimentally Δz=0.8mm is reported.) Because Δz is proportional to $z^2\delta$, a 3-D image in a distant view can not be perceived if a relative motion between the eyes and the object is prohibited.

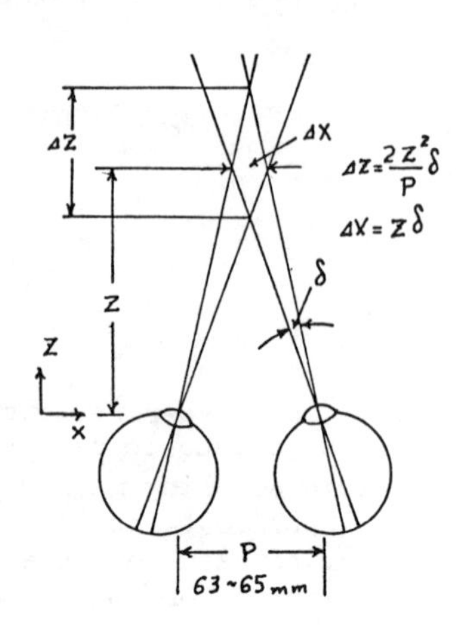

Fig.1 Perception of a 3-D image

2. Methods of recording/reconstruction of 3-D images

Table 1 shows a list of points of view in comparing various recording/reconstruction methods for 3-D images. In Table 1, the resolutions and the signal-to-noise ratio of the reconstructed images(I-a,b) and the technical and economical feasibility (III-a) are very important.

Table 2 shows methods for recording/reconstructing 3-D images. In Table 2, the projection methods(I) record projection images of an object seen from two or many positions in a viewing zone. In other words, it records the position, the direction and the intensity of the rays which reached the viewing zone. In the holography, the ray direction is determined by the orientation and the pitch of the recorded interference fringes.

The methods grouped in I-1 use the observer's eye lenses as optical elements in the 3-D image reconstruction. These methods themselves can not objectively reconstruct a 3-D image without the aid of small aperture lenses. They have a long history[1 2], and still they are the most promising methods even for an autostereoscopic 3-D TV. The parallax panoramagram(I-1-d) can be divided into two subgroups; the one uses a 1-D array of cylindrical lenses (a lenticular sheet), and the other uses a 2-D array of spherical lenses (a fly's eye lens) or crossed 1-D arrays of cylindrical lenses. The integral photography (I-2-a) requires both a very good resolution for recording/reconstructing planes and a very sharply directive lens array. Successful experiments aimed to its realization by using a lens array unexpectedly turned out to be those of parallax panoramagram using a 2-D lens array.

The holography(I-2-b) provides geometrically the most perfect 3-D

Table 1 Points of view in comparing various 3-D methods

I. Reconstructed image qualities
a.Resolutions (lateral, longitudinal)
b.Signal-to-noise ratio
c.Autostereoscopic image
d.No distortions
e.No phantom images
f.No image flippings
II. Recording/reconstruction conditions
a.Ordinary illumination
b.Color images
c.Real-time processings
III.Feasibilities
a.Technical and economical feasibility
b.Compatibility with 2-D methods

Table 2 Methods of recording/reconstructing 3-D images

I. Projection methods
1. Methods using eye lenses for reconstructing a 3-D image
a.Stereoscope
b.Stereoglasses
c.Parallax stereogram
d.Parallax panoramagram
2. Methods without aids of eye lenses
a.Integral photograghy
b.Holography
II. Volume scanning methods
1. Methods using varifocal mirrors
2. Picosecond pulse holography

image, but it requires a laser light at least in the recording stage, and has difficulty in recording color images. By the efforts of many workers, the tremendous amount of information contained in a hologram has been reduced by 10^{-4}, but holography still needs a plane of very high resolution at reconstruction to diffract visible light with the reproduced interference fringes. At present, it is not possible to build real-time-controllable planes of this high resolution, and consequently, holography is not a candidate for a practical 3-D TV.

The volume scanning method (II-1) uses varifocal mirrors [3,4,5] to scan the plane to be recorded(the sectional plane) in the depth direction. It records instantaneously many cross-sectional images or contour images at various depths[6] and at reconstruction these images are sequentially relocated at the original depths. This method needs to handle the smallest amount of information in reconstructing true autostereoscopic images, because the projection methods must handle two degrees of freedom for the directions of the rays while this method needs to handle one degree of freedom for the depth. The volume scanning can be also done by picosecond pulse holography where the time-spatial coincidence at the recording plane determines the cross-sectional surface of the object to be recorded. But this method is not practical in a 3-D TV.

In the following sections, autostereoscopic television experiments by the volume scanning method using varifocal mirrors and those by the parallax panoramagram method will be described.

VOLUME SCANNING METHOD

1. Varifocal mirror arrangements

In a volume scanning method, many cross-sectional images or contour images of an object at various depths must be recorded in an integration time(0.04sec) of a human eye, and they must be reproduced at the original depths during this time at reconstruction. This scanning in the depth direction can be done by varifocal mirror arrangements.

Figure 2 shows two varifocal mirror arrangements, which are representatively used for reconstruction. VM, VM1 and VM2 are varifocal mirrors. A varifocal mirror consists of a dynamic speaker whose opening is covered by an aluminum coated polyethylene film under a uniform tension. The mirror-pole is displaced by the air pressure in the volume surrounded by the film and the paper cone. The inverse focal length is proportional to the driving current of the dynamic speaker, if the driving is done at a low frequency.

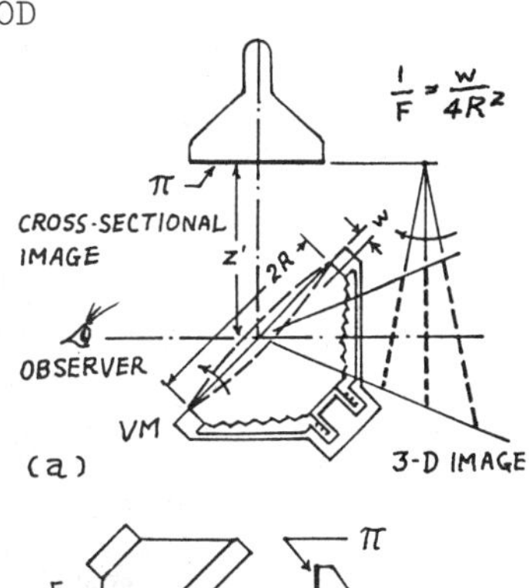

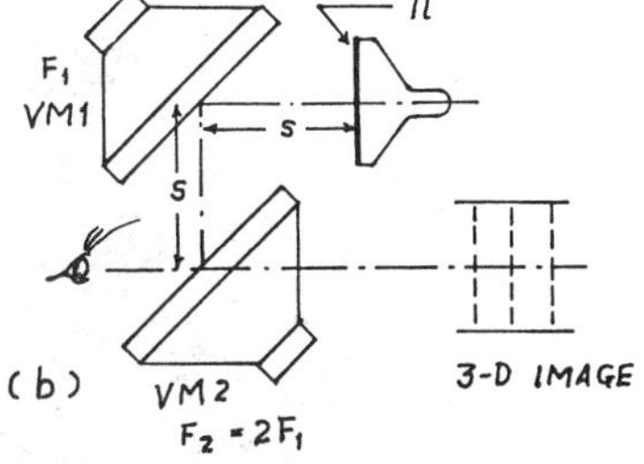

Fig.2 Varifocal mirror arrangements

Figure 2(a) shows the simplest arrangement[7]. When many contour images at various depths are sequentially reproduced on the plane π and the inverse focal length is varied in synchronization with the contour images, an observer sees the contour images at the sectional planes which are shown by the broken lines in Fig.2(a). When the positions of the sectional planes are geometrically similar to those of recording, the observer sees a 3-D image of the object. One assumes that the lateral dimension of the displaying surface at reconstruction is α times larger than that of the image of the recording plane by the camera lens when VM is stationary. To obtain a distortion free 3-D image, (1) the driving current of VM at reconstruction must have the inverted phase with respect to that of recording, and (2) the mirror pole displacement ratio $w_2 w_1^{-1}$ and the distance ratio $z_2' z_1'^{-1}$ must be given by

$$w_2 w_1^{-1} = \alpha^{-1}, \quad z_2' z_1'^{-1} = \alpha R_2^{2} R_1^{-2}, \tag{2}$$

where w and R are the mirror pole displacement and the radius of the varifocal mirror, respectively, z' is the distance between the centers of VM and π, and the subscripts 1 and 2 are assigned to the values of recording and of reconstruction, respectively.

Figure 2(b) shows a two-varifocal-mirror arrangement[8]. By this arrangement, the contour images reproduced on the plane π can be seen with a common magnification on the sectional planes which are mutually parallel as shown by the broken lines. This arrangement is useful when one needs mutually parallel and equally spaced contour images with a common magnification to be recorded, or one needs to reconstruct a 3-D image from many mutually parallel and equally spaced contour images with a common magnification. One assumes that the distance from the center of VM1 to the plane π is the same as the

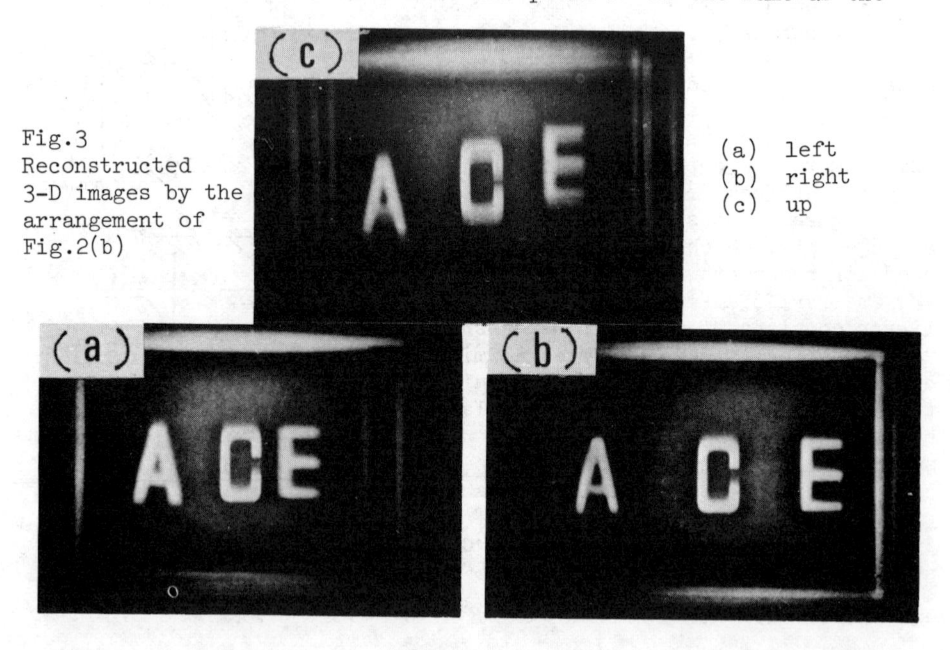

Fig.3 Reconstructed 3-D images by the arrangement of Fig.2(b)

(a) left
(b) right
(c) up

distance s between the centers of VM1 and VM2, and it is much smaller than the minimum focal length of VM1. To obtain an almost distortion free 3-D image, (1) the driving current of VM2 must have an inverted phase with respect to that of VM1, and (2) the focal length of VM2 must be twice that of VM1. Then, the lateral magnification M from the display surface to the sectional plane and the scanning distance Δz are given by

$$M=(1-s^2F_1^{-2}2^{-2})^{-1},\quad \Delta z=-s^2F_1^{-1}2^{-1/2}(1-sF_1^{-1}2^{-1/2}),\quad |sF_1^{-1}|\ll 1, \tag{3}$$

where F_1 is the instantaneous focal length of VM1. When a straight line is displayed on π at a distance t from the plane defined by the centers of VM1, VM2 and π, its image is rotated by θ in this plane and by ψ around the axis of observation, where θ and ψ are given by

$$\theta \simeq s^2F_1^{-2}2^{-2},\quad \psi\simeq -tF_1^{-1}2^{-1}(1+sF_1^{-1}2^{-1/2}). \tag{4}$$

Because $|sF_1^{-1}|\ll 1$ is satisfied, M is approximately unity and Δz is proportional to F_1^{-1}, but the small image rotation ψ for a point at a finite t is inherent to this arrangement.

Figure 3 shows photographs of a reconstructed 3-D image by this arrangement. They are taken from three directions(left, right and up). They show that true autostereoscopic image can be displayed from the cross-sectional images of a common magnification.

2. Phase stabilization of the spatial carrier [7]

In order to extract an object contour image on a moving sectional plane, one must spatially modulate that plane of the object which is illuminated by a spatial carrier source. When the phase and the frequency of the spatial carrier fluctuate at the recording plane during a period of the varifocal mirror, and when a temporally continuous source is used for illumination, a moving object can not be recorded and a contraction of the field of view occurs even for a stationary object, because the storage effect of the recording plane of the camera tube decreases the modulation depth.

Figure 4 shows an illumination and imaging configuration for the frequency and phase stabilization of the spatial carrier in the real time sectioning. In Fig.4 a varifocal mirror is replaced by an equivalent lens VM, the spatial carrier source is a fine grating G illuminated by a tungsten lamp, and a half mirror HM separates the illuminating light and the object light. Using

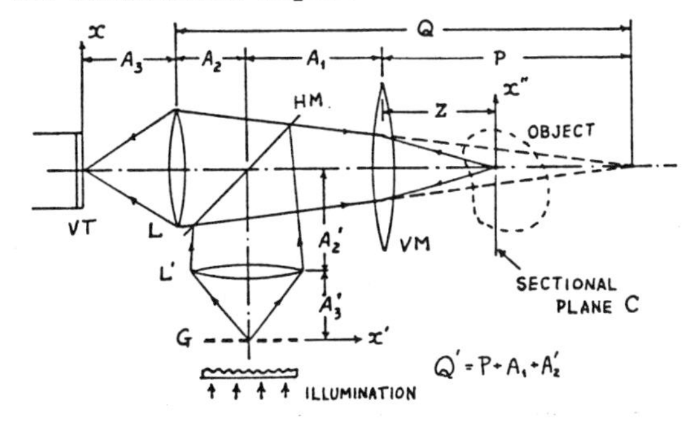

Fig.4 Illumination and imaging configuration

the notations shown in Fig.4, the stabilization condition is given by

$$P=Q-A_1-A_2=Q'-A_1'-A_2',\quad Q=(F^{-1}-A_3)^{-1},\quad Q'=(F'^{-1}-A_3'^{-1})^{-1}, \tag{5}$$

where F and F' are the focal lengths of the imaging lens L and the projection lens L', respectively. Eq.(5) means that the image of G must be focused on a sectional plane C, and that the image of C must be focused on the recording plane VT. Then, the ratio of the lateral coordinates of VT and G, and the longitudinal coordinate of the sectional plane C are given, respectively, by

$$xx'^{-1}=A_3Q'A_3'^{-1}Q^{-1},\qquad z^{-1}=F_v^{-1}+P^{-1}, \tag{6}$$

where F_v is the instantaneous focal length of VM. Because the first equation of Eqs.(6) is independent of F_v, the image of G on VT is stationary.

3. Optical sectioning and longitudinal resolution [7]

Figure 5 shows the geometrical spread of the image of G on VT, when the object surface is on an out-of-focus plane. The spreading process is divided into two steps; the first is the spread of the image of G on the plane C' which is displaced by Δ from the in-focus plane C, and the second is the spread of the image on VT of a point at C'. When the spatial carrier frequency of the grating G is denoted by ξ_c', the spatial carrier frequency of the image on C is given by

$$\xi_c''=-\xi_c'A_3'A'^{-1},\qquad A'=zQ'P'^{-1}, \tag{7}$$

and the spatial frequency on VT for the image of the grating image on C is given by

$$\xi_c=-\xi_c''AA_3^{-1},\qquad A=zQP^{-1} \tag{8}$$

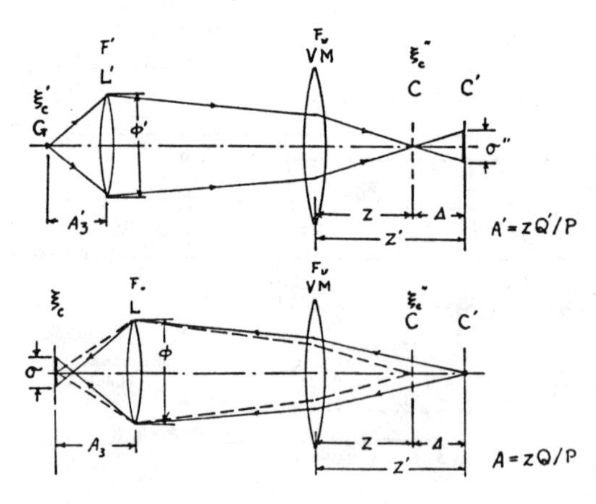

Fig.5 Geometrical spread of the image of G

If the modulated components in the vicinity of the spatial frequency ξ_c are extracted by a filter, the longitudinal resolution curves of this modulated components are given by a product of two Airy functions; they are shown in Fig.6. The abscissa and the parameter are the normalized out-of-focus distance and the normalized diameter ratio of L and L', respectively; they are given by

$$\Delta'\simeq\pi\phi'\xi_c''\Delta A'^{-1},\quad \beta\simeq\phi\phi'^{-1}Q'Q^{-1} \tag{9}$$

where Φ and Φ' are the diameters of L and L', respectively. The longitudinally resolvable distance Δ_Z can be defined by the halfwidth of the resolution curve. It is approximately given by

$$\Delta_Z=1.39|\xi_c''|^{-1}(\Phi'^2A'^{-2}+\Phi^2A^{-2})^{-1/2}. \quad (10)$$

To obtain a high longitudinal resolution, one must use a high spatial carrier frequency in the object space, and large values for both $\Phi'A'^{-1}$ and ΦA^{-1}.

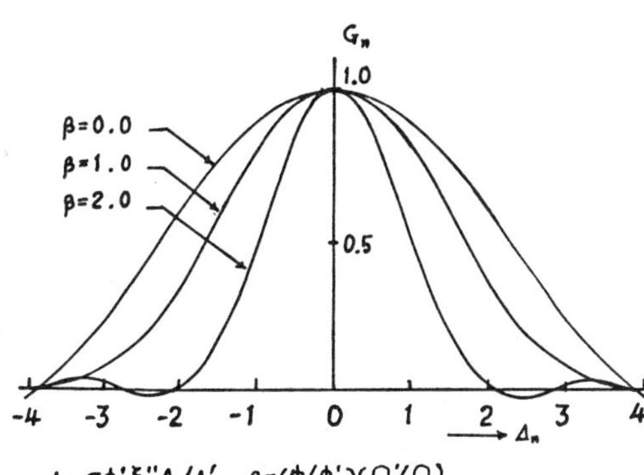

Fig.6 Longitudinal resolution curves

Note that the baseband component to be removed by the filter consists of many out-of-focus images of the object and the low frequency component of the modulated image of the in-focus plane of the object, and therefore, it is generally much larger than the signal to be extracted. Although this baseband component can be removed by the filter, it still contributes to the random noise superimposed on the signal.

Suppose that the varifocal mirror is driven by a symmetrical triangular waveform with a temporal frequency f_v, which is much lower than the field frequency f_f of a camera tube, and it scans a total depth L_d of object space, which is much larger than Δ_Z. Concerning the contour image of the plane at a given depth, its signal is taken in a short time when the moving sectional plane is passing through a slab of the object space with a thickness Δ_Z containing that plane. The mean-square-signal current from the camera tube is given by

$$\overline{i_s^2}=0.072\ I_{dc}^2\epsilon^2f_f\Delta_z^2f_v^{-1}L_d^{-2}, \quad (11)$$

where I_{dc} is the dc current mainly originated by the baseband component, and ϵ ($\lesssim 1$) is the response factor of the combination of the camera and the filter. To obtain a sufficient signal current, one must use high values for I_{dc}, ϵ, and $f_f f_v^{-1}$, and a moderate value for $L_d\Delta_Z^{-1}$.

If the electron beam noise is taken into account, the random noise current $\overline{i_n^2}$ is given by

$$\overline{i_n^2}=4eB(I_{dc}+I_d)+eBN_F/(40R_L),\ \text{(in mks units)}, \quad (12)$$

where B is the bandwidth of the filter, I_d is the dark current, e is the electronic charge, and R_L and N_F are the input resistance and the noise figure of the amplifier, respectively.

4. Lateral resolution [7]

When a continuous source is used for illumination, a sectional image on VT continuously changes by the movement of VM even if the object is stationary. If the object is a simple grating, the spatial frequency of its image on VT changes at the time when the sectional plane is passing through the slab of the thickness containing the object. Because of the combined effect of this spatial frequency change and the storage effect of the recording plane VT, the contrast of the extracted contour image is degraded especially at the rims of the recording plane.

The cutoff spatial frequency ξ_0 is defined as the frequency at which the contrasts of the signal current is reduced by a half. It is given by

$$\xi_0 \begin{cases} =0.075\, A_3 R^2 f_f (PQw_0 f_v D)^{-1}, & \text{if } f_f^{-1} \ll \Delta_z v_z^{-1}, \\ =1.21\, A_3 P (QD_z)^{-1}, & \text{if } f_d^{-1} \gg \Delta_z v_z^{-1}, \end{cases} \tag{13}$$

$$v_z = 16 w_0 f_v P^2 R^{-2},$$

where the dimension of VT is assumed to be D×D, and w_0 and R are the peak mirror-pole displacement and the radius of the varifocal mirror, respectively. Actually, the lateral resolvable distance Δ_x is not simply given by $1/(2\xi_0)$, but also by the low-pass characteristics of the camera and the filter.

5. Filtering ----Frequency interleaving [9] ----

In extracting the modulated component, the response factor ϵ in Eq.(11) is important in determining the signal-to-noise ratio of the reconstracted 3-D image. If the frequency interleaving method is used, this factor can be made to approach unity.

In the frequency interleaving method, the grating G is so oriented that the phase of the grating image on VT is alternatively inverted for every horizontal electron beam scanning. Figure 7 shows the electron beam scanning and the image of the grating image. In Fig.7, p_y is the pitch of the horizontal scanning, ξ_c is the spatial frequency of that image. When the intensity distribution of the object plane focused on VT without spatial modulation is denoted by R(x,y), $R(x,2kp_y)$ has a waveform very similar to $R(x,(2k+1)p_y)$, where k is an integer. When the horizontal scanning period and its inverse are denoted by T and $f_H(=T^{-1})$, respectively, the current from the spatially modulated image has a spectrum peaked at $(n+1/2)f_H$ (n=0,1,2, ...), and this spectrum is interleaved with the spectrum peaked at nf_H of the baseband component.

To extract the current of the spatially modulated image, one can use the comb-filter technique; the current from a horizontal scanning is delayed by T and it is subtracted from the current from the next scanning, and then, the difference current is rectified by a linear full-wave rectifier. The output of the rectifier has a waveform of the sectional image, and it is not contaminated by many blurred images.

In this method, the lateral resolution in x direction is given by the bandwidth of the camera tube and the amplifier, if the cutoff

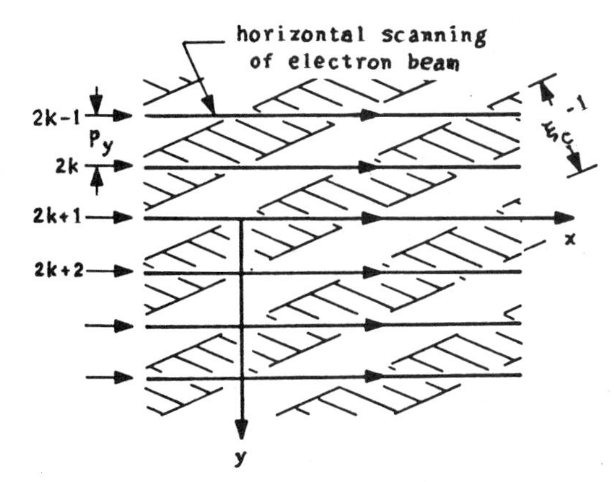

Fig.7 Elecron beam scanning and grating image

frequency described in the previous subsection is high. The lateral resolvable distance Δy in the y direction is given by $2p_y$. When $\partial R/\partial y$ is not zero, it produces an undesired current. This becomes significant when the object has an appreciable component of the spatial frequency higher than $1/(4p_y)$ in the y direction.

6. Experiments [9]

Figure 8 shows a schematic diagram of a 3-D TV. At recording, the lenses, L and L', have large apertures and low f-numbers, and the varifocal mirror VM1, as well as VM2 at reconstruction, has a very large aperture. HM is a half-mirror. A grating G is focused by L' and VM1 onto a sectional plane at some depth. The sectional plane is focused onto the vidicon target VT (the recording plane) by VM1 and L. When VM1 is driven by a triangular wave(VM driver), the modulated images and many blurred images are superimposed, and they are successively formed on VT. The grating G is oriented according to the frequency interleaving method. The current from the camera tube is led into a comb filter, which consists of a delay line(1HDL) and a subtractor. The output of the filter is led to a linear full-wave rectifier(DET). The output of DET is a pure sectional image signal at the instantaneous section-

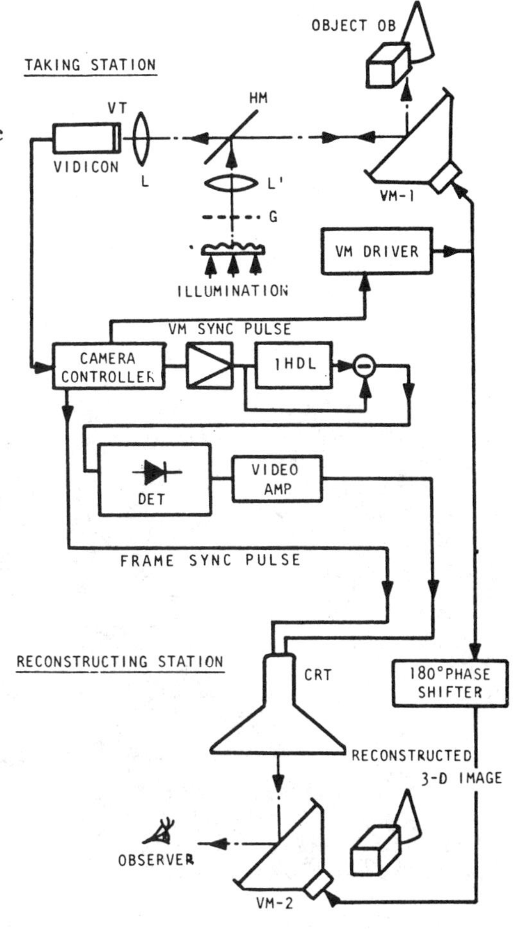

Fig.8 Schematic diagram of 3-D TV by volume scanning

al plane, and it is sent to the receiving station. The synchronization signals for horizontal, the vertical and the depth scannings are also sent to the same station.

At reconstruction, many contour images of the object at various sectional planes are successively displayed on the CRT surface, and VM2 is driven in synchronization with the depth change of the contour images. The driving current of VM2 must have the inverted phase and a proper amplitude with regard to those of VM1. An observer sees virtual images which are geometrically positioned at depths similar to the original depths, and he recognizes a 3-D image.

The theoretical results described above have been experimentally verified, and 3-D TV experiments carried out. Figure 9 shows photographs of a reconstructed 3-D image taken from two directions. The object was a 20cm×15cm×30cm plasticine dog figure. The photographs have a clear parallax, and a reasonably good signal-to-noise ratio and resolutions. In this experiment, a 300W tungsten lamp and a black-and-white grating with 80 cycles/cm were used. The focal lengths of L and L' are 55mm and 105mm, respectively, and their f-numbers are /1.4 and /2.8, respectively. VM1 and VM2 were made from 30cm diameter dynamic speakers, and they were driven at 15cps. An RCA 4532 vidicon tube having a 1cm×1cm silicon diode array was used at 100 fields/sec and at 263 horizontal scannings/field. 1HDL is an ultrasonic delay line having a 31.643μsec delay with a slight trimming. The distance from the center of the object to VM1 was 40cm, and the distance from VM1 to L was 70cm. The distance from VM2 to CRT was 32cm, and the display surface of CRT had a 16cm×16cm area. A 4MHz bandwidth was used for the signal transmission. The experimentally observed value of the response factor ϵ was 0.9. The lateral resolutions observed were 250 TV lines in the horizontal di-

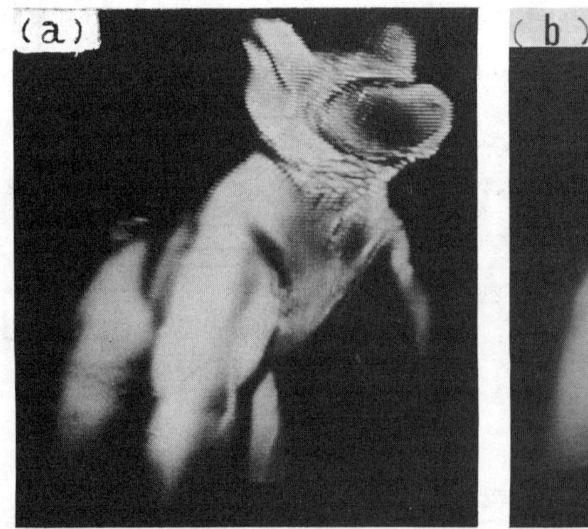

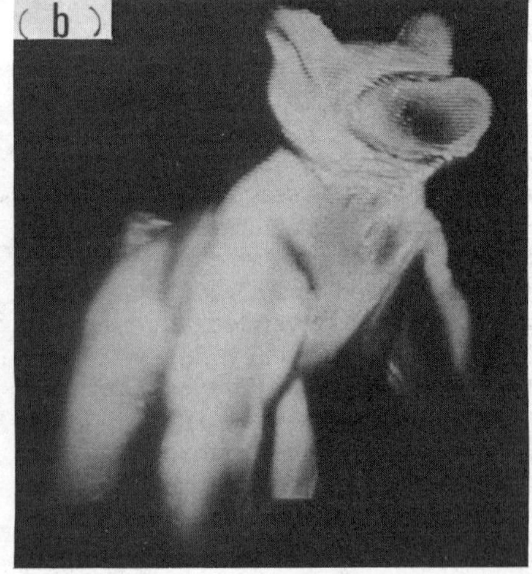

Fig.9 Photographs of a reconstruct image of a 3-D TV by the volume scanning method

rection and 80 TV lines in the vertical direction. The resolvable distance in the depth direction was 4cm (6 sectional planes in the total object depth).

7. Discussions

It has been shown that a 3-D TV of reasonable quality is made from the readily available optical and electronic components with a 4MHz bandwidth by using the volume scanning method. Since the lateral resolutions required for a moving 3-D image are perhaps lower than those for a 2-D image, 10 sectional images with 250 TV line resolutions in both of the lateral directions will be sufficient for a pictorial 3-D TV. When 25 3-D images/sec are needed to avoid flickering of the images, a 16MHz bandwidth is required. And the camera and the monitor must have a 250 fields/sec rate.

Although the volume scanning method can record/reconstruct a true autostereoscopic 3-D image by available components, it has limitations; (1) the object must be illuminated by a spatially modulating light source, (2) the reconstructed image has the phantom image problem which is inherent to the monocular nature of the recording apparatus, (3) the depth of the object is limited by the number of available sectional planes, and (4) this method is not compatible with the presently prevailing 2-D TV systems, because it needs both a high field frequency and a varifocal mirror.

PARALLAX PANORAMAGRAM METHOD

1. 3-D image space formed by parallax

In the projection methods, the position, the direction and the intensity of the rays which reach a viewing zone are recorded. If they are precisely reproduced at reconstruction, they constitute objectly a true autostereoscopic 3-D image.

Figure 10 shows the geometry of the 3-D image space formed by parallax. On the plane S, a left-eye picture and a right-eye picture are simultaneously reproduced; the former and the latter are selectively seen by the left eye and the right eye, respectively, because of the directivity of the plane S. The plane of these characters can be produced by a stereoscope, stereoglasses, a combination of polarizers and a half-mirror, a lens array, a corner-cube array, a hologram and any kind of apparatus used for one of the projection methods. In Fig.10, a point of the left-eye picture and the corresponding point of the right-eye picture are assumed to exist at A_L and A_R on S, respectively. Because the rays coming from these two points are the same as the

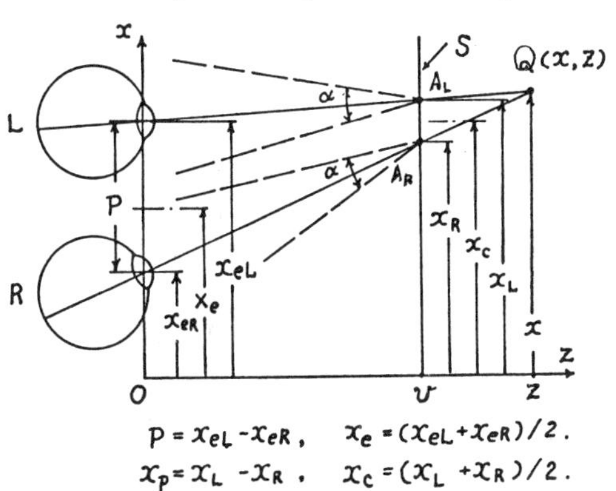

Fig.10 Image space formed by parallax

rays coming from the point Q, which is the intersection of the two rays, the observer see a 3-D image point at Q. The coordinates of Q(x,z) are given by

$$x=(x_c P-x_e x_p)(P-x_p)^{-1}, \quad z=vP(P-x_p)^{-1}, \tag{14}$$

where v is the distance from the eyes to S; the interpupillary distance, the separation between A_L and A_R, the coordinate of the center of the two pupils, and the coordinate of the center between A_L and A_R are denoted by P, x_p, x_e and x_c, respectively; they are given by

$$P=x_{eL}-x_{eR}, \quad x_p=x_L-x_R, \quad x_e=(x_{eL}+x_{eR})/2, \quad x_c=(x_L+x_R)/2. \tag{15}$$

When the observer translates the position of his head along the x and z directions within the area covered by the angle α, which is given by the directivity of S, x and z change proportionally with x_e and v, respectively. Consequently, the reconstructed 3-D image deforms according to the observer's head position. Another observer having a different P recognizes a slightly different image because x and z are dependent on P. For these reasons, it is said that the reconstructed 3-D image by parallax does not objectively exist if α covers an area much larger than the pupil opening. The integral photography requires a very small α to obtain a 3-D image objectively, but it has only been successful when many pin-holes are used. The holography has practically realized a sharply directive surface by recording many fine interference fringes. But both of these methods are not candidates for a 3-D TV for the reasons described in the Preliminaries.

In the methods shown in Table 2 II-1, one needs a moderately directive surface even for an autostereoscopic image, and the amount of information recorded and to be reproduced is tremendously reduced in comparison with holography.

When the coordinates on S are positionally quantized, it is seen from Eqs.(14) that the coordinates of the reconstructed image are also quantized. The sampling distance in z changes with the z value, and the sampling distance in x on the plane having a quantized z value is determined by this z value. The sampled position in the image space can be easily obtained by Eqs.(14).

The y coordinate, whose axis is perpendicular to both the x and z axis, can be readily taken into account in Eqs.(14). When the directivity of S in the y direction is very broad, autostereoscopicity is obtained in the x direction only, and the resolution along the y direction is greatly reduced. If the resolution on S and the bandwidth are limited, autostereoscopicity in the y direction can be ignored because the eyes are horizontally separated.

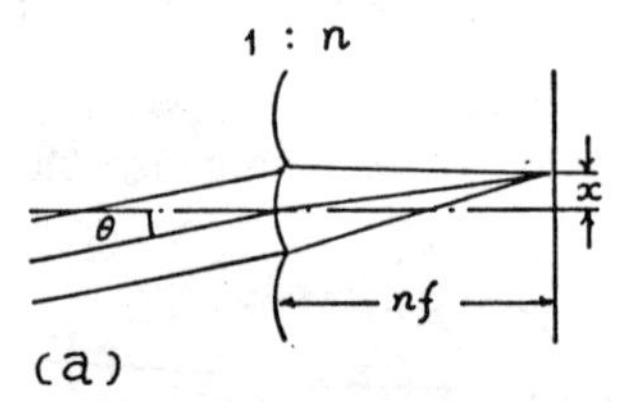

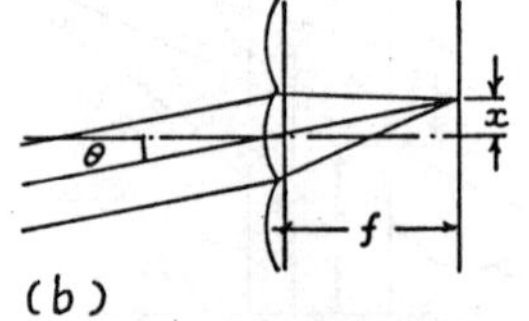

n : REFRACTIVE INDEX
f : FOCAL LENGTH
$f = x/\theta$

Fig.11 Lenticular sheet and a thin-lens array

2. Lenticular sheet

A lenticular sheet is an array of cylindrical lenses. Figure 11 shows the cross-section of a lenticular sheet. Usually the back focal plane coincides with the rear surface of the sheet for convenience of contacting a recording plane. The focal length is defined by the ray displacement at the back focal plane divided by the incident angle of the incoming ray. In Fig.11,an equivalent array of thin lenses is also shown. To obtain a wide viewing zone, one needs a lenticular sheet with a large acceptance angle. Since a lenticular sheet ordinarily has an array-lens surface, it has the spherical aberrations predominantly at a large incident angle. These aberrations cause difficulties in achieving a sharp directivity. They can not be reduced even by a proper shaping of the lens surface[12], but can be reduced if two or many array-lens surfaces are combined into a lenticular sheet.

When a recording plane is fixed on the rear surface of the lenticular sheet, the position of an incident ray is sampled by an elementary lens of the sheet. The direction of the ray is converted to the position at the back focal plane of the elementary lens (Fourier transform by the lens). To record the direction of the ray, the direction must be sampled well in advance because the elementary lens has a moderate directivity; if the sampling is not done, many rays with various directions form overlapped elliptic discs (line images) at the focal plane, and they can not be recorded in the separate forms which are needed for a clear reconstruction of the ray.

3. Image conversion optic[10]

When a reflected light from an object is recorded by using a lenticular sheet, the reconstructed 3-D image is pseudoscopic. To record a 3-D image in real time, one must form in advance a pseudoscopic 3-D image having quantized ray directions, and he must record this image by a lenticular sheet. Although this image conversion can be done in real time by an high speed electronic switching circuit, it is also done by a simple optical arrangement.

Figure 12 shows the principle of the image conversion optic(ICO). It is composed of several lens-mirror sets(LMS), each of which consists of a plane

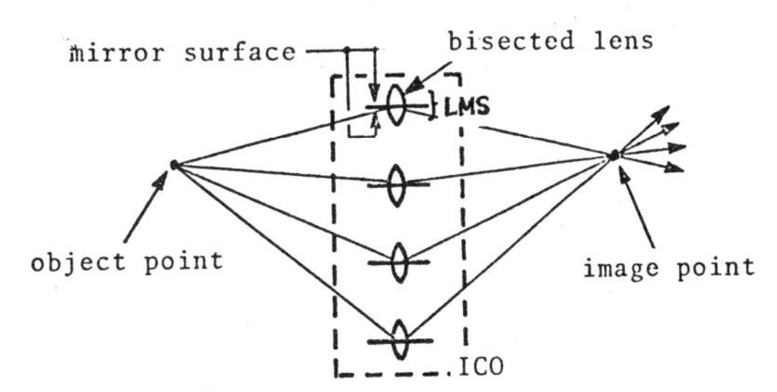

Fig.12 Principle of ICO

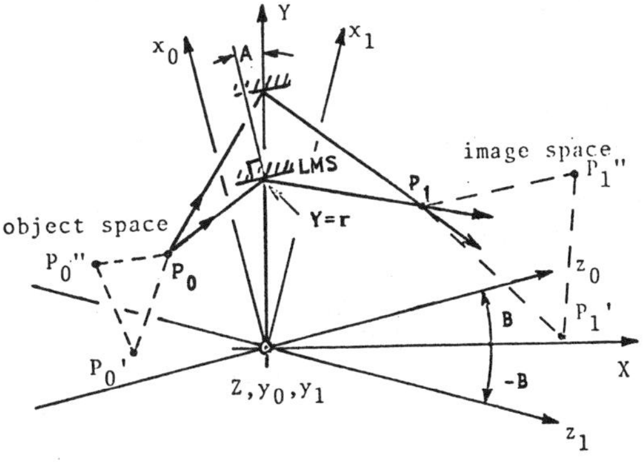

Fig.13 Array of LMS's

mirror and two bisected lenses having a common and small aperture. The sets are separated from one another in the horizontal plane by the average interpupillary distance divided by an integer. The directions of the rays from an object are quantized by the aperture of the lenses. It is assumed that all lenses overlappingly form images of the object center at the image center, and the ray passing through a lens without reflection is blocked by some shields. Because the apertures are small, images in the vicinity of the image center are not blurred too much and they are overlappingly formed in the image space. Then, the intersection of the rays coming from an object point and passing through ICO construct a pseudoscopic mirror image of the object point.

Figure 13 shows an array of LMS's. It is assumed that lenses have infinitesimal apertures, and the mirrors, fixed on the Y axis, have an inclination angle A'. When the y coordinate of a mirror is denoted by r, A is defined by

$$A=-(1/2)rg^{-1}, \tag{16}$$

where g is a constant of ICO, and A and B are assumed to be small. The condition of perfect imaging is given by A=0. When $P_1(x_1,y_1,z_1)$ is the image of $P_0(x_0,y_0,z_0)$ ($z_0<0<z_1$), the image coordinates are given by

$$x_1=x_0, \quad y_1=-y_0, \quad z_1=-z_0. \tag{17}$$

If the paraxial approximations are made, the imaging formulas of ICO are given by

$$x_0z_0^{-1}+x_1z_1^{-1}=0, \; y_0z_0^{-1}-y_1z_1^{-1}=0, \; z_0^{-1}+z_1^{-1}=g^{-1}. \tag{18}$$

The lateral and longitudinal magnifications are given, respectively, by

$$M_1=g(g-z_0)^{-1}, \quad M_2=-M_1^{2}. \tag{19}$$

If a distortion-free 3-D image with magnification is needed, the ratio $M_2M_1^{-1}$ can be compensated by choosing a proper focal length for a lenticular sheet at reconstruction; at this point, the focal length must be M_1^{-1} times that used in the recording stage.

When an object point exists at a position separated from the object center in the z_0 direction, and the lens has a finite aperture, the intersection of rays has a spread. If the image point is formed at the position separated from the image center by z_1' in the z direction, the spread of the image projected on the image center plane is given by

$$S_L(z_1')=D_Lz_1'(v_1+z_1')^{-1}, \tag{20}$$

where D_L is the aperture diameter, and v_1 is the z_1 coordinate of the image center. In recording a 3-D image, the lenticular sheet with the recording plane must be placed in the vicinity of the image center, where the spread is given by Eq. (20).

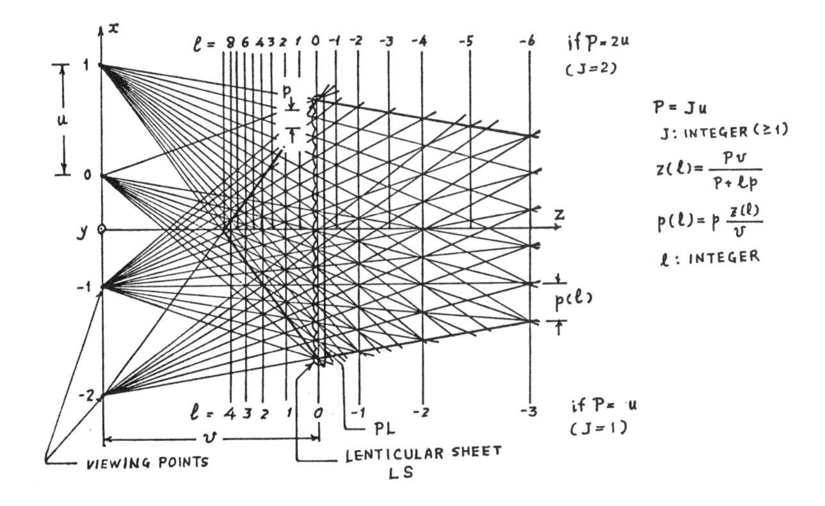

Fig.14 Geometry of the image space

4. Geometry of image space [10,14]

Figure 14 shows the geometry of the image space formed by the viewing points and the elementary lenses of the lenticular sheet(LS). The viewing points are equally spaced by u on x axis, and their total number is denoted by N; the width of the viewing zone is Nu. At recording, an LMS must be placed at the viewing point. At reconstruction, one of the observer's eyes is placed in the vicinity of the viewing point. LS is placed at z=v, and it has the M' elementary lenses with the pitch p; the width of the field of view is M'p. At recording, a ray to be recorded must pass through one of the viewing points and the center of one of the elementary lenses, and then the ray should be recorded in a line image at the back focal plane of LS. At reconstruction, the ray to be reconstructed from a line image passes through the center of the elementary lens under which the line image is reproduced.

If LS has an infinite directivity and the line image has an infinitesimal width, and if LS is perfectly in register with the line images, the ray must pass through the viewing point that was crossed at recording. The intersections of these rays constitute a spatial lattice. In Fig.14, many of these rays are drawn. When u is the interpupillary distance divided by an integer J (J=1,2,3, ..), the lattice point to be seen by an observer exists on one of the quantized planes. The z coordinate of this plane and the pitch of the lattice points on this plane are given, repectively, by

$$z(\ell)=Pv(P+\ell p)^{-1}, \quad p(\ell)=pz(\ell)v^{-1}, \quad (\ell=..,-2,-1,0,1,2,..) \qquad (21)$$

Eqs.(21) have the quantized forms of Eqs.(14); they determine the geometry of the image space. The resolvable distances and the volume of the image space are obtained from Eqs.(21). To obtain the equally spaced line images, the focal length of LS must be given by

$$f=pv(Nu-p)^{-1}. \tag{22}$$

5. Registration, image flippings and resolutions [10,14,15]

Hereafter, we assign the subscripts 1 and 2 to denote the recording and reconstruction processes, if needed.

To obtain a 3-D image by a lenticular sheet, one must overlay LS2 on PL2 which is the display surface at the back focal plane of LS2, in register with the line images on PL2, and he must see the image at an appropriate place. If the magnification from PL1, which is the recording plane, to PL2 is unity, the condition for distortion-free image reconstruction is given by

$$\left|1-p_2p_1^{-1}\right| \lesssim f_1v_1^{-1}N^{-1}, \quad v_1f_1^{-1}=v_2f_2^{-1}. \tag{23}$$

If the pitch error is a systematic one, the observer sees discontinuities in the horizontal direction of the image. If the error is a random one, he sees many random and short vertical lines in the background. The pitch allowance given by Eqs.(23) is approximately satisfied by high-quality lenticular sheets. If the last equation of Eqs.(23) is not satisfied, the longitudinal magnification of the reconstructed 3-D image is not the same as the lateral magnification.

The angle of image flipping seen by an observer is given by

$$F_\ell=\ell pv^{-1}J^{-1}, \tag{24}$$

where ℓ is the index defining the quantized planes in Eqs.(21).

In practical situations, the elementary lenses of LS have a moderate directivity, and the recorded/ reproduced line images have a finite width. For these reasons, the reconstructed rays are not completely sorted into the viewing points, but diverge within a certain angle. The effective line width W of the line images is defined as the width of the line which produces the same diverging angle as the line image when using an ideal elementary lens. To record a still 3-D image, W is mainly determined by the directivity of LS and the resolution of the recording film. In a 3-D TV, W is mainly determined by the horizontal resolution of the 2-D TV system which is used for sending the line images.

To cover the viewing zone without overlapped images, the reconstructed ray from a line image with the finite width W should cover a strip of width u centered at the viewing point. When the line images are equally spaced in this situation, W imposes a restriction on J, and the maximum of J, J_{max}, determines the minimum image-flipping angle F_{min}; they are given by

$$J\lesssim fPv^{-1}W^{-1}, \quad F_{min}=pWf^{-1}P^{-1}. \tag{25}$$

Since the image space is already quantized, the resolutions of the reconstructed image are not degraded until the deviations reach the threshold values. These conditions are given by

$$_{max}D_Lu_1^{-1}\lesssim 1, \quad N(S_{A1}^{\;2}+S_{A2}^{\;2}+S_{T1}^{\;2}+S_{T2}^{\;2}+S_M^{\;2})^{1/2}p_1\lesssim 1, \tag{26}$$

where S_{A1}, S_{A2}, etc. are defined in Table 3, and the unit magnification between PL1 and PL2 is assumed. In Table 3, the definitions of S_{T1} and S_{T2} should be replaced by the resolution of the recording film at recording/reconstruction of a 3-D image if a still image is taken.

When one choose the maximum D_L and N satisfying Eqs.(26), the resolvable angles seen by the observer are given by

$$R_{x2}(\ell)=p_2 v_2^{-1} S_0, \quad R_{y2}(\ell)=(S_{T1}^2+S_{T2}^2+p_1^2)^{1/2} v_2^{-1} S_0. \tag{27}$$

The numbers of pic-cells to be transmitted by the 2-D TV system are given by

$$N_{x1}=NM', \quad N_{y1}=H_1 S_{y1}^{-1}, \tag{28}$$

where H_1 is the vertical dimension of PL1.

6. Experiments on a still camera [11]

Figure 15 shows a sketch of an autostereoscopic 3-D camera with unit magnification. A pseudoscopic image is produced by ICO, and LS decomposes this image into the line images. A color photographic film is fixed on the back of LS.

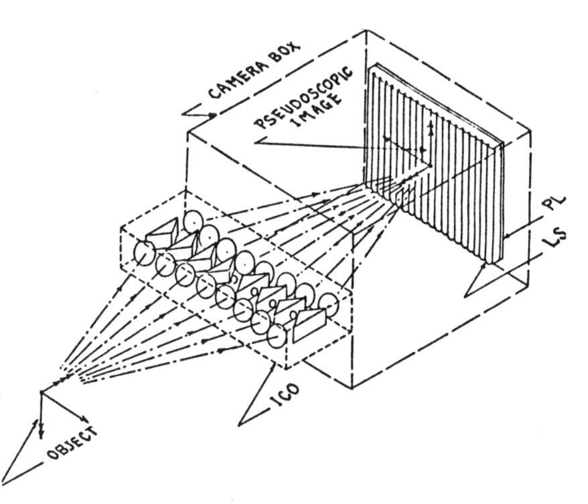

Fig.15 Autostereoscopic 3-D camera

An element of ICO, LMS, is shown in Fig.16. Three plane mirrors, M_1, M_2 and M_3, reflect the incoming ray from the object, similarly to LMS shown in Fig.12. Three barriers, B_1, B_2 and B_3, prevent the ray from passing through ICO without reflections. The outgoing angle θ_2 has the same magnitude and opposite direction as the angle of incidence θ_1. The aperture of ICO is given by the opening of M . Two lenses, L_1 and L_2, form a sharp image of the object center at the image center, where LS is placed. These lenses can be eliminated by replacing M_3 by a convex mirror having a proper focal length. In the experi-

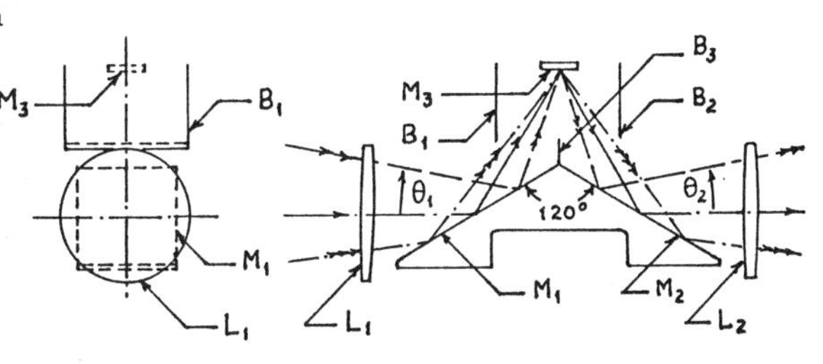

Fig.16 Lens mirror set (LMS). An element of ICO

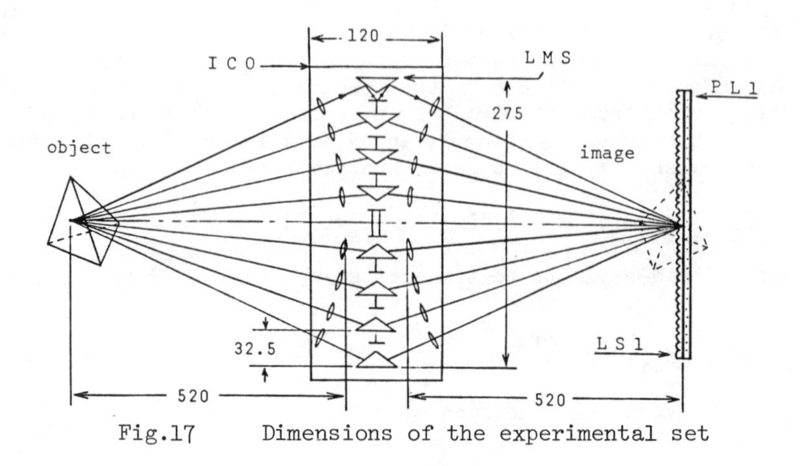

Fig.17 Dimensions of the experimental set

ments, the diameter of M_3, the lateral dimension of M_1 and M_2, the focal length of L_1 and L_2, and the diameter of L_1 and L_2 are 6mm, 16mm, 520mm and 22mm, respectively.

The dimensions of the camera arrangement are shown in Fig.17. Color films (127mm×178mm and 203mm×254mm, ASA 60) were used for the experiments. The object was illuminated by two 300W tungsten lamps with a 1sec exposure time.

After the film was developed and fixed, another lenticular sheet having the same pitch and focal length is fixed on the film in register with the line images. The lenticular sheets have approximately a 0.6mm pitch and a 1.2mm focal length. The registration is easily done by viewing the image with the naked eye.

Figure 18 shows the photographs of a reconstructed color image (the objects are vegetables on a bamboo tray) taken from two directions. If an observer looks at the actual reconstructed image, he will see a pictorial autostereoscopic image with his naked eye.

7. Experiments on a 3-D TV [10]

ICO arrangement used for the 3-D TV experiments is shown in Fig. 19(a). LMS of this ICO consists of a plane mirror and a small aper-

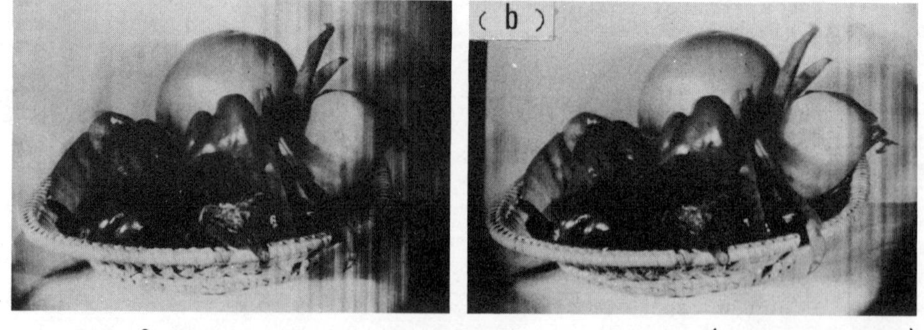

Fig.18 Photographs of a reconstructed 3-D image (left and right)

ture lens. The axis of the object space and the axis of the image space scan an angle of 90 degrees to prevent the direct rays through ICO. Figure 19(b) shows the block diagram of the experimental setup. Since the color TV set used in this experiment has a 350 TV line resolution, arrays of cylindrical lenses having a p=3mm pitch and a f=8.2mm focal length were used for LS1 and LS2, and the number of the cylindrical lenses effectively used in LS1 and LS2 was M'=35. The line images on the back focal plane PL1 were imaged on the target of the camera tube by RL1. The reproduced images on the monitor tube were imaged on PL2 at the back of LS2 by using RL2. The images on PL2 were registered by a zoom lens. The lateral direction of the image space was inverted by reversing the horizontal scan direction of the monitor tube.

Figure 20 shows the line images on PL1. Four lines for each elementary lens are seen in Fig.20. The scan nonlinearity induces an interference pattern between the reproduced line images and the cylindrical lenses, that results in a contraction of the field of view. The spread of the line images was determined by the TV set, and the reproduced line images were barely seen on the display surface of the monitor tube. Figure 21 shows the photographs of the reconstructed 3-D image taken from two directions. The photographs show a clear parallax, but indicate that a 3-D TV needs a 2-D color TV system with the improved resolution and the improved scan linearity. To suppress the interference pattern between the stripe structure of LS2 and the color stripes of the monitor screen, the line images on PL2 were slightly blurred by defocusing the camera lens.

8. Discussion

The experiments with a still camera have revealed that J is not smaller than 2 (and preferably 3) for a good autostereoscopic 3-D image. If 4 scenes are needed for a 3-D image, the theoretical investigations show that a 5 MHz bandwidth is needed to send a 3-D image with lateral resolution of 244 lines in (1/30) sec. The resolutions required for the 2-D TV are 1400 TV lines in the horizontal direction and 250 TV lines in the vertical direction, and the scan-linearity required for the 2-D TV is 0.07%.

In Fig 19(b), two relay lenses, RL1 and RL2, are used. If RL2 is eliminated by combining LS2 and the monitor screen, a large amount of

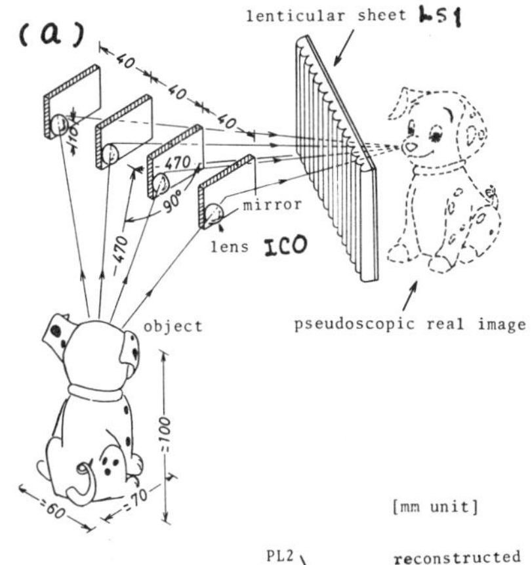

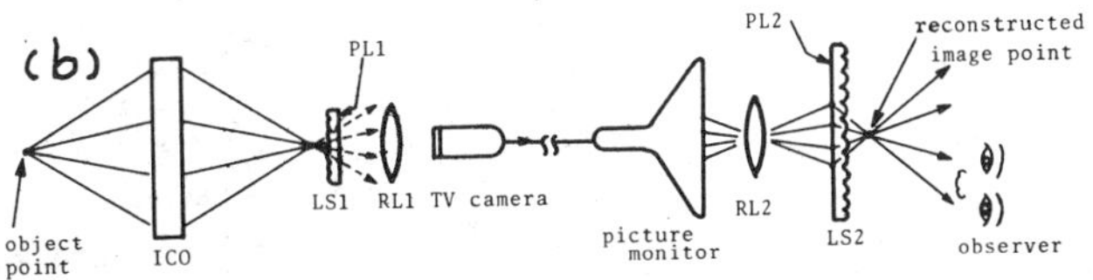

Fig.19 Configuration of the experimental TV
(a) ICO, (b) Block diagram

light is easily saved.

The compatibility between the present TV and an ordinary 2-D TV can be achieved if the following provisions are made. First, at the image-taking station, the line images taken from the central viewing point are separated from other line images by an electronic switching network. The image thus separated is transmitted through an ordinary baseband; other images go through higher frequency bands. Then, a 2-D image can be seen with an ordinary TV monitor . In the 3-D TV monitor, the image from the central viewing point as well as other images are restored at their proper positions to form a 3-D image using another electronic switching network. When an ordinary 2-D signal is received by this 3-D TV monitor, the ordinary 2-D image is seen by repeating the line image to cover a cylindrical lens of LS2.

Fig.20 Line images on PL1. 4 lines per lenslet

To the author's knowledge, an autostereoscopic 3-D TV based on the parallax panoramagram method is most promising in the future, because (1) it can be used under ordinary illumination, (2) it can be compatible with the already prevailing 2-D TV systems and (3) the technically remaining problems will be solved in the near future with some efforts from engineers. The technical efforts to build a 2-D TV having (1) a high resolution in the horizontal direction only, (2) a highly linear scan in the horizontal direction and (3) a screen combined with a lenticular sheet are very important, since the monitor with these features can be used for any 3-D representation of an image. When these problems are solved, a pictorially autostereoscopic 3-D TV having a 15 MHz bandwidth will be widely used in the world.

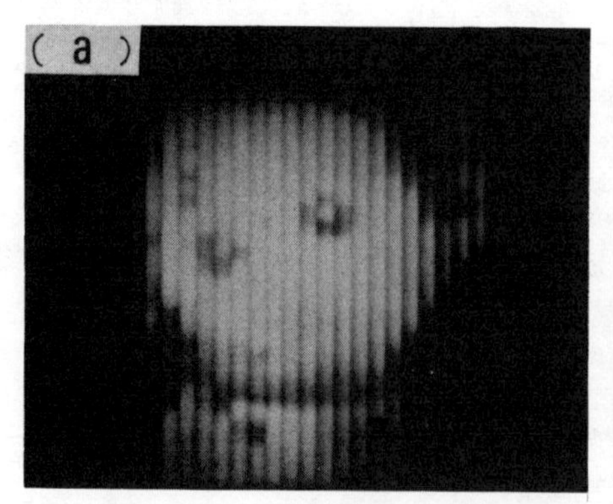

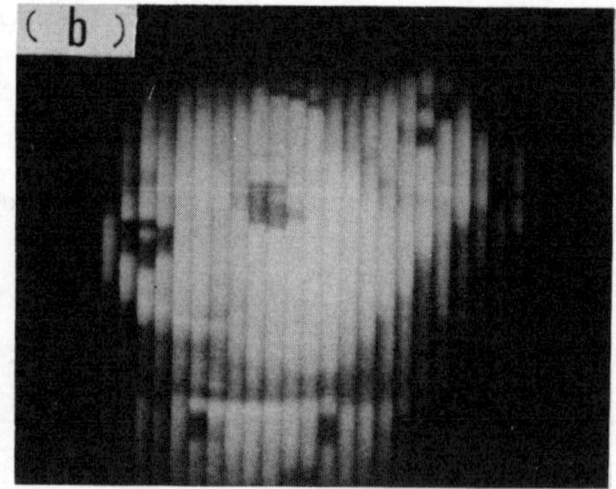

Fig.21 Photographs of a reconstructed image by the experimental 3-D TV (left, and right)

Table 3
Different spreads of line images.
(See Subsec.5)

Different Spreads of Line Images

(1) Aberrations of $LS1$	S_{A1} independent of z_1'
Aberrations of $LS2$	S_{A2} independent of z_2'
(2) Point spreads of the TV camera	S_{T1} independent of z_1'
Point spreads of the TV monitor	S_{T2} independent of z_2'
(3) Line spreads due to finite D_L	$S_M = D_L f_1(\dot{u}_1 + lp_1)u_1^{-1}v_1^{-1}$
(4) Quantization due to $LS1$	$S_Q = 1 + [[\vert l \vert D_L u_1^{-1}]]$
(5) Geometrical defocusing of eyes	$S_O = [1 + (lD_e u_2^{-1})^2]^{1/2}$

The relations used were $S_M = D_L f_1(v_1 + z_1')^{-1}$, $S_Q = 1 + [[S_L(z_1(l)')p_1^{-1}]]$, $S_L(z_1') \doteq D_L z_1'(v_1 + z_1')^{-1}$, and $z_2' = z_2 - v_2$. D_e is the diameter of eye pupils. An ICO using LMSs is assumed.

OTHER EXPERIMENTS ON 3-D IMAGE RECORDING/RECONSTRUCTION

1. Adjustable stereoglasses [13]

The ordinary stereoglasses cause eye strains for an observer, because the position of the image on the retina with stereoglasses is different from that of the usual 3-D vision. The direction of sight and the focal length of the eye lenses are automatically adjusted for 3-D vision, and he has his own positions for the images on the two retinal planes.

In Fig. 22, a scheme for using a pair of adjustable stereoglasses is shown. The stereoglasses consist of a pair of combined lenses, L_1 and L_2, where L_1 is an off-center piece of a convex lens and is continuously adjustable in the horizontal direction. L_2 is the compensating lens, and L_0 is the eye lens. By this scheme, the horizontal position of the image on the retina can be easily adjusted to the position of the usual 3-D vision. The eye strains are greatly eliminated, and also a 3-D vision is obtained in a wide range of positions of the left-eye picture and the right-eye picture.

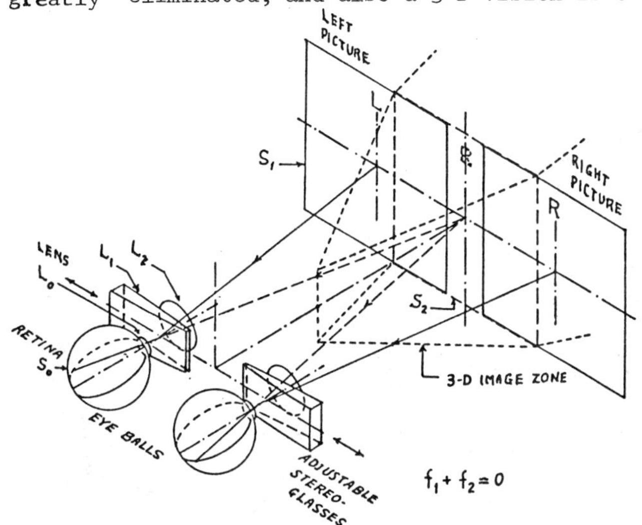

Fig.22 Adjustable stereoglasses

As shown by some experimental and theoretical investigations, this scheme is very useful, because (1) it lightens the eye strain, (2) an ordinary 2-D recording/reproducing instrument (such as a camera, a movie camera and a screen, and a TV camera and a TV monitor, etc.) can be used for a 3-D image, if a stereo-adapter is attached to the lens of the camera, (3) the cost of the glasses can be very low if the adjustable lenses are made of plastics.

2. Direct recording/reconstruction of 3-D x-ray images [14,15]

Using an x-ray grid instead of the lenticular sheet at recording, a 3-D x-ray image is directly recorded and reconstructed by the parallax panoramagram method. No ICO is needed in this case, because direct rays through an object are recorded on the film.

Figure 23 shows the schemes for recording and for reconstruction. T is the x-ray source energized at discrete positions. The grid G is made of tungsten rods and fixed with an x-ray film P at a distance f. The separation c can be zero or positive according to the magnification. The reconstruction procedure is the same as that of the 3-D image recorded by the camera, which is described in the previous section. Figure 24 shows the photographs of a reconstructed image by this method. The details of the internal structure can be seen from the reconstructed 3-D image with the naked eye.

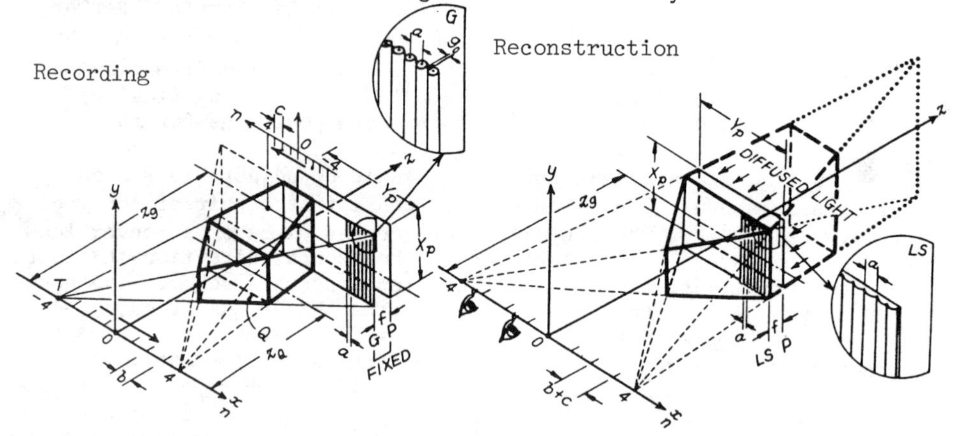

Fig.23 Direct recording/reconstruction of 3-D x-ray images

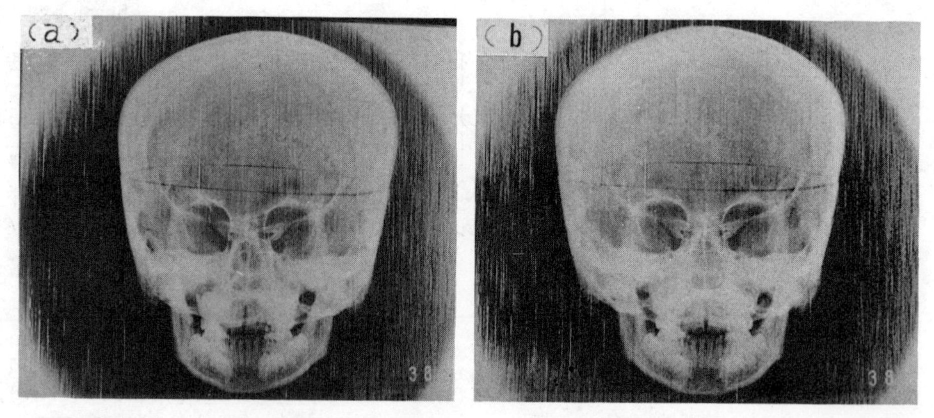

Fig.24 Photographs of a reconstructed 3-D image taken from two directions (left, and right)

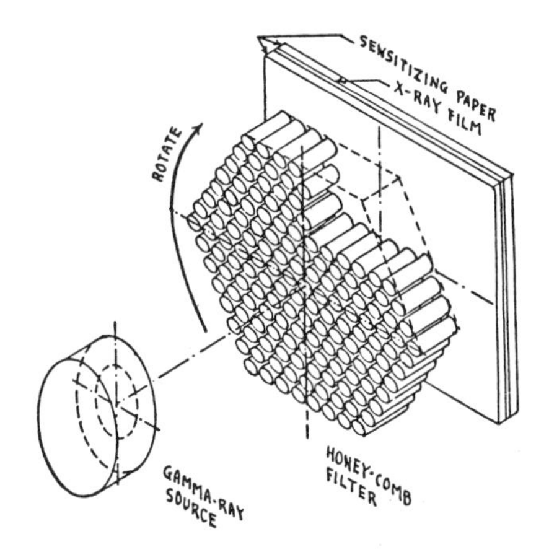

Fig.25 Direct recording of gamma-ray source distribution

Fig.26 Image of gamma-ray source

3. Direct recording of gamma-ray source distributions [16]

Figure 25 shows a method for direct recording of a gamma-ray source distribution by a honey-comb filter. This filter can be looked upon as a collimator or a spatial-frequency filter, and consists of 8000 tungsten rods(diameter:1.24mm, and length:20mm). Since the filter selects the rays coming in the normal direction to the filter, at the back of the filter, an image of parallel projec tion is obtained. When several images of parallel projection are obtained, a 3-D image can be reconstructed with the help of an electronic computer.

As in the case of an ordinary pin-hole type camera, the filter fixed with a recording film can be used in direct contact with an object which contains a distributed gamma-ray source. Furthermore, the dosage of gamma-ray can be made small when an efficient honey-comb filter with many tiny holes of hexagonal shape becomes available.

Figure 26 shows a record of a gamma-ray source. In this experiment, the filter was not rotated. When rotated, the triangular granules will smear out, and the shape of the source is easily seen. The distance between the source and the recording film was 4cm.

4. X-ray tomography from a few projection images [17,18]

Tomographic images can be reconstructed from a few projection images. Figure 27 shows the method to record projection images. It is similar to the method of direct recording a 3-D x-ray image except that no grid is used and an image is recorded on a film.

Using an algorithm of a min-max method and a minimum-mean-square method, applied to the geometry of the parallax panoramagram, a cross-sectional (tomographic) image can be computed from a few projection images. Figure 28 shows a result of a preliminary experiment and a simplified computation. A cross-sectional image of a porcelain cup is seen in the figure.

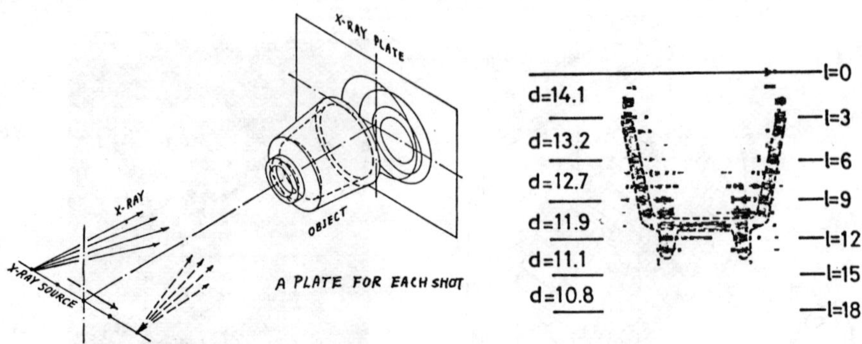

Fig.27 Recording a few projetion images

Fig.28 Reconstructed tomogram from 7 projection images(line shaped)

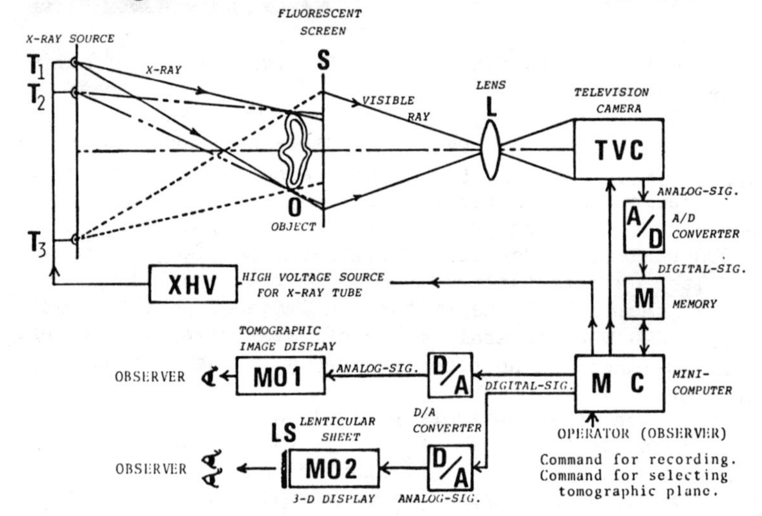

Fig.29 Block diagram of the proposed inspection apparatus

When the tomographic reconstruction from a few x-ray projection images is established, an inspector sees the tomogram of the desired sectional plane (which can be inclined at the desired angle), and at the same time he can see that sectional plane in the whole 3-D image, which is reconstructed on a TV screen with a lenticular sheet. Fig. 29 shows a block diagram of the proposed inspection apparatus.

CONCLUSIONS

The general discussions on various methods for recording/reconstructing 3-D images for the purpose of realization of an autostereoscopic 3-D television, and some details on the volume scanning method and the parallax panoramagram method are reviewed. By some more efforts on the improvement of the 2-D television for the resolution and the linearity in the horizontal direction, a pictorically autostereoscopic 3-D television will be established in the near future.

Some experiments related with 3-D image recording/reconstruction of radiations of various kinds were briefly described. The details of the experiments and the theoretical investigations will be shown in the future.

These works are financially supported mainly by Ministry of Education of Japan, Toray Science Foundations, and Fujitsu Laboratories, Ltd. The author wishes to express his sincere thanks to people who have worked with me for years on the 3-D imaging problem at the Institute of Industrial Science, University of Tokyo.

REFERENCES

1. L. P. Dudley, Stereoptics (McDonald & Company, London, 1951)
2. T. Okoshi, Three dimensional image technology (Sangyo Tosho, Tokyo, 1972)
3. A. C. Taraub, Appl. Opt. 6, 1085 (1967)
4. E. G. Rawson, Appl. Opt. 7, 1505 (1968)
5. M. C. King and D. H. Berry, Appl. Opt. 9, 2035 (1970)
6. H. J. Gerritsen and B. Horwitz, Appl. Opt. 10, 862 (1971)
7. J. Hamasaki, Y. Nagata, H. Higuchi and M. Okada, Appl. Opt. 16, 1675 (1977)
8. K. Miyazawa, J. Hamasaki and M. Okada, National Convention Rd. of IECE Japan, 1013 (1976)
9. H. Higuchi, J. Hamasaki and M. Okada, Appl. Opt. 16, 1777 (1977)
10. H. Higuchi and J. Hamasaki, Appl. Opt. 17, 3895 (1978)
11. J. Hamasaki and M. Okada, National Convention Rd. of IECE Japan, 1133, 5-160 (1980)
12. J. Hamasaki, M. Kawabata and Y. Murakami, National Convention Rd. of IECE Japan, 845, 4-105 (1980)
13. K. Yokota and J. Hamasaki, National Convention Rd. of IECE Japan, 1132, 5-159 (1980)
14. J. Hamasaki and K. Yokota, Appl. Opt. 17, 3125 (1978)
15. J. Hamasaki and K. Yokota, Appl. Opt. 18, 4039 (1979)
16. J. Hamasaki and K. Yokota, in preparation
17. J. Hamasaki, K. Yokota, M. Okada and M. Matsui, National Convention Rd. of IEE Japan, 179, 235 (1979)
18. J. Hamasaki, M. Matsui and K. Yokota, National Convention Rd. Japan, S10-6, 5-405 (1980)
19. R. L. Montebello, Proc. of SPIE, 120, 73 (1977)
20. J. F. Butterfield, The Society of Motion-Picture and Television Engineers, October (1979)

APPENDIX

Montebello's Integram uses a sophisticated three-step method, which is needed for maintaining a very wide viewing angle, as Lippman's pin-hole-array method has, without repetition of viewing zones and without too much loss of incoming light. At the viewing stage, however, Integram uses an array of spherical lenses having two array lens surfaces with f/1.3 and 2.4mm diameter. Actually, this elementary lens quantizes the incoming ray positions, as Stanhope loupe does, and it has considerable spherical aberrations that limit the the directivity of the array. Therefore, Integram does not reconstruct objectively a 3-D image without aids of eye lenses, while the original concept of Lippman's integral photography does. Although Integram does not quantize the ray directions, it is redeemed as one of parallax panoramagrams using a 2-D lens array.

A survey of stereoscopic home TV systems is described in ference 20. An emphasis is on "group viewing". The stereo-pair system with glasses and the autostereoscopic panoramic system are promising. The particular emphasis is placed on the system which employs many TV projectors and a lenticular screen assembly to combine images from many directions into a 3-D image. This method does not need crucial registration accuracy at the expense of the projection assembly consisting of many picture tubes.

REAL TIME OPTICAL RECORDING USING A NEGATIVE PHOTORESIST*

Jaime Frejlich+ & Lucila Cescato++
Instituto de Física - Universidade Estadual de Campinas
13100 Campinas - SP., Brasil

ABSTRACT

We describe a process that allows using a commercial negative photoresist as a real time recording material. The process results in an optical modulation of the film, without using any wet development. We describe its temporal evolution, and the influence of temperature and spatial frequency of the recorded signal on the velocity of the process. We describe the experimental set-up that allows us measuring changes of less than 10^{-4} µm in the optical pathlength of a thin transparent film.

INTRODUCTION

Applications of photoresists to Optics have been since long studied because of their interesting properties as a recording material: low noise, possibility of a wide linear response range[1,2], and the fact of being a "phase material". However, serious handicaps exist that limit their widespread utilization, such as: limited spectral wavelength range response, low energy sensitivity, and wet development required. Photopolymers do exist that need no chemical development; some of them are based on a mass

*This work was partially supported by a grant from FAPESP, São Paulo, Brazil.

+Fellow researcher of CNPq, Brazil.

++ Research fellowship from FAPESP, São Paulo, Brazil.

ISSN:0094-243X/81/650557-18$1.50

diffusion mechanism[3], plus molecular polarizability changes[4], others are based on local mass density changes due to cross-linking chemical reactions[2].

The process we describe in this paper and that we may call "Negative Photoresist Heat Development" (NPHD) is probably originated in a refractive index modulation of the film. Though the molecular mechanism is still unknown, we are able to state that there is no mass diffusion involved in the process.

Description of the Process. We had already seen[5] that when a film of KMR-747 (a negative Kodak photoresist) is exposed to an active light and is subsequently kept at a moderate temperature (40 to 60°C), an optical modulation appears without any wet development. The rapidity of the NPHD process showed to increase strongly with the temperature and does not depend on the registered spatial frequency.

Independent measurements, at zero spatial frequency (that means using an homogeneous light for exposure), showed that the refractive index of the film may decrease by 0.4% when exposed and then heated. However, it should be remarked that the modulation amplitude may behave differently for higher spatial frequencies.

No relevant surface modulation of the film was observed, and this photoresist exhibits a rather low absorption coefficient at the employed wavelength.

EXPERIMENTAL METHOD

We studied the KMR 747 photoresist by recording on it a sinusoidal interference fringe pattern of light, and following the subsequent time evolution of the light diffracted through the film. This method presents some advantages such as: the possibility of large variations of the recorded spatial frequency, and avoiding measuring near the low spatial frequencies where noise predominates due to non uniformities, dust and scratches on all optical interfaces.

Sample Preparation. Our films were prepared as has been described in former papers[6], using a rolling machine. After being coated on a 1.5 mm thick glass plate substrate, they were dried on a hot plate and then finish drying in an oven at 50 - 60°C for 1 or 2 hours. When low irradiance exposure was used, the sample was previously kept for some time in an N_2 atmosphere which increases its sensitivity[7]. The photoresist film need not being carefully prepared because all noise arising from dust, scratches, non uniformity, etc, is low spatial frequency noise. Films obtained were 1 to 2 μm thick.

Sample Chamber. The glass coated film was positioned as a window in a special made chamber (Fig. 1). Its temperature was regulated and as previously explained, a stream of N_2 was sometimes passed through the heated chamber so as to diffuse out the O_2 gas, away from the film, as the latter gas deactivates the photoreaction[7]. The introduction of additional optical interfaces does not affect the measurements as was explained above.

Exposure Set-up. In Fig. 2 the experimental set up is shown. For the interference fringe pattern projection on the film we used the 456 mm line of the Argon ion laser. A polarization rotator sets the correct beam polarization and a variable transmittance metallic beamsplitter plate allows simultaneous splitting and intensity matching of both beams. Because of all interfacesencountered by both beams, depolarizing effects may be corrected[8,9] so as to get reasonable contrast fringes at the film surface.

Time exposures of about 5 minutes totalizing 0.5 J/cm^2 energy density were used. This photoresist has low sensitivity at the wavelength we used but we did not dare using longer exposure times because of lack of confidence

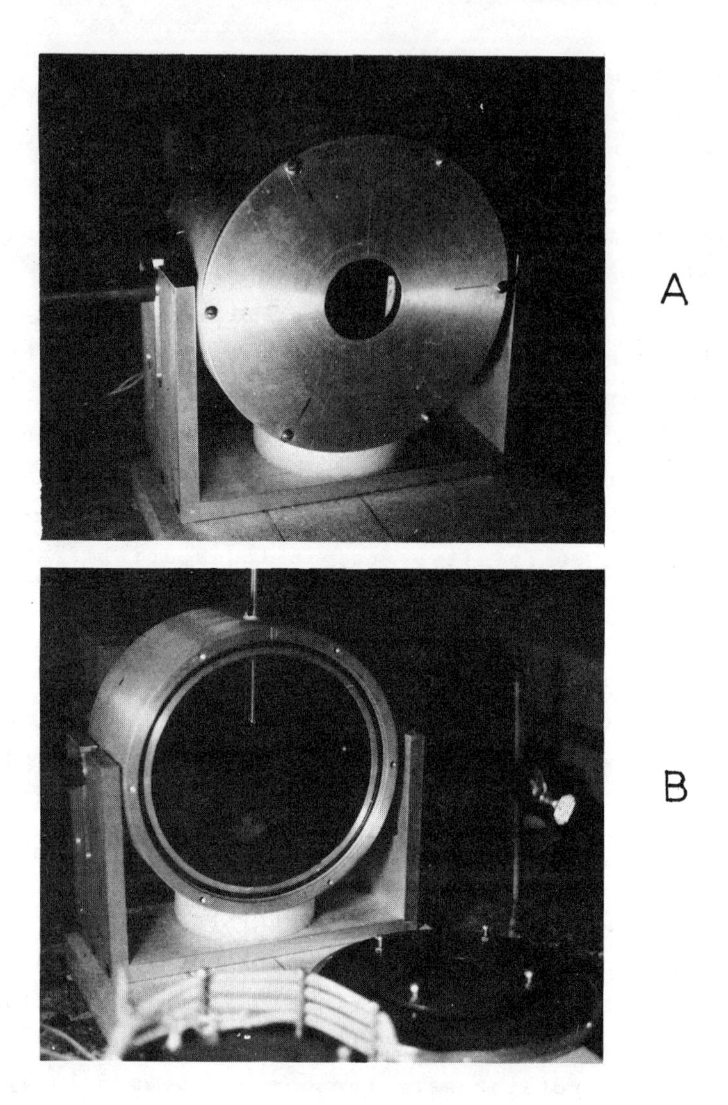

FIGURE 1

Figure A shows an external view of the chamber. The glass coated photoresist film forms the window shown there. Figure B shows the interior and some details.

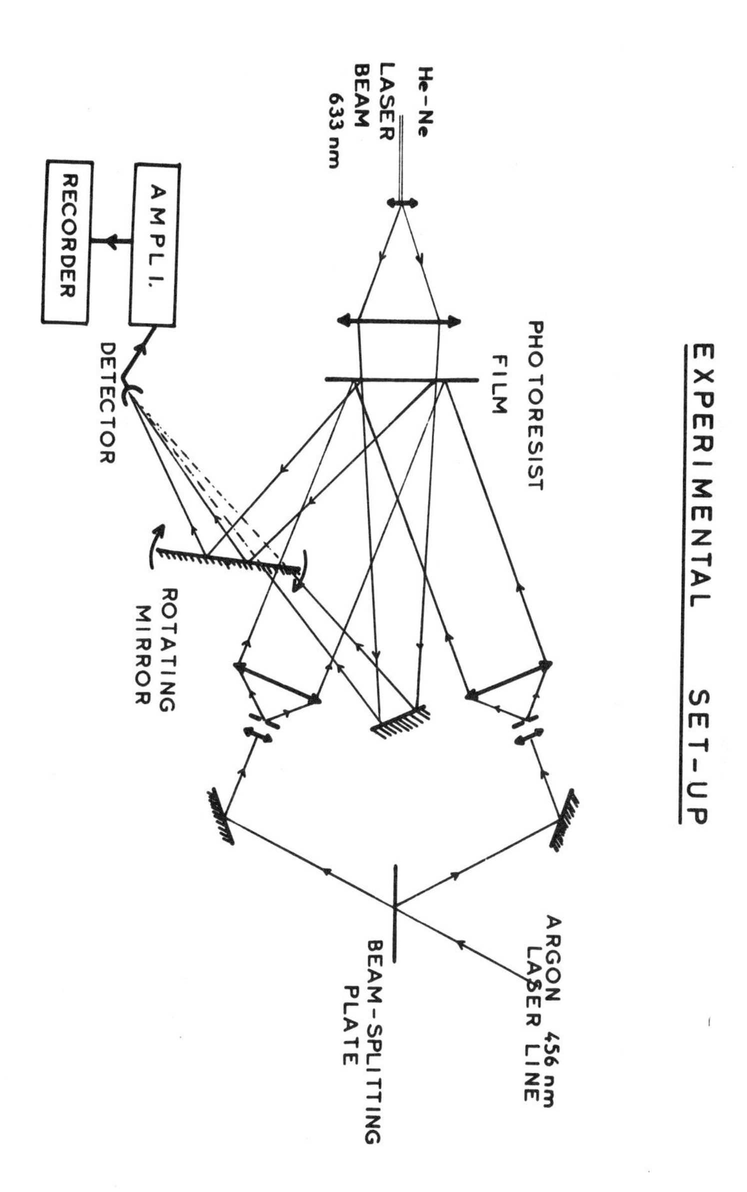

EXPERIMENTAL SET-UP
He-Ne
LASER
BEAM
633 nm
PHOTORESIST
FILM
AMPLI.
RECORDER
DETECTOR
ROTATING
MIRROR
ARGON
LASER LINE
456 nm
BEAM-SPLITTING
PLATE

FIGURE 2

in the mechanical stability of our set-up. The whole is supported on a 200x100x10 cm granite bench on a very massif wood table standing on a sand basement. A foamy anti vibrating board is placed between the granite and the wood.

Measurement Set-up. The measurement set-up is also shown in Fig. 2. Using a 6328 Å He-Ne laser and a 500 mm focal length lens, we follow the evolution of the order one-to-order zero diffraction intensity ratio resulting as a periodical optical modulation develops in the film. A fixed mirror reflects the zero order beam into a fixed detector; a rotating mirror interrupts it periodically, directing the 1st order into the detector, thus allowing alternate measurements of both orders. Two polarizers were used to attenuate the intensity of the zero order and to facilitate registration and comparison of both orders.

Ratios I_1/I_2 as small as 10^{-6} could be measured, corresponding to a phase optical modulation [10] of

$$x \equiv \frac{2\pi}{\lambda} \Delta(ne) = 4x10^{-3} \qquad (1)$$

λ, $\Delta(ne)$ being respectively the 6328 Å He-Ne wavelength, and the peak-to-peak difference in film optical thickness for a sinusoidal modulation. Figures 3 and 4 show photographs of the registered diffracted spectrum, during NPHD process.

EXPERIMENTAL RESULTS

Optical Modulation of the Photoresist Film. The photoresist film received a short exposure and the subsequent evolution of the I_1/I_o ratio was measured. As the modulation x (eq.1) is a better parameter to caracterize any change in the optical structure of the film, we transformed the I_1/I_o ratios into the x modulation using the relation:

$$I_1/I_o = \left[\frac{J_1(x/2)}{J_o(x/2)}\right]^2 \qquad (2)$$

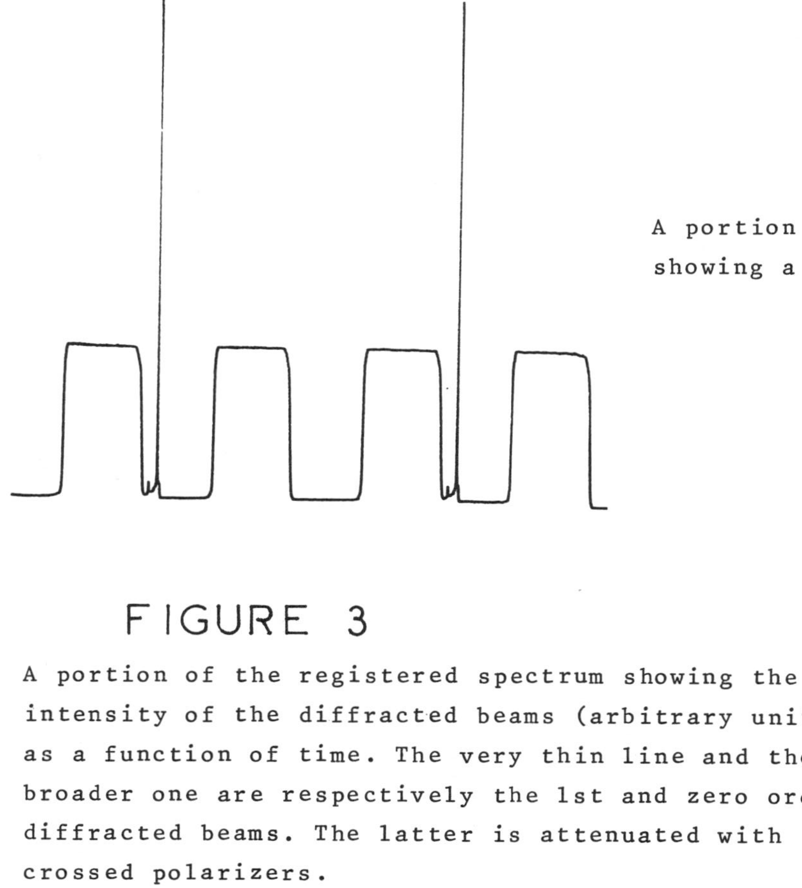

FIGURE 3

A portion of the registered spectrum showing the intensity of the diffracted beams (arbitrary units) as a function of time. The very thin line and the broader one are respectively the 1st and zero order diffracted beams. The latter is attenuated with crossed polarizers.

FIGURE 4

A portion of a registered spectrum as in Figure 3, showing a rapid increase in 1st order intensity.

J_i = i-th order Bessel function.
The above relation is graphically displayed in Fig. 5. In doing this, we had to suppose that our film exhibits a strictly sinusoidal modulation. This is not difficult to accept, as the modulation values involved are quite low ($10^{-2} < x < 10^{-3}$) and at that limit the photoresist has a rather linear response[11]. There is strong experimental evidence that the optical modulation obtained is due to an index of refraction modulation, because of the following facts:

a) liquid gate techniques show that our samples have no relevant surface modulation.
b) the absorption coefficient at the employed 6328 Å wavelength is low enough so as to neglect any influence of its possible modulation on the diffracted spectrum.
c) a 2nd order diffraction beam was sometimes observable, a fact that couldn't appear for a pure amplitude sinusoidal grating (a non-linearity effect being out of question because of the linear response of the photoresist, as explained above).

That is why Eq. 1, above, may be written as:

$$x = \frac{2\pi}{\lambda} e \, \Delta(n) \qquad (3)$$

<u>Temporal Evolution of the NPHD Process in KMR 747</u>. The time evolution of the x variable (eq. 1), fits reasonably well an exponential law:

$$x = x_o \; (1-A.\exp[-K.t])$$

or (4)

$$1 - \frac{x}{x_o} = A.\exp[-K.t]$$

where x_o is the limiting value for x, t being the time and K the time constant of the process (K is a measure of its velocity). The coefficient A is related to the initial

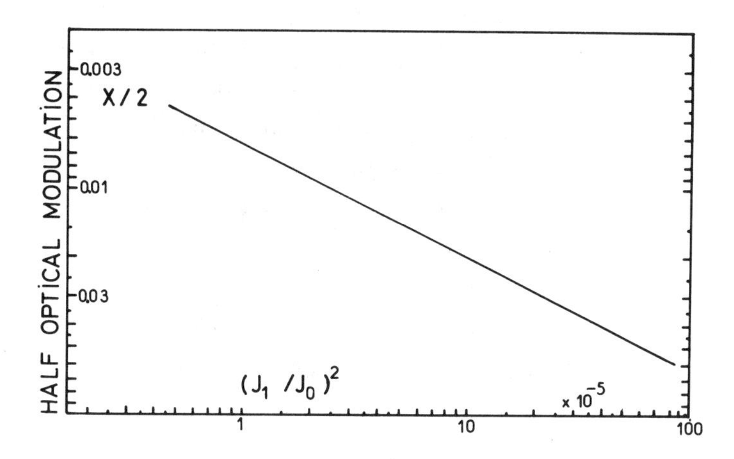

FIGURE 5

Graphical display of the relation "Half optical modulation" (x/2) as described in Eq. 1, versus $[J_1(x/2)/J_o(x/2)]^2$ as described in Eq. 2. For a strictly sinusoidal phase grating, the latter ratio represents the I_1/I_o 1st order-to-zero order diffraction intensity beams ratio.

conditions. Fig. 6 illustrates the fit of the data with eq. 4. Following Figs. 7 and 8 also show good agreement with eq. 4. We should remark that eq. 4 could represent also a mass diffusion process[5], as a 1st order chemical reaction[12], or any other formally analogous phenomenon.

At the beginning the sample temperature is not quite stable yet, so that the data depart from the theoretical curve; at the end of our run the data do not fit either, probably because of some secondary slow decaying process taking place. This "decay" appears to be more relevant the higher the temperature as one can see in Fig. 7. For the reasons above, the first set of points and the last ones (typically for $x > 0.9\ x_o$) are rejected for calculation purposes.

Temperature Influence on the Velocity of the NPHD Process. The velocity of the NPHD process, as caracterized by the time constant parameter K (eq. 4) is highly dependent on temperature, as can be seen in Fig. 8. In Fig. 9 we can see the dependence of the K parameter on the reciprocal absolute temperature of the sample, for spatial frequencies varying from 238 to 1040 1/mm. There it can be seen that there is no sensible spatial frequency dependence on this relation. We can then write:

$$K = a.\exp\left[-b.\frac{1}{T}\right] \qquad (5)$$

T being the absolute temperature and the experimental values for parameters a and b being

$$a = 1.37 \times 10^{34}\ \text{min}^{-1}$$
$$b = 25826^{o}K$$

Spatial Frequency Influence on the Velocity of the NPHD Process: A Non Mass Diffusion Process. If we suppose the NPHD process to be based on a mass diffusion phenomenon, a strong spatial frequency dependence on parameter K should exist, as has already been reported[5,13]. Through data from Fig. 9 we may conclude that no such a dependence exists.

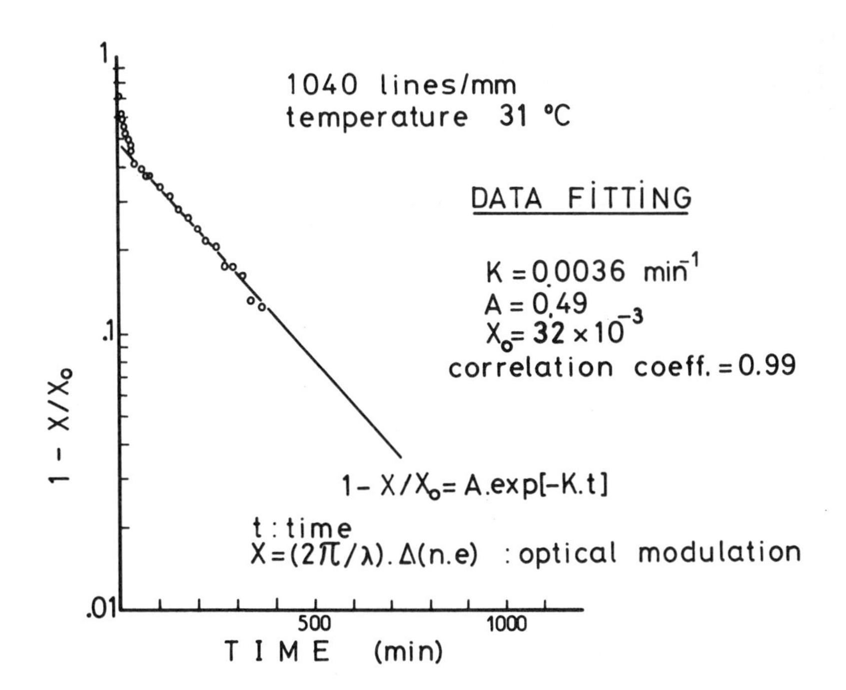
1040 lines/mm
temperature 31 °C
DATA FITTING
K = 0.0036 min⁻¹
A = 0.49
X₀ = 32 × 10⁻³
correlation coeff. = 0.99
1 - X/X₀ = A.exp[-K.t]
t : time
X = (2π/λ). Δ(n.e) : optical modulation
1
.1
.01
1 - X/X₀
500
1000
T I M E (min)

FIGURE 6

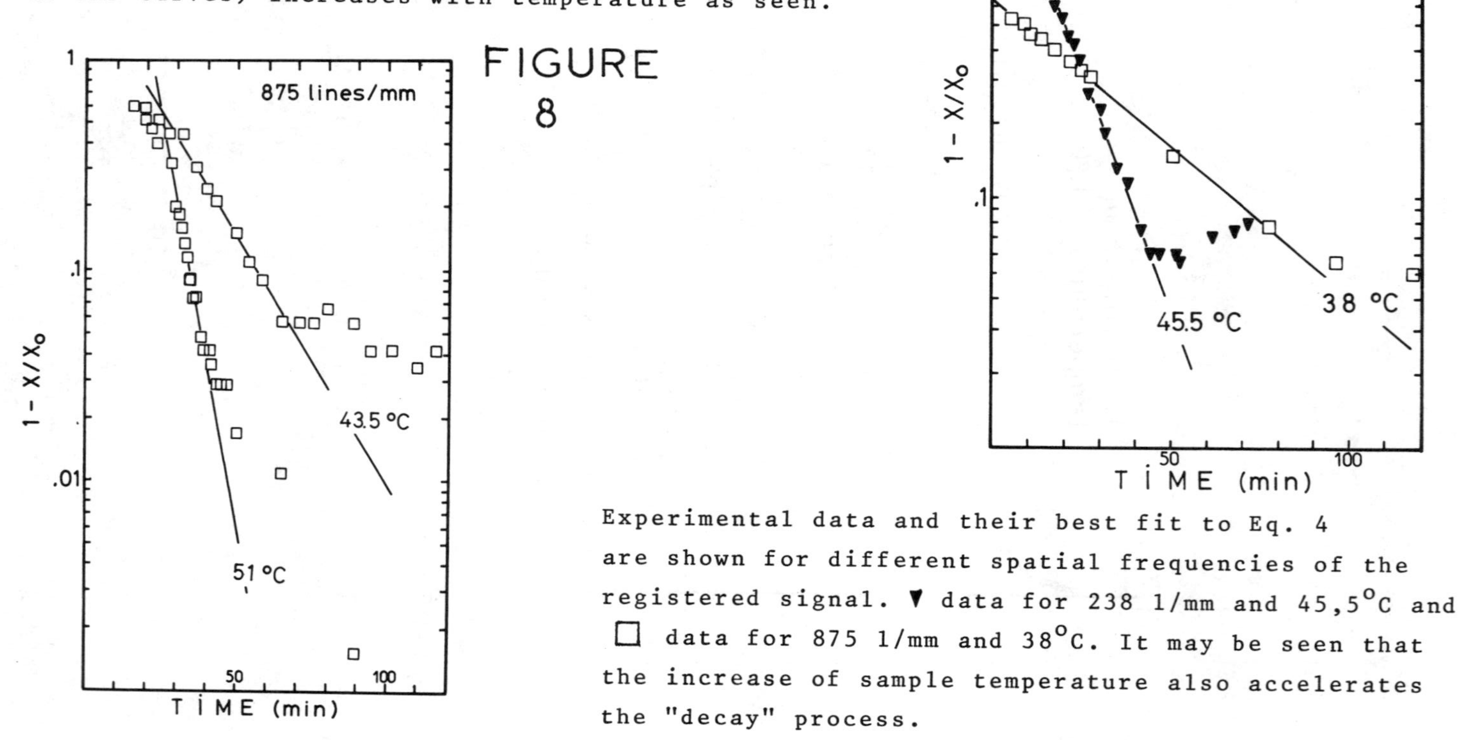

FIGURE 7

Experimental data and their best fit to Eq. 4 are shown for different spatial frequencies of the registered signal. ▼ data for 238 1/mm and 45,5^{o}C and □ data for 875 1/mm and 38°C. It may be seen that the increase of sample temperature also accelerates the "decay" process.

FIGURE 8

Experimental data and their best fit to Eq. 4 are shown for the same spatial frequency of the registered signal (875 1/mm) and different temperatures. The time constant K in eq. 4 (represented by the slope of the curves) increases with temperature as seen.

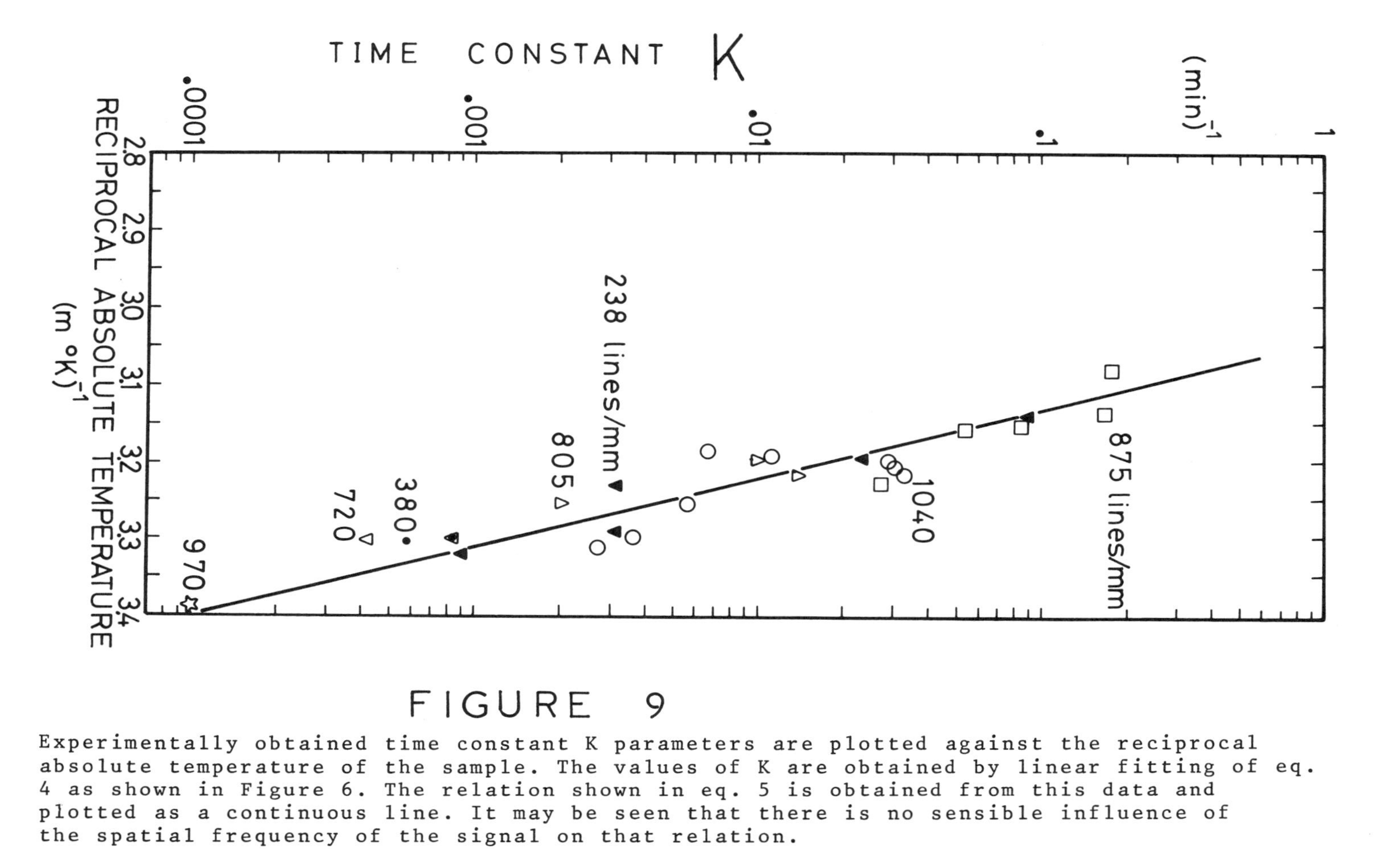

FIGURE 9

Experimentally obtained time constant K parameters are plotted against the reciprocal absolute temperature of the sample. The values of K are obtained by linear fitting of eq. 4 as shown in Figure 6. The relation shown in eq. 5 is obtained from this data and plotted as a continuous line. It may be seen that there is no sensible influence of the spatial frequency of the signal on that relation.

Hence we are forced to believe that the NPHD process, though essentially unknown, must be related with local molecular structure changes or chemical reactions concerning statistically stationary molecules. Any mass diffusion mechanism should result in a spatial frequency square dependence [5,13], which is far from our experimental results.

SOME RELEVANT DERIVATIONS

KMR 747 Photoresist Utilization as a Real Time Recording Material. The NPHD process not requiring any wet development and its velocity being so highly dependent on sample temperature, we propose to use it as a real time recording material, the meaning of this term depending on experimental conditions and user's necessities. Fig. 10 illustrates this fact: at room temperature this material "autodevelops" very slowly, but at 80°C it reaches 90% of its final modulation value in about 0.6 seconds, which may satisfy most common requirements.

When using the NPHD process care must be taken so as to optimize the heating time as there seems to exist a competitive, though slower, "decaying" process which increases also with temperature.

Index of Refraction and Molecular Concentration: An Estimate of the Activation Energy for an Unknown Process. From the general relation[14]

$$\frac{n^2-1}{n^2+2} = \frac{4\pi}{3} Na \tag{6}$$

relating the index of refraction n with the concentration N of a molecular specimen having polarizability a, and from eq. 3 we can obtain

$$x = \frac{2\pi}{\lambda} e\Delta n = Z\ \Delta a(N_o - N), \tag{7}$$

$$K = a \cdot \exp(-b \cdot 1/T)$$
$$a = 1.37 \times 10^{34} \, \text{min}^{-1}$$
$$b = 25826 \, °K$$

KMR-747 as a "REAL TIME" recording material		
TEMP. °C	K min^{-1}	time for reaching 90% of final modulation
20	7.2×10^{-5}	32 060 min
30	1.32×10^{-3}	1 748 min
40	0.02	115 min
50	0.3	9 min
60	2.85	0.8 min
70	27.3	5 sec
80	231	0.6 sec

FIGURE 10

in the limit when $\Delta n/n << 1$. In eq. 7 the parameter Z is defined as

$$Z = \frac{4\pi^2}{9} \frac{e}{\lambda} \frac{(n^2 + 2)^2}{n} ,$$

$\Delta a = a^1 - a$ is the change in polarizability, and $N_o - N$ is the molecular concentration of a specimen whose polarizability is changed by Δa.

Thus it follows that

$$dx = -Z\Delta a \; dN. \tag{8}$$

For a first order chemical reaction involving a specimen with molecular concentration N we expect that[12]

$$dN = -k_a \, Ndt , \tag{9}$$

k_a being the rate constant of the reaction. Also, from eq. 4 we have

$$dx = x_o Ae^{-Kt} \, Kdt , \tag{10}$$

so that after combining eqs. 10, 9 and 8 we can conclude that

$$Z\Delta a \; k_a N = x_o Ae^{-Kt} K , \tag{11}$$

and

$$Z\Delta a \; k_a N = (x_o - x) K. \tag{12}$$

On the other hand, from eq. 7 we see that

$$x_o - x = Z \; \Delta a \; N \tag{13}$$

so that, after substituting eq. 13 into eq. 12, we find

$$K = k_a \tag{14}$$

This enables us to identify the time constant K of our process with the rate constant k_a of the chemical reaction. We are then allowed to evaluate the "activation energy" for our NPHD process, according to the well known relation[12]

$$k_a = k_{ao} \exp(-E_A/RT) \tag{15}$$

where E_A is the activation energy for the reaction. On comparing eq. 15 with eq. 5 we can write

$$E_A = bR \qquad (16)$$

This leads to an activation energy E_A = 51 Kcal/mol for the NPHD process.

CONCLUSIONS

We discuss a process that should allow utilization of a negative commercial photoresist as a real time irreversible recording material. This process should also be utilized as a mean for identifying or positioning the recorded data, prior to the development in the usual wet way.

The velocity of the process is not dependent on the spatial frequency of the recorded signal, but the modulation amplitude may have such a dependence.

An index of refraction modulation seems to be the recording mechanism, though there is no mass diffusion involved.

Whatever the intimate nature of the process should bem we are able to assign an activation energy to it, provided some hypothesis are accepted. This activation energy should eventually contribute to identify the process in question.

Complementary studies concerning the spectral sensitivity, modulation amplitude sensitivity and its dependence on spatial frequency, should be necessary for further caracterizing the NPHD process.

REFERENCES

1. Norman and Singh, Appl. Opt. 14, 818 (1975).
2. A.A. Friesen, Z. Rav-Noy, and S. Reich, Appl. Opt. 16, 427 (1977).
3. W.S. Colburn and K.A. Haines, Appl. Opt. 10, 1636 (1971).
4. W.J. Tomlinson, E.A. Chandross, H.P. Weber and G.D. Aumiller, Appl. Opt. 15, 534 (1976).
5. J. Frejlich and J.J. Clair, J. Opt. Soc. Am. 67, 1644 (1977).
6. J.J. Clair, J. Frejlich, J.M. Jonathan, L.H. Torres, Nouv. Rev. Optique 6, 303 (1975).
7. Kenneth, G. Clark, ARPS, MIRT. Elect. Comp. 29 June 1973.
8. T. Hirschfeld, Appl. Spect. 29, 192 (1975).
9. D.R. Smith and E. Loewenstein, Appl. Spect. 30, 303 (1976).
10. J.W. Goodman, Introduction to Fourier Optics, McGraw Hill (1968).
11. J. Frejlich and J.J. Clair, J. Opt. Soc. Am. 67, 92 (1977).
12. S. Glasstone, Text book of Physical Chemistry, D. Van Nostrand Co. Inc.
13. Wopschall Pampalone, Appl. Opt. 11, 2096 (1972).
14. E. Wolf and Max Born, Principles of Optics, Pergamon Press 1975.

OPTICAL CONTOUR DISPLAY OF SURFACE PATTERNS BY INTERFERENCE OF DIFFRACTED BEAMS

H. Hattori
Nagoya Municipal Industrial Research Institute
Nagoya, 456 Japan

T. Wada & M. Umeno
Nagoya Institute of Technology
Nagoya, 456 Japan

ABSTRACT

This paper describes a method for optical contour display of surface patterns by interference of diffracted beams using an optical filtering technique. Measurements with an adjustable sensitivity ranging from a few to several tens of micrometers have been performed by selecting appropriate diffraction orders. This technique may be applied to the measurement of the surface roughness of metallic and non-metallic materials.

INTRODUCTION

Recently, optical techniques have been used to study the strength of metallic materials against fracture. These techniques provide information about the mechanism leading to the fractures, and lead to the measurement of strength parameters. For this purpose, it is necessary to measure the effect of plastic deformations in three dimensions. Present methods for the measurement of deformations include interferometry by oblique illumination and Moire interferometry. However, the sensitivity of the former is too fine and that of the latter scheme is too coarse.

This paper describes an optical contour display method with adjustable measuring sensitivity ranging from a few to several tens of micrometers. This method uses an optical grating which is placed over, but not in contact with, the test surface.[1-3] By illuminating the optical grating with a laser beam, different diffraction orders are obtained. An optical filter separates a chosen diffraction order from the others for the purpose of forming an image of the interference pattern on a screen. An experiment has been carried out with an optical grating having a grating constant of 25 μm. This technique may be applied to the measurement of the surface roughness of metallic or non-metallic materials.

THEORETICAL FORMULATION

Our method employs a grating placed at a small distance from the surface to be tested. According to diffraction theory,[4] each incident wave at the grating separates into reflected and transmitted diffraction components. The diffracted waves propagate at discrete angles which are defined by

ISSN:0094-243X/81/650575-04$1.50

$$\sin\theta_m = \sin\theta_0 + m\lambda/g,\ m = 0, \pm 1, \pm 2, \cdots, \tag{1}$$

a)

b)

c)

Fig. 1 Optical path of diffraction and reflection.

where θ_0 is the angle of incidence, λ is the illumination wavelength, g is the grating constant and m is the order of diffraction. If a transmission grating is used, the light passing through the grating is reflected from the test surface and interferes with the light diffracted in reflection from the grating. Figure 1 shows the formation of different first diffracted waves as a function of the number of reflections at the test surface. The resultant waves are obtained as follows:

$$\varepsilon_0 \propto E_1 \tag{2}$$

$$\varepsilon_1 \propto E_1\ [\exp\{-i(2\pi(f_1+\alpha)h+\pi)\} + \exp\{-i(2\pi f_2 h+\pi)\}] \tag{3}$$

$$\varepsilon_2 \propto E_1\ [\exp\{-2i(2\pi(f_1+\alpha)h+\pi)\} + \exp\{-i(2\pi(f_1+\alpha+f_2)h+2\pi)\} + \exp\{-2i(2\pi f_2 h+\pi)\}] \tag{4}$$

where, $f_1 = 2\cos\theta/\lambda$, $f_2 = 2\cos\theta_0/\lambda$ and $\alpha = 2(\sin\theta_0 - \sin\theta_0)\tan\theta/\lambda \simeq 0$; The total amplitude of all first order diffracted waves is given by

$$\varepsilon = \Sigma_{n=0}^{n}\ \varepsilon_n = \frac{r_1(1-r)^2}{r} E_1\left[\frac{1}{\{1-(rR)X\}\{1-(rR)Y\}} - \frac{1-2r}{(1-r)^2}\right] \tag{5}$$

where r_1 is the diffraction efficiency of first order diffracted beam, r and R are the reflectivities of the grating and test surface, respectively, $X=\exp\{-i(2\pi(f_1+\alpha)h+\pi)\}$ and $Y = \exp\{-i(2\pi f_2 h+\pi)\}$. After multiplication by its complex conjugate, Eq.(5) yields the following expression for the intensity

$$I = \frac{1}{2}|\varepsilon|^2$$

$$\propto C' - \{C'' + \beta(\cos 2\pi f_1 h + \cos 2\pi f_2 h) + \gamma\cos 2\pi(f_1+f_2)h\}/\{(\gamma' + \cos 2\pi f_1 h)(\gamma' + \cos 2\pi f_2 h)\} \tag{6}$$

where C', C", β, γ, and γ' are constants. This expression describes the intensity variation as a function of the gap h between the grating and the test surface. It corresponds to a

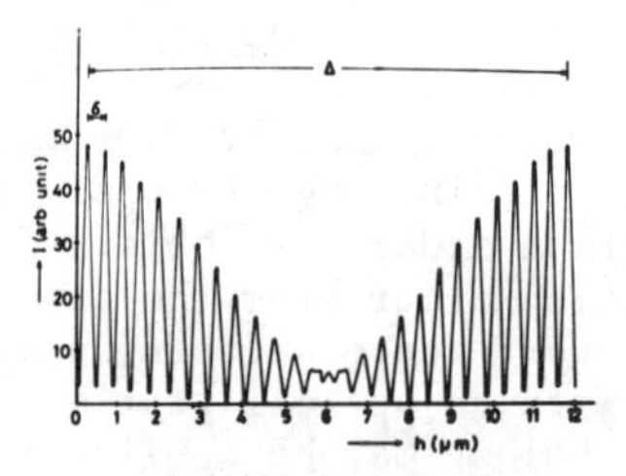

a) Calculated Result

b) Experimental One

Fig. 2 Intensity Distribution

multiplicative type of superposition of two oscillations with frequencies f_1 and f_2. Figure 2a shows the calculated intensity variation as a function of the air gap h between the grating and the test surface for a grating constant g = 25 μm, an illumination wavelength 6328 A°and an angle of incidence $\theta_0 = 45°$, r = 0.2 and R = 0.3. The resulting intensity distribution shows a beat pattern composed of narrow intensity peaks whose amplitudes are modulated by a slowly varying periodic function. Figure 2b shows the experimental results displaying the interference beat pattern at various distances h. The successive maxima of the envelope curve occur at intervals

$$\Delta = \frac{\lambda}{2\left|\cos\theta_m - \cos\theta_0\right|} \tag{7}$$

The separation of the narrow peaks are given by

$$\delta = \frac{\lambda}{4}\left(\frac{1}{\cos\theta_0} + \frac{1}{\cos\theta_m}\right) \tag{8}$$

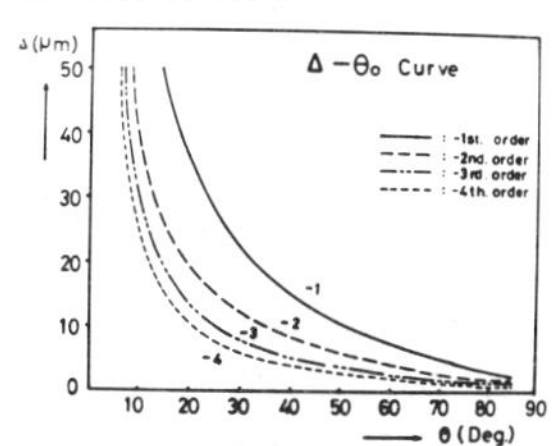

Figure 3

Note that the frequency of occurrence of the narrow peaks at the minima of the envelope curve becomes twice the frequency at the maximum of the macrostructure. Figure 3 shows the dependence of the measuring sensitivity on the angle of incidence and the order of diffraction. The measuring sensitivity can be adjusted by selecting an appropriate diffraction order. Thus, we have analyzed our measurement technique with variable measuring sensitivity by using optical filtering instead of changing the angle of incidence.

EXPERIMENTAL METHOD & RESULTS

The experimental set up is shown in Figure 4. A beam from a single mode He-Ne laser with an output power of 50 mW is expanded and collimated to produce a beam of about 60 mm diameter with a practically perfect plane wavefront and near uniform intensity. This beam illuminates the transmission grating with a grating constant of 25 μm and an area of 50x50 mm^2 which is placed over a reflecting surface.

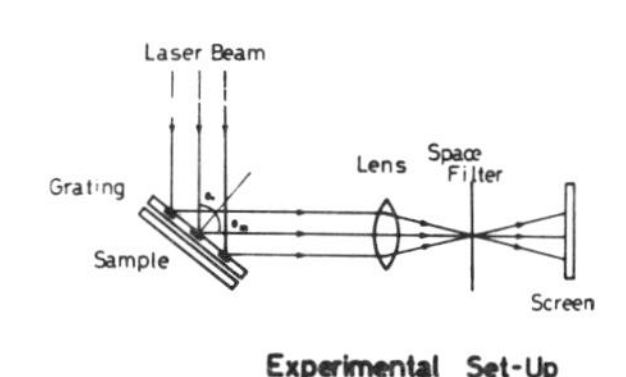

Figure 4

The light diffracted in reflection from the grating and the test surface is collected by the lens. The optical filtering is done by setting a filter at the focal plane of a lens in order to separate diffracted beams of different order.[5] This technique provides a beat frequency pattern with variable fringe separation. One of the experimental results is shown in Fig. 5. The sample was made of stainless steel with a depression of about 100 μm. In this case, the angle of incidence was 45°. The measurement sensitivity was varied by selecting the diffraction orders. Figure 6 shows the measured pattern in the domain of plastic deformation of a compact tension specimen with chevron notch.[6] The slip lines in the ligament can be detected.

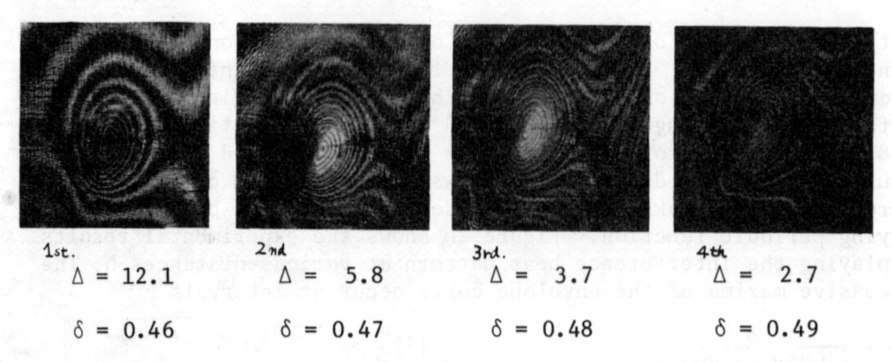

Figure 5. The measured patterns of stainless steel with a concave of 100 μm.

Figure 6. The measured pattern of slip lines of CT specimen.

CONCLUSIONS

Optical contour displays of surface patterns produced by interference of diffracted beams have been obtained by using an optical filtering technique. Interferometry by oblique illumination and Moire interferometry are not usually suitable for measurements requiring sensitivities of the order of a few to several tens of microns. The technique discussed in this paper is appropriate for this sensitivity range and may be applicable for metallic and non-metallic materials.

The authors would like to thank Professor T. Okochi for supplying the samples used in this work.

REFERENCES

1. W. Jaerisch et al. , Applied Optics 12, 1552 (1973).
2. W. Jaerisch et al. , Applied Optics 17, 740 (1978).
3. G. Makosch et al. , Applied Optics 17, 744 (1978).
4. M. Born et al. , Principle of Optics (Pergamon, London, 1970)
5. H. Hattori et al. , ICO at Madrid (1978)
6. for example: American National Standard Z260.2, E399 (1974)

AUTOMATIC MEASUREMENT OF 3-D SHAPES USING SCANNING MOIRÉ METHOD

T. Yatagai and M. Idesawa
Institute of Physical and Chemical Research
2-1, Hirosawa, Wako, Saitama 351, Japan

ABSTRACT

A new type of moiré topography called scanning moiré method is proposed. Moiré fringes are obtained by electronic scanning and sampling techniques, instead of superimposing a reference grating in the conventional projection type moiré topography. The electronic technique enables us to distinguish between a depression and an elevation of an object under measurement, and also interpolate moiré fringes automatically. By controlling the phase of the virtual grating corresponding to the scanning lines of the image sensors, the automatic sign determination of the moiré contour lines is accomplished. We have developed a practical system, in which a 1728-element linear photodiode array on a micro-stage is employed as a scanning image sensor. Some examples of analysis are presented.

INTRODUCTION

Moiré topography is one of the most powerful techniques of describing 3-D shapes, but the application of optics and digital technology to the problem of analyzing moiré fringe patterns has been limited. The job of locating fringes and extracting useful information from a moiré fringe pattern is a tedious and generally unrewarding task. The system to be described in this paper provides for automatic generation and analysis of a moiré fringe pattern.

There are two types of moiré topography. First is the shadow moiré topography, which involves positioning a grating close to an object and observing its shadow through the grating.[1,2] Under a certain condition, the resultant moiré fringes correspond to a contour line system of the object. Second is the projection-type moiré topography, in which a shadow grating produced by projecting a grating on an object is observed through another grating.[3,4] In order to obtain quantitative data, great efforts have been exerted to analyze fringe data by using image processing techniques.[5,6] However, a fully automatic measuring system can not be realized because a moiré contour has no information of the fringe sign, or the fringe order. We have described that the scanning moiré method provides for automated surface measurement.[7] This system eliminates the slope ambiguity problem. Recently, Moore et al discussed the use of phase-lock technique in moiré topography.[8] This method has a high accuracy, but the object size is restricted, because the object must be moved during the measurement.

In this paper, a practical measurement system developed is presented. This system is based on the scanning moiré method, and is useful for medical applications.

ISSN:0094-243X/81/650579-06$1.50

SCANNING MOIRÉ METHOD

The arrangement of the scanning moiré topography is shown in Fig. 1. A grating is projected on an object to be measured. The shadow of the grating on the object is observed with an electronic scanning device, such as a TV camera and a photodiode array. A moiré fringe pattern is observed on a monitor TV. The procedure of scanning and sampling in the image input device corresponds to superposition of a reference grating in the conventional projection-type moiré topography. Thus, we can generate moiré contours of the object in the same manner as the conventional method.

From a single set of moiré contours, we can not distinguish between a depression and an elevation of the object, since the relative order or sign between two adjacent contours is unknown. Additional information is necessary to remove this ambiguity. In the case of the scanning moiré method, we can change the phase, the pitch, and the direction of the scanning lines, and therefore, various contours can be immediately generated. This means that sign determination and interpolation of moiré contours are made, as described below.

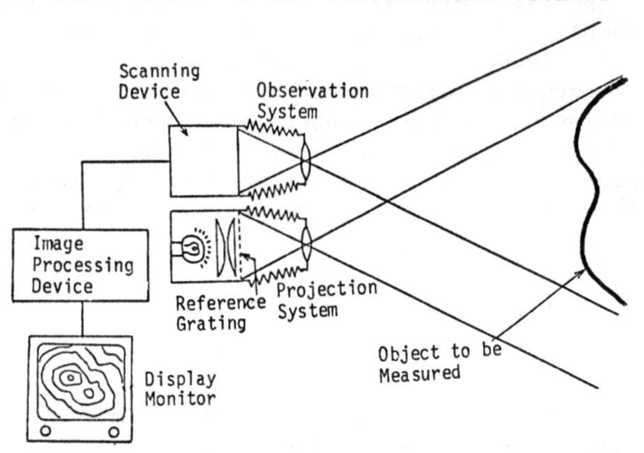

Fig. 1 Schematic of scanning moiré method.

In a typical arrangement of the projection-type moiré topography, the relation between a point (x_n, y_n, z_n) in the object space and a point (x_m, y_m) on the observation plane corresponding to the n-th moiré fringe is given as follows;

$$z_n = - a\ell \,/[\, P\,(\, n + \delta\,)] \qquad x_n = z_n\, x_m \,/\, a \qquad y_n = z_n\, y_m \,/\, a \tag{1}$$

where

$$n = n_m - n_p \qquad \delta = \delta_m - \delta_p \tag{2}$$

In Eq (2), n_m, δ_m, n_p, and δ_p are the index number of the observation grating, its phase factor, the index number of the projection grating, and its phase factor, respectively. The distance between the two optical axes, the image distance from the lens and the grating pitch are a, ℓ, and P, respectively. Also, n denotes the absolute order of the moiré fringe, δ its phase factor, respectively. When the phase factor δ_m of the observation grating varies, the phase factor δ of the moiré fringe also varies, and so moiré contour levels move accordingly. This means that the direction of the contour displacement depends on whether the slope of the object runs up or down. In the scanning moiré method, automatic sign determination of moiré fringes can be made by a computer, because the phase of the observation grating is electronically controllable. In practice, we determined the sign of moiré fringes by sampling a shadow image of the projected grating with 3 different phases.

MOIRÉ CAMERA

We have improved a commercially available moiré camera (FM 77, Fuji Photo Optical Co.) for mass examination of scoliosis. Figure 2 shows the improved moiré camera for the scanning moiré method. An optical system is configured with the projection system and the observation system. Focal lengths of the projection lens L_1 and the observation lens L_2 are 150 mm. The camera distance is 180 cm fixed from the lens front to an object. The measurement area is 99 cm high and 60 cm wide. In the ordinary case of scoliosis mass examination, the pitch of the grating is 0.1 mm, which gives 5 mm of height separation between successive moire fringes. Because the scanning moiré method can interpolate moiré fringes, we employed a coarse grating of 0.5 mm pitch for the projection.

In the observation system, light from the object is divided by a beam splitter. After reflection, about 10 % of the light goes to a TV camera, and the remainder passes through the beam splitter to reach a photodiode array on a precision stage. The signal from the TV camera is used to monitor the position of the object.

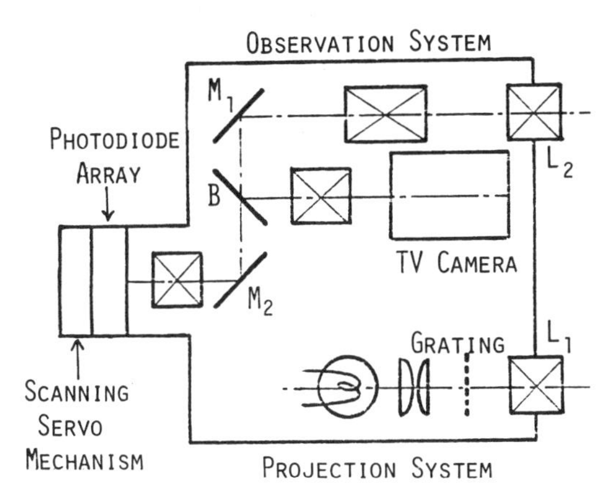

Fig. 2 Optical arrangement of moiré camera.

HARDWARE CONFIGURATION

The block diagram of the automatic measurement system using the scanning moiré method is shown in Fig. 3. To make a stable and reliable measurement in the scanning moiré method, it is necessary that the image scanning device have high positional accuracy. We employed a photodiode array such as the Reticon RL-1728H on a precision stage because of high positional accuracy. This device has 1728 photosensitive elements whose spacing is 15 μm. A scanning

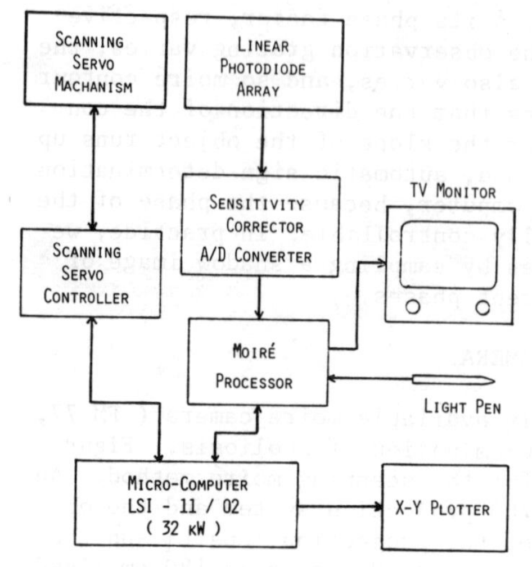

Fig. 3 Block diagram of the automatic measurement system using scanning moiré method.

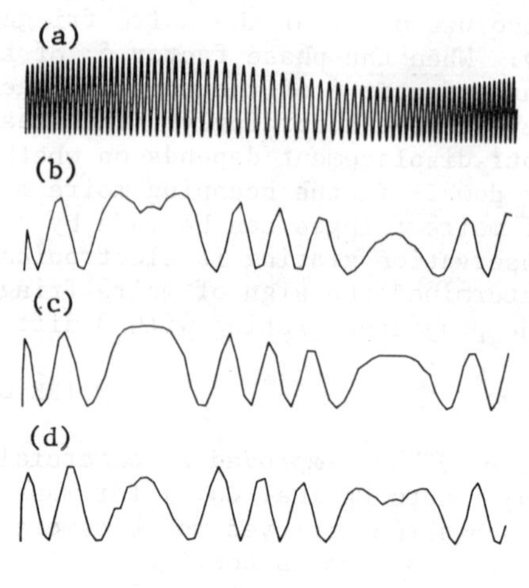

Fig. 4 Fringe generation by sampling. (a):Shadow grating. (b)-(d):Moiré fringes whose sampling phases are different.

servo mechanism moves the photodiode array along the transverse direction, so that all the image area can be covered. Because of non-uniformity in sensitivity and dark levels in the photodiode array, video signals are fed into a sensitivity corrector. An A/D converter converts the corrected signals to 8 bit digital signals, stored in a moiré processor memory. In order to obtain moiré fringes on a TV monitor, video signals from the TV camera are mixed with reference grating signals in the moiré processor. These moiré fringes on the TV monitor are not used for the measurement but for monitoring of object alignment.

FRINGE ANALYSIS

A moiré processor samples image data in the memory along the virtual grating corresponding to the reference grating in the conventional projection-type moiré topography. Sampling is practically made by culling image data from a photodiode array. When no photodiode element exists at the point to be sampled, interpolation of the image data is necessary. The sign of moiré contour lines can be determined by shifting the phase of the virtual grating. As mentioned before, phase shift of the virtual grating causes a displacement of the moiré fringes. Figure 4 shows an example of the fringe displacement. Figure 4 (a) is a shadow grating image obtained by the photodiode array, and Figs 4(b)-(d) are moiré fringes produced by sampling a grating image shown in Fig. 4(a) in a virtual grating. The phases of the virtual grating are shifted by $2\pi/3$ in

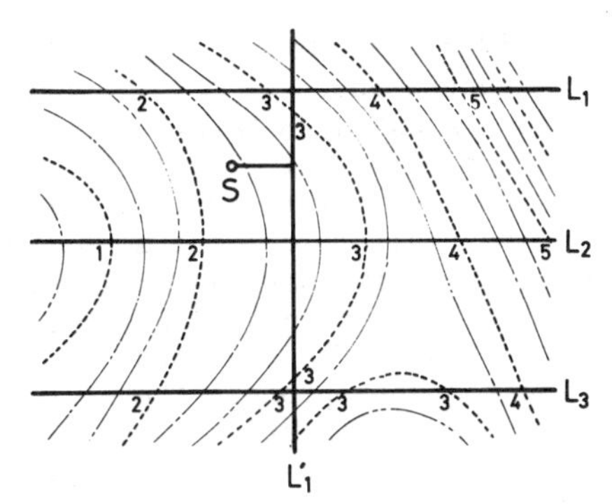

S : standard point
------: phase = 0
—-—: phase = $2\pi/3$
—--—: phase = $4\pi/3$

Fig. 5 Analysis grid. Sectional shapes along the grid lines are measured and relative fringe orders are determined.

Fig. 4(c) with respect to Fig. 4(b) and by $4\pi/3$ in Fig. 4(d), respectively. We can obtain additional information on relative fringe orders from the direction of the fringe displacement.

In the scanning moiré method, contour maps are not always required. For instance, as shown in Fig. 5, it is often enough to measure the sectional shapes only along the grid lines, and interpolate the area other than the grid lines by using data from the measured sectional shapes. In such a case, a shadow image is sampled only in the area along the grid lines, as shown in Fig. 5. Relative fringe orders for each grid line are determined by shifting the virtual grating as mentioned above. All the fringe orders are determined in relation to the relative orders of successive grid lines, that is, the fringe order is determined by counting the order from a reference point along the grid lines and keeping track of the signs of the fringes.
The entire system is controlled by a micro-computer LSI-11/02 with 32 k words of IC memory. The developed software system actually provides two modes: the automatic mode and the man-machine interactive mode. The automatic mode measures all the area. Sectional shapes are measured along the measuring grids which are previously set, and determines relative fringe orders according to the method mentioned above.

In the interactive mode, an operator can use a light pen as an input device to position the area to the line to be measured.

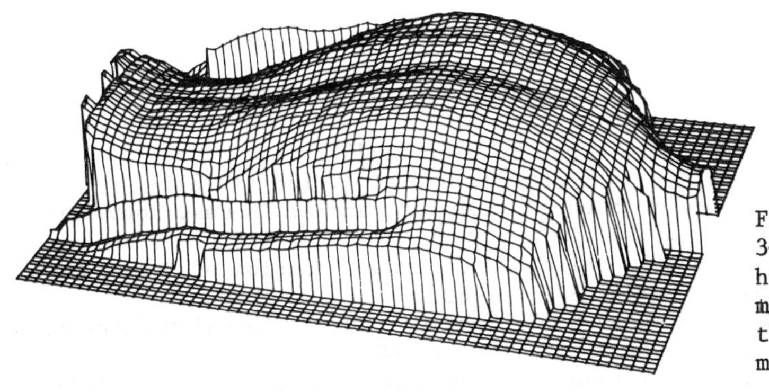

Fig.6 3-D plot of human back measured by the automatic system.

A small rectangular mark, whose longer sides are along either the vertical or the horizontal axes, can be superimposed on the TV monitor. At the start of this mode, operator is asked to locate rectangular marks fixing the boundary of the area or the line to be measured. The interactive mode allows the operator to control the operations of the light pen by locating special areas in the TV monitor, with the help of command keys, e.g.: Move the mark up or down, Move the mark right or left, Speed the mark motion up or down, Exit the procedure, Enter the next procedure, Display the sectional shape on the TV monitor, and so on. The computer prompts the operator to locate each to the command keys. Some application softwares are available for spline interpolation of data, sectional shape plots or 3-D plots on an X-Y plotter, and so on.

Figure 6 shows a 3-D plot of final data. The object is a human body. Some errors of fringe order determination appear in the axillary part, where the visibility of the shadow grating is very low.

CONCLUSION

In this paper, we have presented a method of automating 3-D shape measurements by a new type of moiré topography. Sampling a shadow grating with electronic scanning techniques, instead of superimposing the reference grating in the conventional projection type moiré topography, produces various sets of moiré contours, so that we can obtain automatic sign determination of moiré contours, as well as fringe interpolation. On the base of the scanning moiré method, we have developed a practical system for automatic analysis of 3-D shapes. This system consists of a moiré camera and a moiré processing system with a micro-computer. Softwares and analyzed data are stored in cassette magnetic tapes. Because this system is compact and portable, it is expected that the system will be widely used in the screening of scoliosis.

The authors wish to thank Dr. H. Saito and Dr. E. Goto of the Institute of Physical and Chemical Research for many advices. They wish to thank Mr. H. Ohshima and Mr. M. Suzuki of Fuji Photo Optical Co. for assistance in developing the facility.

REFERENCES

1. D. M. Meadows, W. O. Johnson, and J. B. Allen, Appl. Opt. 9, 942 (1970).
2. H. Takasaki, Appl. Opt. 9, 1467 (1970).
3. Y. Yoshino, Kogaku (Jpn. J. Opt.) 1, 128 (1072).
4. P. Benoit, E. Mathieu, J. Hormiere and A. Thomas, Nouv. Rev. Opt. 6, 67 (1975).
5. T. Yatagai and H. Saito, Proc. Annual Meeting of Jpn. Soc. Appl. Phys. March 1979, p. 83.
6. T. Yatagai, M Idesawa, Y. Yamaashi, Proc. Annual Meeting of Jpn. Soc. Appl. Phys. April 1980, p. 80.
7. M. Idesawa, T. Yatagai, T. Soma, Appl. Opt. 16, 2152 (1977).
8. D. T. Moore and B. E. Truax, Appl. Opt. 18, 91 (1979).

HOLOGRAPHIC INTERFEROMETRY APPLIED TO MEASUREMENTS OF DIFFUSION IN LIQUIDS

Henry Fenichel and L. Gabelmann-Gray
Department of Physics, University of Cincinnati
Cincinnati, Ohio 45221, USA

ABSTRACT

This paper discusses the application of holography to the study of mutual diffusion in transparent liquids. An expression is derived for determining the diffusion constant, D, from measurements of the position and time evolution of fringes on a double exposed hologram. An iterative procedure is developed to account for distortions due to refraction effects. Results of applying the technique to the temperature dependence of D for NaCl solutions are presented.

INTRODUCTION

The advent of holographic interferometry provides a new technique for studying small temporal changes in transparent media. The technique is more accurate than previously known methods and, furthermore, has the advantage that it lends itself to real time analysis. The method has been used successfully to investigate the structure of vibrating systems and flow visualization.[1]

The present paper discusses the application of hologram interferometry to the study of mutual diffusion in transparent liquids. Our technique for determining the diffusion coefficient will be discussed first. The method is a refinement of the approaches of Bochner[2] and Shustin[3]. Next, the effects of refraction, which are present when light traverses a medium with an index of refraction gradient, will be considered. We shall show how the results are modified by the refraction. Finally, results of the application of the technique to the temperature dependence of diffusion in NaCl solutions, will be presented.

THEORY

The method consists of recording on a hologram two beams of laser light which have traversed through the experimental cell at times t_1 and $t_2=t_1+\Delta t$. The main features of the technique are summarized in Fig. 1.

ISSN:0094-243X/81/650585-09$1.50

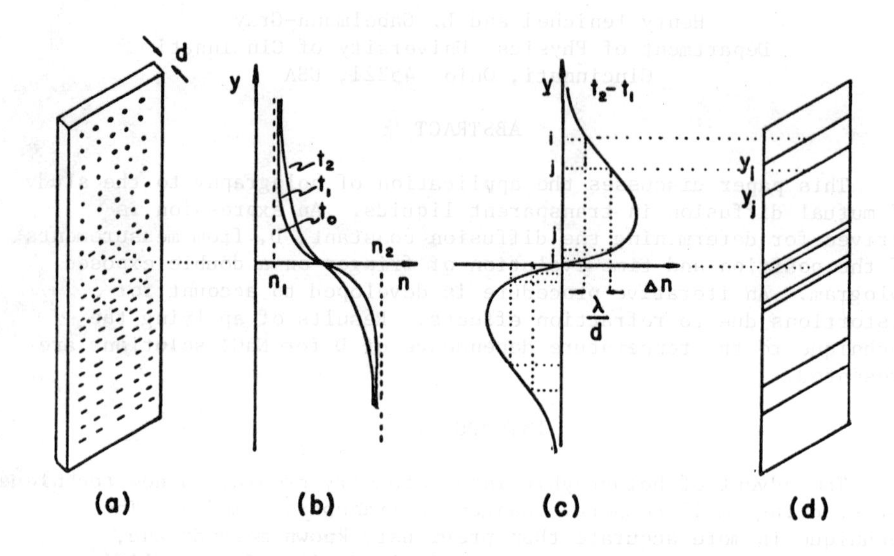

Fig. 1. Main features of the technique for measuring diffusion in liquids. See text for discussion.

Fig. 1(a) is a sketch of the cell, establishing the geometry of the experimental arrangement. At time t=0 the two fluids are separated, with the lighter one, of index of refraction n_1, resting on the liquid with an index of refraction n_2. As diffusion progresses, the concentration of the fluids along direction y will change. This, in turn, will result in a change in the index of refraction along the cell. The index of refraction profiles at two times t_1 and t_2 ($t_1 < t_2$) are shown in Fig. 1(b). The plot shows the variation of the index of refraction along the cell. Subtracting the two curves, (for t_2 and t_1, respectively) from each other gives the change in the index of refraction, Δn, shown in Fig. 1(c). This is the information which is stored on the hologram as a series of horizontal fringes. When the doubly exposed hologram is reconstructed, a series of horizontal interference fringes will be observed as sketched in Fig. 1(d). The fringes appear superimposed on the outline of the cell wherever the following condition is satisfied.

$$\Delta n = (2i+1)\lambda/2d, \tag{1}$$

where d is the thickness of the experimental cell, λ is the wavelength of the laser light and i is an integer representing the order of the fringe. There can be as many as four fringes per order, two above and two below the location of the original interface between the fluids. Fig. 1(c) shows the location of two sets of fringes i, and j respectively.

The location of the fringes will depend on the time, t, elapsed since the start of the diffusion process, and the diffusion constant, D. Thus, by measuring the location of the fringes y_i and y_j, and time, t, one can work backwards and solve for the diffusion constant.

Analytically, this is done as follows. We start with the assumption that, to a first order, the index of refraction n(y,t) varies linearly as the concentration. n(y,t) must then obey Fick's law:

$$\frac{\partial n(y,t)}{\partial t} = D \frac{\partial^2 n(y,t)}{\partial x^2} . \tag{2}$$

The solution for our geometry is an Error function of the form

$$n(y,t) = \frac{n_1 + n_2}{2} + \frac{(n_1 - n_2)\pi}{2\sqrt{2}} \int_o^{z(y)} \exp(-z^2)dz, \tag{3}$$

where $z = y/(2\sqrt{Dt})$. n_1 and n_2 are the initial indecies of refraction of the two solutions. Eq. 3 is the profile shown in Fig. 1(b). Fig. 1(c) is similarly obtained by substituting y_i and y_j into Eq. 3 and subtracting the two. Combining this latter result with the condition for the occurence for the fringes, Eq. 1, we obtain the following expression for the diffusion constant[4]

$$D = \frac{(y_i^2 - y_j^2)}{4t} \quad \ln\left[\frac{(2j + 1)y_i}{(2i + 1)y_j}\right]^{-1} . \tag{4}$$

Note that this result depends only on the location of the ith and j^{th} fringe and the time elapsed since the start of the diffusion. n_1 and n_2 conveniently cancel out in the derivation. Thus Eq. 4 has the advantage that the diffusion constant depends only on a few parameters which are easy to measure. The equation, however, neglects the effects of refraction of the laser light as it traverses the experimental cell. This effect depends critically on the magnitude of the diffusion constant and the index of refraction gradient (dn/dy).

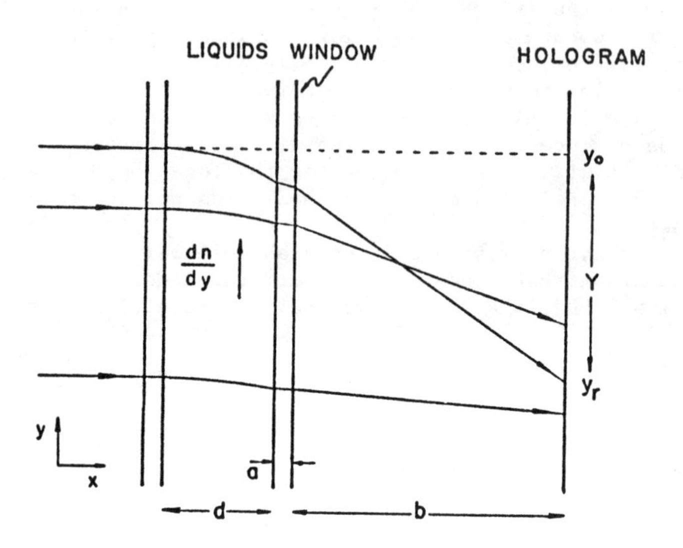

Fig. 2. Non-uniform bending of light passing through a medium with an index of refraction gradient.

Fig. 2 shows the path of light of a number of rays in the beam. During the early phase of diffusion there will be large concentration gradients near the original interface and small ones far from the interface. Correspondingly, refraction will be large near the interface and diminish near the end of the cell.

Consider the path that the terminus of a particular ray (e.g. the one initially reaching y_o) will trace on the photographic plate. As diffusion progresses, the ray will be displaced downward by an amount Y, to location y_r. Y will increase, non-linearly, until the gradient has reached a maximum. As time progresses Y will decrease again and the ray will return to y_o when diffusion ceases and the mixture is uniform. At any arbitrary time, t, the rays emerging from the cell will not be parallel, and some will cross each other.

Thus, the distortion introduced by refraction is non-uniform along the cell. The differential equation which governs refraction in a medium containing an index of refraction gradient is given by[5]

$$\frac{d^2y}{dx^2} = \left[1 + \left(\frac{dy}{dx}\right)^2 \right] \frac{1}{n} \frac{dn}{dy} . \tag{5}$$

Solving this equation for the geometry shown in Fig. 2, and assuming small angles of refraction, one obtains[5]

$$Y = \left(\frac{d}{2n} + \frac{a}{n_g} + \frac{b}{n_a}\right) d \frac{dn}{dy} + \frac{1}{2}\left[\frac{a}{n_g} + \frac{b}{n_a}\right] \frac{d^3}{n^2} \left(\frac{dn}{dy}\right)^3, \tag{6}$$

where a, d, and b, are the distances shown in the Fig. 2, n, n_g, n_a are the indecies of refraction of the sample, glass windows, and air respectively. Y can be evaluated by substuting into Eq. 6 n and its derivative from Eq. 3. We used Simpson's method to evaluate n(y,t) by numerical integration. A computer was used for this purpose and a series of curves were generated showing the dependence of Y on the variables D, t and b.

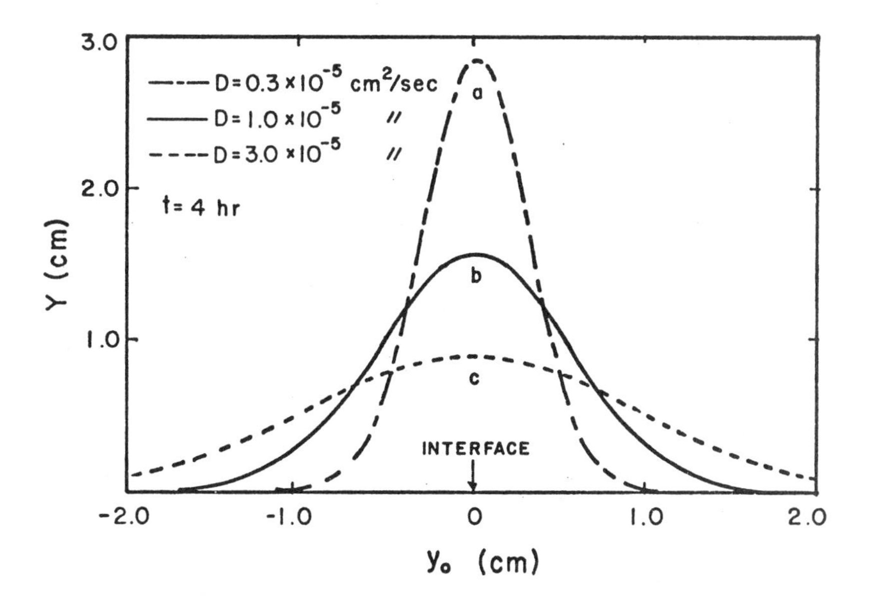

Fig. 3 Deviation of the light beam, Y, as a function of its entry position, y_o, for different diffusion constants, D.

Fig. 3 shows the variation of Y for different values of D for a typical experimental setup. (d=1.2cm, b=50cm, a=0.1cm). Note that the effects of refraction are most pronounced for the smallest diffusion constant. However, beyond y=0.9 cm from the interface it becomes negligible. The magnitude of Y decreases as D increases, however, its effects correspondingly extend over a greater range.

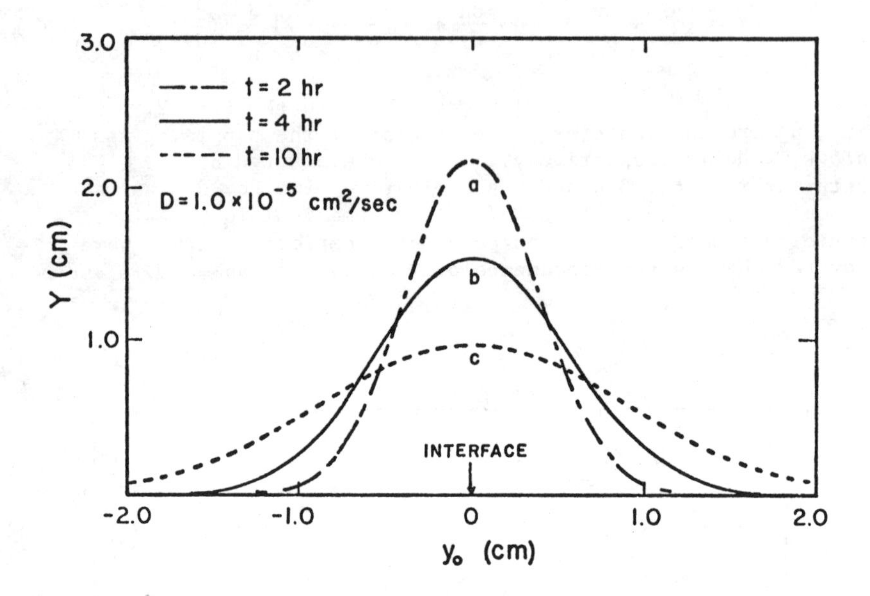

Fig. 4 Deviation of the light beam, Y, as a function of its entry position, y_o, for different times, t.

Similar results are obtained when D is held constant and the time, t, is varied. Fig. 4 shows how critical it is to take the hologram at the appropriate time. In this figure the D is held constant, at the same value as curve (b) in Fig. 3. Here again the effect is largest near the interface at short times and drops after a few hours. Note, however, that if one waits too long, refraction becomes important even at the far ends of the cell.

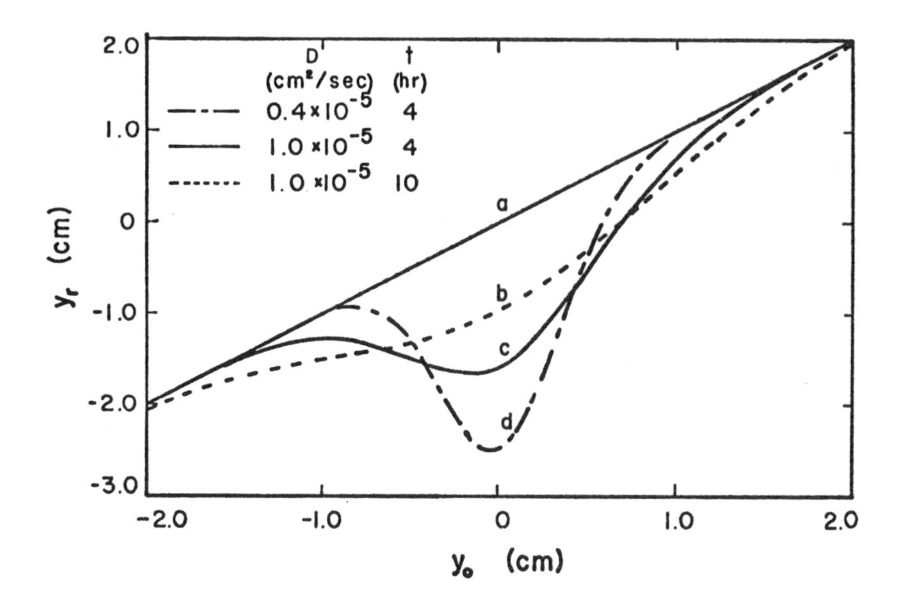

Fig. 5 The observed position of a fringe as a function of its entry position along the cell.

The previous results can be summarized in a different format as seen in Fig. 5. This is a plot of the observed position of a ray (y_r), which has been refracted, vs. the location (y_o) when refraction is absent. The straight line, (a) through the origin is the curve representing no bending. The bending results in an asymmetric curve about the interface. Curve (c) has the parameters $D = 1.0 \times 10^{-5}$ cm^2/sec and t = 4 hr. Curves (b) and (d) shows the changing trend in displacement when t and D are varied, respectively. Note that as many as three rays (and hence three fringes on the film) may fall on the same location below the interface. This points out that only data from the first quadrant, far from the interface, will yield unambiguous results.

RESULTS

As a test of the method we measured the temperature dependence of D for NaCl solutions. The experimental arrangement was similar to the one reported previously.[4] The temperature of the apparatus was controlled by coupling it to a large water bath which surrounded the sample cell. Air shafts containing glass windows permitted the passage of laser light to the cell.

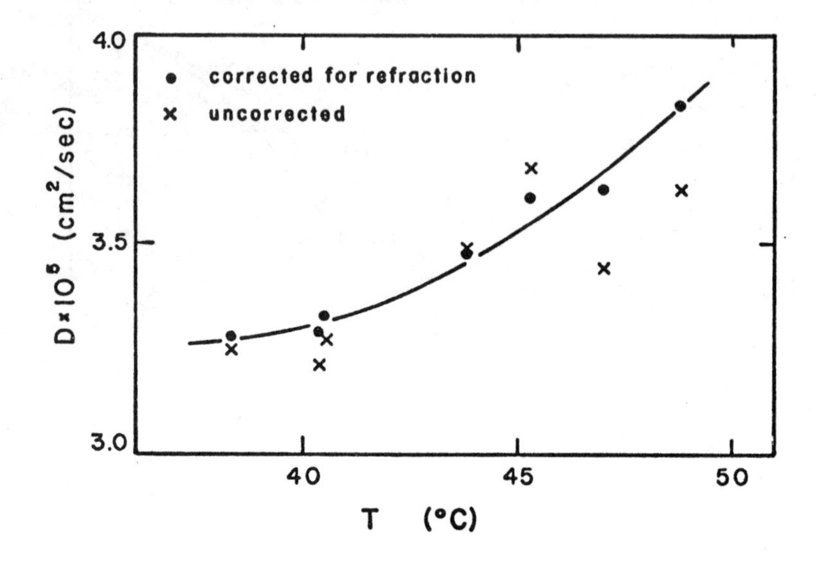

Fig. 6. Diffusion constant of 1.7 M NaCl in H_2O as a function of temperature.

The results are plotted in Fig. 6. The crosses (×) are the values of the diffusion constant obtained by substituting the positions of the fringes y_r on the hologram directly into Eq. 4. This value of D was used in Eq. 6 to calculate a displacement Y for each of the fringes. Y was then added to y_r and a new diffusion constant, D', was calculated. The process was then repeated, iteratively, until successive Y's came within 1% of each other. The final D's obtained in this manner are plotted as circles in Fig. 6. Note that the points, which previously appeared to be scattered at random, are now readjusted to fall on a smooth curve. This clearly points out the care which must be used when applying interferometric techniques to study dynamic processes in fluids.

In conclusion, this paper demonstrates the value of using holographic interferometry as a probe to study diffusion in transparent liquids. The technique does not require an elaborate setup. Our apparatus consists of an inexpensive vibration isolation table, a simple beam expander, a few beam splitters and a 5 mW laser. We must emphasize, however, that caution should be exercised to choose fringes on the hologram that are located sufficiently far above the interface in order to minimize the effects of refraction.

REFERENCES

1. For a review of work in this area see "Hologram Interferometry and Speckle Metrology" a technical digest of papers presented at a meeting of the Optical Society of America, in North Falmouth on June 2-4, 1980.
2. N. Bochner and J. Pipman, J. Phys. D. 9, 1825 (1976).
3. O.A. Shustin, T.S. Velichkina, T.G. Chernevich, and I.A. Yakovlev, JETP Lett. 21, 24 (1975).
4. L. Gabelmann-Gray and Henry Fenichel, Appl. Opt. 18, 343 (1979).
5. F. Andreasi Bassi, G. Avcovito, and G. D'Abramo, J. Phys. E. 10, 249 (1977).

CONDITIONS FOR MAKING AND RECONSTRUCTING MULTIPLEX HOLOGRAMS

J. Tsujiuchi, T. Honda, K. Okada
Imaging Science and Engineering Laboratory
Tokyo Institute of Technology
Nagatsuta, Midori-ku, Yokohama 227

M. Suzuki, T. Saito
Fuji Photo-Optical Co. Ltd.
Uetake-cho, Omiya 330

F. Iwata
Toppan Printing Co. Ltd.
Taito, Taito-ku, Tokyo 110
Japan

ABSTRACT

A machine for making multiplex holograms has been constructed, and optimum conditions for making and reconstructing the holograms are studied by analysing the distortion and resolution of the reconstructed images.

INTRODUCTION

Multiplex holograms, i.e. cylindrical holographic stereograms with white light reconstruction[1)2)], are now very widely used because of their wide viewing angle, white light reconstruction, bright image, and flexibility in the choice of objects.

A machine for making such a multiplex hologram has been built, and the reconstructed images of simple objects have been analysed under various production and reconstruction conditions mainly for the purpose of studying distortion, resolution and optimum configurations.

MAKING AND RECONSTRUCTING HOLOGRAMS

A multiplex hologram is made from a series of photographs of an object taken with different lines of sight in the horizontal direction. Such an original picture is usually taken with a fixed cinecamera by rotating the object on a turntable, or by rotating the cinecamera around the fixed object.

Fig.1 shows a basic optical system by which a multiplex hologram is synthesized. A frame of the original picture OF is projected by a lens L on the pupil of a field lens of large aperture composed of two elements: one is a high power cylindrical lens CL that focuses the illuminating laser beam as a vertical line image on a film H where the hologram is to be recorded; the other is a low power spherical lens SL that focuses the laser beam at a point E_o behind H. The horizontal angular aperture of the field lens with respect to H limits

ISSN:0094-243X/81/650594-10$1.50

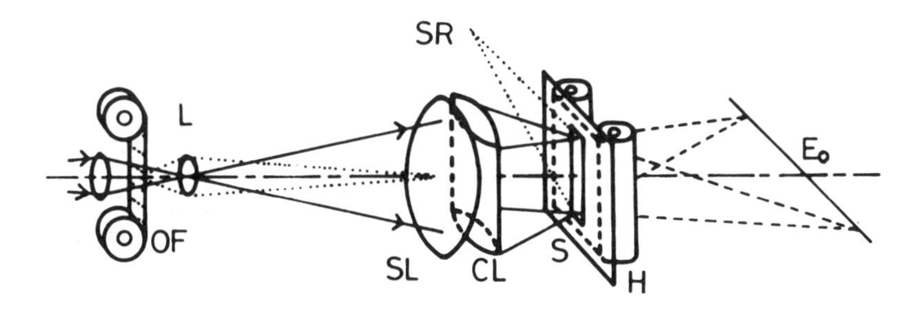

Fig. 1 Basic optical system for making holograms

the field angle of the reconstructed image. A narrow vertical strip hologram of the projected image OF is recorded through a fairly wide slit S using a reference beam emitted from a point source SR positioned above the reference axis, and such that its projection along the optical axis is at a distance r from H. This hologram is a sort of rainbow hologram[3] which becomes an element of the final multiplex hologram. After recording this component, the original picture is moved to the next frame, the hologram film H is advanced by d, the pitch of the component hologram, and the next element is recorded in the same way as the previous one. On repeating this process, a series of component holograms is recorded and the complete multiplex hologram can be synthesized.

In the reconstruction stage, the multiplex hologram is shaped into the form of a cylinder of radius r, and is illuminated by a small white light source positioned on the axis of the cylinder and at the same position as the original reference source. The reconstructed image is observed inside the cylinder in a uniform color if the observer is at a distance b_o from the axis of the hologram, where b_o is the distance selected in making the hologram between E_o and the reference axis.

Fig.2 shows a schematic diagram of the machine that we have constructed for making multiplex holograms. The machine is so arranged that the optical system shown in Fig.1 turns by 90° around the optical axis to avoid the mount of the reference source. A 15 mW He-Ne laser LA is used with an electronic shutter SH as a light source, and a specially designed 35 mm movie projector PR is employed for projecting the original picture on the pupil of the field lens. This projector has a precise mechanism for registering each frame, and the successively projected images on the field lens coincide precisely with each other except for the difference due to parallax.

The field lens is composed of two lenses: one is a large aperture spherical lens SL that focuses the laser beam at b_o = 1500 mm. The other is a high power cylindrical lens CL. Two types of cylindrical lenses have been constructed; one is specially shaped with a plastic sheet and filled with silicon oil; the other is a 4-element

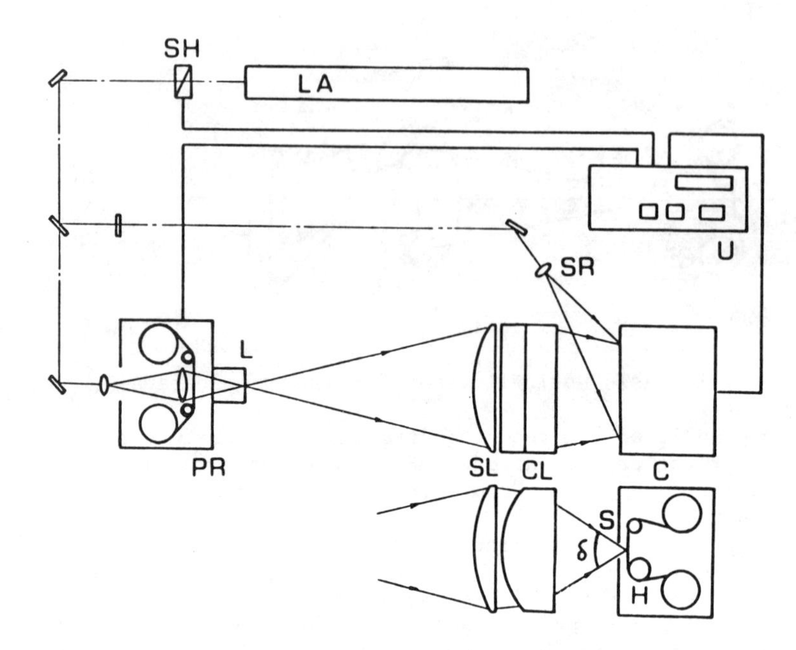

Fig. 2 Machine for making holograms (a top view). A side view of the field lens and the camera of hologram film is shown separately at the bottom.

glass lens system. The angular apertures δ of these cylindrical lenses are 54.3° and 57.7° respectively. The latter type of lens has been used primarily in our work.

The machine is typically adjusted to make hologram with r = 200 mm and a magnification m' = 10.7 of the projected image relative to the original picture. The reference beam is generated by means of a small cylindrical lens SR placed so as to satisfy the condition r = 200 mm. The hologram film H is kept in a camera C equipped with a slit aperture S of a few mm width and a film transporting mechanism that advances the film by a pitch d for every exposure.

These optical components are arranged on a large optical bench, and the sequentical operation for recording the series of elementary holograms is controlled automatically by an electronic control unit U. A Kodak SO-173 film with 300 mm width is used. The exposure time for a component hologram is about 0.5 s and an interval of 5 s is allowed between two successive exposures to avoid the influence of mechanical movement due to the film transportation on recording the hologram. If d = 0.5 mm, it takes about 4 h to make a complete 360° hologram with r = 200 mm.

DISTORTION OF RECONSTRUCTED IMAGES

Distortion of reconstructed images from a multiplex hologram can be studied from the viewpoint of photographic perspective.

Fig.3(a) shows the horizontal section of a multiplex hologram in which an object AB is recorded. If, by any possibility, the reconstructed image is observed by an eye positioned just behind a component hologram H_0, the natural perspective will be obtained if the following law of photographic perspective is satisfied;

$$m'f = a', \tag{1a}$$

where m' is the magnification of the projected image to the original picture, f is the image distance of the cinecamera, and a' is the distance between the hologram and the exit pupil of the field lens which is chosen nearly the same as r. In some cases, the following expression

$$m\,a = a' \tag{1b}$$

derived from eq. (1a) is more convenient, where m is the magnification of the projected image to the object, and a is the distance between the lens of the cinecamera and the center of the turntable. However, as mentioned later, it is impossible to put the eye just behind H_0 in order to observe the entire image.

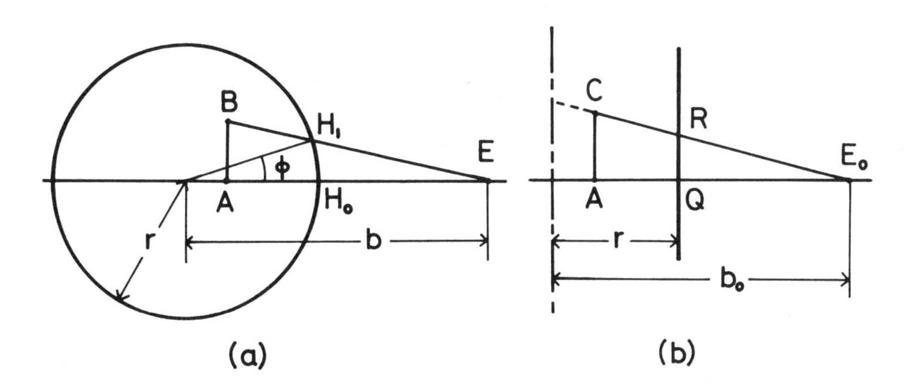

Fig. 3 Distortion in multiplex hologram. (a) horizontal section, (b) vertical section.

If the eye moves to a point E, the point A is reconstructed by another hologram H_1, and if the angle Φ at the circumference between H_0 and H_1 is equal to the angle between the position of the camera when the corresponding original pictures were taken, the reconstructed image of AB is observed in perspective and the natural stereoscopic vision is obtained when both eyes are placed horizontally.

This means that the angle at the circumference between two adjacent component holograms should be the same as that of two adjacent camera positions. This condition can be expressed as

$$M\,N = 2\,\pi\, r/d, \qquad (2)$$

where M is the period in seconds of one turntable revolution, N is the number of frames per second, and d is the pitch of the component hologram. Hence, if both conditions (1) and (2) are satisfied, perspective in the horizontal direction is natural, and a distortion-free image is obtained regardless of the observer's position.

In the vertical direction, however, the field lens focuses at E_0 and an object AC is recorded as QR on the hologram as shown in Fig.3(b). In this case, a reconstructed image with a uniform color and natural perspective is obtained if the eye is at E_0 and if the following equation holds.

$$b_0 = m'f = m\,a. \qquad (3)$$

As the radius r of the hologram is equal or very close to a', the compatibility of both conditions (1) and (3) implies that E_0 should be on or just behind the hologram; however such a condition is impossible to realize. In order to record a component hologram as a rainbow hologram, E_0 should be far behind the hologram, and in this case the observer cannot see good reconstructed images unless he is far from the hologram. So, it is impossible to realize the distortion-free reconstruction satisfying all the conditions (1), (2) and (3), and practically reconstructed images have always distortions.

In addition, it is desirable that the observer be able to move close to or away from the hologram as needed. Then the reconstructed image is no longer uniformly colored if the observer leaves E_0, and

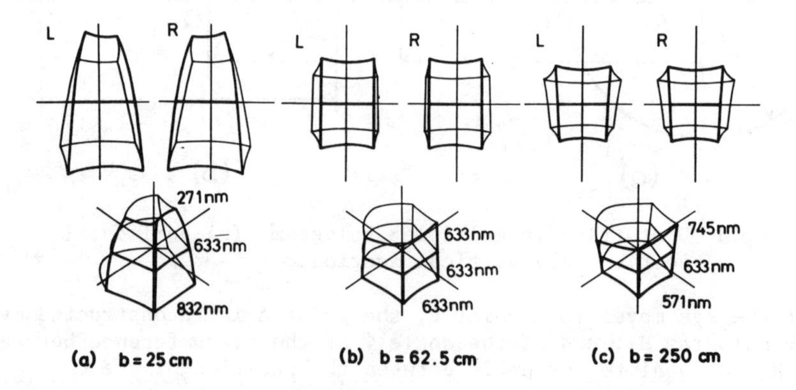

Fig. 4 Some reconstructed images of a cubic framework. Upper images are stereoscopic pairs and lower images are perspective drawings with reconstructed wavelengths.

additional distortions can be observed because of the partial color change of the reconstructed image.

Accordingly, practical conditions for making holograms have been studied by tracing reconstructed rays forming the image of a simple object. In this way we have obtained, for eq.(1b),

$$m\,a = 2\,a' = 2\,r \qquad (4)$$

as the optimum condition for minimization of the RMS distortion of the reconstructed image for any value of b_0 which is selected on the basis of ease of observation[4]. Under this condition, the distortion becomes a minimum if the observer is positioned at E_0, and no considerable change of distortion takes place for observers removed away from E_0, but serious distortion appears if the observer moves too close to the hologram.

Fig.4 shows examples of reconstructed images of a cubic framework of 120 mm side from a hologram with $r = a' = 125$ mm in which the object is recorded under the conditions $a = 250$ mm, $m = 1$ and $b_0 = 625$ mm[4]. The reference beam is projected down from the top.

RESOLUTION OF RECONSTRUCTED IMAGES

The resolution of reconstructed images is mainly influenced by the white light reconstruction process.

Consider a multiplex hologram in which a point object A is recorded as shown in Fig.5. Each component hologram records the interference fringes produced by the object and the reference beams around a point Q where the point A is recorded as a rainbow hologram. If

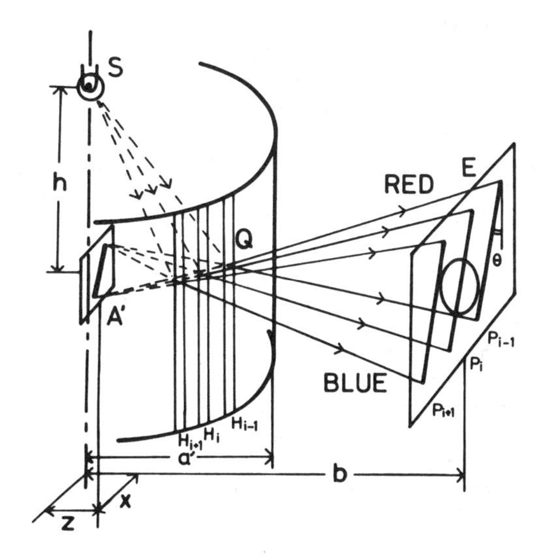

Fig. 5 Reconstruction of a point image

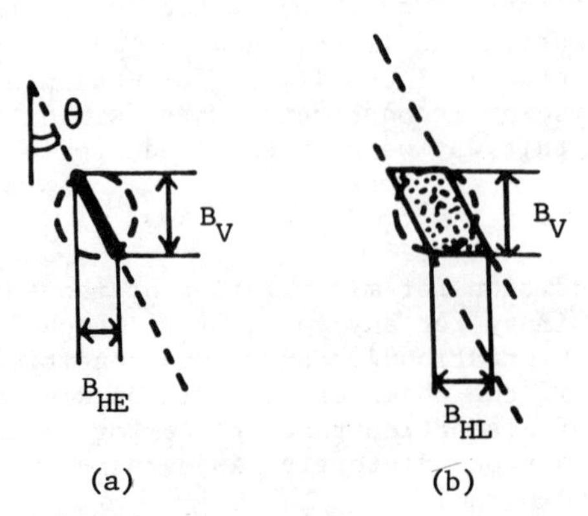

Fig. 6 Blurs of reconstructed point images
(a) Point source, (b) Line source

this hologram is illuminated by a white light point source S placed at the same position as the reference source, a grating around Q in a component hologram H_i diffracts the illuminating beam and produces a linear diffraction pattern P_i in rainbow colors in a plane in which the pupil E of the observer's eye is included. This diffraction pattern is inclined by a small angle θ from the vertical direction if the component hologram H_i is removed from the central component hologram H_o located on the line of sight intersecting the axis of the hologram. Adjacent component holograms to H_i such as H_{i-1} and H_{i+1} produce similar diffraction patterns P_{i-1} and P_{i+1} located side by side. The light beams that produce these diffraction patterns enter the observer's eye E, and produce a narrow linear image A' in rainbow colors as the reconstructed image of A.

If the eye has a small aperture, the diffraction patterns, and consequently the reconstructed image of a point are stopped down only in the vertical direction as shown in Fig.6(a), and the reconstructed image becomes nearly monochromatic with a linear blur which can be evaluated by the vertical blur B_V and the horizontal blur B_{HE}.

In a practical reconstruction, an incandescent lamp with a vertical linear filament is often used, and both the length and the width of the filament will influence the resolution. The length of the filament makes the diffraction pattern P_i spread vertically; in addition the observer's eye stops the pattern in the vertical direction. Thus, the reconstructed image is stopped down, as schematically shown in Fig.6(b), the vertical blur B_V remains unchanged, and the spectrum that contributes to the reconstructed image becomes wider than in the case of a point source. However, the inclination θ of the diffraction pattern generates a new horizontal blur B_{HL}. As for the width of the filament, the observer's eye produces no stopping effect, so that the reconstructed image spreads horizontally in proportion to the width of the filament and causes a horizontal blur B_{HW} regardless of the size of the observer's pupil.

Fig.7 shows a schematic diagram for calculating blurs. If one takes a cartesian coordinate system (x,y,z) with the center 0 of the hologram as the origin, as shown in the figure, $S(0,h,0)$ is an illuminating source of length l and width w, Q is a point that records an object point A is the component hologram H_i, $A'(x,y,z)$ is the reconstructed image observed by the eye $E(0,0,b_o)$ having a pupil of diameter p. In terms of this geometrical set up, the following relations based on diffraction theory can be derived[5];

$$B_V = (r - z)\varepsilon_E \tag{5a}$$

$$B_{HE} = (r - z)\varepsilon_E \tan\theta \tag{5b}$$

$$B_{HL} = (r - z)\varepsilon_L \cos\alpha \tan\theta \tag{5c}$$

$$B_{HW} = (r - z)\varepsilon_W / \cos\beta \tag{5d}$$

$$\tan\theta = \tan\beta / \sin\alpha, \tag{6}$$

In eqs.(5-6), ε_E is the angle subtended by the observer's pupil, ε_L by the length of the illuminating source, and ε_W by the width of the source; α is the elevation angle of the center of the source, and β is the horizontal diffraction angle of the reconstructed beam. All these angles are measured from the point Q in the component hologram H_i. The angles shown in Fig.7 are drawn without considering inclination factors for simplicity.

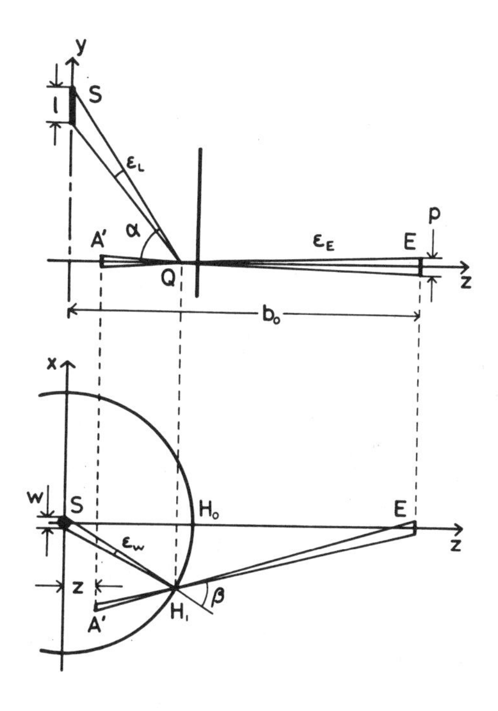

Fig. 7 Geometry of point image reconstruction

Consider, for example, a hologram with r = 200 mm reconstructed by a source with l = 10 mm and w = 1 mm at h = - 200 mm. Furthermore suppose that the reconstructed image A'(x,y,0) is observed by an observer with p = 4 mm at b_o = 1000 mm. The blurs become larger as x and - y increase, and the maximum values of blurs are calculated by choosing x = - y = 100 mm. We have then B_V = 1.0 mm, B_{HE} = 0.9 mm, B_{HL} = 3.3 mm, B_{HW} = 0.8 mm and the total horizontal blur becomes B_H = B_{HE} + B_{HL} + B_{HW} = 5.0 mm. These blurs can be measured by taking a photograph of the reconstructed image with a camera whose aperture is 4 mm, but the practical visual image looks a little more distinct than expected by the calculated blurs. This is so because the pupil of the eye tends to be smaller, the image-forming beam is astigmatic and the blur has a spectral distribution.

The largest blur is B_{HL}, and the length of the filament should be kept within a limit. A reasonable limit can be obtained by making B_{HL} nearly equal to the other blurs, so that the length of the filament should be less than about 3 mm in this case.

CHANGE OF HOLOGRAM FORMAT

As a multiplex hologram needs a special apparatus for reconstruction, the format of a hologram should be standardized in several types. We are now considering the following three types:

	I	II	III
Diameter of hologram ($2r$, mm)	400	300	150
Height of hologram (mm)	300	200	100
Position of reference source (h, mm)	- 250	-190	- 95

Each hologram can be synthesized by an exclusive machine and a suitable original film, but it can also be made by using a versatile machine if the optical system is properly adjusted every time the type is changed.

Suppose that a machine is adjusted to make holograms of Type I, and that other type holograms are to be made. For that purpose, one should first move the field lens or the hologram film so as to make $a' = r$. This would be very difficult in most cases. It is best to leave a' as it is, to put the reference source at the proper position, and to change the size of the projected image on the pupil of the field lens only in the vertical direction so as to keep the horizontal field angle of the hologram constant. An anamorphic lens is very convenient for this purpose. At the same time, the pitch d should be changed to satisfy the condition (2).

CONCLUSION

A machine for making a multiplex hologram from an original movie film has been constructed, and the quality of the reconstructed images is analyzed mainly with regard to the distortion and resolution. It becomes clear that the reconstructed image from the multiplex hologram is always distorted because of the difficulty of satisfying the natural photographic perspective condition and that the horizontal resolution is strongly influenced by the size of the illuminating source used for reconstruction not only in its horizontal width but also in its vertical length. From these results, optimum conditions for making and reconstructing holograms with minimum distortion and best resolution are derived.

One of the purpose of this machine is to make multiplex holograms from a series of medical X-ray photographs, and a spatial filter is employed in the image-projecting beam to enhance the contrast and the resolution of the original pictures.

REFERENCES

1) S. A. Benton, Opt. Eng. 14, 402 (1975)
2) M. Grossmann, P. Meyrueis and J, Fontaine, Holography in Medicine and Biology, Edited by G. von Bally (Springer-Verlag, 1979) p.110
3) S. A. Benton, Applications of Holography and Optical Data Processing, Edited by E. Marom et al. (Pergamon Press, 1977) p.401
4) K. Okada, T, Honda and J. Tsujiuchi, Proceedings of SPIE, Vol.212 Optics and Photonics Applied to 3-D Imagery (SPIE, 1979) p.28
5) K. Okada, T. Honda and J. Tsujiuchi, Opt. Commun. (in preparation)

DYNAMIC HOLOGRAPHY

M. Soskin & S. Odoulov
Institute of Physics, Academy of Sciences of Ukrainian SSR
252 650, Kiev

ABSTRACT

This paper gives a brief review of some practical applications of dynamic holography such as coherent light beams amplification, phase conjunction, and image processing.

INTRODUCTION

Dynamic holography is a branch of modern optics that studies the formation of light induced gratings in nonlinear media and the propagation of electromagnetic beams in the presence of self-diffraction.

One of the main technical problems in dynamic holography is to arrive at an understanding of the key physical processes that govern the nonlinear interaction of light with media which are characterized by many different types of response (instantaneous or inertial, local or non-local, resonant or non-resonant).

The solution of this problem is not simple because of the complicated character of the nonlinear interaction. Dynamic holograms can be regarded as systems with nonlinear optical feedback: the incident light beams create a periodic structure in the medium, they are diffracted by this structure and acquire in the process new field amplitudes and phases; in turn, the new fields alter the character of the diffraction grating.

One can identify two main practical goals in the field of dynamic holography: to develop techniques for handling the image processing in real time, and to provide new schemes for investigating the nonlinear processes that are responsible for the formation of the periodic structures (gratings).

The second class of investigations is well developed at the present time. Some very fine and sensitive grating techniques in nonlinear spectroscopy (1,2) and stimulated Rayleigh scattering (3,4) have been proposed and realized. These have produced many new results in the physics of semiconductors (1,2), dielectrics (3), liquids (5), and liquid crystals (6,7).

In this paper we give a brief review of the application of dynamic holography to the processing of light beams and images. We also consider the questions of holographic amplification, wavefront conjugation, correlation analysis and spectral image processing.

HOLOGRAPHIC AMPLIFICATION

The dynamic hologram amplifier is a device in which a weak signal and a strong pump beam produce the dynamic grating, and where the intensity of the weak beam increases because of self-diffraction.

ISSN:0094-243X/81/650604-09$1.50

Many practical arrangements are possible: with only two beams (a signal and a pump), with three beams (one pump and two signals) and with four beams (two pumps and two signal beams). We shall analyze all of them.

An empirical rule holds for all types of interactions: the transfer of energy caused by the diffraction process occurs only if the holographic grating is shifted with respect to the fringes produced by the interacting beams (8,9). Interactions involving both static (10) and dynamic (8) gratings obey this rule.

The double beam interaction process can be considered as a degenerate four wave mixing; two incident waves produce two diffracted waves that propagate in the same direction. This is the case of Bragg diffraction from a three dimensional grating and corresponds to the wave diagram

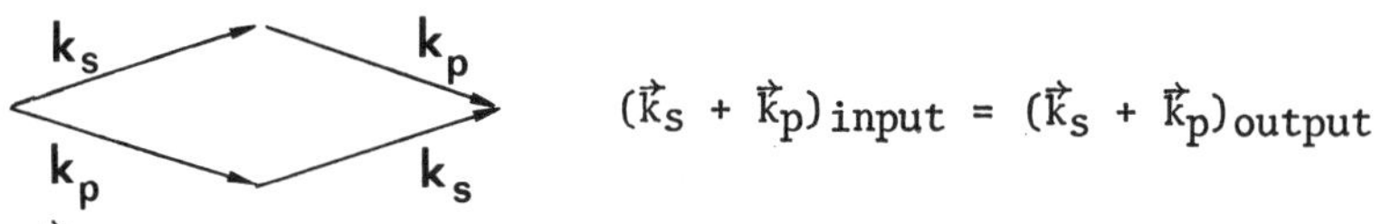

$$(\vec{k}_s + \vec{k}_p)_{input} = (\vec{k}_s + \vec{k}_p)_{output}$$

where $\vec{k}_s$ and $\vec{k}_p$ are the wave vectors of the signal and of the pump waves, respectively.

When the interaction of two coherent light beams occurs in a nonlinear medium with a local response of the type $\Delta n(xy) \sim I(x,y)$ stationary energy transfer between the beams is forbidden and amplification is not possible (11). This conclusion has been tested experimentally in a study involving CdTe crystals (12), where no energy transfer was observed for any intensity ratios of the incident beams and with diffraction efficiencies as high as $\eta = 0.1$. A test of this conclusion for large phase modulation ($\eta \simeq 1$) is very difficult because of the appearance of other nonlinear effects, such as light induced scattering, optical damage or self-focusing (13,14). It has been shown in recent theoretical work that the stationary solution with zero energy transfer is unstable (15).

If the interacting beams have different intensities, their phase difference varies with the thickness of the nonlinear layer, even in steady state (16). This phase variation causes a transient spatial mismatch of the fringes and produces a grating at the beginning of the recording process, which results in a large transient amplification of the weak beam (17). Transient amplification has been obtained in absorbing liquids (18-20), semiconductors (21), and ferroelectric crystals (22,23). It has been shown both theoretically and experimentally that transient amplification occurs over relatively long time intervals, up to ten times longer than the decay time of the nonlinearity.

When the response of the medium is nonlocal and the recorded grating is shifted by $\pi/2$ with respect to the light fringes, the phase variation is not observed, but steady-state energy transfer still occur (24,8). There is no additional transient amplification beside the one occurring in steady-state.

Some materials possess an intrinsic nonlocal response. For example, the diffusion nonlinearity in crystals with a linear electrooptic effect is characterized by the response (24)

$$\Delta n(x) \simeq \frac{d}{dx} \ell n \, I(x).$$

In the same crystals with sufficiently large applied external field, the nonlinear response becomes (25)

$$\Delta \eta(x) \sim \int I(x) \, dx.$$

Both types of nonlocal response have been observed recently in $LiNbO_3$ (24), $KNbO_3$ (26), BSO, BGO (27), SBN (28). Steady-state amplification with an exponential gain in the range 1-50 cm^{-1} has been obtained. In addition, the intensity of the acceptor beam has been observed to increase smoothly without transient amplification.

Nonlocal response can be produced artifically in a nonlinear medium with inertial nonlinearity by continuous displacement of the fringe pattern of the grating in a direction normal to the surface of constant phase (29). Amplification in such systems has been achieved by working with heat gratings in flowing dye cells (30,20), free carrier holograms in semiconductors placed within crossed electric and magnetic field (31) and free carrier holograms in semiconductors which were recorded by moving the interference pattern (32).

Three beam interaction occurs when two waves with wave vectors $\vec{k}_p$ and $\vec{k}_s$ enter the nonlinear medium at appropriate angles, and a third wave $\vec{k}'_s$ is produced subject to the vector condition

k_s k'_s k_p k_p

$$2 \, \vec{k}_p = \vec{k}_s + \vec{k}'_s$$

This interaction can be realized under special synchronism conditions (9,33) and in thin dynamic holograms (9).

In a recording medium with positive local nonlinear response ($\Delta\eta > 0$) the geometric dephasing of the first non-Bragg order of diffraction can be compensated by the nonlinear phase change due to the difference of intensities of the pump and the test beams. The synchronism condition is (34)

$$\Delta\eta \, \eta / \sin^2\theta \geq 1 .$$

An amplification of the weak beam in this case has been obtained using a Q-switched ruby laser in CS_2 (35) and a dye laser in sodium vapor (36) to produce the dynamic grating. In a medium with $\Delta\eta < 0$, or in a medium with a non-local response, this kind of synchronism is impossible.

Another type of synchronism is connected with the existence of anisotropy in some recording materials. For example, in properly oriented $LiNbO_3$, crystals in the interaction of two extraordinary waves can give rise to a third wave with ordinary polarization. The condition for "eeo" synchronous operation with negative uniaxial crystals is approximately given by

$$\theta_s = 0.5 \sqrt{\eta_e (\eta_o - \eta_e)}$$

This kind of interaction and the accompanying amplification of the signal beam has been observed in $LiNbO_3$ with non-local diffusion response (33).

A common drawback of all the above three schemes of amplification is the strict limitation on the spatial spectrum of the beam to be amplified. The tolerance angle of deviation from exact synchronism is approximately equal to the angular selectivity of the dynamic grating. For samples with a thickness of a few millimeters, this angle is only a few minutes of arc. Thus, the three beam interaction is more effective for thin layers of recording material.

Four wave interaction of counterpropagating beams obeys the following vector condition

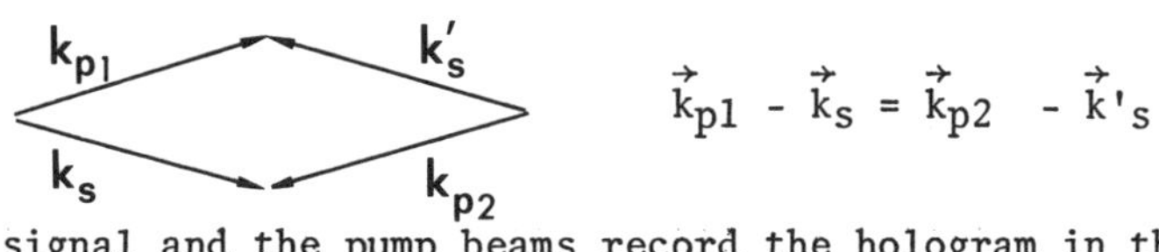

$$\vec{k}_{p1} - \vec{k}_s = \vec{k}_{p2} - \vec{k}'_s$$

If the signal and the pump beams record the hologram in the nonlinear material and the auxiliary third beam (which is counterpropagating with respect to the pump) is used to reconstruct the hologram, a fourth wave is generated which forms a real image of the object. This wave has a phase conjugate wavefront relative to the signal wave (39).

Just as the signal beam, the phase conjugated (PC) beam is also amplified because of the energy transfer from the reference pump and reconstructing beams. The amplification is possible in this case, even for media with local response because the energy transfer for each pair of beams occurs as a result of diffraction from the grating recorded by the two other beams when properly shifted (40).

Signal beam amplification has been observed for many nonlinear materials, including absorbing liquids (39,40), gases (42), metal vapors (43), dielectrics (44), semiconductors (45,46), ferroelectrics (47,48), liquid crystals (49), and others. Amplification of the weak beam, by more than a factor 100, and more than 10% energy transfer into the signal beam have been observed (45). The theory (40,50) predicts that, in principle, total energy transfer of the pump beam into the signal should be possible. In practice, the amplification is always limited by concurrent nonlinear processes, and especially by light induced scattering of the pump.

In order to obtain the highest conversion efficiency in media with a local response, it is necessary to equalize the pump beam intensities as much as possible (51,52). However, violation of the synchronism condition due to unequal pump beam intensities can be compensated by angular detuning of the pump beams (52).

Four wave interaction in media with non-local response is characterized by some unique features. The energy transferred by the grating produced by two of the beams can either be added to or subtracted from the energy which is transferred by diffraction from the grating recorded by the other two beams (47). The signal beam, as a result, can either be amplified or suppressed, depending on the direction of the grating shift.

In materials with non-local response, the phase difference of the interacting beams does not change, and the synchronism can be maintained for arbitrary intensities of all the interacting beams. The highest conversion efficiency is found to correspond to unequal intensities of the pump beams (47,53).

Most holographic amplifiers can provide an exponential gain of the order of 10-100 cm^{-1} and can amplify the weak beam from 2 to 10^4 times (9). The spatial frequency-contrast characteristic may be a step function with a cut-off ranging from 10^2 to 10^4 mm^{-1}. It may also exhibit a linear growth or a decrease of the gain as a function of the spatial frequency depending on the type of nonlinearity (9).

A difficult problem with holographic amplification is the appearance of randomness in the grating recorded by the pump beam, and the consequent light scattering which is a source of considerable noise at large amplification levels.

A comparison between the different types of holographic amplifiers shows that the most efficient one is based on the double beam scheme and operates with a steady state energy transfer in media with non-local response, or with a transient energy transfer in media with local response.

It is apparent that holographic amplifiers provide, at best, the same level of image amplification as, for example, opto-electronic devices. At the same time, they suffer from considerable noise problems, unlike non-holographic amplifiers. However, it must be stressed that a holographic amplifier is a unique device that can be used to increase signal levels without loss of phase information as needed, for example, if one wants to amplify 3D-images (54).

PHASE CONJUGATION

In dynamic, as with the usual static holography, it is possible to reconstruct real images, for example, by generating beams with phase-conjugate wavefronts. Unlike the case of ordinary holography, in dynamic holography this operation can be performed even with time-dependent wavefronts.

The well known technique of phase conjugation is based on the four wave mixing process described in the previous section. Phase conjugation is achieved by reading out the dynamic volume hologram with a beam which has been phase conjugated with respect to the reference beam used in the recording process (39,55). Phase conjugation by four wave mixing has been used for restoration of images distorted by aberration phase plates (38), for compensation of static (38) and time-dependent (56) distortions of regular wavefronts in pattern recognition (57), etc.

On the other hand, real images, as well as phase-conjugate wavefronts, can be generated in different kinds of forward four wave mixing processes (34). These techniques resemble the reconstruction of a real time image by a thin hologram in the "+1" order of diffraction (58). The necessary "+1" order of diffraction can be obtained and amplified in all three-wave interaction schemes, even in thick nonlinear layers. As a result of the anisotropic self-diffraction of the type "eeo" which occurs, for example, in $LiNbO_3$ crystals, generation of phase conjugate wavefronts for simple amplitude transparencies has been achieved (33). However, the image resolution obtained with a 0.3 cm thick crystal has been very poor (of the order of 3-4 mm^{-1}).

The limitations on the spatial frequency content of the image can be successfully overcome if the recorded dynamic hologram is physically thin. An example of a physically thin dynamic hologram is the surface hologram recorded on a mercury mirror by an intense laser pulse (59). If the reference beam is oriented normal to the mirror, the phase conjugate beam propagates strictly in the opposite direction relative to the signal beam, as in the backward wave four wave mixing arrangement. Physically thin gratings and generation of phase conjugate beams in the forward direction has been obtained also in a Na vapor cell (60).

CORRELATION ANALYSIS WITH DYNAMIC HOLOGRAMS

A proposal to use dynamic recording materials for image reconstruction by correlation techniques was first formulated in 1969 (61). Three beams are made to intersect inside a nonlinear material: two beams carry the images to be compared and the third reference beam is in the form of a plane wave. The reference beam and one of the object beams generate a dynamic Vanderlught filter, while the second object beam provides the reconstruction. Under these conditions, the correlation function of the two images appears in certain higher diffraction orders (62). A necessary condition for the success of this scheme is the existence of high diffraction orders. This implies that the hologram must be thin. Using this technique, the comparison of very simple amplitude transparencies and the reconstruction of elementary images have been performed (62,63).

Another way of performing convolution and correlation operations is to use the backward degenerate four wave mixing process (38,64). If two images are introduced as pump and signal beams, and the second pump beam is a plane wave, the correlation function is produced by the fourth beam. If both pump beams are modulated and the signal beam is a plane wave, the convolution function is generated by the fourth wave.

This technique has been applied to produce correlation and convolution of simple bright dot images with a crystal of $LiTaO_3$ using the diffusion nonlinearity as a recording material (57).

IMAGE PROCESSING

In most applications involving holographic amplification, one must select experimental conditions such that the spatial spectrum of the image is not altered after the amplification process. On the other hand, the specific dependence of the gain on the spatial frequency of the recorded grating and on the intensity ratio of the interacting beams can be put to good use for image processing.

For steady state interaction with media having a local response, the phases of selected spatial spectral components can be altered, depending on the intensity ratios of these components relative to the reference beam. In this way, the dynamic holograms can act as pure nonlinear phase filters. On the other hand, dynamic holograms in media with non-local response alter the intensities of the interacting beams, but preserve their phase difference. These holograms can be used under special conditions as pure amplitude spatial filters.

In general, both amplitudes and phases of the interacting beams change when the recording process is a transient one in media with a local response (9).

The first attempts to apply these properties of dynamic holograms to image processing are described in Refs. 16, 17. The subtraction and addition of images due to the stationary energy transfer between two spatially modulated beams has been achieved using crystals of $LiNbO_3$ to record the dynamic hologram (54). In Ref. 66, the transient energy transfer was used for image contrast enhancement.

CONCLUSION

During the last decade, many properties of dynamic holography have been established and investigated (energy transfer between interacting Bragg beams, the appearance of higher orders of diffraction, generation of phase conjugate wavefronts, etc.). The key role played by the phase relation between the light fringes and the dynamic grating has become clear. It has been shown that the nonlinear interaction can be sufficiently strong and that the diffraction efficiency of a dynamic hologram may approach 100%. A wide variety of recording materials with nonlinear time constants ranging from 10^{-12}s to several minutes have been tested and used for dynamic holographic recording. Several schemes for producing light amplification and transformation have been proposed. We hope that the next decade will bring both new ideas and the realization of practical devices in the field of dynamic holography.

REFERENCES

1. A. Maruani, J.L. Oudar, E. Batifol, D.S. Chemla, Phys. Rev. Lett. 41, 1372, 1978.
2. D.S. Chemla, A. Maruani, Opt. & Laser Tech., Dec. 1979, p. 297.
3. D. Pohl, V. Irniger, Phys. Rev. Lett. 36, 480, 1976.
4. D. Pohl, Proc. 4th EPS Conf. (Inst. of Phys., London, 1979), Ch. 2, p. 115.
5. W. Urbach, H. Hervet, F. Rondelez, Mol. Cryst. Liq. Cryst., 46, 209, 1978.
6. K. Thygarajan, P. Lallemand, Opt. Commun. 26, 54, 1978.
7. W. Chan, P.S. Pershan, Phys. Rev. Lett. 39, 1368, 1977.
8. V.G. Sidorovitch, D.I. Stasielko, Sov. Techn. Phys. J, 44, 580, 1974.
9. V.Vinetskii, N. Kukhtarev, S. Odoulov, Sov. Phys. Dokl. 129, 113, 1979.
10. V.P. Kondilenko, V. Markov, S. Odoulov, M.S. Soskin, Optica Acta, 26, 239, 1979.
11. B. Zieldovich, Kratkie Sechscheniya po Fizike, n.5, 20, 1970 (in Russian).
12. V. Kremenitskii, S. Odoulov, M. Soskin, Phys. Stat. Sol. 51, 23 1979.
13. H. Kurz, Optica Acta, 24, 463, 1977.
14. R. Magnussen, T.K. Gaylord, Applied Optics, 13, 1545, 1974.

References Continued

15. A. Khyznyak, to be published in Sov. Techn. Phys. Lett.
16. A.A. Tchaban, Sov. Phys. JETP, 57 1387, 1969.
17. V. Vinetskii, N. Kukhtarev, M. Soskin, Sav. J. Quant. Electron. 7, 1270, 1977.
18. F. Gires, C.R. Ac. Sci. (Paris), Ser. B. 266, 569, 1968.
19. W. Rother, H. Meyer, W. Kaiser, Zs. Naturforsch. 25a, 1136 1970.
20. V. Briskin, A. Grozny, V. Sidorovitch, D. Stasielko, Sov. Phys. Techn. Lett. 2, 219 1976.
21. V. Vinetskii, T. Zaporozhets, N. Kuktarev, S. Odoulov, M. Soskin, Sov. Phys. JETP Lett. 25, 404, 1977.
22. N. Kukhtarev, V. Markov, S. Odoulov, Opt. Commun. 23, 338, 1977.
23. V. Kondilenko, V. Markov, S. Odoulov, M. Soskin, Ukrainian Physical J. 23, 2039, 1978 (in Russian).
24. D.L. Staebler, J.J. Amodei, J. Appl. Phys. 43, 1042, 1973.
25. N. Kukhtarev, V. Markov, S. Odoulov, M. Soskin, V. Vinetskii, Ferroelectrics, 22, 949, 1979.
26. A. Krunins, P. Gunter, Appl. Phys., 19, 153, 1978.
27. J.-P. Huignard, J.-P. Herriau, Appl. Opt. 17, 2671, 1978.
28. I. Dorosh, Yu. Kuzminov, to be published in Sov. J. Quant. Electron.
29. Yu. Anan'ev, Sov. J. Quant. Electron. 4, 929, 1975.
30. A. Grozny, A. Duhovny, A. Leshchov, V. Sidorovich, D. Stasielko, "Optical Holography" ed. Denisiuk, "Nauka" Leningrad, 1979, p. 92.
31. V. Vinetskii, T. Zaporozhets, A. Matveichuk, N. Kukhtarev, M. Soskin, G. Kholodar, Ukrainian Physical J. 22, 1141, 1977 (in Russian).
32. V. Kremenitski, S. Odoulov, M. Soskin, Sov. Techn. Phys. Lett. 6, N 14, 1980.
33. N. Kukhtarev, S. Odoulov, to be published in Sov. Techn. Phys. Lett.
34. A. Khyznyak, M. Soskin, S. Odoulov, Preprint of Inst. of Physics n° 11, Kiev, 1976.
35. R. Chiao, P.L. Kelly, E. Garmire, Phys. Rev. Lett. 17, 1158, 1966.
36. A.M. Bonch-Bruevich, S.G. Przybielski, V.A. Khodovoy, Sov. Phys. JETP 65, 61, 1973.
37. R. Hellwarth, JOSA, 67, 1, 1977.
38. A. Yariv, IEEE J. Quant. Electron. QE-14, 650, 1978.
39. B.I. Stepanov, E.V. Ivakin, A.S. Rubanov, Sov. Phys. Dokl. 19, 46, 1971.
40. M.S. Soskin, A.I. Khyznyak, Sov. J. Quant. Electron. 7, 42, 1980.
41. J.O.Tocho, W. Sibbet, D.L. Bradley, to be published in Opt. Lett.
42. A. Elci, D. Rogovin, Chem. Phys. Lett., 61, 407, 1979.
43. D.M. Bloom, N.P. Economou, Appl. Phys. Lett. 32, 813, 1978.
44. A. Tomita, Appl. Phys. Lett. 34, 460, 1979.
45. V. Kremenitskii, S. Odoulov, M. Soskin, Phys. Stat. Sol. 57, K71, 1980.
46. R.K. Jain, M.B. Klein, R.C. Lind, Opt. Lett. 4, 328, 1979.
47. N. Kukhtarev, S. Odoulov, Opt. Commun. 32, 183, 1980.
48. R. Hellwarth, J. Appl. Phys. 5, 1297, 1980.
49. D. Fekete, A. Yariv, Opt. Lett. 5, 51, 1980.
50. A. Khyznyak, V. Kondilenko, V. Kremenitskii, S. Odoulov, M. Soskin, in "Proc. OPIEM Conf. 1979, Strasbourg". SPIE vol. 213, 1980.
51. B. Zeldovich, V. Chkunov, Preprint of Lebedev Phys. Inst. n 267, Moscow, 1978.
52. A. Khyznyak, Sov. J. Quant. Electron. 7, 685, 1980.

References Continued

53. S. Odoulov, M. Soskin, A. Voroshnin, to be published.
54. V. Markov, S. Odoulov, M. Soskin, Optics & Laser Techn., 11, 95, 1979.
55. A. Yariv, Opt. Commun. 25, 23, 1978.
56. S. Odoulov, M. Soskin, M. Vasnetsov, Opt. Commun. 32, 355, 1980.
57. S. Odoulov, M. Soskin, Sov. Phys. Dokl. 252, 336, 1980.
58. R.J. Collier, C.B. Burckhardt, L. Lin, Optical Holography (Academic Press, 1973), p. 35-40.
59. T. Zaporozhets, S. Odoulov, to be published in Sov. Techn. Phys. Lett.
60. C.V. Heer, N.C. Griffen, Opt. Lett. 4, 239, 1979.
61. E.V. Ivakin, Soviet invention n° 324678, 1.XII, 1969.
62. E. Ivakin, A. Rubanov, I. Petrovich, Sov. J. Quant. Electron. 1, 96, 1973.
63. S.H. Lee, K.T. Stalker, JOSA, 62, 1366, 1972.
64. S.H. Lee, K.T. Stalker, JOSA, 64, 545, 1974.
65. D. Pepper, J. AuYeung, D. Fekete, A. Yariv, Opt. Lett. 3, 7, 1978.
66. V. Markov, S. Odoulov, Sov. J. Quant. Electron. 6, 1436, 1979.

INVERSE SOURCE AND SCATTERING PROBLEMS IN OPTICS

A. J. Devaney
Schlumberger-Doll Research, P.O. Box 307, Ridgefield, CT, 06877, USA

ABSTRACT

Determination of the internal structure of an object from measurements of the field generated by the object is sometimes referred to as an inverse problem - an inverse source problem if the object is a radiator (source) and an inverse scattering problem if the object is a scatterer. In this paper, we formulate and discuss the inverse problem for spatially incoherent, localized sources and localized weak scatterers where the Born approxmation is valid. It is shown for both of these cases that unique reconstructions of the object (either source or scatterer) are not possible from data collected in a finite number of experiments. Approximate reconstructions are, however, possible and two techniques for obtaining such approximate reconstructions are presented. It is shown that these approximate methods will yield accurate renditions of the object (source or scatterer) as long as the object possesses no significant structure on a scale finer that a half-wavelength.

1. INTRODUCTION

The problem of determining the internal structure of three-dimensional objects from measurements of the field generated by such objects in either radiation or scattering experiments is of great importance in a number of practical contexts. Probably the greatest advances in this problem have been in diagnostic radiology where the object structure is determined by means of radiographic imaging techniques in conjunction with computed tomography (CT)[1]. The success of these radiographic techniques depend critically on the validity of certain heuristic models[2] which are used to relate the object to the measured field data. These models are probably adequate for radiographic applications but are questionable in optical and acoustic applications where diffraction effects can become important. Because of this, the use of radiographic imaging techniques in optical and acoustic applications is somewhat limited.

An entirely different approach to the inverse problem was suggested by E. Wolf[3] who showed that semi-transparent objects can, in principle, be approximately reconstructed from measurements (via

ISSN:0094-243X/81/650613-14$1.50

holography) of the scattering amplitude produced in a sequence of experiments using different propagation directions of an incident (illuminating) plane wave. Wolf's work was based on a rigourous solution of the scattering problem (within the Born approximation) and, thus, is applicable, (independent of wavelength), whenever the object being probed is semi-transparent. In addition, Wolf's procedure requires that only far-field measurements be performed, thus offering the promise of being applicable in "in-situ" diagnostics where near field measurements are sometimes not possible. The success of this technique depends on the fact that the scattering amplitude in the Born approximation is a boundary value of the three-fold Fourier transform of the scattering potential. By performing a number of experiments with varying directions of illumination it is possible to determine this transform at a number of sample points from which the scattering potential can then be approximately reconstructed. Wolf did not address the important question of how these sample points should be selected, nor did he provide algorithms for reconstructing the scattering potential from the sample values. Carter[4], who applied Wolf's procedure to some simple objects, noted that the procedure, in general, requires a great amount of data and involves experimental problems in properly registering data collected from different experiments[5].

In the case of radiating sources it is also possible to approximately determine the internal structure of the source from measurements performed in the far field. In particular, it has been shown[6] that the autocorrelation function of the radiation pattern produced by a spatially incoherent source is a boundary value of the three-fold Fourier transform of the intensity distribution of the source. Thus, by performing a number of interference experiments in the far field it is possible to determine this transform at various sample points from which an approximation to the source's intensity distribution can be obtained. As is the case for semi-transparent scatterers, the selection of the sample points at which the transform is determined and the availability of a stable algorithm are crucial to the success of the reconstruction procedure.

In this paper we present a unified treatment of the two inverse problems described above. The underlying theory for these problems is developed in Section 2 while Sections 3 and 4 are devoted to approximate methods for reconstructing weak scatterers or incoherent sources from the available field data. The primary goal of this paper is to provide a review of certain known results in these two problem areas. However, the paper also contains some new results such as the technique for approximately reconstructing line projections of objects from radiation or scattering data[7]. This later procedure, which is described in Section 4, can be used in conjunction with standard algorithms of computed tomography (CT) to yield either planar projections or full three-dimensional reconstructions of the object. As such it bridges the gap, so-to-speak, between the CT method used in diagnostic radiology and the method of inverse scattering.

2.0 INVERSE SOURCE AND SCATTERING PROBLEMS

In this section we review briefly the formulation of the inverse problem for spatially incoherent, localized sources and localized weak scatterers where the Born approximation is valid. It is found for the inverse source problem that measurements of the radiated field's spatial coherence function in the wave zone leads to a partial specfication of the three-fold Fourier transform of the source's intensity profile. Similarly, measurements of the scattering amplitude produced by a weak scattering potential (e.g., semi-transparent object) leads to a partial specification of the three-fold Fourier transform of the scattering potential. Methods for approximately reconstructing an intensity profile or scattering potential from the partial specification of their transforms obtained by such measurements are presented in the following two sections.

We first formulate the inverse source problem which, as explained above, consists of determining the internal structure of a radiating source from measurements of the field generated by the source. We shall restrict our attention to the scalar case where the source $\rho(\underline{r})$ and field $\psi(\underline{r})$ are related according to the inhomogeneous Helmholtz equation

$$(\nabla^2+k_o^2)\psi(\underline{r}) = \rho(\underline{r}) \quad , \tag{2.1}$$

where $k_o = 2\pi/\lambda$ is the wavenumber of the field ψ and λ is its wavelength. The source will be assumed to be localized to within some finite volume τ which, for simplicity, we take to be a sphere of radius R_o centered at the origin. In any practical situation one is limited to measurements of ψ outside the region of localization τ. The inverse source problem then consists of determining the source from such measurements.

The radiated field obeys Sommerfeld's radiation condition so that in the far field

$$\psi(r\underline{s}) \sim -\frac{1}{4\pi}\,\hat{\psi}(\underline{s})\frac{e^{ik_o r}}{r} \quad , \quad k_o r\to\infty \quad , \tag{2.2}$$

where $\hat{\psi}(\underline{s})$ is the radiation pattern of the field in the direction specified by the unit vector $\underline{s} = \underline{r}/r$. A fundamental theorem[8] relating to solutions of the Helmholtz equation obeying Sommerfeld's radiation condition states that the radiation pattern specified for all directions $\underline{s}$, uniquely determines the field everywhere outside the source region τ. It follows that the radiation pattern carries the maximum amount of information about the source that is obtainable from field measurements performed external to τ. Thus, for all practical purposes, the inverse source problem can be defined[9] as being that of determining $\rho(\underline{r})$ from the radiation pattern $\hat{\psi}(\underline{s})$.

In most applications the source will not be perfectly coherent but, rather, will be a realization of a random process characterized by an autocorrelation function[10]

$$\Gamma_\rho(\underline{r}_1,\underline{r}_2) = \langle\rho^*(\underline{r}_1)\rho(\underline{r}_2)\rangle \qquad (2.3)$$

and higher order moments. One important class of random processes are those for which the autocorrelation function is of the form

$$\Gamma_\rho(\underline{r}_1,\underline{r}_2) = I(\underline{r}_1)\delta(\underline{r}_2-\underline{r}_1) \qquad (2.4)$$

where $I(\underline{r})$ is the <u>intensity profile</u> of the source and $\delta(\underline{r}_2-\underline{r}_1)$, is Dirac's delta function. Random sources having autocorrelation functions of the form given in Eq.(2.4) are said to be <u>spatially incoherent</u>[11]. The sources of thermal radiation (e.g., blackbody radiation) and of the gamma radiation given off by nuclear medicine are examples of spatially incoherent sources.

The inverse problem for random sources consists of determining the source autocorrelation function $\Gamma_\rho(\underline{r}_1,\underline{r}_2)$ from field interference experiments performed exterior to the source volume τ. It can be shown[5] that the autocorrelation function of the radiation pattern

$$\mathcal{E}(\underline{s},\underline{s}_o) = \langle\hat{\psi}^*(\underline{s})\hat{\psi}(\underline{s}_o)\rangle \qquad (2.5)$$

carries the maximum amount of information concerning Γ_ρ that can be obtained from such interference experiments. The inverse problem for random sources can thus be defined to be that of determining the source autocorrelation function $\Gamma_\rho(\underline{r}_1,\underline{r}_2)$ from the autocorrelation function of the radiation pattern $\mathcal{E}(\underline{s},\underline{s}_o)$ as measured for some set of observation directions $(\underline{s}, \underline{s}_o)$.

The autocorrelation function $\mathcal{E}(\underline{s},\underline{s}_o)$ can be determined from interference experiments performed in the far field of the source.

For example, the fringes produced in the vicinity of the central point in a Young's interference experiment allow one to determine the spatial coherence function of the field

$$\Gamma_{\psi}(R\hat{\underline{r}}_1, R\hat{\underline{r}}_2) = \langle \psi^*(R\hat{\underline{r}}_1)\psi(R\hat{\underline{r}}_2)\rangle \qquad (2.6)$$

at the pinhole locations $R\hat{\underline{r}}_1$ and $R\hat{\underline{r}}_2$. If the experiment is performed in the far field of the source then according to Eq. (2.2) the field at the pinhole locations is proportional to the radiation pattern evaluated in the directions $\hat{\underline{r}}_1$, $\hat{\underline{r}}_2$ of the pinholes. The autocorrelation function $\mathcal{E}(\hat{\underline{r}}_1, \hat{\underline{r}}_2)$ is then obtained from $\Gamma_{\psi}(R\hat{\underline{r}}_1, R\hat{\underline{r}}_2)$ by means of the equation

$$\mathcal{E}(\hat{\underline{r}}_1, \hat{\underline{r}}_2) = (4\pi)^2 R^2 \Gamma_{\psi}(R\hat{\underline{r}}_1, R\hat{\underline{r}}_2) \quad . \qquad (2.7)$$

It is not difficult to show that for the case of spatially incoherent sources the autocorrelation function of the radiation pattern and the sources intensity profile are related according to the equation[6]

$$\tilde{I}[k_o(\underline{s}-\underline{s}_o)] = \mathcal{E}(\underline{s}, \underline{s}_o) \quad . \qquad (2.8)$$

Here, $\tilde{I}(\underline{K})$ is the three-fold Fourier transform of the source's intensity profile, i.e.,

$$\tilde{I}(\underline{K}) = \int d^3r I(\underline{r}) e^{-i\underline{K}\,\underline{r}} \quad . \qquad (2.9)$$

Eq.(2.8) states that $\mathcal{E}(\underline{s}, \underline{s}_o)$ is equal to the boundary value of the Fourier transform of the source's intensity profile evaluated at spatial frequencies $\underline{K} = k_o(\underline{s}-\underline{s}_o)$. The inverse problem for incoherent sources then reduces to that of determining $\tilde{I}(\underline{K})$ from its value as specified according to Eq.(2.8) for some set of observation directions $\underline{s}$, $\underline{s}_o$.

We turn now to the <u>inverse scattering problem</u> which consists of determining the internal structure of an object from measurements of the field scattered by the object. We shall again restrict our attention to the scalar case and, in addition, assume that the object is characterized by a localized scattering potential $V(\underline{r})$. The total field $\phi(\underline{r})$ generated in a scattering experiment and the scattering potential $V(\underline{r})$ are related according to the time independent Schrodinger equation

$$(\nabla^2+k_o^2)\phi(\underline{r}) = V(\underline{r})\phi(\underline{r}) \quad . \tag{2.10}$$

Here, $k_o = 2\pi/\lambda$ is the wavenumber of the field incident to the scatterer which we shall take to be the plane-wave $\exp[ik_o\underline{s}_o\cdot\underline{r}]$ and $V(\underline{r})$ will be assumed to be localized to within a sphere of radius R_o centered at the origin.

The scattered field obeys Sommerfeld's radiation condition so that

$$\phi(r\underline{s}) \sim e^{ik_o\underline{s}_o\cdot\underline{r}} - \frac{1}{4\pi}\hat{\phi}(\underline{s},\underline{s}_o)\frac{e^{ik_o r}}{r} \quad , \quad (k_o r\to\infty) \quad , \tag{2.11}$$

where $\hat{\phi}(\underline{s},\underline{s}_o)$ is the scattering amplitude in the direction specified by the unit vector $\underline{s}$. The scattering amplitude plays the same role in the inverse scattering problem as that of the radiation pattern in the deterministic inverse source problem. In particular, it follows at once from the theorem[8] alluded to in the discussion of the inverse source problem that the scattering amplitude $\hat{\phi}(\underline{s},\underline{s}_o)$ carries the maximum amount of information about the scattering potential that is obtainable from field measurements performed external to the scattering volume τ[12]. The inverse scattering problem thus reduces to that of determining the scattering potential $V(\underline{r})$ from the scattering amplitude $\hat{\phi}(\underline{s},\underline{s}_o)$ as determined for one or more directions of propagation $\underline{s}_o$ of the incident plane wave $\exp[ik_o\underline{s}_o\cdot\underline{r}]$.

The scattering amplitude may, in some cases (e.g., in acoustic experiments), be measured directly. In optical applications it can be determined from holograms of the scattered field recorded over plane surfaces located in either the near or far field of the scatterer. For example, the two-dimensional Fourier transform

$$A(k_x,k_y,z) = \iint dxdy\ \phi^{(s)}(x,y,z)e^{-i(k_x x+k_y y)} \tag{2.12}$$

of the scattered field $\phi^{(s)}$ evaluated over some plane $z>R_o$ is related to the scattering amplitude $\hat{\phi}(\underline{s},\underline{s}_o)$ according to the equation[13]

$$\hat{\phi}(\underline{s},\underline{s}_o) = 2ik_o s_z e^{-ik_o s_z z} A(k_o s_x, k_o s_y, z) , \qquad (2.13)$$

with s_x, s_y, s_z being the three Cartesian components of the unit observation vector $\underline{s}$. The scattering amplitude can thus be determined by holographically recording the scattered field over a plane surface and then processing (say digitally) the resulting hologram according to Eqs. (2.12) and (2.13).

For semi-transparent objects the scattering amplitude is given approximately by[3]

$$\hat{\phi}(\underline{s},\underline{s}_o) \approx \hat{\phi}_B(\underline{s},\underline{s}_o) = \tilde{V}[k_o(\underline{s}-\underline{s}_o)] , \qquad (2.14)$$

where $\hat{\phi}_B(\underline{s},\underline{s}_o)$ denotes the Born approximation to the scattering amplitude and where

$$\tilde{V}(\underline{K}) = \int d^3 r V(\underline{r}) e^{-i\underline{K}\,\underline{r}} \qquad (2.15)$$

is the three-fold Fourier transform of the scattering potential. Eq. (2.14) states that the scattering amplitude of a semi-transparent object is approximately equal to the boundary value of the Fourier transform of the scattering potential evaluated at spatial frequencies $\underline{K} = k_o(\underline{s}-\underline{s}_o)$. The inverse scattering problem for semi-transparent objects then reduces to determining $\tilde{V}(\underline{K})$ from its value as specified according to Eq.(2.14) for some set of incident plane wave propagation directions $\underline{s}_o$ and observation directions $\underline{s}$[3].

We conclude from the discussion presented above that the inverse problem for incoherent sources and the inverse scattering problem for semi-transparent objects both reduce to the same mathematical problem: namely, that of determining a function (say $F(\underline{r})$) from its three-fold Fourier transform (say $\tilde{F}(\underline{K})$) as specified for those wavevectors $\underline{K}$ which satisfy the equality $\underline{K} = k_o(\underline{s}-\underline{s}_o)$. For the inverse source problem the unit vectors $\underline{s}$ and $\underline{s}_o$ are the two directions at which the autocorrelation of the radiation pattern is measured and $F(\underline{r})$ is the intensity profile of the source. For the inverse scattering problem $\underline{s}_o$ is the direction of propagation of a plane wave incident to the object, $\underline{s}$ is the direction at which the scattering amplitude is measured and $F(\underline{r})$ is the scattering potential. The problem of reconstucting the function $F(\underline{r})$ from the partial specification of its transform $\tilde{F}(\underline{K})$ obtained in radiation and scattering experiments is investigated in the following two sections.

3. APPROXIMATE RECONSTRUCTION USING FOURIER SERIES

We showed in the preceeding section that the problem of reconstructing an incoherent source or weak scatterer from radiation or scattering data reduces to that of determining a function $\tilde{F}(\underline{K})$ from its specification over a set of points in $\underline{K}$ space which satisfy the equation $\underline{K}=k_o(\underline{s}-\underline{s}_o)$. For a fixed value of $k_o\underline{s}_o$ these points lie on the surface of a sphere (called the Ewald sphere[14]) whose radius is k_o and which is centered at $-k_o\underline{s}_o$. As $\underline{s}_o$ varies over the unit sphere the surface of the Ewald sphere sweeps out the volume of $\underline{K}$ space contained within a sphere of radius $2k_o$ which is centered at the origin (the so-called limiting Ewald sphere[14]). It follows that it is possible to find one or more combinations of $\underline{s}_o$ and $\underline{s}$ which will cause the measurement vector $\underline{M}=k_o(\underline{s}-\underline{s}_o)$ to assume any given value within the limiting Ewald sphere (i.e., for which $|\underline{K}|\leq 2k_o$). The transform $\tilde{F}(\underline{K})$ can thus be determined at any $\underline{K}$ value lying within the limiting Ewald sphere by appropriate selection of $\underline{s}_o$ and $\underline{s}$ in a radiation or scattering experiment.

In practice one is limited to performing a finite number of experiments so that it will be possible to determine $\tilde{F}(\underline{K})$ only over a finite number of points within the limiting Ewald sphere. It is, of course, not possible in general to uniquely determine a function from its specification over a finite number of points so that in practice the inverse source and scattering problems discussed here do not possess unique solutions[15].

Although it is not possible to uniquely determine an object from data generated in a finite number of experiments it is possible to obtain approximate reconstructions which will be accurate renditions of the object as long as it possesses no significant structure on a scale finer than a half-wavelength[16]. In particular,we show in this section that it is possible to approximate the function $F(\underline{r})$ by means of a truncated three-dimensional Fourier series whose expansion coefficients (Fourier coefficients) are simply the values of $\tilde{F}(\underline{K})$ at the sample points in a cubic array contained entirely within the limiting Ewald sphere. The mean square error of this approximation is shown to be small as long as $|\tilde{F}(\underline{K})|$ remains small for values of $|\underline{K}| \geq 2k_o$ (i.e., so long as the object possesses no significant structure on a scale finer than $\lambda/2$).

We begin by noting that since $F(\underline{r})$ is assumed to be localized to within a sphere of radius R_o, centered at the origin, it can be expanded into the three-dimensional Fourier series[17]

$$F(\underline{r}) = \frac{1}{8R_o^3} \sum_{\ell=-\infty}^{\infty} \sum_{m=-\infty}^{\infty} \sum_{n=-\infty}^{\infty} C_{\ell,m,n} e^{i\frac{\pi}{R_o}[\ell x+my+nz]} \qquad (3.1)$$

Here, $F(\underline{r})$ is either an intensity profile or scattering potential and the Fourier series converges in the mean square sense to $F(\underline{r})$ for $|\underline{r}| \leq R_o$.

The Fourier coefficients in the above expansion are found to be the values of $F(\underline{K})$ at the sample points of a cubic array; i.e.,

$$C_{\ell,m,n} = \int d^3 r F(\underline{r}) e^{-i\frac{\pi}{R_o}[\ell x+my+nz]} = \tilde{F}[\ell\frac{\pi}{R_o}, m\frac{\pi}{R_o}, n\frac{\pi}{R_o}] \ . \qquad (3.2)$$

As discussed above, the Fourier transform $\tilde{F}(\underline{K})$ can be determined from radiation or scattering data at any finite number of points lying within the limiting Ewald sphere. It follows that the Fourier coefficients $C_{\ell,m,n}$ can be determined from such data for values of the three indicies satisfying the inequality

$$\ell^2+m^2+n^2 \leq (2k_oR_o/\pi)^2 \ . \qquad (3.3)$$

For example, by taking $\underset{\sim}{s}$ and $\underline{s}_o$ such that $\underline{s}-\underline{s}_o$ is parallel to the z axis one can determine $\tilde{F}(0,0,n\pi/R_o)$ for n ranging from $-2k_oR_o/\pi$ to $+2k_oR_o/\pi$. This is accomplished in a total of $4k_oR_o/\pi+1$ measurements where the vector $\underline{s}-\underline{s}_o$ varies in discrete steps from $-2\hat{\underline{z}}$ to $+2\hat{\underline{z}}$ in increments equal to $(\pi/k_oR_o)\hat{\underline{z}}$.

An approximation to $F(\underline{r})$ is given by the truncated Fourier series representation

$$F(\underline{r}) = \frac{1}{8R_o^3} \sum_{\ell,m,n} C_{\ell,m,n} e^{i\frac{\pi}{R_o}[\ell x+my+nz]} \ , \qquad (3.4)$$

where the summation extends over those indicies for which $C_{\ell,m,n}$ has been determined from the available radiation or scattering data. The above approximation to $F(\underline{r})$ is found to possess the mean square error

$$\text{Error} = \frac{1}{8R_o^3} \sum_{\ell,m,n} |\tilde{F}[\ell\frac{\pi}{R_o}, m\frac{\pi}{R_o}, n\frac{\pi}{R_o}]|^2 \qquad (3.5)$$

where the summation extends over all values of ℓ,m,n not included in the truncated Fourier series approximation of $F(\underline{r})$ given in Eq.(3.4).

To keep the mean square error to within an acceptable level will, in general, require a very large number of experiments to be performed It is important to note that even if enough experiments are performed so that $\tilde{F}(\underline{K})$ is sampled out to the maximum value of $|\underline{K}|=2k_o$ the error may still be unacceptably large. This will occur, for example, if $|\tilde{F}(\underline{K})|$ remains large for values of $\underline{K}$ lying outside the limiting Ewald sphere. This is simply another way of saying that the scattered or radiated field is only sensitive to details of the object's structure greater than a half-wavelength. Structure on a finer scale than this simply cannot be determined via radiation or scattering experiments[18]. The unobservable nature of such structure is the underlying reason for the inherent non-uniqueness of inverse source and inverse scattering problems. It is only by knowing apriori that the object does not possess such fine structure that one can be certain that the truncated Fourier series (3.4) yields a good approximation to the object.

4. RECONSTRUCTION OF LINE PROJECTIONS

The Fourier series reconstruction of $F(\underline{r})$ requires that the measurement vector $\underline{M}=k_o(\underline{s}-\underline{s}_o)$ assume sample values on a cubic array. As pointed out by Carter[4] this requirement poses severe problems for the experimentalist; problems which may make this method unfeasable in practice. A considerably less complex series of experiments are required to obtain sample values of $\tilde{F}(\underline{K})$ on a spherical <u>polar</u> sampling array. We show in this section that sample values of $\tilde{F}(\underline{K})$ contained entirely within the limiting Ewald sphere and lying along the polar grid lines of such a sampling array can be used to obtain <u>one-dimensional</u> Fourier series approximations to so-called <u>line projections</u> of the object. Each line projection requires many less samples than is required in the full three dimensional Fourier series reconstruction of the object, and provides valuable partial information about the object's structure. Moreover, approximate reconstructions of the object itself can be obtained from a sufficient number of line projections by use of algorithms used in computed tomography[1]. As in the case of the three-dimensional Fourier series reconstruction technique these later methods will be accurate as long as the object (i.e.,$F(\underline{r})$) is effectively bandlimited to within the limiting Ewald sphere.

The results presented in this section make use of the concept of line projections of three-dimensional functions. A projection of a function onto a line (say the z axis) is simply the integral of this function over planes which are perpendicular to the given line (in this case the x-y planes); e.g.,

$$F_{xy}(z) = \iint_{-\infty}^{\infty} dxdy\ F(\underline{r}) \tag{4.1}$$

is the line projection of $F(\underline{r})$ onto the z axis. More generally,

$$F_{\xi\eta}(\nu) = \iint_{-\infty}^{\infty} d\xi d\eta\ F(\underline{r}) \tag{4.2}$$

is the line projection of $F(\underline{r})$ onto the ν axis. Here, (ξ,η) are the Cartesian coordinates on planes which are perpendicular to the ν axis.

A fundamental theorem having to do with projections is the so-called PROJECTION-SLICE THEOREM[19]. This theorem states that the Fourier transform of a projection of a function is a "slice" through the Fourier transform of this function. As an illustration of this theorem consider the Fourier transform of the line projection $F_{xy}(z)$ of $F(\underline{r})$ defined in Eq.(4.1). This transform is found to be

$$\tilde{F}_{xy}(K_z) = \int_{-\infty}^{\infty} dz F_{xy}(z) e^{-iK_z z} = \tilde{F}(0,0,K_z) \quad . \tag{4.3}$$

Thus, the Fourier transform of the projection $F_{xy}(z)$ of $F(\underline{r})$ is simply the Fourier transform of $F(\underline{r})$ evaluated on the K_z axis; i.e., $F_{xy}(K_z)$ is a "slice" through $\tilde{F}(\underline{K})$.

Because $F(\underline{r})$ is localized to within a sphere of radius R_o, centered at the origin, any line projection of $F(\underline{r})$ is localized within the interval $[-R_o, R_o]$. It follows that the line projections admit one-dimensional Fourier series expansions which converge in the mean square sense to the projection within this interval. For example,

$$F_{xy}(z) = \frac{1}{2R_o} \sum_{n=-\infty}^{\infty} F_n e^{i\frac{\pi}{R_o} nz} \tag{4.4}$$

with

$$F_n = \int_{-\infty}^{\infty} dz\, F_{xy}(z)\, e^{-i\frac{\pi}{R_o} nz} = \tilde{F}(0,0,n\frac{\pi}{R_o}) \tag{4.5}$$

In deriving Eq.(4.5) we have made use of the PROJECTION-SLICE THEOREM [cf, Eq.(4.3)].

Now, $\tilde{F}(0,0,K_z)$ can be determined at points within the limiting Ewald sphere from scattering or radiation experiments by fixing the direction of the measurement vector $\underline{M}=k_o(\underline{s}-\underline{s}_o)$ to lie along the z axis. A truncated Fourier series approximation to $F_{xy}(z)$ is then obtained by choosing $K_z=n\pi/R_o$ and substituting the resulting experimentally determined sample values of $\tilde{F}(0,0,n\pi/R_o)$ into Eq.(4.4) and discarding all terms in the expansion whose Fourier coefficients are unknown. The mean square error of the resulting approximation is given by

$$\text{Error} = \frac{1}{2R_o} \sum_n \left| \tilde{F}(0,0,n\frac{\pi}{R_o}) \right|^2 \tag{4.6}$$

where the summation extends over those values of n not included in the truncated Fourier series approximation.

In a similar fashion the Fourier coefficients required in truncated Fourier series expansions of other line projections of $F(\underline{r})$ are obtained by rotating the direction of the measurement vector and performing sets of measurements at equally spaced points in $\underline{K}$ space along these various directions. Each set of measurements at a fixed direction of the measurement vector results in equally spaced samples of $\tilde{F}(\underline{K})$ along a polar grid line. The samples along any given grid line are then, according to the PROJECTION-SLICE THEOREM, equal to the required Fourier coefficients in a Fourier series approximation to a line projection of the function.

We conclude by noting that a number of algorithms exist for approximating a function from a collection of its projections. We shall not discuss here the various algorithms that have been devised for this purpose, but refer the reader to the rather vast literature on the subject[1].

ACKNOWLEDGEMENT

The author is indebted to Professor Emil Wolf for many helpful discussions on the inverse scattering problem.

FOOTNOTES AND REFERENCES

1. For a survey of computerized tomography see: A.C. Kak, Proc. IEEE 67,1245 (1979).
2. In transmission tomography (scattering experiments) these models are, essentially, ray optics descriptions of the imaging process while in emission tomography (radiation experiments) radiometric type models are employed.
3. E. Wolf, Optics Commun. 1, 153(1969).
4. W.H. Carter, J. Opt. Soc. Am. 60, 306 (1970); W.H. Carter and P.-C. Ho, Appl. Opt. 13, 162 (1974); P.-C. Ho and W.H. Carter, Appl. Opt. 15, 313 (1976).
5. Some recent applications of Wolf's procedure are given in: A.F. Fercher et. al., Appl. Opt. 18, 2427 (1979).
6. A.J. Devaney, J. Math. Phys. 20, 1687 (1979).
7. A more complete account of some of these new results will be presented elsewhere (cf, A.J. Devaney and E Wolf, Optics News A6, 42 (1980)).
8. C. Muller, *Foundations of the Mathematical Theory of Electromagnetic Waves* (Springer, New York,1969), p339.
9. An excellent review of the (non-random) inverse source problem is given by B.J. Hoenders, *Inverse Source Problems in Optics*, edited by H.P. Baltes, (Springer-verlag, Heidelberg, 1978) pp.41-82.
10. Ensemble averages will be denoted by angular brackets $\langle \ \rangle$ enclosing the quantity to be averaged and complex conjugates by a superscript asterisk * on the quantity being conjugated.
11. No natural source of radiation can be completely spatially incoherent due to the interaction of the source with the (partially coherent) radiation field. However, the spatially incoherent model (2.4) provides a good approximation as long as the wavelength is small relative to the internal structure of the source. This later requirement will be found to be necessary in order to obtain accurate reconstructions of the source from radiation field data.
12. A.J. Devaney, J. Math. Phys. 19, 1526 (1978).
13. A.J. Devaney and G.C. Sherman, SIAM Review 15, 784 (1973).
14. R Hosemann and S.H. Baghi, *Direct Analysis of Diffraction of Matter* (North-Holland, Amsterdam, 1962), Chap. I, Sec. 6.
15. For a discussion of the non-uniqueness in the inverse source problem see Ref. 6. The uniqueness question for the inverse scattering problem is investigated in Ref. 12.
16. The lack of uniqueness of solutions to the inverse source and scattering problems implies, of course, that *exact* reconstructions of the object are not possible.
17. This Fourier series expansion of $F(\underline{r})$ can be derived from the Whittaker-Shannon sampling series representation of $\tilde{F}(\underline{K})$.

The use of the sampling series in the reconstruction problem was suggested to the author by E. Wolf and leads to an elegant and mathematically rigourous treatment of the problem (A.J. Devaney and E. Wolf, unpublished paper).

18. Here, we discount entirely any "super-resolution" techniques which attempt to determine $\tilde{F}(\underline{K})$ at points outside the limiting Ewald sphere from direct measurements of the radiated or scattered field. Such techniques rarely work in practice. (See, however E.G. Williams and J.D. Maynard, Phys. Rev. Letters 45, 554, (1980)).
19. R.M. Mersereau and A.V. Oppenheim, Proc. IEEE 62, 1319 (1974).

HOLOGRAPHY, WAVE-LENGTH DIVERSITY AND INVERSE SCATTERING

N.H. FARHAT
University of Pennsylvania
The Moore School Graduate Research Center
Philadelphia, PA 19104

ABSTRACT

The use of wavelength diversity to enhance the performance of thinned coherent imaging apertures is discussed. It is shown that wavelength diversity lensless Fourier transform recording arrangements that utilize a reference point source in the vicinity of the object can be used to access the three-dimensional Fourier space of non-dispersive perfectly reflecting or weakly scattering objects. Hybrid (opto-digital) computing applied to the acquired 3-D Fourier space data is shown to yield tomographic reconstruction of 3-D image detail either in parallel or meridional (central) slices. Because of an inherent ability of converting spectral degrees of freedom into spatial 3-D image detail true super-resolution is achieved together with suppression of coherent noise. The similarity of our key equations to those of inverse scattering theory is pointed out and the feasibility of using other forms of broadband radiation such as impulsive, noise and thermal is discussed. Finally, the possibility of utilizing wavelength diversity imaging in microscopy and telescopy is discussed.

INTRODUCTION

A frequently encountered question in the science of image formation is how to make an available aperture collect more information about the scene or object being imaged in order to enhance its resolving power beyond the classical Rayleigh limit. This process is known as super-resolution and is relevant to all imaging systems whether holographic or conventional. There are five known methods for achieving super-resolution. These include: weighting or apodization of the aperture data[1,2]; analytic continuation of the wavefield measured over the aperture[3,4]; use of evanescent wave illumination[5]; maximum entropy method[6]; and use of the time channel[7]. Weighting and analytical continuation techniques are known to become rapidly ineffective as the signal to noise ratio of the collected data decreases. Maximum entropy techniques are known to be more robust, as far as noise is concerned but involve usually extensive computation. Illumination with evanescent waves is practical in limited situations where full control of the recording arrangement exists as in microscopy, for example.

ISSN:0094-243X/81/650627-16$1.50

This leaves the time channel approach in which one can collect in time more information about the object through the available recording aperture by altering the object aspect relative to the aperture by means of rotation or linear motion[8,9,28] or by altering the parameters of the illumination such as directions of incidence, wavelength and/or polarization as discussed in this paper. These latter operations are known to increase the degrees of freedom of the wavefields impinging on the recording aperture enhancing thereby their ability to convey information about the nature of the scattering object. Sophisticated imaging systems endeavor to convert the nonspatial degrees of freedom of the wavefield, e.g. angular, spectral and polarization to spatial image detail, enhancing thereby the resolution capability beyond the classical Rayleigh limit of the available physical aperture. Obviously such procedures involve more signal processing than that performed by conventional imaging with lens systems or holography.

In this paper we consider generalizing the holographic concept to include wavelength diversity as a means of enhancing resolution. A quick examination of the basic equations of holography reveals that the lensless Fourier transform hologram recording arrangement is amenable to this generalization. This conclusion is used then as a starting point for a Fourier optics formulation of wavelength diversity imaging of 3-D (three dimensional) nondispersive objects. The results show that measurement of the multiaspect or multistatic frequency (or wavelength) response of the 3-D object permits accessing its 3-D Fourier space. The resulting formulas are identical to those obtained from a multistatic generalization of inverse scattering[10,11,12] establishing thus a clear connection between holography and the inverse scattering imaging problem. The inclusion of wavelength diversity in holography is shown to have several important features: (a) the availability of the 3-D Fourier space data permits 3-D image retrieval tomographically in parallel or meridional (central) slices or cross-sectional outlines by the application of Fourier domain projection theorems; (b) suppression of coherent noise and speckle in the retrieved image; (c) removal of several longstanding constraints on longwave (microwave and acoustical) holography such as the impractically high cost of the apertures needed, the inability to view a true 3-D image as in optical holography because of a wavelength scaling problem, and minimization of the effects of resonances on the object.

WAVELENGTH DIVERSITY

We start by inquiring into the conditions under which the data from N holograms of the same nondispersive object recorded over the same aperture, each at a different wavelength, can be combined to yield a single image superior in quality to the image retrieved from any of the individual holograms.

One approach to answering the question posed above would be to determine the conditions under which the well known formulas[13] for the focusing condition, magnification and image location in holography can be made independent of wavelength. This quickly leads to the conclusion that wavelength independence can be met if a reference point source centered on the object is used and proper scaling of the individual holograms by the ratio of recording to the reconstruction wavelength is performed before superposition [15,24]. The former condition is required for recording a lensless Fourier transform hologram[14,29], where the presence of the reference point source in the object plane leads to the recording of a Fraunhofer diffraction pattern of the object rather than its Fresnel diffraction pattern because of the elimination in the recorded hologram of a quadratic phase term in the object wavefield. This is known to result in a highly desirable reduction in the resolution required from the hologram recording medium and is therefore of practical importance especially in nonoptical holography. More detail of the processing involved in combining the data in multi-wavelength hologram can be found elsewhere[15].

Additional insight into the process of attaining super-resolution by wavelength diversity is obtained by considering the concept of wavelength or frequency synthesized aperture [16-20]. The synthesis of a one dimensional aperture by wavelength diversity is based on the simple fact that the Fraunhofer or far field diffraction pattern of a nondispersive planar object changes its scale, i.e. it "breathes", but does not change its shape (functional dependence), as the wavelength is changed. A stationary array of broadband sensors capable of measuring the complex field variations deployed in this breathing diffraction pattern at suitably chosen locations would sense different parts of the diffraction pattern as the wavelength is altered, this allows collecting more information on the nature of the diffraction pattern, and therefore on the object that gave rise to it than if the wavelength was fixed (stationary diffraction pattern). Each stationary sensor in the array is thus able to collect, as the wavelength is changed, and the breathing diffraction pattern sweeps over it, the same set of data or information collected by mechanically scanning a sensor over the diffraction pattern when it is kept stationary by fixing the wavelength. Hence the term wavelength or frequency synthesized aperture.

The orientation and location of the wavelength synthesized aperture for any planar distribution of sensors deployed in the Fraunhofer diffraction pattern of a planar object and the retrieval of an image from the collected data has been treated earlier[16,17]. It was clear, however, that extension of the wavelength diversity concept to the case of 3-D objects is necessary before its generality and practical use could be established.

For this purpose we considered[20] as shown in Fig. 1(a) an isolated planar object of finite extent with reflectivity $D(\bar{\rho}_o)$, where $\bar{\rho}_o$ is a two dimensional position vector in the object plane (x_o, y_o). The object is illuminated by a coherent plane wave of unit-amplitude and of wave vector $\bar{k}_i = k\,\bar{1}_{k_i}$ produced for example by a distant source located at $\bar{R}_T$. The wavefield scattered by the object is monitored at a receiving point designated by the position vector $\bar{R}_R$ belonging to a recording aperture lying in the far field region of the object. The receiving point will henceforth be referred to as the receiver and the source point at the transmitter. The position vectors $\bar{\rho}_o$, $\bar{R}_T$ and $\bar{R}_R$ are measured from the origin of a cartesian coordinate system (x_o, y_o, z_o) centered in the object. The object is assumed to be nondispersive i.e., D is independent of k. However, when the object is dispersive such that $D(\bar{\rho}_o,k) = D_1(\bar{\rho}_1)D_2(k)$ and $D_2(k)$ is known, the analysis presented here can easily be modified to account for such object dispersion by correcting the collected data for $D_2(k)$ as k is varied.

Referring to Figure 1(a) and ignoring polarization effects, the field amplitude at $\bar{R}_R$ caused by the object scattered wavefield may be expressed as,

$$\psi(k,\bar{R}_R) = \frac{jk}{2\pi} \int D(\bar{\rho}_o) e^{-j\bar{k}_i \cdot \bar{r}_T} \frac{e^{-jk\,r_R}}{r_R} d\bar{\rho}_o \tag{1}$$

where $d\bar{\rho}_o$ is an abbreviation for $dx_o dy_o$ and the integration is carried out over the extent of the object. Noting that $\bar{r}_T = \bar{\rho}_o - \bar{R}_T$, $\bar{R}_T = -R_T \bar{1}_{k_i}$ and using the usual approximations valid here: $r_R \simeq R_R + \rho_o^2/2R_R - \bar{1}_R \cdot \bar{\rho}_o$ for the exponential in (1) and $r_R \simeq R_R$ for the denominator in (1) where $\bar{1}_R = \bar{R}_R/R_R$ and $\bar{1}_{k_i} = \bar{k}_i/k$ are unit vectors in the $\bar{R}_R$ and $\bar{k}_i$ directions respectively, one can write eq. (1) as,

$$\psi(k, \bar{R}_R) = \frac{jk}{2\pi R_R} e^{-jk(R_T + R_R)} \int D(\bar{\rho}_o)\, e^{-j\,\bar{p}\,\cdot\,\bar{\rho}_o}\, d\bar{\rho}_o, \tag{2}$$

In eq. (2) we have used the fact that the observation point is in the far field of the object, so that $\exp(-jk\,\rho_o^2/2R_R)$ under the integral sign can be replaced by unity. In addition, $\bar{p} = k(\bar{1}_{k_i} - \bar{1}_R) \triangleq p_x\,\bar{1}_x + p_y\,\bar{1}_y + p_z\,\bar{1}_z$ is a three dimensional vector whose length and orientation depend

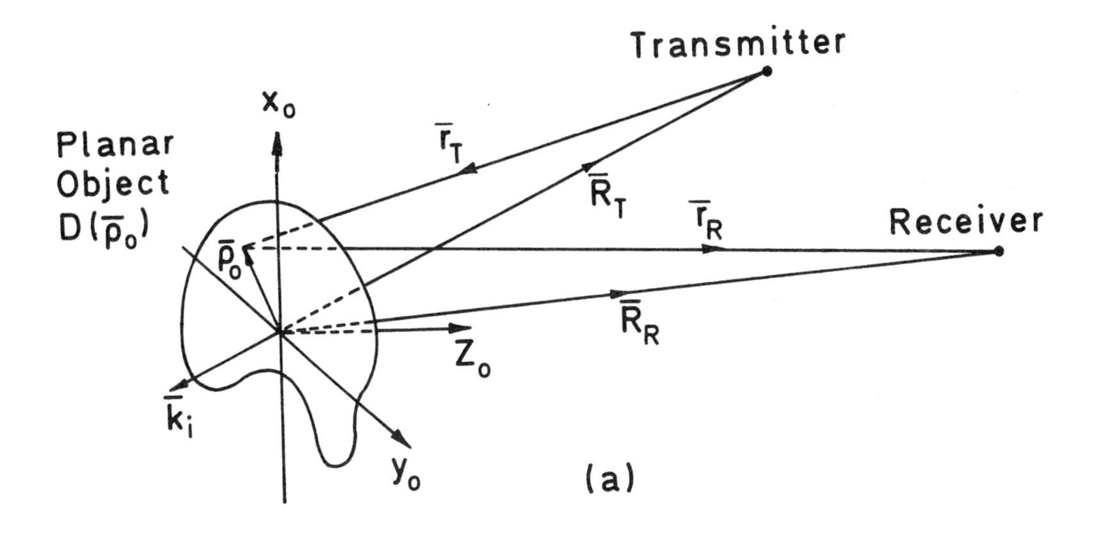

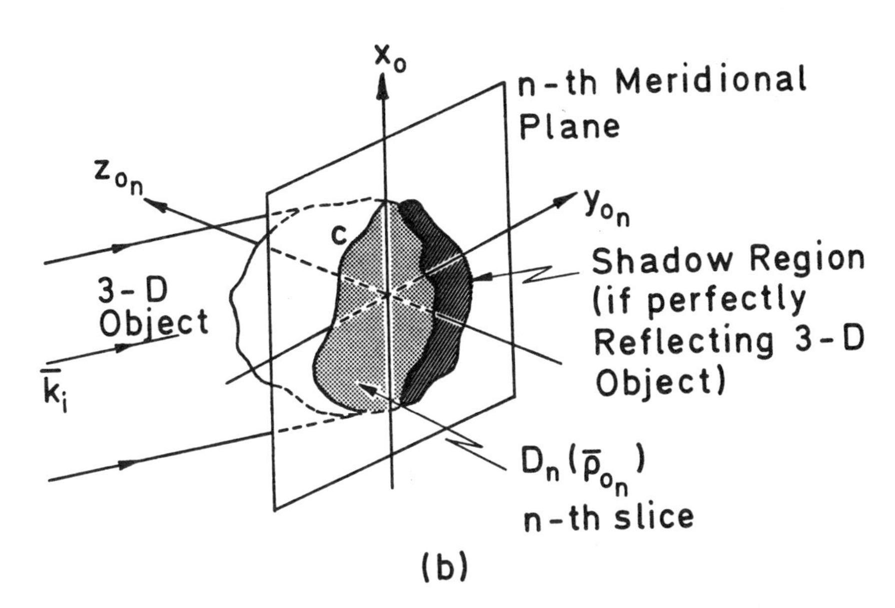

Fig. 1. Geometries for wavelength diversity imaging. (a) Two dimensional object, (b) Three dimensional object with the n-th meridional (central) slice and cross sectional outline c shown.

on the wavenumber k and the angular positions of the transmitter and the receiver. For each receiver and/or transmitter present, $\bar{p}$ indicates the position vector for data storage. An array of receivers for example would yield therefore a 3-D data manifold as k is varied (frequency diversity) or as $\bar{k}$ ($=k\bar{1}_{k_i}$) is varied (wave-vector diversity) The projection of this 3-D data manifold on the object plane yields $\psi(k,R_T)$ because $\bar{p} \cdot \bar{\rho}_o = \bar{p}_t \cdot \bar{\rho}_o = p_x x_o + p_y y_o$ where $p_x = k(\bar{1}_{k_i} - \bar{1}_R)_x$ and $p_y = k(\bar{1}_{k_i} - \bar{1}_R)_y$ are the cartesian components of the projection $\bar{p}_t$ of $\bar{p}$ on the object plane. Accordingly eq. (2) can be expressed as,

$$\psi(k, R_R) = \frac{jk}{2\pi R_R} e^{-jk(R_T + R_R)} \int D(x_o,y_o)\, e^{-j(p_x x_o + p_y y_o)} dx_o dy_o \quad (3)$$

Because of the finite extent of the object, the limits on the integral can be extended to infinity without altering the result. The integral in (3) is recognized then as the two dimensional Fourier transform $\tilde{D}(p_x,p_y)$ of $D(x_o,y_o)$. It is seen to be dependent on the object reflectivity function, the angular positions of the transmitter and the receiver and on the values assumed by the wavenumber k but is entirely independent of range. Information about D can thus be collected by varying these parameters. Note that the range information is contained solely in the factor $F = jk \exp[-jk(R_T + R_R)]/2\pi R_R$ preceeding the integral. The field observed at $\bar{R}_R$ has thus been separated into two factors one of which, the integral, $\tilde{D}$, contains the lateral object information and the other F, contains the range information. The presence of F in eq. (3) hinders the imaging process since it complicates data acquisition and if not removed, gives rise to image distortion because R_R is generally not the same for all receivers. To retrieve an image of the object via a 2-D Fourier transform of eq. (3), the factor F must first be eliminated. Holographic recording of the complex field amplitude given in (3) using a reference point source located at the center of the object will result in the elimination of the factor F and the recording of a Fourier transform hologram. This operation yields $\tilde{D}$ over a two dimensional region in the p_x,p_y plane. The size of this region, which determines the resolution of the retrieved image depends on the angular positions of the transmitter and the receiver and on the values assumed by k, i.e. the extent of the spectral window used. The latter dependence on k implies super-resolution imaging capability because of the frequency synthesized dimension of the 2-D data manifold that is generated. Because of the dependence of resolution on the relative positions of the object, the transmitter, and receiving aperture, the impulse response is clearly not invariant. In fact a receiver point situated at $\bar{R}_R$ for which $\bar{p}$ is normal to the object plane can not collect any lateral object information because for this condition ($\bar{p} \cdot \bar{\rho}_o = 0$) the integrals in (2) and (3) yield a constant. Such receiving point is located in the direction of specular reflection from the object where the diffraction pattern is stationary i.e. does not change with k. In this case the observed field is solely proportional to F containing thus range information only. Obviously this case can easily be avoided through the use of more than one receiver which is required anyway when 2-D or 3-D object resolution is sought[20,21].

The analysis presented above can be extended to three dimensional objects by viewing a 3-D object as a collection of thin meridional or central slices as depicted in Fig. 1(b) each of which representing a two dimensional object of the type analyzed above. With the n-th slice we associate a cartesian coordinate system x_o, y_{o_n}, z_{o_n} that differs from other slices by rotation about the common x_o axis. Since the vectors $\bar{p}$, $\bar{R}_T$ and $\bar{R}_R$ are the same in all n-coordinate systems, eq. (3) holds. $\psi_n(k,\bar{R}_R)$ is then obtained from projection of the three dimensional data manifold collected for the 3-D object on the x_o, y_{on} plane associated with the n-th slice. An image for each slice can then be obtained as described before. An inherent assumption in this argument is that all slices are illuminated by the same plane wave. This is a reasonable approximation when the 3-D object is a weak scatterer and the Born approximation is applicable, or when the 3-D object is perfectly reflecting and does not give rise to multiple reflections between its parts. In the latter case the two dimensional meridional slices $D_n(\bar{\rho}_{o_n})$ deteriorate into contours, such as C in Fig. 1(b) defined by the intersection of the meridonal planes with the illuminated portion of the surface of the object. Accordingly we can write for the n-th meridional slice or contour,

$$\psi_n(k,R_R) = F \int D_n(\bar{\rho}_{o_n})\, e^{-j\bar{p}\cdot\bar{\rho}_{o_n}}\, d\bar{\rho}_{o_n} \qquad (4)$$

We can regard $D_n(\bar{\rho}_{o_n})$ as the n-th meridional slice or contour or a three dimensional object of reflectivity $U(\bar{r})$ where $\bar{r}$ is a three dimensional position vector in object space. This means that $D_n(\bar{\rho}_{o_n}) = U(\bar{r})\ \delta(z_{o_n})$ where δ is the Dirac delta "function". Consequently eq. (4) becomes,

$$\psi_n(k,R_R) = F \int U(\bar{r})\, \delta(z_{o_n})\, e^{-j\bar{p}.\bar{\rho}_{o_n}}\, d\bar{\rho}_{o_n}$$

$$= F \int U(\bar{r})\, \delta(z_{o_n})\, e^{-j\bar{p}.\bar{r}}\, d\bar{r} \qquad (5)$$

where $d\bar{r}$ designates an element of volume in object space and where the last equation is obtained by virtue of the sifting property of the delta function.

After summing up the data from all slices or contours of the object we obtain,

$$\sum_n \psi_n = F \int U(\bar{r})\, e^{-j\bar{p}.\bar{r}}\, d\bar{r} = \psi(\bar{p}) \qquad (6)$$

because

$$\sum_n U(\bar{r})\, \delta(z_{o_n}) = U(\bar{r}).$$

On assuming that the Factor F in eq. (6) is eliminated as before, equation (6) reduces to

$$\psi(\bar{p}) = \int U(\bar{r})\, e^{-j\bar{p}.\bar{r}}\, d\bar{r} \qquad (7)$$

which is the 3-D Fourier transform of the object reflectivity $U(\bar{r})$. Thus, wavelength diversity allows one to access the 3-D Fourier space of a nondispersive object and provides the basis for 3-D Lensless Fourier transform holography. An alternate formulation of super-resolved wave-vector diversity imaging of 3-D perfectly conducting objects is possible[22] by extending the analysis of the inverse scattering imaging problem[10,11] to the multistatic case, along lines that are similar to but somewhat different from those given by Raz[12]. The resulting scalar formulas are identical to (7) thus establishing the connection between the holographic and the inverse scattering approaches to the imaging problem.

THREE DIMENSIONAL IMAGE RETRIEVAL

The above considerations of multiwavelength holography have provided a means by which the 3-D Fourier space of the object can be accessed by employing snychronous detection. It is clear that once the 3-D Fourier space data is available, 3-D image detail can be retrieved by means of an inverse 3-D Fourier transform that can be carried out digitally. Alternately, holographic techniques

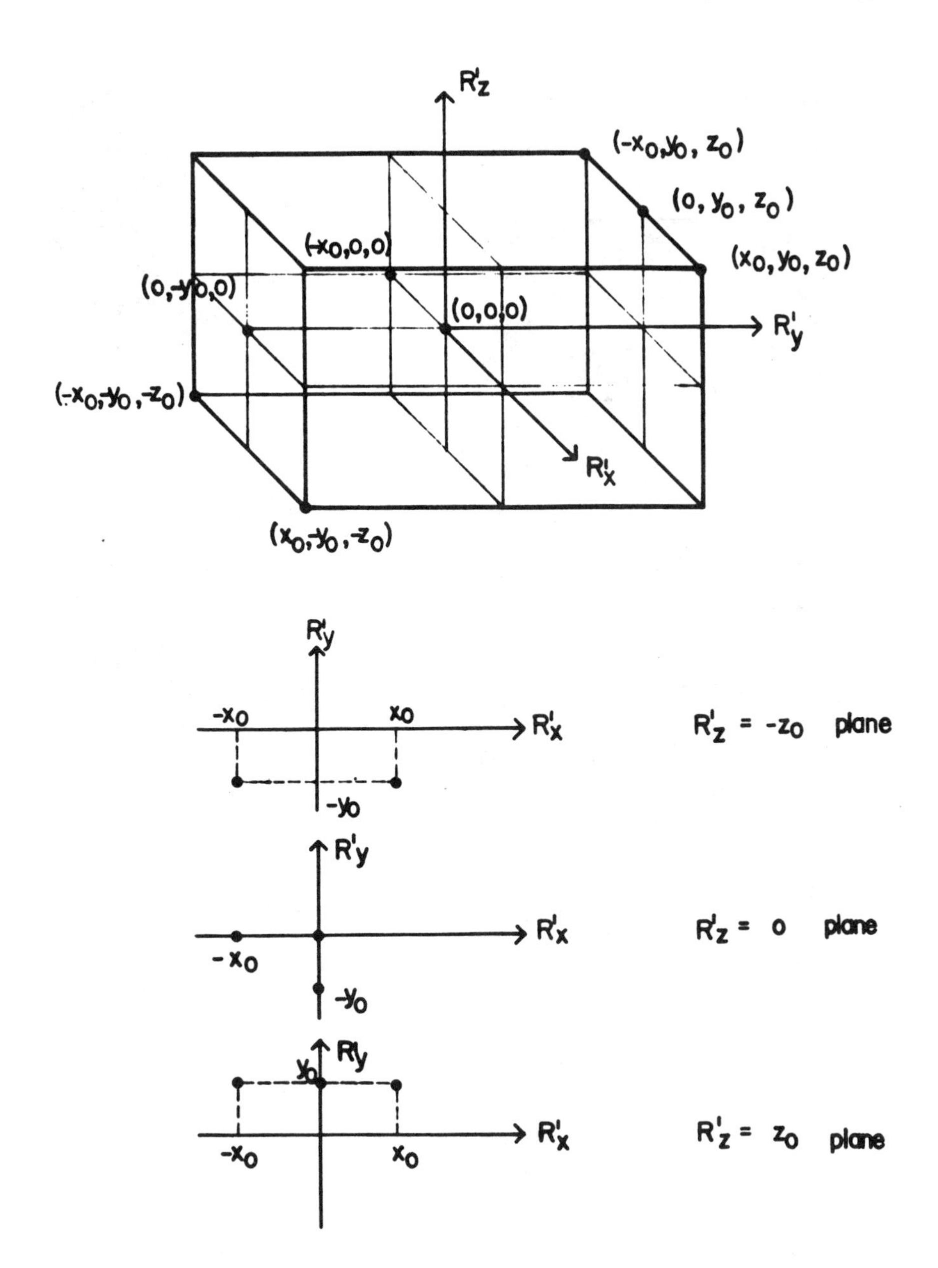

Fig. 2. 3-D object consisting of a set of eight point scatterers shown in isometric and R_x'-R_y' plane views at $R_z'=-z_o,0,z_o$. $x_o=y_o=z_o$ =100 cm.

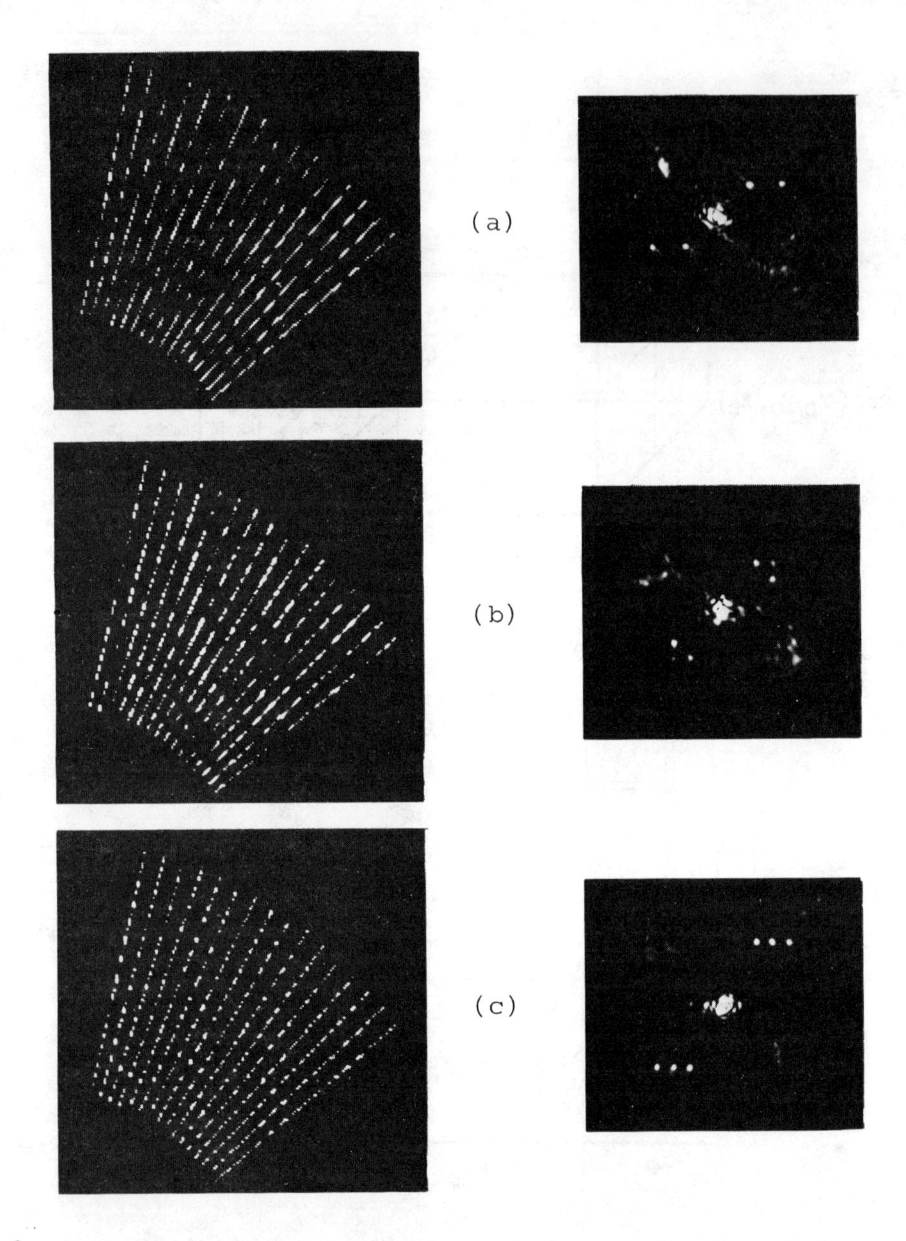

Fig. 3. Projection holograms and their optical reconstructions for the set of point scatterers in Fig. 2 at different R_z' planes. (a) Hologram and reconstructed image of scatterers at $R_z'=-z_o$ plane. (b) Hologram and image at $R_z'=0$ plane. (c) Hologram and image at $R_z'=z_o$ plane. $x_o=y_o=z_o=100$ cm.

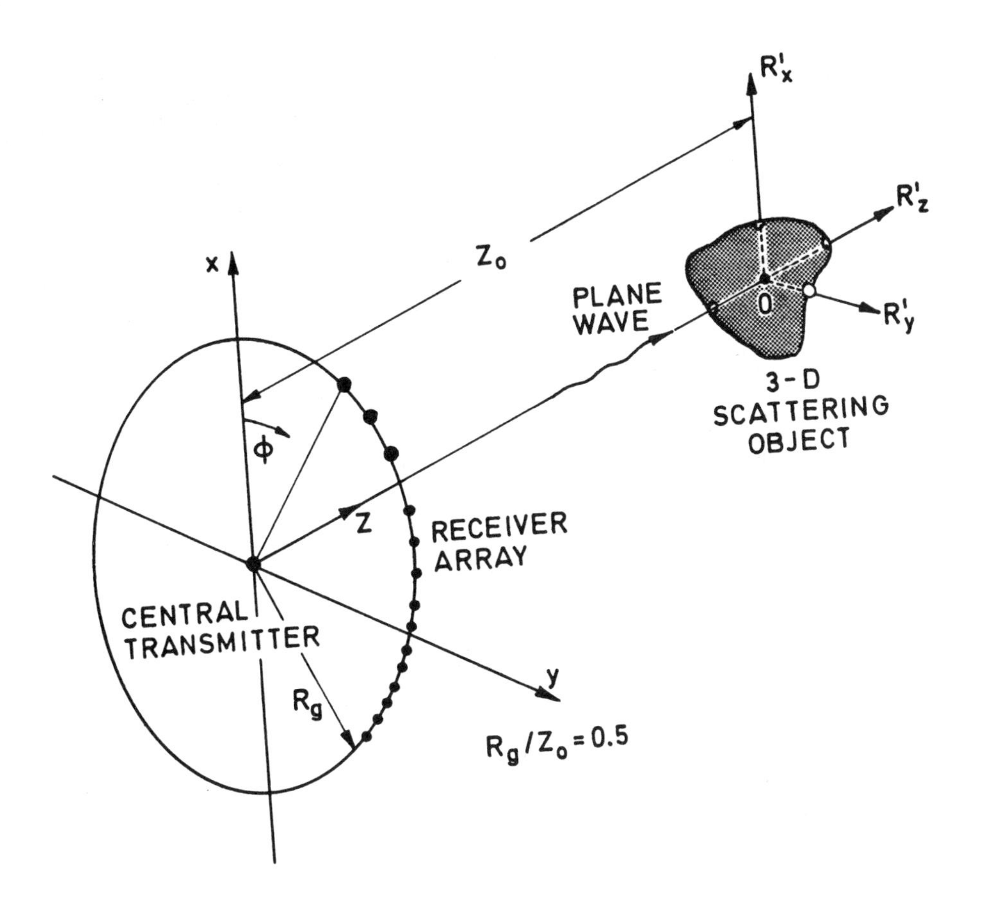

Fig. 4. Arrangement used in computer simulation of wavelength diversity imaging.

can be invoked again. Fourier domain projection theorems[23] that are dual to the spatial domain projection theorem[25,26] can be applied to the Fourier space data to produce a series of projection holograms from which 2-D images of meridional or parallel slices of the object can be retrieved on the optical bench[20]. This procedure does not involve any specific scaling of the size of the optical hologram transparency relative to the size of the original recording aperture by the ratio of the recording to the reconstruction wavelengths as in longwave holography there the scaling necessary for viewing a 3-D image free of longitudinal distortion ususally leads to an impractically minute equivalent hologram transparancy that cannot be readily viewed by an observer. The lateral and longitudinal resolutions in the retrieved image depend now on the dimensions of the volume in Fourier space accessed by wavelength diversity. This volume depends, in turn, on the wavelength range and on the recording geometry. Thus the longitudinal resolution does not deteriorate as rapidly with range as in conventional monochromatic imaging systems.

An example of computer simulations of frequency diversity holographic imaging of a 3-D object consisting of eight point scatterers distributed as shown in Fig. 2 is given in Fig. 3. Shown in Fig. 3 are three weighted Fourier domain projection holograms and the corresponding optically retrieved images for three equally spaced parallel slices of the object containing distinguishable 2-D distributions of scatterers. The simulated recording arrangement shown in Fig. 4 consisted of an array of 16 receivers equally distributed on an arc extending from $\phi = 40^{o}$ to $\phi = 77.5^{o}$ surrounding a central transmitter capable of providing plane wave illumination of the object. The results shown were obtained with microwave imaging in mind assuming a frequency sweep of (2-4)GHz. They clearly indicate a lateral and longitudinal resolution capability of the order of 25 cm. Wider sweep widths yield better resolution. For example a (1-18)GHz sweep would yield a 3-D resolution of the order of 1.5 cm.

DISCUSSION AND CONCLUSIONS

Seeking means by which the information content in a hologram can be increased, for example by wavelength diversity, we have arrived at a formulation of 3-D Lensless Fourier transform holography capable of furnishing 3-D image detail tomographically. This ability of producing 3-D images in slices from coherently detected wavefields enables us to regard the method also as coherent tomography. The Fourier space accessed in the above fashion by wavelength diversity can be viewed as a generalized 3-D hologram in which one dimension has been synthesized by wavelength diversity. Such a generalized hologram contains not only spatial amplitude and phase data as in conventional holography, but also spectral information. Consequently, it can yield better resolution than the classical Rayleigh limit of the available aperture operating at the shortest wavelength of the spectral window used. This super-resolving property is further enhanced through an inherent suppression of the effects of object resonances and speckle in the retrieved image; the latter property is a consequence of the fact that frequency diversity tends to make the impulse response of the system unipolar and to resemble that of a non-coherent imaging system that is free of speckle and coherent noise artifacts[15]. Further enhancement of information content and resolution can be achieved by polarization diversity where the $\bar{p}$ space can be multiply accessed for different nonredundant polarizations of the illumination and the receivers and the resulting polarization diversity images added either coherently or non-coherently in order to achieve a degree of noise averaging as discussed elsewhere[15].

The removal of several longstanding constraints on conventional longwave (microwave and acoustic) holography attained through the use of wavelength diversity as described here leads to a new class of imaging systems capable of converting spectral degrees of freedom into 3-D image detail, and thus of furnishing true super-resolution. Wavelength diversity is applicable to the imaging of two classes of objects: perfectly reflecting objects of the type encountered in radar and sonar, and weakly scattering objects of low or known dispersion of the type encountered in biology and medicine. The practical application of the concepts presented here to optical wavefields is presently under consideration. The availability of tunable dye lasers and electronic imaging devices suggest interesting possibilities of three dimensional wavelength diversity microscopy. Here one can conceive of an arrangement in which a minute semitransparent object with homogeneous or known dispersion is transilluminated by a collimated coherent light beam from a tunable dye laser which can also be made to provide a coherent reference point source in the immediate vicinity of the object. The resulting reference and the object scattered wavefields are intercepted by the photocathode of an electronic imaging device of known spectral response such as a vidicon. Because of the minute size of the object, the photocathode can easily be situated in the far field of the object. Thus nearly a lensless Fourier transform hologram recording arrangement results. The spatial frequency content in the resulting hologram is therefore expected to be sufficiently low to be resolved by a high resolution electronic imaging device. By recording and digitally storing the resulting detected hologram fringe pattern as a function of dye laser wavelength one can gain access to the 3-D Fourier space of the object, since $\bar{1}_{k_i}$ and $\bar{I}_R$ are precisely known for the recording geometry.

A similar recording arrangement can be envisioned in the active coherent imaging of a distant reflecting object (active telescopy) where the object can be made to furnish a reference point source situated on its surface, such as a wavelength independent stationary glint point or an intentionally placed retroreflector. Because in such an arrangement the reference and the object wavefields travel over the same path, atmospheric effects are expected to be minimized. The generation of an object derived reference geometry in longwave (microwave and acoustic) wavelength diversity imaging has been described elsewhere[20,27].

Finally it is worthwhile to note that since the scattering process is linear the multiaspect or multistatic frequency or wavelength response measurements referred to in this paper can be obtained also by measuring the multiaspect impulse response followed by a Fourier transformation of the individual impulse responses that have been measured[19]. This means that impulsive illumination can also be utilized. Because the impulse response of a linear system can be measured by using random noise excitation and by cross-correlating the output with the input[19], the possibility also emerges of using random noise (white light) illumination and cross-correlation detection rechniques as a means for accessing the 3-D Fourier space of the object.

ACKNOWLEDGEMENT

The research presented in this paper was sponsored in part by the U.S. Air Force Office of Scientific Research, Air Force Systems Command, USAF under grant No. AFOSR-77-3256A and in part by U.S. Army Research Office, Durham under grant No. DAAG-76-G-0230.

REFERENCES

1. Schelkunoff, S.A., Bell Syst. Tech. J., 22, 80, (1943).

2. Anderson, A.P. and J.C. Bennet, Proc. IEE (letters), 64, 376, (1976).

3. DiFrancia, G.T., J. Opt. Soc. Am., 45, 497, (1955).

4. DiFrancia, G.T., J. Opt. Soc. Am., 59, 799, (1969).

5. Nassenstein, H., Opt. Comm., 2, 231, (1970).

6. Wernecke, S.J. and L. D'Addario, IEEE Trans. on Computers, C-26, 351, (1977).

7. Lukosz, W., J. Opt. Soc. Am., 56, 1463, (1966).

8. Leith, E.N., Advances in Holography, 2, N. Farhat (Ed,), (M. Decker, New York. 1976).

9. Lukosz, W., J. Opt. Soc. Am., 57, 932, (1967).

10. Bojarski, N.N., Final Report, contract B000-19-73-C-0316, Naval Air Syst. Command, (1974).

11. Lewis, R.M., IEEE Trans. on Ant. and Prop., AP-17, 308, (1969).

12. Raz, S.R., IEEE Trans. on Ant. and Prop., AP-24, 66, (1976).

13. Meier, R.W., J. Opt. Soc. Am., 55, 987, (1965).

14. Stroke, G.W., Appl. Phys. Letters, 6, 272, (1965). Also in: An Introduction to Coherent Optics and Holography, 2nd Edition, (Academic Press, New York, 1969), 114.

15. Farhat, N.H. and C.K. Chan, in Optica Hoy Y Manãna, J. Bescos et. al. (eds.), (Sociedad Espanola De Optica, Madrid, 1978), 399.

16. Farhat, N.H., Ultrasonics Symposium Proceedings, IEEE Cat. No. 75 CHO 944-4SU, (1975).

17. Farhat, N.H., Proc. International Optical Computing Conference, IEEE Cat. No. 76-CH 1100-7C, (1976).

18. Farhat, N.H., Proc. IEEE (letters), 64, 379, (1976).

19. Farhat, N.H., J. Opt. Soc. Am., 67, 1015, (1977).

20. Farhat, N.H. and C.K. Chan, in Acoustical Imaging, A. Metherell (ed.), (Plenum, New York. 1980), 499.

21. Waters, W.M., Proc. IEEE (letters), 66, 609, (1978).

22. Chan C.K., Analytical and numerical studies of frequency swept imagery, Ph.D. Dissertation, Univ. of Pennsylvania, Philadelphia, (1978).

23. Stroke, G.W. and M. Halioua, Trans. Amer. Crystallographic Assoc., 12, 27, (1976).

24. Farhat, N.H., Univ. of Pennsylvania Report No. F1 Annual Report, AFOSR Grant No. 77-3256, July (1978).

25. Bracewell, R.N. and S.J. Wernecke, J. Opt. Soc. Am., 65, 1342, (1975).

26. Bracewell, R.N., Australian Journal of Physicas, 9, 198, (1956).

27. Farhat, N.H., C.K. Chan and T.H. Chu, Proc. AP-S/URSI, Symposium, Quebec, Canada (1980).

28. Stroke, G.W., in Applications of Holography, E.S. Barrakette, W.E. Kock, T. Ose, J. Tsuyiuchi and G.W. Stroke, (Eds.),(Plenum Press, New York, 1971), 35.

29. Ojeida-Castenada and S. Guel Sandoval, App. Optics, 18, 3350, (1979).

THE INVERSE SCATTERING PROBLEM

G. Ross and M.A. Fiddy

Physics Department, Queen Elizabeth College,
University of London, Campden Hill Road,
London W8 7AH, U.K.

M. Nieto-Vesperinas

Instituto de Optica, C.S.I.C., Serrano 121,
Madrid - 6, Spain. (Supported by the Juan March Foundation)

ABSTRACT

Scattered field are described by entire functions of exponential type, characterised everywhere by their distributions of zeros. The paper shows how the inverse scattering problem in one and two dimensions is reduced to establishing procedures for zero location. Several cases are distinguished and suitable methods are put forward for each of them.

INTRODUCTION

The inverse problem in structural determinations is the generic name given to a distinct sequence of tasks, whose ultimate aim is to obtain a representation of an object, in terms of all relevant intensive parameters, from observable data. When dealing with inverse scattering problems, the observable data are associated with the scattered field. The sequence of tasks involved consists

ISSN:0094-243X/81/650643-09$1.50

of

I) determining a spatial distribution, or spatial distributions, of the complex scattered field; this requires the representation of the field from discrete data. At optical frequencies, this implies that phase information is available or retrievable.

II) inverting the scattered field, according to the constraints of the problem, in order to obtain the object wave in one, two or three dimensions.

III) determining the object, i.e. the intensive parameters, from the object wave; this requires knowledge of the incident field.

Determining the object from the object wave, i.e. III, may always be made a trivial operation in one dimension by a judicious choice of polarisation and may be achieved to within a good approximation in those two dimensional situations when multiple scattering can be neglected. In three dimensions this task is again trivial if the object is a weak scatterer. For a strong scatterer, a complete solution has not yet been put forward.

The second task, II, is relatively simple for a one and two dimensional situation, at any rate when the resolution determined by the interval of measurement is acceptable. In three dimensions, obtaining the three dimensional object wave is difficult, and several methods to achieve this have been proposed. This problem is the content of a separate paper in this volume, discussing in detail the three dimensional problem.[1]

In this paper we shall restrict ourselves only to the discussion of I, and only for a one and two dimensional situation, the three dimensional case being investigated in reference (1).

THE COMPLEX SCATTERED FIELD

The first task, namely determining the spatial distribution of the complex scattered field, in one and two dimensions is considered. Two distinct situations can be distinguished:

A) the object or image wave is amenable to processing

B) only scattered data are available.

The approach adopted for this consists of making use of the analytic properties of the scattered field. We assume that the object wave is confined to a finite domain and that a Fourier relation may be assumed between the object wave and the scattered field; i.e. we consider fields only in the Fresnel and Fraunhofer region[2]. This implies that the scattered field is a function of class E , namely entire, of order I and finite type[3]. Moreover, such functions have a denumerable infinity of zeros, regularly distributed. The zero locations determine the function everywhere by a Hadamard product which may be written for one dimension

$$F(z) = cz^m \prod_{j=-\infty}^{\infty} \left(1 - {}^{z}/_{z_j}\right) \qquad (1)$$

or, in two dimensions, locally,

$$F(z_1, z_2) = cz_1^m z_2^n \prod_{j=-\infty}^{\infty} \left(1 - {}^{z_1}/_{\phi_j(z_2)}\right) \qquad (2)$$

where z_j are the isolated complex zeros corresponding to a one dimensional situation, and $\phi_j(z_2)$ defines the complex zero lines in two dimensions. c is a constant and the integers m, n are the orders of possible zeros at the origin. The properties of zero distributions have been described elsewhere[2,4].

The Hadamard product shows how all information in the Fourier space is encoded by zeros. Given the Fourier relationship between the object and Fresnel/Fraunhofer spaces, all information in the object space is also encoded by the zeros. The specific way in which this is done is outlined in reference (5).

One can thus reduce the determination of the spatial distribution of the complex field to the determination of the location of all zeros of the function. There is usually considerable a priori information available about their overall distribution[2].

ZERO LOCATION

A) If the object (or image) wave f(t) can be manipulated, but not measured directly (when the problem becomes trivial), it is convenient to scan the two dimensional function, thus reducing this case to a sequence of one dimensional problems. For one dimension, the problem consists of continuing the scattered field F(x) defined on the real x axis by

$$F(x) = \int_{-a}^{a} f(t)\exp(-i\langle k\rangle tx)\,dt \qquad (3)$$

into the part of the complex plane in which the function's zeros lie. In other words, it is necessary to determine $|F(x+iy)|^2$ where

$$F(x+iy) = \int_{-a}^{a} f(t)\exp(\langle k\rangle yt)\exp(-i\langle k\rangle tx)\,dt \qquad (4)$$

and, for a succession of values of y, find the zeros of F(z). Since f(t) is amenable to processing, the operation indicated in (4), namely multiplying f(t) by an exponential function followed by Fourier transformation can be achieved using optical filters.

One can display a strip of the complex plane - and hence locate the zeros - using a one-dimensional multichannel spectrum analyser[5]. If this procedure is possible, it represents the simplest method for determining directly complex scattered fields, i.e. from geometrical measurements. The disadvantage of sampling functions is thus removed.

B) If only scattered field data are available, several situations may occur.

i) When the field can be measured (which is not possible at optical frequencies), strictly speaking there is no need to locate the zeros. Of course, one can locate them from the sampled field if necessary, for example in information transmission. One way of achieving this is by solving a set of non-linear equations of the form

$$A_p = \prod_{\substack{q \\ q \neq p}} \frac{q}{q-p} \prod_q \left(1 - \frac{\pi p}{\langle k \rangle a} \frac{1}{x_q + i y_q}\right) \cos p\pi \qquad (5)$$

which expresses the complex amplitude A_p at each sample point in terms of zero coordinates. If the field can be oversampled, the advantage of locating the zeros consists of the possibility of achieving superresolution. The zero co-ordinates can be calculated by solving a corresponding larger system of equations generated from the Hadamard product. Locating zeros in this way requires the assumption that there are no more than N complex zeros, when only N values of A_p are available. All zeros beyond this are assumed to be real and equidistant thus constructing a finite interval in t space.

This approach can be easily extended to a two-dimensional situation.

ii) When the field cannot be measured and hence only an intensity is available, the intensity, $I(z)$ has the zeros of the field $F(z)$ and those of its complex conjugate $F^*(Z^*)$, For N complex zeros in $F(z)$ there are 2^N fields compatible with $I(z)$[6]. The zeros of $I(z)$ can be located by essentially the same procedure of analytic continuation as described in section A. More exactly, one must apply the exponential function to the Fourier transform of $I(x)$. This could be achieved optically or on the computer.

The problem remains of determining which of zeros of $I(z)$ belong to $F(z)$.

a) On the face of it, holography avoids this problem by shifting all the zeros of $F(z)$ into one half plane, e.g. the lower half plane. This is realized by the application of a reference wave which satisfies Rouché's theorem[7]; the zero ambiguity in the intensity has been removed and the object is uniquely encoded by $I(x)$. However, the certainty that Rouché's theorem has been satisfied depends upon the knowledge that the reference wave is adequate to displace all zeros into the lower half plane. This is difficult to establish without some a priori knowledge of the locations of the zeros of $F(z)$.

b) Another approach which removes this difficulty relies on the use of optical phase conjugation. Having located the zeros of $I(z)$, knowledge of the zeros of $|\mathrm{Re}F(z)|$, observed on the real axis enables one to distinguish between the zeros of $F(z)$ and those of $F^*(z^*)$. The zeros in one half plane of $I(z)$ which occur between the zeros of $|\mathrm{Re}F(x)|$ belong to one half plane of $F(z)$. This approach, which requires only geometrical measurements, is discussed in detail in reference (8).

c) Although, in principle, one can always manipulate the field, in practise it may not be feasable or convenient. From one distribution of intensity and without a priori information, the zeros cannot be located. The use of a priori information usually requires the calculation of the 2^N possible functions[7,9]. A method has been developed which reduces the number of operations to 2N but which is valid only for weak scatterers[10]. Having located the zeros of I(z), each zero in turn is moved to the two positions compatible with I(z), from an initial equidistant array of real zeros. At each step, a Fresnel intensity is calculated and compared to a measured one. The better fit determines the correct half-plane for each zero.

The methods outlined in a) can be extended to two dimensional situations. However in case b) the full two dimensional problem present entails zero location in a four dimensional space. The difficulties associated with this are that the zeros are no longer isolated points but are aperture dependent lines.
Thus for weak scatterers, the task consist of locating straight lines determined by two complex parameters for a rectangular aperture, or circles determined by three complex parameters for a circular aperture. Therefore, specifying these zeros should not be too difficult. For stronger scatterers, when we expect these zero contours to become distorted by coalescing, characterising them fully may become a problem. The method outlined for the one dimensional case will still work for weak scatterers in two dimensions if these zeros can be located.

CONCLUSIONS

The paper presents an approach to the solution of the inverse scattering problem based on the analytic properties of scattered fields, more specifically on the location of the zeros of functions ϵE. There are several advantages to this description. The first is the characterisation of intensities and complex fields by specifying zero positions rather than non-zero complex magnitudes at sample points. The second concerns the formulation of the ambiguity inherent when only intensity data are available.
The third is the stability offered by a zero description when attempting analytic continuation of the data for superresolution.[11] Furthermore, if the object wave can be addressed, an optical procedure is available for solving the inverse scattering problem entirely from geometrical measurements.

REFERENCES

1. M. Nieto-Vesperinas, M. Gea, G. Ross and M.A. Fiddy. "Solutions to the three dimensional inverse scattering problem." Proc. Optics in Four Dimensions.
2. G. Ross, M.A. Fiddy and M. Nieto-Vesperinas ch. 2 in 'Inverse Scattering Problems in Optics'. 20 of Topics in Current Physics, Ed. H.P. Baltes, Springer-Verlag, Berlin (1980)
3. R.E.A.C. Paley and N. Wiener, 'Fourier transforms in the Complex Domain' Amer. Math. Soc. Colloq. Pub XIX Providence, Rhode Island (1934).
4. B. Ja Levin 'Distributions of zeros of entire functions' Translations of Mathematical Monographs 5 Amer. Math. Soc. Pub., Providence, Rhode Island (1964).
5. M.A. Fiddy, G. Ross and M. Nieto-Vesperinas 'Optical processing for the encoding of information in inverse problems' Proc. Optics in Four Dimensions
6. A. Walther 'The question of phase retrieval in optics' Opt. Acta 10 41 (1963)
7. R.E. Burge, M.A. Fiddy, A.H. Greenaway and G. Ross, 'The phase problem', Proc.Roy.Soc.Lond., A350 191 (1976)
8. R.E. Burge, M.A. Fiddy, T.J. Hall and G. Ross, 'Optical phase conjugation for the study of the analytic properties of optical fields'. Proc. Optics in Four Dimensions.
9. A.H. Greenaway, 'Proposal for phase recovery from a single intensity distribution'. Opt. Lett. I 10 (1977)
10. G. Ross, M.A. Fiddy and H. Moezzi, 'The solution to the inverse scattering prohlem, based on zero location from two measurements. Opt.Acta (in press)
11. J.B. Abbiss, C.De Mol, M.A. Fiddy and G. Ross 'The superresolution problem in Doppler difference laser anemometry' Proc.4th Int.Conf. on Photon Correlation Techniques in Fluid Mechanics. Stanford University, Aug. 25th - 27th 1980.

FIELD CONSTRAINTS ON DISCONTINUOUS SOLUTIONS OF THE MAXWELL EQUATIONS

A. George Lieberman
National Bureau of Standards, Washington, D.C. 20234

ABSTRACT

Relations governing changes in the field vectors of a discontinuous electromagnetic field are formulated in this paper within the framework of special relativity theory. The treatment emphasizes the physical aspects of the problem and for this reason the medium supporting the propagation of the discontinuity is conveniently assumed to be isotropic but otherwise to have arbitary properties. Four field discontinuity conditions and a relation describing electrical charge conservation at the discontinuity are presented.

INTRODUCTION

Present achievements in optical pulse formation and detection have recently stimulated renewed interest in the propagation and transient scattering of discontinuous electromagnetic fields. A review of the literature, dating back to Heaviside's original investigations[1] concerning the transmission of abrupt electrical disturbances, discloses that the field relations at a radiated electromagnetic discontinuity were often either incompletely formulated or else formulated for restricted classes of propagation media. The theory was typically limited to media characterized as isotropic, linear and non-dispersive[4-8], and in the earliest works also homogeneous[1-2]. These restrictions result from the scalar functional forms and dependences assumed for the dielectric permittivity, the magnetic permeability, and the conductivity. In two works[4,10], generality was maintained by never introducing into the theory any constitutive relations between the field vectors. The possible existence of free charges and induced currents in the medium was not addressed in the earlier investigations[1-4] and not always consistently[8] or explicitly[5,6,9] handled in the most recent works.

Two basic approaches have been used to derive the field conditions at a radiated discontinuity. The Maxwell equations were either: directly applied in differential form to a Heaviside step distribution representing the electromagnetic discontinuity surface[6,8]; or else reformulated as integral equations, usually in space-time[3,4,5,7,9]. An entirely different treatment based upon special relativity theory is presented in this paper. The approach is novel, emphasizes the physics and is generally applicable to isotropic matter containing free charges and induced currents.

In order to apply special relativity theory the hypothesis is made that, in the rest frame of the propagation, the field vectors at a radiated electromagnetic discontinuity obey the same boundary conditions as those which are customarily associated with the field vectors at a stationary material interface. By means of the Lorentz transformation these boundary conditions are transformed from the rest frame of the radiated discontinuity into the inertial frame of the fixed observer. The transformed boundary conditions, hereafter referred to as field discontinuity conditions or jump relations, are interpreted as constraints between changes in the field vectors at the discontinuity.

In deriving the field jump relations, the Lorentz transformation is applied locally to each element of the surface comprising the electromagnetic discontinuity. It is assumed that every element of the discontinuity surface moves in the direction of its normal. Since no other assumptions are imposed the derived results appear to be applicable to any isotropic medium of otherwise arbitary properties. The present treatment recovers an apparently overlooked (surface charge) convection term in one of the four field jump relations. The applicability of the field jump relations is further broadened by introducing a generalized charge conservation equation for use at the discontinuity surface.

LORENTZ TRANSFORMATION OF THE FIELD AND SOURCES

As a consequence of special relativity theory, observables describing phenomena in a moving inertial frame of reference may be expressed through a Lorentz transformation in terms of observables belonging to an arbitarily-fixed inertial frame of reference. As customary, observables in the moving frame are denoted by primed symbols to distinguish them from the same observables viewed from the 'fixed' frame. Two widely known consequences of special relativity theory are the Lorentz contraction of lengths paralleling the direction of motion and the dilation of time. These phenomena are governed by the expressions: $\ell = \ell'/\gamma$ and $\delta t = \gamma \delta t'$, where ℓ and δt refer to the length and the time interval, respectively; $\gamma = [1 - (v/c)^2]^{-1/2}$ and $\vec{v}$ is the relative velocity of the moving frame. Also of direct bearing are the field, current and charge transformation equations. In general, the projections of a vector along directions parallel and perpendicular to the relative motion transform differently. It will therefore be convenient to resolve the fields and currents into components longitudinal and transverse to $\vec{v}$, and to symbolically denote these respective projections by the subscripts $//$ and $\perp$. The transformation equations for the magnetic induction $\vec{B}$, the electric displacement $\vec{D}$, and the electric and magnetic field intensities, $\vec{E}$ and $\vec{H}$, follow[10].

$$\vec{B}'_{//} = \vec{B}_{//}, \quad (1a), \qquad \vec{B}'_{\perp} = \gamma(\vec{B} - \frac{1}{c^2}\vec{v} \times \vec{E})_{\perp}, \quad (1b)$$

$$\vec{D}'_{//} = \vec{D}_{//}, \quad (2a), \qquad \vec{D}'_{\perp} = \gamma(\vec{D} + \frac{1}{c^2}\vec{v} \times \vec{H})_{\perp}, \quad (2b)$$

$$\vec{E}'_{//} = \vec{E}_{//}, \quad (3a), \qquad \vec{E}'_{\perp} = \gamma(\vec{E} + \vec{v} \times \vec{B})_{\perp}, \quad (3b)$$

$$\vec{H}'_{//} = \vec{H}_{//}, \quad (4a), \qquad \vec{H}'_{\perp} = \gamma(\vec{H} - \vec{v} \times \vec{D})_{\perp}. \quad (4b)$$

The corresponding transformation equations for the current density $\vec{J}$ in terms of the bulk charge density are expressed by the relations:

$$\vec{J}'_{\perp} = \vec{J}_{\perp}, \quad (5a), \qquad \vec{J}'_{/\!/} = \gamma(\vec{J}_{/\!/} - \vec{v}\rho) \; . \quad (5b)$$

The foregoing relations will be applied directly to the boundary conditions at a stationary discontinuity of the electromagnetic field.

BOUNDARY CONDITIONS AT A STATIONARY SURFACE

At a surface or material interface rapid changes in physical parameters occur and manifest themselves as macroscopic discontinuities of the medium and the electromagnetic field. To treat such cases, formal solutions of the Maxwell equations are obtained for each side of the surface, and boundary conditions are imposed to match the electromagnetic fields across the surface. At each element of the surface the boundary conditions are given by [10]:

$$\hat{n} \cdot (\vec{B}'_2 - \vec{B}'_1) = 0 \quad , \qquad (6)$$

$$\hat{n} \cdot (\vec{D}'_2 - \vec{D}'_1) = \rho'_s \; , \qquad (7)$$

$$\hat{n} \times (\vec{E}'_2 - \vec{E}'_1) = 0 \quad , \qquad (8)$$

$$\hat{n} \times (\vec{H}'_2 - \vec{H}'_1) = \vec{J}'_s \; . \qquad (9)$$

The symbol $\hat{n}$ denotes a unit vector that is directed normal to the surface element in question, from medium 1 into medium 2. Numerical subscripts on the field vectors refer to the medium. The quantities ρ'_s and $\vec{J}'_s$ represent the local surface charge and surface current densities, respectively.

The currents and charges at a boundary surface are related by a current continuity equation expressing charge conservation. To account for the singular nature of the current and charges on the surface the following constraint must be invoked:

$$\nabla'_s \cdot \vec{J}'_s + \hat{n} \cdot (\vec{J}'_2 - \vec{J}'_1) = -\frac{\partial \rho'_s}{\partial t'} \; . \qquad (10)$$

The first term represents the two-dimensional divergence of the current in the surface; the second term corresponds to the net current flowing out of the surface in the direction of the normal. Boundary condition (10) relates the local loss of surface charge with the total current leaving the surface element.

It is important to realize that each boundary condition holds at every point in space independent of the actual presence of a material discontinuity. Boundary conditions (6) through (10) are just as valid for a uniform medium partitioned by a surface charge density ρ'_s and a surface current $\vec{J}'_s$ as for a material interface. If both ρ'_s and $\vec{J}'_s$ are zero, then equations (6) through (10) are statements of electromagnetic field and current continuity. In the following paragraphs, according to the hypothesis made earlier in the text, the above boundary conditions are now regarded as applying point-by-point in the rest frame of some radiated field discontinuity not necessarily caused by an abrupt spatial variation of the medium.

JUMP RELATIONS AT A MOVING SURFACE

Field relations across an arbitrary moving surface are obtained by conceptually dividing the surface into infinitesimal elements and then applying the Lorentz tranformations at some representative element. For translations of this element at velocity $\vec{v} = v\hat{n}$ in the direction of its normal, application of field transformation (1a) to the left side of boundary condition (6) yields:

$$\hat{n} \cdot (\vec{B}_2' - \vec{B}_1') = B_{2/\!/}' - B_{1/\!/}' = B_{2/\!/} - B_{1/\!/} = \hat{n} \cdot (\vec{B}_2 - \vec{B}_1) \ .$$

If the field jump across the surface is now denoted by $\Delta B = \vec{B}_2 - \vec{B}_1$, then

$$\hat{n} \cdot \Delta\vec{B} = 0 \quad . \tag{11}$$

It follows, for each and every point on the moving surface, that the fixed observer sees no discontinuity in the normal component of the magnetic induction.

Transformation of boundary condition (7) is implemented in a similar fashion under field transformation (2a) with due consideration for the surface charge density ρ_s'. This term is invariant ($\rho_s' = \rho_s$) under transformation because electric charge is a scalar invariant of the motion and because surface dimensions transverse to a motion do not suffer Lorentz contractions. As a consequence:

$$\hat{n} \cdot \Delta\vec{D} = \rho_s \quad . \tag{12}$$

Since the direction of surface motion at each point is assumed to coincide with the direction of the surface normal, field jump relations (11) and (12) take the same form as boundary conditions (6) and (7).

Turning to boundary condition (8) it should be recognized that the vector product of $\hat{n}$ or $\vec{v}$ with a field vector selects the transverse field component. The appropriate field transformation is equation (3b) and so:

$$\hat{n} \times \Delta\vec{E}' = \hat{n} \times \Delta\vec{E}_\perp' = \gamma\,[\hat{n} \times \Delta\vec{E}_\perp + \hat{n} \times (\vec{v} \times \Delta\vec{B})].$$

Expansion of the triple vector product, using relation (11) to eliminate $\hat{n} \cdot \Delta\vec{B}$, leads to the third field jump relation:

$$\hat{n} \times \Delta\vec{E} - v\,\Delta\vec{B} = 0 \quad . \tag{13}$$

Repeating this procedure for the left side of boundary condition (9), now using transformation (4b) and jump relation (12) for $\hat{n} \cdot \vec{D}$ gives:

$$\hat{n} \times \Delta\vec{H}' = \gamma[\hat{n} \times \Delta\vec{H}_{\perp} - \rho_s\vec{v} + v\,\Delta\vec{D}] \quad .$$

The surface current density appearing on the right side of equation (9) represents the electric charge flowing across an elemental length located in the surface per unit time interval. In transforming between reference frames only the time interval is not invariant and to account for time dilation it is necessary that $\vec{J}_s' = \gamma\vec{J}_s$. Reconstructing the transformed equation produces the remaining field jump relation:

$$\hat{n} \times \Delta\vec{H} + v\,\Delta\vec{D} = \vec{J}_s + \rho_s\,\vec{v} \quad . \qquad (14)$$

Altogether, relations (11) through (14) restrict the discontinuities which may arise in a physical electromagnetic field. For convenient reference the four field jump relations are repeated here:

$$\hat{n} \cdot \Delta\vec{B} = 0 \qquad (11)$$

$$\hat{n} \cdot \Delta\vec{D} = \rho_s \qquad (12)$$

$$\hat{n} \times \Delta\vec{E} - v\,\Delta\vec{B} = 0 \qquad (13)$$

$$\hat{n} \times \Delta\vec{H} + v\,\Delta\vec{D} = \vec{J}_s + \rho_s\vec{v} \quad . \qquad (14)$$

Jump relations (11) and (12) are not independent of jump relations (13) and (14). This is true because the four boundary conditions on the fields are derived from the four Maxwell equations, and the divergence of either of the two curl equations produces one or the other of the two divergence equations. In the present instance the scalar product $\hat{n}$ with (13) produces (11), while the scalar product of $\hat{n}$ with (14) produces (12), remembering that $\hat{n}$ is parallel to $\vec{v}$ and perpendicular to $\vec{J}_s$. The presence of the surface charge density term in (14) is essential for consistency with (12). It appears that this term has been overlooked in at least one reference publication[8] and not explicitly addressed in the other references[5,6,9].

The field jump relations are most interesting when the surface currents and charges are unspecified but must self-consistently accommodate the field discontinuities at each point on the electromagnetic surface while conserving charge. In the rest frame of a surface element the constraint of charge conservation is administered through boundary condition (10). In the inertial frame of the fixed observer the surface element moves with velocity $\vec{v}$ in the direction of its normal. For such motion the surface element does not appear to contract and the two dimensional differential operator $\nabla_s' = \nabla_s$ is an invariant. Making use of this and (5b) together with the transformation laws for t', $\vec{J}_s'$, and ρ_s' in boundary condition (10) gives for the moving surface:

$$\nabla_s \cdot \vec{J}_s + (\hat{n} \cdot \Delta\vec{J} - v\Delta\rho) = -\frac{\delta\rho_s}{\delta t}, \qquad (15)$$

where the term on the right represents the variation of the surface charge density with time at a position fixed on the surface.

CONCLUSION

Previous workers have limited their analysis to isotropic media possessing linear and non-dispersive properties. The treatment employed in this paper to derive the field jump relations is independent of any description or modeling of the medium and, additionally, allows for the presence of free charges and the flow of currents in the medium. The only assumption made in deriving the field jump conditions concerns the colinearity at each point on the discontinuity surface of the normal vector and the propagation velocity. This is a most useful case and implicit in earlier publications. The extension of the present approach to anisotropic materials is, nevertheless, a straightforward, albeit tedious exercise. The approach used in this paper emphasizes the physical aspects of the problem and, without recourse to any mathematical artifice has resulted in the recovery of an overlooked (surface charge convection) term in one of the field jump relations. A surface charge conservation equation has also been introduced which, together with the field jump relations, describes the structure and transport of a most general electromagnetic field discontinuity.

REFERENCES

1. O. Heaviside, Electrical Papers, Vol. II., (Macmillan and Co., London and New York, 1892), p. 405 et seq.
2. A. E. H. Love, Proc. Lond. Math. Soc. 1, 56 (1904).
3. H. Bateman, Partial Differential Equations of Mathematical Physics (Dover Publications, N.Y., 1944), §2.71, pp. 192-196.
4. R. K. Luneburg, Mathematical Theory of Optics (mimeographed lecture notes, Brown University 1944, printed version published by University of California Press, Berkely and Los Angeles, 1964), §5-7, pp. 18-25.
5. M. Kline, Comm. Pure and Appl. Math. 4, 239 (1951).
6. H. Bremmer, Comm. Pure and Appl. Math. 4, 419 (1951).
7. A. Rubinowicz, Acta Phys. Polonica 14, 209 (1955).
8. M. Born and E. Wolf, Principles of Optics, Fifth Ed., (Pergamon Press, 1975), Appendix VI, pp. 763-766.
9. M. Kline and I. W. Kay, Electromagnetic Theory and Geometrical Optics (Interscience Publishers, John Wiley and Sons, New York, 1965), §3, pp. 37-51.
10. J. A. Stratton, Electromagnetic Theory (McGraw-Hill, New York and London, 1941), §1.23-1.24, pp. 74-82.

OPTICAL PROCESSING FOR THE ENCODING OF INFORMATION IN INVERSE PROBLEMS

M.A. Fiddy, G. Ross and J. Wood
Physics Department, Queen Elizabeth College,
University of London, Campden Hill Road,
London W8 7AH, U.K.

M. Nieto-Vesperinas
Instituto de Optica, C.S.I.C., Serrano 121,
Madrid - 6 Spain
(Supported by the Juan March Foundation)

ABSTRACT

The significance of a description for scattered fields, based upon entire functions is discussed, in particular the way in which object wave information is encoded by zero positions. An optical one dimensional multichannel Fourier transformer is put forward to enable one to visualise the complex plane and thus locate the zeros of the field.

INTRODUCTION

Many inverse problems in optics and related areas are based upon a Fourier relationship between two spaces[1]. Functions expressed by a finite Fourier transform may be analytically continued into an associated complex space; they represent functions of class E,

ISSN:0094-243X/81/650658-10$1.50

characterised everywhere by the location of their complex zeros. If the distribution of zeros could be displayed, only the geometrical measurement of their position is required in order to reconstruct the function, no measurement of amplitude being necessary. An optical procedure for achieving this is put forward in this paper.

These zeros are the key to solving inverse problems and, at the same time, offer an optimum modality for encoding information. Since measurement of zero coordinates enables one to determine a complex field, this is equivalent to associating a phase with a measured modulus.[2] The information thus obtained is important in inverse optical problems such as structural determinations, image analysis, coherence theory as well as in such tasks as the encoding and transmission of information.

ENCODING OF INFORMATION BY ZEROS

Given a field propagating between two spaces related by a Fourier transform, e.g. object and Fraunhofer, and a function $f(t)$ of finite support in one of them, we can write

$$F(\vec{x}) = \int_V f(\vec{t})\exp(-i\langle k\rangle\vec{t}.\vec{x})\,d\vec{t} \qquad (1)$$

where $\langle k\rangle$ is the average wavenumber and V a compact support. $F(\vec{x})$ thus defined is an entire function of exponential type, i.e. a function of class E[3]. Such functions may also be expressed locally in the form

$$F(\vec{x}) = \prod_{j=-\infty}^{\infty}\left(1-\frac{x_1}{\phi_j(x_2,x_3\ldots)}\right) \qquad (2)$$

where $\vec{x} = (x_1,x_2,x_3,\ldots\ldots)$, i.e. $F(\vec{x})$ and hence $f(t)$ is described everywhere by its zeros.

We restrict our treatment to one dimension, the two dimensional case will be discussed later.

In one dimension the zeros of $F(x)$ are a denumerable infinity of isolated points. They are "regularly" distributed and tend to be equidistant, asymptotically, on lines parallel to the real axis. The zero density is determined by the object support.[4] The simplest possible distribution of zeros consists of an equidistant lattice of points defined by zeros at $z_j = \frac{j\pi}{\langle k \rangle a}$ where $f(t) \equiv 0$, $|t| > a$, $j = 0, \pm 1, \pm 2 \ldots.$

This distribution of zeros is non-physical, generating the function $\sin\langle k\rangle az$, which is not square integrable and corresponds to two imaginary Dirac delta functions marking the edges of the interval $(-a,a)$, Square integrability is achieved by the removal of at least one zero. It is convenient to remove the zero at the origin, thus creating $\frac{\sin\langle k\rangle az}{z}$ which defines a d.c. level in t space.

It is interesting to establish how each zero encodes the information in the t space. The removal of the nth zero, results in

$$f(t) = P_o\,(t/a)\left[1-\cos n\pi \exp\left(\frac{in\pi t}{a}\right)\right]. \qquad (3)$$

Moving the nth zero from $\frac{n\pi}{\langle k\rangle a}$ to $x_n + iy_n$ leads to

$$f(t) = Po(t/a)\left[1 - \left[1 - \frac{\frac{n\pi}{\langle k\rangle a}}{x_n + iy_n}\right] \cos n\pi \exp\left(\frac{in\pi t}{a}\right)\right]. \qquad (4)$$

The general expressions, corresponding to any number, N , of zeros are, for zero removal,

$$f(t) = P_o(t/a)\left[1 - \sum_p^N \left(\prod_{\substack{q \\ q\neq p}} \frac{q}{q-p}\right) \cos p\pi \exp\frac{ip\pi t}{a}\right] \qquad (5)$$

and for zero displacement

$$f(t) = P_0(t/a)\left[1 - \sum_p^N \prod_{\substack{q \\ q\neq p}} \frac{q}{q-p} \prod_q \left(1 - \frac{p\pi}{\langle k\rangle a}\,\frac{1}{x_q + iy_q}\right) \cos\pi p \,\exp\frac{it\pi p}{a}\right]. \quad (6)$$

It can be easily seen that moving or removing pairs of zeros, symmetric about the y axis from the $\frac{\sin\langle k\rangle az}{z}$ positions, encodes real harmonics in t space.

The field in the Fraunhofer space associated with the general expression in eq. (6) may be calculated

$$F(x) = \prod_{q=-\infty}^{\infty} (1 - x/z_q)$$

$$= j_0(\langle k\rangle ax)\left[\delta(x) - \sum_p^N \left(\prod_{\substack{q \\ q\neq p}} \frac{q}{q-p}\right) \prod_q \left(1 - \frac{p\pi}{\langle k\rangle a}\,\frac{1}{x_q + iy_q}\right) \cos p\pi\, \delta\!\left(x - \frac{p\pi}{a}\right)\right]$$

$$= \sum_p^N A_p\, j_0\!\left(\langle k\rangle ax - \frac{p\pi}{a}\right) \quad (7)$$

which is the more conventional Shannon description with

$$A_p = \prod_{\substack{q \\ q\neq p}} \frac{q}{q-p} \prod_q \left(1 - \frac{p\pi}{\langle k\rangle a}\,\frac{1}{x_q + iy_q}\right) \cos p\pi \,. \quad (8)$$

DIRECT MEASUREMENTS IN THE COMPLEX PLANE

Given the usefulness of the zero description for propagating fields, the obvious task is to be able to locate them in a simple manner. The procedure is very simple conceptually. Starting from equation (1) written for one dimension

$$F(x) = \int_{-a}^{a} f(t) \exp(-i\langle k\rangle tx)\,dt \qquad (9)$$

it is easy to see that multiplication of f(t) with the function $\exp(\langle k\rangle ty)$ leads to

$$F(x+iy) = F(z) = \int_{-a}^{a} f(t)\exp(-i\langle k\rangle t(x+iy))\,dt \qquad (10)$$

Thus, for a succession of y values, the behaviour of F(z) in the complex plane can be explored. If F(x+iy) has a zero at x_n+iy_n then this will be self evident on the line $y=y_n$.

For an accessible object, or an object with an accessible image, the operations described by eq. (10) may be performed optically. By making a set of exponential filters, each filter encoding a different value of y, one can insert them in succession in the object space; upon Fourier transformation the corresponding strip of the complex plane is displayed. The procedure can be simplified, in the one dimensional case, by repeating f(t) in a direction perpendicular to the t axis and performing one dimensional multichannel Fourier transforms[5,6]. If a two dimensional exponential filter is made, having t and y as axes, a range of y values in the complex plane can be generated in parallel. The observable, $|F(z)|^2$, will obviously display the zeros of F(z) whose co-ordinates could be measured as distances. In the region of the z space thus constructed, the complex F(z) is characterised. The method is illustrated in figure 1.

Preliminary results using a crude version of the optical processor described have been obtained. The major difficulty encountered is the manufacture of an exponential filter having an adequate dynamic range. We have been studying, optically, the behaviour of zero distributions using computer plotted objects, whose zero

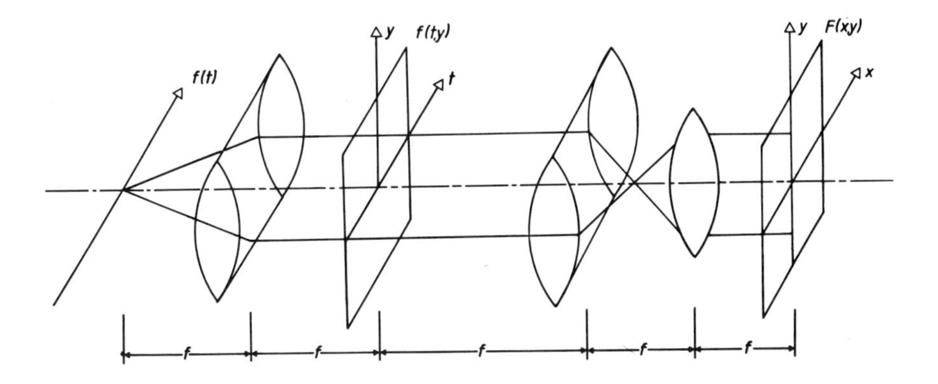

Figure 1

A one dimensional multichannel spectrum analyser (reference[5])

locations are known analytically. An example is given in figure 2 which presents a set of two objects and the associated $|F(x)|^2$. The object wave used is

$$f(t) = 1 \qquad -2a \leqslant t \leqslant 0$$
$$= r \qquad 0 < t \leqslant 2a \qquad (11)$$
$$= 0 \qquad |t| > 2a \quad .$$

The alternate zeros of $F(z)$ are off the real axis and occur at the positions

$$y_{2n+1} = \frac{\ln r}{2a} \qquad n = 0,1,2,3 \; ....$$

The zeros all have $x_n = \frac{n\pi}{2a}$. It can be seen that, as r decreases, alternate zeros are moved further from the real axis, into the lower half of the complex plane, resulting in larger minima between the real zeros.

When a t space is not available, the intensity on the real x axis, $|F(x)|^2$, is normally the observable. This leads to an ambiguity in possible $f(t)$ as outlined in reference[7]. A potential method for resolving this ambiguity is discussed in reference 8.

For a two dimensional situation, there are two possible approaches. The first is when the object wave $f(t_1,t_2)$ may be scanned with a slit and thus the problem is reduced to the one dimensional task discussed above. The second is when a two dimensional exponential filter is applied which can only give the scattered field in one plane in the four dimensional complex space. In order to locate the zeros in this space, a set of such filters could be required. However, for weak scatterers, since the zero lines are aperture dependent only a few points may need locating on these lines in order to characterise them completely.

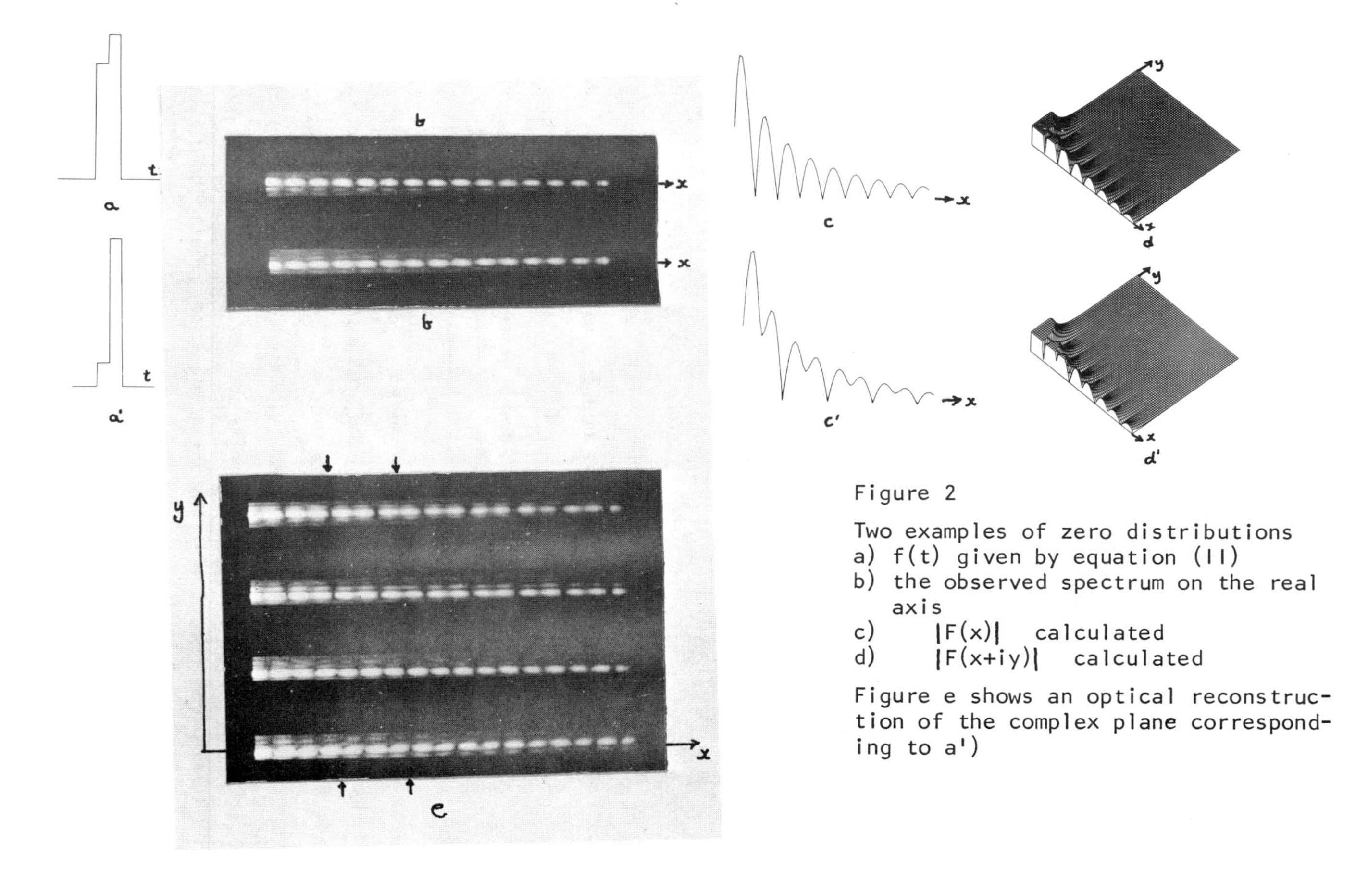

Figure 2

Two examples of zero distributions
a) f(t) given by equation (II)
b) the observed spectrum on the real axis
c) $|F(x)|$ calculated
d) $|F(x+iy)|$ calculated

Figure e shows an optical reconstruction of the complex plane corresponding to a')

CONCLUSION

The significance of zeros for encoding information about objects and propagating fields has been outlined. This description is the most economical one can envisage since only geometrical data are required. Thus it is important to be able to determine the positions of these zeros in the associated complex space. A simple optical method was described enabling one to generate the function in sections of the complex plane or space without any measurement of field or intensity.

Having located the zeros, the way in which each zero encodes a linear phase factor in t space was presented. This description was compared with the Shannon sampling procedure and shown to be equivalent.

REFERENCES

1. G. Ross, M.A. Fiddy and M. Nieto-Vesperinas ch 2 of 'Inverse Scattering Problems in Optics' 20 of Topics in Current Physics, Ed. H.P. Baltes, Springer-Verlag (1980).
2. H.A. Ferwerda, ch.2 of 'Inverse Source Problems in Optics' 9 of Topics in Current Physics Ed.H.P. Baltes, Springer-Verlag (1979)
3. R.E.A.C. Paley and N. Wiener, 'Fourier transforms in the Complex Domain' Amer. Math. Soc. Colloq. Sub. XIX Providence, Rhode Island. (1934)
4. B. Ja Levin, 'Distribution of zeros of Entire functions' Translations of Mathematical Monographs 5 Amer. Math. Soc. Pub., Providence, Rhode Island. (1964)
5. F.T.S. Yu, 'Introduction to Diffraction, Information Processing, and Holography'. M.I.T. Press, Cambridge, Mass., (1973).
6. B.J. Pernick, 'Optical systems for combined I-D image orthogonal Fourier transform processing.' Appl. Opt. 19 754 (1980)
7. G. Ross, M.A. Fiddy and M. Nieto-Vesperinas, 'The inverse scattering problem' Proc. Optics in Four Dimensions.
8. R.E. Burge, M.A. Fiddy, T.J. Hall and G. Ross 'Optical phase conjugation for the study of the analytic properties of optical fields' Proc. Optics in Four Dimensions.

FUNDATION OF SINGLE FRAME IMAGE PROCESSING

H. Harold Szu
Naval Research Laboratory
Washington, D.C. 20375

ABSTRACT

Single frame image processing with noise can be cast into the Fredholm integral equation of the first kind. Its inversion for object reconstruction is known to be increasingly unstable when the ratio of the maximum and minimum eigenvalues (of the square of the point spread function) increases as the number of discrete samples increases. Therefore single-frame image processing belongs to the class of ill-posed inverse problems. The root of instability is identified to be the singularity of the ratio of noise power spectrum and modulation transfer function. Such an over amplification of noise at the high spatial frequency is well known[1] in optics[2,3] as in the super resolution[4,5,6] and the image extrapolation problems.[7,8,9,10] Ad hoc procedures have been casually taken to suppress it. Recently an entirely different approach in a geological inversion problem has been applied to the ad hoc procedure in trading off noise and resolution in image problems. However, the ad hoc combination procedure can be simplified using the Euler-Lagrangian variational approach instead of the conventional gradient method. This will be shown to be rigorously achieved. Qualitative illustrations will be presented which give insight to the present method of dynamical regulation for the illposed inverse problem. Two exact solutions of the Bojarski-Lewis inverse scattering problem will be obtained showing different noise behaviors in the reconstructed object profile. Its impact in the super resolution and the image extrapolation problems will be indicated using an optical implementation of iteration algorithm for image restoration.

1. CONCEPTS OF DYNAMIC IMAGE FILTER

The linear image equation may be considered as stochastic object filtering

$$I(\vec{x}) = \int S(\vec{x} - \vec{x}_0, \vec{x}_0)\, O(\vec{x}_0) d\vec{x}_0 + N(\vec{x}). \tag{1}$$

The image formation through the point spread function (PSF) $S(\vec{x} - \vec{x}_0, \vec{x}_0)$ may depend in realistic cases on the location of the unknown object $O(\vec{x}_0)$. Then S is known as the general space-variant point spread function. It clearly needs more information to solve for the unknown object $O(\vec{x}_0)$ in one equation of three unknowns, O, S and N. A technique of extracting objective information from multiple frame time sequence has been reported[11] elsewhere. A single frame of incoherent image processing requires some subjective condition and a priori information in order to synthesize an image filter $M(\vec{x}_0, \vec{x})$ for an object reconstruction $Q(\vec{x}_0)$ to be as close to the unknown object $O(\vec{x}_0)$ as possible.

$$Q(\vec{x}_0) = \int M(\vec{x}_0, \vec{x}) I(\vec{x}) d\vec{x}. \tag{2}$$

ISSN:0094-243X/81/650668-11 $1.50

Substituting Eq. (2) into Eq. (1) gives the inverse image equation

$$Q(\vec{x}_o) = \int \Psi(\vec{x}_o,\vec{x}_o')O(\vec{x}_o')d\vec{x}_o' + \mathcal{N}(\vec{x}_o) \tag{3}$$

where the restoration point spread function

$$\Psi(\vec{x}_o,\vec{x}_o') = \int M(\vec{x}_o,\vec{x})S(\vec{x} - \vec{x}_o',\vec{x}_o')d\vec{x} \tag{4}$$

plays the central role determining the dynamic image filter M which nullify the restorative noise.

$$\mathcal{N}(\vec{x}_o) = \int M(\vec{x}_o,\vec{x})\, N(\vec{x})d\vec{x}. \tag{5}$$

The ideal goal of single-frame image processing expresses itself explicitly in terms of the inverse image Eq. (3) namely

$$\Psi(\vec{x}_o,\vec{x}_o') = \delta(\vec{x}_o - \vec{x}_o') \tag{6}$$

$$\mathcal{N}(\vec{x}_o) = 0 \tag{7}$$

where δ is the Dirac delta function. If Eq. (6) were true then Eq. (4) defines $M = S^{-1}$, which shows the zero of S to be the singularity of the image filter M. Wherever $S = 0$, M becomes divergent. However, the restoration PSF Ψ can not be realistically as sharp as the Dirac delta function Eq. (6) nor can the restoration noise $\mathcal{N}$ be identically zero. A dynamic trade-off between two Eqs. (6) and (7) has been recently given[12] to synthesize the dynamic image filter M which peaks at the middle frequency similar to the on-axis human visual response. The dynamic trade-off equation of $\tilde{\psi}(\vec{f})$ derived from the Hamilton principle turns out to be the Schrodinger equation of the trade-off potential

$$\tilde{V}(\vec{f}) = v_o < |\tilde{N}(\vec{f})|^2 > / < |\tilde{S}(\vec{f})|^2 > \tag{8a}$$

(cf. Fig. 1). The zero of $\tilde{S}$ makes $\tilde{V}$ infinite where $\tilde{\psi}$ must vanish according to Schrodinger equation

$$-\nabla_f^2 \tilde{\psi} + \tilde{V}(\vec{f})\tilde{\psi} = E\tilde{\psi}. \tag{8b}$$

The kinetic energy is the curvature in the frequency space

$$-\int \tilde{\psi}^*(\vec{f})\, \nabla_f^2 \tilde{\psi}(\vec{f})d\vec{f} = \int |\vec{\nabla}_f \tilde{\psi}(\vec{f})|^2 d\vec{f} = \int |\vec{x}|^2 \tilde{\psi}^2(\vec{x})d\vec{x}. \tag{8c}$$

which by using the integration by parts gives the dynamic extension of the radius of gyration of Backus-Gilbert regulation method.[13,14]

The Gel'fand-Levitan inverse scattering technique becomes straightforward applicable to synthesize the restoration filter $\tilde{\psi}(\vec{f})$. Such a dynamic trade-off scheme (cf. Fig. 2) is a natural regulation method to cure the ill-posed single frame image processing.

2. SMALL EIGENVALUES OF ILL-POSED PROBLEMS

The linear image equation can be cast into the Fredholm integral equation of the first kind.

$$\mathcal{I}(\vec{x}) \equiv I(\vec{x}) - N(\vec{x}) = \int S(\vec{x} - \vec{x}_o,\vec{x}_o)\, O(\vec{x}_o)d\vec{x}_o. \tag{9}$$

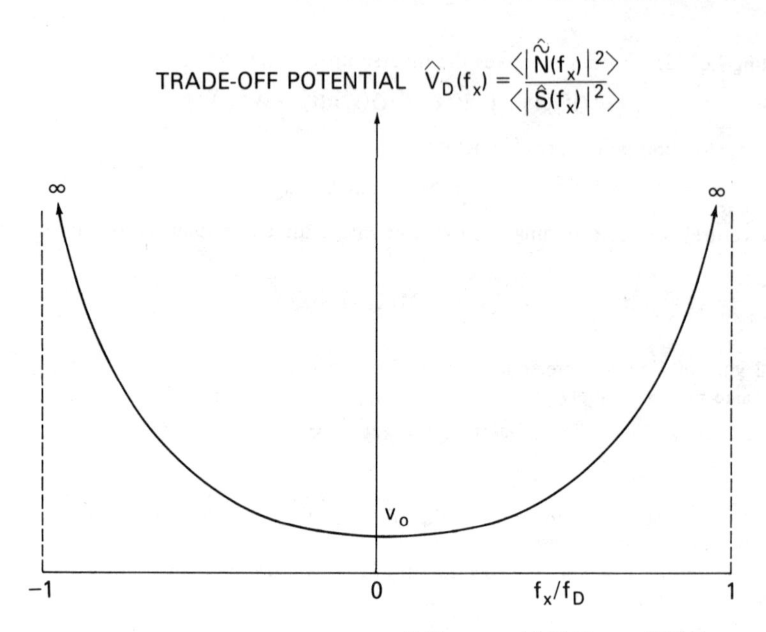

Fig. 1 — The dynamic trade-off potential $\tilde{V}(\vec{f}) = v_o < |\tilde{N}(\vec{f})|^2 > / < |\tilde{S}(\vec{f})|^2 >$, which is derived from the Hamilton principle, shows the high-frequency overamplification of noise power spectrum.

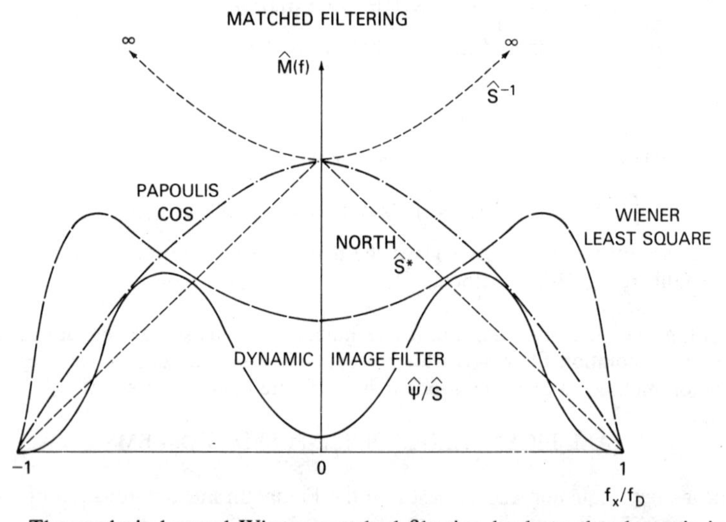

Fig. 2 — The analysis beyond Wiener matched filtering leads to the dynamic image filter, which is compared with the inverse, the North and the Wiener filters. The dynamic image filter, similar to human visual response, may be useful for the optical pattern recognition, which was oversensitive in differentiating high frequency within-class objects (the same letters B, ***B***) but was insensitive in differentiating low frequency between-class objects (different letters B and E) (due to excessive low pass filtering of the dominant low frequency object content.)

The inversion for object reconstruction is known[15,16] to be increasingly unstable when the ratio of the maximum and minimum eigenvalues (of the square of the point spread function) increases as the number of discrete samples increases. This is easy to prove as follows.

$$S\ O = I. \tag{10}$$

Let the adjoint of S be S^+. Then from Eq. (10) follows

$$S^+SO = S^+I \quad , \quad SS^+SO = SS^+I. \tag{11a,b}$$

Let the eigenfunction u_l of S^+S be different from the eigenfunction v_l of SS^+.

$$S^+S\ u_l = \mu_l\, u_l \quad , \quad SS^+\ v_l = \mu_l\, v_l. \tag{12a,b}$$

Then expanding O in u_l and I in v_l

$$O = \sum o_l\, u_l \quad , \quad I = \sum i_l\, v_l \tag{13a,b}$$

reveals the root of ill-condition

$$o_l = i_l \,/\, \sqrt{\mu_l}. \tag{14}$$

Wherever the eigenvalue μ_l becomes small, the error in image data i_l leads to large error in object reconstruction o_l. The proof of Eq. (14) is based on the following Ansatz of adjoint eigenfunctions

$$v_l = S\ u_l \,/\, \sqrt{\mu_l}. \tag{15}$$

If the Ansatz is taken to be correct, then it gives the conclusion Eq. (14) by substituting Eq. (15) into Eq. (10).

$$\text{LHS} = \sum o_l\, S u_l = \sum o_l \sqrt{\mu_l}\, v_l = \text{RHS} = \sum i_l\, v_l. \tag{16}$$

Using the fact that S^+ is the adjoint associated with the real eigenvalue μ_l, the Ansatz of eigenfunction, Eq. (15), can be verified by checking the orthonormality condition

$$\delta_{l,m} = (v_l, v_m) = \left(\frac{1}{\sqrt{\mu_m}}\ S\, u_l,\ \frac{1}{\sqrt{\mu_m}}\ S\, u_m \right) = \frac{1}{\sqrt{\mu_l \mu_m}}\ (u_l, S^+S\, u_m) = \frac{\sqrt{\mu_m}}{\sqrt{\mu_l}}\ (u_l, u_m), \tag{17}$$

and by checking the consistency of double S eigenfunction expansions Eq. (12b)

$$\text{LHS} = \underline{SS^+S\, u_l} \,/\, \sqrt{\mu_l} = S\underline{\mu_l u_l} \,/\, \sqrt{\mu_l} = \mu_l\, v_l = \text{RHS}, \tag{18}$$

where use is made of Eq. (15) and Eq. (12a).

If we define for a single S

$$S\ w_l = \lambda_l\ w_l \tag{19}$$

then for symmetric S

$$\mu_l = \lambda_l^2. \tag{20}$$

Therefore the object expansion will amplify the noisy image expansion i_l of small eigenvalue λ_l.

$$o_l = i_l / \lambda_l. \tag{21}$$

If the PSF has a constant trace[17,18]

$$T_r S = \sum \lambda_l = \text{constant} \tag{22}$$

then the increase of discrete samples produces more small eigenvalues λ_l's which create sources of instability in the object reconstruction from noisy image data. Therefore, single frame image processing belongs to the class of ill-posed inverse problem.

If the expansion in terms of these dual set of eignefunctions is terminated at the desired number of terms, then this is called the singular value decomposition used by Huang and Narendra (1975)[19] to restore the image and extended by Sahasrabudhe and Kulkarni (1979)[20] to the space variant image restoration.

3. AN ALLEGED ILL-POSED PROBLEM OF BOJARSKI-LEWIS HIGH PASS FILTER

Perry (1974)[21] concluded that the ill-posed nature of Bojarski[22]-Lewis[23] inverse scattering problems arises from the inaccessibility of low frequency information, i.e., it is a high pass filter (1-rect(ν/m)). Obtaining from optical inverse diffraction data (or radar back scattering data) for all frequency $\tilde{f}(\nu)$,

$$\tilde{f}(\nu) = [1 - \text{rect}(\nu/m)]\,\tilde{\gamma}(\nu). \tag{23}$$

Lewis and Bojarski can determine the body shape function $\tilde{\gamma}(f)$. The shape function $\gamma(x)$ equals to one if x belongs to the (unknown) body and otherwise equals zero. It satisfies the following Fredholm integral equation of the second kind, obtained as the Fourier inverse transform of Eq. (23).

$$\gamma(x) = f(x) + \epsilon \int_{-\infty}^{\infty} \frac{\sin(C(x-y))}{\pi(x-y)}\,\gamma(y)\,dy. \tag{24}$$

An order parameter ϵ is inserted for scaling convenience, and the space bandwidth product $C = \pi$ Lm where L is the unit of length. Perry's conclusion seems to be in contradiction to the general belief that the Fredholm integral equation of the second kind Eq. (24) is correctly posed. It will be shown that the singularity arises from the truncation of infinite spatial domain

$$\gamma(x) = f(x) + \epsilon \int_{-1}^{+1} \frac{\sin(C(x-y))}{\pi(x-y)}\,\gamma(y)\,dy. \tag{25}$$

These two integral equations are exactly solved as follows. The noise behaviors are entirely different in two cases. In the original Bojarski-Lewis equation the noise is smoothed and directly added, while in the truncated case of Perry is unduely amplified by $(1 - \lambda_l\epsilon)\lambda_l$. These will be exactly proved as follows. Since rect function is a projection

$$\text{rect}(\nu/m)^2 = \text{rect}(\nu/m) = 1,\ |\nu| \leqslant \frac{m}{2},\ \text{otherwise zero} \tag{26}$$

then the Fourier transform of a projection function is self convoluting

$$\int_{-\infty}^{\infty} S(x-x')\,S(x'-x_o)\,dx' = S(x-x_o) \tag{27a}$$

where

$$S(x) = FT^{-1}\,(\text{rect}(\nu/m)) = \frac{\sin(Cx)}{\pi x} \tag{27b}$$

is the kernel of angular prolate spheroidal functions (APS).[24]

$$\int_{-1}^{+1} S(x-y)\,\phi_i(y)dy = \lambda_i\,\phi_i(x) \tag{27c}$$

The basic difference in two results lies in the eigenvalues of APS functions which only enters into the orthonormality condition of a finite domain, but not for the infinite domain. Using Eqs. (27c) and (27a) one finds the basic difference in the orthonormality conditions

$$\delta_{i,j} \equiv \int_{-\infty}^{\infty} \phi_i(x)\phi_j(x)dx = \tag{28}$$

$$\frac{1}{\lambda_i\lambda_j}\int_{-\infty}^{\infty}\iint_{-1}^{+1} S(x-x')\phi_i(x')\,S(x-x'')\,\phi_j(x'')dx'dx''dx = \frac{1}{\lambda}\int_{-1}^{+1}\phi_i(x)\phi_j(x)dx.$$

The von Neuman-Born iteration gives Eq. (24) the following solution

$$\gamma(x) = f(x) + \epsilon\int_{-\infty}^{\infty} S(x-y)\,f(y)dy + \epsilon^2\iint_{-\infty}^{\infty} S(x-y)S(y-z)\,\gamma(z)dy\,dz. \tag{29}$$

which can be exactly sum up, using Eq. (27a),

$$\gamma(x) = f(x) + (\epsilon + \epsilon^2 + \dots)\int_{-\infty}^{\infty} S(x-y)f(y)dy = f(x) + \frac{\epsilon}{1-\epsilon}\int_{-\infty}^{\infty} S(x-y)f(y)dy \tag{30}$$

Therefore, the noise in $\gamma(x)$ equals the noise in $f(x)$ and a convolution-smoothed noise.

An eigenfunction expansion yields, using Eq. (28), the spectral decomposition of S

$$S(x-y) = \Sigma\,\phi_i(x)\phi_i(y). \tag{32}$$

The finite domain solution of Eq. (25) can be obtained using the following expansions

$$\gamma = \Sigma\, a_i\phi_i, \qquad f = \Sigma\, b_i\phi_i$$

and solving a_i in terms of b_i

$$a_i\lambda_i = (\phi_i, \gamma)_1, \qquad b_i\lambda_i = (\phi_i, f)_1$$

from Eq. (25)

$$\Sigma\, a_i\phi_i = \Sigma\, b_j\phi_j + \epsilon\,\Sigma\,\lambda_k a_k\phi_k.$$

Thus

$$\gamma(x) = \Sigma_l\,\frac{1}{(1-\lambda_l\epsilon)\lambda_l}\,(\phi_l, f)_1\phi_l(x). \tag{33}$$

Since multiplying Eq. (27c) with $S(y'-x)dx$ and integrating dx over $(-\infty, \infty)$ one obtains, using Eqs. (27a) and (27c),

$$\int_{-\infty}^{\infty} S(x-y)\phi_i(y)dy = \phi_i(x) \tag{34}$$

then Eq. (30) can be expanded in the infinite domain, using Eq. (34),

$$\gamma(x) = f(x) + \frac{\epsilon}{1-\epsilon}\,\Sigma^i(\phi_i, f)_\infty\phi_i(x) = \frac{1}{1-\epsilon}\,\Sigma_l(\phi_l, f)_\infty\phi_l(x), \tag{35}$$

i.e., the spectral decomposition, Eq. (32), is valid for the infinite and finite domains. If one compares the finite domain result Eq. (33) with the infinite domain result Eq. (35), the Fredholm integral equation of the second kind is not the real reason of the alleged ill-posed Bojarski-Lewis inverse scattering method. Using the step function behavior of eigenvalues

$$\lambda_l \cong \begin{cases} 1 - \dfrac{l}{C} & l \leqq C \\ 0 & l > C \end{cases} \tag{36}$$

the noise in data f(x) is over amplified in Eq. (33) for mode l larger than the space-bandwidth product C, whenever the cut-off of the integration limits gives a finite space-bandwidth product C. Several comments are in order.

(i) Tikhonor (1963)[25]-Miller (2970)[26] method of regulation amounts to insert a nonzero parameter in Eq. (33) to make the denominator nonzero.

(ii) The Fredholm integral equation of the second kind amounts to the subtraction-regulated-kernel of the Fredholm integral equation of the first kind.

$$f(x) = \int_{-\infty}^{\infty} [S(x-y)\epsilon - \delta(x-y)]\gamma(y)dy \tag{37}$$

(iii) the Gel'fand-Levitan integral equation for quantum mechanical inverse scattering problem is the Fredholm integral equation of the second kind.

$$K(x, x') = -\,\Omega(x, x') - \int K(x, x'')\Omega(x'', x')dx'', \tag{38}$$

which has never been shown to be ill posed.

4. OPTICAL ITERATION ALGORITHM FOR SINGLE SPACE-VARIANT COHERENCE-VARIANT IMAGE RESTORATION

Iteration algorithms are efficient. They can be carried out with a limited memory space in a digital computer. They can also be optically implemented in parallel. Earlier success were mainly in one dimensional images studied by Frieden (1967)[27], but the signal-to-noise ratio was shown by Harris (1966)[27] to limit the practicality of the system. The extension of iteration schemes to 2-D was done digitally by Kawata and Ichioka (1980)[28,29] and optically by Marks (1980)[30]. Two non-iterative methods of signal frame image processing are given to set the stage, then the optical implementation of regulated iterative method follows. Given the matrix equation

$$SO = I \tag{39}$$

one can always introduce an arbitrary gauge vector P

$$SO = I - SP + SP, \tag{40}$$

and then invert it using a generalized (pseudo) inverse S^{-1} of Moore-Penrose type

$$O = S^{-1}(I - SP) + P \tag{41a}$$

$$= S^{-1}I + (I - S^{-1}S)P \tag{41b}$$

Reducing the extra P on Eq. (41a) sequentially is the purpose of the so-called gradient method (cf. Angel and Jain (1978)[31]).

The distance between the image data I and the unknown object O is bounded by the distance between the true object O and the reconstructed one Q.

$$||O - I||^2 = ||O - Q + Q - I||^2 = ||O - Q||^2 + 2(O - Q, Q - I) + ||Q - I||^2$$
$$\leqq ||O - Q||^2, \text{ if and only if,} \tag{42}$$
$$(O - Q, Q - I) = 0 \tag{43}$$

This Eq. (43) is a nonlinear generalization[12] of Wiener matched filtering a la Pugachev's projection. If the linear image filter is required

$$Q \equiv MI \tag{44}$$

then the necessary and sufficient condition Eq. (43) is reduced

$$(O - Q, I) = 0 \tag{45}$$

which implies

$$(O, I) \equiv (O, SO) = (Q, I) \equiv (MI, I) \equiv (MSO, SO) \tag{46}$$

the familiar inverse filter

$$M = S^{+}(SS^{+})^{-1} \tag{47}$$

for arbitary object vector. This is known as the least-norm solution for the inverse filter.

The convergence of the inversion

$$O = S^{-1}I \tag{48}$$

is determined in cases of iterative algorithms by the "acceleration" (or "relaxation") parameter c_o

$$c_o \equiv -\frac{\delta O}{\delta I} \cong -\frac{O^{(new)} - O^{(old)}}{I^{(new)} - I^{(old)}} \tag{49}$$

It was shown by Kawata and Ichioka (1980) in the Fourier space that the fastest convergence for a space-invariant PSF S is bounded by

$$c_o \leqq 2$$

Such an iteration algorithm is extended to space variant coherent imagery in the coordinate space and implemented optically in parallel. From the definition of c_o follows

$$O^{(new)} = O^{(old)} + c_o(I^{(old)} - I^{(new)}) \tag{50}$$

Now the iteration method means that Eq. (39) begins with

$$SO^{(old)} = I^{(new)} \tag{51}$$

Since the single frame image processing has only one input image, therefore,

$$I^{(old)} = I \tag{52}$$

Consequently the following formula of Eq. (50), using Eqs. (51) and (52), is useful to derive the iteration

$$O^{(n)} = O^{(n-1)} + c_o(I - SO^{(n-1)}) \tag{53}$$

The initial condition is the first guess of the unkown object 0, e.g.

$$O^{(o)} = I \tag{54}$$

Then the first order iteration follows

$$O^{(1)} = (1 - c_o S)I + c_o I \tag{55}$$

$$= [1 - (1 - S)(1 - c_o S)]S^{-1}I \tag{56}$$

and the second order iteration gives

$$O^{(2)} = (1 - c_o S)O^{(1)} + c_o I \tag{57}$$

$$= [1 - (1 - S)(1 - c_o S)^2]S^{-1}I \tag{58}$$

Thus, the n-th iteration synthesizes the following image filter

$$M^{(n)} = [1 - (1 - S)(1 - c_o S)^n]S^{-1} \tag{59}$$

The reconstruction of the unknown object is given by

$$Q = M^{(n)} I \tag{60}$$

From Eq. (59) the condition of convergence follows

$$|1 - c_o S| < 1 \tag{61}$$

which implies for the space variant case

$$|c_o S| < 2 \tag{62}$$

There exists no noise over amplification because for a finite iteration

$$M^{(n)} = 0, \qquad S = 0 \tag{63}$$

The point spread function is positive, real and space-variant because it is the modulus of impulse response function

$$S(\vec{x} - \vec{x}_o, \vec{x}_o) = |h(\vec{x} - \vec{x}_o, \vec{x}_o)|^2 \tag{64}$$

Therefore we obtain naturally the following results in the space coordinate for the general space-variant case

$$M^{(n)} = [1 - (1 - |h|^2)(1 - c_o |h|^2)^n]|h|^{-2}, \tag{65}$$

where the impulse response function can be written in terms of the complex pupil function $P = |P| \exp(iW)$

$$h = (\lambda_o R)^{-2} \int |P| e^{iw} \exp\left(-i \frac{2\pi}{\lambda_o R} (\vec{x} - \vec{x}_o) \cdot \vec{\xi}\right) d\vec{\xi}, \tag{66}$$

according to the Huygens-Fresnel principle.

The present iteration scheme differs from Kawata and Ichioka's digital processing in that no space-invariant assumption and no artificial reblurring in the Fourier space are assumed. It provides the physical explanation of why the iteration when filtering through the incoherent point spread function $S = |h|^2$ is better than that of the coherent impulse response function, h.

The usefulness of such a regulated $M^{(n)}$, Eq. (65), can be optically demonstrated and parallel implemented. Eqs. (55, 57) are rederived for a coherent image, which is iterated back and forth, through (1) a matched filtering $c_o|h|^2$ and (2) the input image hologram and (3) a phase plate, by a semi-transparent retro-reflector mirror placed at the output end.

The hologram is shining with the laser plane wave and then blocking off, and the other end is the output. The irradiance of reconstructed object, that passes back and forth between the 4-wave mixing and amplifying phase-conjugate mirror and the semi-transparent retroreflector at the output plan, may be time integrated on a film at the semi-transparent output plan (Fig. 3).

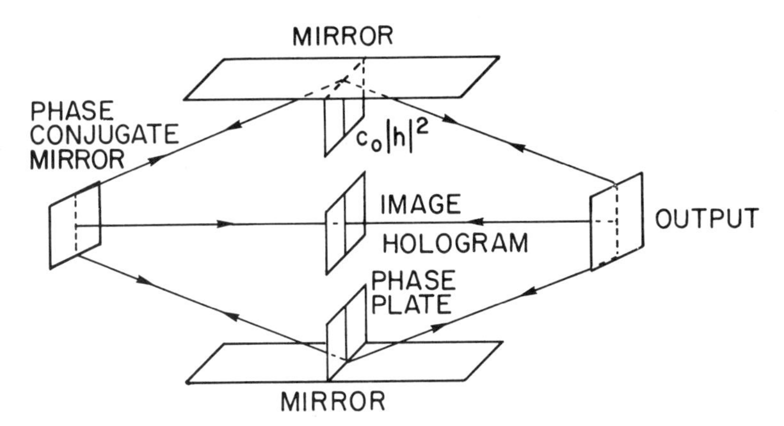

Fig. 3 — Using a beam splitter (not shown) to enter the input image hologram of the impulse response h one can use a 4-wave mixing phase-conjugate mirror to amplify the light for image restoration beyond the classical resolution limit. Such an increase of the signal to noise ratio through amplifications and filtering with $c_0|h|^2$ will be iteratively carried out in two dimensions.

REFERENCES

1. G. Toraldo di Francia, J. Opt. Sci. Am. 59, 799-804 (1969)
2. C. K. Rushforth and R. W. Harris, J. Opt. Soc. Am. 58, 539-545 (1968)
3. B. E. A. Saleh, J. Opt. Soc. Am. 67, 71-76 (1977)
4. R. W. Gerchberg, Opt. Acta 21, 709-720 (1974)
5. W. Lukosz, J. Opt. Soc. Am. 57, 932-941 (1967)
6. P. de Santis and F. Gori, Opt. Acta 22, 691-695 (1975)
7. M. Bertero, C. De Mol, and G. A. Viano, J. Math. Phys. 20, 509-521 (1979)
8. H. Wolter, in: Prog. in Optics, 1, 155-210 (1961)
9. C. W. Barnes, J. Opt. Soc. Am. 56, 575-578 (1966)
10. C. L. Rino, J. Opt. Soc. Am. 59, 547-553 (1969)
11. H. H. Szu, J. A. Blodgett, and L. Sica, "Local instances of good seeing," (submitted for publication)
12. H. H. Szu, "Analysis beyond matched filtering," (submitted for publication)
    H. H. Szu, "Basis of optical computing," (in prepagation)
13. G. Backus and F. Gilbert, Roy. Soc. London Phil. Trans 266, 123-192 (1970)
14. B. E. A. Saleh, Appl. Opt. 13, 1833-1838 (1974)

15. D. K. Faddeyev and V. N. Faddeyeva, USSR compu. Math. and Math. Phys. 1, 447-454 (1961), original in Zh. Vych. Mat. i Mat. Fiz 1, 412-417 (1961)
    B. R. Hunt, IEEE Trans. Audio Electroacoust., AU–20, 94-95 (1972).
16. J. Todd, Proc. Cambridge Philos. Soc. 46, 116-118 (1950); I. Quart. J. Mech. and Appl. Math. 2, 469-472 (1949)
17. F. Gori and G. Guatari, Opt. Comm. 7, 163 (1973)
18. H. H. Szu, Phys. Rev. *A* 11, 350 (1975)
19. T. S. Huang and P. M. Narendra, Appl. Opt. 14, 2213-2216 (1975)
20. S. C. Sahasrabudhe and A. D. Kulkarni, Compt. Graphics Image Proc. 9, 203-212 (1979)
21. W. L. Perry, IEEE Trans. Ant. and Prop., Ap–22, 826-829 (1974)
22. N. N. Bojarski, Report of Syracuse Univ. Res. Corp., Syracuse, N. Y. (1967)
23. R. M. Lewis, IEEE Trans. Antennas Propagat., Ap–17, 308-314 (1969)
24. B. R. Frieden, Prog. in Opt., Ed. E. Wolf, Vol.9, Ch. 8, 311-407 (1971) (North Holland Pub., Amsterdam, 1971)
25. V. F. Turchin, V. P. Kozlow and M. S. Malkevich, Sov. Phys. Uspekhi 13, 681-840 (1971)
    A. N. Tihonov, Sov. Math. 4, 1624-1627 (1963)
26. K. Miller, SIAM J. Math. Anal. 1, 52-74 (1970)
27. J. L. Harris, J. Opt. Soc. Am. 56, 569-574 (1966)
    R. B. Frieden, J. Opt. Soc. Am. 57, 1003-1019 (1967)
28. S. Kawata and Y. Ichioka, J. Opt. Soc. Am. 70, 768-772 (1980)
29. S. Kawata and Y. Ichioka, J. Opt. Soc. Am. 70, 762-768 (1980)
30. R. J. Marks II, Appl. Opt. 19, 1670-1672 (1980)
31. E. S. Angel and A. K. Jain, Appl. Opt. 17, 2186-2190 (1978)

SOLUTIONS TO THE THREE DIMENSIONAL INVERSE SCATTERING PROBLEM

M. Nieto-Vesperinas[1] and M. Gea
Instituto de Optica, Serrano 121,
Madrid 6, Spain.

G.Ross and M.A.Fiddy
Physics Department, Queen Elizabeth
College, University of London,
Campden Hill Road, London W8 7AH, U.K.

1) Supported by the Juan March Foundation.

ABSTRACT

We discuss methods for solving the three dimensional scattering problem based on the reconstruction of the scattering object from projections. First, the reconstruction of 2-D functions is **studied** and later the 3-D problem is discussed .

INTRODUCTION

In two other papers of this volume[1,2] a solution to the inverse scattering problem in 1-D situations was put forward, (see also references [3, 4, 5]). The methods of solution were shown to reduce essentially to zero location of the wavefunction extended into the complex plane.

Of great interest is the generalization of these studies to scattered fields from 2-D and 3-D objects. However, a straightforward application of the theory would imply handling a 4-D complex space which constitutes a formidable task.

ISSN:0094-243X/81/650679-09$1.50

In this paper we present a method for solving 3-D inverse scattering problems. We shall start with the reconstruction of 2-D functions by means of 1-D strips. We shall see later that the 3-D problem may be reduced to a 2-D one.

RECONSTRUCTION OF 2-D FUNCTIONS

Let $f(t_1, t_2)$ be a 2-D object wave, its spectrum expressing the scattered field in the Fraunhofer region, being:

$$F(x_1, x_2) = \int_{-\infty}^{\infty} f(t_1, t_2) \exp(-i\langle k\rangle(x_1 t_1 + x_2 t_2)) dt_1 dt_2 \tag{1}$$

where an obliquity factor has been neglected and $\langle k\rangle$ being the mean wavenumber. In general, the following analysis may be applied to any problem of information transmission between two spaces related by a Fourier transform. The problem consists of continuing the function $F(x_1, x_2)$ into the complex plane in order to encode it.

a) Let us assume first that the function $f(t_1, t_2)$ or its image is amenable to processing.

In this case $f(t_1, t_2)$ can be divided into 1-D strips $f(t_1, K) \times \delta(t_2-K)$ for each value K of t_2 and then the multi-channel one-dimensional coherent Fourier transformer of reference[2] may be used for each of those strips. For each K, a strip $f(t_1, K) \times \delta(t_2-K)$ is obtained and then the whole function $f(t_1, t_2)$ is reconstructed by juxtaposing all those strips.

b) An alternative procedure, avoiding processing the object or image relies on using only information in the (x_1, x_2)-space. We define in this case the projection $G(x_1)$ of $F(x_1, x_2)$:

$$G(x_1) = \int_{-\infty}^{\infty} F(x_1, x_2)\, dx_2 = F(x_1, x_2) ** \delta(x_1) \tag{2}$$

where the double asterisk denotes a double convolution. This corresponds in the (t_1, t_2)-space to a strip of $f(t_1, t_2)$:

$$f(t_1, t_2)\ \delta(t_2) = f(t_1, o)\,\delta(t_2) \tag{3}$$

In general, the projection;

$$G(x_1, x_2) = \int_{-\infty}^{\infty} F(x_1, u_2) \exp(iKu_2)\, du_2 \,.\, \exp(-iKx_2) = F(x_1, x_2) ** \delta(x_1) . \exp(-iKx_2) \tag{4}$$

corresponds, in the (t_1, t_2)-space, to the strip:

$$f(t_1, t_2)\ \delta(t_2 - K) = f(t_1, K)\ \delta(t_2 - K) \tag{5}$$

Obviously, the functions $G(x_1, x_2)$ and $G(x_1)$, i.e. for K=0, are 2-D functions, constant along the x_2-direction. The operation (2) and (4) is made by using a cylindrical lens in front of $F(x_1, x_2)$ with the axis along the x_1-direction. Applying this treatment one can determine each projection $G(x_1, x_2)$ and for each K one reconstructs a 1-D strip of the function $f(t_1, t_2)$.

Another possibility is to successively rotate the integration direction, i.e. by successively rotating the axes through angles α according to the equations:

$$x_1' = x_1 \cos\alpha + x_2 \sin\alpha$$

$$x_2' = -x_1 \sin\alpha + x_2 \cos\alpha$$

Two situations may be distinguished according to the determination of $G(x_1,x_2)$:

i) The function $G(x_1,x_2)$ is amenable of processing.

Since the function $G(x_1,x_2)$ is equal to the 1-D **strip** $G(x_1,x_2)$ **X** $\delta(x_2-K)$ uniformly repeated along the x_2-axis one can insert in the plane of this function the exponential filter $\exp(-\langle k\rangle x_1 s_1)$ and display the zeros of its 1-D Fourier Transform in the complex plane t_1+is_1 as indicated in reference [2] thus obtaining $G(x_1,x_2)$ X $\delta(x_2-K)$ from the zeros in the conjugate space.

ii) Only the intensity $|G(x_1,x_2)|^2$ is accessible to processing.

In this case one must use the 1-D algorithm described in references [4,5]. This implies the measurement also of a projection of a Fresnel transform $F_d(x_1, x_2)$ obtained at a certain distance d of the (t_1, t_2)-plane:

$$F_d(x_1, x_2) = \int_{-\infty}^{\infty} f(t_1, t_2) \exp(\frac{i\langle k\rangle}{2d}(t_1^2 + t_2^2)) \exp(-i\langle k\rangle(x_1t_1+x_2t_2)) \quad \text{X} dt_1 dt_2 \tag{6}$$

i.e. a projection given by:

$$G_d(x_1,x_2) = \int_{-\infty}^{\infty} F_d(x_1, u_2) \exp(i\, Ku_2)\, du_2 \,.\, \exp(-iKx_2) = F_d(x_1, x_2) ** \delta(x_1).\, \exp(-iKx_2) \tag{7}$$

which yields in the (t_1, t_2)-space the strip:

$$f(t_1, t_2) \exp(\frac{i\langle k\rangle}{2\,d}(t_1^2 + t_2^2))\; \delta(t_2-K) = f(t_1,K) \exp(\frac{i\langle k\rangle}{2d}(t_1^2 + K^2))\; \delta(t_2 - K) \tag{8}$$

Thus, by comparing the reconstructed Fresnel intensity $|G_d(x_1 x_2)|^2$ with the measured Fresnel intensity (7) one may determine the function $G(x_1, x_2)$ and hence the successive 1-D strips $f(t_1, K) \times \delta(t_2 - K)$ with which one reconstructs the 2-D function $f(t_1, t_2)$.

c) If the field is not amenable to processing and only intensities $|F(x_1, x_2)|^2$ are available, the full treatment using the theory of functions of two complex variables is required, as well as additional a priori information. The latter consists of knowledge of the aperture shape and a Fresnel intensity. Assuming weak scatterers and choosing, for exemple, a circular aperture, the zero circles could be located using an extension of the method presented in references[4,5].

RECONSTRUCTION OF THREE DIMENSIONAL OBJECTS

Two important situations can be distinguished.

A) Geometrical optics. In this limit two cases may occur:

i) Straight line propagation, which is the instance considered by conventional tomography, (e.g. references [6-14]), this is well documented and will not be discussed here.

ii) Propagation described by the eikonal equation[12]. As far as we are aware this case has not been treated yet.

Strictly speaking, section A) is a subset of section B), but owing to its importance it can be considered separately.

B) Scattering process taken into account.

The relationship between the object wave $f(\vec{t})$ and the scattered field $F(\vec{x})$ in the Fraunhofer region, again neglecting an obliquity factor, is given by[3]:

$$F(\vec{x}) = \int_{-\infty}^{\infty} d^3\vec{t}\ f(\vec{t})\ \exp(-i\langle k\rangle \vec{t}.\vec{x}) \tag{9}$$

The task of obtaining $f(\vec{t})$ is considerably more difficult in three dimensions, as compared to one or two dimensional problems, owing to the apparent impossibility of inverting Eq.(9). It is well-known[15,16] that the three components of $\vec{x}$ are not independent, being cosine directors. Thus the scattered field $F(\vec{x})$ is essentially two-dimensional. This is not surprising since the wavefunction satisfies the wave equation and hence its 3-D structure is completely determined by giving its value at a surface. Thus, from 2-D information in $\vec{x}$-space , obtained from one single measurement it is not possible to retrieve a three- dimensional object wave $f(\vec{t})$.

The method put forward here for reconstructing the object wave relies on the possibility of operating on the object.

Let us assume that the measurement is performed in the equatorial plane, $x_3 = 0$. With this it is easy to obtain from Eq.(9):

$$F(x_1) = \int_{-\infty}^{\infty} dt_1 dt_2\ \exp(-i\langle k\rangle(t_1x_1+t_2x_2))\ [\int_{-\infty}^{\infty} dt_3\ f(\vec{t})] \tag{10}$$

In writing Eq.(10) we have considered $F(x_1, x_2) = F(x_1)$ since $x_1^2 + x_2^2 = 1$.

In order to obtain the projection $\int_{-\infty}^{\infty} f(\vec{t})\ dt_3$ from $F(x_1)$ two possibilities arise. The first consists of taking one slice of $f(\vec{t})$ by performing the operation:

$$\delta(t_2-t_2^o)\int_{-\infty}^{\infty} f(\vec{t})\ dt_3 = \delta(t_2-t_2^o)\int_{-\infty}^{\infty} f(t_1,t_2^o,t_3)\ dt_3 \tag{11}$$

which suggests illuminating the object -assumed a weak scatterer- with a plane wave of one dimensional wavefront, oriented in a direction parallel to the t_3-axis, propagating in a direction parallel to t_1 , at t_2^o .

Thus we obtain a one dimensional situation given by:

$$F(x_1) = \exp(-i\langle k\rangle x_2\ t_2^o)\int_{-\infty}^{\infty} dt_1\ [\int_{-\infty}^{\infty} dt_3\ f(t_1,t_2^o,t_3)]\ \exp(-i\langle k\rangle x_1 t_1) \quad (12)$$

from which the 1-D function $\int_{-\infty}^{\infty} dt_3\ f(t_1,t_2^o,t_3)$ can be found. The problem remains of course to obtain $f(t_1,t_2^o,t_3)$ from the above projection. This can be achieved by using the Radon transform[13,14]. In order to perform it,it is necessary to carry out the measurements in a succession of planes, for a range of values of x_3, i.e. by rotating the circle of measurement about the t_2 axis. This procedure must be repeated for a range of t_2 values, in order to obtain the 3-D reconstruction.

The second possibility of solving the two dimensional problem, expressed by equation (10), corresponds to illuminating the object with a plane wave having a two dimensional wavefront; in this case operation (11) is not realised. This is obvious since now $\int_{-\infty}^{\infty} f(\vec{t})dt_3$ in equation (10) depends upon two variables, namely t_1 and t_2. As such, the inversion of equation (10) cannot be carried out since $F(x_1,x_2)$ is a function of only one independent variable. In order, to render F a function of two independent variables, it is necessary to vary the direction or frequency of the incident beam. This generates a family of one dimensional functions equivalent to a single function of two independent variables. This approach is equivalent to the method put forward by Wolf[15] and Dandliker and Weiss[16].

The first approach presented above is the direct generalisation of the two dimensional problem. We may regard the second approach as a further generalisation of the same procedure. It is possible to devise another approach, based on Wolf's[15] procedure and consisting of measuring F not on a circle but on a two dimensional surface.

However, instead of using a holographic procedure, in order to determine the complex field, we suggest as an alternative, the method outlined in section b) in the two dimensional case discussed earlier.

CONCLUSIONS

The specific difficulties of the three dimensional inverse problem are largely due to the fact that the scattered field is a function of two independent angular variables. The means for overcoming this have been examined and methods for reducing the three dimensional problem to a sequence of soluble two dimensional and even one dimensional problems have been investigated.

REFERENCES

1 G.Ross, M.A.Fiddy and M.Nieto-Vesperinas: Proc Optics in Four Dimensions

2 M.A.Fiddy, G.Ross, J.Wood and M.Nieto-Vesperinas: Proc Optics in Four Dimensions.

3 G.Ross, M.A.Fiddy and M.Nieto-Vesperinas: ch 2 of 'Inverse Scattering Problems in Optics',Ed. H.P.Baltes, Topics in Current Physics 20 Springer Verlag, (1980).

4 G.Ross, M.A.Fiddy and H.Moezzi: Opt. Acta (in press)

5 G.Ross, M.A.Fiddy, M.Nieto-Vesperinas and M.W.L.Wheeler: Proc. Roy.Soc.Lond. A360, 25 (1978).

6 B.J.Hoenders: ch.3 of 'Inverse Source Problems in Optics', Ed. H.P.Baltes, Topics in Current Physics 9 , Springer Verlag, (1978)

7 P.D.Rowley: J.Opt.Soc.Amer. 59, 1496, (1969).

8 M.V.Berry and D.F.Gibbs: Proc.Roy.Soc.Lond.,A314, 143, (1970).

9 D.W.Sweeney and C.M.Vest: Appl.Opt., 12 ,2649, (1973).

10 G.T.Herman (Ed.):'Image Reconstructions from Projections', Topics in Applied Physics 32 , Springer Verlag, (1979).

11 A.M.Cormack: J.Appl.Phys., 34, 2722, (1963).

12 M.Born and E.Wolf: 'Principles of Optics', Pergamon Press, Oxford (1970).

13 H.H.Barrett and W.Swindell: Proc I.E.E.E., 65, 89, (1977).

14 R.Gordon and G.T.Herman:Int.Rev.Cytol., 38, 111, (1974).

15 E.Wolf: Opt.Comm., 1, 153, (1969).

16 R.Dandliker and K.Weiss: Opt.Comm., 1, 323, (1970).

THE DETECTION OF MOVING OPTICAL STRUCTURES WHEN USING DETECTOR ARRAYS SUFFERING FROM SPATIAL NOISE

R. Röhler
Institut für medizinische Optik der Universität München
D 8000 München 2

ABSTRACT

Detector arrays usually suffer from spatial noise, i.e. unpredictable variations of the sensitivity of individual detectors. In the most common class of such arrays, the retinae of men and higher vertebrates, sensitivity variations are due to local adaptation. Technical devices, such as CCDs, show similar effects due to space charge shielding and charge leakage. It has been shown earlier that a recuction of spatial noise effects may be obtained for stationary images by small controlled lateral shifts between image and detector array, e.g. saccadic eye movements.

In general it is not possible, however, to evaluate temporally varying structures with such arrays. An important exception are translationally moving structures, the shape of which undergoes at most limited and slow changes during movement. Such structures may be detected and evaluated when it is possible to get additional information on their velocity (defined by magnitude and direction). With such information the hypothesis may be tested whether some spatio-temporal signal pattern is generated by a moving structure or not. Usually several structures with different velocities are present in complex scenes. Therefore a local velocity analysis over the image field is necessary. Such local analysis may be obtained from sets of spatial and temporal masks or filters which analyse the spatial and temporal surrounding of selected signal structures. By this procedure, a four-dimensional representation of the spatial-temporal signal manifold may be obtained. It displays the spatial arrangement of image elements with a velocity vector attached to each element. There is considerable evidence, that the visual system uses this type of signal evaluation. For technical image analysis, this has advantages even when spatial noise is no limiting factor.

INTRODUCTION

In a recent paper [1], attention was drawn to the problem that under normal visual conditions the sensitivity of receptive units of the visual system is subject to spatial and temporal variations, and that consequently in performing pattern recognition the visual cortex has to discriminate between external luminance structure and internal sensitivity

ISSN:0094-243X/81/650688-13$1.50

structure. It was suggested that this discrimination problem is solved with the help of eye movements. Similar conditions arrise with photodetector arrays, such as CCDs [2]. In the cited papers the analysis was limited to stationary objects. In the present paper the case of non stationary objects will be considered. It is shown that the general case of a spatially and temporally changing object cannot be solved, i.e. it is not possible to recognize such a structure with an array of detectors of variable sensitivity. But an important special case will be considered in more detail, namely objects which undergo translational movements in the visual field but do not change their shape. This case can be extended to slow deformations of the objects during their motion.

In the visual perception of a moving object two modes have to be distinguished:
a) the detection of the object moving in the field of view, i.e. the retinal image of the object moves relative to the receptor array,
b) the examination of a moving object under visual fixation, i.e. the eye performs a smooth movement to maintain a stationary retinal image of the object.

From the point of view of image analysis, the mode b) is essentially equal to the case of a stationary object. It must be borne in mind however, that this mode must be preceded by mode a) since otherwise the visual system has no cue for initiating an appropriate eye movement. In this paper only the mode a) of detecting a moving object is considered.

It is evident from experience of every day life that the visual system is able to recognize moving objects in the visual field, i.e. to discriminate these objects from spatially and temporally varying random pattern as being objects which maintain their shape. It is even possible to detect slow deformations of these objects during their course within the field of view.

At first sight there is little difference between the case of a stationary object which is scanned by eye movements and the case of a moving object within a stationary field of view. In both cases a relative motion between the receptor array and the retinal image occurs, though in the case of a stationary object the motion is saccadic, in the other case smooth. This latter point is considered in the discussion. Yet another important difference has to be considered: in the case of a stationary object, the scanning eye movements are controlled, so that the system has an internal knowledge of size and direction of the movements. This information is not automatically available to the visual system in the case of moving objects. Therefore, it is essential for the visual system, to extract velocity information from the visual input.

Possibilities of the visual system to solve this problem and implications for automatic pattern recognition are discussed in later sections.

In order to analyse the similarities and differences between the two cases just mentioned, it is helpful to reconsider the case of a stationary object. This is done in the next section. In the third section the general case of a spatially - temporally variable object is analysed, which is followed by the main subject under discussion, the case of a moving, invariable object.

ANALYSIS OF A STATIONARY OBJECT WITH AN ARRAY OF PHOTODETECTORS WITH VARIABLE SENSITIVITY

It has been shown [1,2] that with the help of eye movements it is possible to evaluate the luminance distribution of a stationary object, even when the sensitivity of the individual detectors are not equal and undergo temporal changes in an unpredictable way. A presupposition is, that the change of sensitivity be slow compared to the eye movements, so that the sensitivity of an individual detector does not change appreciably during the time required for a few eye movements. The subject will be reconsidered here from a different point of view. In order to simplify calculations, it is assumed that the detectors have linear transducer characteristics and that the outcoming signals undergo logarithmic transformations.

Consider a narrow section of the retinal image, represented by a one-dimensional row x_i $(1 \leq i \leq n)$ of sampling points with illuminances $f(x_i)$, and a corresponding row of detector elements i with sensitivities e_i. The neuronal signals generated by the detectors are then

$$S_1(x_i) = \log e_i f(x_i) \qquad (1)$$

After an eye movement which shifts the detector array by one sampling interval with respect to the illuminance distribution, detector number 2 "sees" the illuminance $f(x_1)$, number 3 "sees" $f(x_2)$ etc., which gives rise to a series of signals

$$S_2(x_i) = \log e_{i+1} f(x_i) \ , \quad (1 \leq i \leq n-1)$$

Since $f(x_i)$ remains constant, the signal difference

$$S_2(x_i) - S_1(x_i) = \log (e_{i+1} / e_i)$$

gives the ratio of the sensitivities of two neighbouring detectors. Hence one detector is calibrated by its neighbour, provided that the sensitivity of one detector is fixed arbitrarily. The process of successive calibration leads to the knowledge of the relative sensitivity distribution of the detector array and, in consequence, the relative illuminance distribution.

More than one eye movement may be necessary in more complicated situations, e.g. a two dimensional array, high noise level, large field, or nonlinear detector characteristic.

It is worthwhile to consider in more detail the spatial and temporal signal structure of the detector array. Fig. 1 shows in the upper part an example of a spatial illuminance structure over six sampling points, corresponding to a linear row of detectors numbered from 1 to 6. Consider then two successive shifts, each by one sampling distance, so that e.g. receptor 1 yields in temporal succession the signals

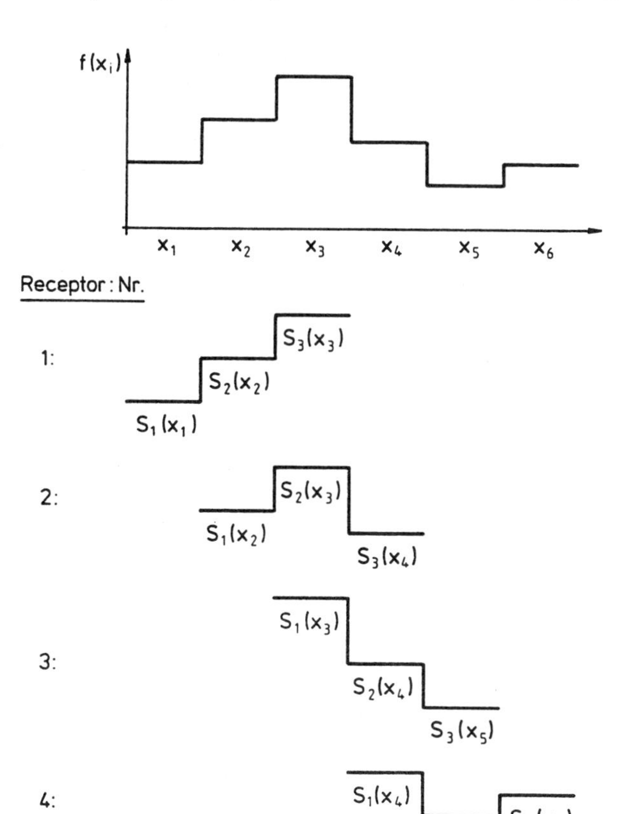

Fig. 1. Composition of a stationary illuminance distribution from the signals of individual photodetectors of a linear detector array. Upper part: Stationary illuminance distribution $f(x_i)$ at a series of spatial sampling points x_i. Lower part: Temporal succession of signals from individual detectors, when the array is shifted to the right two times for one sampling interval. S_1, S_2, S_3 indicate signals before the first, after the first, and after the second shift respectively.

$$\left.\begin{aligned} S_1(x_1) &= \log e_1\ f(x_1) \\ S_2(x_2) &= \log e_1\ f(x_2) \\ S_3(x_3) &= \log e_1\ f(x_3) \end{aligned}\right\} \qquad (2)$$

These signals and those of the next three detectors are indicated qualitatively in the lower part of fig. 1. These detector signals are the basic information from which the visual cortex has to reconstruct the illuminance distribution by fitting together the picture elements delivered from the single detectors. This fitting together is possible if there is an overlap of the individual picture elements and if the temporal order of the signals of each detector can be associated with a spatial order. This assiciation is rendered possible if the size of the single eye movements is known. Hence the eye movements have to be controlled.

SPATIALLY AND TEMPORALLY VARIABLE OBJECTS

It is easy to see in a qualitative way, that the mechanism of the preceding section does not work, when the object undergoes unpredictable changes in shape between two successive recordings. For if two detectors scan in temporal succession the same sampling point and thereby give different signals, this may be due to a difference in their sensitivities or due to a change of illuminance. This will be treated in a more quantitative way in the following.

Contrary to the discrete signal structure considered before, smooth functions of space and time are adopted here, since they allow simple methods of analysis to be applied. Formally it is straightforward to change from smooth to discrete functions by sampling methods. The physiological aspect of this point is discussed later.

Consider an illuminance distribution $f(x, t)$, depending on one spatial coordinate x and time t, and a sensitivity function $e(y)$ depending on a retinal coordinate y, the direction of y being parallel to that of x, so that $y = x - \delta$. δ measures shifts between spatial and retinal coordinates and is changed by eye movements. Corresponding to (1), it is

$$S(y, x, t) = \log f(x, t) + \log e(y) \qquad (3)$$

At $t = 0$ the origins of the x- and y-axis may be adjusted so that $y = x$ and

$$S(x,x,0) = \log f(x,0) + \log e(x) \qquad (4)$$

At $t = t_1$ an eye movement may have taken place, so that $y = x - \delta_1$ and

$$S(y,x,t_1) = \log f(x,t_1) + \log e(y) \tag{5}$$

Assume that t_1, δ_1 are small and $f(x,t)$, $e(y)$ are sufficient smooth, so that

$$\log f(x,t_1) = \log f(x,0) + t_1 \frac{f_t(x,0)}{f(x,0)} \tag{6}$$

$$\log e(y) = \log e(x) - \delta_1 \frac{e'(x)}{e(x)} \tag{7}$$

where $f_t(x,0)$ denotes the partial derivative of f with respect to t. From (4,5) and (6,7) it follows that

$$S = S(y,x,t_1) - S(x,x,0)$$

$$S = t_1 \frac{f_t(x,0)}{f(x,0)} - \delta_1 \frac{e'(x)}{e(x)} \tag{8}$$

If $e(x)$ and $e'(x)$ are not known, it is not possible to give a unique solution of this differential equation for $f(x,0)$.

When no eye movement is introduced, we have $\delta_1 = 0$. In this case one can solve the equation for $f(x,0)$, but here x is considered as parameter, and for each value a separate initial condition would be necessary. Each detector measures, then, a time function at a separate location, and there is no communication between different detectors. Hence no spatial picture is gained.

TRANSLATIONALLY MOVING OBJECTS

Consider a visual object which moves with velocity v along the x-direction, but does not change its shape. This object gives rise to an illuminance distribution

$$f(x,0) = f(x + \Delta x, \Delta t) \tag{9}$$

where $\Delta x / \Delta t = v$. If no eye movement takes place, it is not necessary to discriminate between spatial and retinal coordinates (x and y) and (4) may be written

$$S(x,x,0) \equiv S(x,0) = \log f(x,0) + \log e(x) \tag{10}$$

$$S(x+\Delta x, \Delta t) = \log f(x+\Delta x, \Delta t) + \log e(x+\Delta x)$$

$$S(x+\Delta x, \Delta t) = \log f(x,0) + \log e(x) + \Delta x \frac{e'(x)}{e(x)} \quad (11)$$

Thus two different equations for $f(x,0)$ exist, and $f(x,0)$ may be eliminated. This gives an expression for the sensitivity:

$$\frac{e'(x)}{e(x)} = \frac{S(x+\Delta x, \Delta t) - S(x,0)}{\Delta x} = \frac{S(x+v\Delta t, \Delta t) - S(x,0)}{v\Delta t} \quad (12)$$

$$\log \frac{e(x)}{e(x_0)} = \int_{x_0}^{x} \frac{S(x'+v\Delta t, \Delta t) - S(x',0)}{v\Delta t} dx' \quad (13)$$

Herewith the illuminance may be evaluated:

$$\log \frac{f(x,0)}{f(x_0,0)} = S(x,0) - S(x_0,0) - \int_{x_0}^{x} \frac{S(x'+v\Delta t, \Delta t) - S(x',0)}{v\Delta t} dx' \quad (14)$$

x_0 is a reference point.

This formula shows that the illuminance distribution $f(x,0)$ can be calculated, if the signal difference

$$S(x+v\Delta t, \Delta t) - S(x,0) \quad (15)$$

is measured along the x-axis, beginning at a suitable reference point x_0. It is also seen from (12) that the correcting term in (14), i.e. the integral, is zero if $e(x)$ is constant over the field. The difference (15) can only be evaluated, if the velocity v of the translational motion is known, since otherwise no corresponding values Δx, Δt are known.

This latter point elucidates an important difference with the case of a stationary object, which is evaluated by eye movement. Since the eye movement is controlled, the system has an internal knowledge of the size of the lateral shift that links the two signal types S_1 and S_2. An equivalent knowledge is not available in the case of a moving object. Without such knowledge, the single detectors yield temporal series of signals quite similar to the situation described in fig. 1, but since the relation between temporal succession and spatial shift is not known, the individual picture elements cannot be glued together.

If the visual system can estimate the value of v, it can easily test whether the hypothesis of a moving object is correct. It can repeat the calculation of the illuminance distribution with a second set of data measured at two later temporal sampling moments. The illuminance distribution derived from these data should be identical to the one derived from the first two sampling moments, if the object has not changed shape. Furthermore, if the evaluated illuminance structure changes systematically for successive sets of data, gradual deformations of the object can be detected.

THE PROBLEM OF VELOCITY MEASUREMENT

It has been shown in the preceding section that the detection of a moving object is possible when independent information about the velocity of the object and its retinal image is available to the system. Since the visual system can detect moving objects very well, it obviously has developed a method to measure or estimate velocity, including magnitude and direction.

The technical standard method for velocity measurements is sketched in fig. 2a. A suitable detail, e.g. an edge, is selected and the time Δt is measured, which is needed for

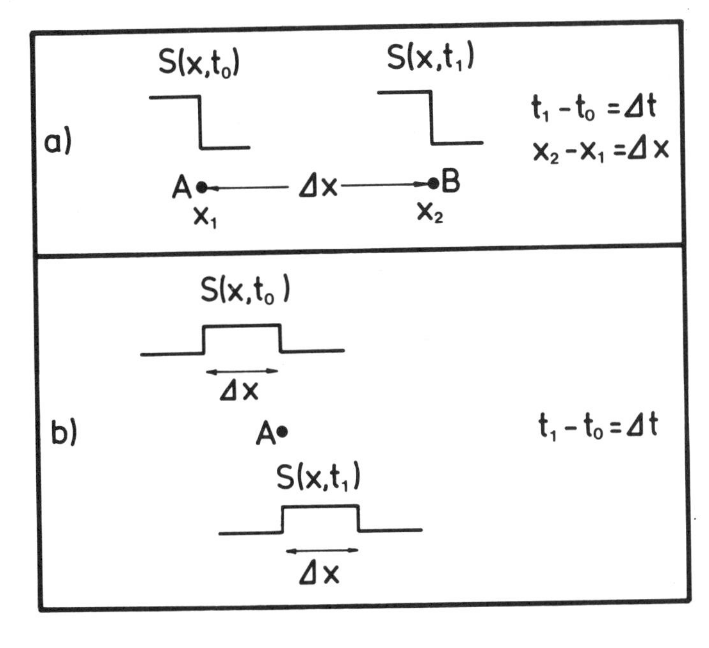

Fig. 2. Two possibilities for measuring the velocity of moving structures in a stationary visual field.

a) Bilocal method: Two locations A,B with distance Δx in the visual field are passed by the same detail in temporal succession. The velocity is calculated as $v = \Delta x/\Delta t$

b) Local method: A detail, whose width Δx is measured locally passes a location A within a time interval Δt. The velocity is $v = \Delta x/\Delta t$.

the detail to traverse the distance Δx between two suitable positions. This is a bilocal method, it requires to measure the temporal order of events at two different loci. The visual system has extreme difficulties to perform this type of measurement, since the perceived temporal order of light- (and dark-) signals is largely influenced by masking phenomena [3,4,5,6], which depend on luminance, stimulus time, adaptation etc.. So this method is probably not used by the visual system.

A second, local, method for the measurement of the amount of velocity is indicated in fig. 2b. Assume that a suitable detail, e.g. a bar or the period of a sine wave, moves across a detector A. The time for its passage is indicated by a flicker signal from the detector. If it is possible to measure locally the width Δx of the detail, its local velocity can be calculated. This local measurement of width can be accomplished with the help of a series of receptive fields of different sizes, the center of which is the detector A. The fields should have the usual spatial organisation with an excitatory center and an inhibitory surround (or vice versa)[7]. That field which responds optimally during the passage of the detail is adapted best to the detail width[8]. Of course the optimal response can be distinguished only by evaluating the signals from all receptive fields, since the excitation of a single field depends both on size and luminance of the detail. But in comparing the signals from fields of various sizes, the influence of luminance can be eliminated and a local measuremant of width is possible.

Evidently the visual system does possess just these receptive fields which have been postulated from the requirements of velocity measurement. However, in a two-dimensional image analysis not only the absolute value but also the direction of local velocity is needed. This is also in accordance with the performance of the visual system, since receptive fields are known[9] which respond maximally to bars moving over the field with a specific orientation. These fields may be generated by selective lateral inhibition or excitation between adjacent detector elements[10].

CONCLUSIONS FOR SIGNAL PROCESSING AND AUTOMATICAL PATTERN RECOGNITION

The momentary signals from a two-dimensional detector array fill a two-dimensional field. In order to detect moving structures, the signals from successive sampling times have to be stored. They fill a three-dimensional storage memory. A two-dimensional section of this data structure which is generated by an one-dimensional detector array, is represented by the following scheme:

spatial sampling points / temporal sampling moments	x_o	x_1	x_2
t_o	$S(x_o,t_o)$ ↘	$S(x_1,t_o)$ ↘	$S(x_2,t_o)$
t_1	$S(x_o,t_1)$	$S(x_1,t_1)$ ↘	$S(x_2,t_1)$ ↘
t_2	$S(x_o,t_2)$	$S(x_1,t_2)$	$S(x_2,t_2)$

In cases of simple data structure, a correlation analysis may reveal, that e.g. the signals $S(x_o,t_o)$, $S(x_1,t_1)$, $S(x_2,t_2)$ are consistently larger than the rest of the data. It may be concluded then, that a bright spot has moved from x_o over x_1 to x_2 with the velocity

$$v = (x_1-x_o)/(t_1-t_o) = (x_2-x_1)/(t_2-t_1)$$

In complex scenes, this type of data analysis is not very effective, especially in the presence of noise[11]. Therefore it is very helpful for data analysis if additional information of local velocities is available. This information may be represented by arrows indicating the proper $\Delta x - \Delta t$ connection for each element in the above data scheme. In the given example the arrows suggest the hypothesis, that two image elements which at the time t_o are located at x_o, x_1 have been shifted during the first time interval to the positions x_1, x_2. During the second time interval, the left image element has shifted to x_2, while the other element has left the field. The hypo - thesis has to be checked by data analysis. If it is confirmed, much calculation time has been saved, compared to the correlation analysis mentioned. If it is disproved, something more complicated has happened which cannot be extracted from the available data.

This consideration shows that additional information on the velocity distribution in the image plane is very helpful even in cases when spatial noise is not the limiting factor.

For a proper allowance of the velocity information, the three-dimensional data field has to be transformed into a four-dimensional space with two spatial and two velocity components. Such a data structure may easily be dissected into equal velocity compartments. Even in visual pattern recognition, such a dissection is necessary, as can be seen from every day experience.

A procedure of facilitating pattern analysis by dissecting along velocity conturs is possible only when the local velocity distribution is available. A bilocal velocity analysis presupposes that pattern analysis has been finished already.

In automatical pattern recognition, local velocity evaluation may be performed along similar lines as has been suggested in the preceding section as a model for visual signal processing. Appropriate receptive fields may be generated by masks or filtering processes. Two or more successive spatial image distributions have to be evaluated to generate a local velocity distribution.

DISCUSSION

As has been mentioned in the foregoing, a similarity exists between the perception of moving objects in a steady field of view and the perception of stationary objects with the help of eye movements, since in both ceses a movement is involved between the retinal image and the detector array. However the movement is smooth in the case of a moving object and saccadic in the case of a stationary object. The type of signal processing which is postulated in this paper for the detection of moving objects requires some discontinuation of the detector signals, since the illuminance distribution at one instant of time t_o has to be compared with that at another definite time t_1, with $t_1 - t_o = \Delta t$ being related to a spatial shift Δx. This requires a temporal sampling process, which has some interesting aspects. In the first place the illuminance has to be integrated over a time interval long enough that sufficient radiation energy is absorbed to give a reliable signal. Then this signal has to be sampled, i.e. to be discontinuated from further integrative contributions of the detectors. Furthermore the signal has to be stored long enough to allow a comparison with the value from the next sampling interval.

There is evidence that this type of signal processing is indeed performed by the visual system. A momentary visual impression persists for some time[12]. This is easily demonstrated by the experiment of d'Arcy[13], which is performed by looking at a light source that is moved on a circular path. The faster the movement, the longer is the perceived "tail" of the light point, until with fast enough movement a closed luminous circle is seen. Hence a sample and hold operation is performed by the system. The time of persistence is influenced by movement[14], which indicates that the sampling process is adapted to the requirements of the visual task[15], in the present model to the comparison of signals from different retinal loci. Masking experiments[16, 17, 18, 19] as well as physiological findings[20] revealed a complicated system of mutual inhibition and facilitation between transient and sustained channels, which may very well be responsible for this sampling process.

According to the present model, the measurement of velocity is necessary for the detection of moving objects, and this measurement is done with the help of receptive fields of different size and of specific directional sensitivity. There is ample electro-physiological evidence for the existence of such receptive fields in the visual neuronal pathway of vertebrates. Moreover in psychophysical experiments spatial frequency channels for moving targets and with different peak frequencies have been found [21]. By Fourier transform, the peak frequency of a spatial frequency channel is related to the width of the central part of a corresponding receptive field, while the bandwith is related to size and spatial organization of the field [22].

If spatial frequency channels are but another description of local receptive fields, they should be organized locally, e.g. adaptation to a sine wave grid which is confined to a certain part of the visual field should not be transfered to the complementary part of the field. This has been found indeed for stationary grids [23]. On the other hand it has been shown [6], that there exists a masking interaction over large distances in the visual field, which seems to be in contradiction with the assumption of a local organization of spatial frequency channels. However it has to be borne in mind that in this experiment the target was presented for a very short time and was above threshold. Thus the assumptions of the present model are not fulfilled. There was not enough time for an eye movement to detect sensitivity variations and furthermore the detector characteristics may be nonlinear. It has been discussed [2] that with nonlinear detectors contrast is relative and needs calibration. The conditioning stimulus in Weisstein's experiment may serve as a calibration normal.

When the moving structure to be detected is an edge, the proposed mechanism of local velocity measurement seems to be impossible. Firstly, however, this is a limiting case and is not directly related to the detection of moving objects. Also Fourier transform gives problems with this structure. Furthermore a luminance step is differentiated by the visual system resulting in Mach band, which in turn has a definite detail width. Some caution is required, however, with this explanation, since the creation of Mach bands is usually assumed to be effected by receptive fields of a special size. These fields then should precede in the visual path the family of fields which serve for width detection.

For insects, mechanisms of movement detection have been found[24] which are based on other principles. But these investigations are based on flight orientation behavior, which is a task quite different from the detection and recognition of small objects in an otherwise stationary world. From a logical point of view it is no surprise, that correlation methods are quite efficient for flight orientation, since the whole visual world is affected consistently by this type of motion. These results will not be considered here in detail.

ACKNOWLEDGEMENT

This work has been supported financially as a part of the Schwerpunktprogramm Optik und Information by the Deutsche Forschungsgemeinschaft.

REFERENCES

1. R. Röhler, Biol. Cybernetics 32, 101 (1979).
2. R. Röhler, Optica Acta 26, 1407 (1979).
3. R. M. Boynton and J. Siegfried, J. Opt. Soc. Am. 52, 720 (1962)
4. R. M. Boynton, Discrimination of Homogeneous Double Pulses of Light. Handbook of Sensory Physiology VII/4, (Springer, Berlin, Heidelberg, New York, 1972), p. 202 ff.
5. N. Weisstein, Psychol. Rev. 75, 494 (1968).
6. N. Weisstein and Ch. S. Harris, Vis. Research 17, 341 (1977).
7. S. W. Kuffler, Cold Spring Harbor Sympos. Quant. Biol. 17, 281 (1952).
8. W. von Seelen, Kybernetik 7, 43 (1970).
9. D.H. Hubel and T. N. Wiesel, J. Physiol. (Lond.) 148, 574 (1954).
10. H. B. Barlow and W. R. Levick, J. Physiol. 178, 477 (1965).
11. J. P. Foith (Verf.), Angewandte Systemanalyse, DAGM Symp. Karlsruhe 1979,(Springer, Berlin, Heidelberg, New York 1979).
12. J. Plateau, Dissertation sur quelque propriétés des impressions produites par la lumière sur l'organe de la vue. Liège, Dessain 1829.
13. C. d'Arcy, Mém. Acad. Roy Sci. 37, 700 (1773).
14. W. T. Pollack,J. exp. Psychol. 45, 449 (1953).
15. G. van den Brink, Retinal Summation and the Visibility of Moving Objects, Thesis, Utrecht 1957.
16. B. G. Breitmeyer, Vision Research 17, 861 (1977).
17. B. G. Breitmeyer, Vision Research 18, 1401 (1978).
18. B. G. Breitmeyer, Vision Research 18, 1443 (1978).
19. M. W. von Grünau, Vision Research 18, 197 (1978).
20. W. Singer and N. Bedworth, Brain Res. 49, 291 (1973).
21. D. J. Tolhurst, J. Physiol. 231, 385 (1973).
22. I. D. G. MacLeod and A. Rosenfeld, Vision Res. 14, 909 (1974)
23. S. Klein, C. F. Stromeyer and L. Ganz, Vision Res. 14, 1421 (1974)
24. W. Reichardt and T. Poggio, Quarterly Rev. Biophys. 9, 311 (1976).

VARIANCE OF A TWO FREQUENCY SPECKLE FIELD AFTER PROPAGATION THROUGH THE TURBULENT ATMOSPHERE

J. Fred Holmes
V. S. Rao Gudimetla
Oregon Graduate Center, Beaverton, OR 97006

ABSTRACT

The extended Huygens-Fresnel principle is used to develop a formulation for the variance of the received intensity for a two frequency speckle field after propagation through the turbulent atmosphere. The formulation includes the effects of the log-amplitude covariance as well as the wave structure function. It takes into account the effects of the turbulent atmosphere on the laser beams as they propagate to the target and on the two frequency speckle pattern as it propagates back to the receiver.

INTRODUCTION

The source illuminating the target is taken as a TEM_{oo} laser beam consisting of two discrete frequencies with wave numbers k_1 and k_2. It is assumed for the development that the difference frequency between k_1 and k_2 is large enough to be beyond the pass band of the receiver electronics; but small enough that the perturbation of the intensities due to the atmosphere are almost perfectly correlated for the two wave numbers. From a practical standpoint this is not a severe restraint and does not exclude many problems of practical interest. In addition, the target is assumed to act like a perfectly diffuse surface at both wave numbers.

The variance of the received intensity can be expressed as

$$\begin{aligned}\sigma_I^2 &= \langle[(u_1 + u_2)(u_1^* + u_2^*)]^2\rangle - \langle(u_1 + u_2)(u_1^* + u_2^*)\rangle^2 \\ &= \langle(u_1u_1^* + u_2u_1^* + u_2u_2^* + u_1u_2^*)^2\rangle - \langle u_1u_1^* + u_2u_1^* \\ &\quad + u_1u_2^* + u_2^*u_2\rangle^2 \qquad (1)\end{aligned}$$

where u_1 and u_2 are the fields at the receiver due to transmitter wave numbers k_1 and k_2 respectively. Since the mutual intensities $u_2u_1^*$ and $u_1u_2^*$ will be beyond the pass band of the receiver, this reduces to

$$\sigma_I^2 = \langle I_1^2\rangle + \langle I_2^2\rangle + 2\,\langle I_1I_2\rangle - \langle I_1\rangle^2 - \langle I_2\rangle^2 - 2\,\langle I_1\rangle\langle I_2\rangle\ . \qquad (2)$$

Using the extended Huygens-Fresnel principle, a formulation for equation (2) that includes the effects of atmospheric turbulence will be developed.

ISSN:0094-243X/81/650701-07$1.50

ANALYSIS

In order to evaluate the quantity $<I_1I_2>$ analytically, it needs to be written in terms of the fields as shown in (3)

$$<I_1I_2> = <u_1u_1^* \, u_2u_2^*> \; . \tag{3}$$

The fields at the receiver in (3) can be evaluated in terms of the speckle field by using the extended Huygens-Fresnel formulation given by[1]

$$U(\overline{p}) = \frac{k}{2\pi iL} \exp\left(ik(L + p^2)/2L\right) \times \int U(\overline{\rho}) \exp\left[ik(\rho^2 - 2\overline{p} \cdot \overline{\rho})/2L + \psi_2(\overline{p},\overline{\rho})\right] d\overline{\rho} \tag{4}$$

where $\exp[\psi_2(\overline{p},\overline{\rho})]$ represents the effects of the turbulent atmosphere on the fields from a point source located at $\overline{\rho}$ in the speckle source plane as they propagate to a point $\overline{p}$ in the receiver plane, k is the wave number, and L is the path length. The complex, random quantity ψ is usually expressed as

$$\psi = \chi + i\phi$$

where χ represents the log-amplitude perturbation of a spherical wave due to atmospheric turbulence and ϕ the phase perturbation.

If (4) is now used to generate the four field quantities in (3) the resultant expression for $<I_1I_2>$ is given by

$$\begin{aligned} <I_1I_2> = & \left(\frac{k}{2\pi L}\right)^4 \iiiint d\overline{\rho}_1 \, d\overline{\rho}_2 \, d\overline{\rho}_3 \, d\overline{\rho}_4 \\ & \times <u_1(\overline{\rho}_1)u_1^*(\overline{\rho}_2)u_2(\overline{\rho}_3)u_2^*(\overline{\rho}_4)> \\ & \times \exp\left[\frac{ik}{2L} (\rho_1{}^2 - \rho_2{}^2 + \rho_3{}^2 - \rho_4{}^2 - 2\overline{p}_1 \cdot (\overline{\rho}_1 - \overline{\rho}_2) - 2\overline{p}_1 \cdot (\overline{\rho}_3 - \overline{\rho}_4)\right] \\ & \times < \exp\left[\psi_2(\overline{\rho}_1) + \psi_2^*(\overline{\rho}_2) + \psi_2(\overline{\rho}_3) + \psi_2^*(\overline{\rho}_4)> \; . \right. \end{aligned} \tag{5}$$

For the case where the speckle field is perfectly diffuse (spatially incoherent) then[2,3]

$$<u_1(\overline{\rho}_1)u_1^*(\overline{\rho}_2)u_2(\overline{\rho}_3)u_2^*(\overline{\rho}_4)> = \left(\frac{4\pi}{k^2}\right)^2 <I_1(\overline{\rho}_1)><I_2(\overline{\rho}_3)> \delta(\overline{\rho}_1 - \overline{\rho}_2) \, \delta(\overline{\rho}_3 - \overline{\rho}_4) \; . \tag{6}$$

Using (6) in (5) and performing the $d\overline{\rho}_1$ and $d\overline{\rho}_3$ integrations, it becomes

$$\langle I_1 I_2 \rangle = \frac{1}{\pi^2 L^2} \iint d\overline{\rho}_2 \, d\overline{\rho}_4 \, \langle I_1(\overline{\rho}_2)\rangle \langle I_2(\rho_4)\rangle \, e^{4C_\chi(\overline{\rho}_4 - \overline{\rho}_2)} \tag{7}$$

where use has been made of[4]

$$\left. \langle \exp[\psi_2(\overline{\rho}_1) + \psi_2^*(\overline{\rho}_2) + \psi_2(\overline{\rho}_3) + \psi_2^*(\overline{\rho}_4)\rangle \right|_{\substack{\overline{\rho}_1 = \overline{\rho}_2 \\ \overline{\rho}_3 = \overline{\rho}_4}} = e^{4C_\chi(\overline{\rho}_4 - \overline{\rho}_2)} \tag{8}$$

since $k_1 \approx k_2 \approx k$.

The log-amplitude covariance function C_χ is given to first order by[5]

$$C_\chi = 0.132\, \pi^2 k^2 L \int_0^1 dt\, C_n^{\,2}(t) \int_0^\infty du\, u^{-8/3} \operatorname{Sin}^2\left[\frac{u^2 t(1 - t) L}{k}\right] \times J_o\left[u\left|(\overline{\rho}_4 - \overline{\rho}_3)(1 - t)\right|\right] . \tag{9}$$

Before Eq.(7) can be carried any further, the quantity $\langle I(\overline{\rho}_i)\rangle$ which partially defines the spatially incoherent speckle field will have to be specified. One method of generating the speckle field that has wide applicability is to illuminate a diffuse target with a laser source. The source will be taken as a TEM_{oo} laser beam with the laser transmitter located at the same end of the propagation path as the receiver and adjacent to it. As indicated earlier, the transmitter radiates at two discrete wavelengths with amplitudes u_{01} and u_{02}.

The field distribution for a laser beam at the transmitter is given by

$$U_o(\overline{r}) = U_o \exp\left(- \frac{r^2}{2\alpha_o^{\,2}} - \frac{ikr^2}{2F}\right) \tag{10}$$

where α_o and F are the characteristic beam radius and focal length, respectively. The field at the target before scattering from the target then is written from the extended Huygens-Fresnel principle as

$$U(\overline{\rho}) = \frac{k e^{ik\left[L + \frac{\rho^2}{2L}\right]}}{2\pi i L} \int U_o(\overline{r}) \exp\left[\frac{ik(r^2 - 2\overline{r} \cdot \overline{\rho})}{2L} + \psi_1(\overline{\rho},\overline{r})\right] d\overline{r} \tag{11}$$

where ψ_1 describes the effects of the random medium on the propagation of a spherical wave from the source to the target. The effect of the target has been already taken into account in Eq.(7). Again utilizing the extended Huygens-Fresnel principle, the intensity distribution on the target is given by

$$<I(\bar{\rho}_i)> = <u(\bar{\rho}_i)u^*(\bar{\rho}_i)>$$

$$= \left(\frac{k}{2\pi L}\right)^2 U_o^2 \iint d\bar{r}_1\, d\bar{r}_2 \exp\left[-\left(\frac{r_1^2 + r_2^2}{2\alpha_o^2}\right)\right.$$

$$\left. + i\frac{k}{2L}\left(1 - \frac{L}{F}\right)(r_1^2 - r_2^2) - i\frac{k}{L}\bar{\rho}_i \cdot (\bar{r}_1 - \bar{r}_2)\right]$$

$$<\exp\left[\psi(\bar{\rho}_i, \bar{r}_1) + \psi^*(\bar{\rho}_i, \bar{r}_2)\right]> \tag{12}$$

where for uniform turbulence

$$<\exp \psi(\bar{\rho}_i, \bar{r}_1) + \psi^*(\bar{\rho}_i, \bar{r}_2)> = e^{-\left(\frac{|\bar{r}_2 - \bar{r}_1|}{\rho_o}\right)^{5/3}}, \tag{13}$$

$$\rho_o = (0.545625\, C_n^2\, Lk^2)^{-3/5},$$

and C_n^2 is the refractive index structure constant. Now performing several of the integrations in (12) it reduces to[2,3]

$$<I(\rho_i)> = \left(\frac{k}{L}\right)^2 \frac{U_o^2\alpha_o^2}{2} \int r dr\, J_o\left(\frac{k}{L}\rho_i r\right) \exp\left[-\frac{r^2}{4\alpha_o^2} - \left(\frac{r}{\rho_o}\right)^{5/3}\right.$$

$$\left. - \left[\frac{k}{L}\frac{\alpha_o}{2}\left(1 - \frac{L}{F}\right)\right]^2 r^2\right] \tag{14}$$

where $r = |\bar{r}_2 - \bar{r}_1|$.
Using (14) in (7) and using previously developed techniques to accomplish several of the integrations,[2,3] it becomes

$$<I_1 I_2> = <I_1><I_2> \left(\frac{k}{L}\right)^2 \iint r_2 dr_2\, d\rho\, J_o\left(\frac{k}{L} r_2\rho\right)\left(e^{4C_\chi(\rho)}\right)$$

$$\times \exp\left[-\frac{r_2^2}{2\alpha_o^2} - 2\left(\frac{r}{\rho_o}\right)^{5/3} - 2\left(\frac{k}{L}\frac{\alpha_o}{2}\left(1 - \frac{L}{F}\right)\right)^2 r_2^2\right] \tag{15}$$

where $\langle I_1 \rangle$ and $\langle I_2 \rangle$ are the expected values of the intensity at the receiver due to each of the discrete frequencies given by

$$\langle I_i \rangle = \frac{U_{oi}^{\,2} \, \alpha_o^{\,2}}{L^2} \quad . \tag{16}$$

It is customary to normalize the variance to the mean received intensity given by

$$\langle I \rangle = \langle I_1 \rangle^2 + \langle I_2 \rangle^2 + 2 \, \langle I_1 \rangle \langle I_2 \rangle \quad . \tag{17}$$

In addition it is convenient to let

$$\langle I_2 \rangle = G \langle I_1 \rangle$$

and recognizing by comparing (15) to previous work[3] that

$$\frac{\langle I_1^{\,2} \rangle}{\langle I_1 \rangle^2} = \frac{\langle I_2^{\,2} \rangle}{\langle I_2 \rangle^2} = \frac{2 \, \langle I_1 I_2 \rangle}{\langle I_1 \rangle \langle I_2 \rangle} \tag{18}$$

then

$$\sigma_{I_N}^{\;2} = 2 \, \langle I_1 I_2 \rangle_N \, \frac{[1 + G^2 + G]}{[1 + G]^2} - 1 \tag{19}$$

where

$$\langle I_1 I_2 \rangle_N = \frac{\langle I_1 I_2 \rangle}{\langle I_1 \rangle \langle I_2 \rangle} \quad . \tag{20}$$

NUMERICAL RESULTS

The quantity $\langle I_1 I_2 \rangle_N$ has the following form for the case of uniform turbulence:

$$\langle I_1 I_2 \rangle_N = \left(\frac{k}{L}\right)^2 \int_0^\infty \!\! \int_0^\infty r dr \, \rho d\rho \, e^{4C_\chi(\rho)} \, J_o\!\left(\frac{k}{L} \rho r\right) f_2(r) \tag{21}$$

where

$$f_2(r) = \exp\left[- r^2 \left(\frac{1}{2\alpha_o^{\,2}} - 2\left(\frac{k}{L} \frac{\alpha_o}{2} \left(1 - \frac{L}{F}\right)\right)^2 \right) - 2\left(\frac{r}{\rho_o}\right)^{5/3} \right] \quad . \tag{22}$$

The integrand contains the term $rJ_o(kr\rho/L) f_2(r)$. Our approach[3] is to expand $f_2(r)$ in a Fourier-Bessel series and then make use of

$$\int_0^\infty R dR \; J_m(r_1 R) J_m(r_2 R) = \frac{2 \, \delta(r_1 - r_2)}{r_1 + r_2}$$

to simplify the integration. The series is given by

$$f_2(r) = \sum_m b_m J_o\left(\frac{P_m}{A} r\right) \tag{23}$$

where

$$b_m = \frac{2}{A^2 J_1^2(P_m)} \int_o^A x f(x)\, J_o\left(\frac{P_m}{A} X\right) dx \tag{24}$$

and

$$J_o(P_m) = 0 \quad .$$

This expansion can be done because $f_2(r)$ becomes negligible for some value of r that is easily calculated and does not depend on the other integration variables. In addition, the shape of $f_2(r)$ allows it to be represented by only a few terms in the series.

Using this result, the normalized variance becomes

$$\sigma_{I_N}^{\ 2} = \left[2 \sum_m b_m\, e^{4C_\chi\left(\frac{L}{k} \frac{P_m}{A}\right)} \right] \frac{[1 + G + G^2]}{[1 + G]^2} - 1 \quad . \tag{25}$$

Equation (25) is plotted versus the log-amplitude variance, $\sigma_\chi^{\ 2}$, for G = 0 (single line) and G = 1 (two lines) in Figure 1. It can be seen that the effect of the atmosphere is to increase the normalized variance of the speckle field above its vacuum value. For only two modes the effect is slight. However, extension of this work to larger numbers of modes shows that the effect of the atmosphere can be substantial.

It should be noted that since C_χ plays a major role in determining the effect of the atmosphere on the variance, a form of C_χ that accounts for saturation should be used for $\sigma_\chi^{\ 2} \geqslant 0.3$.

A formulation valid for all levels of turbulence is given by[6]

$$C_\chi(\bar{\rho},o) = 2.95\, \sigma_T^{\ 2} \int_o^1 du\, [u(1 - u)]^{5/6} \int_o^\infty dy\, \frac{\sin^2 y}{y^{11/6}}$$

$$\times \exp\left\{-\sigma_T^{\ 2}\, [u(1 - u)]^{5/6}\, f(y)\right\} J_o\left[\left(\frac{4\pi y u}{1 - u}\right)^{1/2} \frac{\rho}{\sqrt{\lambda L}}\right] \tag{26}$$

where

$$\sigma_T^{\ 2} = .124\, k^{7/6}\, L^{11/6}\, C_n^{\ 2} \ ,$$

$$f(y) = 7.02\, y^{5/6} \int_{.7y}^\infty dx\, x^{-8/3} \left[1 - J_o(x)\right] \ ,$$

and J_o is the zero-order Bessel function. Direct evaluation of Eq.(26) is rather difficult. However, an efficient method for reducing Eq.(26) to numbers for use in Eq.(25) is described in Ref. 7.

REFERENCES

1. Lutomirski, R. F., and Yura, H. T., Appl. Opt., 10, 1652 (1971).
2. Lee, M. H., Holmes, J. Fred, and Kerr, J. R., J. Opt. Soc. Am., 66, 1164 (1976).
3. Holmes, J. Fred, Lee, Myung Hun, and Kerr, J. Richard, J. Opt. Soc. Am., 70, 355 (1980).
4. Lee, M. H., Holmes, J. F., and Kerr, J. R., J. Opt. Soc. Am., 67, 1279 (1977).
5. Wang, Ting-i, Clifford, S. F., and Ochs, G. R., Appl. Opt., 13, 2602 (1974).
6. Clifford, S. F., and Yura, H. T., J. Opt. Soc. Am., 64, 1641 (1974).
7. Holmes, J. Fred, Kerr, J. Richard, Elliott, Richard A., Lee, Myung H., Pincus, Philip A., and Fossey, Michael E., U.S. Army Armament Research and Development Command report ARSCD-CR-79-007, September 1978.

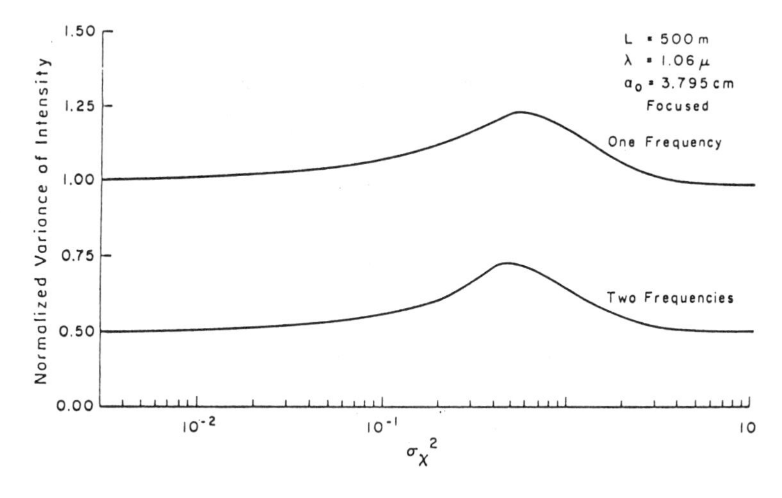

Figure 1 - Comparison of the normalized variance of the intensity for a single frequency and a two frequency speckle field after propagation through the turbulent atmosphere.

AN ELONGATED STRUCTURE OF SPECKLE PATTERNS IN POLYCHROMATIC LASER LIGHT

K. Nakagawa, Y. Tomita, and T. Asakura

Research Institute of Applied Electricity, Hokkaido University
Sapporo, Hokkaido, Japan

A fibrous radial structure of speckle patterns produced at the far-field diffraction plane by diffuse objects under illumination with polychromatic laser light has been studied experimentally by means of the spatial correlation function of the speckle intensity. It is shown that the fibrous structure of speckle patterns strongly depends on both the surface roughness of diffuse objects and the radial distance from the center of the patterns. The elongated polychromatic speckle patterns produced in the region far from their center are next studied in connection with the surface roughness of diffuse objects under illumination. For this purpose, a new method is proposed by which the one-dimensionally elongated speckle patterns are produced and their structure is adequately processed.

INTRODUCTION

In recent years, a fibrous radial structure of speckle patterns produced at the far-field diffraction plane under illumination with polychromatic laser light or white light has received much attention by many investigators. Fujii and Asakura[1,2] have given a qualitative explanation for the fibrous radial structure of speckle patterns from the viewpoint of the temporal coherence of illuminating light. Parry[3] and Pedersen[4] analyzed this structure by taking into account the statistical properties of diffuse objects. Very recently, Stansberg[5] has theoretically and experimentally studied the roughness-dependent fibrous structure of polychromatic speckle patterns by a Fourier transforming technique and has proposed a new method for measurements of surface roughness of diffuse objects. Except for the work of Stansberg, no quantitative experimental study has been done so far to verify the fibrous radial structure of speckle patterns in connection with the surface roughness properties of diffuse objects.

In this paper, a fibrous radial structure of speckle patterns produced at the far-field diffraction plane by the polychromatic laser light with four lines is first investigated experimentally by using the spatial correlation function of the speckle intensity. The elongated polychromatic speckle patterns produced in the region far from their center and outside the fibrous area are also investigated in connection with the surface roughness of diffuse objects under illumination. For this purpose, a new method is proposed for

ISSN:0094-243X/81/650708-06$1.50

producing and processing the one-dimensionally elongated speckle patterns.

FIBROUS RADIAL STRUCTURE OF POLYCHROMATIC SPECKLE PATTERNS [6]

To investigate the fibrous radial structure of polychromatic speckle patterns, a novel technique is devised to measure the spatial correlation function of the speckle intensity. Figure 1 shows a schematic diagram of the experimental arrangement. An argon-ion laser operating multimode with wavelengths of 476.5, 488.0, 496.5 and 514.5 nm is employed as a polychromatic light source. A ground glass plate used as a sample diffuse object is mounted on a rotating stage drived by an electric motor and is actually made to rotate in such a way that the center of rotation coincides with the center of illumination over the object plane. If the object is situated at an accurate optimum position, the speckle pattern produced at the observation plane rotates around the center of the optical axis without the boiling motion. The light scattered from the rotating object is divided by a beam splitter and detected by two photomultipliers Ph_1 and Ph_2 at the far-field diffraction planes of F_1 and F_2 with the same speckle pattern. Therefore, this situation is equivalent to the two-point detection study of one polychromatic speckle pattern by using the two same patterns. The output signals $I(\mathbf{r})$ from the photomultipliers at the positions $\mathbf{r}$ are amplified and brought into a correlator by which their dc components are first cut off and the correlation of the resultant ac components, expressed as $\Delta I(\mathbf{r}) = I(\mathbf{r}) - \langle I(\mathbf{r}) \rangle$, is calculated next as a function of the two detecting points. The spatial correlation function of the speckle intensity fluctuations at the positions $\mathbf{r}$ and $\mathbf{r}+\delta\mathbf{r}$ is given by

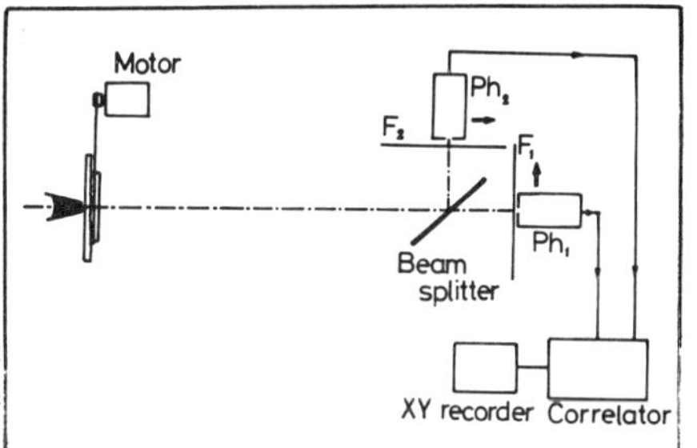

Fig.1 Experimental arrangement for measuring the two-dimensional spatial correlation function.

$$g_{\Delta I}(\mathbf{r},\mathbf{r}+\delta\mathbf{r}) = \frac{\langle \Delta I(\mathbf{r})\ \Delta I(\mathbf{r}+\delta\mathbf{r}) \rangle}{[\langle \Delta I^2(\mathbf{r}) \rangle \langle \Delta I^2(\mathbf{r}+\delta\mathbf{r}) \rangle]^{1/2}} \tag{1}$$

and can be measured as a function of the two detecting points by rotating the speckle patterns with the rotating motion of the diffuse object under illumination. The measuring method proposed and used here to obtain the two-dimensional spatial correlation area of the speckle pattern may be explained by using Fig.2. In this figure, the optic axis corresponds to the center of the fibrous radial speckle pattern. We now investigate the two-dimensional correlation area around a certain fixed point A, separated by a distance $\mathbf{r}$ from the center of the speckle pattern, by means of the spatial correlation function $g_{\Delta I}(\mathbf{r},\mathbf{r}+\delta\mathbf{r})$ of the speckle intensity fluctua-

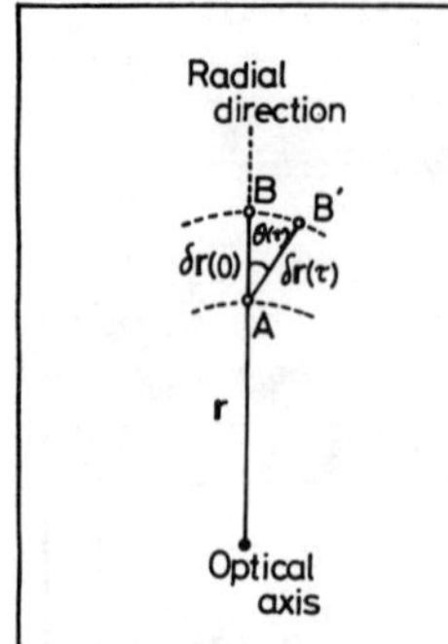

Fig.2 Geometry for measuring the two-dimensional spatial correlation area.

tions. The correlation area is defined as the area within a contour of the point B' where the correlation function becomes 1/e times its maximum value of unity obtained when the point A coincides with the point B. In order to follow the point B' two-dimensionally, the continuously time-delayed correlation function $\langle I_A(t) I_B(t+\tau) \rangle = \langle I_A(t) I_{B'}(t) \rangle$ is calculated from the time-varying intensities of $I_A(t)$ and $I_B(t)$ from the photomultipliers Ph_1 and Ph_2 at the points A and B. The time delay of τ_o at which the correlation function decays to 1/e is determined.

By using the time delay τ_o, and with the assumption that the radial distance r is large enough compared with the correlation area of the speckle pattern, the distance $\delta r(\tau_o)$ and the angle $\theta(\tau_o)$ which identify the point B' are given by

$$\left.\begin{aligned} |\delta \mathbf{r}(\tau_o)| &= \left[|\delta \mathbf{r}(0)|^2 + |\mathbf{r} + \delta \mathbf{r}(0)|^2 \omega^2 \tau_o^2\right]^{1/2}, \\ \theta(\tau_o) &= \cos^{-1}\left[\delta \mathbf{r}(0) \,/\, \delta \mathbf{r}(\tau_o)\right], \end{aligned}\right\} \quad (2)$$

where ω is the angular velocity of the diffuse object. In this case, $\delta \mathbf{r}(\tau_o)$ satisfies the equality of

$$g_{\Delta I}(\mathbf{r}, \mathbf{r} + \delta \mathbf{r}(\tau_o)) = 1/e \quad (3)$$

and corresponds to the correlation length measured from the fixed point A. Since $\mathbf{r}$ and $\delta \mathbf{r}(0)$ are known from the experimental setting of the points A and B, ω is known from the rotation velocity of the diffuse object, and τ_o is determined from the correlation function of the speckle intensity fluctuations, $\delta \mathbf{r}(\tau_o)$ and $\theta(\tau_o)$ are obtained from Eq.(2) and the point B' is correctly designated with respect to the fixed point A. This experimental procedure is repeated by changing the point B along the same radial direction until the central value of the correlation function reaches the value of 1/e.

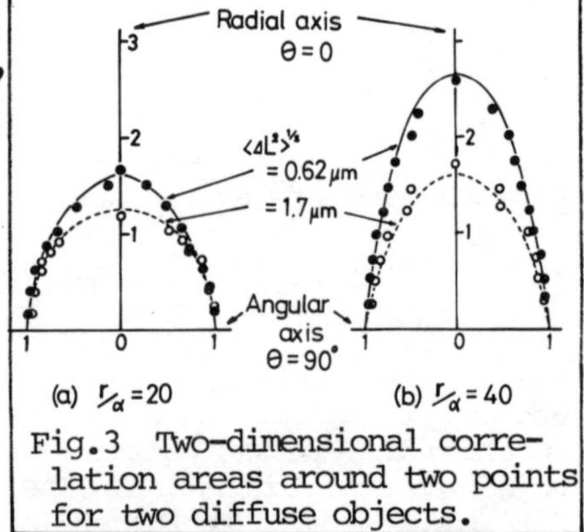

Fig.3 Two-dimensional correlation areas around two points for two diffuse objects.

Figure 3 shows the correlation areas obtained experimentally with respect to two values of $\mathbf{r}/\alpha = 20$ and 40 for two diffuse objects having rms surface roughnesses of $\langle \Delta L^2 \rangle^{1/2} = 0.62$ and $1.7\,\mu m$. The measuring points and the correlation lengths are normalized by the average speckle size α for the wavelength of 514.5 nm. This figure shows that a fibrous structure in polychromatic speckle patterns appears. As is clearly seen from this figure, the radial elongation of the speckle patterns increases with an increase of the radial distance from the center of the patterns and decreases with an increase of surface roughness of diffuse objects. The phenomenon of increasing elongation is due to the angular dispersion of light by which the radial decorrelation is produced. On the other hand, a decrease in elongation is due to an increase of surface roughness by which the wavelength decorrelation is enhanced.

ELONGATED POLYCHROMATIC SPECKLE PATTERNS IN THE REGION FAR FROM THEIR CENTER

A fibrous structure of speckle patterns produced at the far-field diffraction plane by the plychromatic light with discrete spectral lines, for example, from an argon-ion laser may differ from that of speckle patterns produced by the white light with continuous spectra. This difference can be observed in the region far from the center of the elongated speckle patterns and outside the area where the fibrous radial structure is produced. In this region, each of the spectral components of speckle patterns produced by the polychromatic light with discrete spectra are separated and, consequently, their fibrous structure is collapsed. As a result, we have produced random speckle patterns which are usually observed with monochromatic light to have a somewhat radial structure. The condition under which speckle patterns are produced far from their center is expected to have a certain relation whith the surface roughness of diffuse objects. This section investigates this relation by using one-dimensionally elongated speckle patterns instead of the two-dimensionally elongated radial ones.

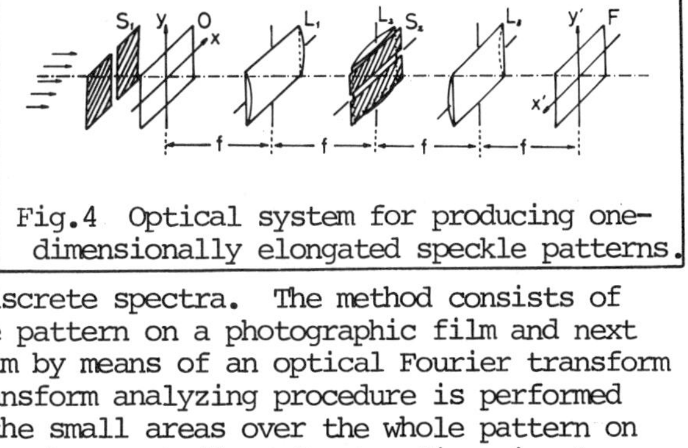

Fig.4 Optical system for producing one-dimensionally elongated speckle patterns.

We now propose a new method of obtaining some information about the roughness dependent characteristics of speckle patterns, in the region far from their center and outside the fibrous structure, produced by the polychromatic light with discrete spectra. The method consists of recording first the speckle pattern on a photographic film and next analyzing the developed film by means of an optical Fourier transform technique. The Fourier transform analyzing procedure is performed independently for each of the small areas over the whole pattern on the film. This procedure may become tedious for two-dimensional radial speckle patterns. To overcome this tedious procedure and to make easy the processing of patterns, the one-dimensionally elongated speckle patterns are effectively used and can be obtained by the optical system of Fig.4. A diffuse object O is illuminated by the collimated polychromatic light through a slit S_1. An image of the diffuse object along the direction of the y axis is formed at the plane F by a double diffraction imaging system consisting of two cylindrical lenses L_1 and L_3 together with a slit S_2. The x component of the light scattered from the diffuse object is Fourier transformed by a cylindrical lens L_2 onto the plane F. In this way, the one-dimensionally elongated speckle pattern is produced at the plane F. In this system, the average grain size of speckles produced with one wavelength is actually limited by the widths of two slits S_1 and S_2 for the directions of x' and y', respectively.

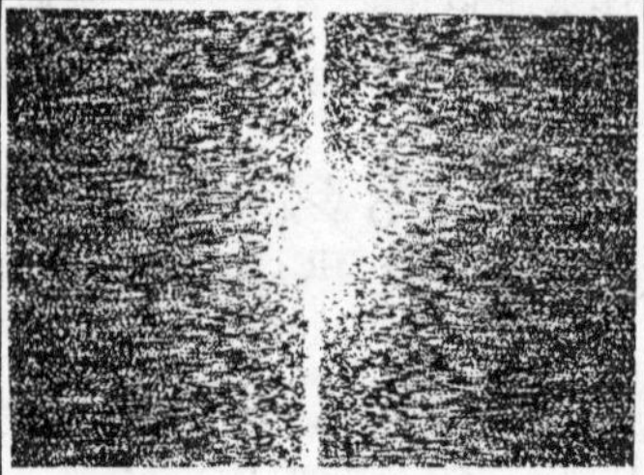

Figure 5 shows one typical photograph of one-dimensionally elongated speckle patterns obtained experimentally by using the optical system of Fig.4. A central bright line is produced by the specular component of the scattered light. For this pattern, a ground glass plate with rms roughness of $<\Delta L^2>^{1/2} = 0.14$ m was used as a diffused object, and the same argon-ion laser operating as in the previous experiment in a multiline mode with four wavelengths was employed again as a polychromatic light source. The developed film F is analyzed next by the optical Fourier transoming system of Fig. 6. A slit S_1 is made to move along the direction of the x' axis and actually successively moved along the direction of the x' axis in order for a rectangle area of the developed film along the direction of the y' axis to be Fourier transformed. Then, the Fourier transformed pattern is produced at the plane D.

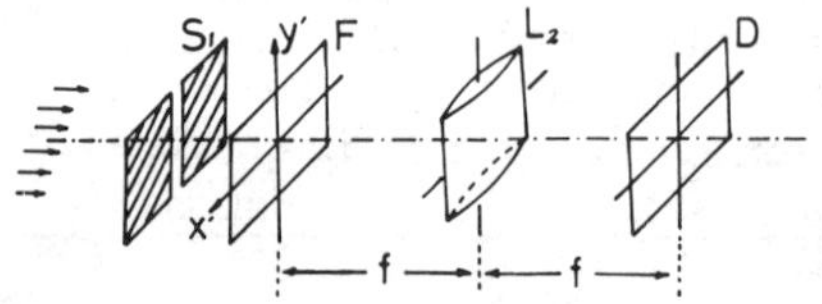

Fig.5 One-dimensionally elongated speckle pattern produced by the polychromatic light.

Fig.6 Fourier transforming system

Figure 7 shows the Fourier transformed patterns obtained for two different diffuse objects of $<\Delta L^2>^{1/2} = 0.14$ and $1.7\ \mu m$ by moving the position of the slit S_1 from the center of the developed film. The position of the slit is normalized the average speckle size α. In this case, the original elongated speckle patterns recorded on the films were obtained under illumination of the dichromatic light with two wavelengths of 488.0 and 496.5 nm. As is clearly seen from Fig.7, Young's fringes are observed for large values of x'/α whose spacing becomes narrower with an increase of the distance x'/α of the slit from the center of the patterns.

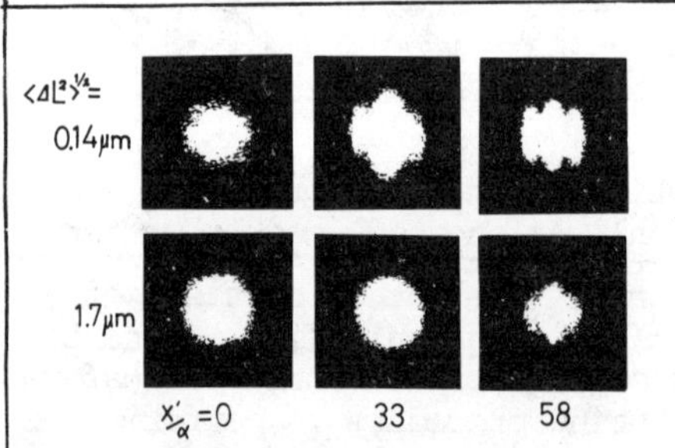

Fig.7 Fourier transformed patterns obtained from the partial slit areas of speckle patterns with three different positions.

This effect results from the wavelength dispersion. When the slit moves from the center of the patterns, the position x'/α of the slit where Young's fringes are clearly observed is changed by the surface roughness of diffuse objects. This position moves away from the center of the patterns with an increase of surface roughness of the diffuse objects. This is caused by decorrelation between the speckle patterns produced by different wavelengths in the illuminating light. Dependence of the Fourier transforming position x'/α on the rms surface roughness of diffuse objects is shown in Fig.8 for five different diffuse objects of $\langle \Delta L^2 \rangle^{1/2} = 0.14$, 0.36, 0.67, 0.99 and 1.7 μm. The position x'/α does not vary from an almost constant value for very smooth surfaces, but it increases gradually with an increase of surface roughness of the diffuse objects. This tendency can be interpreted by considering the decorrelation phenomenon of speckle patterns due to different wavelengths. The varying tendency of the increasing distance x'/α from the center is produced from a certain value of surface roughness and is enhanced gradually with an increase of surface roughness. A varying feature of the roughness dependent curve in Fig.8 may depend on the number and combination of wavelengths involved in the illuminating light.

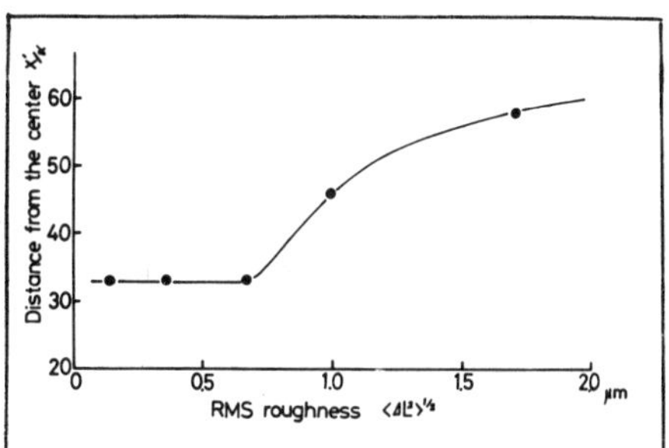

Fig.8 Dependence of the Fourier transforming position x'/α on the rms roughness of diffuse objects.

REFERENCES

1. H. Fujii and T. Asakura, Optik 39, 99 (1973).
2. H. Fujii and T. Asakura, Optik 39, 284 (1974).
3. G. Parry, Opt. Commun. 12, 75 (1974).
4. H.M. Pedersen, Opt. Acta 22, 523 (1975).
5. C.T. Stansberg, Appl. Opt. 18, 4051 (1979).
6. Y. Tomita, K. Nakagawa and T. Asakura, Appl. Opt. 19 (1980), to be published.

IMAGE FORMATION BY A CENTRALLY OBSTRUCTED CIRCULAR APERTURE: SPECKLE CONSIDERATIONS

M. Devi & K. Singh
Physics Department
Indian Institute of Technology, Delhi
New Delhi - 110 016

ABSTRACT

In this paper we present the results of some theoretical investigations on the aperture shape dependence of the speckle statistics in and near the far field of an optical system. Making use of the known results from diffraction theory, contrast and related quantities are obtained in terms of the statistical parameters of a rough surface and optical performance characteristics of the imaging system. Detailed results are presented for the oscillatory behavior of speckle contrast both in the focal plane and along the axis of the optical system.

INTRODUCTION

Speckle pheonomena in coherent optics have been, and still are, a subject of extensive study[1,2]. Recent work on speckle statistics, particularly in the far field, led to some useful applications in estimating the statistical parameters pertaining to surface roughness[3]. The speckle statistics along the axis of a conventional optical system with Gaussian apodization was studied by Jakeman and Welford[4]. Ouchi, in a subsequent paper[5], pointed out that with an optical system having a hard aperture, the contrast passes through maxima and minima; specifically, for a circular hard aperture, he derived[6] expressions for the contrast in the far field and out of focus axial points.

In addition to circular apertures, optical systems with central obstructions are by far most common. The analytical description of performance characteristics of an optical system in cassegranian and coaxial trans-receiver-lidar configuration neglecting the central obstruction represents an idealization that cannot be realized in practice. Masters and Rye investigated the efficiency of a heterodyne lidar[7] by taking into consideration the central obstruction of the receiver. Chelli, et al.[8] and Aime et al.[9], took into account the central obstruction of the telescope in their study of the M.T.F. for speckle interferometry. On the other hand, optical systems with deliberately large central obstructions of the pupil are used because of the narrowing of the central portion of the impulse response and the increased depth of focus[10]. It is noteworthy that interest in the performance of centrally and non-centrally obstructed pupils continues to grow[11-15].

In the present paper we derive expressions for the speckle contrast at an arbitrary point in the far field and an axial point in out of focus planes of an optical system with an annular pupil. The diffuser responsible for speckle is assumed to be a thin screen introducing a Gaussian distributed random phase fluctuations with a Gaussian

ISSN:0094-243X/81/650714-17$1.50

correlation function. We thus extend the analysis of Ouchi[6] to the case of an obstructed aperture. The lens is assumed to be aberration free and the illumination perfectly coherent, with plane phase front normal to the optic axis. The mathematical formulation of Stoffregen[16] for the average intensity and contrast is used to derive these quantities in our case. It is found that Ouchi's results[6] can be derived more readily by using this formulation instead of that due to Jakeman and Welford[4] (equation 17a and 17b). The equivalence of the expressions derived by Stoffregen[16] and by Jakeman and Welford[4] for contrast and average intensity can also be established easily.

THEORY

Following Ouchi we consider the optical system illustrated in Fig. 1. A normally incident plane wave with wavelength λ illuminates the diffuser at the plane $r_0 \equiv (x_0, y_0)$. The near field of the scattered wave is collected by an optical system with an annular pupil close to a lens of focal length f. The pupil function is unity for for the range $w_1 \leq r_0 \leq w_2$ and zero elsewhere; w_1 and w_2 being the inner and outer radii of the annular pupil. The diffuser as mentioned in the Introduction produces Gaussian distributed random phase fluctuations $\phi(r_0)$ whose correlation has a Gaussian form given by

$$\langle\phi(r_{01})\,\phi(r_{02})\rangle / \langle\phi^2\rangle = \exp\left[-|r_{01}-r_{02}|^2/\xi^2\right] \tag{1}$$

where ξ is the characteristic correlation length and $\langle\phi^2\rangle = \sigma^2$ is the variance.

The complex amplitude after the diffuser is given by $A_0 \exp[i\phi(r_0)]$ where A_0 is a constant with appropriate dimensions. The complex amplitude at an arbitrary point $(r,z) \equiv (x,y,z)$ to the right of the optical system is given by Fresnel diffraction theory as

$$A(r,\xi) = \frac{A_0}{\pi(w_2^2-w_1^2)} \iint \exp\left[i\,\phi(r_0)\right] \cdot \exp\left[-ik\left(\frac{r_0 \cdot r}{f} + \frac{r_0^2 z}{2f^2}\right)\right] \tag{2}$$

where $k = 2\pi/\lambda$ and the integration extends over the annular pupil A'.

We work under the usual assumption that many correlation cells contribute to the field at a point of observation so that the integral on the right hand side of eq. 2 may be considered as composed of many independent contributions. Then by the central limit theorem, the real and imaginary parts of the complex amplitude will behave in general as a correlated joint Gaussian process. The ensemble average (denoted by the angular brackets < >) and the variance of the intensity distribution in terms of those of A, AA* and A^2 are given by (Ref. [16])

$$\langle I\rangle = \langle A_r^2\rangle + \langle A_i^2\rangle = \langle|A|^2\rangle \tag{3}$$

where A_r and A_i are the real and imaginary parts of the complex amplitude A and

$$\sigma_I^2 = |\langle A^2 \rangle|^2 - 2|\langle A \rangle|^4 + \langle |A|^2 \rangle^2 \quad (4)$$

The variances of the real and imaginary components of the complex amplitude σ_r^2 and σ_i^2 and the correlation coefficient C can be also expressed in terms of $\langle A^2 \rangle$, $\langle |A|^2 \rangle$ and $\langle A \rangle$

$$\sigma_r^2 = \frac{1}{2}\langle |A|^2 \rangle + \frac{1}{2} \text{Re}\,[\langle A^2 \rangle] - \text{Re}^2[\langle A \rangle] \quad (5)$$

$$\sigma_i^2 = \frac{1}{2}\langle |A|^2 \rangle - \frac{1}{2} \text{Re}\,[\langle A^2 \rangle] - \text{Im}^2[\langle A \rangle] \quad (6)$$

$$C = \text{Im}\,[\langle A^2 \rangle - \langle A \rangle]/(2\,\sigma_r\,\sigma_i) \quad (7)$$

The argument of A is omitted for simplicity. We shall calculate the various averages and moduli of A in equations 3 and 4, and obtain the speckle contrast in the far field and in out of focus planes. The variances of A_r and A_i, and their correlation are also calculated.

Far Field Speckle Contrast

From equation 2 the complex amplitude in the far field is given as

$$A(r,0) = \frac{A_0}{\pi(w_2^2 - w_1^2)} \iint_{A'} \exp[i\phi(r_0)] \cdot \exp -\left[\frac{ikr \cdot r_0}{f}\right] d^2 r_0 \quad (8)$$

The ensemble average of A is given by

$$\langle A \rangle = A_0 \exp\left(-\frac{\sigma^2}{2}\right)\left[2\,\frac{J_1(U_2)}{U_2} - 2\,\frac{J_1(U_1)}{U_1}\,\varepsilon^2\right]/(1-\varepsilon^2)$$

$$= A_0 \exp\left(\frac{-\sigma^2}{2}\right) \cdot \mu'_1 \quad (9)$$

where

$$U_j = \frac{kw_j}{f}\,r \qquad (j = 1,2)$$

$$\varepsilon = w_1/w_2$$

and

$$\mu'_1 = \left[2\,\frac{J_1(U_2)}{U_2} - 2\,\frac{J_1(U_1)}{U_1}\,\varepsilon^2\right]/(1-\varepsilon^2)$$

The average value of the intensity[17] is given by

$$\langle |A|^2 \rangle = \langle I \rangle = A_0^2 \exp(-\sigma^2) \left[\left\{ 2 \frac{J_1(U_2)}{U_2} - 2 \frac{J_1(U_1)}{U_1} \varepsilon^2 \right\}^2 / (1-\varepsilon^2) + \frac{\xi^2}{w_2^2 - w_1^2} \sum_{m=1}^{\infty} \frac{(\sigma^2)^n}{n!} \frac{1}{n} \exp\left(- \frac{k^2 r^2 \xi^2}{4nf^2} \right) \right]$$

$$= A_0 \exp(-\sigma^2) \left[\mu'^2_1 + \alpha' T_f \right] \qquad (10)$$

where

$$\alpha' = \xi^2 / (w_2^2 - w_1^2)$$

and

$$T_f = \sum_{n=1}^{\infty} \frac{(\sigma^2)^n}{n!} \frac{1}{n} \exp\left(- \frac{k^2 r^2 \xi^2}{4nf^2} \right)$$

Returning again to $\langle A^2 \rangle$, this is found from equation 8 to be given by (Ref. [17])

$$\langle A^2 \rangle = A_0^2 \exp(-\sigma^2) \left[\mu'^2_1 + \alpha' \left\{ \frac{J_1(2U_2)}{U_2} - \frac{J_1(2U_1)}{U_1} \varepsilon^2 \right\} \frac{1}{(1-\varepsilon^2)} \sum \frac{(-\sigma^2)^n}{n!} \frac{1}{n} \exp\left(- \frac{k^2 r^2 \xi^2}{4nf^2} \right) \right]$$

$$\equiv A_0 \exp(-\sigma^2) \left[\mu'^2_1 + \mu'_2 \alpha' S_f \right] \qquad (11)$$

After substituting for the various terms in equation 4 we obtain

$$\sigma_I^2 = A_0^4 \exp(-2\sigma^2)[\alpha'^2 \mu'^2_2 S_f^2 + 2\alpha' \mu'^2_1 \mu'_2 S_f - \mu'^4_1] + A_0^4 \exp(-2\sigma^2)[\mu'^2_1 + \alpha T_f]^2 \qquad (12)$$

and

$$\frac{\sigma_I^2}{\langle I \rangle^2} = 1 + \frac{\alpha'^2 \mu'^2_2 S_f^2 + 2\alpha' \mu'^2_1 \mu'_2 S_f - \mu'^4_1}{(\mu'^2_1 + \alpha' T_f)^2} \qquad (13)$$

On substituting the values of $\langle A \rangle$, $\langle |A|^2 \rangle$ and $\langle A^2 \rangle$ from equation 9, 10, 11 in equations 5, 6 and 7 we obtain

$$\sigma_r^2 = \frac{1}{2} A_0 \exp(-\sigma^2) \alpha' [T_f + \mu'_2 S_f] \qquad (14)$$

$$\sigma_i^2 = \frac{1}{2} A_o \exp(-\sigma^2)\, \alpha' \,[T_f + \mu'_2 S_f] \tag{15}$$

$$C = 0 \tag{16}$$

Speckle Contrast in out of focus planes.

Coming to the contrast of the speckle pattern at an arbitrary axial point near the far field (distance z from the focal plane), the complex amplitude at a point (0,0,z) is given by

$$A(0,0,z) = \frac{A_o}{\pi\left(w_2^2 - w_1^2\right)} \iint_{A'} \exp[i\phi(r_o)] \cdot \exp\left[-i\frac{kz}{2f^2} \cdot r_o^2\right] d^2 r_o \tag{17}$$

where A' is the region of the annular pupil. If we write $z_e = f^2/3$ in equation 17, the integral represents the complex amplitude at an axial point z_e behind a diffuser illuminated by a plane wave truncated by a centrally obscured aperture A'.

As mentioned in the beginning, the central limit theorem is invoked, and equation 17 then represents constribution to the amplitude by many independent correlation elements. The ensemble average of A is given by

$$\langle A \rangle = A_o \exp(-\sigma^2)\, \mathrm{sinc}\left(\frac{a_2 - a_1}{2}\right) \exp\left(-i\,\frac{a_1 + a_2}{2}\right) \tag{18}$$

where

$$a_j = s w_j^2 \qquad (j = 1,2)$$

and

$$s = k/(2|z_e|)$$

The average values of the intensity $\langle AA^* \rangle$ and of $\langle A^2 \rangle$ are given by

$$\langle AA^* \rangle = \langle I \rangle = A_o \exp(-\sigma^2) \left[\mathrm{sinc}^2\left(\frac{a_2 - a_1}{2}\right) + \frac{1}{s^2(w_2^2 - w_1^2)} \sum_{n=1}^{\infty} \frac{(\sigma^2)^n}{n!} \frac{1}{n} \left\{ \exp\left(-\frac{b_1^2}{n}\right) - \exp\left(-\frac{b_2^2}{n}\right) \right\} \right] \tag{19}$$

where

$$b_j = s w_j \xi \qquad (j = 1,2)$$

and

$$\langle A^2 \rangle = A_0^{\,2} \exp(-\sigma^2) \cdot \left[\mathrm{sinc}^2 \left(\frac{a_2-a_1}{2}\right) \cdot \exp\left(-i(a_1+a_2)\right) \right.$$

$$+ \frac{\xi^2}{2s(w_2^{\,2}-w_1^{\,2})} \left\{ (s_2 \sin 2a_2 - s_1 \sin 2a_1) \right.$$

$$\left.\left. + \; i\,(s_2 \cos 2a_2 - s_1 \cos 2a_1) \right\} \right]$$

Also

$$|\langle A^2 \rangle|^2 = A_0^{\,4} \exp(-2\sigma^2) \left[\mathrm{sinc}^4 \left(\frac{a_2-a_1}{2}\right) + \frac{\xi^2}{4s^2(w_2^{\,2}-w_1^{\,2})^4} \right.$$

$$(s_2^{\,2} + s_1^{\,2} - 2s_1s_2 \cos\{2(a_2-a_1)\}) +$$

$$\frac{\xi^2}{s(w_2^{\,2}-w_1^{\,2})^2} \mathrm{sinc}\left(\frac{a_2-a_1}{2}\right) \left\{ \cos(a_1+a_2)(s_2 \sin 2a_2 - \right.$$

$$\left.\left. s_1 \sin 2a_1) - \sin(a_1+a_2)(s_2 \cos 2a_2 - s_1 \cos 2a_1) \right\} \right] \tag{20}$$

where

$$s_j = \sum_{n=1}^{\infty} \frac{(\sigma^2)^n}{n!} \frac{1}{n} \exp\left(-\frac{b_j^{\,2}}{n}\right)$$

In addition, if we let

$$T_j = \sum_{n=1}^{\infty} \frac{(-\sigma^2)^n}{n!} \exp\left(-\frac{b_j^{\,2}}{n}\right)$$

and substitute for the various terms in equation 4 we obtain the following expression for the contrast

$$\frac{\sigma_I^{\,2}}{\langle I \rangle^2} = 1 + \left[\frac{\xi^2}{4s^2(w_2^{\,2}-w_1^{\,2})^4} \left\{ s_2^{\,2}+s_1^{\,2}-2s_1s_2 \cos 2(a_2-a_1) \right\} \right.$$

$$+ \frac{\xi^2 \,\mathrm{sinc}^2 \left(\frac{a_2-a_1}{2}\right)}{s\,(w_2^{\,2}-w_1^{\,2})^2} \left\{ \cos(a_1+a_2)\,(s_2 \sin 2a_2 - s_1 \sin 2a_1) \right.$$

$$\left.\left. - \sin(a_1+a_2)\,(s_2 \cos 2a_2 - s_1 \cos 2a_1) \right\} \right] \Big/$$

$$\left[\mathrm{sinc}^2 \left(\frac{a_2-a_1}{2}\right) + \frac{1}{s^2(w_2^{\,2}-w_1^{\,2})} (T_1-T_2) \right]^2 \tag{21}$$

TABLE 1. Some Features of the Amplitude Point Spread Function of an Annular Aperture

Obscuration Ratio	Position of the First Minima	First Secondary Maxima Position/Relative Amplitude		Position of the Second Maxima	Second Secondary Maxima Position/Relative Amplitude	
0.1	3.70	5.13	-0.1433	7.13	8.42	0.0559
0.2	3.66	5.12	-0.1742	7.41	8.46	0.03867
0.3	3.50	5.07	-0.2179	7.62	8.57	0.03321
0.4	3.32	4.97	-0.2659	7.51	8.70	0.0578
0.5	3.14	4.82	-0.3103	7.11	8.67	0.0111
0.6	2.97	4.64	-0.3468	6.82	8.45	0.1749
0.7	2.81	4.44	-0.3735	6.46	8.11	0.2308
0.8	2.66	4.23	-0.3908	6.13	7.75	0.2709

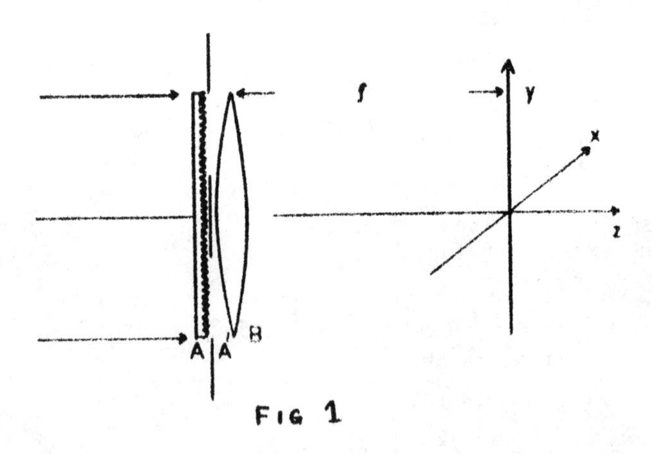

Fig. 1

The Optical System Forming Speckles. A-diffuser A'-annular aperture. B-Lens

Again the variance of the real and imaginary parts of the complex amplitude and their correlation follow from equations 5,6,7

$$\sigma_r^2 = \frac{1}{2} A_o e^{-\sigma^2} \left[\frac{T_1 - T_2}{s^2 (w_2^2 - w_1^2)^2} + \frac{\xi^2}{2s(w_2^2 - w_1^2)^2} (s_2 \sin 2a_2 - s_1 \sin 2a_1) \right]$$

$$\sigma_i^2 = \frac{1}{2} A_o e^{-\sigma^2} \left[\frac{T_1 - T_2}{s^2 (w_2^2 - w_1^2)^2} - \frac{\xi^2}{2s(w_2^2 - w_1^2)^2} (s_2 \sin 2a_2 - s_1 \sin 2a_1) \right]$$

$$C = \frac{1}{2\sigma_r \sigma_i} \frac{\xi^2}{2s(w_2^2 - w_1^2)^2} (s_2 \cos 2a_2 - s_1 \cos 2a_1) \qquad (22)$$

DISCUSSION OF RESULTS

For the light field in the focal region of our optical system, we have derived expressions for the average intensity, contrast and variance of the real and imaginary parts of the complex amplitude in addition to their correlation coefficient. The equations 13 and 21 giving the average intensity are the sum of two terms: the first represents the specular component, while the second describes the speckle pattern. As known[6] this superposition leads to contrast variations. The peaks of the contrast are located at the zeros of the first term (which is the intensity point spread function) and the minima of the contrast at its peaks. Also, when $\sigma_r^2 = \sigma_i^2$ the amplitude statistics are circular Gaussian. When $\langle A_r \rangle = 0$, the intensity statistics are negative exponential. In all other cases, the intensity is described by a Rice distribution. In the first case, the contrast reaches a maximum of unity while in the latter case, the maximum is less than one. From equations 14,15, and 9, it can be seen that the first situation prevails at the even zeros of μ'_2 (which also correspond to the zeros of μ'_1). The second situation occurs at the odd zeros of μ'_2 where μ'_1 is not equal to zero. From Fig. 2, one can see that the contrast peaks are located at the zeros of μ'_1. It is known that the first zero of μ'_1 shifts continuously toward the center as the size of the obstruction ε increases. The first contrast maximum as seen from Fig. 2 shifts to the center for higher values. However, the second zero of μ'_1 is known to move away from the center when ε increases from 0 to approximately 0.3, and then returns to the starting point for ε of the order of 0.5. A further increase in ε takes it nearer to the center (Table 1 gives some features of the amplitude point spread function for different values of ε). The second contrast peak indeed follows this variation (Fig. 2). At points other than the zeros of μ'_2, the amplitude statistics is non-circular.

In addition, for $r = 0$ (from equation 10) and $z = 0$ (from equation 9), we find the same value of $\langle I \rangle$. The variances σ_r^2 (and σ_i^2) from equations 14, 22 (and 15, 23) turn out to be equal in this case. When $\sigma_r^2 \neq \sigma_i^2$, it is clear that the statistics are not circular at the focus. The axial intensity variation follows a sinc^2 pattern whose argument is proportional to $(1-\varepsilon^2)$. The plots of contrast (Fig. 3) show this clearly.

It is seen that our plot for $\varepsilon = 0.1$ is very close to that of Ouchi[6] for a circular pupil. In addition, for $\varepsilon = 0$, all the expressions derived here reduce to the corresponding expressions given by Ouchi[6], which can be derived more readily by using the formulation of Stoffregen[16] instead of the equations 17a and 17b of Jakeman and Welford[4], and the defining relations for σ_r^2, σ_i^2 and C.

ACKNOWLEDGEMENT

The authors wish to thank Professor M.S. Sodha for his interest in the work.

REFERENCES

1. K. Singh, J. Optics (India) 8, 51 (1975)
2. J.C. Dainty, (Ed) Laser Speckles and Related Phenomena (Springer Verlag 1975).
3. W.T. Welford, Opt. & Quant. Electron. 9 (1977).
4. E. Jakeman & W.T. Welford, Opt. Commun. 21, 72 (1977).
5. K. Ouchi, Opt. Commun. 24, 273 (1978).
6. K. Ouchi, Opt. & Quant. Electron 11, 345 (1979).
7. D. Masters & B.J. Rye (unpublished).
8. A. Chelli, P. Lena, C. Roddier, F. Roddier, F. Sibille, Opt. Acta 26, 583 (1979).
9. C. Aime, S. Kadiri, C. Ricort, C. Roddier, J. Vernin, J. Opt. Acta 26, 575 (1979).
10. G.C. Steward, The symmetrical optical system (Cambridge Univ. Press, 1958).
11. V.N. Mahajan, Appl. Opt. 17, 964 (1978).
12. V.N. Mahajan, J. Opt. Soc. Am. 68, 742 (1978).
13. J.Y. Wang, Appl. Opt. 17, 2580 (1978).
14. M.L. Belov & V.M. Orlov, Opt. Spectrs 44, 231 (1978).
15. H.F.A. Tschunko, Appl. Opt. 13, 1820 (1974).
16. B. Stoffregen, Optik 52, 305 (1978-79).
17. P.J. Chandley, W.T. Welford, Opt. & Quant. Electron 7, 393 (1975).

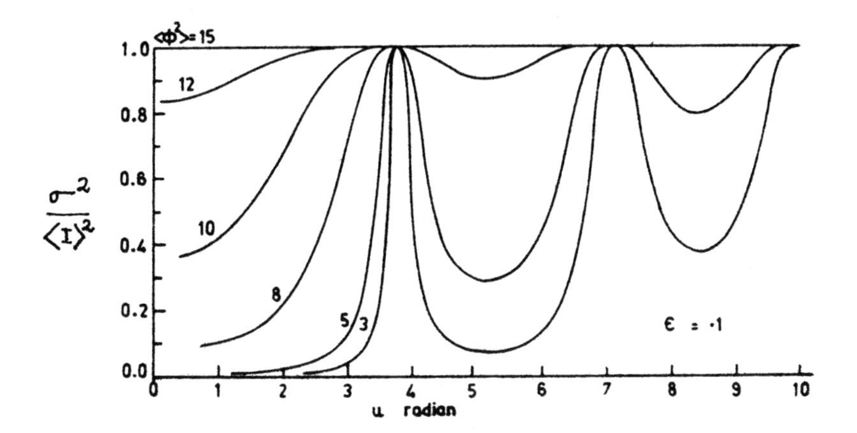

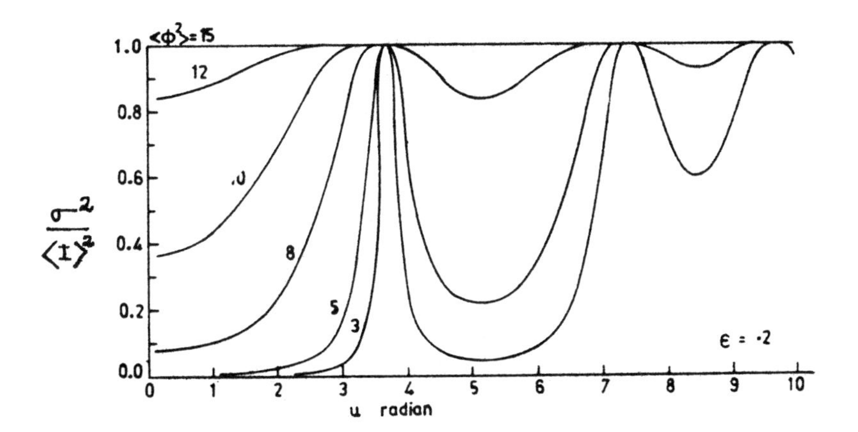

Fig. 2(a)

Computed speckle contrast at the focal plane
W = 1 mm $\xi = 10\ \mu m$, $u = (k\,W_{2/f})\,r$

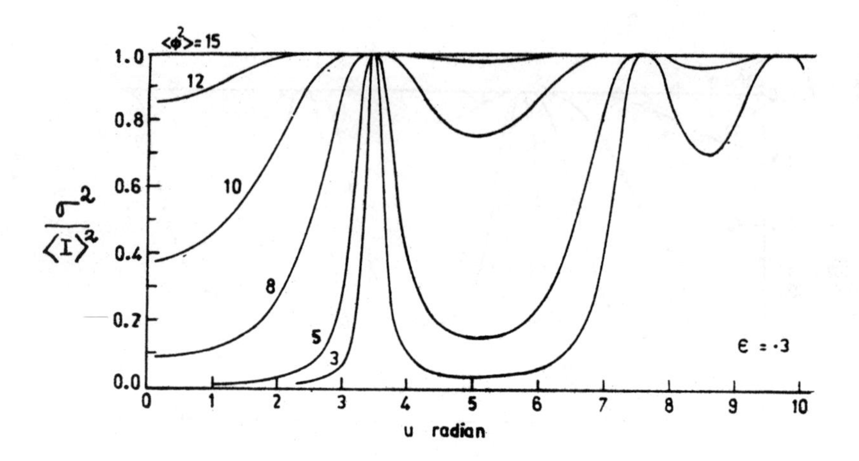

$<\phi^2>=15$
12
10
8
5
3
1.0
0.8
0.6
0.4
0.2
0.0
$\frac{\sigma^2}{\langle I \rangle^2}$
0
1
2
3
4
5
6
7
8
9
10
u radian
$\epsilon = \cdot 3$

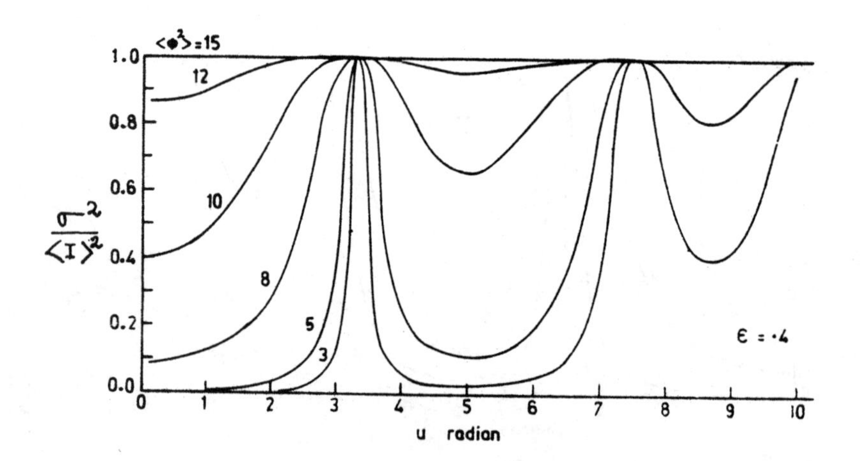

$<\phi^2>=15$
12
10
8
5
3
1.0
0.8
0.6
0.4
0.2
0.0
$\frac{\sigma^2}{\langle I \rangle^2}$
0
1
2
3
4
5
6
7
8
9
10
u radian
$\epsilon = \cdot 4$

Fig. 2(b)

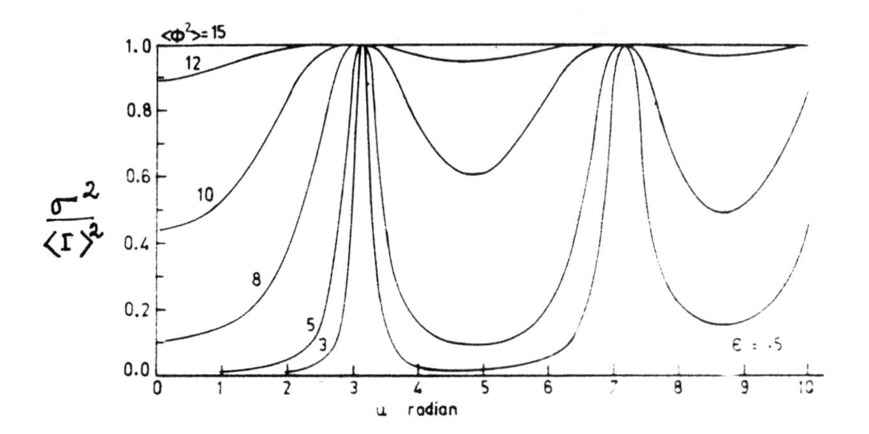
$\langle\phi^2\rangle = 15$
1.0
0.8
0.6
0.4
0.2
0.0
12
10
8
5
3
$\frac{\sigma^2}{\langle I\rangle^2}$
ε = .5
0
1
2
3
4
5
6
7
8
9
10
u radian

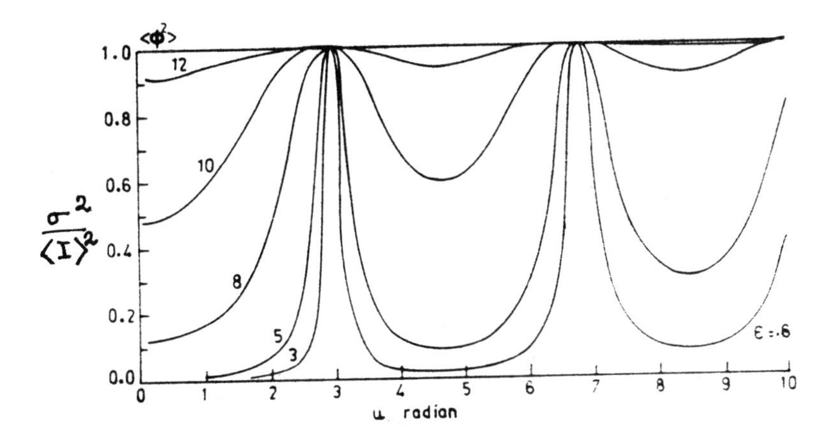
$\langle\phi^2\rangle$
1.0
0.8
0.6
0.4
0.2
0.0
12
10
8
5
3
$\frac{\sigma^2}{\langle I\rangle^2}$
ε =.6
0
1
2
3
4
5
6
7
8
9
10
u radian

Fig. 2(c)

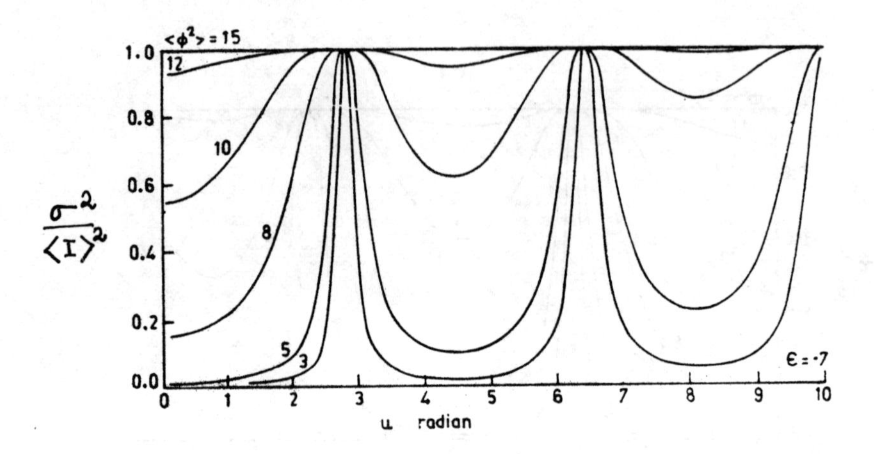

<φ²> = 15
12
10
8
5
3
1.0
0.8
0.6
0.4
0.2
0.0
σ²/⟨I⟩²
0
1
2
3
4
5
6
7
8
9
10
u radian
ε = ·7

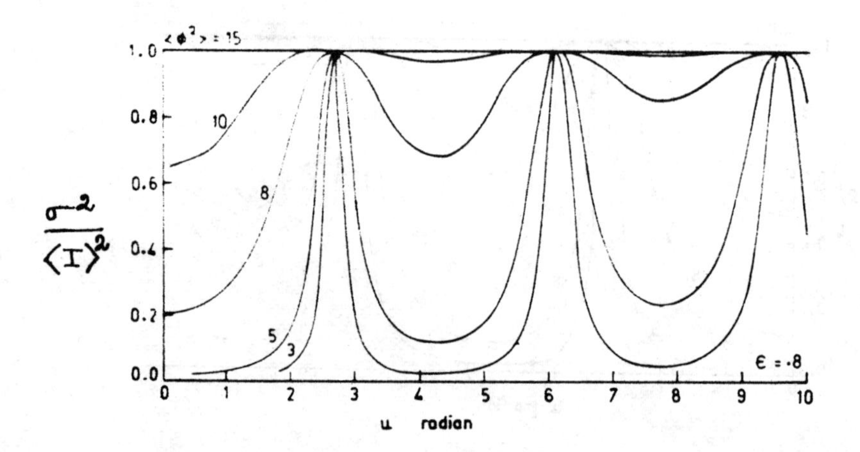

<φ²> = 15
10
8
5
3
1.0
0.8
0.6
0.4
0.2
0.0
σ²/⟨I⟩²
0
1
2
3
4
5
6
7
8
9
10
u radian
ε = ·8

Fig. 2(d)

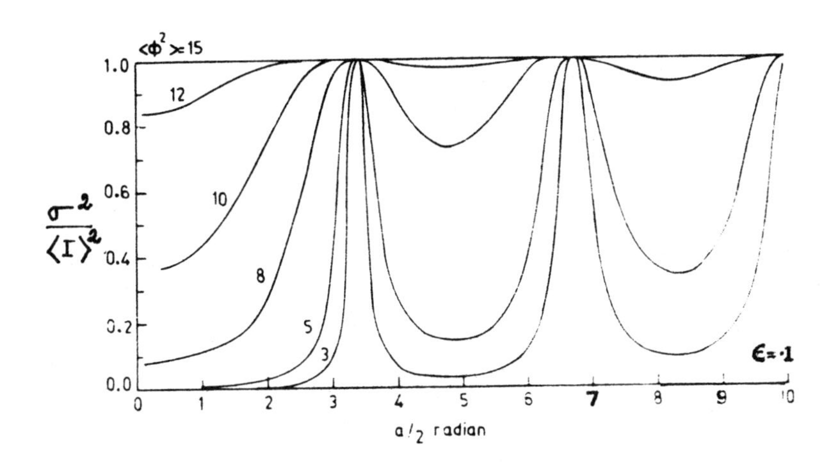

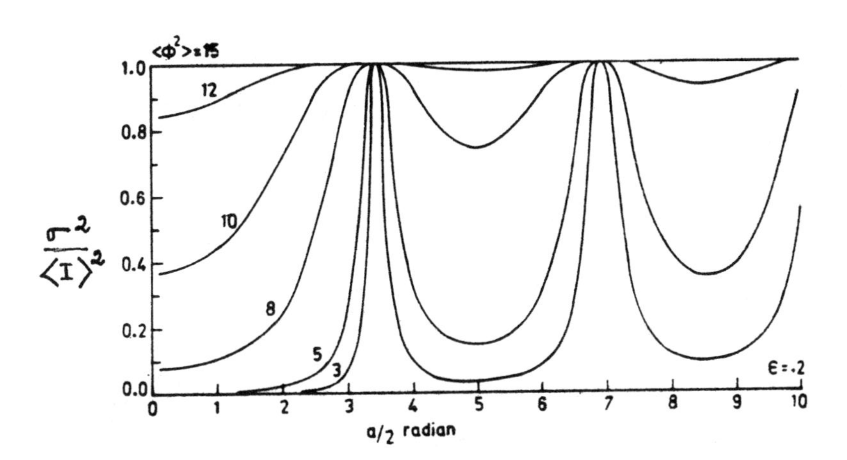

Fig. 3(a)

Computed speckle contrast along the axis $a = 2k(W_{2}/2f)^{2}z$. Other parameters same as in Fig.2.

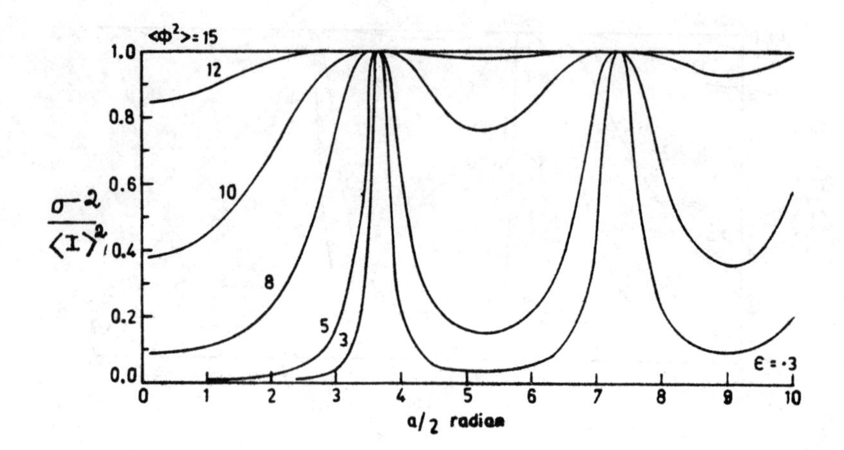

<<Φ²>= 15
1.0
0.8
0.6
0.4
0.2
0.0
12
10
8
5
3
ε = ·3
0
1
2
3
4
5
6
7
8
9
10
a/2 radian
σ²/⟨I⟩²

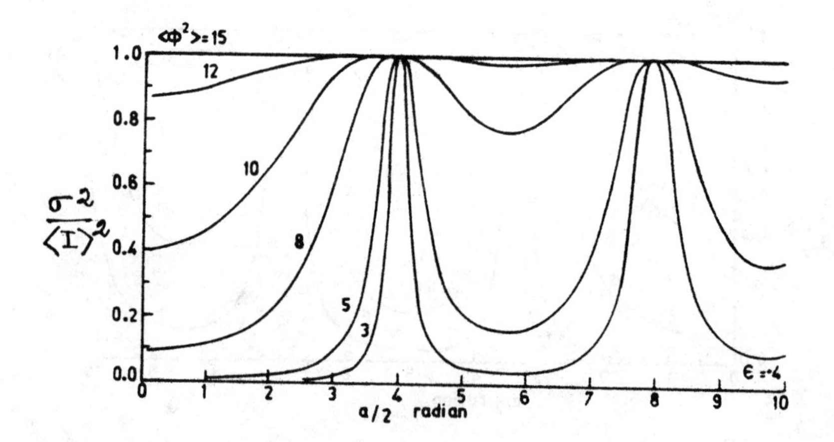

<<Φ²>= 15
1.0
0.8
0.6
0.4
0.2
0.0
12
10
8
5
3
ε = ·4
0
1
2
3
4
5
6
7
8
9
10
a/2 radian
σ²/⟨I⟩²

Fig. 3(b)

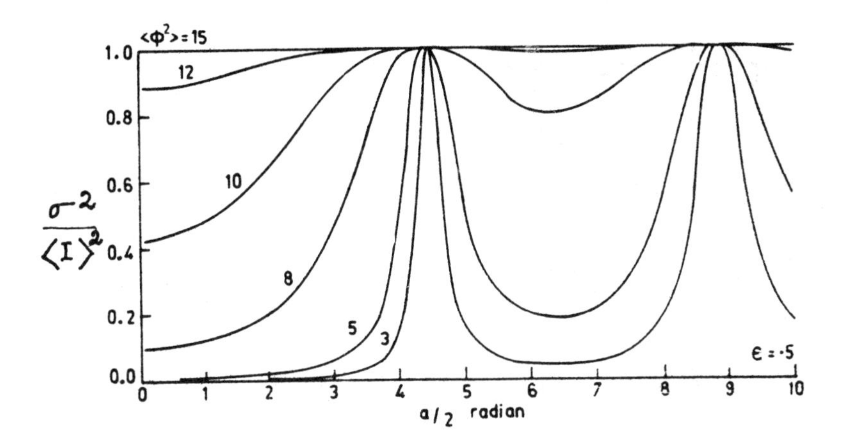

<φ²> = 15
1.0
0.8
0.6
0.4
0.2
0.0
12
10
8
5
3
σ²/⟨I⟩²
ε = ·5
0
1
2
3
4
5
6
7
8
9
10
a/2 radian

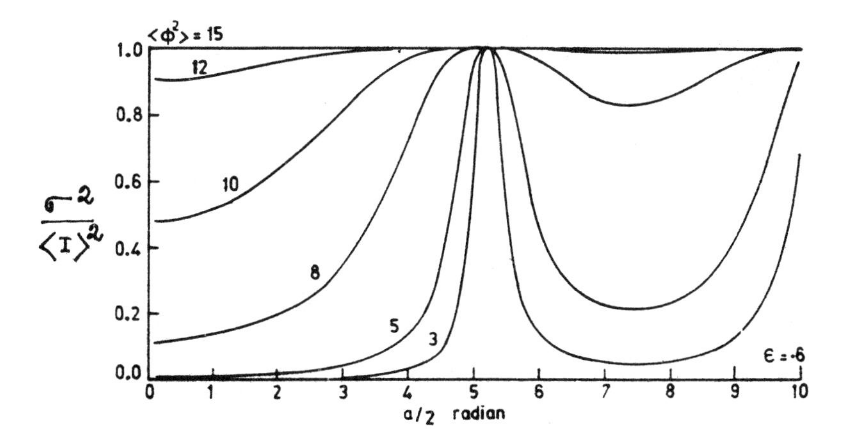

<φ²> = 15
1.0
0.8
0.6
0.4
0.2
0.0
12
10
8
5
3
σ²/⟨I⟩²
ε = ·6
0
1
2
3
4
5
6
7
8
9
10
a/2 radian

Fig. 3(c)

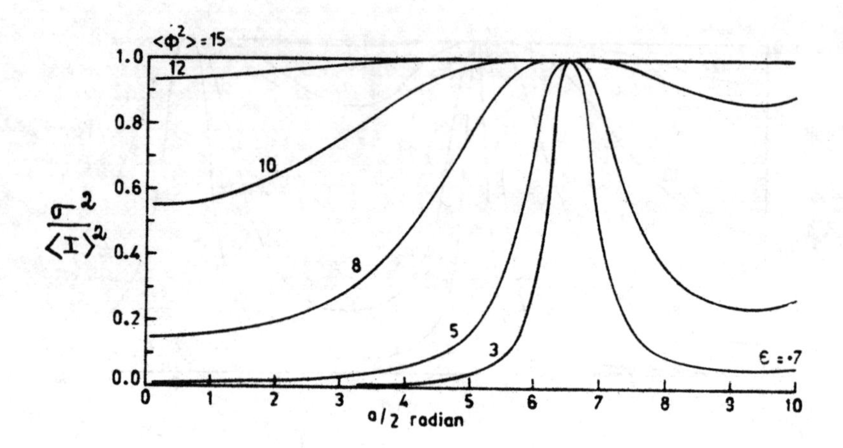

$<\phi^2> = 15$
12
10
8
5
3
$\epsilon = \cdot 7$
$\frac{\sigma^2}{\langle I \rangle^2}$
1.0
0.8
0.6
0.4
0.2
0.0
0 1 2 3 4 5 6 7 8 9 10
a/2 radian

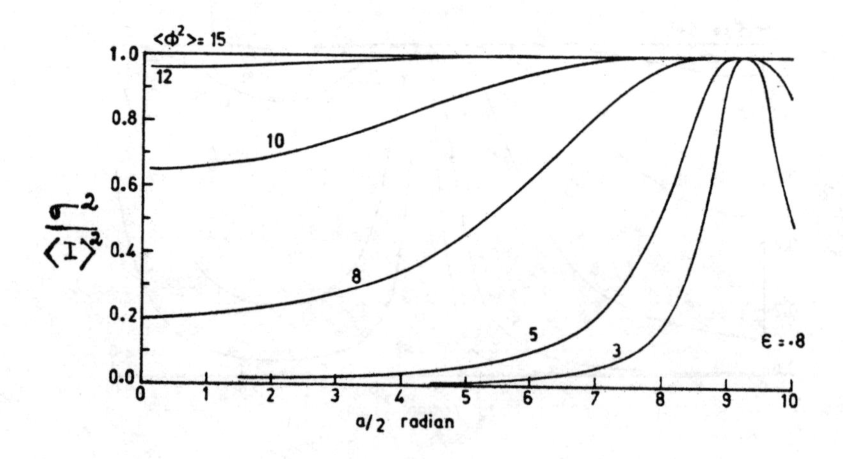

$<\phi^2> = 15$
12
10
8
5
3
$\epsilon = \cdot 8$
$\frac{\sigma^2}{\langle I \rangle^2}$
1.0
0.8
0.6
0.4
0.2
0.0
0 1 2 3 4 5 6 7 8 9 10
a/2 radian

Fig. 3(d)

SPATIAL AND TEMPORAL PROPERTIES OF SPECKLE INTERFEROGRAMS

G. P. Weigelt
Physikalisches Institut, Rommelstraße 1, 8520 Erlangen, Germany West

ABSTRACT

In order to investigate the spatial structure of a three-dimensional speckle intensity distribution, we have studied the intensity correlation function of two arbitrary points in a 3-D-speckle field. The result is an extension of the lateral autocorrelation function of speckle interferograms first derived by L. I. Goldfischer.

Speckle interferograms will fluctuate temporally if the random object (rough surface or air turbulence) varies in time. The temporal properties of speckle interferograms are important for various applications. Therefore we have developed a method for measuring the "life time" of speckles. We record the temporal behaviour of the speckles with a 16mm-movie camera. The life time is determined by measuring the average crosscorrelation of adjacent frames on the 16mm-motion picture films. We used this method to measure the life time of stellar speckle interferograms. Based on these studies it was possible to perform speckle interferometry measurements of rather faint astronomical objects.

THE 3-D-CORRELATION FUNCTION OF A 3D-SPECKLE INTENSITY DISTRIBUTION

A three-dimensional speckle intensity distribution is produced if, for example, a rough surface is illuminated by laser light as shown in Fig. 1.

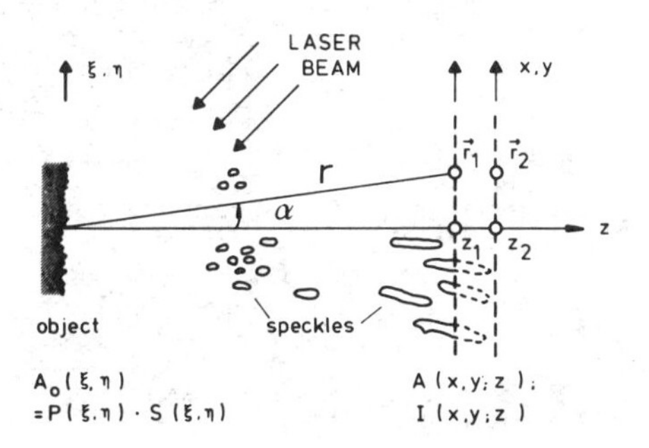

Fig. 1. Three-dimensional speckle intensity distribution in front of a rough surface ($P(\xi)$ is the illumination function). From Ref.2.

ISSN:0094-243X/81/650731-03$1.50

The lateral autocorrelation function $\langle I(x+\Delta x,y+\Delta y)I(x,y)\rangle$ of a lateral speckle intensity distribution $I(x,y;z=\text{const.})$ was first derived by Goldfischer [1]. In order to investigate the <u>three-dimensional</u> properties of 3-D-speckle interferograms we extended Goldfischer's study. We derived the average crosscorrelation function of speckle interferograms $I(x,y;z_i)$ in different z-planes z_1 and z_2 (with B. Stoffregen, Ref. 2). From the obtained results we derived

(1) the point correlation $\langle I(x+\Delta x,y+\Delta y;z_1)I(x,y;z_2)\rangle$ of two arbitrary points in space, and

(2) the longitudinal correlation function (in r-direction; see Fig. 1)

$$\langle I(r-\Delta r/2)I(r+\Delta r/2)\rangle =$$

$$\text{const.} + \left| \int |P(\xi)|^2 \exp(-ik\xi^2\Delta r/2r^2)d\xi \right|^2 \quad (1)$$

where r is the radius of polar coordinates and $P(\xi)$ is the illumination function of the object (see Ref. 3, chapter 2.5.1).

Fig. 2 shows for illustration a part of a longitudinal speckle intensity distribution function $I(y,z;x\approx 0)$ of a three-dimensional speckle intensity distribution $I(x,y,z)$.

Fig. 2 A longitudinal speckle intensity distribution $I(y,z;x\approx 0)$. From Ref. 2.

The motivation for this study was the application of various methods for measuring longitudinal motions (parallel to the z-axis)[4].

(2) MEASUREMENT OF THE LIFE TIME OF SPECKLE INTERFEROGRAMS

In order to study the temporal properties of speckle interferograms we developed a method for measuring the "life time" of speckle interferograms [5]. We record the temporal behaviour of the speckles with a 16mm-movie camera. The life time is determined by measuring the average crosscorrelation of adjacent, next-to-adjacent,... speckle interferograms recorded on 16mm-motion picture film. We produce

$$(1/N) \sum_n I(x,y;t_n) \otimes I(x-mx_o,y;t_{n+m}) \qquad (2)$$

where $I(x,y;t_n)$ denotes the intensity distribution of the speckle interferogram recorded at the time t_n and x_o is the distance between the centers of two adjacent frames.

The life time of speckle interferograms plays an important role in stellar speckle interferometry [6]. Therefore we used the described method for measuring the life time of stellar speckle interferograms [4]. Based on these studies it was possible to perform speckle interferometry measurements of rather faint objects. Astronomical objects up to 14 th magnitude were resolved with high SNR [7].

ACKNOWLEDGEMENT

I would like to thank A. W. Lohmann and B. Stoffregen for many helpful discussions.

REFERENCES

1. L. I. Goldfischer, J. Opt. Soc. Am. 55, 247 (1965)
2. G. P. Weigelt, B. Stoffregen, OPTIK 48, 399 (1977)
3. J. W. Goodman, in Laser Speckle and Related Phenomena (Ed. J. C. Dainty), Topics in Applied Physics series, Springer Verlag, 1975
4. A. W. Lohmann and G. P. Weigelt, J. Opt. Soc. Am. 66, 1271 (1976)
5. A. W. Lohmann, G. P. Weigelt, OPTIK 53, 167 (1979)
6. A. Labeyrie, Astron. Astrophys. 6, 85 (1970)
7. G. P. Weigelt, "Speckle interferometry and Speckle Holography at Low Light Levels", in the Proc. of the conference "Speckle and Related Phenomena", San Diego, 1980

TEMPORAL AND SPATIAL BEHAVIOR OF ELECTROMAGNETIC FIELDS IN DISSIPATIVE OR CONDUCTIVE MEDIA

P. Halevi and A. Mendoza-Hernández

Instituto de Ciencias de la Universidad Autónoma de Puebla
Apdo. Post. J-48, Puebla, Pue, México

ABSTRACT

We study the temporal and spatial behavior of electromagnetic waves and of Poynting's vector in material media. The waves are excited by direct incidence of light at the surface of the medium, and we consider the cases of p- and s-polarized light. The material properties are described by a complex permittivity ϵ and real permeability μ. We find that the real electromagnetic fields are not perpendicular to the real propagation vector $Re\,\vec{q}$. The tips of the vectors $Re\,\vec{E}_p(\vec{r},t)$ and $Re\,\vec{B}_s(\vec{r},t)$ describe ellipses whose planes are parallel to the plane of incidence ("longelliptical polarization"). None of the axes of these ellipses is parallel to the surface normal. The tip of the real, instantaneous Poynting's vector, $\vec{S}(\vec{r},t)$, also describes an ellipse as a function of time t at any point $\vec{r}$. The plane of this "Poynting's ellipse" is also parallel to the plane of incidence and its axes are parallel to those of Poynting's ellipse. The center of Poynting's ellipse is given by the time-averaged Poynting's vector $\langle \vec{S} \rangle$.For s-polarized light $\langle \vec{S} \rangle \parallel Re\,\vec{q}$ .For p-polarized light, however, these two vectors are not parallel. Thus, for the same angle of incidence p- and s-polarized light are refracted at different angles by a lossy medium. The effect should be observable, and certainly important for frequencies approaching absorption lines.

INTRODUCTION

The optical properties of dissipative and/or conductive media are frequently described in terms of a complex index of refraction. Electromagnetic waves which propagate in such media are "inhomogeneous", i.e. the planes of constant phase and the planes of constant amplitude are in general not parallel. An exception occurs in the case when the waves in the material medium are excited by light which is normally incident at the surface of the medium. For oblique incidence, however, the propagation vector $Re\,\vec{q}$ is not parallel to the attenuation vector $Im\,\vec{q}$. In fact, the former vector gives the direction of the refracted ray, while the latter vector is perpendicular to the surface.

ISSN:0094-243X/81/650734-12$1.50

In a recent publication[1] one of the authors has raised the question whether such inhomogeneous waves are transverse, i.e. do the equations $Re\,\vec{E}\cdot Re\,\vec{q} = Re\,\vec{B}\cdot Re\,\vec{q} = 0$ hold for the real electromagnetic fields? The answer was an unequivocal "no" — contrary to the popular opinion that "electromagnetic waves are transverse".

Consider the simple case that the electric field of the incident light is either parallel to the plane of incidence ("p-polarization") or perpendicular to this plane ("s-polarization"). For sure, the magnetic field $Re\,\vec{B}$ in the first case, and the electric field $Re\,\vec{E}$ in the second case are transverse in the medium. However, the other two fields - $Re\,\vec{E}_p$ for p-polarization, and $Re\,\vec{B}_s$ for s-polarization - are not transverse. They are not longitudinal, either. Their polarization is best described by the term (Ref. 1), "longelliptical". It is distinct from ordinary "transelliptical" polarization in which case the plane of the polarization ellipse is perpendicular to the propagation vector $Re\,\vec{q}$. For longelliptical polarization the plane of the polarization ellipse is parallel to the plane of incidence and, therefore, to $Re\,\vec{q}$. Moreover, the size of the polarization ellipse diminishes with depth into the material medium.

All this raises an intriguing question. If the direction of $Re\,\vec{E}(\vec{r},t)$ (for p-polarization) or $Re\,\vec{B}(\vec{r},t)$ (for s-polarization) varies with time, then what happens to Poynting's vector? Is it possible that the direction of energy flux changes with time?

THE POLARIZATION ELLIPSE

The medium is assumed to be homogeneous and isotropic and characterized by a relative permittivity or dielectric "constant" $\epsilon = \epsilon_1 + i\epsilon_2$ and a relative permeability $\mu = 1$. A plane wave of wave vector $q_0 = \omega/c$ is incident at an angle θ_0 at the surface of the medium. The x axis is taken parallel to the surface and the z axis normal to it; thus the plane of incidence is the x-z plane. The incident wave excites in the material medium an inhomogeneous wave whose wave vector is[1]

$$\vec{q} = \hat{x}\, q_0 \sin\theta_0 + \hat{z}\,(q_{z1} + i\,q_{z2}) \qquad (1)$$

where the real and imaginary parts of q_z are given by

$$2(q_{z1,2}/q_0)^2 = \left[(\epsilon_1 - \sin^2\theta_0)^2 + \epsilon_2^2\right]^{1/2} \pm (\epsilon_1 - \sin^2\theta_0) \qquad (2)$$

The meaning of eq.(1) is born out by considering the factor

$$e^{i(\vec{q}\cdot\vec{r} - \omega t)} = e^{-q_{z2}z}\, e^{i(q_x x + q_{z1}z - \omega t)}$$

$$= e^{-q_{z2}z}\, e^{i(Re\,\vec{q}\cdot\vec{r} - \omega t)} \equiv e^{-q_{z2}z}\, e^{i\phi} \tag{3}$$

All the components of the electromagnetic fields are proportional to this factor. Then we see that the wave propagates in the direction of $Re\,\vec{q}$ and the angle of refraction is given by

$$\tan\theta = q_x/q_{z1} = q_o \sin\theta_o / q_{z1} \tag{4}$$

Thus the planes of constant phase form an angle θ with the surface, while the planes of constant amplitude are parallel to it. All fields decay as $\exp(-q_{z2}z)$ away from the surface. In eq.(3) we have introduced the phase angle

$$\phi = q_x x + q_{z1}z - \omega t \tag{5}$$

This is a real angle that depends on the point of observation (x,z) in the plane of incidence and on the time of observation t.

We shall first consider the case of p-polarization. We may write the x component of the electric field in the complex representation as

$$E_x(\vec{r},t) = E_x \exp(i\phi) \tag{6}$$

with a real amplitude E_x , which includes the factor $\exp(-q_{z2}z)$ .Then, the real electric field is[1]

$$Re\,\vec{E}_p(\vec{r},t) = \hat{x}\,E_x\cos\phi - \hat{z}\,E_x(q_x/|q_z|^2)(q_{z1}\cos\phi + q_{z2}\sin\phi) \tag{7}$$

Faraday's Law for plane waves is

$$\omega\vec{B}(\vec{r},t) = \vec{q}\times\vec{E}(\vec{r},t) \tag{8}$$

In component form, we find that $B_x = B_z = 0$ and

$$\omega\,Re\,B_y(\vec{r},t) = q_{z1}\,Re\,E_x(\vec{r},t) - q_{z2}\,Im\,E_x(\vec{r},t) - q_x\,Re\,E_z(\vec{r},t)$$

Substitution of eqs.(6) and (7) gives

$$Re\,\vec{B}_p(\vec{r},t) =$$

$$\hat{y}\,E_x(\omega|q_z|^2)^{-1}\left[q_{z1}(q_x^2 + |q_z|^2)\cos\phi + q_{z2}(q_x^2 - |q_z|^2)\sin\phi\right] \tag{9}$$

Note that the three nonvanishing components of the electromagnetic fields, $Re\ E_x$, $Re\ E_z$, and $Re\ B_y$ are not in phase because of dissipative effects ($q_{z2} \neq 0$).

In the case of s-polarization the x component of the magnetic field in the complex representation is taken as

$$B_x(\vec{r}, t) = B_x \exp(i\phi) \tag{10}$$

with a real amplitude B_x, which again includes the factor $\exp(-q_{z2}z)$. Then the real magnetic field is[1]

$$Re\ \vec{B}_s(\vec{r}, t) = \hat{x} B_x \cos\phi - \hat{z} B_x (q_x/|q_z|^2)(q_{z1}\cos\phi + q_{z2}\sin\phi) \tag{11}$$

The electric field is found from the Ampere-Maxwell Law, $\vec{\nabla} \times \vec{H} = \partial\vec{D}/\partial t$, where the current has been "absorbed" in an effective displacement vector $\vec{D}$. For plane wave solutions we have

$$\omega\mu_0\epsilon_0\epsilon\,\vec{E}(\vec{r}, t) = \vec{B}(\vec{r}, t) \times \vec{q} \tag{12}$$

Because $q^2 = (\omega/c)^2\epsilon$ this equation may be rewritten as

$$\vec{E}_s(\vec{r}, t) = (\omega/q^2)[\vec{B}(\vec{r}, t) \times \vec{q}\,] \tag{13}$$

The only nonvanishing component of $\vec{E}_s$ is E_y. Using the equation $\vec{q}\cdot\vec{B}(\vec{r}, t) = 0$ we find that

$$E_y(\vec{r}, t) = \omega B_z(\vec{r}, t)/q_x \tag{14}$$

With the help of eq.(11) we get the real electric field:

$$Re\ \vec{E}_s(\vec{r}, t) = -\hat{y} B_x(\omega/|q_z|^2)(q_{z1}\cos\phi + q_{z2}\sin\phi) \tag{15}$$

Comparing eqs.(15) and (11)(or directly from eq.(14)) we see that $Re\ E_y$ is in phase with $Re\ B_z$, however these fields are not in phase with $Re\ B_x$.

If we ignore dissipative or conductive effects then $\vec{q}$ is a real vector and, by eq.(2), $q_{z2} = 0$. In this case we see from eqs.(7) and (11) that both components of $Re\ \vec{E}_P(\vec{r}, t)$ and of $Re\ \vec{B}_s(\vec{r}, t)$ are proportional to $\cos\phi$, i.e. they are in phase. Thus both fields are linearly polarized, as it should be for a dissipationless and nonconductive dielectric. If, however, $q_{z2} \neq 0$ then the x and z components of the electromagnetic fields which are parallel to the plane of incidence ($\vec{E}_P$ and $\vec{B}_s$) are not in phase. We expect then that the tips of these vectors describe ellipses which are parallel to the plane of incidence, as is born out by a detailed calculation. As we have suggested in the Introduction a convenient term for this type of polarization is "longelliptical". It turns out[1] that the polarization ellipses for both $Re\,\vec{E}_p(\vec{r},t)$ and $Re\,\vec{B}_s(\vec{r},t)$ have the same characteristics. The minor half-axis B forms an angle δ with the surface normal, where

$$\tan\delta = \frac{2\tan\theta}{1-\tan^2\theta + q_{z2}^2/q_{z1}^2} \tag{16}$$

The ratio of the axes is given by

$$\frac{A^2}{B^2} = \frac{1+\tan\theta\,\tan\delta + q_{z2}^2/q_{z1}^2}{1-\tan\theta\,\cot\delta + q_{z2}^2/q_{z1}^2} \tag{17}$$

The polarization ellipse and the relevant geometry are shown in Fig.1. The semiaxes A and B are proportional to the factor exp $(-q_{z2}z)$. Hence the polarization ellipse decreases exponentially with depth Z into the material medium. This is illustred in the inset of Fig.1.

Here we have considered only the simple cases of p- or s-polarized light. The case of arbitrary polarization has been studied in reference 1. Then the tips of both $Re\,\vec{E}(\vec{r},t)$ and $Re\,\vec{B}(\vec{r},t)$ describe ellipses as a function of time. The planes of these ellipses are not parallel in general, and neither polarization ellipse is parallel to the plane of incidence or perpendicular to the propagation vector $Re\,\vec{q}$. Moreover, the real electric and magnetic fields are not in general perpendicular to each other.

THE INSTANTANEOUS POYNTING'S VECTOR: POYNTING'S ELLIPSE

We wish to study the spatial and temporal behavior of the real, instantaneous Poynting's vector,

$$\vec{S}(\vec{r},t) = Re\,\vec{E}(\vec{r},t) \times Re\,\vec{H}(\vec{r},t) \tag{18}$$

If the electric field is parallel to the plane of incidence then $Re\,\vec{E}$ and $Re\,\vec{H}$ are given by eqs.(7) and (9). If the electric field is normal to the plane of incidence then these fields are given by eqs.(15) and (11). Note that $Re\,\vec{E}_p$ and $Re\,\vec{B}_s$ have the same form, and that $Re\,\vec{B}_p$ and $Re\,\vec{E}_s$ are both parallel to y and have a linear dependence on cos ϕ and sin ϕ .

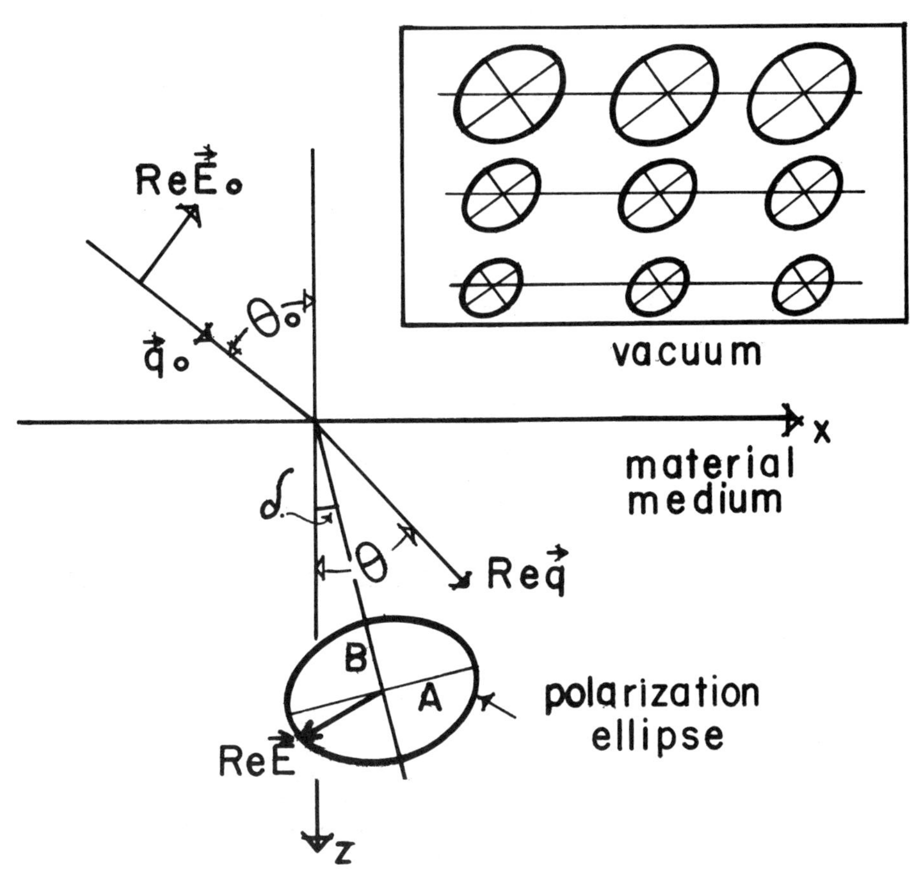

Fig. 1. Polarization ellipse traced by electric field

Therefore we may express Poynting's vector in a notation which is valid for both polarizations:

$$\vec{S}(\vec{r},t) =$$

$$(M\cos\phi + N\sin\phi)\left[\hat{x}(q_{z1}\cos\phi + q_{z2}\sin\phi)q_x/|q_z|^2 + \hat{z}\cos\phi\right] \qquad (19)$$

Only M and N assume different values for the two polarizations considered, and they are given in Table I. We note that $\vec{S}(\vec{r},t)$ vanishes whenever and wherever tan ϕ $= -M/N$.

Table I. The parameters M and N for p- and s- polarized light.

	P	S
M	$\frac{E_x^2}{\mu_0 c}\frac{q_{z1}}{q_0}\frac{q_x^2+q_{z1}^2+q_{z2}^2}{q_{z1}^2+q_{z2}^2}$	$\frac{B_x^2 c}{\mu_0}\frac{q_0 q_{z1}}{q_{z1}^2+q_{z2}^2}$
N	$\frac{E_x^2}{\mu_0 c}\frac{q_{z2}}{q_0}\frac{q_x^2-q_{z1}^2-q_{z2}^2}{q_{z1}^2+q_{z2}^2}$	$\frac{B_x^2 c}{\mu_0}\frac{q_0 q_{z2}}{q_{z1}^2+q_{z2}^2}$

It is already clear at this point that the direction of Poynting's vector is not constant in time. This is because its x and z components have different phases. What is the locus in time of the tip of $\vec{S}(\vec{r},t)$? We may write out eq.(19) for S_x and S_z , then calculate cos ϕ and sin ϕ as a function of these components, and finally sum the expressions for $\cos^2\phi$ and $\sin^2\phi$. This eliminates the phase ϕ and we find that $\vec{S}(\vec{r},t)$ satisfies the following equation — for any $\vec{r}$ and t .

$$|q_z|^4 S_x^2 + q_x^2|q_z|^2 S_z^2 - 2q_x q_{z1}|q_z|^2 S_x S_z$$
$$-q_x q_{z2}|q_z|^2 N S_x + q_x^2 q_{z2}(q_{z1}N - q_{z2}M)S_z = 0 \tag{20}$$

Here $S_{x,z}$ is an abbreviated notation for $S_{x,z}(\vec{r},t)$. Eq. (20) is the equation of an ellipse, which we call "Poynting's ellipse".

The characteristics of Poynting's ellipse are readily calculated by a translation and a rotation of the "coordinates" S_x and S_z .[2] The plane of the ellipse is parallel to the plane of incidence and its center is given by the vector

$$\langle \vec{S} \rangle = \hat{x}(q_{z1}M + q_{z2}N)q_x/(2|q_z|^2) + \hat{z}M/2 \tag{21}$$

Is is not difficult to see from eq. (19) that this is just the time-averaged Poynting's vector $\langle \vec{S}(\vec{r},t) \rangle$ Thus Poynting's vector may be rewritten as the sum of the usual, average Poynting's vector and a fluctuating part $\vec{S}'$:

$$\vec{S}(\vec{r},t) = \langle \vec{S} \rangle + \vec{S}'(\vec{r},t) \tag{22}$$

This is illustrated in Fig. 2, which gives the temporal behavior of Poynting's vector at a given point $\vec{r}$. We show a complete cycle, whose period is π/ω, i.e. one half of the cycle of the fields.

It turns out [2] that the major semiaxis A_1 of Poynting's ellipse forms an angle δ with the surface normal, i.e. the axes of Poynting's ellipse and those of the polarization ellipse are parallel to each other. The ratio of the axes is given by

$$\frac{A_1^2}{B_1^2} = \frac{(\tan\theta - \tan\mu)^2 + \tan^2\mu\,(q_{z2}/q_{z1})^2}{(1 + \tan\theta\,\tan\mu)^2 + (q_{z2}/q_{z1})^2} \tag{23}$$

The shape of Poynting's ellipse is thus determined by the material parameters ϵ_1 and ϵ_2 and the angle of incidence θ_0 ; it is the same shape for both polarizations.

While the orientations of Poynting's ellipse and the polarization ellipse are the same, their eccentricities are different. This may be seen by comparing eqs. (23) and (17).

The semiaxes A_1 and B_1 are proportional to - - - $(M^2 + N^2)^{1/2}$ and therefore-depending on the polarization-to E_x^2 or B_x^2 . These factors, in turn, are proportional to $\exp(-2q_{z2}z)$. Therefore, the size of Poynting's ellipse decreases exponentially with the depth z into the material medium. We show this in Fig.3, which gives the spatial behavior of Poynting's vector at a given instant of time t. There is a periodicity in the direction x parallel to the surface, the period being given by π/q_x . We also note that Poynting's ellipse passes through the origin, as is clear from the fact that $\vec{S}(\vec{r},t)$ vanishes for $\tan\phi = -M/N$.

REFRACTION: HOW MANY ANGLES?

From eq. (21) we easily find that the average Poynting's vector forms an angle η with the surface normal, where η is given by

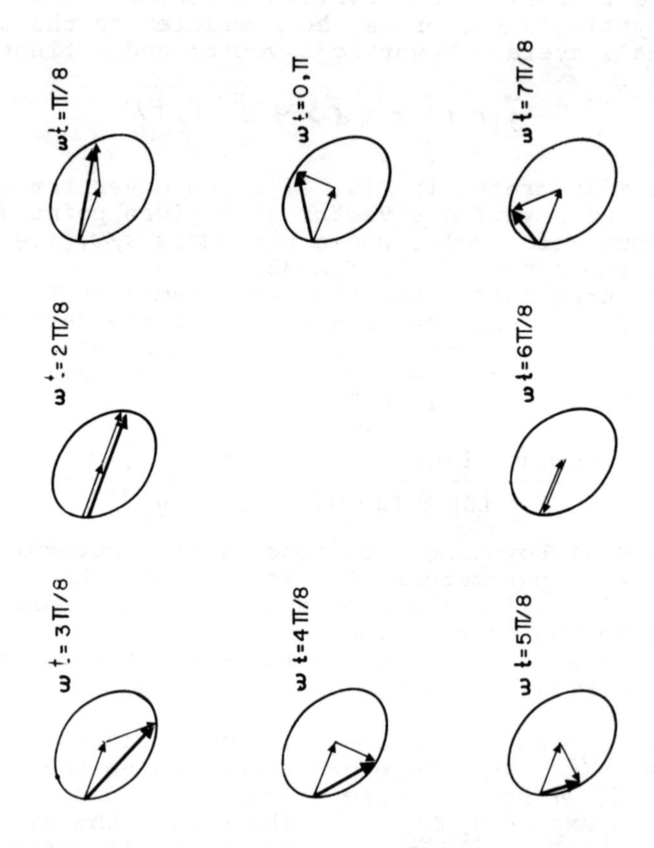

Fig. 2 Temporal behavior of Poynting's vector at a point $\vec{r}$. The sum of the constant part $\langle \vec{S} \rangle$ and the fluctuating part is the instantaneous Poynting's vector $\vec{S}(\vec{r},t)$. "Poynting's ellipse" is parallel to the plane of incidence.

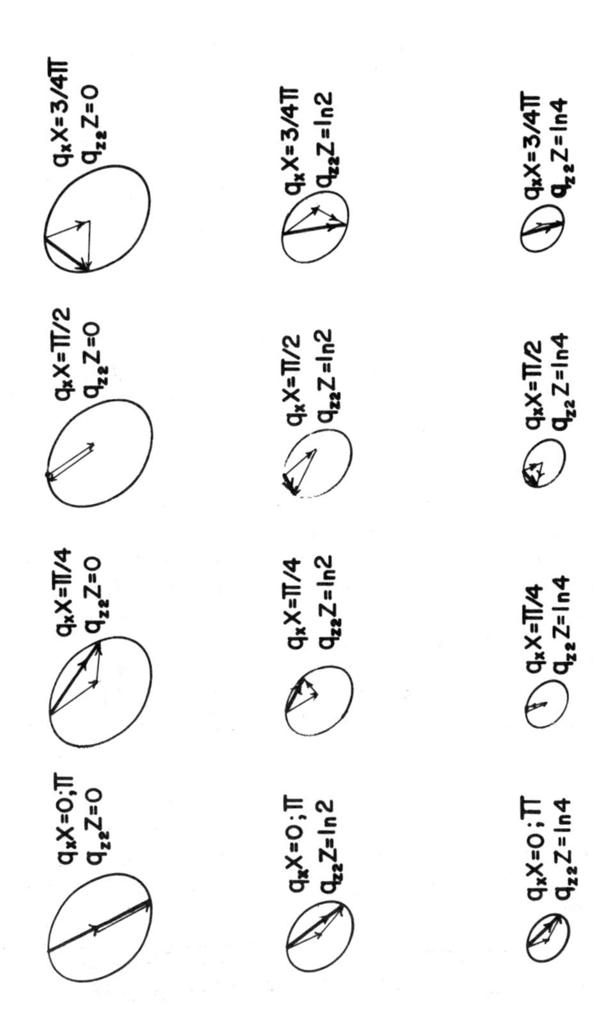

Fig. 3 Spatial behavior of Poynting's vector at a given time t. It exhibits periodicity in the direction x parallel to the surface and exponential decay in the normal direction z.

$$\tan\eta = \frac{\langle S_x\rangle}{\langle S_z\rangle} = \frac{q_x(q_{z1}M + q_{z2}N)}{|q_z|^2 M} \tag{24}$$

Using Table I, we evaluate this expression for the two polarizations considered. We find that

$$\eta_s = \theta \tag{25a}$$

$$\tan\eta_p = \tan\theta\,\frac{1+\tan^2\theta - q_{z2}^2/q_{z1}^2}{1+\tan^2\theta + q_{z2}^2/q_{z1}^2} \tag{25b}$$

This is rather surprising result. If the electric field is perpendicular to the plane of incidence (s-polarization) then eq. (25a) tells us that $\langle \vec{S} \rangle \parallel \mathrm{Re}\,\vec{q}$, i.e. there is one well-defined angle of refraction. However, for an electric field parallel to the plane of incidence (p-polarization) these vectors are not parallel. From eq. (25b) se see that $\eta_p < \theta$. Thus there are two meaningful angles of refraction, θ and η_p, however only one ray! Ordinarily, the direction of a ray coincides with that of energy transport, i.e. one measures η_p. An interesting question arises: is it possible to measure θ and thus distinguish experimentally between θ and η_p for p-polarized light?

The energy-refraction angle η_p is obtained explicitly by substituting eqs. (4) and (2) in eq. (25b). This gives η_p as a function of the angle of incidence θ_0 and $\epsilon = \epsilon_1 + i\epsilon_2$. Thus eq. (25b) is the refraction law for p-polarized light in "ordinary" optical experiments, in lieu of Snell's law for nonconducting, dissipationless media. The generalization is straightforward to the the case that the vacuum side of the interface is replaced by a medium with a real and positive dielectric constant.

We have the remarkable conclusion that, for the same angle of incidence, a p-polarized ray and s-polarized ray are refracted at different angles ($\eta_p < \eta_s$) !

This difference in the refractive behavior of p-and s-polarized light may be explained by our choice of complex ϵ, but real μ. If we reverse the choice, i.e. $\mu = \mu_1 + i\mu_2$, $\epsilon = 1$ we find that the average Poynting's vector is paralell to the propagation vector for p-polarization, however not for s polarization. In the general case of a medium with dissipative electric and magnetic properties we expect that the time-averaged Poynting's vector is not parallel to the propagation vector for any polarization of the light. In this case the algebra becomes somewhat messy.

The fact that the direction of the time averaged Poynting's vector depends on the polarization (p or s) is particularly noteworthy. It means that, for the same angle of incidence, p- and s-polarized light will be refracted at different angles by a lossy medium. Therefore the generalized law of refraction, which replaces Snell's law, is no longer purely "kinematic": it also depends on "dynamic" properties of the light. The possibility of experimental verification of this effect will be investigated in a future publication.

REFERENCES

1. P. Halevi, Amer. J. Phys., to be published.

2. A. Mendoza-Hernández and P. Halevi, to be published.

AIP Conference Proceedings

No.	Title	L.C. Number	ISBN
No.1	Feedback and Dynamic Control of Plasmas	70-141596	0-88318-100-2
No.2	Particles and Fields - 1971 (Rochester)	71-184662	0-88318-101-0
No.3	Thermal Expansion - 1971 (Corning)	72-76970	0-88318-102-9
No.4	Superconductivity in d-and f-Band Metals (Rochester, 1971)	74-18879	0-88318-103-7
No.5	Magnetism and Magnetic Materials - 1971 (2 parts) (Chicago)	59-2468	0-88318-104-5
No.6	Particle Physics (Irvine, 1971)	72-81239	0-88318-105-3
No.7	Exploring the History of Nuclear Physics	72-81883	0-88318-106-1
No.8	Experimental Meson Spectroscopy - 1972	72-88226	0-88318-107-X
No.9	Cyclotrons - 1972 (Vancouver)	72-92798	0-88318-108-8
No.10	Magnetism and Magnetic Materials - 1972	72-623469	0-88318-109-6
No.11	Transport Phenomena - 1973 (Brown University Conference)	73-80682	0-88318-110-X
No.12	Experiments on High Energy Particle Collisions - 1973 (Vanderbilt Conference)	73-81705	0-88318-111-8
No.13	π-π Scattering - 1973 (Tallahassee Conference)	73-81704	0-88318-112-6
No.14	Particles and Fields - 1973 (APS/DPF Berkeley)	73-91923	0-88318-113-4
No.15	High Energy Collisions - 1973 (Stony Brook)	73-92324	0-88318-114-2
No.16	Causality and Physical Theories (Wayne State University, 1973)	73-93420	0-88318-115-0
No.17	Thermal Expansion - 1973 (lake of the Ozarks)	73-94415	0-88318-116-9
No.18	Magnetism and Magnetic Materials - 1973 (2 parts) (Boston)	59-2468	0-88318-117-7
No.19	Physics and the Energy Problem - 1974 (APS Chicago)	73-94416	0-88318-118-5
No.20	Tetrahedrally Bonded Amorphous Semiconductors (Yorktown Heights, 1974)	74-80145	0-88318-119-3
No.21	Experimental Meson Spectroscopy - 1974 (Boston)	74-82628	0-88318-120-7
No.22	Neutrinos - 1974 (Philadelphia)	74-82413	0-88318-121-5
No.23	Particles and Fields - 1974 (APS/DPF Williamsburg)	74-27575	0-88318-122-3

No.24	Magnetism and Magnetic Materials - 1974 (20th Annual Conference, San Francisco)	75-2647	0-88318-123-1
No.25	Efficient Use of Energy (The APS Studies on the Technical Aspects of the More Efficient Use of Energy)	75-18227	0-88318-124-X
No.26	High-Energy Physics and Nuclear Structure - 1975 (Santa Fe and Los Alamos)	75-26411	0-88318-125-8
No.27	Topics in Statistical Mechanics and Biophysics: A Memorial to Julius L. Jackson (Wayne State University, 1975)	75-36309	0-88318-126-6
No.28	Physics and Our World: A Symposium in Honor of Victor F. Weisskopf (M.I.T., 1974)	76-7207	0-88318-127-4
No.29	Magnetism and Magnetic Materials - 1975 (21st Annual Conference, Philadelphia)	76-10931	0-88318-128-2
No.30	Particle Searches and Discoveries - 1976 (Vanderbilt Conference)	76-19949	0-88318-129-0
No.31	Structure and Excitations of Amorphous Solids (Williamsburg, VA., 1976)	76-22279	0-88318-130-4
No.32	Materials Technology - 1975 (APS New York Meeting)	76-27967	0-88318-131-2
No.33	Meson-Nuclear Physics - 1976 (Carnegie-Mellon Conference)	76-26811	0-88318-132-0
No.34	Magnetism and Magnetic Materials - 1976 (Joint MMM-Intermag Conference, Pittsburgh)	76-47106	0-88318-133-9
No.35	High Energy Physics with Polarized Beams and Targets (Argonne, 1976)	76-50181	0-88318-134-7
No.36	Momentum Wave Functions - 1976 (Indiana University)	77-82145	0-88318-135-5
No.37	Weak Interaction Physics - 1977 (Indiana University)	77-83344	0-88318-136-3
No.38	Workshop on New Directions in Mossbauer Spectroscopy (Argonne, 1977)	77-90635	0-88318-137-1
No.39	Physics Careers, Employment and Education (Penn State, 1977)	77-94053	0-88318-138-X
No.40	Electrical Transport and Optical Properties of Inhomogeneous Media (Ohio State University, 1977)	78-54319	0-88318-139-8
No.41	Nucleon-Nucleon Interactions - 1977 (Vancouver)	78-54249	0-88318-140-1
No.42	Higher Energy Polarized Proton Beams (Ann Arbor, 1977)	78-55682	0-88318-141-X
No.43	Particles and Fields - 1977 (APS/DPF, Argonne)	78-55683	0-88318-142-8
No.44	Future Trends in Superconductive Electronics (Charlottesville, 1978)	77-9240	0-88318-143-6

No.45	New Results in High Energy Physics - 1978 (Vanderbilt Conference)	78-67196	0-88318-144-4
No.46	Topics in Nonlinear Dynamics (La Jolla Institute)	78-057870	0-88318-145-2
No.47	Clustering Aspects of Nuclear Structure and Nuclear Reactions (Winnepeg, 1978)	78-64942	0-88318-146-0
No.48	Current Trends in the Theory of Fields (Tallahassee, 1978)	78-72948	0-88318-147-9
No.49	Cosmic Rays and Particle Physics - 1978 (Bartol Conference)	79-50489	0-88318-148-7
No.50	Laser-Solid Interactions and Laser Processing - 1978 (Boston)	79-51564	0-88318-149-5
No.51	High Energy Physics with Polarized Beams and Polarized Targets (Argonne, 1978)	79-64565	0-88318-150-9
No.52	Long-Distance Neutrino Detection - 1978 (C.L. Cowan Memorial Symposium)	79-52078	0-88318-151-7
No.53	Modulated Structures - 1979 (Kailua Kona, Hawaii)	79-53846	0-88318-152-5
No.54	Meson-Nuclear Physics - 1979 (Houston)	79-53978	0-88318-153-3
No.55	Quantum Chromodynamics (La Jolla, 1978)	79-54969	0-88318-154-1
No.56	Particle Acceleration Mechanisms in Astrophysics (La Jolla, 1979)	79-55844	0-88318-155-X
No. 57	Nonlinear Dynamics and the Beam-Beam Interaction (Brookhaven, 1979)	79-57341	0-88318-156-8
No. 58	Inhomogeneous Superconductors - 1979 (Berkeley Springs, W.V.)	79-57620	0-88318-157-6
No. 59	Particles and Fields - 1979 (APS/DPF Montreal)	80-66631	0-88318-158-4
No. 60	History of the ZGS (Argonne, 1979)	80-67694	0-88318-159-2
No. 61	Aspects of the Kinetics and Dynamics of Surface Reactions (La Jolla Institute, 1979)	80-68004	0-88318-160-6
No. 62	High Energy e^+e^- Interactions (Vanderbilt , 1980)	80-53377	0-88318-161-4
No. 63	Supernovae Spectra (La Jolla, 1980)	80-70019	0-88318-162-2
No. 64	Laboratory EXAFS Facilities - 1980 (Univ. of Washington)	80-70579	0-88318-163-0
No. 65	Optics in Four Dimensions - 1980 (ICO, Ensenada)	80-70771	0-88318-164-9